Biology *of the*
Invertebrates

Biology *of the*
Invertebrates

Third Edition

Jan A. Pechenik
Tufts University

WCB **Wm. C. Brown Publishers**

Dubuque, IA Bogota Boston Buenos Aires Caracas Chicago
Guilford, CT London Madrid Mexico City Sydney Toronto

Book Team

Editor *Margaret J. Kemp*
Developmental Editor *Kathleen R. Loewenberg*
Production Editor *Kay Driscoll*
Designer *Kathleen F. Theis*
Art Editor *Tina Flanagan*
Photo Editor *Janice Hancock*
Permissions Coordinator *Mavis M. Oeth*

Wm. C. Brown Publishers

President and Chief Executive Officer *Beverly Kolz*
Vice President, Publisher *Kevin Kane*
Vice President, Director of Sales and Marketing *Virginia S. Moffat*
Vice President, Director of Production *Colleen A. Yonda*
National Sales Manager *Douglas J. DiNardo*
Marketing Manager *Thomas C. Lyon*
Advertising Manager *Janelle Keeffer*
Production Editorial Manager *Renée Menne*
Publishing Services Manager *Karen J. Slaght*
Royalty/Permissions Manager *Connie Allendorf*

Copyedited by Sarah Lane

Cover photo © Kevin Schafer/Peter Arnold, Inc.

Design elements (Chapter Openers, Research Focus Boxes, End of
Chapters): Undersea Life CD, Digital Stock

Illustrators: Arthur Ciccone: Figures 2.7*a*, 2.9, 3.1, 3.8*a*, focus 3.1, focus 3.2,
3.22, 3.26*a*, 3.27, 3.34, 3.35*a*, 4.4, 4.5, 4.6, 5.4, 5.13*a*, 5.16*c*, 5.21, focus 5.1, focus 5.2,
6.2, 7.3, 7.11, 7.15*c,d,* 7.16*a*, 7.17, 7.19, 7.20, 7.21, opening illustrations for
chapters 7, 8, 8.2, 8.3, 10.1, 11.4*a,* 11.5*a,* 11.12, 12.3, box 13.1, 14.3, 14.4,
14.5*d,e,* 14.5*f,* 14.6*e,* 14.7*b,c,* 14.8, 14.10, 14.11*d,* 14.15, 14.16, 14.18*b,c,* 14.19,
14.21, 14.24, 14.29, 14.30, focus 14.2*a,b,e,* 14.37*b,* 14.38*a,* 14.40*a,* 14.42*c,*
14.44*a,* 14.44*d,* 14.46, 14.47, 15.1, 15.2, 15.5, 15.6, 15.8, 15.9*b,* 15.18*d,* 16.6,
17.6*a,b,* 18.5, 18.9*a,b,c,e,* 18.10, 18.16, 18.18*b,* 18.20, 18.22, 18.23, 18.26, 18.28,
18.34*a,* 18.36, 18.40, TA18.3, TA18.5, pg. 397, TA18.6, pg. 398, TA18.11,
pg. 405, TA18.16, pg. 410, 20.1*a,d,* 20.6*d,* 20.7, 20.18, 21.2, 21.12, 21.15*a,c,d,*
23.4, 24.4*c,* 24.6*c,* 24.9

Eric Vincent: Opening illustrations for chapters 1, 2, 9, 10, 12. Margaret
Soltysik-Espanola: Figures 14.26, 14.27*a,* 14.28

To Oliver
favorite son and l.w.

Brief Contents

Expanded Contents

List of Research Focus Boxes

Preface

Invertebrate zoology is a fascinating but enormous field. More than 98% of all known animal species are invertebrates, and that proportion will undoubtedly increase with time as more species are described. Invertebrates are distributed among more than 40 phyla and a mind-boggling number of classes, subclasses, orders, and families. The degree of morphological and functional diversity found within some groups, even within single orders, can overwhelm the beginning student. The enormity of the field and the great range of potential approaches to the subject make invertebrate biology challenging to teach and to learn. In preparing the third edition of this book I have endeavored to make the tasks of both teaching and learning easier and more interesting.

About This Book

Like previous editions, the third is designed as a nonintimidating, readable introduction to the biology of each group, emphasizing those characteristics that set each group apart from all others. Even so, a considerable vocabulary must be mastered. Mastering that vocabulary opens the door to a great and growing literature; in this respect, learning invertebrate biology is much like learning a foreign language: Mastering the vocabulary and grammar of the French language pays off once one can read books by Molière or Camus as the authors wrote them. This book serves as the foundation for further learning, in lecture, laboratory, field, and library, a foundation that is largely manageable by the student. Because the book presents the basic vocabulary and grammar of invertebrate zoology, the instructor is free to embellish and expand on that foundation to suit any desired focus: phylogeny, behavior, diversity of form and function, physiology, ecology, or current research in any of those areas. The book whets the student's appetite and provides the required background, buying instructors the time to discuss more fully whatever they feel are the most interesting aspects of the field. This is the guiding principle in my decisions about the level of detail to provide: I generalize wherever possible to build a firm foundation without intimidation or confusion. Given a chance, the animals themselves soon win most students over.

The most difficult part of writing this book has been deciding what to leave out. These decisions are based largely on reading many dozens of research articles on the biology of each group and determining the specific terminology and level of background information students need to read those papers. Although all phyla are now covered, I have aimed for conciseness, not exhaustiveness, and have emphasized unifying principles rather than the diversity found within each group. Students are best prepared to encounter the diversity of form and function in lecture, laboratory, and field, once they have mastered basic concepts and terminology.

Too many people think of invertebrate zoology as an exercise in memorizing terms and, perhaps, interesting but trivial stories about interesting but irrelevant animals. Too many people think of invertebrate zoology as outdated, a field in which everything is already known. I hope that this book alters those perceptions, by presenting invertebrate zoology as a lively and modern area of ongoing and worthwhile biological inquiry.

Contented users of the previous edition will be pleased to find that the book retains its readability and manageability, despite its greater length. The text remains somewhat biased toward functional morphology, bringing animals to life for students and preparing readers to make careful observations of living animals in the laboratory and in the field. As before, most chapters contain a section entitled *Topics for Further Discussion and Investigation,* which highlights research issues encapsulating many of the major questions that have been and are being addressed for the animals covered in each chapter. For each topic, I have selected references from the primary literature that are likely to be found in the typical college library and that should be intellectually accessible to any interested, beginning student once he or she has read the relevant textbook chapter. I have had to exclude many excellent papers because they were too advanced, were published in less widely distributed journals, or were review papers rather than primary journal articles. The topics I have chosen, along with the accompanying references, could be used as a basis for lectures, class discussion, term papers, research proposals, or short essays, or simply as a convenient way of easing students into the original literature, by having them investigate topics that excite their curiosity. Students gain little by reading predigested summaries of the primary literature; they gain much by reading and discussing that literature.

The *Research Focus Boxes* are based on recent papers from the primary literature to illustrate the range of questions that biologists are asking about invertebrates and the variety of approaches that are being taken to address those questions. My goal here is to prepare students to read the primary literature by focusing on how questions are formulated, how data are collected and interpreted, and how each study typically leads to further questions. Students interested in the topic of a particular *Research Focus* might wish to read the original paper on which that *Focus* was based and then use the *Focus* as a model for writing about other papers on related topics. The excitement of invertebrate biology is found in the primary literature, and a major goal of this book is to motivate and prepare students to read that literature. The *Research Focus Boxes* also dispel the notion that invertebrate zoology is a dead field.

As with previous editions, chapters are self-contained and can be assigned in whatever order best suits the organization of any particular course, once the introductory chapters have been covered. Within each chapter, the material has been arranged in manageable, readable units for the convenience of both student and instructor. For example, a section entitled *Introduction and General Characteristics* might be assigned prior to a lecture on a particular group of organisms, while a section called *Feeding and Digestion* might best be assigned prior to the accompanying laboratory session; it is always easier to assign additional sections of a text than to tell students what *not* to read within a larger section. The final chapter, concerning general principles of invertebrate reproduction and development, brings together all of the major phyla in a consideration of basic problems shared by the members of all groups covered in the text, providing students an opportunity to reminisce about all the animals they have encountered during the term.

This edition has a somewhat greater phylogenetic orientation than its predecessors (see especially chapter 2), but it still avoids prolonged phylogenetic discussion. There is no consensus concerning many invertebrate interrelationships, and any statements to the contrary are misleading. Some argue that deuterostomes evolved directly from protostome ancestors, while others argue that coelomic cavities were independently evolved in the two groups. Some doubt that the various protostome groups are closely related. Some view the molluscs and arthropods as acoelomate, derived from ancestors that never had coelomic cavities. Certain groups traditionally placed among the protozoans may be highly degenerate bilateral animals. Priapulids are probably pseudocoelomates but may be coelomates. Gastrotrichs are probably acoelomates but may be pseudocoelomates. Molluscs may have no direct evolutionary relationship with annelids. Arthro-

pods and annelids may have no close evolutionary connections. Nemertine worms, long considered acoelomates, may actually be unusual coelomates with no direct flatworm affinities. Lophophorates may be deuterostomes, or they may be protostomes. Molecular studies are aiding and abetting these and other controversies. Although the phylogenetic atmosphere is certainly charged with excitement, beginning students typically find such controversies unsettling, pointless, or both, and they view textbook discussions of such issues as simply one more presentation of facts to be memorized. For this reason, such issues are best treated in lecture, where they can be used to animate class discussion. Indeed, the controversies surrounding phylogenetic speculation are what make phylogeny interesting. The textbook provides a foundation upon which the instructor can build.

Changes in the Third Edition

The third edition contains more information than its predecessor. The publisher has managed to again accomodate this expansion without increasing the number of pages (and, correspondingly, the cost), by increasing the page dimensions. All chapters have been revised to reflect new discoveries and expanding research areas, and each now concludes with a listing of general references and appropriate review papers. Chapter 2, on **classification schemes, has been expanded and updated** considerably. Chapter 3, on **protozoans, has been completely rewritten** and reorganized to reflect the recent upheavals that have taken place since the second edition. Molluscs now precede annelids, which, in turn, precede the arthropods, by popular request, and the **chapter on molluscs has been substantially reorganized** to facilitate phylogenetic discussion.

I have added **new sections on social insects, on the evolution of trematode life cycles, on gastropod shell coiling, and on cephalopod bioluminescence,** and I have expanded the number and scope of the *Research Focus Boxes;* one new *Research Focus Box* (see Chapter 10) **highlights key issues in molecular phylogeny.** The new edition also **incorporates results of recent physiological studies.** For example, we now know that insect prothoracic glands do not secrete ecdysone, but rather secrete an ecdysone precursor that is converted to ecdysone by hemolymph enzymes. Similarly, recent evidence suggests that duogland systems may play more important roles than suction in the locomotory activities of echinoderm tube feet, recent endoscopic investigations have altered our understanding of bivalve gill function, and recent studies have clarified how zooxanthellae-containing corals are protected from damage by toxic forms of oxygen.

I also have **incorporated recent shifts in our understanding about evolutionary relationships.** For example, orthonectids and rhombozoans are now given independent phylum status, myxozoans are no longer believed to be protozoans, pentastomids are now viewed as highly modified crustaceans (and *bona fide* arthropods), and chaetognaths are now regarded as probable coelomates.

In preparing this edition I have relied on the following references, which interested instructors and students are urged to consult:

Backeljau et al. 1993. Cladistic analysis of metazoan relationships: A reappraisal. *Cladistics* 9:167–81.

Eernisse, D. J., et al. 1992. Annelida and Arthropoda are not sister taxa: A phylogenetic analysis of spiralian metazoan morphology. *Syst. Biol.* 41:305–30.

Erwin, D. H. 1991. Metazoan phylogeny and the Cambrian radiation. *TREE* 6:131–34.

Halanych, K. M., J. Bacheller, A. M. Aguinaldo, S. Liva, D. M. Hillis, and J. A. Lake. 1995. 18S rDNA evidence that the lophophorates are protostome animals. *Science* 267:1641–43.

Morris, S. C. 1993. The fossil record and the early evolution of the Metazoa. *Nature* 361:219–25.

Shinn, G. 1994. Epithelial origin of mesodermal structures in arrowworms (Phylum Chaetognatha). *Amer. Zool.* 34:523–32.

Turbeville, J. M., et al. 1992. Phylogenetic position of phylum Nemertini, inferred from 18S rRNA sequences: Molecular data as a test of morphological character homology. *Molec. Biol. Evol.* 9:235–49.

Wada, H., and N. Satoh. 1994. Details of the evolutionary history from invertebrates to vertebrates, as deduced from the sequences of 18S rDNA. *Proc. Nat. Acad. Sci. USA* 91:1801–4.

Willmer, P. 1990. *Invertebrate Relationships.* New York: Cambridge University Press.

Many **new figures have been added,** drawn largely from the recent research literature, and **more than 30 figures from the previous edition have been redrawn,** all thanks to the keen eye and willing pen of Art Ciccone. Most of the minor arthropod groups (e.g., Remipedia, Mystacocarida, and Cephalocarida) are now illustrated. I also have added two new invertebrate riddles, bringing the total to four; readers are invited to submit their answers (and their own riddles) to me directly.

As always, I welcome constructive criticism from all readers.

JPecheni @ emerald.tufts.edu

Acknowledgments

It is a great pleasure to thank the many colleagues who have helped to make this a better text.

Charles Biernbaum
College of Charleston

Amelie H. Scheltema
Woods Hole Oceanographic Institution

Ron Campbell
University of Massachusetts at Dartmouth

John O. Corliss

Bernard Fried
Lafayette College

Steve Palumbi
University of Hawaii

Nancy Milburn
Tufts University

Robert Robertson
The Academy of Natural Sciences

J. Malcolm Shick
University of Maine

Steve Vogel
Duke University

J. Evan Ward
Salisbury State University

Edward O. Wilson
Harvard University

Robert Woollacott
Harvard University

William J. Pohley
Franklin College of Indiana

Audrey Gabel
Black Hills State University

William W. Kirby-Smith
Duke University

Robert Knowlton
George Washington University

Ben Foote
Kent State University

Other colleagues answered specific queries by telephone and e-mail, and many more contributed their expertise to previous editions. I am grateful to all; their devotion to, enthusiasm for, and knowledge about invertebrate biology is both inspiring and humbling. I also wish to thank all who allowed me to reproduce their graphs and photographs and those who commented on my new *Research Focus* sections based on their original papers. Lastly, my favorite neighbor and colleague, Nancy Milburn (Tufts University), kept me going through thick and thin with friendship, good sense, and good humor.

Introduction

Much fascinating and important research has been conducted and continues to be conducted using invertebrates. Many diseases of humans and of the animals and plants upon which we depend are caused by invertebrates, either directly or indirectly, and invertebrates are near the base of most food webs in all habitats. Much of what we presently know about the control of gene expression, mitosis, and meiosis; programmed cell death and the mechanisms of pattern formation during embryonic development; the transmission of nerve impulses; the biochemical basis of learning and memory; and the biology of vision come from studies on various invertebrate species. Much of what we know about the mechanisms by which genetic diversity originates, is maintained, and is transmitted to succeeding generations comes to us through the study of invertebrates, as do many basic principles of animal behavior, physiology, ecology, and evolution.

In addition, present research on invertebrates is helping to unravel the story of how immune recognition systems evolved and how they work. Interest in certain invertebrates as biological agents for controlling various agricultural pests and as sources of unique chemicals of potential biomedical importance also is increasing. Some of the substances isolated from marine sponges, for example, promise to be potent antitumor agents, and others isolated from certain spiders and venomous snails are providing neurobiologists with highly specific chemical probes for studying key aspects of nerve and muscle function, such as how ion channels are opened and closed. Still other substances derived from invertebrates show considerable promise as instant adhesives (glues produced by onychophorans and barnacles and by some spider and bivalve species, for example) and anticorrosion agents (e.g., barnacle cements). Invertebrates also have become widely used to evaluate and monitor pollutant stress in aquatic environments, and the rapid loss of invertebrate species from both terrestrial and aquatic habitats is gaining increasing attention in biodiversity studies.

This book, together with lectures and laboratory sessions, opens the door to the great and growing literature of invertebrate zoology.

1

Environmental Considerations

The organisms considered in this book are grouped into 45 phyla. The members of almost half of these phyla are entirely marine, and the members of the remaining phyla are found primarily in marine and, to a lesser extent, freshwater habitats. Excluding the arthropods, invertebrates have generally been far less successful in invading terrestrial environments. Even those invertebrate species that are terrestrial as adults often have aquatic developmental stages. It is worthwhile, therefore, to consider some of the physical properties of water—both fresh and salt—to discover why so many species are aquatic for all or part of their lives. The physical properties of salt water, freshwater, and air play major roles in determining the structural, physiological, and behavioral characteristics displayed by animals living in various habitats.

Air is dry, whereas water is wet. As trivial as this statement may seem, the repercussions with respect to morphology, respiratory physiology, nitrogen metabolism, and reproductive biology are tremendous, as seen in Table 1.1.

Gases diffuse only across moist surfaces. Since aquatic organisms are in no danger of drying out, gas exchange can be accomplished across the general body surface. Thus, the body walls of aquatic invertebrates are generally thin and water permeable, and any specialized respiratory structures that exist may be external and in direct contact with the surrounding medium. Gills, which can be structurally quite complex, are simply vascularized extensions of the outer body wall. These extensions increase the surface area available for gas exchange and, if they are especially thin walled, may also increase the efficiency of respiration (measured as the volume of gas exchanged per unit time per unit area).

In contrast to the minimal complexity required for aquatic respiratory systems, terrestrial organisms must cope with potential desiccation (dehydration). Terrestrial species relying on simple diffusion of gases through unspecialized body surfaces must have some means of maintaining a moist outer body surface, as by the secre-

Table 1.1   Summary of the Different Lifestyles Possible in the Two Major Environments, Aquatic and Terrestrial, As They Reflect Differences in the Physical Properties of Water and Air

Property	Water	Air
Humidity	High: Exposed respiratory surfaces; external fertilization*; external development; excretion of ammonia	Low: Internalized respiratory surfaces; internal fertilization; protected development; excretion of urea and uric acid
Density	High: Rigid skeletal supports unnecessary; filter-feeding lifestyle possible; external fertilization; dispersing developmental stages*	Low: Rigid skeletal supports necessary; must move to find food; internal fertilization; sedentary developmental stages
Compressibility	Low: Transmits pressure changes uniformly and effectively	High: Less effective at transmitting pressure changes
Specific heat	High: Temperature stability	Low: Wide fluctuations in ambient temperature
Oxygen solubility	Low: 5–6 ml O_2/liter of water	High: 210 ml O_2/liter of air
Viscosity	High: Organisms sink slowly; greater frictional resistance to movement	Low: Faster rates of falling; less frictional resistance to movement
Rate of oxygen diffusion	Low: Animal must move (or must move water) for gas exchange	High: (about 10,000 times higher than in water)
Nutrient content	High: Salts and nutrients available through absorption directly from water for all life stages*; adults may make minimal nutrient investment per egg*	Low: No nutrients available via direct absorption from air; adults supply eggs with all nutrients and salts needed for development
Light-extinction coefficient	High: Animals may be far removed from sites of surface-water primary production	Low: Animals never far from sites of primary production

*Signifies features that are especially characteristic of marine invertebrates and uncommon among freshwater invertebrates.

tion of mucus in earthworms. Truly terrestrial invertebrates generally have a water-impermeable outer body covering that prevents rapid dehydration. Gas exchange in such species must be accomplished through specialized, internal respiratory structures.

The union of sperm and egg, and the subsequent development of a zygote, can be achieved far more simply by aquatic invertebrates than by terrestrial species. Marine organisms, in particular, may shed sperm and eggs freely into the environment. Because the gametes and embryos of marine species are not subject to dehydration or to osmotic stress, fertilization and development can be completed entirely in the water. Fertilization in the terrestrial environment, on the other hand, must be internal to avoid dehydration of gametes; terrestrial invertebrates therefore require more complex reproductive systems

than do their marine counterparts. Successful fertilization of terrestrial eggs often involves complex reproductive behaviors as well.

Ammonia is the basic end product of amino acid metabolism in all organisms, regardless of habitat. Ammonia is usually very toxic, largely through its effects on cellular respiration. Even a small accumulation of ammonia in the tissues and blood is detrimental to individuals of most species. However, few terrestrial organisms can afford the luxury of constantly eliminating ammonia as it is produced, since the water required to flush out the ammonia is in short supply. As an adaptation to life on land, terrestrial organisms usually incorporate ammonia into less toxic compounds (urea and uric acid), which can then be excreted in a smaller amount of water. This detoxification of ammonia requires additional biochemical pathways and an increased expenditure of energy. Aquatic invertebrates, on the other hand, can simply use the surrounding water to dilute away metabolic ammonia as it is produced. Moreover, because water is wet, ammonia excretion may occur by simple diffusion across the general body surface of many aquatic invertebrates. In contrast, all terrestrial animals require complex excretory systems.

Water is a remarkably versatile solvent. The benefits to aquatic invertebrates are both direct and indirect. First of all, aquatic animals can potentially take up dissolved nutrients (including amino acids, carbohydrates, and salts) directly from the surrounding water by diffusion or by active uptake. In particular, dissolved salts and organic water-soluble nutrients may be taken up directly from water by developing embryos and larvae. Embryos of terrestrial organisms must be supplied (by their parents) with all food and salts needed for development and must be protected from desiccation as well. Second, as an indirect benefit to aquatic invertebrates, suspension in a nutrient-containing, wet medium permits primary producers to take the form of small (typically less than 25 μm [micrometers]), suspended, single-celled organisms (**phytoplankton**); roots are not mandatory. Phytoplankton cells can attain high concentrations in water, and can easily be harvested and ingested by many suspension-feeding aquatic herbivores; including the developmental stages of many invertebrate species.

Water is far denser than air, a fact that has profound consequences for invertebrates. For example, a rigid skeletal support system is not required in water, since the medium itself is supportive. For the same reason, animals often move with greater efficiency in water than in air, expending less energy to progress a given distance. Indeed, many aquatic invertebrate species expend virtually no energy at all for movement—they simply don't move. How do such animals feed without the ability to move? Because water is wet and dense, microscopic free-floating "plants" and animals (phytoplankton and **zooplankton,** respectively) live in suspension; this enables many other aquatic animals to make their livings "sitting down," capturing food particles directly from the medium as it flows past the stationary animal. Often, some energy must be expended to move water past the animal's feeding structures, but the animal need not use energy in a search for food. Such a suspension-feeding existence, quite commonly encountered in aquatic environments, seems to have been exploited only by web-building spiders in the terrestrial habitat. Potential food particles simply do not occur in high concentrations in the dry, unsupportive air.

External fertilization and external development of embryos, so commonly encountered among marine invertebrates, are made possible as much by water's high density as by its wetness; the water supports both sperm and eggs and the embryo itself as it develops. In many groups of marine invertebrates, external fertilization and/or external larval development is the rule rather than the exception. Because little energy may be required to remain afloat in the aquatic medium, developmental stages (e.g., embryos, larvae) of aquatic invertebrates often serve as the dispersal stages for sedentary adults—exactly the opposite of the situation encountered among most terrestrial animals.

One additional advantage of water as a biological environment is its relatively high temperature stability with respect to air. Water has a high specific heat; that is, the number of calories required to heat 1 g (gram) of water 1°C is considerably greater than that required to raise the temperature of 1 g of most other substances by the same 1°C. Because of its high specific heat, water is slow to cool and slow to heat up; water temperature is relatively insensitive to short-term fluctuations in air temperature. Over a 24-hour period, air temperatures at midlatitudes may vary by 20°C or more. For reasonably large volumes of water, local surface temperatures will probably not vary by more than 1°C to 2°C over the same time interval.

Differences in seasonal temperature fluctuations are even more striking. Near Cape Cod, Massachusetts, for example, local seawater temperature may vary between approximately 5°C during the winter and 20°C during the summer: a seasonal range of about 15°C. Air temperatures, on the other hand, fluctuate between approximately −25°C and 40°C: a seasonal range of 65°C. Even in small lakes and ponds, the annual range of water temperatures is much smaller than that of air temperatures in the same geographic area. Because the rates of all chemical reactions, including those associated with organismal metabolism, are altered by temperature, wide fluctuations in temperature (especially those occurring over short time intervals) are highly stressful to most invertebrates. Invertebrates living in thermally variable environments require biochemical, physiological, and/or behavioral adaptations not required by organisms living in more stable, aquatic habitats.

Life in water does pose some problems. Light is extinguished over a much shorter distance in water than in air, so most aquatic **primary production** (fixation of carbon from carbon dioxide into carbohydrates, generally by photosynthesizing plants, algae, and phytoplankton) is limited to the upper 20 m to 50 m (meters) or so. Moreover, water's oxygen-carrying capacity, volume for volume, is only about 2.5% that of air. An additional problem for aquatic organisms is that the time required for a given molecule to diffuse across a given distance in water is much, much greater than the time required for the same molecule to diffuse across the same distance in air. An organism sitting completely still in motionless water would have a severe gas-exchange problem once the fluid immediately in contact with the respiratory surface had given up all available oxygen (and/or had become saturated with carbon dioxide). On the other hand, even the slightest movement of the water surrounding an animal's respiratory surface enhances gas exchange significantly. **Sessile** (nonmotile) organisms living in areas of significant water-current velocity thus benefit in terms of gas exchange as well as nutrient replenishment. Sessile animals living in still water invariably have some means of creating water flow over their respiratory surfaces.

Potential difficulties are also created by the greater density and viscosity of water. Water is about 800 times more dense and about 50 times more viscous than air. **Viscosity** essentially measures the extent to which the molecules of a fluid stick to each other. In contrast, density, which I referred to earlier in discussing the *benefits* of life in water, is a measure of mass per unit volume. Because of water's greater density and viscosity, animals swimming through it or facing a current experience far more frictional resistance (called **drag**) than they would experience in air. For large animals moving quickly (or facing a fast-moving current), the greater drag is due primarily to the greater density of water, whereas small animals moving slowly (or facing a slow-moving current) are affected mainly by water's greater viscosity. Because viscosity increases much more dramatically in water than in air for any given decline in temperature, small, slow-moving aquatic organisms experience noticeably greater frictional resistance in swimming as temperature falls.

Indeed, small organisms—which really can't help but swim slowly because of their small size—live in a world dominated by viscous forces, a world in which **Reynolds numbers** (essentially a ratio of inertial to viscous forces) are very low. It is difficult for us to imagine what life is like in such a world, a world in which there is really no such thing as "gliding" to a stop; as soon as propulsion stops, the animal stops. Moreover, in a world of low Reynolds numbers, water tends to move primarily around rather than through bristly appendages; in such a world, rake-shaped objects behave much like solid paddles, so that they cannot readily filter food particles from the water. Clearly, animals operating at low Reynolds numbers are subjected to some physical selective pressures quite unlike those acting on larger, faster-moving organisms; even such basic biological functions as locomotion and suspension-feeding may require specialized physiological and behavioral adaptations, adaptations that often seem somewhat peculiar and counterintuitive.[1]

The fact that water is the so-called universal solvent creates another problem that should be particularly acute for aquatic invertebrates. Many of our industrial and agricultural waste products are water soluble, and we insert fantastic amounts of such pollutants into aquatic ecosystems each year. Aquatic animals must live in particularly intimate contact with these pollutants. Consider, for example, that the gas exchange surfaces of aquatic invertebrates are always in direct contact with the surrounding fluid. Consider also that many aquatic invertebrates are small, so that the surface area across which pollutants can diffuse is high relative to the animal's body volume. Free-living embryonic and larval stages, so common in aquatic invertebrate life cycles, would seem especially vulnerable to pollutant insult, partly because of their high surface area:volume ratios and partly because they are undergoing such complex and critical developmental processes. Indeed, for any given toxicant, developmental stages typically suffer adverse effects at only one-tenth to one-hundredth the concentration required to affect adults of the same species to the same degree.

Organisms living in freshwater face several difficulties unique to the freshwater environment. Marine invertebrates are approximately in **osmotic equilibrium** with the medium in which they live; that is, the concentration of solutes in their body fluids matches that of the surrounding seawater. In contrast, the internal body fluids of freshwater organisms are always higher in osmotic concentration than is the surrounding medium; that is, freshwater organisms are **hyperosmotic** to their surroundings, and water tends to diffuse inward along the osmotic concentration gradient. Some freshwater animals have reduced surface permeability to water, reducing the magnitude of this inflow. Complete impermeability to water is not possible, however, because respiratory surfaces must remain permeable for gas exchange to occur. Thus, all freshwater animals must be capable of constantly expelling large volumes of incoming fresh water. Also, because salts are relatively rare in the freshwater medium (by definition of freshwater), most of the salts necessary for embryonic development must be supplied to the egg by the mother. By contrast, all salts required for the differentiation and growth of marine embryos are readily available in the surrounding medium.

The relative paucity of salts in freshwater has additional ramifications for animals living in it. Freshwater organisms, which must constantly expel incoming water, often possess sophisticated physiological mechanisms for reclaiming precious salts from the urine before the urine leaves the body; they also must possess mechanisms for making good any salt loss that does occur.

Finally, the pH of salt water is far less variable, both spatially and temporally, than is the pH of freshwater. The pH of salt water is maintained near 8.1 by the tremendous buffering capacity of the bicarbonate ion. Most freshwater environments lack such a high buffering

---

1. See *Topic for Further Discussion and Investigation.*

capacity, and the pH of the water is therefore far more sensitive to local fluctuation of acid and base content.

It should be noted that most bodies of freshwater are ultimately ephemeral, with smaller ponds and lakes being subject to drying up at yearly or even more frequent intervals. Most marine invertebrates are not faced with such a high degree of habitat unreliability.

From such considerations of the properties of air, salt water, and freshwater, it is easy to understand why life must have originated in the ocean. The specialized physiological and/or morphological adaptations essential for existence on land or in freshwater are not required for the relatively simple and generally less stressful existence possible in the marine environment. Once life arose, various **preadaptations** could eventually evolve that made a transition from saltwater environments to other habitats possible. Such preadaptations for terrestrial and freshwater life apparently arose rarely in many groups of animals and not at all in others. Not surprisingly, most phyla are still best represented in the ocean, both in terms of species numbers and in terms of the diversity of body plans and lifestyles.

Topic for Further Discussion and Investigation

In what ways are small invertebrates adapted to life at low Reynolds numbers?

Vogel, S. 1994. *Life in Moving Fluids,* 2d ed. Princeton, N.J.: Princeton University Press.

2

Invertebrate Classification

"Classifications are theories about the basic natural order, not dull categories compiled only to avoid chaos." *Gould, J. S. 1989. Wonderful Life. W. W. Norton & Company.*

Introduction

At least five million animal species (of which only about one million have been described and named so far), presently exist on Earth, and probably several hundred million other species were here previously but are now extinct. At one time, presumably, there were no animal species. The marvelous variety of animal life-forms seen today and in the fossil record evolved gradually, beginning over three billion years ago; the Earth itself is over 4.5 billion years old.

Fossils of the earliest known unicellular eukaryotes (protists—see Chapter 3) are about 1.8 to 1.9 billion years old, but the oldest known fossils of multicellular animals (called *metazoans*) are a relatively young 580 to 600 million years old; multicellular life seems to have taken quite a long time to evolve from its single-celled ancestral forms. Moreover, none of those first recorded multicellular animals from 580 or so million years ago had shells, bones, or other hard parts, and most were quite small, rarely more than a few millimeters long. The first sizable metazoans with hard parts appear abruptly in the Burgess shale of British Columbia, formed some 530 million years ago in the mid-Cambrian period and discovered in 1909; all of today's major animal groups (phyla) are already represented among those fossils. This amazingly rapid proliferation and diversification of multicellular life about 600 million years ago leaves us no clues about how various groups of animals are related to each other. Studying the fossil record can tell us something about evolution *since* the Cambrian explosion, but not about where these animals came from. However, if we make the very reasonable assumption that animals

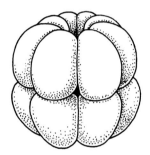

have ancestral forms in common we can *infer* evolutionary relationships, with varying degrees of certainty, based on morphological, developmental, physiological, biochemical, and genetic similarities and differences among animal groups.

Before we can consider the evolutionary interrelationships among different groups of organisms, we must first sort the millions of animal species into categories, which can be done only after determining the degrees of similarity and difference that will define each category. It is important to keep in mind that all classification schemes are, at least in part, artificial attempts to impose order for convenience. Humans decide what characteristics will separate one family from another within the same order, what characteristics will separate one order from another within the same class, and so forth. As we will see throughout this book, many organisms do not fit cleanly into any one group; it is relatively simple to decide upon the categories to be used but often far more difficult to determine the category to which a given organism belongs. Once the organisms are assigned to reasonable taxonomic categories, it becomes possible to consider the evolutionary relationships among and within those categories. In this chapter, we will consider some of the schemes that have been developed to sort animals into groups.

Classification by Cell Number, Embryology, and Body Symmetry

Invertebrates have been categorized in several ways. The most basic division is based upon the number of cells composing any given individual. True animals are multicellular, diploid organisms that develop from a blastula; these organisms are referred to collectively as the Metazoa, or as **metazoans.** Other invertebrates are considered either **unicellular** (single celled) or **acellular** (without cells)—a distinction discussed further in Chapter 3—and do not develop from anything resembling a metazoan embryo. As we will see in the next several chapters, the point at which an association of cells can be viewed as composing a multicellular organism is not always clear-cut.

Animals may be classified according to their general body form. Most metazoans show one of two types of body symmetry (Fig. 2.1a,b). Animals like ourselves are **bilaterally symmetrical,** possessing right and left sides that are approximate mirror images of each other. Bilat-

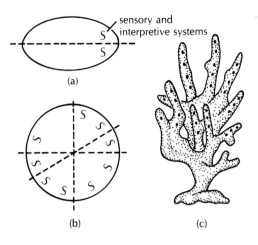

Figure 2.1

Various types of body symmetry. (a) Bilateral symmetry. (b) Radial symmetry. (c) Asymmetrical body plan of a marine sponge.

eral symmetry is highly correlated with **cephalization,** which is the concentration of nervous and sensory tissues and organs at one end of an animal, resulting in distinct anterior and posterior ends. For an animal that shows cephalization, two mirror images can be produced only when a slice is made parallel to the animal's long (anterior-posterior) axis, with the cut passing down the midline. Any cut perpendicular to this midline, even when passing through the center of the animal, creates two dissimilar pieces. This is not so for a **radially symmetrical** organism. Such an animal can be divided into two reasonably equal halves by any cut that passes through the center of the organism. Thus, most animals belong to either the Radiata or the Bilateria, groupings that are often convenient for discussion but that have no formal taxonomic significance. Invertebrates that are asymmetrical— that is, that have no ordered pattern to their gross morphology—are encountered rarely (Fig. 2.1c).

Once again, what seems to be straightforward on the surface is never quite so simple when dealing with actual animals. Many species whose external appearances are the epitome of uncontroversial, radial symmetry have asymmetrical internal anatomies. For such organisms, all knife cuts are thus created equal only superficially. Perhaps it would have been better to group animals based on degree of cephalization rather than on the basis of body symmetry. I bow, however, to historical precedent.

Classification by Developmental Pattern

Multicellular invertebrates generally can be classified into two groups based upon the number of distinguishable germ layers formed during embryogenesis. **Germ layers** may be defined as groups of cells behaving as a unit during the early stages of embryonic development and giving rise to distinctly different tissue and/or organ systems in the adult. In **diploblastic** animals (*diplo* = G: double), only two distinct germ layers form during or following the movement of cells into the interior of the embryo. The outermost layer of cells is called the **ectoderm** (*ecto* = G: outer; *derm* = G: skin), and the innermost layer of cells is called the **endoderm** (*endo* = G: inner). Members of only two phyla are generally considered to be diploblastic (Fig. 2.8). Most metazoans are **triploblastic** (*triplo* = G: triple). During the ontogeny of triploblastic animals, cells of either the ectoderm or, more usually, the endoderm give rise to a third germ layer, the **mesoderm** (*meso* = G: middle). The mesodermal layer of tissue always lies between the outer ectodermal tissue and the inner endodermal tissue.

The absence of a distinct, embryonic, third tissue layer does not imply that the adult of a diploblastic species will lack the derivatives of this layer that are found in adults of a triploblastic species. Muscular elements, for example, derive from the mesodermal layer in triploblastic animals. Diploblastic adults also have musculature, despite the absence of a morphologically or behaviorally distinct group of cells that can be termed mesoderm in the early embryo.

Triploblastic animals can be further classified into three basic plans of body construction, based on whether an organism has an internal body cavity independent of the gut and on how this cavity, if present, is formed during embryogenesis. The most **primitive** triploblastic animals (i.e., those believed to resemble most closely the ancestral metazoans) are generally believed to lack an internal body cavity. These are the **acoelomates** (*a* = G: without; *coelom* = G: a hollow space). Characteristically, the area lying between the outer body wall and the gut of acoelomates is solid, being occupied by mesoderm (Fig. 2.2a).

In a second group of animals, this area between the outer body wall musculature and the endoderm of the gut is a fluid-filled cavity (Fig. 2.2b); in some species, this cavity is derived from the **blastocoel,** an internal space that

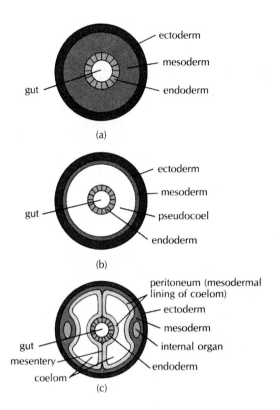

Figure 2.2

(a) Diagrammatic cross section through the body of an acoelomate. The space between the gut and the outer body wall musculature is completely filled with tissue derived from embryonic mesoderm. (b) Cross section through the body of a pseudocoelomate. The gut derives entirely from endoderm and is therefore not lined with mesoderm. (c) Cross section through the body of a coelomate. The entire coelomic space is bordered by tissue derived from embryonic mesoderm.

develops in the embryo prior to gastrulation. Selective pressures that might account for the evolution of such a body cavity from an acoelomate ancestral form are not difficult to imagine. For example, the gut is now somewhat independent from muscular, locomotory activities of the body wall. Also, the animal now has internal space into which can bulge digestive organs, gonads, and developing embryos, and an internal fluid that can serve to distribute oxygen, nutrients, and hormones or neurosecretory substances throughout the body, facilitating the evolution of larger body sizes. Perhaps most significantly, the animal gains the mechanical advantages associated with a more effective locomotory system, to be discussed in Chapter 9.

This fluid-filled cavity sometimes derived from the embryonic blastocoel is termed a **pseudocoel,** and the organism housing it is said to be **pseudocoelomate.** The name is a bit misleading. The *pseudo* is not intended to disparage the *coel;* the body cavity is genuine. The *pseudo* prefix merely draws attention to the fact that this body cavity is not to be confused with a true coelom, which, as we will see, is a precisely defined internal cavity formed through one of several quite different processes and always lined completely with tissue derived from embryonic mesoderm.

This brings us to the third group of triploblastic animals, those with a true **coelom:** an internal, fluid-filled body cavity lying between the gut and the outer body wall musculature and lined with tissue derived from embryonic mesoderm. The animals possessing such a body cavity are **coelomates** (or **eucoelomates;** *eu* = G: real, true). Coelom formation may occur by either of two quite dissimilar mechanisms; mode of coelom formation traditionally is used to assign coelomates to one of two major subgroups: protostomes or deuterostomes. In the **protostomes,** coelom formation occurs by gradual enlargement of a split in the mesoderm (Fig. 2.3). This process is termed **schizocoely** (*schizo* = G: split). In the **deuterostomes,** on the other hand, the coelom typically forms through evagination of the archenteron into the blastocoel of the embryo (Figs. 2.3 and 2.4). Because the coelom of deuterostomes is formed from a part of what eventually becomes the gut, coelom formation in this group of animals is termed **enterocoely** (*entero* = G: gut).

Whether coelom formation occurs by schizocoely or enterocoely, the end result is similar. The organism is left with a fluid-filled internal body cavity lying between the gut and the outer body wall musculature, and unlike the cavity of pseudocoelomates, this cavity is lined entirely by a mesodermally derived epithelium. The fact that internal cavities develop by any of three distinctly different mechanisms (enterocoely, schizocoely, or persistence of embryonic blastocoel) suggests that such cavities have been independently evolved at least three times. If so, the selective pressures favoring the evolution of internal body cavities clearly must have been substantial.

To summarize, triploblastic animals can be divided into acoelomates, pseudocoelomates, and coelomates, depending on whether they possess an internal, fluid-filled body cavity and on whether this body cavity is lined entirely by mesodermally derived tissue. Also, coelomates can be further classified into protostomes and deuterostomes, depending, in part, on how the coelom is created during development. But mode of coelom formation is only one of several characteristics distinguishing protostomes from deuterostomes.

The terms *protostome* and *deuterostome* were actually coined to reflect differences in the origin of the mouth (*stoma* = G: mouth). Among the protostomes, the mouth is formed from or near the blastopore (the opening from the outside into the archenteron)—hence the term *protostome,* meaning "first mouth", since the mouth forms from the first opening that appears during embryonic development. Among the deuterostomes, the mouth never develops from the blastopore. Indeed, the blastopore often gives rise to the anus, the mouth always forming as a second, novel opening elsewhere on the embryo—hence the term *deuterostome,* meaning "second mouth."

What other characteristics distinguish protostomes from deuterostomes? In addition to differing in the mode of coelom formation and in the embryological origin of the mouth, protostomes and deuterostomes differ with respect to the orientation of the spindle axes of the cells during cleavage, the point in development at which cell fates become irrevocably fixed, and how the mesoderm originates.

Cleavage is usually either radial or spiral, depending on the orientation of the mitotic spindles relative to the egg axis. Generally, yolk is asymmetrically distributed within an egg, and the nucleus occurs in, or moves to, the region of lowest yolk density. This is the **animal pole,** and it is here that the polar bodies are given off during meiosis. The opposite end of the egg is termed the **vegetal** (not vegetable!) **pole.**

In **radial cleavage** (deuterostomes), the spindles of a given cell, and thus the cleavage planes, are oriented either parallel or perpendicular to the animal-vegetal axis. Thus, daughter cells derived from a division in which the cleavage plane is parallel to the animal-vegetal axis end up lying in the same plane as the original mother cell (Figs. 2.5a,b). The two daughter cells resulting from a division perpendicular to the animal-vegetal axis come to lie directly one atop the other, with the center of the upper cell lying directly over the center of the underlying cell (Fig. 2.5c–f and 2.6).

In contrast, the spindle axes of cells undergoing **spiral cleavage** (protostomes) are oriented (after the first two cleavages) at 45° angles to the animal-vegetal axis (Fig. 2.5j–k). Moreover, the division line may not pass through the center of the dividing cell. As a result, by the eight-cell stage we often see a group of smaller cells (**micromeres**) lying in the spaces between the underlying larger cells (**macromeres**) (Fig. 2.5k–m). Cell division continues in this fashion, with the cleavage planes always oblique to the polar axis of the embryo.

Cleaving embryos of protostomes and deuterostomes also differ with respect to when their cells become fully committed to a particular fate. Among

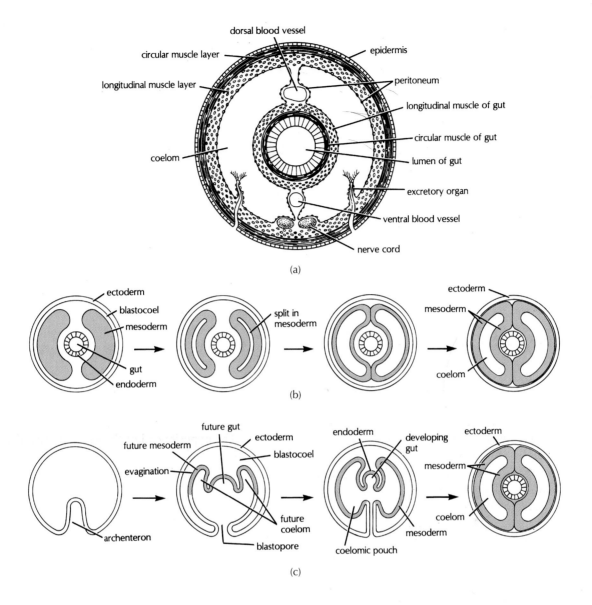

Figure 2.3
(a) Detailed cross section through the body of a coelomate. The tissues bordering the coelomic space include the musculature of the gut; the mesenteries, which suspend the various organs in the coelom; and the peritoneum, which lines the coelomic cavity. (b) Coelom formation by schizocoely—that is, by an actual split, or schism, in the mesodermal tissue. (c) Coelom formation by enterocoely, in which the archenteron evaginates into the embryonic blastocoel. *(a) From Hyman,* The Invertebrates, *Vol. III. Copyright © 1951 McGraw-Hill Book Company, New York. Reprinted by permission.*

deuterostomes, one can separate the cells of a two-cell or four-cell embryo, and each cell will develop into a small but complete and fully functional animal. Thus, deuterostomes are said to show **indeterminate** (or **regulative**) cleavage; each cell retains—sometimes as late as the eight-cell stage—the capacity to differentiate the entire organism if that cell loses contact with its associates. Among protostomes, in contrast, the developmental po-

tential of each cell is irrevocably determined at the first cleavage; separate the blastomeres of a two-celled protostome embryo and each cell will give rise only to short-lived, malformed monsters. Protostome cleavage is therefore said to be **determinate** or **mosaic.** Protostomes can never produce identical twins, which in deuterostomes arise from the natural separation of blastomeres during early cleavage.

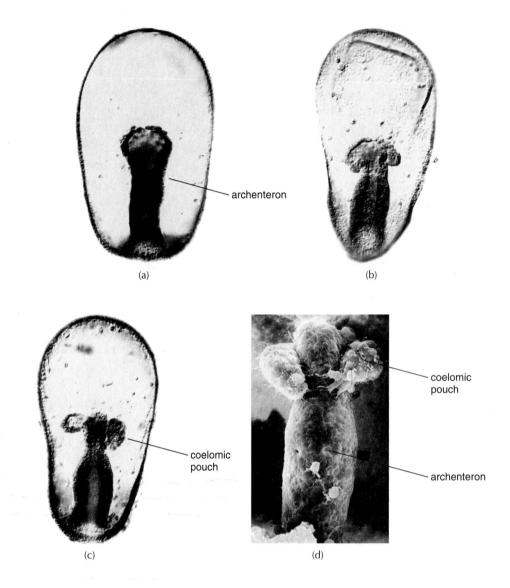

(a)

(b)

archenteron

coelomic
pouch

(c)

coelomic
pouch

archenteron

(d)

Figure 2.4
(a–c) Enterocoely in a sea star, showing the coelomic pouches forming from the sides of the archenteron and splitting off. (d) The coelomic pouches are clearly shown in this scanning electron micrograph. The pouches gradually enlarge to form the coelom.

(a-d) *Courtesy of B.J. Crawford, from Crawford and Chia, 1978* Journal of Morphology. *157:99.*

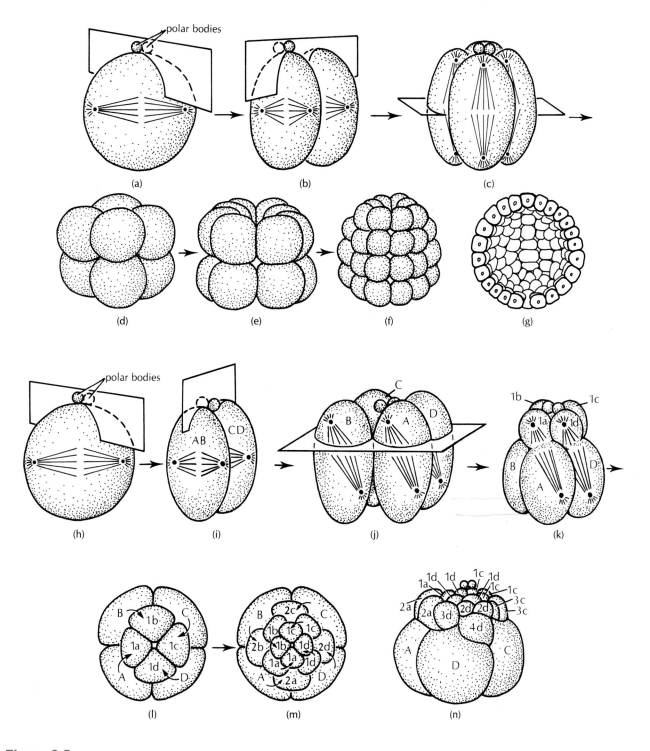

Figure 2.5

(a–g) Radial cleavage, as seen in the sea cucumber *Synapta digitata*. In (g), part of the embryo has been removed to reveal the blastocoel. (h–n) Spiral cleavage. The first two cleavages (h,i) are identical with those seen in radially cleaving embryos, forming four large blastomeres (j). The cleavage plane during the next cleavage, however, is oblique to the animal-vegetal axis of the embryo and does not pass through the center of a given cell (k). This produces a ring of smaller cells (micromeres) lying between the underlying larger cells (macromeres), as shown in (l). The lettering system illustrated was devised by embryologist E. B. Wilson in the late 1800s to make possible a discussion of particular cell origins and fates. The number preceding a letter indicates the cleavage in which a particular micromere was formed. Capital letters refer to macromeres, while lowercase letters refer to micromeres. With each subsequent cleavage, the macromeres divide to form one daughter macromere and one daughter micromere, while the micromeres divide to form two daughter micromeres. The 32-cell embryo of the marine snail *Crepidula fornicata* is shown in (n). Note the 4d cell, from which all of the mesodermal tissue of protostomes will ultimately derive.   *After Richards.*

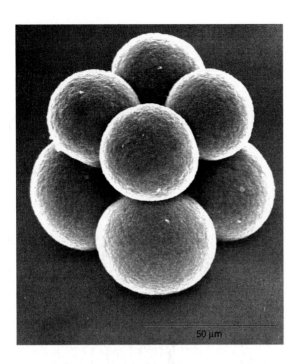

Figure 2.6
Scanning electron micrograph of radial cleavage (8-cell stage) in the cephalochordate *Branchiostoma belcheri* (chapter 24). *Courtesy of N. Kajita. From Hirakow, R. and N. Kajita. 1990. J. Morphol. 203:331–44.*

A further difference between the two groups of coelomates concerns the source of mesoderm. Among protostomes, all mesodermal tissue derives from a single cell of the 64-cell embryo. This is not true of deuterostomes.

During their first one or two cleavages, some protostomes form **polar lobes** (not to be confused with polar bodies, which arise during egg maturation). A polar lobe is a conspicuous bulge of cytoplasm that forms prior to cell division. The lobe contains no nuclear material. After cell division is complete, the bulge is resorbed into the single daughter cell to which it is still attached (Fig. 2.7). Although the functional significance of this phenomenon for the embryo is still not fully understood, polar-lobe formation has provided developmental biologists with an intriguing system through which to study the role of cytoplasmic factors in determining cell fate. In the basic experiment, the fully formed polar lobe is detached from an embryo, and the development of the lobeless embryo is subsequently monitored. Polar-lobe formation is characteristic of only some protostome species (some annelids and some molluscs) but is never encountered among deuterostomes.

The developmental features distinguishing protostomes from deuterostomes are summarized in Table 2.1. Making the distinction between protostomes and deuterostomes can sometimes be difficult. Flatworms (phylum Platyhelminthes) and ribbon worms (phylum Rhynchocoela) are often grouped with the protostomes (under the general term *Spiralia*) even though these worms lack a coelom or even a pseudocoelom—they are acoelomates. They are nevertheless often lumped with the protostomes because in most other respects, flatworms and ribbon worms develop as perfectly good protostomes, showing, for example, spiral cleavage and development of the mouth from the blastopore. For our purposes, it seems best to view the flatworms as being neither protostomes nor deuterostomes but to simply recognize that their development can be characterized as being protostome-like. In this text, the terms *protostome* and *deuterostome* are used only in reference to coelomate animals.

Classification schemes in biology always seem tidy. Unfortunately, biologists often find it far simpler to construct a logical classification system than to neatly distribute animals within it. I have already hinted at some of the difficulties. Many animals show other developmental departures from the characteristics previously outlined. In particular, some triploblastic invertebrate species show characteristics that are part deuterostome and part protostome, while others display certain features that are characteristic of neither group. Because all animal groups have had ancestral forms in common at some time during their evolution, because evolution is an ongoing process, and because embryos as well as adults are subject to the modifying forces of natural selection, some species are likely to have developmental characteristics that fall outside the mainstream. The affinities of the "misfits" must remain uncertain, at least until additional embryological and molecular studies can be completed. In any event, those species exhibiting entirely or primarily protostome characteristics are most likely to be more closely related to each other than to those species exhibiting purely deuterostome characteristics.

Classification by Evolutionary Relationship

Probably the most familiar classification scheme is the taxonomic framework established several hundred years ago (1758) by Carolus Linnaeus. The system is hierarchical; that is, one taxon contains groups of lesser taxa, which in turn contain still more groups of lesser taxa, and so on:

Kingdom
  Phylum
    Class
      Order
        Family
          Genus
            Species

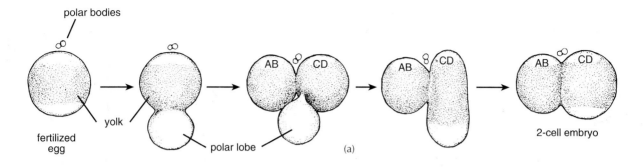

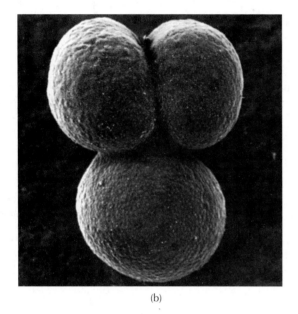

(b)

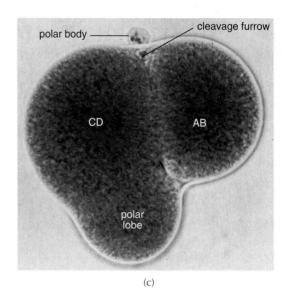

(c)

Figure 2.7

(a) Polar lobe formation during the development of a protostome. Following resorption of the polar lobe, the two blastomeres are clearly unequal in size because the cytoplasm held within the polar lobe does not participate directly in the process of cleavage. (b) Scanning electron micrograph of the two-celled embryo of the marine snail *Nassarius reticulatus.* The polar lobe (at bottom) is nearly equal in size to the blastomere into which it will be resorbed. (c) Polar-lobe formation during first cleavage in the blue mussel, *Mytillus edulis,* as seen with light microscopy. The newly formed cleavage furrow is visible between the AB and CD blastomeres. The polar lobe is clearly affiliated with only one of the daughter cells (the CD blastomere). A polar body (a product of meiotic division prior to cleavage) can be seen at the animal pole of the embryo.

(b) *Courtesy of M. R. Dohmen.* (c) *Carolina Biological Supply Company/Phototake.*

Table 2.1   Summary of the Basic Characteristics of Protostomous and Deuterostomous Coelomates

Developmental characteristic	Protostomes	Deuterostomes
Mouth origin	From blastopore	Never from blastopore
Coelom formation	Schizocoely	Enterocoely
Mesoderm origin	4d cell (Fig. 2.5)	Other
Cleavage pattern	Spiral, determinate	Radial, indeterminate
Polar-lobe formation	Present in some species	Not present in any species

The members of a given phylum show a high degree of morphological and developmental similarity and are presumed to be more closely related to each other than to the members of any other phylum. Indeed, all the members of any particular phylum are presumed to have evolved from a single ancestral form. The evolutionary implications are similar for the other taxonomic categories as well: All members of a given taxonomic group should be derived from a single ancestral form; such a group is said to be *monophyletic* (G: single-tribed). At present, there is considerable debate about whether each category should also include all descendants of that ancestor. A group that does not do so is said to be *paraphyletic.*

The category of **species** has additional biological significance. Theoretically, the members of one species are reproductively isolated from members of all other species. The species, therefore, forms a pool of genetic material that only members of that species have access to and that is isolated from the gene pool of all other species.

The scientific name of a species is binomial (has two parts): the **generic name** and the **specific name.** The generic and specific names (i.e., the **species name**) are usually italicized in print and underlined in writing. The generic name begins with a capital letter, but the specific name does not. For example, the proper scientific name for one of the common shallow-water marine snails found off Cape Cod, Massachusetts, is *Crepidula fornicata.* Related species are *Crepidula plana* and *Crepidula convexa.* Once the generic name is spelled out, it may be abbreviated when used subsequently. Thus, *Crepidula fornicata, C. plana,* and *C. convexa* are common shallow-water marine gastropods found near Woods Hole, Massachusetts. They all belong to the phylum Mollusca and are contained within the class Gastropoda, family Calyptraeidae. The family Calyptraeidae contains other genera besides *Crepidula;* the class Gastropoda contains other families besides the Calyptraeidae; and the phylum Mollusca contains other classes besides the Gastropoda. The taxonomic classification system is indeed hierarchical.

An additional name often follows the species name of an organism. This additional name is capitalized but is not italicized and may be contained in parentheses. This is the name of the person who first described the organism. A barnacle common along the coast of the southeastern United States, for example, is *Balanus amphitrite* Darwin, first described by Charles Darwin. Linnaeus's name is often abbreviated as L., since he is associated with the descriptions of so many species. If the organism was originally described as being in a different genus than the one in which it is currently placed, the describer's name is enclosed within parentheses. Thus, the snail *Ilyanassa obsoleta* (Say) was described by a man named Say, who originally assigned the species to another genus (the genus *Nassa*). This snail was later determined to be sufficiently dissimilar from other members of this genus to warrant its removal. Occasionally, a person's name is followed by a date, identifying the year in which the species was first described. For example, the shrimplike animal known as *Euphausia superba* Dana 1858 was first described by Dana in 1858, and it has remained in the genus *Euphausia* since it was originally named.

A list of all groups containing invertebrates is presented in Figure 2.8, showing where each phylum fits into the classification systems discussed so far in this section. The number following each listing gives the page on which the group is first discussed. The distribution of described species among the various animal phyla is summarized in Figure 2.9. Note that the percentage of species contained within our own phylum—the phylum Chordata—is quite small (only about 5% of all described species) and that this phylum contains invertebrates as well as vertebrates.

Ideally, a taxonomic classification scheme reflects degrees of phylogenetic relatedness; that is, all members of a given taxonomic group should be derived from a single ancestral form. Biologists have long made logical, reasoned guesses about the origins of various animal groups, based upon studies of developmental patterns, studies of morphological and biochemical characteristics, and careful examination of animals preserved in the fossil record. Comparative molecular analyses of protein structure and of DNA and ribosomal RNA sequences among species are altering some of these views substantially.[1] Ferreting out probable relationships is no easy task. Indeed, there is no universally accepted procedure for deducing evolutionary relationships; considerable disagreement on this point is well reflected in the appropriate literature of the past 25 to 35 years.[2] In part, the controversy concerns the relative importance of phenotypic similarities among taxa, phenotypic differences among taxa, and the degree to which one is willing to admit (and deal with the fact) that phenotype may be a very misleading indicator of underlying genetic similarities and differences. Through the process of **convergence,** distantly related animals may come to resemble each other rather closely. For example, the eye of an octopus (a cephalopod mollusc) is remarkably like that of a human, but these visual organs are believed to be analogues, not homologues, and not to indicate any close evolutionary relationship between vertebrates and molluscs. Which features indicate evolutionary closeness and which do not? Should we try to make this distinction? How can we know if we've decided correctly?

---

1. See *Topics for Further Discussion and Investigation,* no. 5, at the end of the chapter.

2. See *Topics for Further Discussion and Investigation,* nos. 1 and 2.

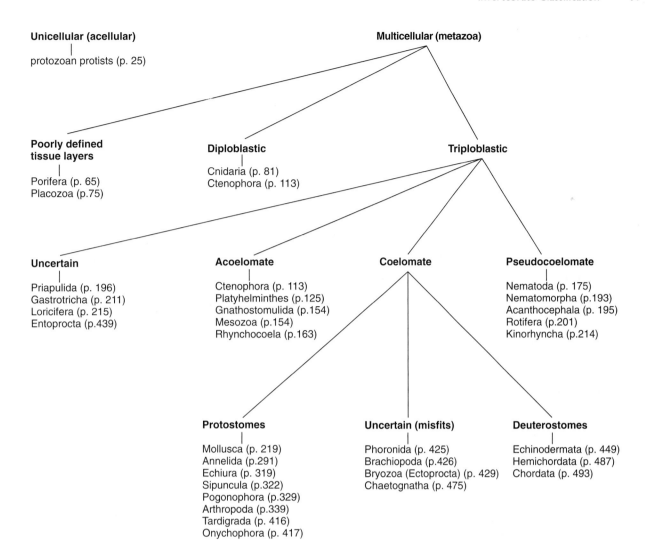

Figure 2.8
Arrangement of the major groups according to the factors discussed in this chapter. Note that the placement of some groups within this framework is uncertain at present.

Moreover, in the evolutionary process, structures sometimes become less complex rather than more complex. Suppose, for example, you discover a new species of wingless insect. How can you tell whether this species evolved before insect wings evolved or whether it instead descended from a winged ancestor and simply lost the wings over time? It is often very difficult to determine which of two character states is the original (*primitive*, or *plesiomorphic*) condition and which is the advanced (*derived*, or *apomorphic*) condition.

Until very recently, evolutionary relationships have been deduced entirely through anatomical and ultrastructural studies, with phenotypes serving as reflections of the underlying genotypes. More recently, particularly during the past 10 years or so, biochemical and molecular studies have allowed us to examine genotypic diver-

sity directly. Particularly remarkable are recent interspecific comparisons of nucleotide sequences of genes coding for ribosomal RNA (rRNA), comparisons made feasible through development of the polymerase chain reaction (PCR) in the mid-1980s. The PCR permits biologists to generate very quickly and inexpensively many copies of specific DNA sequences; several million copies of a single DNA molecule can be obtained in a few hours, producing sufficient material for analysis.

Although molecular studies often confirm the conclusions of prior workers, they also have produced some remarkable and surprising results, results that differ considerably from those of earlier, organismal studies. In some cases, there is considerable disagreement among workers about how the results of molecular studies should be interpreted and about the mathematical

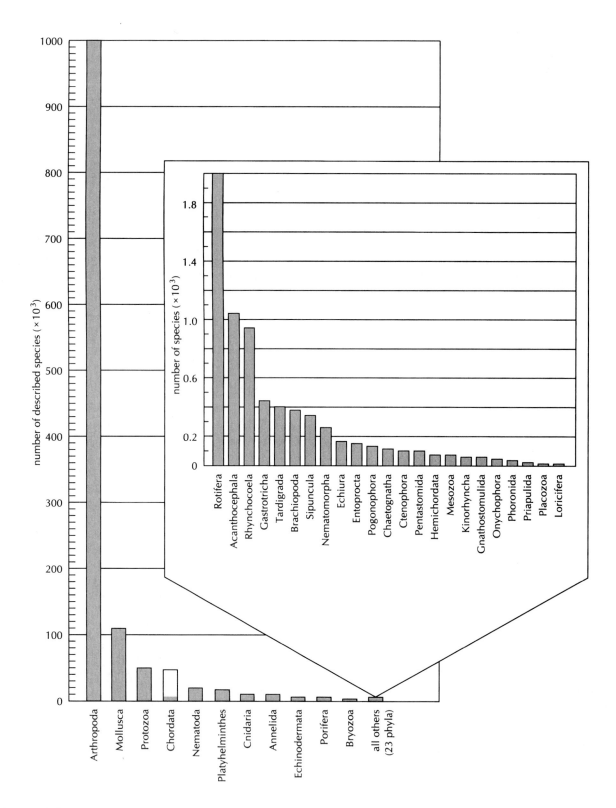

Figure 2.9
Graphic representation of the distribution of described species among the 34 major groups of invertebrates. In this text, protozoans are divided among more than one dozen phyla. Metazoan phyla containing fewer than 2,000 described species are presented in the inset. Note the different scale on the Y-axis of the inset. The open (unshaded) area of the bar labeled *Chordata* represents vertebrate species. All other species in all other phyla are invertebrates.

procedures that should be used to deduce evolutionary relationships from such data. But even before molecular biologists joined the fray, proposed phylogenetic relationships were controversial and a variety of phylogenetic trees have been proposed over the years. Seven of these are illustrated in Figure 2.10. None of the proposed schemes represents idle speculation; all reflect hard work and detailed and careful reasoning. The two oldest schemes (Fig. 2.10a,b) assume that all multicellular animals descended from some form of single-celled protist, most likely a colonial flagellate, and they view sponges (phylum Porifera) as the earliest experiments in multicellularity with no close relationship to any other existing phyla. One recent proposal, based upon molecular comparisons of 18S ribosomal RNA sequences, derives all other animals from a different protist, a slime mold (see p. 55), and separates the cnidarians (jellyfish and sea anemones, for example) markedly from all other animal groups. Another recent scheme (Fig. 2.10f) portrays many different early metazoans as independently evolved from many different protozoan ancestors and many more advanced animals as independently evolved from many different flatworm ancestors. The hypothesized relationships among annelids, arthropods, and molluscs also differ considerably among the different viewpoints. The more closely you look at the different schemes, the more fascinating the comparisons become; it is well worth returning to Figure 2.10 after you have read the rest of this book.

At least some of the differences among the various schemes may be attributed to insufficient data. As additional information about the various groups is gradually obtained, the evidence in favor of one scheme over some others may become more compelling, or additional modifications may be proposed.

Classification schemes reflecting evolutionary relationships are certainly not static, and the assignment of a given animal or group of animals to a particular position within the taxonomic hierarchy is not an irrevocable event. Studies of an animal's early development, for example, can reveal new information about the nature of the organism's internal body cavity, information that may affiliate that organism with an entirely different group of animals from those with which it was previously categorized. Classifications also can change when data from the fossil record are added to data from extant species. Or a detailed study might call the usefulness of particular characters into question. If, for example, a certain embryonic cleavage pattern arose only once in evolution, then those animals that develop in this particular way must be closely related. But if one can convincingly argue that this particular pattern probably evolved independently in several groups of animals, then that trait conveys little, if any, phylogenetic information.

Classifications also change when biologists discover organisms having characteristics not shared with any existing groups. For example, two arthropod classes (the Remipedia and Tantulocarida, p. 413), one echinoderm subclass (the Concentricycloidea, p. 459), and a small but remarkably distinct phylum of recently discovered marine animals called loriciferans (p. 215) have been established in the last 15 years or so.

Finally, molecular studies comparing selected gene sequences among representatives of different groups are quickly altering our understanding of many invertebrate relationships. Where molecular data produce phylogenies very different from those based on morphological criteria, decisions will have to be made about which evidence is more likely to be correct. And it is worth noting that molecular studies, as powerful as they are, will never resolve all phylogenetic issues, no matter how sophisticated these studies become. For one thing, when species diversified too rapidly, molecular studies are unable to resolve the order of divergence. Moreover, molecular studies will never be able to tell us the precise sequence of steps that took place as one form gave rise to another or what selective pressures brought about these morphological changes. Perhaps molecular, paleontological, ecological, and morphological evidence can be used in concert to deduce relationships, but we will still need to decide how much weight to give each line of evidence.

Phylogenetic relationships have been argued about for over 100 years. Such arguments will likely continue long into the future.

Classification by Habitat and Lifestyle

Animals also may be categorized on the basis of habitat or lifestyle; such categories reflect degrees of ecological similarity rather than any evolutionary relationships. For example, one group of animals may be **terrestrial,** living on land, while another is **marine,** living in the ocean. Marine animals, in turn, may be **intertidal** (living between the physical limits of high and low tides and thus exposed to air periodically); **subtidal** (living below the low-tide line and thus exposed to air only under extreme conditions, if ever); or **open ocean** creatures. In addition, animals may or may not be capable of locomotion, either **free-living** (mobile), **sessile** (immobile), or, perhaps, **sedentary** (exhibiting only limited locomotory capabilities). Some aquatic organisms may be able to move but have negligible locomotory powers with respect to the movement of the medium in which they live; such individuals are said to be **planktonic** (G: forced to drift or wander).

Animals are often categorized as to how they feed or what they eat. For example, some species are

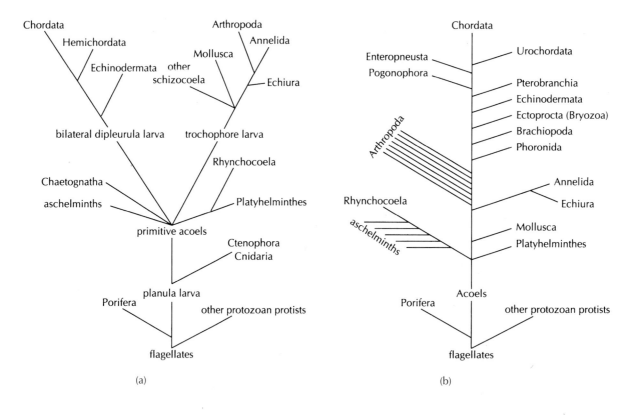

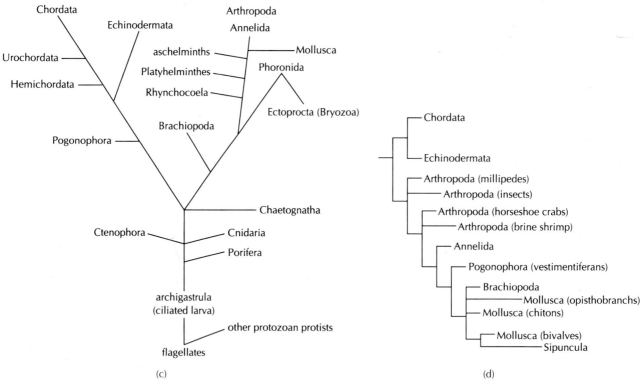

Figure 2.10

(a–g) Seven schemes proposed to illustrate presumed phylogenetic relationships among animals. (a) *According to Hyman, 1940.* (b) *According to Marcus, 1958.* (c) *R. B. Clark, Dynamics in Metazoan Evolution. New York: Oxford University Press, 1964.* (d) *Based on molecular analyses of Lake.* Proceedings of the National Academy of Science, USA 87:763-66, 1990. (e) *From* *Backeljau, et al., in Cladistics, 9:167-181, 1993. Reprinted by permission.* (f) *P. Willmer, Invertebrate Relationships. New York: Cambridge University Press, 1990.* (g) *From Claus Nielsen, in* American Zoologist, *34:492-501, 1994. Reprinted by permission.*

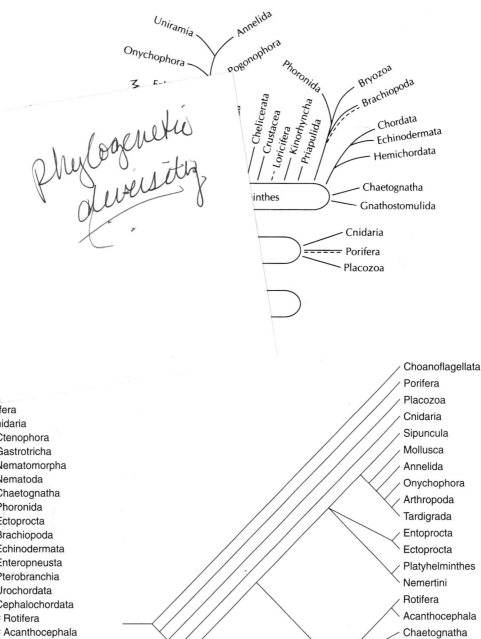

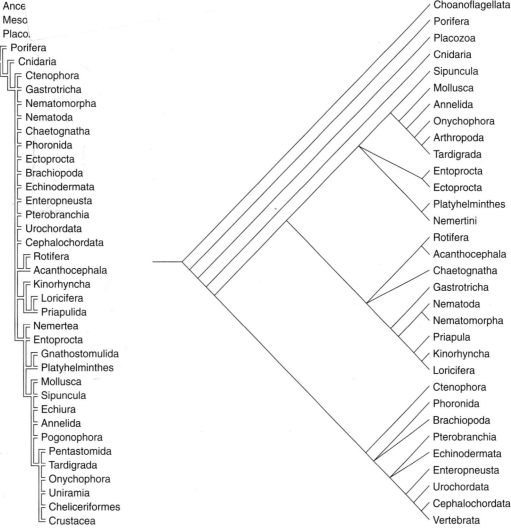

(e)   (g)

(a)

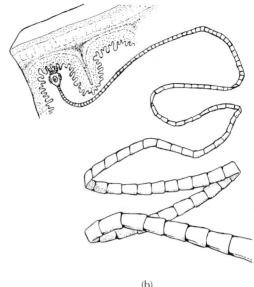

(b)

Figure 2.11
(a) A symbiotic relationship between a sea anemone, *Calliactis parasitica,* and a hermit crab, *Eupagurus bernhardus.* The crab deliberately places the anemones on its shell. (b) A tapeworm,

Taenia solium, shown attached to the intestinal wall of its vertebrate host.   (a) *After Hardy.* (b) *After Villee.*

herbivores (plant eaters), while others are **carnivores** (flesh eaters). Some species remove small food particles from the surrounding medium (**suspension feeders**), while others ingest sediment, digesting the organic component as the sediment moves through the digestive tract (**deposit feeders**).

Members of one species frequently live in intimate association with those of another species. These **symbiotic associations,** or **symbioses,** frequently relate to the feeding biology of one or both of the participants (**symbionts**) in the association (Fig. 2.11). **Ectosymbionts** live near or on the body of the other participant, while **endosymbionts** live within the body of the other participant. When both symbionts benefit, the relationship is said to be **mutualistic,** or an example of **mutualism.** When the benefit accrues to only one of the symbionts and the other is neither benefited nor harmed, the relationship is one of **commensalism,** and the benefiting member is the **commensal.** Lastly, some animals are **par-**

asites; that is, they are utterly dependent upon their **host** for continuation of the species, generally subsisting on either the blood or the tissues of the host. A parasite may or may not substantially impair the host's activities. The essence of parasitism is that the parasite is metabolically dependent upon the host and that the association is obligate for the parasite.

The boundaries between parasitism, mutualism, commensalism, and predation are not always distinct. For example, a parasite that eventually kills its host essentially becomes a predator. A parasite that produces a metabolic end product from which the host benefits borders on being mutualistic. Indeed, transitional forms in the process of evolving from one type of relationship to another are not uncommon. Such transitional forms once again make tidy categorization of animals into human-made schemes difficult; definitions of some categories have been modified by various workers in an attempt to improve the fit, but every rule seems to have an exception.

Topics for Further Discussion and Investigation

1.   There are presently three major theories of animal classification: the theory of **phenetics** (based entirely on degree of overall anatomical and biochemical similarity, without regard to whether the similarities reflect homology or convergence), the theory of **cladistics** (based entirely upon recency of common descent inferred from the mutual possession of particular specialized, derived morphological traits called **synapomorphies**), and the theory of **evolutionary classification** (which attempts to consider both ancestry and the degree to which organisms have subsequently diverged from the ancestral form). Discuss the advantages and disadvantages inherent in any two of these three approaches to the inferring of phylogenetic relationships.

Bock, W. 1965. Review of Hennig, *Phylogenetic Systematics. Evolution* 22:646.

Bock, W. J. 1982. Biological classification. In *Synopsis and Classification of Living Organisms*, vol. 2, edited by S. P. Parker. New York: McGraw-Hill, 1068–71.

Brusca, R. C., and G. J. Brusca. 1990. *The Invertebrates.* Sunderland, Mass.: Sinauer Assoc., 28–39.

Estabrook, G. F. 1986. Evolutionary classification using convex phenetics. *Syst. Zool.* 35:560.

Farris, J. 1967. The meaning of relationship and taxonomic procedure. *Syst. Zool.* 16:44.

Mayr, E. 1965. Numerical phenetics and taxonomic theory. *Syst. Zool.* 14:73.

Mayr, E. 1974. Cladistic analysis or cladistic classification? *Zool. Syst. Evol.-forsch.* 12:94. (Reprinted in E. Mayr. 1976. *Evolution and the Diversity of Life—Selected Essays.* Cambridge, Mass.: Harvard University Press, 433–76.)

Mayr, E. 1991. *Principles of Systematic Zoology*, 2d ed. New York: McGraw-Hill.

Panchen, A. L. 1992. *Classification, Evolution, and the Nature of Biology.* New York: Cambridge University Press.

Smith, A. B. 1993. *Systematics and the Fossil Record: Documenting Evolutionary Patterns.* Cambridge, Mass.: Blackwell Scientific Publications.

2.   What are the characteristics of the "ideal" classification system, and why is this ideal so difficult to attain?

Archie, J. W. 1984. A new look at the predictive value of numerical classifications. *Syst. Zool.* 33:30.

Erwin, D. H. 1991. Metazoan phylogeny and the Cambrian radiation. *Trends Ecol. Evol.* 6:131.

Farris, J. 1967. The meaning of relationship and taxonomic procedure. *Syst. Zool.* 16:44.

Farris, J. 1982. Simplicity and informativeness in systematics and phylogeny. *Syst. Zool.* 31:413.

Mayr, E. 1974. Cladistic analysis or cladistic classification? *Zool. Syst. Evol.-forsch.* 12:94. (Reprinted in E. Mayr. 1976. *Evolution and the Diversity of Life—Selected Essays.* Cambridge, Mass.: Harvard University Press, 433–76.)

Page, M. D., and P. H. Harvey. 1988. Recent developments in the analysis of comparative data. *Q. Rev. Biol.* 63:413.

3.   The fossil record shows a remarkable radiation of animal body plans between about 500 and 570 million years ago in a geologic time period termed the Cambrian. The earliest known metazoans—from the so-called Ediacaran fauna collected in the Ediacara hills of South Australia in the 1940s—are somewhat older. To what extent do the Ediacaran and Cambrian faunas differ, and what might account for those differences?

McMenamin, M. A. S. 1987. The emergence of animals. *Sci. Amer.* 256(4):94.

Moore, J. A. 1990. A conceptual framework for Biology, Part II. *Amer. Zool.* 30:752–71.

Morris, S. C. 1989. Burgess shale faunas and the Cambrian explosion. *Science* 246:339.

4.   Molecular analyses allow us a direct look at differences in the genetic structure among species, potentially overcoming some of the problems (such as the likelihood of convergence) associated with morphological studies. Moreover, if the chosen molecules are not subject to selection and if mutation rates are constant, degrees of difference between the molecules of two different species should indicate the amount of time elapsed since the species diverged from their common ancestor. Why haven't molecular data been able to resolve all previous phylogenetic controversies?

Backeljau, T., B. Winnepenninckx, and L. De Bruyn. 1993. Cladistic analysis of metazoan relationships: A reappraisal. *Cladistics* 9:167.

Hoelzer, G. A., and D. J. Melnick. 1992. Patterns of speciation and limits to phylogenetic resolution. *Trends Ecol. Evol.* 9:104.

Martin, A. P., G. J. P. Naylor, and S. R. Palumbi. 1992. Rates of mitochondrial DNA evolution in sharks are slow compared with mammals. *Nature* 357:153.

Patterson, C. 1990. Reassessing relationships. *Nature* 344:199.

Rand, D. M. 1994. Thermal habit, metabolic rate and the evolution of mitochondrial DNA. *Trends Ecol. Evol.* 9:125.

5.   Compare and contrast the relationships among phyla depicted in any three parts of Figure 2.10. For example, note that molluscs (such as snails, clams, and squid) are shown as having an immediate ancestor in common with annelids (such as earthworms and leeches) in (c), (e), and (g) but as evolving independently of annelids in (f).

3

The Protozoans

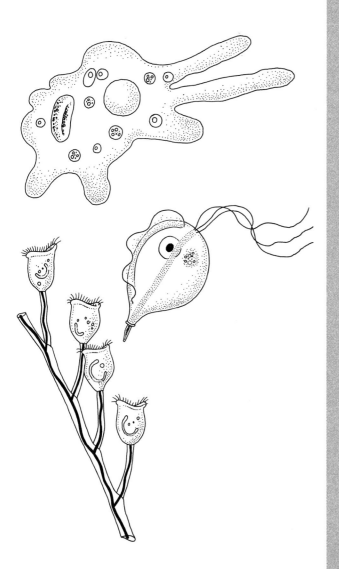

Introduction

The distinction between plant-like and animal-like organisms is not always easily made, particularly among unicellular life forms. This difficulty reflects what is surely a distant but very real evolutionary relationship between plants and animals. As with the mitochondria and other membrane-bound organelles that characterize eukaryotes (*eu* = G: true; *karyo* = G: nuclei), chloroplasts have been acquired by many unicellular organisms through ancient symbiotic relationships. Thus, some unicellular organisms ingest solid food particles, while others photosynthesize. Still others live amidst decaying plant and animal matter—or live as parasites—and feed by taking up dissolved organic material and other nutrients across their body surfaces. Some species are capable of two or even all three nutritional modes, either simultaneously or at different times. For this reason, about 130 years ago, the great German scientist Ernst Haeckel suggested placing all of the troublesome organisms that weren't clearly plants or animals within a separate kingdom, the Protista (*protist* = G: the very first), now sometimes called the Protoctista (*ktistos* = G: to establish). Eventually, the Protista came to include everything except plants, animals, fungi, and bacteria. But the diversity of protist ultrastructure, life cycles, lifestyles, and evolutionary trajectories seems too extreme for a single kingdom to support. In consequence, in the classification scheme adopted for this edition[1] there is no longer a formal taxonomic category called the Protista, although "protist" still serves as a useful general term for referring to this remarkable collection of eukaryotes. The various protist groups have been elevated to kingdom status.

---

1. Corliss, J. O. 1994. An interim utilitarian ("user-friendly") hierarchical classification and characterization of the protists. *Acta Protozool.* 33:1–51. For simplification I follow the suggestions of other protozoologists and include the primitive(?) protists, which lack mitochondria, with the Protozoa.

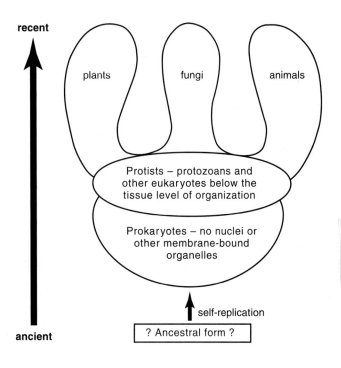

recent

plants　fungi　animals

Protists – protozoans and other eukaryotes below the tissue level of organization

Prokaryotes – no nuclei or other membrane-bound organelles

self-replication

? Ancestral form ?

ancient

Figure 3.1
Likely evolutionary relationships between major groups of organisms. The plant, fungal, and animal kingdoms all have their origins somewhere among the largely unicellular protists. Animals probably evolved from flagellated protozoans. Much further back in time, protists probably evolved from symbiotic associations among various prokaryotic organisms.

This chapter is concerned primarily with the animal-like members of one of those kingdoms, the Kingdom Protozoa (*proto* = G: first; *zoa* = G: animals). The group includes some 38,000 extant species, with hundreds of new species being described each year.

> *Kingdom Proto · zoa*
> (G: the very first · animals)
> prō-tō-zō-ah

Protozoans are clearly eukaryotes, since they harbor distinct nuclei and other membrane-bound organelles, but unlike members of the animal kingdom, protozoans never develop from a blastula embryonic stage. Although protozoans are not animals, animals most likely evolved from protozoan ancestors (Fig. 3.1). Indeed, all animals ultimately seem to have descended from certain flagellated protozoans. This suggestion, first proposed in part because most animals produce flagellated sperm, is well supported by current molecular data. Protozoans, then, bridge the gap between unicellular and multicellular organisms.

Protozoan classification has long been anything but certain, and nucleotide sequence data in particular are changing the scenarios of protozoan interrelationships at a rapid rate. Precise evolutionary relationships are difficult to determine for any group of organisms, but this is particularly true for protozoans. The fossil record for protozoans is sparse, and what does exist is not particularly helpful in deducing relationships; members of some rather sophisticated protozoan groups possessing hard parts are encountered among the fossils of about 600 million years ago and have forms that are virtually identical to those found in some genera today. The small size of most individuals makes structural studies slow and difficult (but not insurmountably so). As with all organisms, structural similarities need not imply close evolutionary relationships; similar structures often arise independently in different, unrelated groups of organisms in response to similar selective pressures. This phenomenon is known as **convergent evolution.** As a result of these factors, the classification of protozoans has been a source of particular difficulty and controversy.

Ongoing molecular studies should soon indicate the extent to which protozoans are a *monophyletic* (G: single-tribed) group of organisms evolved from a single ancestral form or the extent to which they instead represent a *polyphyletic* (G: many-tribed) group, a collection of organisms descended from several different ancestral forms. The next few years will surely contain many surprises. The information presented here should be read with the understanding that the evolutionary relationships among and within the various protozoan groups are still far from clear. This seems a truly exciting time to be a protozoologist!

Although most people never see protozoans, few other groups of organisms rival them in economic and scientific importance. Protozoans play major roles both in primary production and in decomposition, and may serve as a major food source for many invertebrates and, indirectly, for many vertebrates as well.[2] In addition, protozoans cause a number of human ailments, including malaria, African sleeping sickness, and dysentery, and a variety of devastating diseases of poultry, sheep, cattle, cabbage, and other human food sources. Protozoans also have provided biologists with outstanding material for genetic, physiological, developmental, and ecological studies. Nevertheless, these ubiquitous and important organisms have been known for only about 320 years, their discovery awaiting the invention of the microscope and the patience of the Dutch draper and amateur scientist Antony van Leeuwenhoek, who discovered protozoans in 1674.

Protozoans occur wherever there is moisture. Free-living protozoans are found in both marine and freshwater habitats, and in moist soil. Many protozoans live in close association with other protozoans, with animals, or with plants, either as commensals or as parasites. Indeed, every major group of protozoans contains at least some parasitic species, and the members of several major groups are exclusively parasitic. Most protozoans are small, usually 5μm to 250 μm (micrometers) long, and

---

2. See *Topics for Further Discussion and Investigation,* no. 7, at the end of the chapter.

are therefore difficult to work with. The largest protozoans rarely exceed 6 mm to 7 mm (millimeters). About 80,000 protozoan species have been described so far, more than half of which are known only as fossils. Probably, most extant protozoan species have yet to be described.

General Characteristics

It is difficult to generalize about protozoans; they absolutely defy tidy categorization. The members of this kingdom demonstrate a tremendous range of sizes, morphologies, nutritional modes, behavioral and physiological diversity, and genetic diversity. Indeed, in many respects, the diversity of form and function encountered among protozoans rivals that encountered among all invertebrate animals combined. Discussing all of the protozoans within one chapter is analogous to attempting a single coherent discussion of the Annelida, Mollusca, Arthropoda, Echinodermata, and several other groups as well.

Protozoans are clearly among the simplest of life-forms. Nevertheless, in many respects protozoans are as complex as any multicellular animal (**metazoan**), and a great deal of protozoan biology remains poorly understood despite much sophisticated research. Protozoans are not composed of individual cells; in essence, each protozoan *is* an individual cell. This is a major characteristic that sets protozoans apart from animals, and is one of the few characteristics applicable to nearly all protozoans. Remarkably, this single cell functions as a complete organism, doing all of the basic things—feeding, digesting, locomoting, behaving, and reproducing—that multicellular animals do. Complexity arises from the specialization of organelles rather than from the specialization of individual cells and tissues. In addition, most protozoans lack specialized circulatory, respiratory, and excretory (waste-removal) structures. The surface area of their bodies is high relative to body volume, so that gas exchange and removal of soluble wastes can occur by simple diffusion across the entire exposed body surface. The cytoplasm of a few species has been shown to contain hemoglobin, although its role in gas exchange has not yet been demonstrated. The relatively high surface area of protozoans may also facilitate active uptake of dissolved nutrients from the surrounding fluids.

The entire protozoan body is bounded by a **plasmalemma** (cell membrane) that is structurally and chemically identical to that of multicellular organisms. The cytoplasm bounded by the plasmalemma resembles that of animal cells, except that it is often differentiated into a clear, gelatinous outer region, the **ectoplasm**, and an inner, more fluid region, the **endoplasm**. Within the cytoplasm are organelles and other components typical of metazoan cells, including nuclei, nucleoli, chromosomes,

Golgi bodies, endoplasmic reticulum (with and without ribosomes), lysosomes, centrioles, mitochondria, and in some individuals, chloroplasts. In a sense, then, the typical protozoan can be regarded as a single-celled organism, although this cell is functionally more versatile than any single cell found within any multicellular animal.

In addition to the organelles encountered typically in multicellular animals, many protozoans contain organelles not generally found among metazoans. These organelles are, most notably, contractile vacuoles, trichocysts, and toxicysts. Many protozoans are also characterized by highly complex arrays of microtubules and microfilaments. Still other highly specialized organelles are restricted to various small groups of protozoans.

Contractile vacuoles are organelles involved in expelling excess water from the cytoplasm. Apparently, fluid is collected from the cytoplasm by a system of membranous vesicles and tubules called the **spongiome**.[3] The collected fluid is transferred to a contractile vacuole and is subsequently discharged to the outside through a pore in the plasmalemma (Fig. 3.2). Contractile vacuoles are most commonly seen among freshwater protozoan species because concentrations of dissolved solutes in the cytoplasm of freshwater protozoa are much higher than solute concentrations in the surrounding medium. Thus, water continuously diffuses into the cytoplasm across the plasmalemma and does so in proportion to the magnitude of the solute concentration gradient (i.e., the **osmotic gradient**). Without some form of compensating mechanism, water would continue to flow into the protozoan until either the osmotic gradient across the plasmalemma was reduced to zero or the protozoan burst. Moreover, cells can function over only a narrow range of internal solute concentrations, and protozoans are no exception; even a small dilution of the cytoplasmic solute concentration could be disabling.

This discussion indicates that the contractile vacuole must function not only to prevent swelling (by ridding the body of fluid as fast as water crosses the plasmalemma), but also to maintain a physiologically acceptable solute concentration within the cell (Fig. 3.3). In other words, the contractile vacuole must function both in **volume regulation** (maintaining a constant body volume) and in **osmotic regulation** (maintaining a constant intracellular solute concentration).[4] The contractile vacuole achieves the latter result by pumping out of the cell a fluid that is dilute relative to the surrounding cytoplasm in the cell; that is, the solute is essentially separated from the water and retained within the cell before the remaining fluid is expelled. The mechanism by which water is separated from the cytoplasm is not known, nor is the mechanism by which the vacuole fluid is discharged to the outside, although researchers are making some progress in these studies. Contractile vacuoles typically fill

---

3. See *Topics for Further Discussion and Investigation*, no. 3.

4. See *Topics for Further Discussion and Investigation*, no. 4.

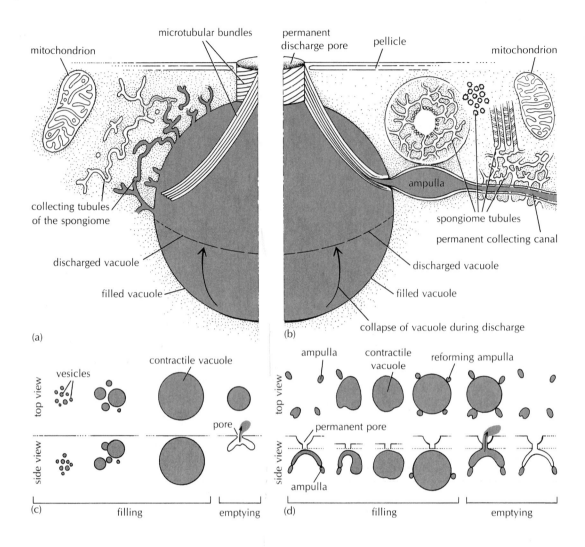

Figure 3.2

(a,b) Diagrammatic illustration of two types of contractile vacuole system encountered among protozoans. The two differ largely in the complexity of the spongiome and in other aspects of the fluid collection system. Other contractile vacuole complexes (not illustrated) may lack ampullae, permanent pores, and the associated bundles of microtubules. (c,d) The behavior of two types of contractile vacuole complexes during filling and emptying; arrows represent the flow of liquid out of the emptying vacuole. The vacuole is viewed from above in the upper series (discharge emerges perpendicular to the plane of the page) and from the side in the lower series. In (c), a permanent pore is lacking, and the vacuole forms through the fusion of many small, fluid-filled vesicles; the vacuole disappears completely following discharge. In (d), the contractile vacuole is filled by conspicuous ampullae and discharges through a permanent pore. The ampullae may begin to refill before the vacuole discharges its fluid to the outside.

From Patterson, in Biological Review of the Philosophical Society, *55:1, 1980.*

Copyright © 1980 Cambridge University Press, New York. Reprinted by permission.

and discharge several to many times per minute, and a single individual may possess several contractile vacuoles.

Other interesting organelles called **trichocysts** develop within membrane-bound vesicles in the cytoplasm and eventually come to lie along the periphery of the protozoan. The trichocysts themselves are elongated capsules that can be triggered by a variety of mechanical and/or chemical stimuli to discharge a long, thin filament (Fig. 3.4). This discharge occurs within several thousandths of a second and is thought to be initiated through an osmotic mechanism involving a rapid influx of water. The adaptive significance of trichocysts is unknown; likely possibilities are that they protect against predation or anchor the animal during feeding. Related structures called **toxicysts** are clearly involved in predation; filaments discharged from toxicysts paralyze prey and initiate digestion. At least 10 other types of **extrusomes** (organelles capable of ejectability) have so far been described from protozoans.

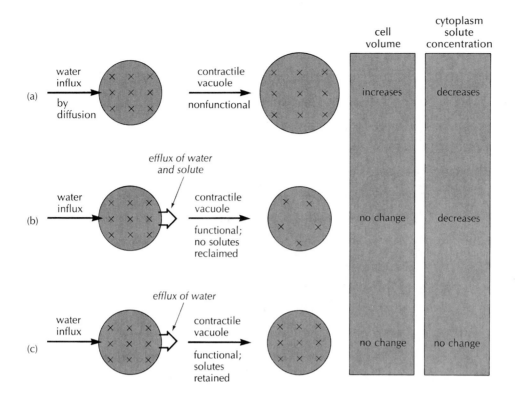

Figure 3.3

Diagrammatic illustration of the functional significance of contractile vacuoles in freshwater protozoans. The vacuole system functions both to maintain body volume and to maintain proper solute concentrations within the cell (X represents solute). (a) No functioning contractile vacuole; water diffuses inward, swelling the cell and diluting internal solute. (b) The contractile vacuole pumps out cytoplasmic fluid, including dissolved solute; cell volume is maintained, but solute is continuously lost. (c) The contractile vacuole pumps out fluid containing little solute, maintaining both cell volume and osmotic concentration.

As single-celled organisms, protozoans necessarily lack gonads. Although sexual reproduction occurs in most groups, asexual (*a* = G: without) reproduction is commonly encountered among all groups of protozoans and is the only form of reproduction reported for many species. By definition, asexual reproduction does not generate new genotypes. Protozoans reproduce asexually through **fission,** a controlled mitotic replication of chromosomes and splitting of the parent into two or more parts. Indeed, asexual reproduction among protozoans has long been exploited by biologists as a general model for the study of mitosis. **Binary fission** occurs when the protozoan splits into two individuals (Fig. 3.5). In **multiple fission,** many nuclear divisions precede the rapid differentiation of the cytoplasm into many distinct individuals. In **budding,** a portion of the parent breaks off and differentiates to form a new, complete individual. In some multinucleate species, the parent simply divides in two in the absence of any mitotic division, the original nuclei being distributed between the two daughter cells. This process is termed **plasmotomy** (*tomy* = G: to cut).

In keeping with the widespread occurrence of asexual reproductive capabilities, many protozoans possess great capacity for regeneration. One example of this capacity is the phenomenon of encystment and excystment exhibited by many freshwater and parasitic species. During **encystment,** substantial dedifferentiation of the organism takes place. The individual loses its distinctive surface features, including cilia and flagella, and becomes rounded. The contractile vacuole(s) pumps out all excess water from the cytoplasm, and a gelatinous covering (or some other type of covering) is secreted. This covering soon hardens to form a protective **cyst.** In this encysted form, the quiescent individual can withstand long periods (several to many years in some species) of exposure to what would otherwise be intolerable environmental conditions of acidity, dryness, thermal stress, and food or oxygen deprivation. Winds, animals, and other vectors may widely disperse individuals in the encysted state. Once conditions improve, **excystment** quickly ensues, with the regeneration of all former internal and external structures.

Patterns of sexual reproduction will be discussed as appropriate on a group-by-group basis, as few generalizations are possible. It is, however, possible to generalize about the digestion of particulate food.

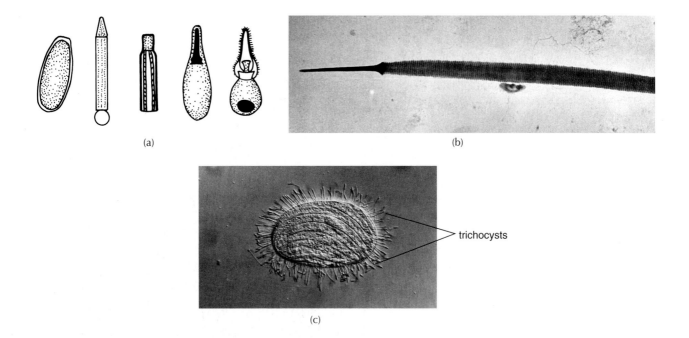

(a)

(b)

trichocysts

(c)

Figure 3.4

(a) Undischarged extrusomes of different types. At least 12 morphologically distinct types of extrusome have been described. Some expel mucus (left); others eject filaments of varying length. Paralytic toxins are injected by one class of extrusome (toxicysts, not shown). Others (far right) are used primarily for adhering to prey during food capture. (b) Transmission electron micrograph of a trichocyst that has been discharged from *Paramecium* sp. (c) A ciliate, *Pseudomicrothorax dubis,* with its trichocysts extended. (a) *From Corliss, in American Zoologist 19:573, 1979. Copyright © 1979 American Society of Zoologists, Thousand Oaks, California.* (b) *Courtesy of M. A. Jakus, National Institutes of Health.* (c) *From S. Eperon and Peck. The Journal of Protozoology 35:280-86, fig. 1, p. 282, Allen Press, Inc. © 1988 Dr. Edna Kaneshiro, editor.*

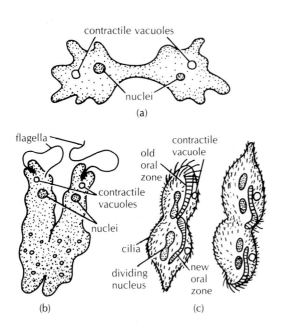

Figure 3.5

Binary fission among protozoans. (a) In an amoeba. (b) In a flagellate. (c) In a ciliate. (a) *After Pennak.* (b,c) *After Hyman; after Gregory.*

Food particles are digested internally among protozoans[5]; since they are single-celled organisms, digestion is, of necessity, entirely intracellular. Ingested food particles generally become surrounded by membrane (similar to the plasmalemma), forming a distinct **food vacuole.** These vacuoles move about in the fluid cytoplasm of the body as the vacuolar contents are digested enzymatically. By feeding protozoans food stained with pH-sensitive chemicals, researchers have determined that the contents of the vacuoles first become quite acidic and later become strongly basic. As in other animals, including humans, digestion requires exposing the food to a series of enzymes, each with a specific role to play and each with a narrowly defined pH optimum. The controlled changes of pH that occur within the food vacuoles of protozoans allow for the sequential disassembly of foods by a series of different enzymes, despite the absence of a digestive tract *per se.* Once solubilized, nutrients move across the vacuole wall and into the endoplasm of the cell. Indigestible solid wastes are commonly discharged to the outside through an opening in the plasma membrane.

Clearly, protozoans are unlike animals in many features of their biology. Protozoans are problematic—and intriguing—in another important respect as well: Even

5. See *Topics for Further Discussion and Investigation,* no. 6.

the distinction between unicellular and multicellular is not always easily made. Although most protozoan species occur as single, one-celled individuals, many other species are **colonial;** that is, a single individual divides asexually to form a colony of attached, genetically identical individuals. Usually, the individuals of a protozoan colony are morphologically and functionally identical. However, the individuals comprising the colonies of several species show a degree of structural and functional differentiation that is very reminiscent of that encountered within primitive groups of metazoans. Protozoans thus bridge the gap not only between plants and animals but between unicellular and multicellular life-forms as well. As the protozoologist John Farmer has said, such "transitional forms play havoc with attempts to keep taxonomy tidy." But they certainly add interest to studies of invertebrate zoology and evolution.

Phylum Ciliophora

Phylum Cilio · phora
(G: cilia bearing)
sil´-ē-ō-for´-ah

Ciliates show the highest degree of subcellular specialization encountered among protozoans and thus are considerably advanced from the primitive protozoan condition. Nevertheless, we will begin with the ciliates because they are predominately free-living and, compared with other protozoan groups, are relatively uniform in basic body plan. The phylum contains about 9,000 described species. Several features are unique to the ciliates, and several other features characterize this group above all others.

First among the unique features of this phylum is the presence of external ciliation in at least some stage of the life cycle. The ultrastructure of cilia is remarkably uniform throughout the Ciliophora and, in fact, throughout the animal (and plant) kingdom(s) as well, although some modifications to the basic pattern do occur sporadically.

The following description of ciliary structure and function is generally applicable to ciliary organelles encountered among all organisms in all phyla.

Structure and Function of Cilia

Each cilium is cylindrical and arises from a **basal body (kinetosome).** Within the cilium are a number of long rods called **microtubules,** composed of a protein known as **tubulin.** Tubulin is extremely similar to the actin of metazoan striated muscle. A cross section of the kineto-

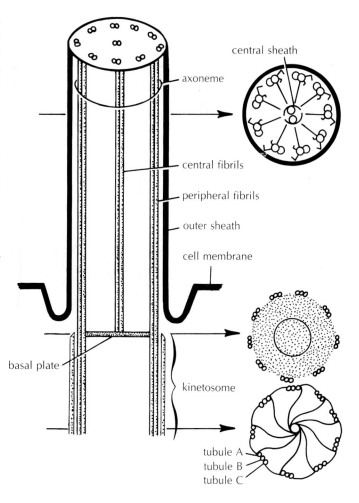

Figure 3.6
Ciliary ultrastructure. The appearance of the cilium in cross section changes along its length as indicated. *After Sherman and Sherman; after Wells.*

some of a cilium in the region beneath the outer body surface shows a ring of nine groups of microtubules, with three microtubules to a group (Fig. 3.6, lower right). The A, or innermost microtubule of each group, is physically connected to the C, or outermost microtubule of an adjacent group, via a thin filament. Additional filaments connect the A microtubule to a central tubule, like the spokes of a wheel. This configuration of microtubules changes somewhat near the distal end of the cilium.

A cross section through a cilium external to the body surface shows a ring of nine groups of microtubules, with only two microtubules per group; the C microtubule is not found (Fig. 3.6, upper right). Instead, a pair of **dynein arms** projects outward from each A microtubule toward the B tubule of the neighboring pair of microtubules. The primary protein component of these arms (**dynein**) is similar in some respects to muscle myosin. Like myosin, dynein possesses the ability to cleave ATP, releasing chemical energy. One pair of microtubules is located centrally within the cilium. These two microtubules form the central shaft of the cilium, and they are often surrounded by a membrane constituting a **central sheath.** Distinct filaments extend from the A microtubule of each outer doublet in toward this central sheath. The entire microtubular complex, consisting of the nine doublet microtubules and the inner pair of single microtubules, is termed the **axoneme.**

Although many details of ciliary operation remain to be elucidated, present evidence indicates that ciliary bending is achieved through the differential sliding of some groups of adjacent microtubules relative to others within the same cilium. The energy for this sliding appears to derive from the ATPase activity of the dynein arms, and the sliding itself appears to involve an interaction between tubulin and dynein, in a manner highly reminiscent of the interaction between the actin and myosin filaments of metazoan striated muscle. As we will see later in this chapter, microtubules play a variety of roles, both direct and indirect, in the locomotion of most cells—even in those lacking cilia.

The rhythmic and coordinated bending and recovery of cilia provide the means of locomotion and often aid in food collection for members of this phylum. Ciliates are the fastest moving of all protozoans, achieving speeds of up to 2 mm per second. In addition, ciliary activity is probably important in removing wastes from the immediate vicinity of the animal and in continually bringing oxygenated water into contact with the general body surface.

To be effective—that is, to create unidirectional movement of the animal in water or unidirectional flow of water past the animal—the **power stroke** and the **recovery stroke** of a cilium must not be identical in form. By analogy, if you are swimming the breaststroke and move your arm forward to the front of your head by the same path and with the same force that you used to bring the arm back to your side, you will move backward by about the same distance that you moved forward during your power stroke. The cilium faces a similar dilemma. During ciliary locomotion, the power stroke, as the name implies, does work against the environment. The cilium is outstretched for maximum resistance as it bends downward toward the body (Fig. 3.7a). During the recovery stroke, the cilium bends in such a manner as to considerably reduce resistance, doing less work against the environment (Fig. 3.7b). Thus, the recovery stroke does not undo the work of the power stroke. In addition, the cilium moves fastest in the power stroke, increasing the force of the power stroke relative to that of the recovery stroke. By analogy, more force is required to move your arm through the water quickly than slowly.

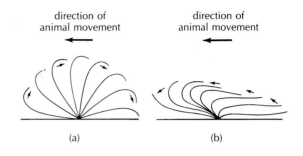

Figure 3.7
(*a*) Power and (*b*) recovery strokes in isolated cilia. Arrows indicate the direction of motion of the cilium. Note that more of the ciliary surface area is involved in pushing against the water in (*a*) than in (*b*). Thus, the recovery stroke does less work and does not undo all the work of the power stroke.

Generally, a large part of the body surface of ciliates is covered by distinct rows of cilia, and movement is achieved by coordinated, **metachronal** beating of the cilia in each row (Fig. 3.8). In metachronal beating, the power and recovery strokes of a cilium are begun immediately following the initiation of the comparable strokes of an adjacent cilium. The direction of metachronal beating may be quickly reversed in response to chemical, mechanical, or other stimuli, permitting mobile organisms to quickly reverse direction and perhaps escape undesirable situations.

Patterns of Ciliation

Individual cilia, seen external to the cell body, are associated with each other through a complex **infraciliature** below the body surface (Figs. 3.9 and 3.10). A striated fibril, called a **kinetodesmos,** extends from each kinetosome (**basal body**) in the direction of an adjacent cilium of the same row. Thus, running along the right side of each row of basal bodies is a cord of fibers, termed the **kinetodesmata.** Direct microtubular connections between adjacent kinetosomes have rarely been demonstrated. Each of the kinetosomes has its own array of microtubular, microfibrillar, and other organelles, giving ciliate bodies a more complex cytoarchitecture than is found in many other protozoans and in any metazoan.

This infraciliature is found only within the Ciliophora and is encountered in the adults of all ciliate species, even in those lacking external ciliation as adults. The structure of this infraciliature is one of the primary tools used to distinguish the different ciliate species and to assess the degree to which different species are related. The functional significance of the complex structure of the infraciliature has not yet been conclusively demonstrated.[6]

---

6. See *Topics for Further Discussion and Investigation*, no. 2.

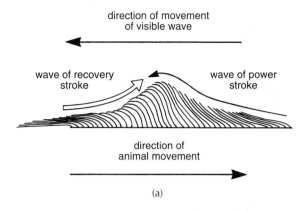

direction of movement
of visible wave

wave of recovery
stroke

wave of power
stroke

direction of
animal movement

(a)

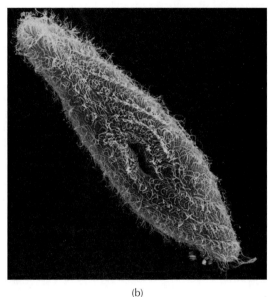

(b)

Figure 3.8

(a) A metachronal wave passing along a row of cilia. (b) A ciliate (*Paramecium sonneborni*), showing metachronal waves of ciliary activity. Note the oral opening near the middle of the body. The paramecium is 40μm in length.   (a) After Wells; after Sleigh. (b) From K. J. Aufderheide et al., The Journal of Protozoology 30:128, Allen Press, Inc. © 1983 Dr. Edna Kaneshiro, editor.

Cilia cover virtually the entire body of some species, but they are reduced or otherwise modified to varying degrees in others. In some species, groups of cilia are functionally associated in such a way as to form discrete organelles. One such organelle is the so-called **undulating membrane,** a flattened sheet of cilia that moves as a single unit (Fig. 3.11*a*). A second commonly encountered ciliary organelle is termed a **membranelle;** here, a smaller number of cilia in several adjacent rows appear to lean toward each other, forming, in effect, a two-dimensional triangular tooth (Fig. 3.11*a,b*). In addition, cilia may form a discrete bundle (**cirrus**), which tapers to

a point toward the tip (Figs. 3.11*a,c* and 3.12). The cilia comprising such organelles are structurally identical to those that function as individuals, and no permanent physical attachments between the cilia comprising undulatory membranes, membranelles, or cirri have been observed. The mechanism by which their activities are so closely coordinated remains uncertain.

Besides the infraciliature and general placement and pattern of body ciliation, other characteristics of major taxonomic significance are the position, ultrastructure, and pattern of ciliation of the oral region. The mouth opening, called the **cytostome** (*cyto* = G: cell; *stoma* = G: mouth), may be located anteriorly, laterally, or ventrally on the body. Often, the cytostome is preceded by one or more preoral chambers, whose complexity varies greatly among major groups of ciliate species.

Other Morphological Features

One other morphological feature particularly characteristic of ciliates is the covering of the body by an often complex series of membranes, forming a **pellicle** (Fig. 3.13). The inner membranes, lying beneath the single plasmalemma enveloping the body, form a series of elongated, flattened vesicles called **alveoli** (*alveolus* = L: a cavity). Cilia project to the outside between adjacent alveoli. The pellicle may be rigid or highly flexible, depending upon how the membranes are organized, and may serve a supportive function in some species, helping to maintain shape. Trichocysts are characteristically associated with the ciliate pellicle.

In keeping with the unusually high level of structural complexity characteristic of ciliates, only the members of this group have permanent excretory pores associated with their contractile vacuoles. These excretory openings are maintained by arrays of microtubules. The system for collecting cytoplasmic fluid and transferring it to the contractile vacuole is also particularly complex among ciliates. Some species also maintain a permanent opening to the outside (a **cytoproct;** *procto* = G: the anus) for expelling undigested wastes.

Reproductive Characteristics

In addition to the presence of cilia and an infraciliature, a second unique characteristic of all ciliates is the possession by each individual of two types of nuclei; that is, the nuclei of every ciliate are **dimorphic** (of two kinds). The ciliates are thus **heterokaryotic** (*hetero* = G: different; *karyo* = G: nucleus), while all other cells, including other protozoans, are **monomorphic** (one kind) or **homokaryotic** (*homo* = G: alike). Every ciliate contains one or more large nuclei, termed **macronuclei,** and one

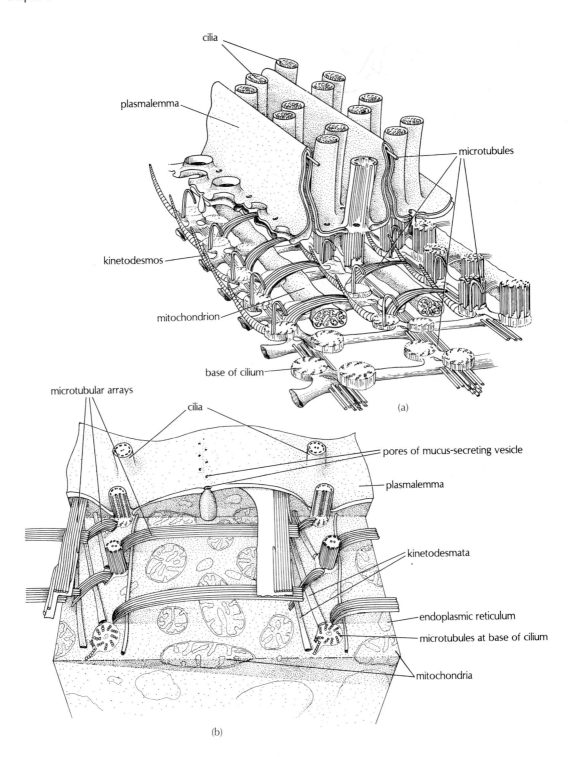

cilia

plasmalemma

microtubules

kinetodesmos

mitochondrion

base of cilium

(a)

microtubular arrays

cilia

pores of mucus-secreting vesicle

plasmalemma

kinetodesmata

endoplasmic reticulum

microtubules at base of cilium

mitochondria

(b)

Figure 3.9

The complex infraciliature of ciliates. (*a*) *Conchophthirus* sp. (*b*) *Tetrahymena pyriformis*. These ultrastructural details were unknown before the advent of the electron microscope.

(a) *From Corliss, in* American Zoologist, *19:573, 1979. Copyright © 1979 American Society of Zoologists. Reprinted by permission.* (b) *From R. D. Allen, "Fine Structure, Reconstruction and Possible Functions of Components of the Cortex of* Tetrahymena pyriformis*" in* Journal of Protozoology, *14:553, 1967. Copyright © 1967 Society of Protozoologists, Lawrence, Kansas. Reprinted by permission.*

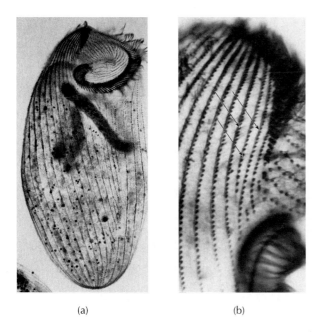

(a)   (b)

Figure 3.10

(a) *Climacostomum* sp., stained to reveal the infraciliature.
(b) Detail of three rows of kinetosomes (basal bodies). A row of kinetosomes, with their associated cilia and kinetodesmata, is termed a kinety.   (a,b) *From C. F. Dubochet et al., The Journal of Protozoology 26:218, Allen Press, Inc. © 1979 Dr. Edna Kaneshiro, ed.*

or more smaller nuclei, termed **micronuclei.** Micronuclei are often more abundant than macronuclei within a given individual; ciliates in some species possess more than 80 micronuclei per individual.

The macronucleus is polyploid, contains both DNA and RNA, and is involved both in the day-to-day operations of the protozoan and in differentiation and regeneration. Macronuclear shape varies markedly among species. Ciliates cannot live without their macronuclei, but they can live without their micronuclei. However, micronuclei are essential for sexual reproduction, while the macronucleus plays no direct role in the sexual activities of ciliates.

Sexuality among ciliates never involves the formation of gametes. Rather, their primary sexual activity invariably involves a process called **conjugation** (Fig. 3.14), typically a temporary physical association between two "consenting" individuals during which genetic material is exchanged. The exchange takes place through a tube connecting the cytoplasm of the two individuals.

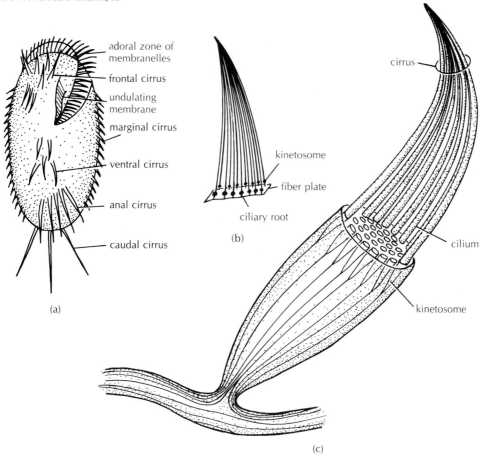

Figure 3.11

(a) *Stylonychia* sp., showing several ciliary organelles: undulating membranes, membranelles, and cirri. (b) Diagrammatic illustration of membranelle structure. (c) Diagrammatic representation of the structure of a single cirrus. Cirri typically contain between 24 and 36 individual cilia.   (a,b) *After Kudo.*
(c) *Sherman/Sherman, The Invertebrates: Function and Form, 2/e, © 1976, pp. 111, 112, 236, 172, 169, 45, 9, 15. Reprinted by permission of Prentice Hall, Upper Saddle River, New Jersey.*

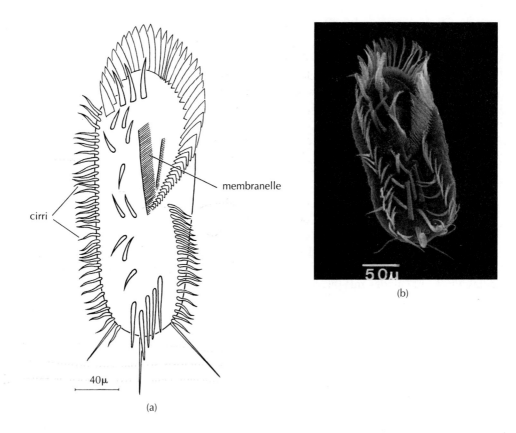

Figure 3.12
(*a*) Illustration of *Stylonychia lemnae*.   (*b*) The same organism,
seen with the scanning electron microscope.   (a) *From Ammermann
and Schlegel, in* Journal of Protozoology, *30:290, 1983. Copyright © 1983 Society of*

*Protozoologists, Lawrence, Kansas. Reprinted by permission. (b) The Journal of
Protozoology from Ammermann and Schlegel, 30:290, Allen Press, Inc. © 1983 Dr. Edna
Kaneshiro, editor.*

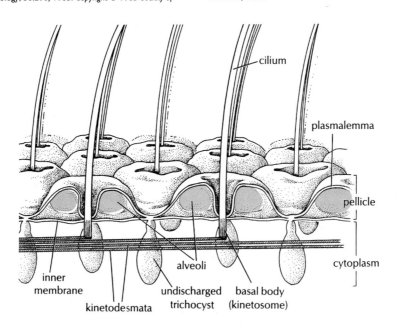

Figure 3.13
Illustration of pellicle, showing alveoli.
After Corliss; after Sherman and Sherman.

During conjugation, the macronuclei disintegrate, and the micronuclei, which are diploid, divide by meiosis so that four haploid **pronuclei** are formed from each micronucleus. Typically, all but one of these pronuclei degenerate, and the remaining pronucleus undergoes mitosis to form two identical haploid pronuclei. One of these two pronuclei somehow migrates through the cytoplasmic tube into the other individual. Exchange of pronuclei is reciprocal so that each individual winds up with one migratory pronucleus. Each migratory pronucleus subsequently fuses with the stay-at-home, stationary pronucleus, restoring the diploid condition through the formation of a **synkaryon** (i.e., a nucleus formed by the fusion of the pronucleus from one individual with its partner's pronucleus). Once the two conjugants separate, following the transfer and fusion of micronuclei, the synkaryon of each exconjugant divides mitotically from one to several times. Some of the products form micronuclei, while others give rise to macronuclei. Cytoplasmic divisions (i.e., actual reproduction of individuals) may follow, resulting in several individual offspring that are genetically dissimilar to the parental conjugants. All but one of the micronuclei disintegrate preceding these "distributive" divisions. The remaining micronucleus then divides mitotically, so that one is received by each of the offspring.

Although sexual dimorphism per se appears to be lacking in ciliates, the species that have been well studied do exhibit different **mating types.** These mating types, in turn, belong to separate varieties called **syngens.** For example, *Paramecium aurelia* has 16 to 18 known syngens, each containing several different mating types. Conjugation occurs only between individuals of different mating types and only between individuals within one syngen. In the case of *P. aurelia*, the various syngens have now been formally recognized as separate taxonomic species, each with its own set of mating types.

All ciliates are capable of asexual reproduction, as are all other protozoans. Among ciliates, asexual reproduction takes the form of transverse binary fission. The animal becomes bisected perpendicular to its long axis, so that complete individuals result from the anterior and posterior halves of the parent (Fig. 3.15). In contrast, binary fission in other groups of protozoa is longitudinal (i.e., parallel to the long axis), producing unicellular offspring that are mirror images of each other. By convention, offspring produced by binary fission are called daughters; this should not be taken as a reference to sexuality.

During binary fission in ciliates, the micronuclei divide mitotically and redistribute throughout the cytoplasm. The macronuclei elongate but do not undergo mitosis. In some cases, all macronuclei fuse prior to elongation, forming one very large macronucleus. A cleavage furrow gradually forms, dividing the macronucleus and the body itself into anterior and posterior halves. The detached, posterior half of the body must then regenerate all external and internal structures that have been appropriated by the anterior half and vice versa. In many species, much of this differentiation is completed prior to cell division. Commonly, all parental organelles are eventually replaced in both daughters. Such differentiation is, as always, under the control of the macronucleus. Current studies of pattern formation during ciliate reproduction may increase our understanding of pattern formation mechanisms in metazoan development.

One other form of nuclear reorganization commonly encountered among ciliates remains to be discussed. In this process, called **autogamy,** a form of sexual activity (in which new genotypes may be formed) takes place with the involvement of only a single individual! In some respects, events are reminiscent of the preliminaries to conjugation. The macronucleus (or macronuclei) degenerates while the micronucleus (or micronuclei) undergoes meiosis so that two pronuclei form from each micronucleus (Fig. 3.16). Meiosis is followed by several mitotic divisions of the pronuclei. Two of these pronuclei fuse, forming a zygote nucleus (a **synkaryon**), while the remaining division products disintegrate. Subsequent events, including the formation of a new macronucleus, resemble those of conjugation following separation of the conjugants.

Periodic renewal of the macronucleus appears to be essential for a number of protozoan species. Laboratory cultures will eventually die out after many generations of asexual reproduction if some form of nuclear reorganization is prevented from occurring periodically. Macronuclear regeneration thus appears to have a rejuvenating effect on the animals, which would otherwise apparently exhibit senescence and, ultimately, death.

Ciliate Lifestyles

A variety of lifestyles and surprisingly sophisticated behaviors are exhibited within the phylum Ciliophora (Research Focus Box 3.1). About 65% of ciliate species are free-living, and most of these are motile. Some other species form temporary attachments to living or nonliving substrates for feeding purposes, while others are permanently attached (i.e., **sessile**) and may form colonies. Although all ciliates have a distinct pellicle, some sessile species also produce a rigid, protective encasement termed a **test** or **lorica** (*lorica* = L: armor). Probably the best-known examples of such test-forming species are

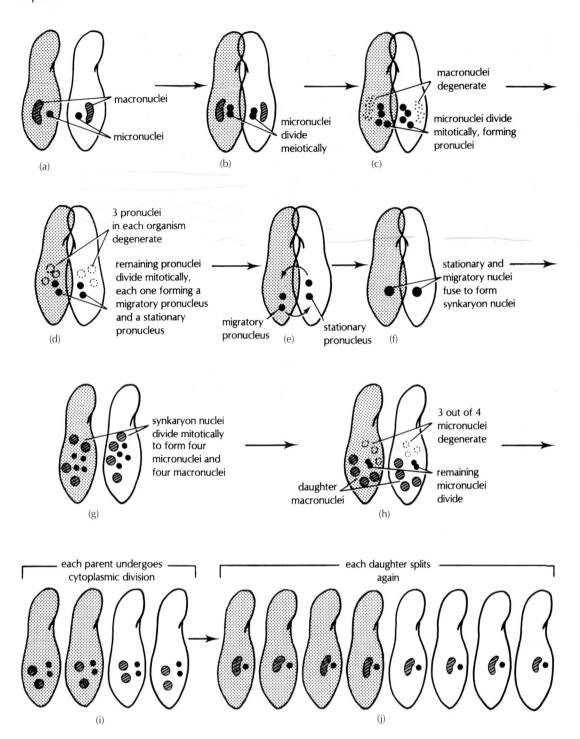

Figure 3.14

Conjugation in *Paramecium caudatum*. (*a*) Two individuals of different mating types come together. (*b*) A tubular cytoplasmic bridge forms between the two conjugants, and the micronuclei divide meiotically. (*c*) The macronuclei degenerate and the micronuclei divide mitotically, forming four pronuclei in each individual. (*d*) In each organism, three pronuclei degenerate and the remaining pronucleus undergoes another mitotic division, forming one migratory pronucleus and one stationary pronucleus. (*e*) The migratory pronuclei are exchanged through a cytoplasmic bridge and fuse with the stationary pronuclei (*f*). (*g,h*) The exconjugants separate, while a series of nuclear divisions and degenerations produce four macronuclei and a pair of micronuclei. (*i,j*) Cytoplasmic divisions produce four daughter individuals from each parent.

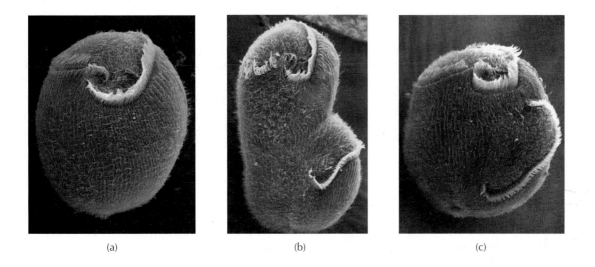

(a)   (b)   (c)

Figure 3.15
Binary fission in *Stentor coeruleus*.   *From D. R. Diener et al., The Journal of Protozoology 30:84, fig. 2A-D, Allen Press, Inc. © 1979 Dr. Edna Kaneshiro, editor.*

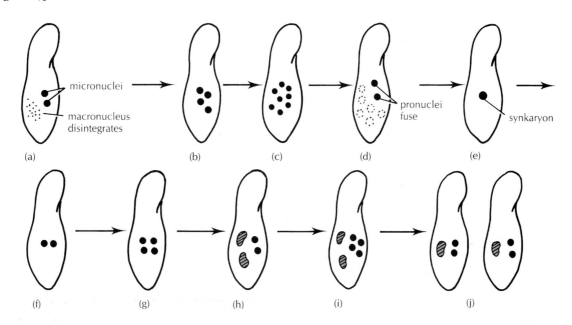

Figure 3.16
Diagrammatic illustration of autogamy. (*a*) Degeneration of macronucleus. (*b–d*) Micronuclei replicate meiotically and then mitotically; selective degeneration of haploid pronuclei follows. (*e*) Two pronuclei fuse to form a synkaryon. (*f–i*) A series of mitotic divisions produces macronuclei and micronuclei. (*j*) Cytoplasmic division produces two individuals, each with the proper number of micronuclei and macronuclei.

Research Focus Box 3.1

Ciliate Behavior

Malvin, G. M., and S. C. Wood. 1992. Behavioral hypothermia and survival of hypoxic protozoans *Paramecium caudatum*. *Science* 255:1423–25.

People tend to think of unicellular organisms as being particularly "simple." Yet protozoans often show remarkably complex behaviors and physiological responses whose mechanisms are not well understood despite many years of study. This paper concerns the behavioral response of a ciliated protozoan to low oxygen stress, or hypoxia (*hypo* = G: under; in this case, below normal). Multicellular animals faced with hypoxia tend to lower their body temperatures; that is, they exhibit hypothermia, a response that is generally believed to be adaptive in reducing oxygen demand by reducing metabolic rate. In this study, Malvin and Wood asked whether the ciliate *Paramecium caudatum* would migrate to lower temperature water under hypoxic conditions and whether such behavior prolonged its survival.

To test the effects of reduced oxygen partial pressure (PO_2, measured in torr) on ciliate behavior, the researchers pipetted large numbers of *P. caudatum* into a glass trough across which they established a temperature gradient ranging from about 0.4°C on one end of the trough to about 35°C on the opposite end of the trough. They monitored water temperature electronically at 12 points along the thermal gradient and monitored positions of the ciliates using a microscope coupled to a videorecorder. Gradually, they decreased PO_2 in the medium by gently flowing increasing amounts of nitrogen gas through the medium, monitoring changes in PO_2 continuously along with changes in average ciliate positions.

Under normal oxygen conditions the paramecia clustered at about 28°C, but they clustered at significantly ($P < 0.05$) lower temperatures when PO_2 was reduced to 1.3 torr or less (Focus Fig. 3.1). Moreover, the behavioral shift

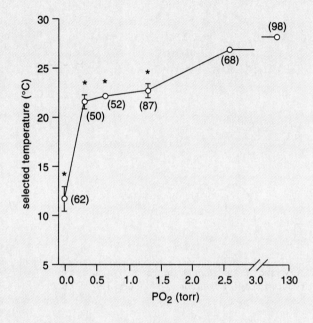

Focus Figure 3.1

The influence of PO_2 on the water temperature selected by the ciliate *Paramecium caudatum*. Each point represents the mean response (± one standard error about the mean) of 50–90 individuals as indicated in parentheses. * indicates means that are significantly different ($p < 0.05$) from the mean control response at 127 torr.

to lower temperature water had a clear survival advantage, as shown in a separate experiment (Focus Fig. 3.2). Groups of paramecia were maintained for up to 7.5 hours in either normal (127 torr) PO_2 or greatly reduced (0.3 torr) PO_2 at

encountered among three groups of ciliates known as tintinnids, folliculinids (Fig. 3.17), and peritrichs (not illustrated). Folliculinids are quite remarkable organisms in that, although they are permanently attached to solid substrate as adults, they are capable of dedifferentiating to a "larval" free-living form, abandoning their encasement, relocating at a distance from the original site, secreting a new case, and resuming the adult lifestyle. In fact, some species seem to dedifferentiate and then undergo fission, with the posterior half of the organism remaining behind to redifferentiate at the original location

while the anterior half of the animal leaves to establish itself elsewhere!

About one-third of all ciliate species are symbiotic in or on a variety of other invertebrates (including crustaceans, molluscs, bryozoans, and annelids) and vertebrates (including the digestive tracts of humans, producing an ulcerative condition of the intestine, and the skin of fishes, causing the disease known to freshwater fish aficionados as "ick"). Some species are commensals, attaching to the outer surface of organisms such as crabs, or within the body of such diverse hosts as cattle, horses, sheep, frogs, and cockroaches.

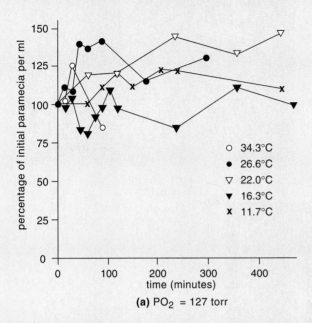

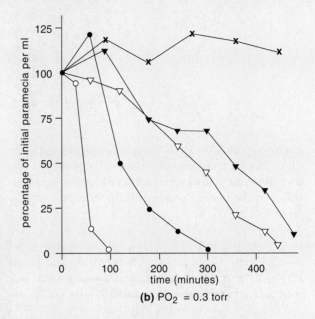

(a) PO$_2$ = 127 torr

(b) PO$_2$ = 0.3 torr

Legend for (a):
- ○ 34.3°C
- ● 26.6°C
- ▽ 22.0°C
- ▼ 16.3°C
- ✕ 11.7°C

Focus Figure 3.2

The influence of different constant water temperatures on survival of *P. caudatum* at control PO$_2$ (a) and at greatly reduced PO$_2$ (b) over about 7.5 hours. Ten 150 μl subsamples were removed from each culture at each indicated time period to determine changes in paramecium concentrations.

six temperatures (11.7–34.3°C); survival was monitored in subsamples of water up to six times during this period. The ciliates survived equally well at all temperatures under normal oxygen conditions (Focus Fig. 3.2a), but under hypoxia they survived well only at the lowest temperatures (Focus Fig. 3.2b). The tendency to seek out lower temperatures under the stress of reduced PO$_2$ thus has obvious adaptive value.

The extent to which the paramecia actually sought a lower temperature rather than the higher oxygen concentration associated with water of lower temperature remains to be explored, although the authors argue effectively that the organisms were responding to water temperature directly.

How would you design an experiment to distinguish between these two possibilities? The extent to which *P. caudatum* actually experiences such remarkably low oxygen conditions in the field also was not addressed in this paper, although many other ciliates are known to live in anoxic or hypoxic environments. In any event, this study demonstrates a clearly adaptive behavior mirroring that shown by multicellular animals with sophisticated nervous systems.

Figures from G. M. Malvin and S. C. Wood, "Behavioral hypothermia and survival of hypoxic protozoans Paramecium caudatum" in Science, 255:1423–25, 1992. Copyright © 1992 Association for the Advancement of Science, Washington DC. Reprinted by permission.

Most free-living ciliates are **holozoic;** that is, they ingest particulate foods. Some species may be **raptorial;** that is, they hunt and ingest animal prey. Still others may be passive suspension feeders, primarily ingesting clumps of bacteria, unicellular algae, other protozoans, or small metazoans.

The mouths of some raptorial species can expand to body width or larger, permitting the ingestion of prey that are large with respect to the ingestor. Members of one species, *Didinium nastum,* about 125 μm in length, can ingest prey (exclusively members of the ciliate genus *Paramecium*) two to three times larger than themselves (Fig. 3.18).

Sedentary suspension feeders, in contrast, are relatively passive. Food particles are carried to the mouth by water currents generated by ciliary activity in the mouth region. Commonly, the suspension-feeding species have stalks, as in members of the genus *Vorticella.* The stalks of such species may contain a **spasmoneme**—a helical, membrane-bound bundle of contractile fibers (Fig. 3.19f,g). Contraction of the spasmoneme fibers results in a very rapid shortening (and coiling) of the stalk and constitutes the organism's only escape response. Other sessile species, such as *Stentor* spp. (Fig. 3.19a), lack

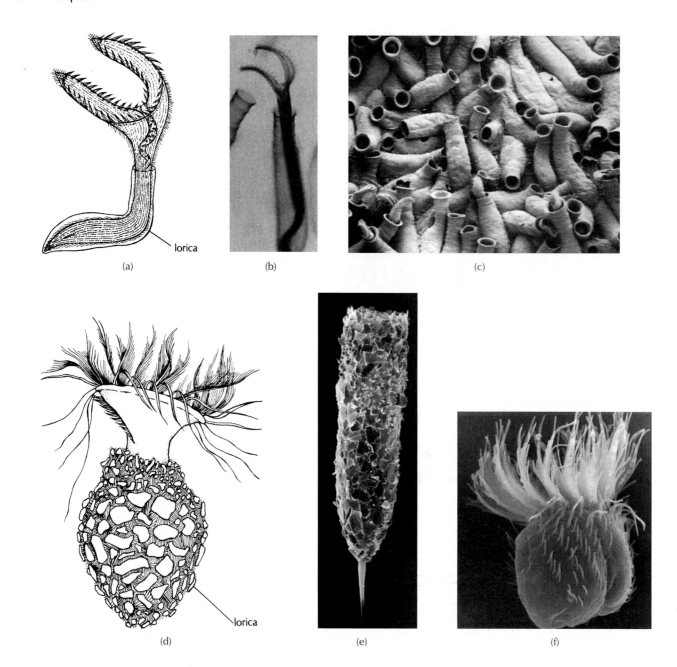

Figure 3.17
(a–c) Folliculinids: animals and loricas. (d) A tintinnid, *Stenosemella* sp. (e) Lorica of a tintinnid, *Tintinopsis platensis,* as seen with a scanning electron microscope. (f) *T. parva* without its test.
(a) *After Jahn, 1949.* (b,c) *From M. Mulisch and Hausmann, The Journal of*

Protozoology 30:97-104, Allen Press. © 1983 Dr. Edna Kaneshiro, editor. (d) From Corliss, in American Zoologist 19:573, 1979. Copyright © 1979 American Society of Zoologists, Reprinted by permission (e,f) From K. Gold, The Journal of Protozoology 28:415, Allen Press, Inc. © 1979 Dr. Edna Kaneshiro, editor.

stalks, but contractile fibers in the body proper often permit extensive shape changes to occur.

One group of usually sessile ciliates, the suctorians, merit special attention in that they, of all free-living ciliates, seem to have diverged farthest from the basic body plan. The body may attach to a substrate directly or by means of a long stalk (Fig. 3.19h). This stalk is never

contractile. Cilia are totally absent in adult suctorians, although the immature stages are ciliated and free-swimming. The systems of kinetodesmata of the immature dispersive stage are retained even after external ciliation is lost during maturation. A cytostome is never found, even in the immature stage. Besides the infraciliature, the only clear indication that adult suctorians are indeed

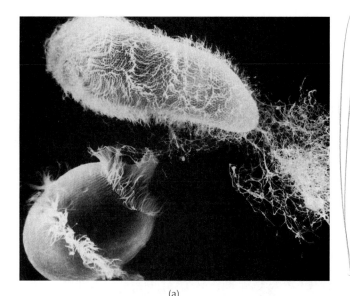

(a)

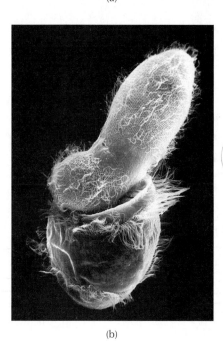

(b)

Figure 3.18
(*a*) This *Paramecium multimicronucleatum* has discharged numerous trichocysts in response to attack by another ciliate, *Didinium nastum*. (*b*) The persistent *D. nastum* begins to ingest its prey.
(a,b) *Courtesy of G.A. Antipa, San Francisco State University.*

ciliates is the presence of both macronuclei and micronuclei. Unlike with most other ciliates, asexual reproduction occurs not by binary fission, but by simple budding off of small parts of the adult. These buds differentiate into ciliated swimming forms that disperse, attach elsewhere, and differentiate to adulthood. Also, sexual reproduction involves the total fusion of the two conjugants rather than the temporary union described previously. One individual typically differentiates into a

microconjugant, which then swims to a partner. Conjugants also fuse in other sessile ciliates.

Since suctorians lack both a conventional cytostome and cilia, food collection and ingestion clearly must take place by methods atypical of ciliates in general. Small prey are captured by tentacles, which stud the general body surface (Fig. 3.19*h*). Some of these tentacles are hollow and bear knobs with individual "mouth" openings at their ends. Numerous membrane-bound organelles called **haptocysts** stud these knobs. When an appropriate prey organism contacts the tentacles, it triggers discharge of the haptocysts. The haptocysts then penetrate the prey's body surface, presumably by enzymatic secretion, and help maintain contact between prey and suctorian tentacle. Prey cytoplasm is then drawn through the lumen of the hollow tentacles into the suctorian body (Fig. 3.19*i*). These are indeed ingenious little suckers.

Amoeboid Protozoans— The Sarcodinids

Members of the next three phyla to be discussed (Rhizopoda, Heliozoa, and Radiozoa) were formerly grouped together as the Sarcodina and are still referred to here for convenience by the general term "sarcodinids." The sarcodinids were themselves formerly grouped with the flagellates and some others (the opalinids) in the phylum Sarcomastigophora (G: bearers of flesh or whips), a grouping that is no longer believed to be monophyletic.

Some 56,000 sarcodinid species have been described to date, with about 44,000 of those species known only as fossils. None of the sarcodinids have cilia or infraciliature, none have a pellicle, and none conjugate; where sexual reproduction is known to occur, gamete formation is usually involved. Some species produce flagellated gametes, which has intriguing phylogenetic implications, although no adult stages possess flagella. In addition, sarcodinids generally contain a single type of nucleus. A few species are multinucleate, but the nuclei are always monomorphic, in contrast to the distinctly dimorphic macronuclei and micronuclei found among ciliates.

Contractile vacuoles are present, particularly among freshwater species, although they are not fixed in position as they are in ciliates but move freely within the cytoplasm. As in the Ciliophora, specialized sensory organelles are absent. The typical sarcodinid is characterized above all by cytoplasmic extensions of the body called "pseudopodia," or "pseudopods" (*pseudo* = G: false; *pod* = G: a foot), used for locomotion, food collection, or both. Body shapes range from completely amorphous and ever changing to highly structured, with elaborate, rigid skeletal supports. Members of most species feed on small particulate material exclusively, although some sarcodinids house photosynthesizing endosymbiotic algae.

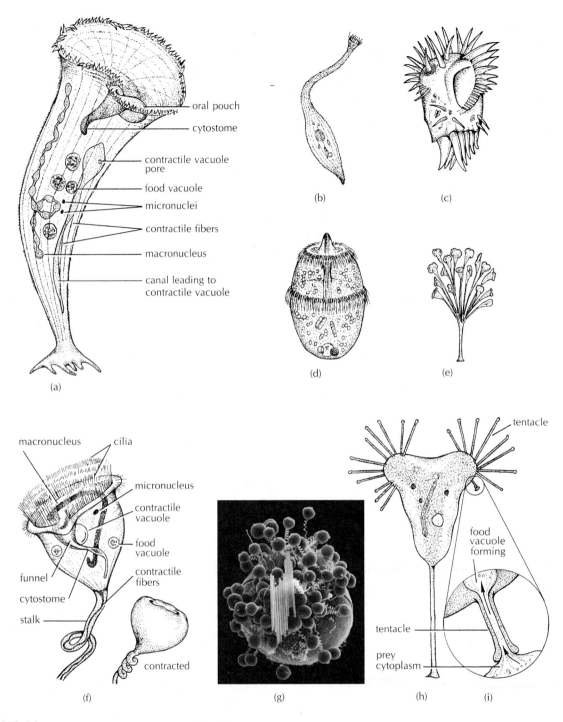

Figure 3.19

Ciliate diversity. (*a*) *Stentor.* (*b*) *Tracheloraphis kahli.* (*c*) *Diophyrs scutum.* (*d*) *Didinium nastum.* (*e*) A colony of *Epistylis* sp. (*f*) *Vorticella.* (*g*) Vorticellids covering the shell of a larval marine bivalve. Note the coiled, springlike stalks. The larval shell is approximately 200 μm in length. (*h*) A suctorian, *Tokophyra quadripartita.* The body is approximately 100–175 μm long. (*i*) Prey cytoplasm being moved through a hollow tentacle of the predaceous suctorian. Haptocysts aid in maintaining contact between prey and predator. (*a*) *After*

Sherman and Sherman. (*b,c*) *From Bayard H. McConnaughey and Robert Zottoli, Introduction to Marine Biology, 4th ed. Copyright © 1983 The C. V. Mosby Company, St. Louis, Missouri; redrawn from T. Fenchel, 1969, "Ecology of Marine Microbenthos IV" in Ophelia 6:1-182, July. Reprinted by permission.* (*d*) *From Hyman after Blochmann.* (*e*) *From Hyman.* (*f*) *After Sherman and Sherman.* (*g*) *Courtesy of C. Bradford Calloway and R.D. Turner.* (*h*) *From Hyman; after Kent.*

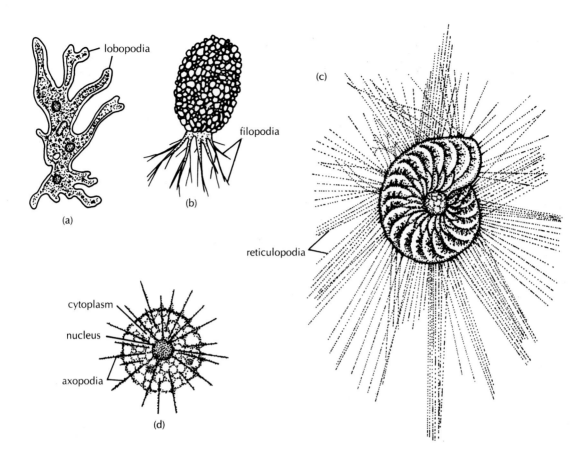

Figure 3.20
(*a*) A naked sarcodinid, *Amoeba proteus*, with lobopodia.
(*b*) *Pseudodifflugia* sp., with filopodia. (*c*) A foraminiferan,
Polystomella strigillata, with reticulopodia. (*d*) *Actinophrys sol*, with
axopodia.   (a) *From Hyman, The Invertebrates, Vol. III. Copyright © 1951 McGraw-Hill Book Company, New York. Reprinted by permission. (b) After Pennak. (c) After Kingsley; after Whiteley. (d) After Beck and Braithwaite.*

Most sarcodinids are free-living, although about 2% of known species parasitize vertebrate and invertebrate hosts, including other protozoans. Amoebic dysentery is probably the best-known human ailment caused by a member of this group. Some sarcodinid species are parasites within the bodies of other parasites; that is, they are **hyperparasites.**

All sarcodinid species reproduce asexually, chiefly by binary fission or multiple fission. In addition, sexual reproduction has been described for many species. In some cases, a single individual essentially becomes a gamete, which then fuses with another individual. In other cases, flagellated gametes are formed, pairs of which fuse to form zygotes. Encystment is common among the Sarcodina, especially in freshwater and parasitic species, providing a means of withstanding unfavorable environmental circumstances.

Phylum Rhizopoda

Phylum Rhizo • poda
(G: root foot)
rī-zō-pō′-dah

The approximately 4,000 members of this phylum are called *amoebae*, referring to their constantly changing body shapes (*ameba* = G: change). Rhizopods are common in freshwater, moist soil, and the sea. Some species are *naked*, with the body surrounded only by a plasma membrane (Figs. 3.20 and 3.21), but the bodies of most are partially surrounded with a single-chambered protective covering, called a **test** or shell. The test may be a secretion product of the cell itself or may consist of a structured agglutination of fine sand particles or detritus,

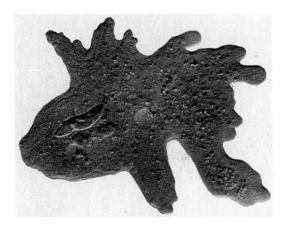

Figure 3.21

Chaos sp., a naked sarcodinid that can attain lengths of up to 5 mm. Note the numerous pseudopodia extending in various directions about the periphery of the organism. Also note the ingested *Paramecium* to the left of the (central) contractile vacuole.   *Supplied by Carolina Biological Supply Company.*

In such cases, the body is given form, and the pseudopodia project through one or more openings in the test (Fig. 3.20). Fossilized tests of shelled amoebae have been found in sedimentary rock deposited some 500 million years ago.

Members of this phylum exhibit several forms of pseudopodia, as illustrated in Figure 3.20. **Lobopodia** are typical of the familiar amoeba. These pseudopodia are broad with rounded tips, like *fingers* (*lobo* = G: a lobe), and bear a distinctly clear ectoplasmic area, called the **hyaline cap,** near each tip. In contrast, all other forms of pseudopod lack a hyaline cap and are very slender. **Filopodia** (*filo* = L: a thread) often branch, while **reticulopodia** always branch extensively and anastomose, forming dense pseudopodial networks (*reticul* = L: a network).

Most rhizopods move by using lobopodia or filopodia.[7] Typically, rhizopods flow into the advancing pseudopodium, a process called *cytoplasmic streaming.* The rhizopod body is thus truly formless, lacking permanent anterior, posterior, or lateral surfaces (Fig. 3.21); pseudopodia can generally form at practically any point on the body surface. This is also true of the parasitic species, including the ubiquitous agent of amoebic dysentery, *Entamoeba histolytica.* The mechanism by which pseudopodia form and change shape is not certain, although it seems clear that movement involves a controlled transition of cytoplasm between the gelatinous, ectoplasmic form (**gel**) and the more fluid, endoplasmic form (**sol**). The factors coordinating this transformation in different parts of the body are not fully understood, although the hypotheses have become wonderfully complex during the past decade. The transformation may in-

volve the interaction of actin and myosin molecules, both abundant in amoeba cytoplasm. A model of sliding actin filaments has been proposed, in which actin and myosin interact in a way that resembles their interaction in the muscle tissue of multicellular animals. Other hypotheses minimize the role of myosin in the gel-sol-gel transitions and instead emphasize the potential role of selective actin polymerization and depolymerization. In one model (Fig. 3.22), localized actin disassembly creates an area of increased osmotic pressure, which draws water from the more central region of the body to the periphery, forming a pseudopod. The actin is then repolymerized to form a bracing network, fixing the pseudopod's shape until the next bout of depolymerization. Whatever the mechanism, pseudopodial locomotion is extremely slow, usually less than 300 μm per minute.

All free-living rhizopods are particle feeders, and all use their pseudopodia to capture food.[8] Commonly, ingestion involves **phagocytosis** (*phago* = G: to eat). In this process, lobopodia or filopodia advance on both sides of, and often on top of, the intended prey, forming a **food cup.** The tips of the pseudopodia soon come together and fuse, thereby internalizing the prey item. Sarcodinids typically feed on bacteria, other protozoans, and unicellular algae (e.g., diatoms). In **pinocytosis** (*pino* = G: to drink), much smaller pseudopods are formed, capturing extremely small particulates or fluids rich in dissolved organic matter. Reticulopodia function as a food-catching net, as well as in locomotion, whereas axopodia are used almost exclusively for capturing food particles. Species with axopodia are incapable of active locomotion, most species being free-floating in fresh- or salt water.

About 50% of all extant free-living sarcodinids are foraminiferans (*foramin* = L: an opening; *fera* = L: to bear). These organisms are primarily marine. There are no parasitic species, although some are ectocommensal. Foraminiferans secrete multichambered tests, largely of calcium carbonate (Fig. 3.23). The individual chambers of these tests are often demarcated from each other by perforated septa. Reticulopodia project from minute openings in the test (Fig. 3.20c) and function primarily in food capture. Repeated extension and shortening of the pseudopodia also permit slow crawling over the ocean bottom. Foraminiferans feed on a remarkable variety of food, including other protists, small metazoans, fungi, bacteria, and organic detritus. In addition, some species house photosynthesizing symbionts, including dinoflagellates, and others can take up dissolved organic material from seawater.

Foraminiferan tests (representing some 30,000 species!) abound in the fossil record. The oldest specimens date from the Early Cambrian, about 550 million years ago. Although tests of present-day species rarely exceed 9 mm in diameter, with those of most species being

---

7. See *Topics for Further Discussion and Investigation*, no. 1.

8. See *Topics for Further Discussion and Investigation*, no. 6.

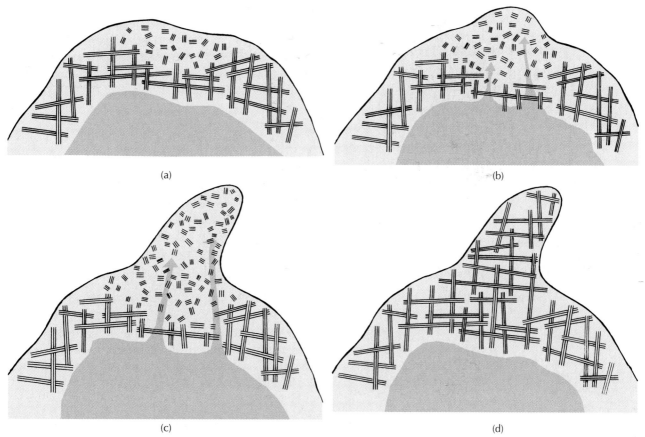

(a)

(b)

(c)

(d)

Figure 3.22
Hypothesized mechanism of pseudopod formation during amoeboid locomotion. (*a*) Localized breakdown of the actin network increases osmotic concentration in that part of the cytoplasm. (*b*) Fluid from the interior of the cell moves toward the periphery along the osmotic gradient, forming a pseudopod (*c*). (*d*) Actin is repolymerized, reforming a stabilizing network of filaments.   *From T. P. Stossel, "How Cells Crawl" in* American Scientist, *78:408-23, 1990. © American Scientist. Reprinted by permission.*

less than 1 mm, fossilized tests of up to 15 cm in diameter have been found—quite a remarkable size for a protozoan. The extensive fossilized remains of foraminiferans have taken on considerable economic importance; for example, cement and blackboard chalk are foraminiferan-containing products. In addition, geologists have become quite interested in certain foraminiferan fossils as fairly reliable indicators of likely places to drill for oil.

Phylum Radiozoa

Radio • zoa
(L: ray animals)
rā´-dē-ō-zō´-ah

The members of this phylum, called radiolarians and acantharians, are spherical, supporting their pseudopodia with thin, radiating microtubules that give a spiny, rayed appearance to many species (Fig. 3.24a,b). The complexity and symmetry of their rigid infrastructure make many radiolarians exceedingly beautiful. The pseudopodia and their microtubular supports are called *axopodia* (*axo* = G: an axle). Radiozoans also possess a rigid endoskeleton composed either of silica (in radiolarians) or strontium sulfate (in acantharians). Because of their siliceous (silica-containing) skeletons, radiolarians, like the foraminiferans and shelled amoebae, are prominent in the fossil record; of the 11,000 described radiolarian species, about 3,300 are extant and 7,700 are known only as fossils. In contrast, acantharians have left no fossil record, as the strontium sulfate supports degrade soon after the organism dies. Only about 500 acantharian species have been described.

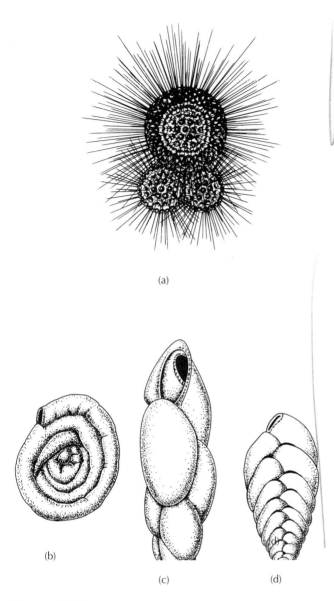

(a)

(b)

(c)   (d)

Figure 3.23
Foraminiferan tests. Probably 80% of all described foraminiferan species are extinct and are known only from the fossil record.
(*a*) *Globigerina bulloides.* (*b*) *Glomospira gordialis* (×140).
(*c*) *Stainforthia concava* (×200). (*d*) *Brizalina spathulata* (×100).
(a) *After Kingsley.* (b-d) *From Haynes, Foraminifera. Copyright © 1981 Macmillan Accounts and Administration, Ltd., Hampshire, England.*

In contrast to the mostly bottom-dwelling Foraminifera, all radiolarians and most acantharians are planktonic organisms, passively carried about by ocean currents. Many species possess symbiotic algae and so meet at least some of their nutritional requirements through photosynthesis. They also feed as carnivores, capturing microscopic prey with the cytoplasm that flows along their axopods.

The radiolarian body is generally spherical and is divided into an **intracapsular zone** and an **extracapsular zone** by a perforated, spherical membrane or capsule. Food vacuole formation and digestion occur in the extracapsular region. The nucleus is contained in the intracapsular zone. The acantharian body is also composed of distinct layers, with the innermost layer containing a number of small nuclei.

Phylum Heliozoa

Phylum Helio • zoa
(G: sun animals)
Hēl´-ē-ō-zō´-ah

Like the radiolarians and acantharians, heliozoans are primarily floating organisms with axopodia (Fig. 3.24c), but they are largely restricted to freshwater habitats. As with the radiozoans, heliozoan bodies are demarcated into (1) a frothy outer region of ectoplasm in which digestion occurs and (2) a less highly vacuolated inner region of endoplasm containing the nucleus. However, heliozoans do not show a distinct physical boundary (i.e., a capsular membrane) between the two regions. The axopodia serve primarily to capture food items, although they function in locomotion in some species, too. In such cases, a coordinated retraction and reextension of axopods permit individuals to roll slowly over a surface. Microtubules located in the cores of the axopods are apparently involved in mediating the changes in axopod length. Only about 200 heliozoan species have been described.

Flagellated Protozoans

Flagellates are quite unlike the sarcodinids, both structurally and functionally. As we will see shortly, however, the dramatic morphological differences between adult flagellates and adult amoebae are somewhat misleading; some very intriguing transitional forms suggest a close evolutionary relationship between the two groups.

Flagellates (or mastigophorans, depending upon whether one's classical affinities are Latin or Greek, respectively) are characterized by the possession of a pellicle, giving the body a definite shape, and most especially by the possession of one or more flagella. Most species are free-living and motile. A cytostome is present in some species, but its morphology is never as complex as that encountered within the Ciliophora. Contractile vacuoles may be present, particularly in freshwater species; if present, their position is fixed within the cell cytoplasm, in contrast to the contractile vacuoles of sarcodinid species.

Flagella are structurally much like cilia. In cross section, flagella exhibit the same characteristic arrangement of nine pairs of microtubules ringing a pair of central

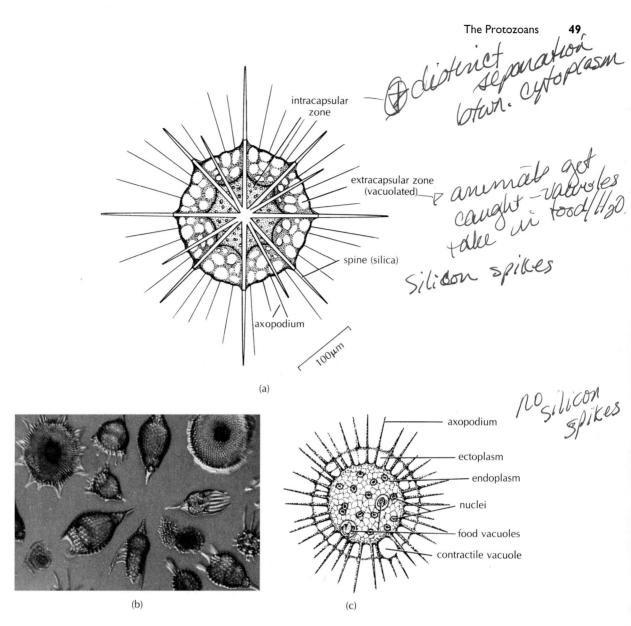

(handwritten notes:)
① distinct separation btwn. cytoplasm

animals get caught – vacuoles take in food/H₂O.

Silicon spikes

no silicon spikes

Figure 3.24
(*a*) A radiolarian, *Acanthometra elasticum.* (*b*) Radiolarian
skeletons, composed of silica. (*c*) A heliozoan, *Actinosphaerium* sp.
(a) *After Farmer.* (b) © *Eric V. Grave.* (c) *After Beck and Braithwaite.*

microtubules over most of their length. Like cilia, flagella are produced from basal bodies, and the movement of flagella is believed to involve the sliding of microtubules in relation to each other. The cross-sectional appearance of the basal body of a flagellum is modified as previously described for cilia. Indeed, the structural and functional similarities between cilia and flagella are so striking—and differ so much from those of bacterial flagella—that eukaryote cilia and flagella are sometimes grouped together under the descriptive heading **undulipodia.**

Unlike cilia, however, flagella often bear numerous, external hairlike projections (**mastigonemes**) along the length of the organelle. Presumably, these mastigonemes increase the effective surface area of the flagellum, thus increasing the power that it can generate when it is moved through the water. Flagella are usually longer than cilia, and the typical flagellate bears many fewer flagella than the typical ciliate has cilia. Unlike with cilia, several waves of movement may be in progress simultaneously along a single flagellum, and the wave may be initiated at the tip of the flagellum rather than at the base, pulling the organism forward. Flagellate locomotion can be fairly rapid, up to about 200 µm per second. This is only about one-tenth the speed attained by many ciliates but is about 40 times that reached by the fastest amoebae.

More than 7,500 flagellated protozoan species have been described and named to date. Flagellates include photosynthesizing, particle-feeding, and parasitic species, as described next. To simplify discussion, I have grouped flagellates according to lifestyle. The protozoan

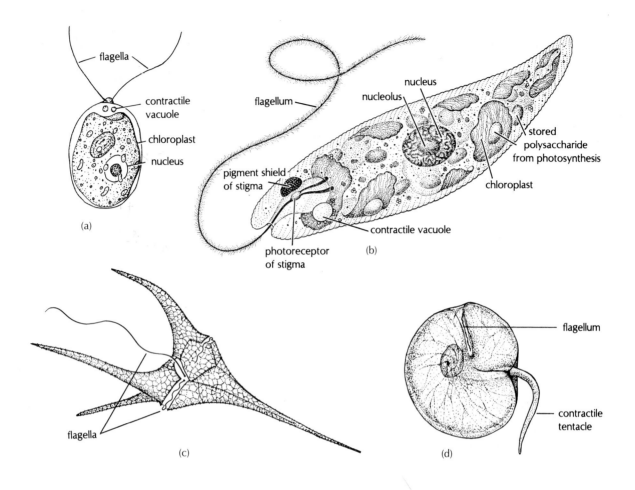

Figure 3.25

Flagellate diversity: phytoflagellates. (*a*) *Chlamydomonas.*
(*b*) *Euglena.* (*c*) *Ceratium hirundinella,* an armored dinoflagellate.
Like many other dinoflagellates, the members of this genus are
encased in sturdy cellulose plates. (*d*) *Noctiluca scintillans,* an
unarmored dinoflagellate. Unlike other dinoflagellates, these
possess only a single, very short flagellum and a long, sticky,
contractile tentacle that catches particulate food. Members of
this genus are strikingly bioluminescent.   (a & c) *After Pennak.* (b) *From
Purves and Orians, Life: The Science of Biology, 2d ed. Copyright © 1983 Sinauer
Associates, sunderland MA. Reprinted by permission.*

flagellates discussed here are distributed among five different phyla, as indicated in the *Taxonomic Summary* at
the end of this chapter. Many other flagellated species
have been assigned to other kingdoms and are therefore
omitted from this presentation; farewell, dear *Volvox*
(now placed in the kingdom Plantae).

The Phytoflagellated Protozoans

Some flagellated protozoans contain chlorophyll, obtain
their energy directly from sunlight, and rely exclusively
on carbon dioxide as a carbon source. Other photosynthesizing species use light as an energy source but require
various dissolved organic compounds as well. Neither
type of organism has a mouth or forms food vacuoles.
Together, these two groups comprise the flagellated
plant-like protozoans (Fig. 3.25). Both freshwater forms,
such as *Euglena,* and marine forms, including the di-

noflagellates, are common. Dinoflagellates are best
known for commonly exhibiting **bioluminescence** (biochemical production of light) and for occasionally producing highly toxic "red tides." These are dense aggregations of certain dinoflagellate species whose neurotoxins
(saxitoxin and related compounds) kill fish and crustaceans and accumulate in the tissues of otherwise edible
clams, mussels, and oysters. Various benthic dinoflagellates produce a similarly potent neurotoxin that accumulates in certain warm-water fish. This "Ciguatera poisoning" can kill people that eat the fish, although the fish
themselves are somehow unaffected.

Many—but not all—dinoflagellates contain chlorophyll and photosynthesize. Individuals of some of these
autotrophic species live intracellularly as important symbionts in the tissues of various multicellular marine invertebrates, including reef-forming corals. In such cases,
the photosynthesizing activities of the dinoflagellates
(now referred to as **zooxanthellae**) contribute significantly

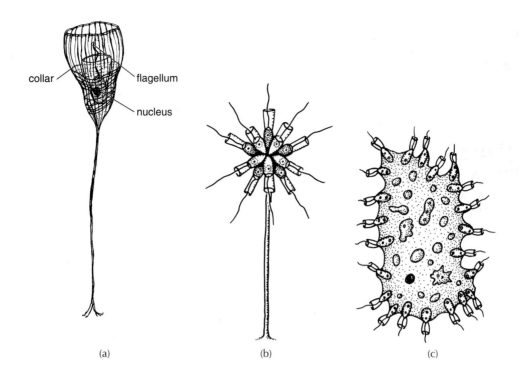

Figure 3.26
Flagellate diversity; zooflagellates. (*a–c*) Choanoflagellates:
(*a*) *Stephanoeca campanula* (*b*) *Codosiga botrytis*, a colonial species.
(*c*) *Proterospongia*, another colonial species, with the individuals
embedded in a thick, gelatinous matrix.   (a) *From H. A. Thomsen,*

*K. R. Buck, and F. P. Chavez, in Ophelia 33:131-164, 1991. Copyright © 1991 Ophelia
Publications, Helsingor, Denmark. Reprinted by permission. (b) After Kingsley. (c) From
Hyman; after Kent.*

to the nutritional needs of their host. Other species, such
as the bioluminescent *Noctiluca* (Fig. 3.25*d*), lack chlorophyll and feed only as heterotrophs, ingesting particulate
foods.

Many phytoflagellate species bear a red, cup-shaped,
photosensitive organelle called a **stigma** (Fig. 3.25*b*).
This is one of the few true sensory organelles known
among protozoans. Both flagella and stigma are clearly
adaptive to the extent that they help individuals maintain
themselves in the narrow region of the water column
where sufficient light is available for net photosynthesis
to occur.

Although some species are completely **autotrophic**
(i.e., self-nourishing through photosynthesis), many can
feed on particulate foods if necessary. Some euglenids,
for example, become holozoic if maintained in darkness
for a sufficient time. These individuals then ingest solid
food in perfectly good animal-like fashion.

The Zooflagellated Protozoans—Free-Living Forms

Most strictly animal-like flagellate species (i.e., the
zooflagellates) are free-living in either fresh or salt water.
One of the most interesting groups of free-living species
are the choanoflagellates, a group found primarily in
freshwater. Many choanoflagellate species are sessile,

being permanently attached to a substrate (Fig. 3.26).
Each individual bears a single flagellum, which extends
for part of its length through a cylindrical network (collar) of closely spaced protoplasmic strands called **microvilli.** Flagellar movements create feeding currents;
small food particles stick to the microvillar collar and are
ingested. Commonly, individuals are stalked and/or embedded in a gelatinous secretion. Most species are colonial and immobile. Members of the genus *Proterospongia*
form (planktonic) colonies of up to several hundred
cells; these colonies may bear a striking resemblance to
primitive sponges, as discussed in the next chapter.
Whether this similarity reflects a true phylogenetic relationship between the unicellular flagellates and the multicellular sponges or whether the similarities are a product of independent, convergent evolution is not certain.

The Zooflagellated Protozoans—Parasitic Forms

About 25% of zooflagellate species are parasitic or commensal with plants, invertebrates, and vertebrates, including humans. Species in these parasitic groups often exhibit
levels of structural and functional complexity not observed
in any other sarcomastigophoran. Life cycles are especially
complex, as they include adaptations for perpetuating the
species from host to host. **Intermediate hosts** commonly

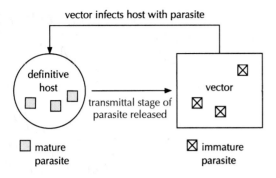

vector infects host with parasite

definitive host

transmittal stage of parasite released

vector

☐ mature parasite

☒ immature parasite

Figure 3.27
Diagram of a parasitic life cycle in which an invertebrate vector transfers the parasite from one definitive host to another. The parasite cannot attain sexual maturity within the body of the vector.

serve as **vectors,** which transport developing individuals from one **definitive host** to another; sexual maturity of the protozoan parasite can be attained only in the definitive host (Fig. 3.27). The problems implied in maintaining a parasitic lifestyle—such as locating or attracting proper intermediate and definitive hosts, adapting to the physiological requirements of different hosts, and evading host immune responses—are common to parasites of all phyla; these issues are discussed in a subsequent section of this chapter and in Chapter 7.

Trypanosomes are parasites of special interest to humans (Fig. 3.28). Various species are pathogens in flowering plants, in people, and in cattle, sheep, goats, horses, and other domesticated animals. A key morphological characteristic is the possession of a single, very large mitochondrion, called a **kinetoplast** (Fig. 3.28a). Trypanosomes, with the tsetse fly as vector, are responsible for the well-known African sleeping sickness and for many other human ailments. *Leishmania donovani*, for example, causes extreme disfigurement and death (with a mortality rate approaching 95%) in many areas of the world. About 12 million people are affected. Sand flies serve as vectors for the transmission of leishmanial parasites from human to human. Dogs also serve as suitable definitive hosts, functioning, in a sense, as reservoirs for later human infestation. The human defense system recognizes these trypanosomes as foreign beings, and the trypanosomes are quickly phagocytozed by appropriate cells (macrophages) of the human reticuloendothelial system. Remarkably, however, *L. donovani* is not then digested within these cells but instead succeeds in greatly increasing in numbers through repeated intracellular binary fission. How the host defense system is circumvented in this way is not known. Other trypanosomes are equally successful at evading the host's humoral immune response, either by changing the chemical composition of

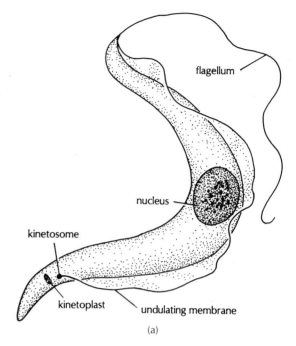

flagellum

nucleus

kinetosome

kinetoplast

undulating membrane

(a)

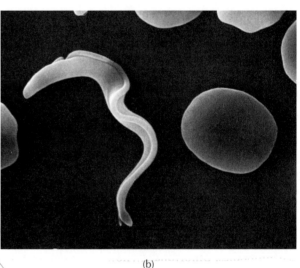

(b)

Figure 3.28
Flagellate diversity: trypanosomes. (*a*) *Trypanosoma lewisi.* Note the undulating membrane (not to be confused with the ciliate structure of the same name). (*b*) Scanning electron micrograph (5,500 × magnification) of an African trypanosome (*Trypanosoma brucei brucei*) among host red blood cells. This trypanosome is a close relative of the subspecies causing African sleeping sickness. (a) *Sherman/Sherman, The Invertebrates: Function and Form, 2/e, © 1976, pp. 111, 12, 236, 172, 169, 45, 9, 15. Reprinted by permission of Prentice Hall, Upper Saddle River, New Jersey. (b) From Donelson, J. E. 1988. The Biology of Parasitism. New York: Wiley-Liss, 371-400, fig. 1, 372.*

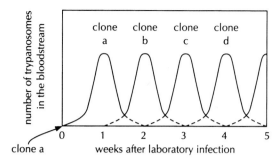

Figure 3.29

The cyclic appearance of trypanosomes in the host bloodstream. The test animal was initially injected with a single trypanosome, which gave rise to other individuals (clone *a*) by asexual replication; all of these individuals express a single surface antigen. Most of these individuals are destroyed by the host immune system within a few days, by which time a few trypanosomes have begun producing a new surface antigen, one not recognized by the circulating antibody produced in response to clone *a*. Clones *b*, *c*, and *d* represent trypanosomes expressing three distinctly different surface antigens. With the periodic expression of new surface antigens, the host can never eliminate the infection.   *From J. E. Donelson, in The Biology of Parasitism. Copyright © 1988 Alan R. Liss, Inc., New York. Reprinted by permission of John Wiley & Sons, Inc., New York.*

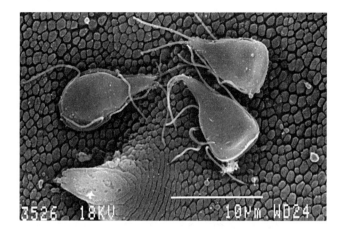

Figure 3.30

Scanning electron micrograph of *Giardia* sp. (three individuals) attached to the outer body wall (tegument) of the parasitic flatworm *Echinostoma caproni*. Both parasites were taken from the small intestine of an experimentally infected golden hamster.   *Courtesy of S.W.B. Irwin, University of Ulster at Jordanstown, Northern Ireland.*

their antigenic surface coats frequently enough to prevent the host from producing specific antibody in effective concentrations or by quickly invading specific tissues that are reasonably well isolated from circulating antibody.

The trick of changing surface antigens has been especially well studied in African trypanosomes, *Trypanosoma* spp. Each individual is covered with a single protein. One would think that the human immune system could easily deal with such a target, and that it should be a simple matter for scientists to produce antibodies against that protein or to help the host produce such antibodies. Unfortunately, however, the biochemical nature of the surface antigen changes as the parasite population multiplies within the host. Initially, all the individuals in an infective population have the same external protein composition. The host immune system turns on and begins producing antibodies against that surface protein; within a few days, most of the invaders are, in fact, destroyed. But for a few individuals in the population, a new gene is activated during a division cycle so that a new population begins to develop, one composed of individuals that present a new antigenic surface—one not targeted by the circulating antibody (Fig. 3.29). By the time the host immune system has produced sufficient antibodies to handle the second parasite population, a new variant with a third type of surface coating has appeared, and yet another population of trypanosomes quickly grows (Fig. 3.29). In this way, the trypanosomes stay one step ahead of the humoral immune system. One

trypanosome has been estimated to hold over 100 different genes coding for surface proteins. Since each gene can be activated more than once, the parasite has more than enough gene combinations to maintain an infection for the life of the host. The molecular mechanisms responsible for turning the various genes on and off are currently under intensive study; preventing the parasite from changing its coat would seem an excellent way of combating the disease. At least 15 million people—and innumerable livestock—become infected by various trypanosomes yearly.

Assuming that natural selection ultimately promotes associations that do not kill the host (an obvious detriment to continued existence of the parasite population), the most highly evolved associations between parasitic zooflagellates and their hosts must be nonpathogenic in nature. The trichomonads (phylum Parabasala) represent one such group of primarily commensal flagellates. Nonpathogenic species generally attract less research attention (and funding) than do pathogenic species, but some trichomonads wreak enough havoc upon humans to have merited careful study. *Trichomonas vaginalis*, for example, is a typically small protozoan parasitizing the human vagina, prostate, and urethra. *T. vaginalis* generally causes little or no discomfort in men, but it often produces considerable inflammation and irritation in women. The disease is readily transmittable, sexually and otherwise (through contact with toilet seats and towels, for example).

Other unpleasant flagellates are contained in the genus *Giardia* (Fig. 3.30). One species, *G. lamblia*, causes intense nausea, cramps, and diarrhea in its human hosts, a condition that may take several months to dissipate. The infection is readily transmitted through drinking

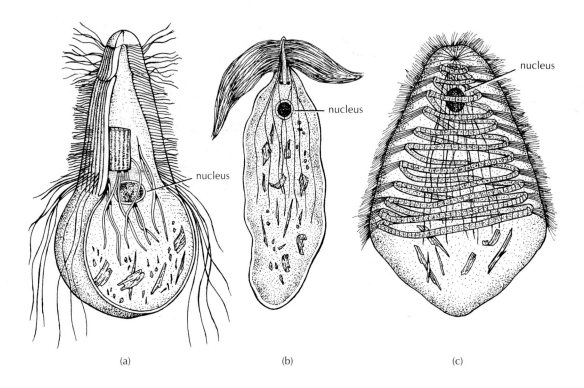

(a)   (b)   (c)

Figure 3.31

Symbiotic flagellates, the most morphologically complex of all flagellate species. (*a*) *Trichonympha collaris* (about 150 μm), taken from the intestine of a termite. (*b*) *Rhynchonympha tarda*, from the gut of a wood roach. (*c*) *Macrospironympha xylopletha*, from the gut of a wood roach.   (a) *From Hyman; after Kirby.* (b & c) *From Hyman; after Cleveland.*

water contaminated with feces. Many sadder-but-wiser hikers have discovered that even water in pristine wilderness areas may be so contaminated.

Probably the most morphologically complex flagellate species are the hypermastigotes (phylum Parabasala) living commensally in the guts of termites, cockroaches, and wood roaches (Fig. 3.31). A great many individuals can easily be obtained for study by squeezing out some of the fluid from the guts of termites. One species alone (*Trichonympha campanula*) may account for up to one-third of the biomass of an individual termite! The species in this group are generally large (for flagellates), typically attaining body lengths of several hundred micrometers. The bodies are commonly divisible into several morphologically and functionally distinct regions and typically bear numerous (up to several thousand!) flagella.

Perhaps the most intriguing feature of the biology of these hypermastigid protozoans is the mutual dependence between host and flagellate. Several species inhabiting termite guts are capable of digesting cellulose, something that most animals, including termites, cannot do. If the termite is deprived of its flagellates, it soon dies, even though it never stops ingesting wood. Similarly, wood-eating roaches soon die if deprived of their flagellate gut fauna. The flagellates, on the other hand, gain from their hosts protection from desiccation and predation and rely upon the host for a source of cellulose to digest; cellulose is the primary carbon source for these protozoans.

Some Transitional Forms and Misfits

I mentioned earlier (p. 43) that some sarcodinids form flagellated gametes in the course of sexual reproduction. The presence of flagellated gametes in the amoeboid life cycle has intriguing phylogenetic implications. Indeed, there are many protozoans that offer additional tantalizing links between amoeboid and flagellated species; the actual phylogenetic implications of these transitional forms are far from being resolved.

One transitional species is *Naegleria gruberi*, an apparently primitive protist that lacks mitochondria and typical Golgi bodies and whose ribosomes exhibit a number of prokaryotic features. *N. gruberi* and its relatives are commonly referred to as *amoebomastigotes* or *amoeboflagellates* because they are clearly neither full-fledged amoebae nor full-fledged flagellates. Under normal laboratory conditions, *N. gruberi* is a completely convincing, amorphous,

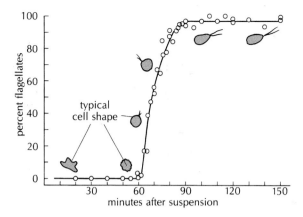

Figure 3.32

The transformation of *Naegleria gruberi* from an amoeboid form to a flagellated form. The transformation was initiated at 25° C by moving the amoebae from the surfaces on which they had been living, into suspension in a test tube. The total number of cells in the culture did not change appreciably during the experiment.

From Fulton and Dingle, in Developmental Biology, 15:165, 1967. Copyright © 1967 Academic Press, Inc. Reprinted by permission.

crawling amoeba. Lobopodia form anywhere on the cell surface for locomotion and phagocytosis, and the single contractile vacuole has no fixed position within the body. However, a variety of stimuli, including salinity changes or simply maintaining the animals in suspension, quickly induce a complete transformation of the amoebae into equally convincing flagellates.[9] A pellicle forms, the contractile vacuole becomes fixed in one position, and functional flagella appear. The entire transformation, which takes place within about one and one-half hours, has been the subject of considerable study by developmental biologists (Fig. 3.32). At least one species of *Naegleria* recently has been shown to be a facultative parasite of humans. It probably enters through the nasal passages while the host is swimming and then migrates to the brain, killing the host within a week.

A second transitional form consists of a major group of organisms, the Mycetozoa (*myceto* = G: a fungus; i.e., the "fungus animals") or the cellular and acellular slime molds. For reasons that will soon become apparent, the slime molds are often classified as plants or as fungi. Nevertheless, for much of their lives, the slime molds exist as individual amoeboid cells, moving and feeding on particulate matter by means of lobopodia. Most species are harmless and are commonly found among decaying vegetation in terrestrial habitats, although some species are plant pathogens of considerable commercial importance. In all cases, the amoeboid individuals eventually aggregate (Fig. 3.33). For the cellular slime molds, aggregation clearly results from chemical communication between cells; probably in response to dwindling

food supplies, some of the amoebae secrete cAMP, which attracts other amoebae to form a large multinuclear mass—essentially a giant, multinucleate amoeba. The acellular slime molds form a true *plasmodium,* in which cell membranes are lost to form a large, immobile syncytial mass. In contrast, the cellular slime molds form what is termed a *pseudoplasmodium,* in which individual cell membranes persist. The pseudoplasmodium is motile, and it continues to move and feed in standard amoeboid fashion for some time (Fig. 3.33).

If two pseudoplasmodia or plasmodia of the same species come into contact with each other, they often fuse, creating an even larger "individual." A decrease in nutrient availability causes the pseudoplasmodium or plasmodium to develop funguslike **sporangia,** or fruiting bodies. Highly resistant spores are released from these sporangia, and only appropriate environmental conditions will bring about germination of these spores. An amoeboid form or a flagellated form may emerge from a spore, depending upon the species. In some cases, the amoeboid individual subsequently transforms into a flagellated individual, which proceeds to eat and undergo repeated binary fission. Eventually, the flagellates give rise to amoebae, which feed and ultimately aggregate to form a multinucleate plasmodium or pseudoplasmodium that then differentiates to produce another round of resistant spores.

Another intriguing group of transitional forms is the archamoebae, such as *Mastigamoeba* (Fig. 3.33b). These organisms possess both locomotory flagella and pseudopodia, either simultaneously or in succession. These species are clearly at the crossroads of flagellate-amoeba organization and provide further evidence of a close evolutionary relationship between the two forms. Archamoebae are most commonly encountered in freshwater environments.

Members of the phylum Labyrinthomorpha are strange, free-living organisms forming colonies of cells that move exclusively within a secreted slime network (Fig. 3.34), giving rise to the common name for the group: slime nets. Some labyrinthomorphs are terrestrial, but most live on aquatic algae and grasses in both freshwater and marine environments. Individual cells are only about 10 μm long, but a colony may measure several centimeters across. The cells can move only if completely surrounded by slime. They lack cilia and flagella and do not produce pseudopods. Nevertheless, they travel along their slime tubes at 10 cm per minute; the locomotory mechanism remains uncertain. The cells apparently feed on particulate foods through extracellular digestion.

Under adverse conditions, groups of cells aggregate to form resistant cysts that can later rupture to reestablish a colony. Sexual reproduction involves meiosis to form flagellated gametes, which is why I mention these

---

9. See *Topics for Further Discussion and Investigation,* no. 5.

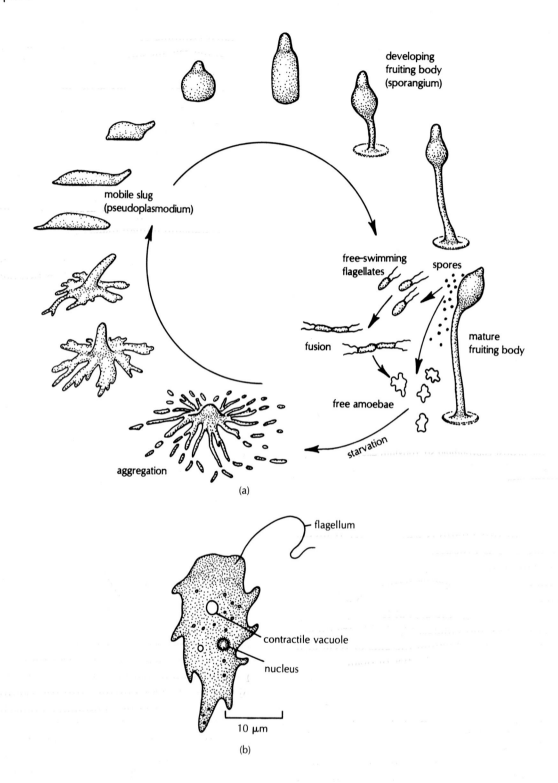

developing
fruiting body
(sporangium)

mobile slug
(pseudoplasmodium)

free-swimming
flagellates

spores

mature
fruiting body

fusion

free amoebae

aggregation

starvation

(a)

flagellum

contractile vacuole

nucleus

10 μm

(b)

Figure 3.33

The life cycle of a cellular slime mold. *Dictyostelium* is perhaps one of the best-studied slime molds. Note that the spores may give rise to swimming flagellates, which eventually fuse and form amoebae, or may germinate amoebae directly, depending upon the species. The slug stage is typically formed from about 100,000 individual amoebae. (b) *Mastigamoeba longifilum.* (a) *Adapted from Biological Science, Third Edition, by William T. Keeton, illustrated by Paula De Santo Bensadoun, by permission of W. W. Norton Company, Inc. Copyright © 1980, 1979, 1972, 1967 by W. W. Norton & Company, Inc. (b) After Farmer.*

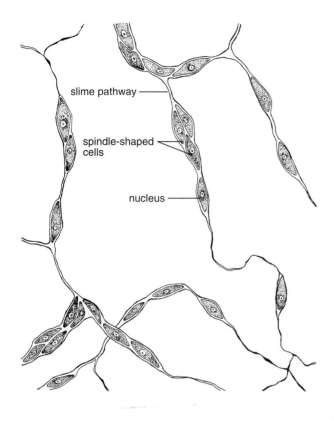

Figure 3.34
Cells of the protist *Labyrinthula* sp. (phylum Labyrinthomorpha) traveling within their slime pathways.   *Source: Margulis and Schwartz, Five Kingdoms, 1988.*

organisms in this chapter; but in other respects the labyrinthomorphs probably do not belong in the kingdom Protozoa. Cells also replicate asexually, by mitotic division. Only about 40 species have been described.

Opalinids are also of uncertain taxonomic status, but these organisms bridge the gap, at least superficially, between flagellates and ciliates. Opalinids are exclusively parasitic or commensal within the guts or recta of frogs, toads, and a few fishes. The bodies of opalinids bear numerous rows of cilia (with an associated infraciliature). However, opalinids lack a mouth and dimorphic nuclei and exhibit sexual reproduction by gamete formation rather than by conjugation. On balance, then, they seem to be more flagellate than ciliate in nature, although they clearly fit comfortably in neither group. Once again, we see the difficulties caused by attempts to wedge living organisms into necessarily artificial categorization schemes.

Spore-Forming Protozoans

Phylum Apicomplexa

The over 5,500 described members of phylum Apicomplexa are all endoparasites in animals. At two stages in the life cycle, all individuals possess a structurally com-

plicated grouping of microtubules and organelles at one end of the cell; the phylum name formally acknowledges this "apical complex." The phylum includes two distinct groups of protozoans: the gregarines and the coccidians. Flagellated gametes are found among the members of both groups. The gregarines parasitize insects and other invertebrates, while the coccidians parasitize both vertebrates and invertebrates, with the invertebrate serving as an intermediate host in the life cycle. A number of coccidian species are blood-cell parasites of humans. Most conspicuous among these are members of the genus *Plasmodium,* the agent of malaria. Four *Plasmodium* species transmit human malaria, with *P. falciparum* being the most deadly of the four.

Malaria has been called the most important disease in the world today: Probably more than two billion people living in (and visiting) tropical and semitropical areas are at risk. Probably over 300 million people presently have malaria, and one to two million people—mostly young children—die of malaria each year. (For comparison, cancer affects only about 10 million people worldwide.) Adults generally survive infection but are severely disabled.

Adult apicomplexans are highly specialized for existence as endoparasites and completely lack locomotory organelles. In typical parasite fashion, apicomplexan life cycles are complicated affairs, including sexual reproduction by means of gamete fusion, asexual reproduction through fission, and the production of resistant or infective spores. The life cycle of the malarial parasite is diagrammed in Figure 3.35. Female mosquitoes (three species in the genus *Anopheles*) are excellent vectors of malaria, easily transmitting it from one person to another, because they must continually ingest blood from vertebrate hosts to nourish their developing embryos.

To begin the cycle, haploid **gametocytes,** quiescent within the red blood cells of their human host, are removed from the human bloodstream through the bite of a female mosquito. Within the mosquito, gametocytes quickly emerge from the red blood cells and mature into either male or female gametes. The female gametes are fertilized within about one-half hour of the mosquito's blood meal. The motile, diploid zygotes (called **ookinetes**) move to the mosquito's stomach, encyst in the stomach wall, and grow. Each encysted ookinete (**oocyst**) now undergoes meiosis and a large number of fission events, the products of which mature into infective haploid **sporozoites.** As many as 10,000 sporozoites are produced asexually from an individual zygote.

Mature sporozoites migrate to the insect's salivary glands, to be injected into a human host at the next blood meal. Within minutes after injection, sporozoites begin infecting human liver cells (hepatocytes); becoming intracellular removes them from direct attack by the host immune system. The sporozoite forms a resistant cyst, which is followed by a remarkable number of asexual divisions; a single fissioning sporozoite (now called a

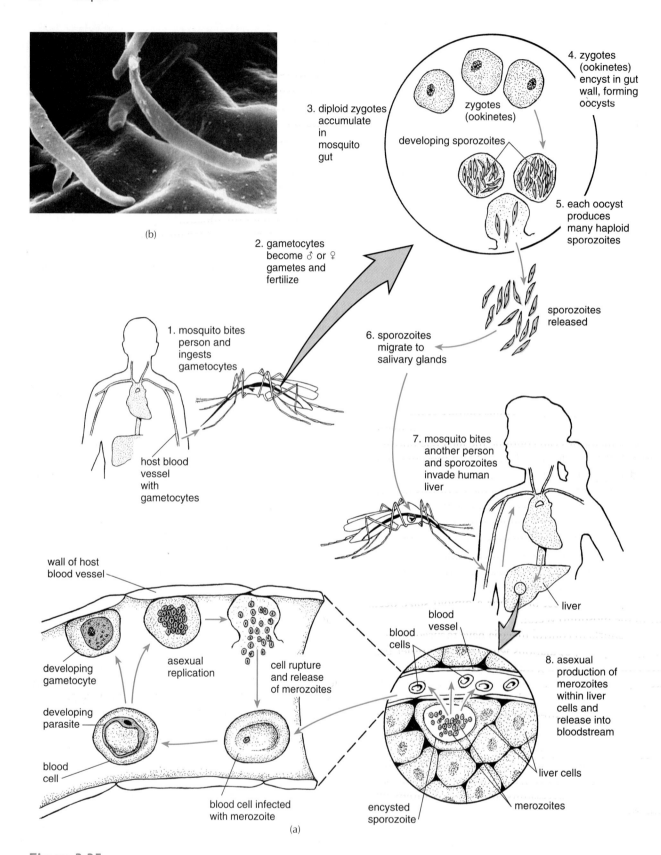

Figure 3.35
(a) The life cycle of the malarial parasite, *Plasmodium* spp. (phylum Apicomplexa). (b) Sporozoites escaping from an oocyst on the gut wall of the mosquito vector *Anopheles gambiae.*   (b) *Courtesy of R. E. Sinden. From Cox, F. E. G. 1992 Nature 359:361–62.*

schizont) can give rise, within about one week, to nearly 40,000 genetically identical offspring called **merozoites.** The merozoites rupture the liver cell housing them, enter the bloodstream, and immediately invade the host's red blood cells. Soon after entering a blood cell, the parasite somehow alters the cell's surface chemistry so that the cell sticks to the tissue lining the blood vessel. This removes the red blood cell (and its parasite) from general circulation, preventing destruction of that blood cell by the immune surveillance system within the host's spleen: The cells simply do not circulate through the spleen. Accumulations of infected cells in the blood vessels of the host's brain can lead to severe neurological impairment or death.

Within the red blood cell, merozoite development follows one of two pathways: Either the merozoite multiplies asexually to form 10 to 20 new merozoites over the next 48 hours, or it differentiates into a gametocyte. The new merozoites lyse the red blood cell and quickly invade new red blood cells for another round of asexual replication and red blood cell lysis (leading to debilitating anemia in the host) or for differentiation into a gametocyte. The fever and chills characterizing malarial infection result from toxins released during the synchronous rupture of infected red blood cells, once every 48 hours for most human malarial species. Gametocytes can continue the life cycle only if the blood cells they reside in are ingested by a mosquito; if not ingested, they perish within several weeks.

Let's summarize some key elements of this life cycle. Only the haploid gametocytes, living within human red blood cells, infect the mosquito vector. Only the haploid sporozoites, which accumulate in certain of the mosquito's salivary gland cells, can infect humans. Pathological effects in humans are caused only by the red blood cell (merozoite) stage of the life cycle. Note that the parasite is haploid for most of its life; a diploid stage occurs only in the mosquito vector.

Past efforts to control malaria have focused on eliminating mosquito populations with DDT and other pesticides and on controlling human infection with the drug chloroquine. Over the past 30 years or so, the treated mosquito populations have become resistant to the pesticides, and the parasites have become resistant to the chloroquine; in consequence, the incidence of malarial infection is increasing rapidly.

Here are some of the other potential control strategies that are currently being investigated: (1) Produce antibodies that will successfully attack the complex antigenic surfaces of the parasites (merozoite stage) during their very limited period of exposure in the bloodstream or that will attack the modified antigenic surfaces of infected red blood cells. This approach will be exceedingly difficult, as the merozoites vary the protein components of their surface coats with great frequency, as do the trypanosomes discussed earlier in this chapter, and vary the proteins expressed on the surfaces of the red blood cells they infect. (2) Prevent the parasite from invading host liver cells (hepatocytes), by somehow disarming the specific sporozoite surface proteins that recognize and bind to receptors on the liver cells. (3) Prevent the diploid zygotes (ookinetes) from producing the chitinase that enables them to penetrate the chitinous membrane lining the mosquito gut and invade the stomach wall. (4) Genetically engineer mosquitoes that will prevent gametocytes from maturing in the mosquito vector or that will produce antibodies against the surface proteins expressed on zygotes and oocysts, and then develop the means of spreading those genes throughout all populations of all relevant mosquito species. (5) Genetically alter the bacteria or protozoans living symbiotically with the mosquito vector, so that the symbiont will kill or disable the parasite. (6) Prevent sporozoites from invading the mosquito's salivary gland cells, by disrupting the binding sites on the parasite that interact with mosquito salivary gland receptors. (7) Disable the vasodilators and anticlotting and antiplatelet compounds found in mosquito saliva, which prevent normal vertebrate defenses against blood loss. This could be done by engineering mosquitos that can't secrete saliva and then developing ways to spread those genes through mosquito populations.

Additional efforts to control the disease may focus on reversing resistance of the parasites to chloroquine and developing new antimalarials. A promising substitute compound (artemisinin), produced naturally by sweet wormwood plants, is now being synthesized successfully in the laboratory; once in the mosquito vector, the chemical breaks down to form free radicals that seem to destroy the parasite's DNA directly.

The phylum Apicomplexa contains many other commercially and medically noteworthy species. For example, parasites in the genus *Perkinsus* are common in oysters and other marine molluscs; they are generally associated with molluscan mortality in both field and aquaculture situations. Of more direct interest to us, *Toxoplasma gondii* is a particularly widespread and common parasite of warm-blooded mammals. Humans typically become contaminated from cats, which release tremendous numbers of encysted eggs in their feces. As the parasite can be deadly to developing fetuses, pregnant women should avoid contact with cats and litter boxes. The parasite *Pneumocystis carinii* may also be a member of this group, although many now believe it is more likely a fungus, belonging to a separate kingdom of eukaryotes, the Fungi. It causes a fatal form of pneumonia in people with suppressed immune systems and is especially common among patients with AIDS (acquired immune deficiency syndrome).

Phylum Microspora

Microsporans superficially resemble the apicomplexans in several respects. As with apicomplexans, all microsporan

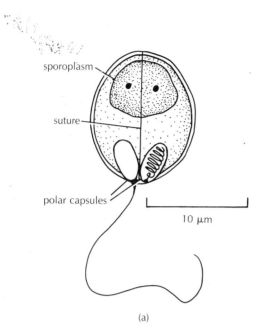

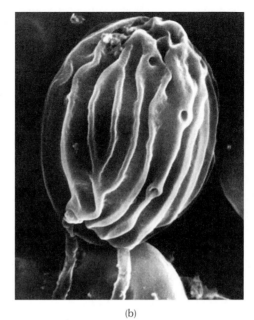

(a)

(b)

Figure 3.36

(a) Myxosporan spore with two polar capsules, one with an extruded filament. Infective sporozoites emerge from the spore following rupture along the suture. (b) Scanning electron micrograph of a spore of *Chloromyxum catostomi*.   (a) *From John N.*

Farmer, The Protozoa. Copyright © 1980 The C. V. Mosby Company, St. Louis, Missouri. Reprinted by permission. (b) From Listebarger, J.K., and Mitchell. 1980. J. Protozool: 27: 155-59, fig. 15, 158.

species are endoparasites, and the life cycle includes production of dispersive or resistant spores. In addition, microsporans lack mitochondria, which is also true of some apicomplexans. Until recently, microsporans were grouped with apicomplexans and other spore-forming protozoans in the now-abandoned phylum Sporozoa.

Unlike apicomplexans, members of the Microspora lack flagella at all stages of the life cycle and possess a distinctive organelle called the **polar filament.** If a spore is ingested by (or in some other way comes into contact with) a suitable host, the polar filament discharges from its capsule and anchors the spore to the host tissues while further events take place. A single amoeboid **sporoplasm** soon exits through the everted polar filament of the attached spore and invades the host tissues. The sporoplasm replicates itself many times by asexual and often complex means, so that the location of a suitable host by a single spore may result in the eventual production of thousands of new spores within that host. About 850 species have been described.

Phylum Myxozoa (= Myxospora)

Myxozoans are some 1,250 species of extracellular, spore-forming parasites formerly grouped with the apicomplexans and microsporans in the now-abandoned phylum

Sporozoa. The infective spore serves to disperse the parasite to a new host. Myxozoans mostly infect fish and aquatic annelids, although the complete life cycle has not been worked out for any myxozoan species. As with the microsporans, myxosporan spores possess polar filaments, and the two groups in the past have sometimes been united in a single phylum, the Cnidosporidia (*cnido* = G: nettle, filament). But each myxozoan often houses several polar filaments rather than only one, and the filaments are always coiled tightly within specialized **polar capsules;** these capsules may occupy much of the space within the spore (Fig. 3.36). Moreover, each spore is often multicellular, containing two or more amoeboid sporoplasms. The sporoplasms are released by rupture of the spore, rather than by discharge through an everted polar filament.

A number of protozoologists have argued for some time that myxozoans have evolved by extreme degeneration from multicellular animals and are not protozoans at all. Recent molecular data[10] support this contention; myxozoans should thus be removed from the kingdom Protozoa and be recognized as true multicellular animals. But it is not yet clear with what metazoan phylum to affiliate them! Although polar capsules are structurally and developmentally similar to the stinging structures found among jellyfish and other cnidarians (Chapter 5), present molecular data argue against a close evolutionary relationship between myxozoans and cnidarians.

---

10. Smothers et al. 1994. *Science* 265:1719–21.

Taxonomic Summary

Kingdom Protozoa

✓Phylum Ciliophora—the ciliates.
(e.g., *Blepharisma, Didinium, Euplotes, Folliculina, Paramecium, Stentor, Stylonychia, Tetrahymena, Tintinopsis, Tokophyra, Vorticella*)

✓Phylum Rhizopoda.
✓Class Lobosea—the naked and shelled amoebae (e.g., *Amoeba, Acanthamoeba, Arcella, Chaos, Difflugia, Entamoeba*)
✓Class Granuloreticulosea—foraminifera, (forams) (e.g., *Allogromia, Elphidium, Globigerina, Spirillina*)

✓Phylum Radiozoa.
Class Acantharea—acantharians (e.g., *Acanthometra, Acantharia, Lithoptera, Pseudolithium*)
Class Radiolaria—radiolarians (e.g., *Atlanticella, Coccodiscus, Collosphaera, Phaeodina*)
✓ Phylum Heliozoa—heliozoans (e.g., *Actinosphaerium, Sticholonche, Gymnosphaera*)
Phylum Archamoebae (e.g., *Mastigamoeba, Mastigella, Pelomyxa*)

The Flagellated Protozoans

✓Phylum Parabasala.
Class Trichomonadea (e.g., *Trichomonas*)
Class Hypermastigotea (e.g., *Macrospironympha, Trichonympha*)

✓Phylum Euglenozoa
Class Euglenoidea (e.g., *Euglena, Ploeotia*)
Class Kinetoplastidea (e.g., *Bodo, Leishmania, Leptomonas, Trypanosoma*)

✓Phylum Metamonada (e.g., *Giardia, Oxymonas, Pyrsonympha*)

✓Phylum Dinozoa—dinoflagellates (e.g., Amphidimium, Ceratium, Dinophysis, Gonyaulax, Gymnodinium, Noctiluca, Prorocentrum, Alexandrium [= Protogonyaulax], Pyrocycstis, Zooanthella)

✓Phylum Choanozoa—choanoflagellates (e.g., *Acanthoeca, Codosiga, Proterospongia*)

Misfits

✓Phylum Mycetozoa—the cellular and acellular slime molds (e.g., *Dictyostelium, Echinostelium, Protostelium*)
✓Phylum Labyrinthomorpha—slime nets (e.g., *Labyrinthula*)
✓Phylum Opalozoa—opalinids (e.g., *Opalina, Zelleriella, Proteromonas*)

Spore-Forming Protozoans

Phylum Apicomplexa
Class Perkinsidae (e.g., *Perkinsus*)
Class Gregarinidea (e.g., *Gregarina*)
Class Coccidea (e.g., *Diplospora, Eimeria, Plasmodium, Sarcocystis, Toxoplasma*)
Phylum Microspora (e.g., *Buxtehudea, Loma, Microfilum*)
Phylum Myxozoa (e.g., *Chloromyxum, Myxidium*)

*[handwritten marginalia: Phytoflagellates; Phytoflagellates; free living zooflagellates; *most btm them Break them into 8 phyla]*

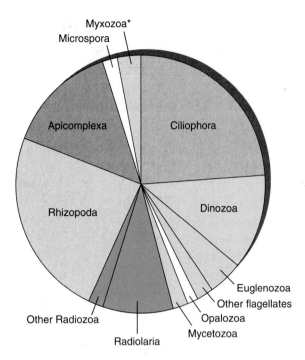

About 38,000 species total. Fossil species not included.
*Myxozoans, long believed to be protozoans, may actually
be highly degenerate multicellular animals.

Taylor, D. L., J. S. Condeelis, P. L. Moore, and R. D. Allen. 1973. The contractile basis of amoeboid movement. I. The chemical control of motility in isolated cytoplasm. *J. Cell Biol.* 59:378.

Travis, J. L., J. F. X. Kenealy, and R. D. Allen. 1983. Studies on the motility of the Foraminifera. II. The dynamic microtubular cytoskeleton of the reticulopodial network of *Allogromia laticollaris. J. Cell Biol.* 97:1668.

2.   Discuss the coordination of ciliary beating in free-living ciliates. What is the evidence that the infraciliature is not involved in this coordination?

Naitoh, Y., and R. Eckert. 1969. Ciliary orientation: Controlled by cell membrane or by intracellular fibrils? *Science* 166:1633.

Tamm, S. L. 1972. Ciliary motion in *Paramecium:* A scanning electron microscope study. *J. Cell Biol.* 55:250.

3.   Discuss what we do and do not understand about how contractile vacuoles function.

Ahmad, M. 1979. The contractile vacuole of *Amoeba proteus.* III. Effects of inhibitors. *Canadian J. Zool.* 57:2083.

Organ, A. E., E. C. Bovee, and T. L. Jahn. 1972. The mechanisms of the water expulsion vesicle of the ciliate *Tetrahymena pyriformis. J. Cell Biol.* 55:644.

Patterson, D. J., and M. A. Sleigh. 1976. Behavior of the contractile vacuole of *Tetrahymena pyriformis* W: A redescription with comments on the terminology. *J. Protozool.* 23:410.

Riddick, D. H. 1968. Contractile vacuole in the amoeba, *Pelomyxa carolinensis. Amer. J. Physiol.* 215:736.

Wigg, D., E. C. Bovee, and T. L. Jahn. 1967. The evacuation mechanism of the water expulsion vesicle ("contractile vacuole") of *Amoeba proteus. J. Protozool.* 14:104.

Topics for Further Discussion and Investigation

1.   The past several decades have seen considerable advances in our understanding of amoeboid locomotory mechanisms. What aspects of their locomotory biology remain puzzling?

Allen, R. D., D. Francis, and R. Zeh. 1971. Direct test of the positive pressure gradient theory of pseudopod extension and retraction in amoebae. *Science* 174:1237.

Cullen, K. J., and R. D. Allen. 1980. A laser microbeam study of amoeboid movement. *Exp. Cell Res.* 128:353.

Edds, K. T. 1975. Motility in *Echinosphaerium nucleofilum.* II. Cytoplasmic contractility and its molecular basis. *J. Cell Biol.* 66:156.

Fukui, Y., T. J. Lynch, H. Brzeska, and E. D. Korn. 1989. Myosin I is located at the leading edges of locomoting *Dictyostelium* amoebae. *Nature* 341:328.

Heath, J., and B. Holifield. 1991. Actin alone in lamellipodia. *Nature* 352:107.

Lazowski, K., and L. Kuznicki. 1991. Influence of light of different colors on motile behavior and cytoplasmic streaming in *Amoeba proteus. Acta Protozool.* 30:73.

Martin, R. E. 1987. Adhesion, morphology, and locomotion of *Paramoeba pemaquidensis* Page (Amoebida, Paramoebidae): Effects of substrate charge density and external cations. *J. Protozool.* 34:345.

Stossel, T. P. 1989. From signal to pseudopod: How cells control cytoplasmic actin assembly. *J. Biol. Chem.* 264:18261.

4.   What is the evidence that contractile vacuoles play a role in osmotic regulation?

Cronkite, D. L., J. Neuman, D. Walker, and S. K. Pierce. 1991. The response of contractile and non-contractile vacuoles of *Paramecium calkinsi* to widely varying salinities. *J. Protozool.* 38:565.

Hampton, J. R., and J. R. L. Schwartz. 1976. Contractile vacuole function in *Pseudocohnilembus persalinus:* Responses to variation in ion and total solute concentration. *Comp. Biochem. Physiol.* 55A:1.

Schmidt-Nielsen, B., and C. R. Schrauger. 1963. *Amoeba proteus:* Studying the contractile vacuole by micropuncture. *Science* 139:606.

Stoner, L. C., and P. B. Dunham. 1970. Regulation of cellular osmolarity and volume in *Tetrahymena. J. Exp. Biol.* 53:391.

5.   Several protozoans can alter their morphology considerably, quickly transforming from a convincing member of one taxon to an equally convincing member of another. Describe the morphological changes involved and discuss the environmental factors that initiate these dramatic transformations.

Fulton, C., and A. D. Dingle. 1967. Appearance of the flagellate phenotype in populations of *Naegleria* amebae. *Devel. Biol.* 15:165.

Nelson, E. M. 1978. Transformation in *Tetrahymena thermophila*. Development of an inducible phenotype. *Devel. Biol.* 66:17.

Willmer, E. N. 1956. Factors which influence the acquisition of flagella by the amoeba, *Naegleria gruberi. J. Exp. Biol.* 33:583.

6. Investigate some of the morphological, behavioral, and physiological adaptations for nutrient acquisition among free-living protozoans.

Alexander, S. P., and T. E. DeLaca. 1987. Feeding adaptations of the foraminiferan *Cibicides refulgens* living epizoically and parasitically on the antarctic scallop *Adamussium colbecki. Biol. Bull.* 173:136.

Bowers, B., and T. E. Olszewski. 1983. *Acanthamoeba* discriminates internally between digestible and indigestible particles. *J. Cell Biol.* 97:317.

Bowser, S. S., T. E. DeLaca, and C. L. Rieder. 1986. Novel extracellular matrix and microtubule cables associated with pseudopodia of *Astrammina rara*, a carnivorous antarctic foraminifer. *J. Ultrastr. Mol. Res.* 94:149.

Mast, S. O., and F. M. Root. 1916. Observations on ameba feeding on rotifers, nematodes and ciliates, and their bearing on the surface tension theory. *J. Exp. Zool.* 21:33.

Salt, G. W. 1968. The feeding of *Amoeba proteus* on *Paramecium aurelia. J. Protozool.* 15:275.

Stoecker, D. K., A. E. Michaels, and L. H. Davis. 1987. Large proportion of marine planktonic ciliates found to contain functional chloroplasts. *Nature* 326:790.

Sundermann, C. A., J. J. Paulin, and H. W. Dickerson. 1986. Recognition of prey by suctoria: The role of cilia. *J. Protozool.* 33:473.

7. How does one document the relative importance of microscopic protozoans in aquatic food webs?

Banse, K. 1982. Cell volumes, maximal growth rates of unicellular algae and ciliates and the role of ciliates in the marine pelagial. *Limnol. Oceanogr.* 27:1059.

Carlough, L. A., and J. L. Meyer. 1989. Protozoans in two southeastern blackwater rivers and their importance to trophic transfer. *Limnol. Oceanogr.* 34:163.

Michaels, A. F. 1988. Vertical distribution and abundance of *Acantharia* and their symbionts. *Marine Biol.* 97:559.

Porter, K. G., E. B. Sherr, B. F. Sherr, M. Pace, and R. W. Sanders. 1985. Protozoa in planktonic food-webs. *J. Protozool.* 32:409.

8. This chapter presented several examples of ways that parasitic protozoans confound the sophisticated immune systems of their hosts. Investigate some of the ways that free-living ciliates outfox their would-be predators.

Kuhlmann, H. W., and K. Heckmann. 1985. Interspecific morphogens regulating prey-predator relationships in protozoa. *Science* 227:1347.

Kusch, J. 1993. Predator-induced morphological changes in *Euplotes* (Ciliata): Isolation of the inducing substance released from *Stenostomum sphagnetorum* (Turbellaria). *J. Exp. Zool.* 265:613.

Washburn, J. O., M. E. Gross, D. R. Mercer, and J. R. Anderson. 1988. Predator-induced trophic shift of a free-living ciliate: Parasitism of mosquito larvae by their prey. *Science* 240:1193.

9. Despite their lack of organs and tissues, free-living protozoans often show surprisingly complex behaviors. Discuss the evidence that free-living amoebae can orient to chemical signals.

Bailey, G. B., G. J. Leitch, and D. B. Day. 1985. Chemotaxis by *Entamoeba histolytica. J. Protozool.* 32:341.

Fisher, P. R., R. Merkl, and G. Gerisch. 1989. Quantitative analysis of cell motility and chemotaxis in *Dictyostelium discoideum* by using an image processing system and a novel chemotaxis chamber providing stationary chemical gradients. *J. Cell Biol.* 108:973.

Wang, M., P. J. M. Van Haastert, P. N. Devreotes, and P. Schaap. 1988. Localization of chemoattractant receptors on *Dictyostelium discoideum* cells during aggregation and down-regulation. *Devel. Biol.* 128:72.

Some General References About the Protozoans

Boardman, R. S., A. H. Cheetham, and A. J. Rowell, eds. 1987. *Fossil Invertebrates.* Palo Alto, Calif.: Blackwell Scientific, 67–91.

Cheng, T. C. 1986. *General Parasitology,* 2d ed. New York: Academic Press.

Fenchel, T. 1987. *Ecology of Protozoa: The Biology of Free-Living Phagotrophic Protists.* New York: Springer-Verlag.

Hyman, L. H. 1949. *The Invertebrates, Vol. 1. Protozoa Through Ctenophora.* New York: McGraw-Hill.

Noble, E. R., and G. A. Noble. 1982. *Parasitology: The Biology of Animal Parasites,* 5th ed. Philadelphia: Lea & Febiger.

Parker, S. P., ed. 1982. *Classification and Synopsis of Living Organisms,* vol. 1. New York: McGraw-Hill, 491–637.

4

The Poriferans and Placozoans

Sponges and placozoans are not closely related. They are discussed together in this chapter only because they constitute the simplest of multicellular animals (metazoans).

Phylum Porifera

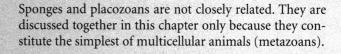

Phylum Pori · fera
(L: pore bearing)
por-i´-fer-ah

Introduction

Less than 2% of all sponge species are found in freshwater, the remaining 98% of the species being marine. There are no terrestrial sponges. The approximately 5,000 to 10,000 species comprising this group of animals (depending on who is doing the counting) are perhaps more remarkable for the characteristics they lack than for those they possess. Unlike almost any other metazoan, sponges lack a nervous system and have no true musculature. Thus, locomotion is generally beyond them, and all species feed on food particles suspended in the water. No specialized reproductive, digestive, respiratory, sensory, or excretory organs are found in this group, either; indeed, no organs are found at all. Often, sponges are amorphous, asymmetrical creatures, although there are some very beautiful exceptions to this generalization. No sponge has anything corresponding to anterior, posterior, or oral surfaces. Moreover, only a few different cell types are encountered within any given individual. These cells are functionally independent to the extent that an entire sponge can be dissociated into its constituent cells in the laboratory, with no long-term impact. The cells dedifferentiate to an amoeboid form, reaggregate, and at least in some species, can redifferentiate to reform the

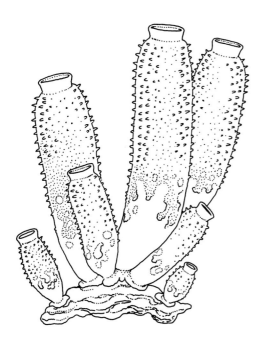

Research Focus Box 4.1

Histoincompatibility in Sponges

Hildemann, W. H., I. S. Johnson, and P. L. Jokiel. 1979. Immunocompetence in the lowest metazoan phylum: Transplantation immunity in sponges. *Science* 204:420–22. Reprinted by Permission.

To qualify as having an immune system, (1) the organism (or the cells of that organism) must evidence some form of antagonism toward foreign substances, (2) the antagonism must be specific toward that substance, and (3) future responses should be altered by the first response—that is, the system must "remember" the prior encounter. Twenty years ago, invertebrates were thought to lack immune systems. This is no longer the case: The ability to distinguish between self and nonself has been clearly demonstrated for many nonvertebrate animal groups, including sea urchins and sea stars, insects, crustaceans, annelid worms, clams and snails, sea anemones, and corals. Sponge cells of any given species have long been known to avoid fusing with cells of other sponge species, but can the cells of a sponge discriminate between sponge cells of another individual of the *same* species? Can those cells show a true immune response?

Hildemann et al. (1979) studied intraspecific (among individuals of one species) tissue incompatibility of the tropical sponge *Callyspongia diffusa*, a large, purple sponge that grows upward like long fingers arising from a common base on coral rock in Hawaii. They chose to study this species because, although they frequently observed tissue fusion between adjacent fingers or pieces of a single sponge colony in the field, they never observed such grafts between adjacent *C. diffusa* colonies, suggesting that the sponge tissues might be capable of intraspecific self-recognition.

To examine this possibility in the laboratory, they wired pieces of sponge (approximately 2cm by 8cm each) to heavy plastic plates, placing two pieces of sponge from one colony (isogeneic treatments) or pieces from two different colonies collected from different locations (allogeneic treatments) close together on each plate. The researchers then kept the sponges in the laboratory at about 27°C and monitored the fate of each sponge daily.

Sponge fragments in the isogeneic treatments (paired from the same colony) always fused together within a few days, showing complete tissue compatibility. In marked contrast, tissues in the allogeneic treatments never fused. In fact, the researchers observed pronounced toxic responses between the tissues of these incompatible sponges; within two weeks, tissues in contact with the neighboring sponge fragment lost their purple coloration and died (Focus Table 4.1). The sponges showed no such reaction to the plastic holders or to the wire they were tied down with; the response was clearly one of histoincompatibility: specific, intense tissue disharmony.

Even more remarkably, when the incompatible sponge fragments were permitted to "rest" apart for 12 days and were then reexposed to each other at new tissue

sponge.[1] Just as remarkably, for such a primitive animal, the cells of one sponge can often distinguish between other cells of the same individual and those of different individuals (Research Focus Box 4.1). Such non-self-recognition (i.e., alloincompatibility) has now been reported from at least 25 sponge species.[2]

Probably the most surprising aspect of sponge biology is that sponges work so well with so little. They are often major components of aquatic communities and compete impressively with other sedentary metazoans for food and, especially, space.

Considering the lack of organs and systems in sponges,[3] one can wonder whether sponges are truly multicellular animals or whether they represent a highly evolved form of colonial living by individual cells. In fact, some freshwater colonial protozoans (the choanoflagellates, p. 51) bear very definite morphological similarities to the simplest sponges, and as recently as 100 years ago it was suggested that sponges be classified as colonial protozoans. Many zoologists have long felt that members of the Porifera are sufficiently different from all other animals to warrant placement in a sepa-

---

1. See *Topics for Further Discussion and Investigation*, no. 5, at the end of the chapter.

2. See *Topics for Further Discussion and Investigation*, no. 7.

3. See *Topics for Further Discussion and Investigation*, no. 4.

Focus Table 4.1   Reaction Times of *Callyspongia diffusa* Fragments to Each Other

1 Source of individuals tested	2 Days to react in first test (median ± one standard deviation)	3 Number of pairs tested	4 Days to react in second test (median ± one standard deviation)	5 Number of pairs tested
A & B	9.0 ± 1.9	24	3.8 ± 0.9	10
A & C	8.9 ± 6.9	30	1.2 ± 1.3	13
B & C	7.2 ± 2.2	21	4.0 ± 1.2	11

Note: For each test, sponges collected from three different locations (A, B, or C) were grafted together (in pairs) intimate soft tissue contact. Tissue death at points of contact was clearly visible in a little over one week, on average (column 2). When the pairs were regrafted after a 12-day separation, the tissue-toxic responses occurred even more rapidly (column 4). For example, the median response for individuals collected at points A and B required about nine days in the first encounter but less than four days in the second encounter. The faster second response clearly was a consequence of prior experience, indicating that these sponges have an immune memory system.

locations, the toxic interactions recurred significantly more rapidly than before (Focus Table 4.1), suggesting an immune memory: The second response was clearly modified by the first.

All 200 intercolony challenges were marked by distinct tissue incompatibility responses in the first trial and accelerated responses in the second. The tissue of each sponge is clearly capable of recognizing the differences between many subtly different surface molecules characterizing individual sponges of the same species. This clearly indicates that the genetic loci coding for histocompatibility molecules (i.e., those responsible for making the self/nonself distinction) at the cell surface of sponges are highly polymorphic, coded for by perhaps hundreds of different genes at particular loci. This ability to discriminate so finely between self and nonself most certainly plays a key role in helping an individual sponge genotype compete successfully for space.

Clearly, even sponges have a spectacularly sensitive immunorecognition system capable of distinguishing even minor chemical differences among allogeneic surface markers. Further studies of this system in sponges may tell us something about the evolutionary origin of cell-mediated immunity in all animals and, perhaps, about the functioning of the major histocompatibility complex of the human immune system.

rate subkingdom, the subkingdom Parazoa ("almost animals"). More recently, a number of authorities have urged that sponges be readmitted as full-fledged members of the animal kingdom. Recent molecular data support this decision: Sponges are indeed animals, although they clearly diverged very early from the evolutionary pathways travelled by other metazoans and gave rise to no other lineages.

General Characteristics

In its simplest form, the typical sponge is a fairly rigid, perforated bag, whose inner surface is lined with flagellated cells. The empty space of this bag is called the **spongocoel**. The flagellated cells lining the spongocoel are called **choanocytes** (literally, "funnel cells"), or **collar**

cells in recognition of the cylindrical arrangement of the cytoplasmic extensions (collars) surrounding the proximal portion of each flagellum (Fig. 4.1). The collars of sponge choanocytes are almost certainly homologous with those of protozoan choanoflagellates. These collar cells perform the following functions:

1. They generate currents that help maintain circulation of seawater within and through the sponge.
2. They capture small food particles.
3. They capture incoming sperm for fertilization.

Adjacent to the choanocyte layer is a gelatinous, nonliving layer of material called the **mesohyl** layer (*meso* = G: middle; *hyl* = G: stuff, matter). Although the mesohyl is acellular and nonliving, it contains live cells.

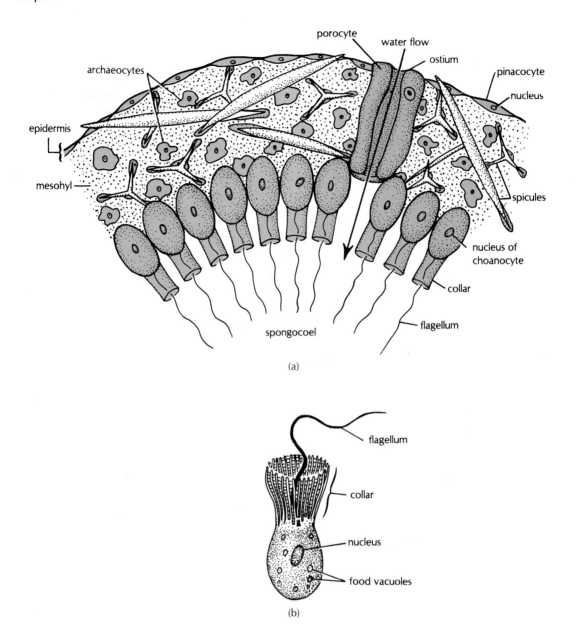

Figure 4.1
(a) Diagrammatic illustration of the body wall of a sponge.
(b) Detail of a choanocyte.   (a) *From Hyman, The Invertebrates, Vol. III.*

Amorphous, amoeboid cells called **archaeocytes** wander throughout the mesohyl by typical cytoplasmic streaming, which involves the formation of pseudopodia as in amoeboid protozoans. Archaeocytes perform a number of essential functions within sponges and can develop into more specialized cell types when necessary. Archaeocytes are responsible for digesting food particles captured by the choanocytes, so that digestion is entirely intracellular. Some archaeocytes also store digested food material. In addition, archaeocytes may give rise to both sperm (which are flagellated) and eggs, although gametes may also arise through morphological modification of exist-

ing choanocytes. They also probably play an active role in non-self-recognition reactions in response to contact with other sponges. Finally, archaeocytes play a role in eliminating wastes and, in addition, can become specialized to secrete the supporting elements located in the mesohyl layer. These support elements may be calcareous or siliceous **spicules,** or they may be fibers composed of a collagenous protein called **spongin.** The cells secreting spicules are termed **sclerocytes,** and those producing spongin fibers are termed spongocytes[4] (Fig. 4.2). Both of these cell types are derived from archaeocytes. Clearly, archaeocyte cells are quite versatile. The spicules and

---

4. See *Topics for Further Discussion and Investigation,* no. 1.

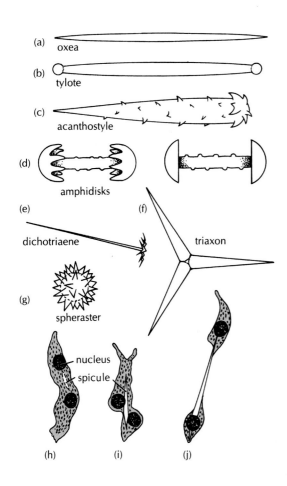

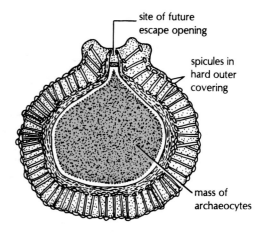

Figure 4.3
The gemmule of a freshwater sponge, *Ephydatia*, as seen in cross section. *From Hyman; after Evans.*

Figure 4.2
(*a–g*) Representative sponge spicule morphologies. (*h–j*) The production of a sponge spicule by a binucleate sclerocyte. The spicule is initiated between the two nuclei. After the spicule is completed, the cells will wander off into thle mesoglea.
Beck/Braithwaite, Invertebrate Zoology, Laboratory Workbook, 3/e, © 1968. Reprinted by permission of Prentice Hall, Upper Saddle River, New Jersey.

fibers secreted by the sclerocytes and spongocytes are of great importance (1) to systematists, as an indispensable factor in species identification, and (2) to sponges, which depend upon the support elements for maintaining shape and, probably, for discouraging predation.

At certain times of the year, many freshwater sponge species (and a few marine species) produce dormant structures called **gemmules.** To begin the process, archaeocytes accumulate nutrients by phagocytizing other cells and then cluster together within the sponge. Certain cells surrounding each cluster secrete a thick, protective covering; the gemmule consists of the cluster plus its surrounding capsule (Fig. 4.3). Gemmules are typically far more resistant to desiccation and freezing than are the sponges that produce them. In fact, the gemmules of many species must spend several months at low temperature before the gemmules are capable of hatching; that

is, they require a period of **vernalization.** Under appropriate environmental conditions, the living cells leave the gemmule (hatch) through a narrow opening and differentiate to form a functional sponge.[5] Gemmule formation thus appears to be a mechanism through which sponges can withstand unfavorable environmental conditions, by entering a stage of developmental arrest, a period of dormancy. In light of some of the differences between freshwater and marine environments mentioned in Chapter 1, the adaptive significance of gemmule formation by freshwater sponges should be particularly apparent. Also, since many gemmules are produced by an individual sponge, gemmule formation can be an effective means of asexual reproduction, resulting in numerous genetically identical offspring.

Instead of forming gemmules, some freshwater and marine species undergo pronounced tissue regression during unfavorable periods, with the sponge becoming reduced to little more than a compact cellular mass with an outer protective covering. The sponge reactivates when environmental conditions improve.

The outer layer of a sponge is composed of flattened contractile cells called **pinacocytes** (Fig. 4.4). These cells also line the incurrent canals and the spongocoel in places where choanocytes are lacking. Contraction of pinacocytes enables the sponge to undergo minor shape changes and also may play a role in regulating water flow through the sponge by varying the diameter of the incurrent openings.

Because they lack muscles, nerves, and deformable bodies, sponges are utterly dependent upon water flow for food, gas exchange, dissemination and collection of sperm, and removal of wastes. Partly as a consequence of choanocyte activity and partly as a consequence of sponge

5. See *Topics for Further Discussion and Investigation*, no. 2.

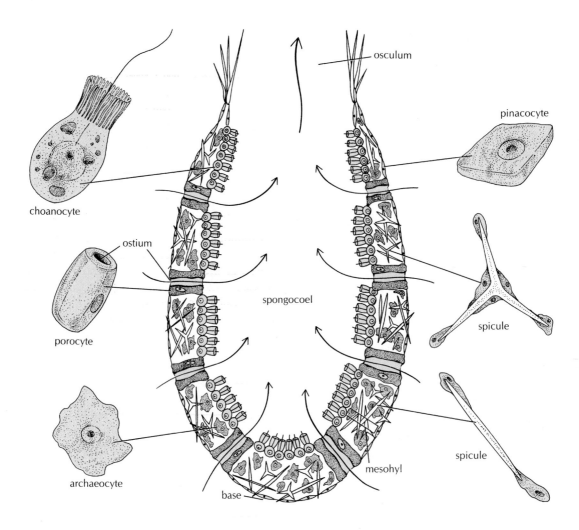

Figure 4.4
Diagrammatic representation of a simple (asconoid) sponge, illustrating its various cellular and structural components. In asconoids, the incurrent canal is simply a tube passing through a modified pinacocyte, called a *porocyte*. Note that six cells are involved in producing a triradiate spicule. *Sherman/Sherman*, The Invertebrates: Function and Form, 2/e, © 1976, pp. 111, 112, 236, 172, 169, 45, 9, 15. Reprinted by permission of Prentice Hall, Upper Saddle River, New Jersey.

architecture (Fig. 4.5),[6] water flows into the spongocoel through narrow openings (**ostia**) and exits the spongocoel through larger openings (**oscula**). Ostia are always numerous on the body of a sponge, but there may be as few as one osculum present per individual. The name Porifera attests to the supreme importance of all these openings in sponge morphology and physiology.

The large size of many sponges—the members of some species exceed 1m (meter) in height or diameter—attests to the large volumes of water filtered for food: probably dozens of liters per sponge per day. However, rapid flow past a food capturing surface is incompatible with efficient particle capture. Ideally, the water would slow down within the choanocyte chambers, allowing time for efficient particle removal, and then increase

velocity on the way out of the sponge to dissipate wastes (and sperm) effectively. In fact, this is exactly what happens, not because of differences in flagellar activity in different parts of the sponge, but simply because of internal sponge architecture. Table 4.1 presents water flow data for a sponge that filters 0.18 cm³ of water per second, which works out to a respectable 0.65 liters of water per hour. This sponge bears a large number of ostia (940,000!); because incoming water is divided among all these available openings, water enters the sponge at a leisurely 0.057 cm per second. However, the ostia lead into a mind-boggling number of flagellated chambers, nearly 29 million in this case. Although these chambers are each very small, multiply their individual surface area by 29 million and you get a surprisingly large total area, as

6. See *Topics for Further Discussion and Investigation*, nos. 3 and 4.

Table 4.1 Water Transport Characteristics for a Marine Leuconoid Sponge.

The sponge on which the data are based had a total volume of 2.4 cm³. *From LaBarbera, M., and S. Vogel. 1982. Amer. Scient. 70:54–60.*

Anatomical feature	Individual surface area (cm²)	Approximate no. per sponge	Total area (cm²)	Water velocity (cm/sec)
ostia	3.33×10^{-6}	940,000	3.14	0.057
flagellated chambers	7.06×10^{-6}	2.88×10^{7}	203.0	8.69×10^{-4}
osculum	0.034	1.0	0.034	5.1

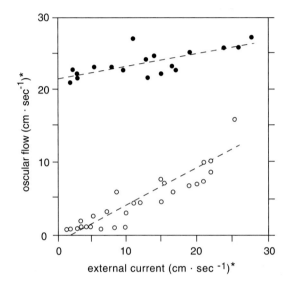

Figure 4.5

Influence of morphology on water flow through the marine sponge *Haliclona viridis*. (•) Velocity of water leaving sponge oscula for undisturbed sponges. (o) Data for sponges whose choanocytes were inactivated by immersing sponges in fresh water for several minutes. The data show that, in a current, water flows through the sponge with or without the help of choanocyte activity. In a current of 15 cm/sec., for example, water moves through an inactivated sponge at nearly 30% of the rate measured in a fully functional sponge, simply because of the way the sponge's architecture takes advantage of surrounding water flow. All flow measurements were made in the field, by positioning one tiny flow-sensitive probe next to a sponge and another in the center of its osculum.   *From Steven Vogel, in* Proceedings of the National Academy of Science, 74:2069-71, 1977. Reprinted by permission of the author.

indicated in the table. Since the volume of water entering the sponge must also be the volume moving through and leaving the sponge in the same amount of time (note that for each row of the table, total area multiplied by water velocity always equals about 0.18 cm³/sec), water velocity must decline dramatically inside the choanocyte chambers, as indicated, increasing the opportunity for food particle or gamete capture. Our own blood capillary system operates similarly in increasing the efficiency of

gas exchange between blood and tissues, but sponges were clearly there first, a nice example of convergent morphological evolution in the face of similar selective pressures. Once particles (and oxygen) have been removed in the flagellated chambers, water velocity again increases, simply due to the decreased total surface area across which the water now flows; to move the same volume of water through a smaller cross-sectional area of the sponge per time, velocity must increase.

Poriferan Diversity

Advances in sponge morphology imply, over evolutionary time, selection for maximizing current flow through the spongocoel and increasing the amount of surface area available for food collection. There are three basic levels of sponge construction: asconoid, syconoid, and leuconoid, in order of increasing complexity. Each form simply reflects an increased degree of evagination of the choanocyte layer away from the spongocoel, increasing the extent of flagellated surface area enclosed by the sponge (Fig. 4.6). Most sponge species are of leuconoid construction.

Representative sponges are illustrated in Figure 4.7. Sponges exist in a tremendous variety of colors and shapes, ranging from encrusting forms only a few millimeters wide to elaborate upright forms over 1 m high. Although most sponges are immobile, members of a few sponge species can move at speeds of several millimeters per day, presumably by the coordinated cytoplasmic movements of individual, amoeboid cells.

Sponges are distributed among the following four classes, based largely upon the chemical composition and morphology of the support elements: Calcarea, Demospongiae, Sclerospongiae, and Hexactinellida.

Class Calcarea

Members of the class Calcarea bear spicules composed only of calcium carbonate ($CaCO_3$). Representatives of all three types of construction occur in this class. Indeed, the only living asconoid forms are found among the Calcarea.

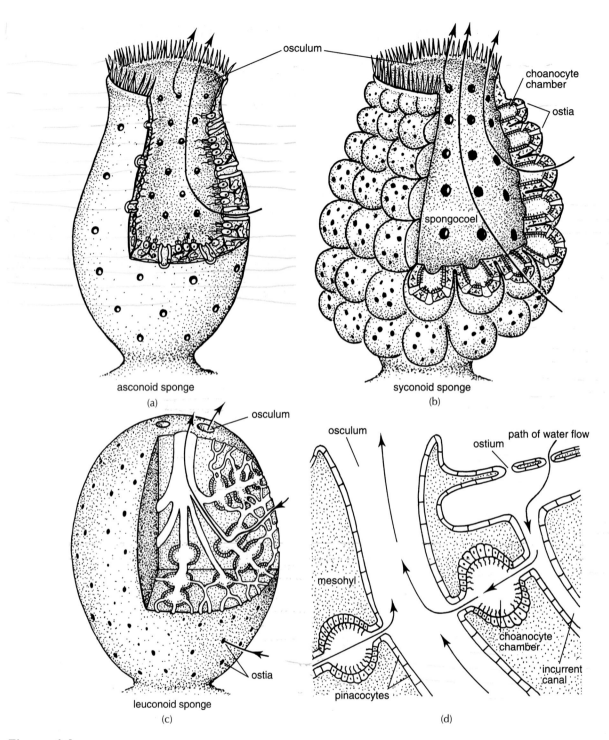

Figure 4.6

Diagrammatic illustrations of the different levels of complexity of sponge architecture. Arrows indicate the direction of water flow: in at the ostia and out at the oscula. (*a*) Asconoid sponge. (*b*) Syconoid sponge. (*c*) Leuconoid sponge. (*d*) Detail of water circulation in a leuconoid sponge.   (a-c) *From Bayer and Owre, The Free-*

Living Lower Invertebrates. Reprinted by permission of the authors. From Johnston and Hildemann, in The Reticuloendothelial System, *Cohen and Siegel, eds. Copyright* © *1982 Plenum Publishing Corporation, New York. Reprinted by permission.* (d) *From Johnston and Hildemann, in* The Reticuloendothelial System, *Cohen and Siegel, eds. Copyright* © *1982 Plenum Publishing Corporation, New York. Reprinted by permission.*

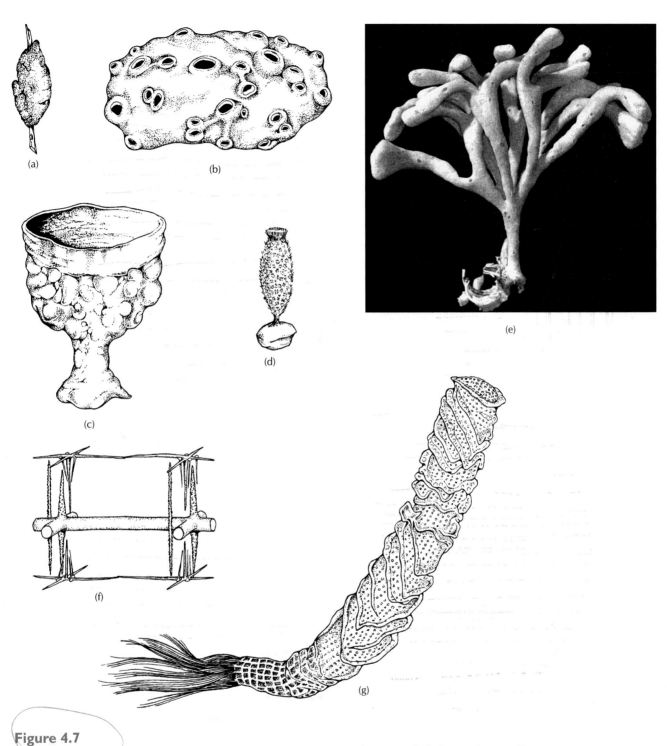

Figure 4.7

Sponge diversity. (a) A freshwater species encrusting a twig.
(b) An encrusting marine sponge. (c) Neptune's goblet sponge,
Poterion neptuni. (d) A syconoid sponge, *Sycon* sp. (e) A fingered
sponge, *Haliclona oculata* (g) *Euplectella* sp., Venus's flower basket.
This is a member of the Hexactinellida. The spicules are six-rayed
and composed of silica, as shown in (f).   (a) *After Hyman.* (b,c) *After
Pimentel.* (d) *After MacGinitie and MacGinitie.* (e) *Courtesy of J.A. Kaandorp, from* Fractal
Modelling: Growth and Form in Biology *Springer-Verlag, Berlin, New York, 1994.*
(f) *Barnes,* Invertebrate Zoology, *4th ed. Orlando: W. B. Saunders Company, 1980.*
(g) *After Bayer and Owre.*

Class Demospongiae

Members of the largest class (containing at least 80% of all sponge species), the Demospongiae, are nearly all of leuconoid construction. The supporting spicules and fibers of the Demospongiae may be composed of spongin and/or silica but never of CaCO₃. All freshwater sponges (fewer than 300 species) are found in this class. Interestingly, these freshwater species possess contractile vacuoles, which are organelles specialized for eliminating water from cytoplasm; they are found elsewhere only among protozoans.

A bizarre group of deep sea sponges was described recently; their spicules place them convincingly within the Demospongiae, although their morphology and feeding biology are so aberrant, one wonders if they really are sponges! They possess no ostia and no oscula—no internal canal system at all, in fact—and no choanocytes.[7] Remarkably, they feed as carnivores, entrapping small (mostly less than 1 mm) swimming crustaceans on numerous long, thin filaments that cover much of the body surface. The captured prey are gradually surrounded by adjacent epithelial cells and the growth of new filaments, and are digested externally within a few days. Additional nutrition may be provided by symbiotic bacteria, as described elsewhere in this book for bivalved molluscs (p. 250) and pogonophorans (p. 332).

Class Sclerospongiae

A third class of sponges, the Sclerospongiae (*sclero* = G: hard), contains only a few species, all restricted to obscure caves and crevices in coral reefs. Like most members of the Demospongiae, all members of the Sclerospongiae are of leuconoid construction. But unlike that of most other sponges, the sclerosponge body is supported by all three of the skeletal materials encountered within the Porifera: CaCO₃, silica, and spongin. However, the calcium carbonate in these sponges forms a solid supporting mass rather than spicules (Fig. 4.8). Some biologists feel that this may represent the primitive sponge condition, and that the solid calcareous framework has been mostly lost in subsequent sponge evolution. Indeed, a solid calcareous skeleton has now been described in at least one living member of the Calcarea and at least one living member of the Demospongiae. There is, thus, a growing feeling that the sclerosponges should be distributed among the Calcarea and Demospongiae rather than segregated in a separate class.

Class Hexactinellida

Finally, sponges whose bodies are supported entirely by six-rayed siliceous spicules are placed in the class Hexactinellida. These sponges, known as the glass sponges, are marvels of structural complexity and symmetry (Fig. 4.7g). Their canal systems may be either syconoid or leuconoid, but hexactinellids stand apart from all other sponges in that the outer layer is syncytial (having many nuclei contained within a single plasma membrane) rather than cellular and lacks contractile elements; thus, there is no pinacoderm layer. The inner, flagellated layer is also syncytial, again setting the glass sponges apart from all others. Some workers argue that the hexactinellids, therefore, belong in a separate phylum.

Other Features of Poriferan Biology: Reproduction and Development

Sponges reproduce asexually, either by fragmentation or through the production of gemmules or buds, and sexually, through the production of eggs and sperm. Many sponge species are **hermaphroditic,** with a single individual producing both types of gametes. Fertilization and early development are typically internal, a surprising elaboration for what is generally considered a rather unsophisticated animal. In at least some species, choanocytes capture the incoming sperm, dedifferentiate to amoeboid form, and then transport the sperm to the mesohyl, where the eggs are fertilized.

Development of the sponge embryo is unlike that of other metazoan embryos. In some calcareous sponge species, the embryo develops into a hollow blastula. Upon attaining the 16-cell stage, one group of eight cells continues to divide rapidly and to become flagellated at one end. The eight cells of the second group divide far more slowly and remain unflagellated. The small internal cavity opens to the outside in the middle of this group of relatively large, slowly dividing cells. In most species, the embryo is said to turn inside out through this opening so that the flagella of the rapidly dividing cells come to lie on the outer surface of the embryo, where they can propel the animal through the water once the larva is discharged from the parent. Even after turning inside out, the embryo remains only one cell thick; this process, called **inversion,** is therefore not to be confused with the process of gastrulation.

The hollow, swimming sponge larva, ciliated at one end only, is termed an **amphiblastula** (Fig. 4.9*a,b*). The

---

7. Vacelet, J., and N. Boury-Esnault. 1995. *Nature* 373:333–35.

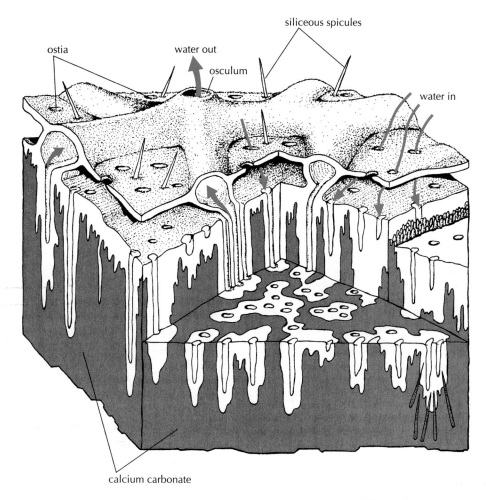

ostia | water out | siliceous spicules

osculum

water in

calcium carbonate

Figure 4.8

Schematic representation of a sclerosponge, collected off the coast of Jamaica. Note the unusually narrow choanocyte chambers and the massive calcareous skeleton.   *From Willenz and*

Hartman in Marine Biology, 103:387-401, 1989. Copyright © 1989 Springer-Verlag, Heidelberg, Germany. Reprinted by permission.

embryos of other sponge species form solid blastulae, each of which generally becomes a solid **parenchymula** (or **parenchymella**) larva, with flagella (one flagellum per cell) covering nearly the entire outer body surface (Fig. 4.9c).

Sponge larvae are typically incapable of feeding, and metamorphose within a few hours of their release from the parent. Before losing the ability to swim, sponge larvae attach to a substrate. During the ensuing process of **gastrulation,** in which the embryonic body wall becomes two cells thick, the flagellated cells apparently are shed or phagocytized. Cells from various parts of the embryo undergo extensive migrations and begin to, or continue to, differentiate to form the future adult sponge. In particular, the choanocytes typically differentiate *de novo* from archaeocytes, not from the flagellated cells of the larva as was previously thought, although the direct

development of choanocytes from larval flagella recently has been reported for one calcareous sponge species.[8]

With respect to their mode of reproduction, sponges are unusual in that both marine and freshwater species have a free-swimming larval stage in their development. Such a free-swimming dispersal stage is often suppressed in the life histories of other freshwater invertebrates, as noted in Chapter 1.

Phylum Placozoa

Phylum Placo · zoa
(G: flat plate animal)
plack-ō-zō-´-ah

8. Amano and Hori, 1993. *Invert. Rep. Devel.* 24:13–26.

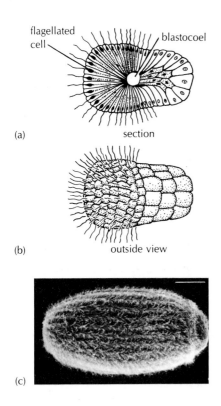

(a) flagellated cell — blastocoel — section

(b) outside view

(c)

Figure 4.9

(a,b) Typical amphiblastula larva. This free-swimming larva will attach to a substrate before undergoing further development. (c) Scanning electron micrograph of a sponge larva, *Halichondria* sp. The larva is about 300 μm long.   (a,b) *From Hyman; after Hammer.* (c) *Courtesy of Claus Nielson, from C. Nielson 1987, "Acta Zoologica" (Stockholm) 68:205-262, Permission of Royal Swedish Academy of Sciences.*

Introduction

Sponges are quite distinct from most other metazoans in their general lack of symmetry and in their lack of true tissue layers and organs. In some respects, the most primitive sponges resemble certain colonial flagellated protozoans more than they resemble other metazoans. Similarly troublesome animals are contained in a separate phylum, the Placozoa. Only a single placozoan species has been described, *Trichoplax adhaerens* (Fig. 4.10), and it has been collected only from marine habitats, including marine aquariums. Although placozoans have been known for about 100 years, their biology is poorly explored.

General Characteristics

Like sponges, placozoans have no front or back and no right or left, and they lack organs and tissues. There is no digestive system, no nervous system, and no true

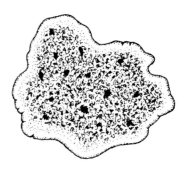

Figure 4.10

The placozoan *Trichoplax adhaerens*. The individual is approximately 0.5 mm in its longest dimension.

musculature. Placozoans, like sponges, additionally lack specialized sensory structures. However, unlike sponges, placozoans are fully mobile. They are apparently planktonic for part of their lives but are most often seen gliding across hard substrates on thousands of motile flagella, changing shape markedly, amoeba-style, as they travel. The animals rarely get much larger than a few millimeters across.

Placozoans are flat, with two distinct cell layers, each containing perhaps a thousand or so cells. The ventral layer is composed of columnar cells, each bearing a single flagellum. Associated glandular cells apparently secrete digestive enzymes beneath the animal as it sits atop the algae and protozoans on which it apparently feeds; digestion seems to be entirely extracellular, as there is no mouth and no sign of phagocytosis. The much thinner, upper layer of the animal bears flagellated cells but no gland cells. In a sense, the upper layer is ectodermal, while the lower layer, because of its involvement in digesting food and absorbing nutrients, is endodermal. Between the upper and lower cell layers is a fluid-filled space containing a dense network of fibrous cells that may be contractile.

The details of sexual reproduction are unknown. While fertilization has never been seen, some individual placozoans have produced embryos; unfortunately, the embryos have never developed very far in the laboratory. Asexual reproduction—by budding, fragmentation, or fission—occurs commonly in the laboratory. Also, individuals have no difficulty regenerating pieces that are cut off.

The relationship between placozoans and other metazoans is obscure. Some suggest they are related in some way to sponges, while others feel they may be more closely related to anemones and their relatives. Placozoans are very simple but clearly multicellular animals, perhaps representing the earliest metazoan condition.

Taxonomic Summary

Phylum Porifera—the sponges
    Class Calcarea
    Class Demospongiae
    Class Sclerospongiae
    Class Hexactinellida—the glass sponges
Phylum Placozoa

Topics for Further Discussion and Investigation

1. How are sponge spicules secreted by sclerocytes?

Dendy, A. 1926. Origin, growth, and arrangement of sponge spicules. *Q. J. Microsc. Sci.* 70:1.

Elvin, D. 1971. Growth rates of the siliceous spicules of the freshwater sponge *Ephydatia mülleri* (Lieberkuhn). *Trans. Amer. Microsc. Sci.* 90:219.

Ledger, P. W., and W. C. Jones. 1977. Spicule formation in the calcareous sponge *Sycon ciliatum. Cell Tissue Res.* 181:553.

2. Investigate the environmental control of hatching from sponge gemmules.

Benfey, T. J., and H. M. Reiswig. 1982. Temperature, pH, and photoperiod effects upon gemmule hatching in the freshwater sponge, *Ephydatia mülleri* (Porifera, Spongillidae). *J. Exp. Zool.* 221:13.

Fell, P. E. 1974. Diapause in the gemmules of the marine sponge, *Haliclona loosanoffi*, with a note on the gemmules of *Haliclona oculata. Biol. Bull.* 147:333.

Jetton, T. L., P. E. Fell, and F. W. Harrison. 1987. Cytological and cytochemical investigations of development from dormant gemmules of the marine sponge, *Haliclona loosanoffi. J. Morphol.* 193:99.

3. How does sponge architecture contribute to increased flow of water through a sponge and to increased feeding efficiency?

LaBarbera, M., and S. Vogel. 1982. The design of fluid transport systems in organisms. *Amer. Scient.* 70:54.

Reiswig, H. M. 1975. The aquiferous systems of three marine Demospongiae. *J. Morphol.* 145:493.

Vogel, S. 1974. Current-induced flow through the sponge, *Halichondria. Biol. Bull.* 147:443.

Vogel, S. 1978. Organisms that capture currents. *Sci. Amer.* 234:108.

4. No one has ever demonstrated nerve tissue in any poriferan. Nevertheless, in some species there is evidence of cooperation among different areas of the sponge, resulting in the regulation of water flow through the animal. What is the evidence for such internal coordination, and through what mechanisms might it be accomplished in the absence of nerve cells?

Lawn, I. D., G. O. Mackie, and G. Silver. 1981. Conduction system in a sponge. *Science* 211:1169.

Mackie, G. O., I. D. Lawn, and M. Pavans de Deccaty. 1983. Studies on hexactinellid sponges. II. Excitability conduction and coordinated responses in *Rhabdocalyptus dawsoni* (Lambe, 1873). *Phil. Trans. Royal Soc. London* B 301:401.

5. How do dissociated sponge cells recognize each other in the reaggregation process?

Galtsoff, P. S. 1925. Regeneration after dissociation (an experimental study on sponges). I. Behavior of dissociated cells of *Microciona prolifera* under normal and altered conditions. *J. Exp. Zool.* 42:183.

Humphreys, T. 1963. Chemical dissolution and in vitro reconstruction of sponge cell adhesions. I. Isolation of and functional demonstration of the components involved. *Devel. Biol.* 8:27.

McClay, D. R. 1974. Cell aggregation: Properties of cell surface factors from five species of sponge. *J. Exp. Zool.* 188:89.

Spiegel, M. 1954. The role of specific surface antigens in cell adhesion. I. The reaggregation of sponge cells. *Biol. Bull.* 107:130.

Wilson, H. V., and J. T. Penney. 1930. The regeneration of sponges (*Microciona*) from dissociated cells. *J. Exp. Zool.* 56:73.

6. Sponges must compete for space with a variety of organisms, including macroalgae, corals, bryozoans, and other sponges. In addition, sponges must be subject to considerable predation, especially by gastropods and fishes. Finally, the great surface area of sponges would seem to make them ideal for other sedentary organisms to settle and grow on. Lacking specialized organs and behaviors, sponges instead rely on chemical means to compete and to protect themselves from these threats. Discuss the evidence for chemical defense among sponges.

Becerro, M. A., N. I. Lopez, X. Turon, and M. J. Uriz. 1994. Antimicrobial activity and surface bacterial film in marine sponges. *J. Exp. Marine Biol. Ecol.* 179:195.

McClintock, J. B. 1987. Investigation of the relationship between invertebrate predation and biochemical composition, energy content, spicule armament and toxicity of benthic sponges at McMurdo Sound, Antarctica. *Marine Biol.* 94:479.

Porter, J. W., and N. M. Targett. 1988. Allelochemical interactions between sponges and corals. *Biol. Bull.* 175:230.

Thompson, J. E. 1985. Exudation of biologically-active metabolites in the sponge *Aplysina fistularis*. I. Biological evidence. *Marine Biol.* 88:23.

Thompson, J. E., R. P. Walker, and D. J. Faulkner. 1985. Screening and bioassays for biologically-active substances from forty marine sponge species from San Diego, California, USA. *Marine Biol.* 88:11.

7. Describe the immune recognition responses of sponges that come into contact with other sponges.

Amano, S. 1990. Self and non-self recognition in a calcareous sponge, *Leucandra abratsbo. Biol. Bull.* 179:272.

Ilan, M., and Y. Loya. 1990. Ontogenetic variation in sponge histocompatibility responses. *Biol. Bull.* 179:279.

Jokiel, P. L., and C. M. Bigger. 1994. Aspects of histocompatibility and regeneration in the solitary reef coral *Fungia scutaria. Biol. Bull.* 186:72.

Van der Vyver, G., S. Holvoet, and P. DeWint. 1990. Variability of the immune response in freshwater sponges. *J. Exp. Zool.* 254:215.

8. Based upon your knowledge of sponge biology and the properties of air and water, why are there no terrestrial sponges?

Taxonomic Detail

Subkingdom Parazoa
Phylum Porifera

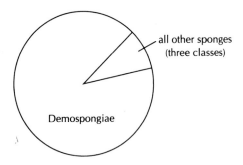

Class Demospongiae.

This class contains more than 90% of all existing sponge species, including all freshwater species. Support elements are never calcareous; they are instead composed of silica, spongin, or both. All members are of leuconoid construction. Species show a variety of forms—thin and encrusting, erect and branching, multilobed, spherical, tubular—and a single individual may exceed 2 m in diameter. Sixty-five families.

Family Clionidae. *Cliona.* Marine sponges that excavate burrows in calcareous material, such as mollusc shells and coral. Etching chemicals are released at the tips of specialized surface cells; these secretions dissolve only the edges of calcareous chips, which then fall to the seafloor, contributing up to 40% of the reef sediment in some places. Species in this family are distributed from shallow water to depths exceeding 2,100 m.

Family Spongiidae. *Spongia, Hippospongia.* All commercial sponges come from these two genera. All species in this family are marine. They live in all waters, from the tropics to the arctic and antarctic.

Family Haliclonidae. *Haliclona.* Members of this family are among the most commonly found and widely distributed of all shallow-water sponges, although some species reach depths of nearly 2,500 m.

Family Halichondriidae. *Halichondria.* These encrusting, marine sponges are common in shallow water.

Family Clathriidae. *Microciona.* The first experiments on sponge regeneration were performed on *Microciona prolifera* during the early 1900s.

Family Callyspongiidae. *Callyspongia.* The species in this group are common in shallow tropical oceans. Species may be upright, encrusting, branching, tubular, or vase-shaped; some are massive.

Family Spongillidae. *Spongilla, Eunapius, Ephydatia, Heteromeyenia.* This group contains most of the approximately 300 freshwater sponge species, which usually encrust solid surfaces in ponds, streams, rivers, and lakes. An individual sponge may exceed 1 m in diameter. A few species are found in brackish (i.e., slightly salty) waters.

Family Lubomirskiidae. This group contains freshwater species restricted to Lake Baikal in Siberia. None of these species produce gemmules.

Family Cladorhizidae. *Asbestopluma, Cladorhiza.* These are mostly deep sea sponges, living to depths as great as 8,840 m, although one species has now been found in a shallow-water Mediterranean cave. Unlike other sponges, these animals lack choanocytes and internal water canal systems. They apparently feed as carnivores, passively entrapping small crustaceans that swim by, and may also gain nutrients from the activities of symbiotic bacteria.

Class Calcarea.

All species are marine and have spicules exclusively composed of calcium carbonate. Species in this class may have asconoid, syconoid, or leuconoid grades of construction. Sixteen families.

Family Leucosoleniidae. *Leucosolenia.* All members of this family are marine, asconoid, and contained in a single genus. Species are found from the intertidal zone to depths exceeding 2,400 m.

Family Grantiidae. *Grantia* (= *Scypha*). Species are all marine. Members of this family are distributed from the intertidal zone to depths of about 2,200 m.

Class Sclerospongiae.

Stromatospongia. Most of the species in this group live in deep water on coral reefs or in caves, crevices, or tunnels within the reefs. All species are of leuconoid complexity. The body is supported by a thick layer of calcium carbonate in addition to spicules of silica and fibers of spongin. Five families.

Class Hexactinellida—the glass sponges.

Skeletal supports of all species are composed entirely of six-sided spicules of silica. Sixteen families.

Family Euplectellidae. *Euplectella*—Venus's flower basket. Individual sponges often harbor a single pair of shrimp, one of each sex, that enter the sponge when small and reach reproductive adulthood imprisoned together within the spongocoel. The species in this family are found at depths ranging from 100 m to over 5,200 m.

Some General References About the Sponges and Placozoans

Bergquist, P. R. 1978. *Sponges.* Berkeley, Calif.: Univ. Calif. Press.

Harrison, F. W., and J. A. Westfall, eds. 1991. *Microscopic Anatomy of the Invertebrates, Vol. 2. Placozoa, Porifera, Cnidaria, and Ctenophora.* New York: Wiley-Liss.

Hyman, L. 1940. *The Invertebrates, Vol. 1. Protozoa Through Ctenophora.* New York: McGraw-Hill.

Morris, S. C., et al., eds. 1985. *The Origins and Relationships of Lower Invertebrates. Systematics Association, Special Vol. 28.* Oxford: Clarendon Press.

Parker, S. P., ed. 1982. *Classification and Synopsis of Living Organisms,* vol. 1. New York: McGraw-Hill, 639 (placozoans), 641–66 (sponges).

Simpson, T. L. 1984. *The Cell Biology of Sponges.* New York: Springer-Verlag.

Thorpe, J. H., and A. P. Covich. 1991. *Ecology and Classification of North American Freshwater Invertebrates.* New York: Academic Press.

5

The Cnidarians

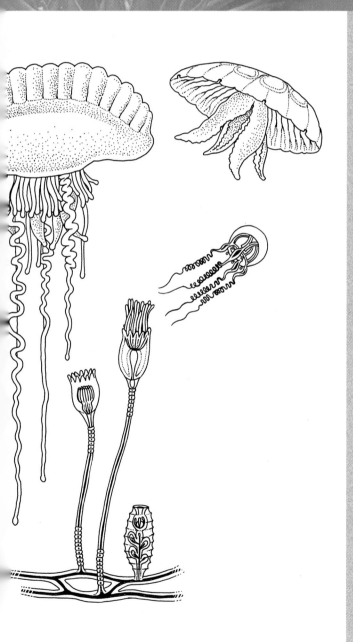

Introduction and General Characteristics

Phylum Cnidaria (= Coelenterata)
(G: a stinging thread [G: hollow gut])
nī-dare´-ē-ah (sih-len´-ter-ah´-tah)

The phylum Cnidaria contains some 10,000 species, including such animals as the sea anemones, corals, jellyfish, freshwater *Hydra,* and the Portuguese man-of-war. Despite these organisms' structural and functional diversity, there is no doubt regarding their membership in a single phylum. All cnidarians have a basic radial symmetry and possess only two layers of living tissue (the epidermis and the gastrodermis). All cnidarians possess a gelatinous layer, the **mesoglea,** located between the epidermis and the gastrodermis. Although the mesoglea is itself nonliving, it may contain living cells derived from embryonic ectoderm. All cnidarians have tentacles surrounding the mouth and only a single opening to the digestive system. All cnidarians possess nematocysts.

Nematocysts (literally, "thread bags"), unique to the members of this phylum, are small organelles formed within cells called **cnidoblasts** (or **cnidocytes**). Each nematocyst consists of a rounded, proteinaceous capsule, with an opening at one end that is often occluded by a hinged operculum. Within the sac is a long, hollow, coiled tube. During nematocyst discharge, the hollow tube shoots out explosively from the sac, turning inside out as it goes (Fig. 5.1). The entire process requires only about 3 ms (milliseconds). Discharge is triggered by a combination of chemical and tactile stimulation, perceived, presumably, through a cluster of modified cilia

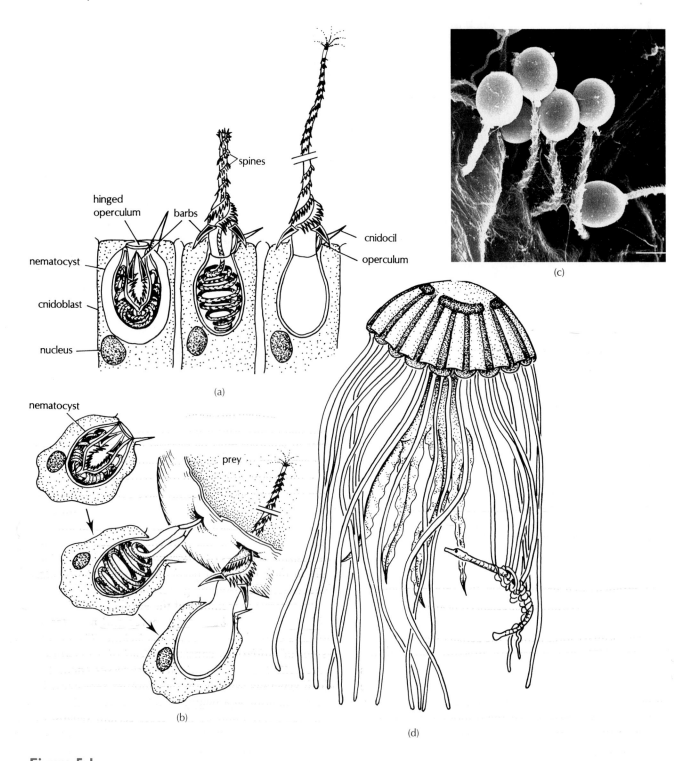

Figure 5.1

(a) Stages in the discharge of a nematocyst, stimulated by chemical and/or physical contact with the cnidocil. The mechanism of discharge remains unclear, although it is apparently mediated by an inflow of water along a concentration gradient. (b) Penetration of the nematocyst filament into another animal. Spines on the filament are exposed as the filament emerges (see part [a]), cutting through the prey tissues. (c) Six nematocysts fired by the scyphozoan jellyfish *Cyanea capillata* penetrating human skin. (Length of scale bar = 10 mm.) (d) Jellyfish (medusa) capturing prey. Nematocysts commonly inject toxins that paralyze the prey prior to ingestion.   (a) From Wells, *Lower Animals.* Copyright © 1968 McGraw-Hill, Inc., New York. Reprinted by permission. (b) *Hardy, The Open Sea: Its Natural History. Boston: Houghton Mifflin Company, 1965. (c) Courtesy of Thomas Heeger. From T. Heeger et al 1992. Marine Biology 113:669-678.*

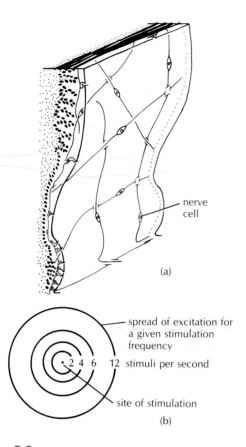

(a)

- spread of excitation for a given stimulation frequency

2 4 6 12 stimuli per second

- site of stimulation

(b)

Figure 5.2

(*a*) Diagrammatic illustration of a cnidarian nerve net. The nerve cells synapse with each other repeatedly. Nerve impulses may cross the synapses in both directions. (*b*) The surface area affected by stimulation of a cnidarian nerve cell varies directly with the frequency of stimulation; that is, the greater the frequency of stimulation, the greater the surface area affected.

(a) *After Bullock and Horridge.*

(the **cnidocil**) that projects from the cnidoblast and by surface chemoreceptors on specialized nearby cells.[1] Each nematocyst can be discharged only once.

The primary force behind the actual expulsion of the nematocyst filament is osmotic pressure, although the exact mechanism remains uncertain. One hypothesis is that discharge results from a sudden, dramatic increase in osmotic concentration within the capsular fluid. An alternative hypothesis is that the osmotic concentration is high at all times, the filament simply discharging when the capsule operculum is somehow opened. Many different morphological and functional types of nematocyst are found within the phylum and even within an individual cnidarian; it may turn out that different types of nematocysts are operated by different mechanisms.

Within a given individual, nematocysts may be specialized for wrapping around small objects, sticking to surfaces, penetrating surfaces, or secreting proteinaceous

toxins, some of which are among the most deadly toxins known. Nematocysts function in food collection, defense, and, to some extent, locomotion. They are especially abundant on the feeding tentacles of all species and within the digestive cavity of some species. The functional significance of nematocysts lies in their great number per square millimeter of body surface, rather than in their individual size; nematocyst capsules rarely exceed 50 μm (micrometers) in diameter and none exceed 100 μm. Nematocyst morphology is often important in making species identifications.

For those contemplating reincarnation, a major drawback to life as a cnidarian would seem to be the absence of an anus. All undigested food material passes out through the same opening through which the food enters: the mouth. This is not particularly appetizing, from the human point of view, but the shortcomings of life without an anus are not merely aesthetic. The sequential disassembly of particulate food material that occurs in an open-ended tubular gut is not possible in the coelenterate digestive system. Moreover, movements of the coelenterate are generally accompanied by physical distortion of the digestive cavity, including partial or complete expulsion of the contained fluid. Extensive movement is therefore not conducive to leisurely, thorough digestion. Finally, gonadal development often takes place within the digestive cavity, and the gametes or embryos must be released into this cavity before their expulsion to the exterior through the mouth.

Cnidarians are primarily carnivorous, although some soft-coral species recently have been found to feed also on phytoplankton. Some species in each of the three main cnidarian classes (Scyphozoa, Hydrozoa, and Anthozoa) obtain additional nutrients through the photosynthesizing activities of unicellular algae living symbiotically in their tissues. In particular, endosymbiotic algae characterize all reef-building (**hermatypic**) corals.[2]

In contrast to members of the Porifera and Placozoa, cnidarians possess bona fide nerves and muscles, although a central nervous system is lacking. The nervous system of cnidarians consists instead of a network of nerve cells (neurons) and their processes (neurites), which generally synapse on one another repeatedly before terminating at a neuromuscular junction (Fig. 5.2). Although nerve impulses may cross certain synapses in one direction only, many synapses permit impulses to pass in both directions. Moreover, a given cell body may give rise to two or more neurites, radiating in different directions. Thus, a nerve impulse received by one neuron may proceed in several directions at once.

With such a nerve network, stimulation of a given sensory cell in the epithelium results in an outward spread of excitation over the animal's entire body (Fig. 5.2b). The amount of surface area of the cnidarian that is

nerve cell

---

1. See *Topics for Further Discussion and Investigation*, nos. 6–8, at the end of the chapter.

2. See *Topics for Further Discussion and Investigation*, no. 3.

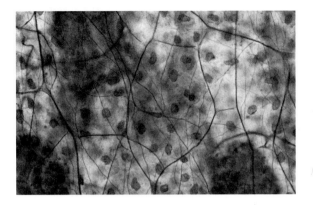

Figure 5.3

The cnidarian nerve net. Nerve cells in epithelial tissue of *Velella,* the by-the-wind sailor (class Hydrozoa), were impregnated with silver to make their outlines stand out when viewed with a light microscope. The network of smaller nerves (less than 1 µm in diameter) seems to be synaptic, and the additional network of larger nerves (up to 5 µm in diameter) may be syncytial, owing to the fusion of smaller nerve cells during development.   *Courtesy of G.O. Mackie. From Mackie, Singla and Arkett, 1988. J. Morphol 198:15–23, Permission of Wiley-Liss Publishers.*

affected by stimulating a given nerve cell increases in proportion to the frequency of stimulation.

In addition to this slow-conducting nerve network, a second, fast-conducting nerve network generally underlies the epithelium. These cells are less branched (bipolar as opposed to multipolar) than are the cells in the slow-conducting network, so that signal transmission is more directed. Furthermore, neurites in the two nets differ in size: Nerves in the fast-conducting network are of greater diameter, allowing more rapid conduction of nerve impulses. Some recent ultrastructural studies[3] indicate that these "giant" fibers (perhaps 1 µm to 5 µm in diameter) may arise through the fusion of smaller fibers during development (Fig. 5.3).

Because cnidarians are diploblastic animals, their musculature cannot (by definition of *diploblastic*) have its origins in embryonic mesoderm. Instead, the muscle layers are composed of numerous epithelial or gastrodermal cells that possess elongated, contractile bases anchored in the mesoglea; these cells are termed *epitheliomuscular cells* or *nutritive-muscular cells,* respectively (Fig. 5.4). Depending upon the orientation of the contractile bases, the cells can form either longitudinal musculature (running from the base to the tip of a tentacle, for example) or circular musculature (running around the circumference of a sea anemone's body column, for example). In addition to their contractile function, nutritive-muscular cells also capture small food particles and then digest them intracellularly. Although epitheliomuscular cells characterize all cnidarians, they are not a uniquely cnidarian feature; they are found sporadically among members of several other phyla.

Class Scyphozoa

Class Scypho • zoa
(G: cup animals)
sky-fō-zoˊ-ah

The Scyphozoa contains only a few hundred species, all of which are marine and many of which are quite large (up to about 2 m across). The mesoglea layer of scyphozoans is very thick and has the consistency of firm gelatin. For this reason, scyphozoans are known collectively as *jellyfish.*

Jellyfish morphology is described as **medusoid.** The body is in the form of an inverted cup, with nematocyst-studded tentacles extending downward from the cup, or **bell** (Fig. 5.5). The mouth is borne at the end of a muscular cylinder known as the **manubrium.**

Members of most scyphozoan species can swim actively, by contracting muscles and exploiting the mechanical properties of the mesoglea. When the muscle fibers of the swimming bell contract, the volume of fluid enclosed under the bell decreases. Water is forcefully expelled from under the bell as a consequence, and the animal is propelled in the opposite direction (Fig. 5.6). The muscular contraction deforms the elastic mesoglea, so that when the musculature is relaxed, the mesoglea "pops" back to its normal shape. This, of course, pulls the jellyfish downward as the volume enclosed by the swimming bell increases. Net forward movement of the animal occurs primarily because the speed with which the bell contracts exceeds the speed with which the bell recoils to its resting state. Note that water is forced out from under the swimming bell, not through the manubrium. Also note that the mesoglea functions here as a skeletal system, stretching the muscles following their contraction.

Scyphozoans are characterized by a well-developed system of fluid-filled **gastrovascular canals,** ultimately connecting to the mouth through the manubrium (Fig. 5.7). Food particles captured by nematocysts on the tentacles and/or oral arms are ingested at the mouth and conveyed to the stomach through the manubrium. Food is then distributed among four **gastric pouches,** which contain short, nematocyst-bearing tentacles (**gastric filaments**) that secrete an array of digestive enzymes. The partially digested food particles are then phagocytized, and digestion is completed intracellularly, a typical feature of cnidarian biology. Fluid within the gut is circulated by cilia lining the walls of the gastrovascular canals. The gastrovascular canals are believed to function in the circulation of oxygen and carbon dioxide (the "vascular" part of *gastrovascular*), as well as in the distribution of nutrients (the "gastro" part of the term).

3. G. O. Mackie, et al. 1988. On the nervous system of *Velella. J. Morphol.* 198: 15–23.

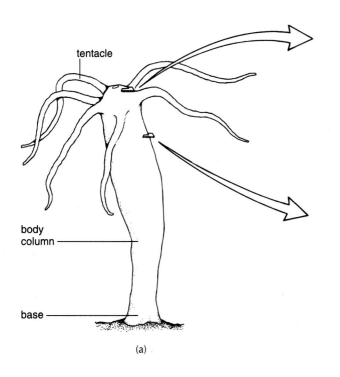

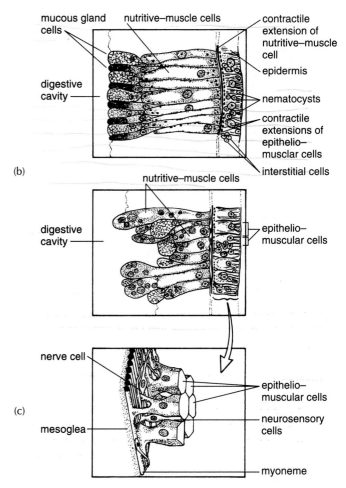

Figure 5.4

(a) Body wall of a freshwater *Hydra* seen in cross section from two regions of the body. Note that, as illustrated, the contractile bases of epitheliomuscular cells form a layer of longitudinal musculature extending up and down the body column, whereas the contractile bases of the nutritive muscle cells form a layer of circular muscles extending around the body column (b).
(c) Epitheliomuscular cells of a cnidarian. The columnar (or sometimes cylindrical) upper portions of the cells form epithelial tissue, while the elongated contractile bases form the musculature. A portion of the nerve net is also shown.

(a, b) *From L. Hyman,* The Invertebrates, *Vol. 1. Copyright © 1940 McGraw-Hill, Inc., New York. Reprinted by permission.* (c) *From G. O. Mackie and L. M. Pussano, in* Journal of General Physiology, *52:600-608, 1968. Copyright © 1968 Rockefeller University Press, New York. Reprinted by permission.*

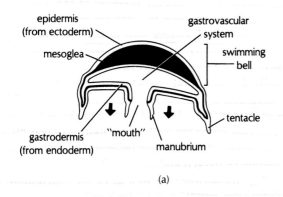

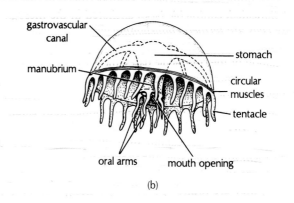

Figure 5.5

(a) A medusa, seen in longitudinal section. Note the very thick layer of mesoglea and the single opening to the gastrovascular cavity; this opening serves as both mouth and anus. Arrows indicate water leaving when bell musculature contracts.

(b) Lateral view of a medusa, showing the gastrovascular canal system and the arrangement of tentacles, oral arms, and musculature of the swimming bell.   (a) *After Russell-Hunter.*

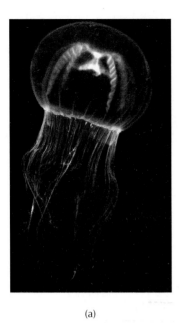

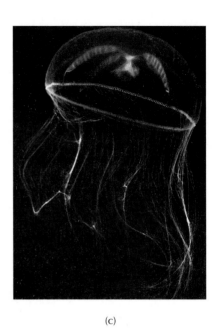

(a)   (b)   (c)

Figure 5.6

Locomotion of the medusa stage of *Mitrocoma cellularia*. The bell has a diameter of about 70 mm. As muscle contractions force water out from under the swimming bell, the animal is propelled in the opposite direction (upward, in photographs). Efficient jet propulsion relies on the incompressibility of water; fluid must leave the bell as it cannot "hide" under the bell by compression. The animal shown is actually a hydrozoan rather than a scyphozoan; the velum is seen clearly as a sheet protruding from the bell in (*a*). (See page 90 for a discussion of hydrozoan medusae.) (*a*) Power stroke nearly completed; note that a bolus of expelled water can be seen pushing the tentacles outward about midway along their length. (*b*) Beginning of recovery period; swimming bell is expanding; note that the bolus of ejected water has moved farther down the tentacles. (*c*) Bell fully relaxed, ready for the next contraction.   (a-c) *Courtesy of Claudia E. Mills.*

Some scyphozoans also obtain nutrients from certain unicellular algae (**zooxanthellae**) that live symbiotically in the jellyfish tissues.

Although neurologically unsophisticated in comparison with most other metazoans, scyphozoans nevertheless show a variety of fairly sophisticated behaviors, including periodic formation of breeding aggregations and periodic vertical migrations from surface waters to deeper waters and back again. As befits any mobile organism, scyphozoan medusae are equipped with fairly sophisticated sensory receptors. These include balance organs (**statocysts**), simple light receptors (**ocelli**), and, in some species, touch receptors (**sensory lappets**). The statocysts and ocelli are contained within club-shaped structures called **rhopalia,** which are distributed along the margins of the swimming bell (Figs. 5.7*b* and 5.8). Dense aggregations of nerve tissue are found associated with the rhopalia. These ganglia act as pacemakers, triggering the rhythmic contraction of the swimming bell.[4]

Statocysts operate on a beautifully simple principle. Tubular pieces of tissue (the **rhopalia**) hang freely at several locations around the margins of the swimming bell. Each of these rhopalia is adjacent to (but not in continuous contact with) sensory cilia. Also, each tube is weighted at the free end with a spherical calcareous mass (the **statolith**). If the animal tilts in a particular direction, those statocysts on the lowermost margin press against their respective cilia (Fig. 5.8*b*), causing the associated nerve cells to generate action potentials. The rhopalium/statocyst system thus provides a mechanism through which the animal can be informed of its physical orientation—that is, whether the body is horizontal or tilted—and the jellyfish can alter its posture accordingly, through stronger contractions of the musculature on one side of the bell.

The non-image-forming ocelli (light receptors) are also found along the bell margin. An **ocellus** is simply a small area, often cup shaped, backed by light-sensitive pigment.

The life cycle of a scyphozoan is a diagnostic feature of its biology. Gonads develop within gastrodermal tissue and are closely associated with the gastric pouches (see Fig. 5.7). With few exceptions, individual medusae are either male or female; that is, the sexes are generally separate, and the species is said to be **dioecious** (G: two houses). This contrasts with the situation frequently encountered among other invertebrates, in which a given individual may be both male and female, either simultaneously or in sequence. Such species are said to be **hermaphroditic.**

A **planula** larva eventually results from the union between sperm and egg. This larva typically has the form of a heavily ciliated, microscopic sausage. The nonfeeding planula larva soon settles on a substrate and transforms into a small polypoid individual called a **scyphistoma**

---

4. See *Topics for Further Discussion and Investigation*, no. 11.

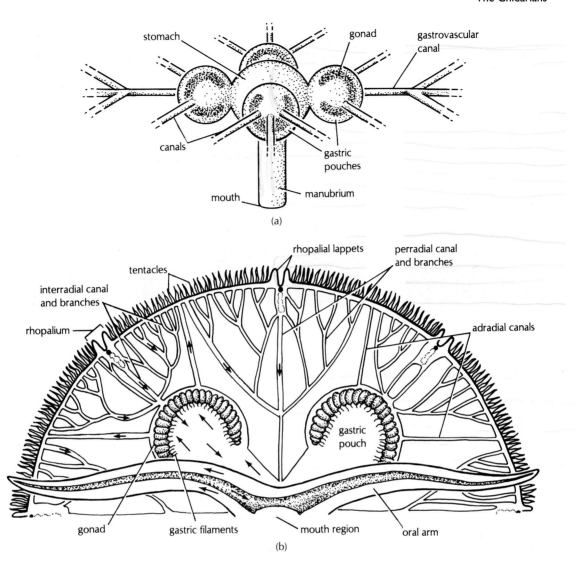

Figure 5.7

Detail of the scyphozoan gastrovascular canal system. (*a*) Lateral view of the moon jelly, *Aurelia* sp. (*b*) Oral view. Cilia lining the canal system draw water in through the mouth to the gastric pouches. From the pouches, water circulates to the periphery of the bell through a complex series of narrow canals, as shown by the arrows. Details of the rhopalia, which are sensory organs, are shown in Figure 5.8.   (*b*) *After Hickman.*

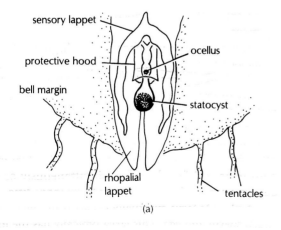

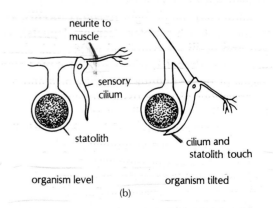

Figure 5.8

(*a*) Detail of one rhopalium from the scyphozoan *Aurelia aurita.* The species illustrated has a statocyst, a simple ocellus, and a pair of specialized sensory lappets associated with each rhopalium. The sensory lappets are believed to function as chemical receptors. (*b*) The principle of statocyst operation. When the animal tilts sideways, the statolith swings against a sensory cilium, initiating a nerve signal to the appropriate muscles; these muscle contractions then restore the animal to proper orientation.

(*a*) *After Hyman.* (*b*) *After Wells.*

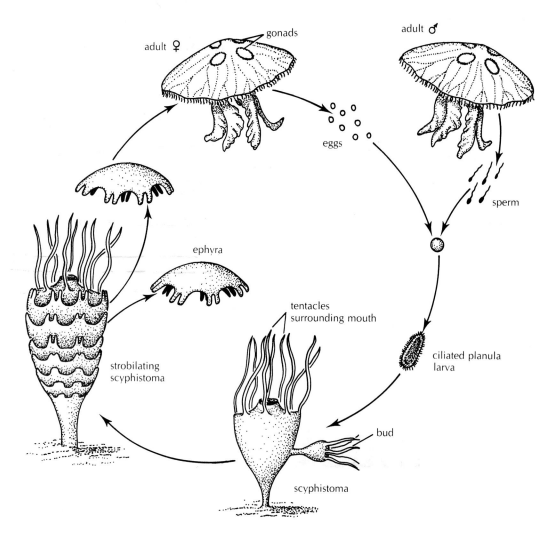

Figure 5.9
Life cycle of the moon jelly *Aurelia aurita*, a typical scyphozoan. The polyp stage (scyphistoma) is small—often only a few

millimeters in length—and is often found hanging downward from the undersides of underwater rock ledges.

(Fig. 5.9). This **polyp** form has the same two-layered construction (plus mesoglea layer) as the medusa, but the mesoglea layer is substantially thinner in the polyp than in the medusa morph. The scyphistoma is sessile and lacks ocelli and statocysts. It is a feeding individual, with the mouth oriented away from the substrate.

As the scyphistoma grows, it may produce additional scyphistomae by asexual budding. Eventually, a process called **strobilation** takes place in most species. During strobilation, the body column of a scyphistoma subdivides transversely, forming numerous discs that are stacked on top of each other like hotel ashtrays (Fig. 5.9). Each "ashtray" eventually breaks away from the stack as a swimming **ephyra**. As it swims, each ephyra gradually grows and changes in physical appearance, becoming an adult scyphozoan.

Reflect for a moment upon this life cycle. Scyphozoans use the relatively inconspicuous polyp morph to achieve

something quite remarkable: From a single fertilized egg, a large number of genetically identical, sexually reproducing medusae are generated. A similar phenomenon occurs among the parasitic flatworms, as discussed in Chapter 7.

Class Cubozoa

The members of the small but interesting class Cubozoa (cube-ō-zo´-ah) are called *cubomedusae*. As in scyphozoans, the medusa (jellyfish) stage dominates the cubozoan life cycle. The cubomedusa is typically only a few centimeters in its largest dimension and extremely transparent. The "cubo" part of the name refers to the fact that all members of the class have a cuboidal swimming bell: The bell is actually square in transverse section. Each individual bears four tentacles, or four clusters of tentacles, emerging from the four corners of the

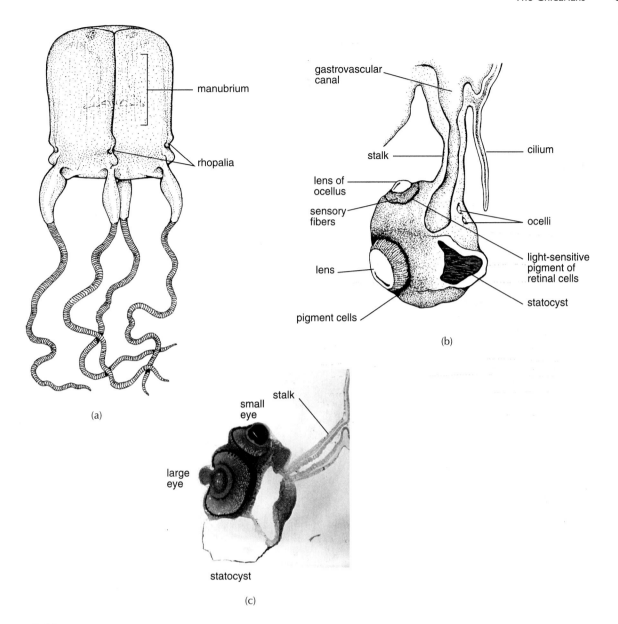

(a)

manubrium

rhopalia

gastrovascular canal

stalk

lens of ocellus

sensory fibers

lens

pigment cells

cilium

ocelli

light-sensitive pigment of retinal cells

statocyst

(b)

small eye

stalk

large eye

statocyst

(c)

Figure 5.10

(*a*) The cubomedusan *Carybdea* sp. Note the distinctly cuboidal shape of the swimming bell. This individual is only about 3 cm tall, excluding the tentacles. (*b*) Longitudinal section through a single rhopalium of *Carybdea* sp., showing the structure of the ocelli. In addition to several simple light receptors, note the more complex eyes, complete with lenses. These are among the most complex eyes found among invertebrates. (*c*) Histological section through a rhopalium of the cubomedusan *Tripedalia cystophora*. (b) *After Bayre and Owre: after Mayer, and after Conant.* (c) *From J. Piatigorsky, et al., in Journal of Comparative Physiology* (A), *164:577-87, 1989. Copyright © 1989 Springer Verlag, New York. Reprinted by permission.*

bell, near the four rhopalia (Fig. 5.10*a*). An animal only 2 cm to 3 cm in diameter can have tentacles over 30 cm long. These tentacles boast highly virulent nematocysts that have earned cubomedusae the well-deserved epithet "sea wasp." Some cubomedusan species can kill humans, inflicting considerable pain along the way. Cubomedusae mostly eat small fish, often killing individuals much larger than themselves.

Cubomedusae are unusually active and strong swimmers for jellyfish and possess an unusually well-developed nervous system and remarkably complex eyes (Fig. 5.10*b,c*). Unlike that of scyphozoans, the swimming bell of sea wasps curves inward at the lower edge, restricting the size of the opening through which water is expelled when the bell contracts. This increases the force with which the water exits the bell, and thus the amount of jet propulsion obtained. The effect is similar to that obtained by putting a nozzle on a garden hose: Greater propulsive force results from pushing the same volume of water out through a smaller opening in the same amount of time.

Cubozoans also differ from the true jellyfish (scyphozoans) in that the polyp stage does not strobilate. Rather, the polyp stage resulting from a single planula larva buds off more polyps, each of which develops into a single medusa. The asexual production of genetically identical individuals characterizing scyphozoans is again achieved in the cubozoan life cycle, but through a different vehicle.

Cubozoans are restricted to tropical and subtropical areas, and they can be quite common in those waters at certain times of the year.

Class Hydrozoa

Class Hydro • zoa
(G: water animals)
hī-drō-zō´-ah

Members of the Hydrozoa are characterized by generally greater representation of the polyp morph in the life cycle than is the case for scyphozoans, although the polyp and medusa morphs are about equally prominent in a number of hydrozoan species, and the medusa morph is dominant in a few others. In contrast to other cnidarians, the gastrodermal tissue of hydrozoans lacks nematocysts, and no cells are found within the mesoglea; nematocysts are restricted to the epidermis. The class Hydrozoa comprises three major orders (and several smaller ones not discussed here) containing fewer than 3,000 species. Most species are marine.

Order Hydroida

Although most members of the Hydroida (hī-droid´-ah) are marine, a number of freshwater species also exist—*Hydra,* for example (Fig. 5.13*a*). Hydroids are generally medusoid as adults; that is, the sexual stage of the life cycle resembles that found among the Scyphozoa, and like scyphozoan medusae, those of hydrozoans are commonly called jellyfish. While the best-known species live in surface waters, recent studies using submersibles and underwater cameras have revealed a number of jellyfish, both hydromedusae and scyphomedusae, living on or just above the ocean bottom several hundred to perhaps 1,000 m or more below the water's surface.

As in scyphozoans, the mesoglea layer of the hydrozoan medusa is thick, the mouth is borne at the end of a manubrium, and ocelli and statocysts are present. The sense organs may be found at the base of the tentacles, as in scyphozoans, or between the tentacles. All medusae are dioecious, a given individual being either male or female but never both. However, hydrozoan medusae tend to be much smaller than scyphozoan medusae (generally only a few centimeters or less across), and they usually possess a shelf of tissue (the **velum**) extending inward from the edge of the swimming bell toward the manubrium (Fig. 5.11). The presence of the velum causes water to be ejected from under the swimming bell through a narrower opening, and thus with greater velocity, when the musculature contracts. The effect is virtually identical to that achieved by the inwardly directed edge of the swimming bell of cubomedusae (p. 89). Scyphozoan medusae lack a velum.

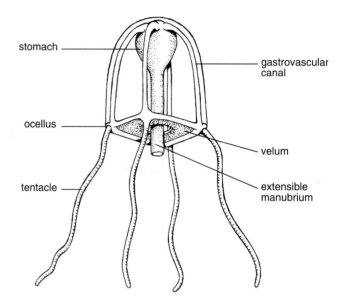

Figure 5.11
Diagrammatic illustration of a typical hydrozoan medusa. Note the conspicuous velum, through which water is forcefully expelled when the musculature of the swimming bell contracts. The water is expelled from under the swimming bell, as in scyphozoan medusae. The velum extends toward, but does not fuse with, the manubrium.

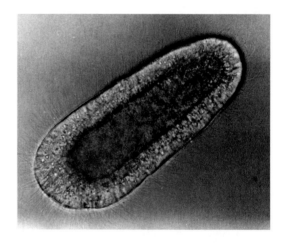

Figure 5.12
The planula larva of a hydrozoan, *Mitrocomella polydiademata.*
Courtesy of V. J. Martin.

As in the Scyphozoa, a planula larva (Fig. 5.12) develops from a fertilized hydrozoan egg, and this planula typically metamorphoses into a sessile polypoid individual lacking both statocysts and ocelli. The polyps of the Hydrozoa are structurally and functionally more complex than are the scyphistomae of the scyphozoan life cycle. The hydrozoan genus most familiar to the reader is probably the freshwater *Hydra.* In *Hydra,* each polyp is a separate, distinct being, completely responsible for its own welfare (Fig. 5.13*a*). Some species of *Hydra* (and some other

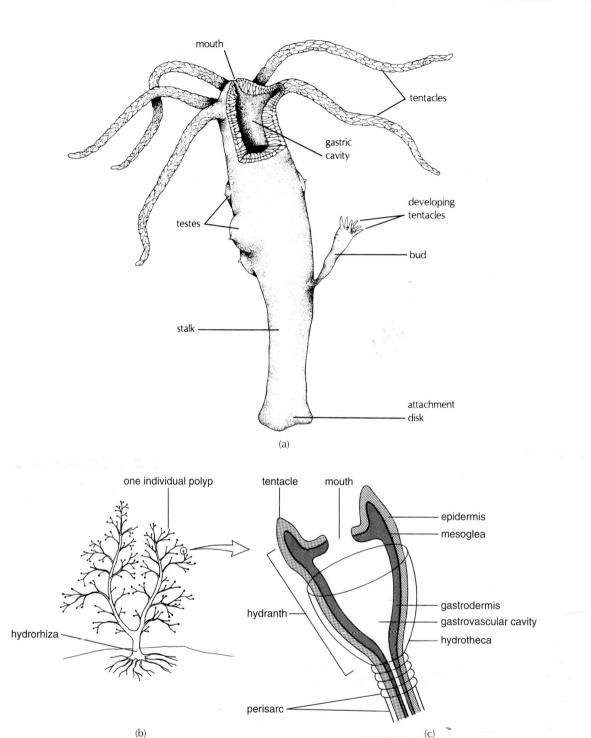

(a)

(b)

(c)

Figure 5.13

(a) A member of the freshwater genus *Hydra*. In most species, a single polyp is dioecious, making either sperm or eggs but not both, as illustrated; a few species are simultaneous hermaphrodites. All can also bud off new individuals asexually, as shown. Each bud will eventually break free and establish itself as a separate individual. (b) A typical colonial hydroid, *Campanularia* sp. The entire colony is only a few centimeters in height. (c) Diagrammatic illustration of a single hydrozoan polyp. Note that the mesoglea layer is much thinner than in the medusoid morph illustrated in Figure 5.5a.   (b) *After Hyman, Vol. I.*

Research Focus Box 5.1

Control of Nutrient Transfer in Symbiotic Associations

From A. E. Douglas, "The Influence Host Contamination on Maltose Release by Symbiotic Chlorella" in *Limnology and Oceanography*, 32:1363–1365, 1987. Copyright © 1987. American Society of Limnology and Oceanography, Inc., Seattle, Washington. Reprinted by permission.

Symbiotic algae supplement the diets of their invertebrate hosts with substantial quantities of carbohydrates and other energy-rich compounds. Although the relationship between cnidarians and their symbiotic algae has been studied intently for more than 25 years, much is still not understood about the process of nutrient exchange between algae and host. One especially intriguing issue is whether the cnidarian host can regulate the release of nutrients from the algal cells. After all, free-living algae retain most of their photosynthetic end products for their own use in respiration, growth, and cell division. When algal cells (*Chlorella* sp.) are isolated from the tissues of freshwater hydrozoans (*Hydra* spp.) and cultured in the laboratory, the algal cells release little carbohydrate to the surrounding medium. Yet, when living symbiotically, the algae release to the host about 60% of the carbohydrate they manufacture, usually as maltose. These observations imply that host tissue somehow encourages symbiotic algae to release carbohydrate that they would normally retain for their own use.

To examine this tantalizing prospect, Angela Douglas (1988) concentrated algal cells from cnidarian tissue by grinding up (**homogenizing**) green hydra and centrifuging the suspension; the preparation at the bottom of the centrifuge tube contained algal cells and host tissue. Douglas hypothesized that, if host tissue stimulates algae to release the products of their photosynthetic activity, such release should occur in the centrifuge tube, and that removing host tissue from the preparation should decrease the rate at which photosynthates are released.

When the crude preparation (algae and host cells) was incubated in the presence of radioactive carbon dioxide ($^{14}CO_2$ provided as ^{14}C-sodium bicarbonate) and adequate lighting, algal cells incorporated the ^{14}C into carbohydrates (Focus Table 5.1, row 1, column 1), indicating that the cells were actively photosynthesizing. More importantly, the algae released maltose into the medium at high rates (Focus Table 5.1, row 1, column 2), as predicted. In contrast, when host tissue in the hydra homogenates was solubilized with a detergent (SDS: sodium dodecyl sulfate), the algae released little maltose, even though they continued to photosynthesize at high rates (Focus Table 5.1, row 2). This comparison strongly suggests that host cells regulate the release of carbohydrates from the algae.

hydrozoans) harbor unicellular green algae, called zoochlorellae, in their tissues. Both the host and the algae benefit from the relationship, but the details of the interaction are still being discovered (Research Focus Box 5.1).[5] Note that, in many respects, *Hydra* is an atypical hydrozoan. For one thing, its life cycle lacks a medusa stage. Also, most other hydrozoans in the polyp stage of the life cycle are colonial; that is, a single planula often gives rise to a large number of polyps, called **zooids,** all of which are interconnected and share a continuous gastrovascular cavity (Fig. 5.13b). The zooids are often connected to each other, or to a substrate, by means of a rootlike **stolon,** termed the **hydrorhiza.** The oral end of a polyp (i.e., the end bearing the mouth and tentacles) is called the **hydranth.**

The stolon and stalks of the colony are commonly encased in a transparent protective tube, known as the **perisarc,** composed of polysaccharide and protein (Fig. 5.14a). The perisarc may or may not extend upward to encase the hydranth of a polyp, depending on the species. The perisarc surrounding the hydranth is known as a **hydrotheca** (Figs. 5.13c and 5.14a), and the hydroid is said to be **thecate** (as opposed to **athecate,** the Greek prefix *a* meaning "not," or "without"; see Fig. 5.14b).

Several structurally and functionally distinct individuals are often present in each hydroid colony; such colonies are **polymorphic,** consisting of two or more types of individuals. The feeding individuals are called **gastrozooids.** Gastrozooids collect small animals using the tentacles

---

Focus Table 5.1 — Effect of Several Treatments on Rates of Photosynthesis and Maltose Release by Symbiotic Zoochlorellae

Treatment	1 Rate of ^{14}C Incorporation (isotope counts per minute per 10,000 cells per hour)	2 Rate of Maltose Release (fmol per cell per hour)
untreated	2,250	1.30
detergent	2,150	undetectable
density gradation centrifugation	2,380	undetectable

Note: Zoochlorellae in this study were concentrated by homogenizing green hydra in a physiologically suitable solution, centrifuging the resulting suspension, and collecting the sedimented cells. (fmol = femtomole [10^{-18} moles]).

As further support for the hypothesis of regulation by hosts, Douglas was able to decrease the rate of maltose release when she removed host tissue from her preparation by an entirely different method: centrifuging the homogenate in the presence of a density gradient, to cleanly separate components differing in density—in this case, separating the algal cells from the host tissue. Once again, maltose release by algae in the absence of host tissue was barely detectable, even though the algae were clearly photosynthesizing (Focus Table 5.1, row 3). In both treatments, purified algal cells incorporated ^{14}C into carbohydrates as fast as nonpurified preparations (Focus Table 5.1, column 1), indicating that the treatments did not harm the algae.

These data suggest that something in host tissue regulates the release of carbohydrates from the symbiotic algae in green hydra. Similar findings have been reported for marine cnidarians. A reasonable next goal is to understand *how* hosts exert these regulatory effects on their algal guests. How would you proceed?

From A. E. Douglas, "The Influence Host Contamination on Maltose Release by Symbiotic Chlorella" in Limnology and Oceanography, *32:1363-1365, 1987. Copyright © 1987 American Society of Limnology and Oceanography, Inc., Seattle, Washington. Reprinted by permission.*

(which are densely clothed in nematocysts) and ingest the prey through the single opening into the gastrovascular cavity. Digestion is extracellular in the gastrovascular cavity, and then becomes intracellular as the partially digested food is distributed elsewhere in the colony, largely through rhythmic contractions of the muscular polyps.

Medusoids are produced asexually by budding from various regions of the polyp colony. A single hydrozoan colony typically produces either male or female medusae, but in some species a colony can apparently produce medusae of both sexes. Often, the medusoids derive from a particular type of individual, called a **gonozooid** (Fig. 5.14*a*). Some gonozooids lack tentacles and so are incapable of feeding; they are specialized for producing medusoids and must depend upon other members of the colony for nutrition. In some hydrozoans, the medusae eventually break free of the polyp colony, swim off, and commingle gametes with other medusae as already discussed. More commonly, however, the gamete-producing medusoid morph remains attached to the hydrozoan colony. No free-swimming medusa stage exists in such species. In fact, the medusa morph may be reduced to little more than a mass of gonadal tissue. Even so, it is best to think of the medusa as the adult. The polyp stage, no matter how conspicuous, is then best thought of as a prepubescent juvenile, which

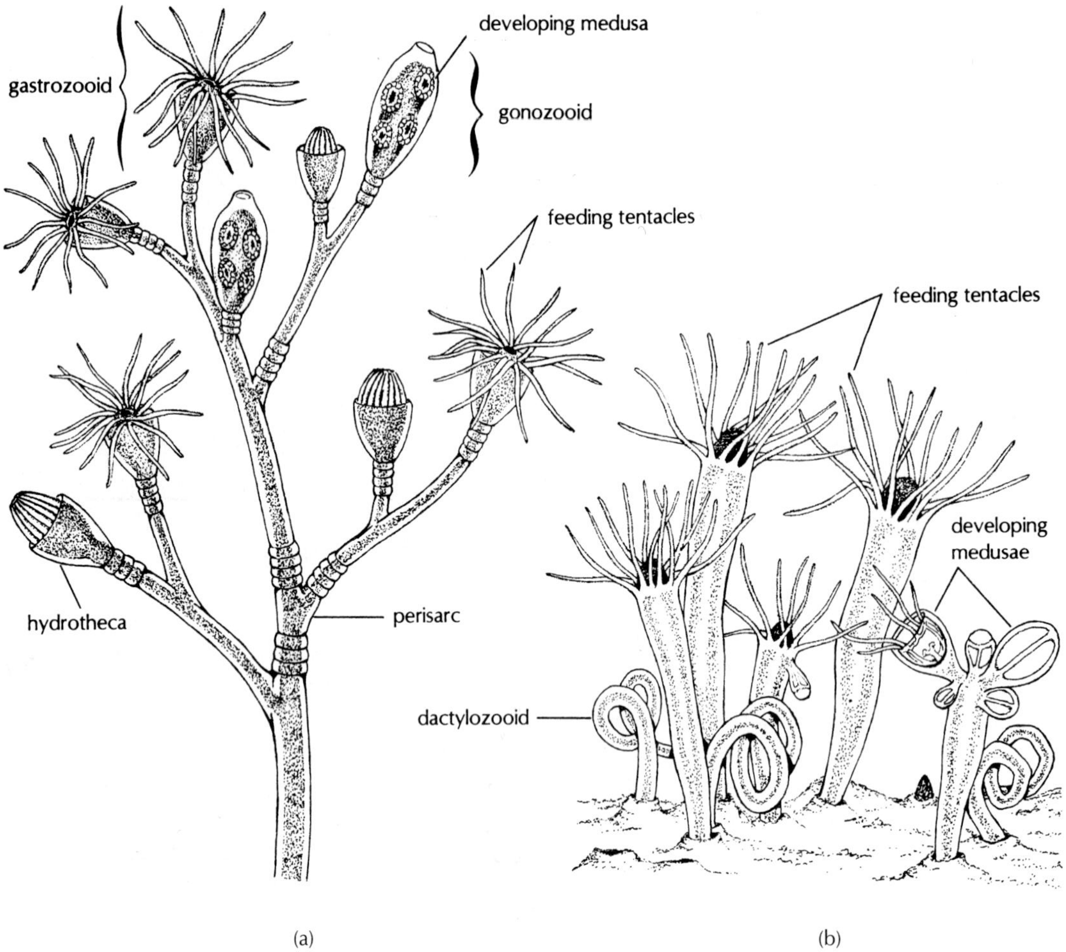

Figure 5.14
(a) A thecate, marine hydroid, *Obelia commissuralis*, showing specialized reproductive and feeding polyps (gonozooids and gastrozooids, respectively). This species commonly forms a pale, bushy covering on pilings and floats in protected harbors. (b) An athecate hydroid, *Podocoryne carnea*. This species possesses individuals specialized for protection (dactylozooids). *P. carnea* is commonly encountered inside the openings of snail shells occupied by marine hermit crabs.   (a) *After McConnaughey and Zottoli; after Nutting.* (b) *After McConnaughey and Zottoli; after Fraser.*

just happens to have a dramatically different morphology and lifestyle than does the sexually mature adult. In a few hydrozoan species, free-living medusae, asexually bud off more, genetically identical medusae either from the manubrium or from the tentacular bulbs, before getting on with the business at hand.

Hydrozoan colonies commonly contain fingerlike members specialized for defense (**dactylozooids;** *dactylus* = G: finger), as well as members specialized for feeding and reproduction (Figs. 5.14b and 5.15). The dactylozooids are heavily studded with nematocysts. Dactylozooids never possess mouths and thus, like some of the highly specialized gonozooids, depend upon the gastrozooids for food collection. All individuals in a given colony, no matter how polymorphic the colony, are originally derived from a single planula larva.

Many hydrozoans form species-specific symbiotic associations with other animals, including fish, sea urchins, sea squirts (ascidians), polychaete annelids, gastropods, bivalves, crustaceans, sponges, and even other hydrozoans (Figs. 5.14b, 5.15a).

Order Siphonophora

The epitome of hydrozoan polymorphism is reached within the order Siphonophora (sī-fon´-ō-for´-ah). The siphonophores, which include the Portuguese man-of-war, are free-floating hydrozoans in which medusoid and polypoid morphs are present simultaneously in a number of different incarnations. Modified medusae serve as gas-filled floats (**pneumatophores**), as individuals modified to propel the colony through the water by jet propulsion

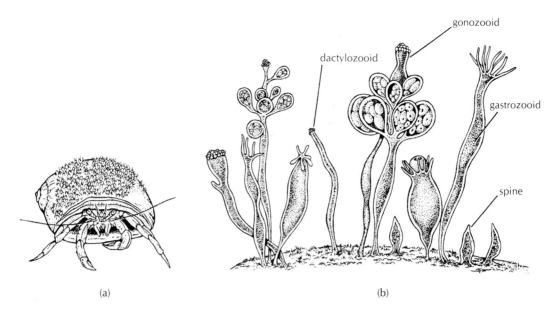

(a)   (b)

Figure 5.15

(a) Colony of *Hydractinia echinata* encrusting the outside surface of a snail shell inhabited by a hermit crab. (b) Detail of *H. echinata* colony, illustrating gastrozooids, gonozooids, dactylozooids, and individuals modified to form sharp protective spines.   (a) From R. I. Smith, Key to Marine Invertebrates of the Woods Hole Region Copyright © 1964 Marine Biological Laboratory, Woods Hole, MA. Reprinted by permission. (b) After Bayer and Owre.

(**nectophores**), or as leaflike defensive individuals (**bracts, or phyllozooids;** *phyllo* = G: leaf). The mesoglea layer is much reduced or entirely absent in pneumatophores. Nectophores lack both mouth and tentacles. Bracts are well-endowed with nematocysts. The polyp morph is represented by gastrozooids, gonozooids, and dactylozooids (Fig. 5.16).

Each gastrozooid has a single tentacle associated with it; elongated, nematocyst-bearing structures (**tentillae**) may project from these tentacles. Dactylozooids may also have associated tentacles. All siphonophore tentacles are highly retractile, and the nematocysts are often very toxic, even to humans.

Individuals within a colony often occur in clusters, called **cormidia,** arranged upon a long stem. Each cormidium typically contains gonozooids, dactylozooids, phyllozooids (bracts), and gastrozooids.

One or more morphs may be absent in some groups of siphonophores. For example, the Portuguese man-of-war lacks nectophores (Fig. 5.16c); movement of the animal depends entirely on wind and water currents. Other species (Fig. 5.16b) may lack pneumatophores.

All siphonophores, like other hydrozoans, are voracious carnivores.

The Hydrocorals

The hydrocorals are sometimes placed in a single order, the Hydrocorallina (hī-drō-cor-ah-līn´-ah), and sometimes distributed among two separate orders, as in the *Taxonomic Detail* at the end of this chapter. Either way, we are talking about a small number of species, all of which are colonial and secrete a substantial calcareous skeleton. Hydrocorals are largely restricted to warm waters. The dactylozooids are especially abundant and potent in many species; the common name "fire coral" is well deserved. These animals are not true corals, however. The true corals are contained within a different class of cnidarians, the Anthozoa.

Class Anthozoa

Class Antho · zoa
(G: flower animals)
an-thō-zoō´-ah

Anthozoans (including the sea anemones and the corals) consist of about 6,000 species, all of them marine. Anthozoans exploit the polyp body form and lifestyle exclusively; no trace of the medusa morph appears in the life cycle. Gametes are produced directly by the anthozoan polyp. A planula larva develops from the fertilized egg and metamorphoses to form another polyp. Many anthozoan species also reproduce asexually, often through longitudinal or transverse **fission** or through a process of **pedal laceration,** in which parts of the pedal disc (foot) detach from the rest of the animal and gradually differentiate to form another anemone.[6]

As with most hydrozoans and scyphozoans, anthozoans are primarily carnivorous; they capture food using

6. See *Topics for Further Discussion and Investigation,* no. 10.

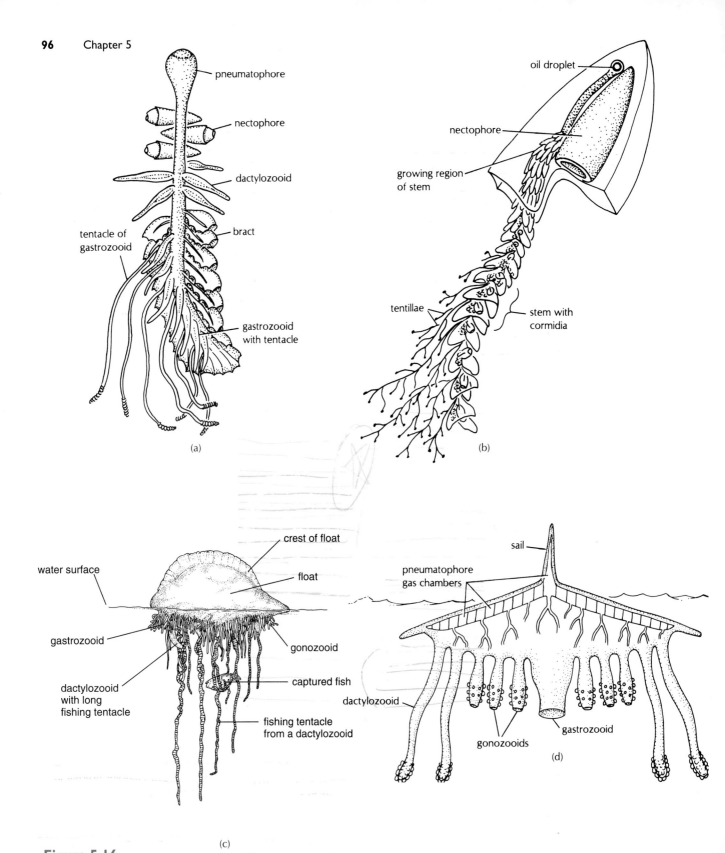

Figure 5.16

Typical pelagic (free-floating) hydrozoans. (*a*) In this form, typical of *Nectalia* sp., the pneumatophore is relatively small. Nectophore contraction provides jet propulsion. (*b*) *Muggiaea* sp., a siphonophore that lacks a pneumatophore. The largest member of the colony is a nectophore. Note the long stem with cormidia (clusters of gonozooids, bracts, gastrozooids, and dactylozooids). (*c*) *Physalia*, the Portuguese Man-of-War. This siphonophore has no nectophores and so moves wherever it is blown by the wind in its "sail." (*d*) In this hydrozoan (*Velella* sp.), the pneumatophore is divided into a series of chambers, and only a single gastrozooid is present. The dactylozooids are studded with nematocysts for prey capture. As with *Physalia*, locomotion is wind and current driven. Some workers consider *Velella* to be a single, elaborate polyp rather than a colony of individuals. It is no longer considered to be a siphonophore but has instead been placed in a separate order, the Chondrophora. (a,b,d) *From Hyman, The Invertebrates, Vol. III. Copyright © 1951 McGraw-Hill Book Company, New York. Reprinted by permission.*

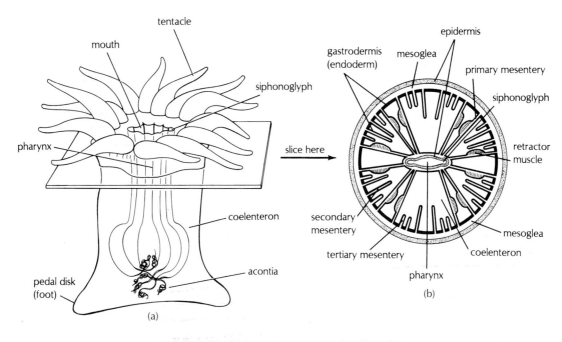

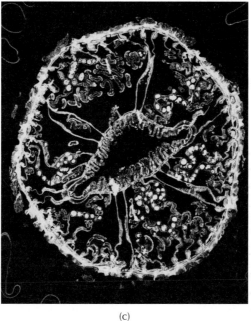

(c)

Figure 5.17

(a) Schematic of a typical anthozoan. The acontia connect to the lower edge of mesenteries (not shown here). (b) Cross section made through the polyp in the area indicated in (a).

(c) Photograph of a cross section through an anthozoan. Note the two siphonoglyphs and the developing embryos present in some of the mesenteries.   (c) *Copyright L.S. Eyster.*

nematocyst-studded tentacles and transfer it to a central mouth opening. However, the polyps of anthozoans differ from those of hydrozoans in several respects. The anthozoan mouth opens into a tubular pharynx rather than directly into the gastrovascular cavity. One or two discrete, ciliated grooves, called **siphonoglyphs**, typically extend down the pharynx from the mouth (Fig. 5.17). The gastrovascular cavity of an anthozoan is partitioned by numerous sheets of tissue called **mesenteries**, whereas no mesenteries are found in the gastrovascular cavity of hy-

drozoan polyps. These infoldings of gastroderm and mesoglea greatly increase the surface area available for secreting digestive enzymes and absorbing nutrients. Mesenteries that extend far enough from the body wall into the gastrovascular cavity to actually attach to the pharynx are called **primary mesenteries.** Those extending only partway into the gastrovascular cavity are termed **secondary** and **tertiary mesenteries.** Along their free edge, these shorter mesenteries typically form numerous short filaments studded with nematocysts and cells

Figure 5.18

A sea anemone, *Anthopleura krebsi*. The individual on the left is displaying numerous stubby acrorhagi below its pointy tentacles. The acrorhagi are used by the anemone to defend territory against invasion by neighboring anemones.   *Courtesy of C.H. Bigger, from Bigger 1980 Biological Bulletin 159:117 fig.1, pg. 120. Permission of Biological Bulletin.*

secreting digestive enzymes. Additionally, the mesenteries contain thick, longitudinal retractor muscles (Fig. 5.17*b*) and bear the gonads. Anthozoans are usually dioecious, but some species are sequential hermaphrodites.

Internally, near the base of an anthozoan, thin filaments called **acontia** extend from the mesenteries (Fig. 5.17*a*). These acontia are loaded with nematocysts and secretory cells and may be used for both defensive and offensive purposes. They can be extended outside the body through small pores in the body wall. The acontia may also function in digestion.

A number of anemone species possess rings of small, spherical bulges extending around the circumference of the body column just below the tentacles. These hollow **acrorhagi** can be extended a substantial distance from the body column, possibly by forcing fluid into them from the gastrovascular cavity, with which they are continuous. Acrorhagi are covered with very potent nematocysts and are used in defending a territory against invasion by other anemones (Fig. 5.18). Some anemone species that lack acrorhagi have, instead, tentacles that are specialized for fighting. These **catch tentacles** are analogous to acrorhagi, functioning in aggressive encounters among individuals.[7]

The tissues of an anthozoan contain both circular and longitudinal muscle fibers (Fig. 5.19). Provided that the animal keeps its mouth closed (by contracting appropriate sphincter muscles), the seawater in the gastrovascular cavity can serve as a hydrostatic skeleton. For example, by closing the mouth, relaxing the longitudinal musculature, and contracting the circular muscles of the body wall, the animal becomes long and thin, as the longitudinal muscles are stretched by the elevated pressure within the gastrovascular cavity (Fig. 5.20). Then, by contracting the longitudinal muscles on one side of the body and relaxing those on the other, the animal can bend to one side, provided that the circular muscles are not permitted to stretch. In an emergency, all the longitudinal muscles can be contracted while the mouth is open, causing the animal to flatten considerably as the fluid of the gastrovascular cavity is expelled. The resulting shape has been referred to as the "bubble gum on a rock" disguise (Fig. 5.20*c*). Reinflation is rather slow, being dependent upon the activity of the cilia lining the siphonoglyphs; these cilia "pump" water back into the gastrovascular cavity.

Many anthozoan species can move from place to place under their own power, although usually very slowly. A fast-moving anemone might achieve a speed of several millimeters per minute. Some anemones are commonly found on the backs of more mobile invertebrates and are transported adventitiously. A few species

7. See *Topics for Further Discussion and Investigation*, no. 9.

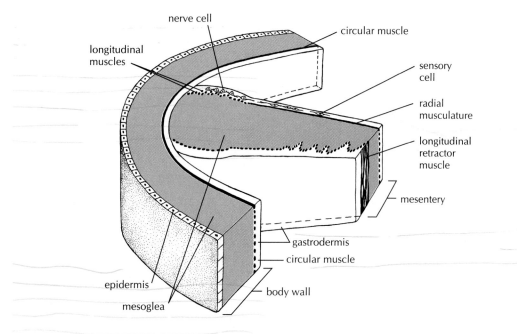

Figure 5.19
Diagrammatic illustration of the musculature and tissue layers in
the body wall and on the mesentery of a typical anthozoan.
After Bullock and Horridge.

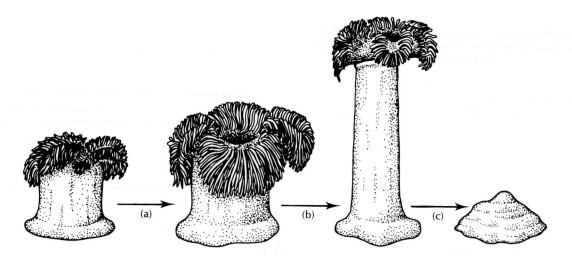

Figure 5.20
Shape changes in the sea anemone *Metridium senile.* (*a*) The
animal is slowly inflating, using the cilia on the siphonoglyph to
drive water into the gastrovascular cavity. (*b*) By closing the
mouth, relaxing the longitudinal muscles, and contracting the
circular muscles, the anemone increases in height but decreases
in body width. (*c*) By opening the mouth and contracting the
longitudinal muscles, most of the fluid in the gastrovascular cavity
is rapidly expelled. To regain its original shape, the animal must
pump water back into the coelenteron by ciliary action, a much
slower process.   *After Batham and Pantin.*

can "swim" for short distances in response to predators,[8] but most anemones are basically stay-at-homes.

Subclass Hexacorallia (= Zoantharia)

Subclass Hexa · corallia (= Zoantharia)
(G: six)
hex-ah-cor-ahl′-ē-ah (zō-an-thair′-ē-ah)

Hexacorallians possess numerous tentacles around the mouth opening (usually in some multiple of six; *hexa* =G: six) and six pairs of primary mesenteries (Fig. 5.17*b*). One pair of siphonoglyphs is associated with the pharynx. Many species in this subclass are **solitary** (i.e., they are independent individuals rather than colonies of connected individuals) and lack any specialized protective covering. These species are the sea anemones, as described in the class description. The other species in this subclass tend to be colonial; unlike the colonial hydrozoans, however, these anthozoans are never polymorphic. The best known of these colonial species are the true (or stony) corals, which secrete substantial external calcium carbonate skeletons. These species are also called scleractinian corals (*sclero* =G: hard), named for the order in which they are placed, the order Scleractinia (Figs. 5.21 and 5.23*a*,*b*). Scleractinian corals may be reef-building (**hermatypic**) or not (**ahermatypic**).

Most hermatypic corals are restricted to clear, warm waters. Coral reefs are especially abundant in tropical areas of the Indo-Pacific, forming chains of islands and other structures of massive proportions. The Great Barrier Reef along Australia's northeast shoreline is more than 2,000 km (kilometers) long and 145 km wide. This reef is one of the world's most diverse, most complex ecosystems. Although other organisms (particularly calcareous red algae, foraminiferans, shelled molluscs, certain tube-dwelling polychaetes, and bryozoans) make significant contributions to reef structure, anthozoans typically play the major role in constructing these spectacular habitats.[9]

Even though almost all anthozoans are carnivores, coral reefs flourish only in areas of low planktonic productivity. This is surprising because zooplankton production and abundance in the surrounding waters are insufficient to support such luxuriant coral growth; yet, the reefs are there! Years of research show that coral growth is aided here by a complex relationship with endosymbiotic, unicellular photosynthesizers (modified dinoflagellates, actually) called **zooxanthellae** (Fig. 5.22).[10] The zooxanthellae are photosynthetic and provide the anthozoan host with energy-rich, small molecular-weight organic compounds; recent measurements show 20% to 95% of the material fixed through photosynthesis being released to the host, principally as fatty acids, lipid droplets, amino acids, glucose, or glycerol. In return, the zooxanthellae have intimate access to the metabolic wastes of the coral, including carbon dioxide for photosynthesis and nitrogenous wastes essential for algal growth. The zooxanthellae also benefit from this symbiotic relationship, by being protected from herbivores. Zooxanthellae typically occur in coral tissue at concentrations of about one million cells per cm^2.

The relationship between the anthozoan and its zooxanthellae has an additional effect on reef growth, quite apart from that mediated by nutritional considerations. Some 30 years ago, scientists determined that rates of calcification by hermatypic corals are considerably higher in the light than in the dark. The implication is that activities of the symbiotic algae play a role in determining the rate at which the anthozoan deposits calcium carbonate, although the mechanism through which this effect is mediated remains uncertain. The most likely possibilities are

1. through an effect on availability of bicarbonate ion (HCO_3^-), an essential constituent of the calcification process;
2. through a contribution by the algae of a critical component of the organic matrix serving as the nucleating site for calcium carbonate deposition;
3. through localized removal of soluble phosphate (PO_4^{3-}) by the algae during photosynthesis (phosphates are known inhibitors of carbonate calcification); and
4. indirectly, through the influence of elevated concentrations of dissolved oxygen on rates of coral metabolism.

Several ahermatypic coral species also contain zooxanthellae in their tissues.

An intimate association with zooxanthellae incurs certain risks to the anthozoan host, particularly the exposure to hydrogen peroxide, oxyradicals, and other potentially toxic forms of oxygen; such active oxygen species are produced from oxygen evolved during photosynthesis, partly in response to ultraviolet radiation from sunlight. Oxyradicals may inactivate enzymes and degrade other proteins, damage nucleic acids, and oxidize membrane lipids. Accordingly, the tissues of anthozoan species harboring zooxanthellae tend to exhibit high activities of superoxide dismutase, catalase, and other enzymes that quickly detoxify free radicals and other reactive oxygen species; these same detoxifying enzymes are also present in the zooxanthellae themselves. In addition, the anthozoans concentrate UV-absorbing compounds

---

8. See *Topics for Further Discussion and Investigation*, no. 2.
9. See *Topics for Further Discussion and Investigation*, no. 13.
10. See *Topics for Further Discussion and Investigation*, nos. 3–5.

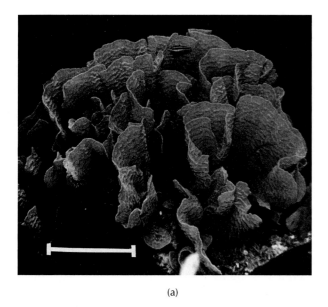

(a)

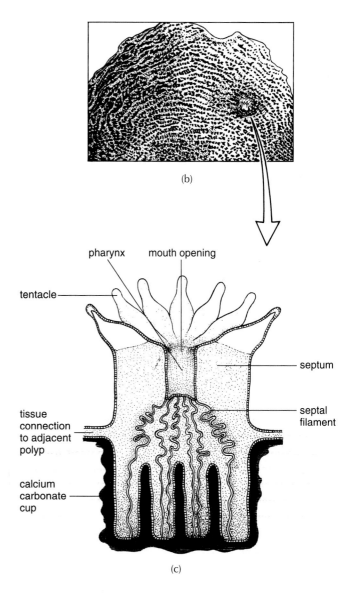

(b)

(c)

Figure 5.21

(a) *Agaricia tenuifolia*, a scleractinian coral from the waters off Belize. Scale bar about 15 cm. Unlike many hard coral species that produce massive, sturdy colonies, A. tenuifolia produces colonies that are fragile and leaflike. (b) Detail of one dead "leaf" of an *Agaricia tenuifolia* colony, showing the calcium carbonate cups, or thecae, in which the individual polyps sat. A single living polyp is illustrated at the right side of the leaf. (c) Detail of polyp sitting in its calcium carbonate cup, as seen in cross section. The cup is secreted by polyp epidermal cells and lies completely external to the polyp.

(a) *Courtesy of E.A. Chornesky. From Chornesky, E.A. 1991. Marine Biology 109:41–51.*
(c) *From Hyman, The Invertebrates, Vol. I. Copyright © 1940 McGraw-Hill, Inc., New York. Reprinted by permission.*

in their tissues and tend to contract during the sunniest part of the day, which may substantially reduce production of active oxygen.

In addition to their importance as reef builders and objects of aesthetic enjoyment, a few stony coral species are being used in certain surgical procedures. Because their system of appropriately sized, interconnected pores is rapidly infiltrated by human blood capillaries and osteoblasts, for example, small pieces of coral are now being used for bone grafts, particularly in face and jaw reconstruction, and in arm and leg surgery. The grafts become firmly anchored to the adjacent bone and, eventually, become completely replaced by normal bone tissue. Even more remarkably, small coral implants are being used to improve the degree to which artificial eyes exhibit natural movement. A piece of coral is implanted into the patient's eye socket and soon becomes invaded by blood vessels and eye muscles. Once connected to the coral implant, the

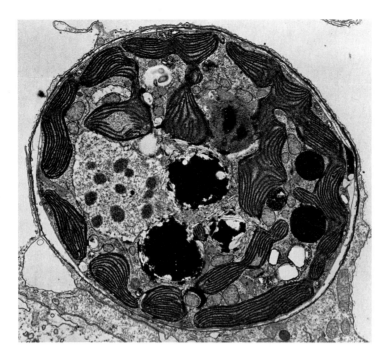

Figure 5.22
A transmission electron micrograph of a zooxanthella in the tissues of the stony coral Pocillopora damicornis. *Courtesy of E. H. Newcomb and T. D. Pugh, University of Wisconsin/BPS.*

artificial eye can be moved almost as naturally as a real eye. In all cases, the coral grafts are chemically converted from calcium carbonate to a form of calcium phosphate (hydroxyapatite) compatible with human bone tissue before implantation.

Subclass Octocorallia (= Alcyonaria)

Subclass Octo • corallia (= Alcyonaria)
(G: eight)
ok-to-cor-ahl´-ē-ah (ahl-sē-on-air´-ē-ah)

Members of the subclass Octocorallia possess eight tentacles (and eight primary septa) and generally have only a single siphonoglyph. The tentacles of octocorallians are **pinnate;** that is, they bear numerous lateral outfoldings called **pinnules** (Fig. 5.23*c*). A few octocoral species recently have been found to ingest phytoplankton, which probably are captured by the pinnules. Pinnate tentacles are only rarely encountered among other anthozoans.

All octocorallian species are colonial, and the colonies are often polymorphic. In species with polymorphic polyps, some individuals are incapable of feeding and function solely in driving water through the gastrovascular spaces of the colony. The polyps of octocorals may be embedded in a thick matrix of mesoglea; these are the soft corals. In other species of octocorals (colonial species with skeletal supports), the polyps are supported by proteinaceous or calcareous internal skeletons secreted by cells in the mesoglea. This latter group of octocorals includes the sea fans and sea whips (known collectively as gorgonians, or horny corals) and the pipe corals. Figure 5.23 shows some typical hexacorals and octocorals.

The tissues of octocorallians have long been known to accumulate a variety of unusual biochemicals derived from fatty-acid metabolism.[11] These compounds are not directly involved in the metabolic processes of the corals, but seem instead to protect the coral from predation and overgrowth by other organisms (Research Focus Box 5.2). Recent research indicates that some of these chemicals kill certain mammalian tumors, increasing interest in cnidarians as potential sources of effective anticancer agents.

---

11. See *Topics for Further Discussion and Investigation,* no. 12.

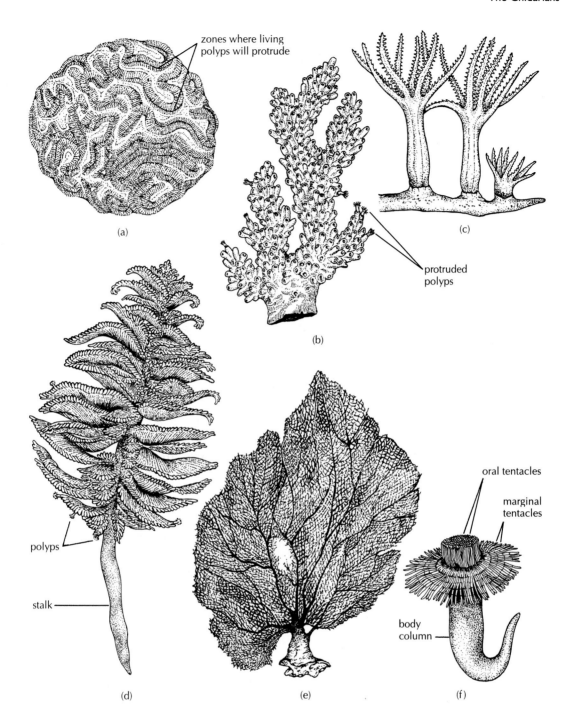

Figure 5.23

Anthozoan diversity. (*a*) The hexacoral, *Heliastra heliopora,* a brain coral. (*b*) Another hexacoral, *Astraea pallida,* with some of its polyps expanded and others retracted into the protective calcareous base of the colony. (*c*) An octocoral, *Clavularia,* showing the pinnate tentacles characteristic of the subclass Octocorallia. (*d*) *Pennatula* sp., the sea pen, another octocoral. (*e*) The sea fan *Gorgonia* sp., another octocoral. (*f*) A burrowing hexacoral anthozoan, *Cerianthus* sp., taken from the muddy substrate in which it lives. These animals (*f*) are structurally similar to sea anemones except that they lack a basal disc and have the tentacles arranged in two rings, as shown. Longitudinal muscles are especially well developed, permitting rapid withdrawal into mucus-lined burrows. Only the tentacles are generally visible at the surface of the sediment.   (a,b) *After Kingsley.* (c) *After Gohar.* (d) *After Kolliker.* (e) *After Bayer and Owre.*

Research Focus Box 5.2

Chemical Defenses on Coral Reefs

Van Alstyne, K. L., C. R. Wylie, and V. J. Paul. 1994. Antipredator defenses in tropical Pacific soft corals (Coelenterata: Alcyonacea). II. The relative importance of chemical and structural defenses in three species of *Sinularia*. *J. Exp. Marine Biol. Ecol.* 178:17–34.

Many organisms, including plants and macroalgae, manufacture chemical defenses against predation. Such **allelopathic** interactions, in which one organism makes a chemical inhibiting some activity in another, have been especially well studied among terrestrial plants and insects; they also occur commonly on coral reefs. Recent detailed biochemical studies show that gorgonians and other octocorals contain many lipid-soluble chemical compounds not found in most other organisms; at least some of these compounds might protect octocorals from predation. In addition, although most octocorals lack a hard, external, protective covering, gorgonians and soft corals do have fine calcareous spicules in their tissues. Do these spicules, which can account for up to 90% of the dry weight in some species, also deter predators? And if so, what is the relative importance of spicules and defensive chemicals in deterring predators?

Kathryn Van Alstyne and her colleagues addressed these questions for three soft-coral species in the genus *Sinularia,* collected from a reef off the coast of Guam. For all three species, the concentration of lipid-soluble chemicals was greater near the tips of the colonies than near the base, raising the possibility that the relative importance of chemicals and spicules might vary spatially within a colony.

The basic design of the experiments was to collect representative coral colonies in the field; extract organic chemicals and calcareous spicules from tips or bases of those colonies in the laboratory; add the extracted substances to a measured amount of homogenized squid; add nontoxic stiffening agents to the mixtures, forming water-insoluble cubes that contained the extracted chemical and/or spicules in concentrations matching those occurring naturally in the corals; and then test the relative attractiveness of the different cubes to predatory fish by suspending the cubes on ropes over a natural reef. The relative attractiveness of each diet was determined from the number of food cubes of each type that were eaten by fish within the 5 minute to 10 minute test periods. Each rope held four cubes of identical composition. The researchers always suspended the ropes of cubes in sets, with cubes of

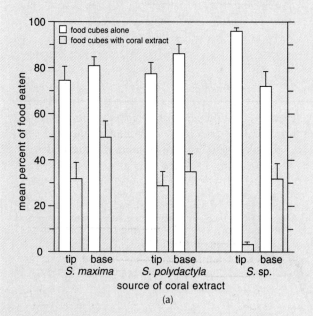

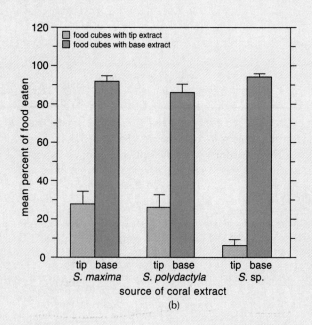

Focus Figure 5.1

(*a*) Effect of coral extract (blue bars) on attractiveness of food cubes to coral reef fish in the field. Extracts were prepared separately from tips and bases of three soft-coral species in the genus *Sinularia.* Each bar represents the mean number of food cubes eaten (error bars show one standard error above the mean) (N = 18–20 replicates per treatment.) (*b*) Direct comparison of relative effectiveness of tip and base extracts in discouraging feeding by coral reef fish in the field (N = 19–20 replicates per treatment.)

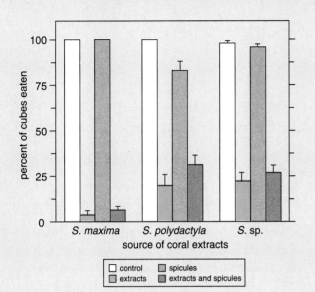

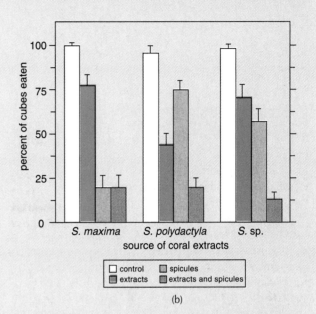

Focus Figure 5.2

Relative effectiveness of soft-coral chemical extracts and spicules in discouraging predation by coral reef fish in the field. Data are mean percentages of food cubes eaten (error bars show one standard error above the mean). (*a*) Extracts prepared from coral tips. (*b*) Extracts prepared from coral bases (N = 17–20 replicates per treatment.)

one type (containing extracts from coral tips, for example) on one rope and cubes of another type (containing extracts from coral bases, for example) on the second rope.

The first question to ask is whether fish will feed readily on the cubed squid homogenate. Thus, in some experiments the cubes of one rope of a pair contained only the squid and stiffening agent mixture as a control; after all, if fish wouldn't eat this control mixture, which is intended to be tasty, then one wouldn't expect them to eat the test mixtures either. As it turned out, fish fed actively on the control cubes (Focus Fig. 5.1a, open bars) but didn't eat nearly as many of the food cubes containing organic extracts from any of the three soft-coral species. (Focus Fig. 5.1a, filled bars). Note that for two of the soft-coral species, extracts from the coral tips were more effective deterrents than extracts from the coral bases when fish were given a choice of feeding on treated or control cubes. When given a choice between feeding only on cubes with extracts from the coral tips and cubes with extracts from the coral bases, the results were even more dramatic: Fish fed substantially less on cubes containing extracts from the coral tips (Focus Fig. 5.1b), probably reflecting the greater concentration of deterrent chemicals at the tips.

Now the researchers could study the relative effectiveness of chemicals and spicules as feeding deterrents. This time, ropes were suspended over the reef in sets of the following four treatments: control cubes, cubes containing only coral extract, cubes containing only spicules, and cubes containing both spicules and extract. In one series of experiments, extracts were made from coral tips (Focus Fig. 5.2a), and in another series of experiments, extracts were made from coral bases (Focus Fig. 5.2b). The results differed among coral species, but in general the spicules taken from coral tips were relatively ineffective in deterring predation compared with the effects of the extracts made from coral tips. For example, fish consumed essentially all of the control cubes (Focus Fig. 5.2a open bars) and essentially all cubes containing spicules (Focus Fig. 5.2a, light gray bars) obtained from *S. maxima*, but they ate less than 5% to 6% of the cubes containing extract from the tips of this species (Focus Fig. 5.2a, light blue bars). Moreover, for all three coral species, cubes containing spicules and extracts from the coral tips were no more effective in deterring predators than were cubes containing only the extract.

In contrast, spicules apparently play a more important role in deterring predation near the coral bases for at least two of the species (*S. maxima* and *S.* sp.; Focus Fig. 5.2b). For *S. maxima*, for example, fish fed somewhat less on cubes containing extract (dark blue bars) than on control cubes (open bars), but they fed even less on cubes containing spicules (light gray bars). Thus, chemicals seem to play the major defensive role near the tips of the colony, while spicules play the major defensive role near the base of the colony in these species.

Figures from K. L. Van Alstyne, et al., in Journal of Experimental Marine Biology and Ecology, *178:17–34, 1994. Copyright © 1994 Elsevier Science Publishers B. V., Netherlands. Reprinted by permission.*

Taxonomic Summary

Phylum Cnidaria (= Coelenterata)
  Class Scyphozoa—the true jellyfish
  Class Cubozoa—the sea wasps
  Class Hydrozoa
    Order Hydroida
    Order Siphonophora
    Order Hydrocorallina
  Class Anthozoa—the sea anemones, corals, sea whips, sea pens, sea fans, and sea pansies
    Subclass Hexacorallia (= Zoantharia)
    Subclass Octocorallia (= Alcyonaria)

Topics for Further Discussion and Investigation

1. Investigate the evolutionary origin of the phylum Cnidaria (Coelenterata). Which class probably represents the most primitive body plan and life history? Justify your answer.

Hyman, L. H. 1940. *The invertebrates,* vol. I. New York: McGraw-Hill, 632–41.

Rees, W. J., ed. 1966. *The Cnidaria and their evolution.* New York: Academic Press.

Willmer, P. 1990. *Invertebrate relationships.* New York: Cambridge University Press, 188–91.

2. Investigate the morphological and functional adaptations for swimming or burrowing encountered among some members of the Hexacorallia and Octocorallia, respectively.

Dalby, J. E. Jr., J. K. Elliott, and D. M. Ross. 1988. The swim response of the actinian *Stomphia didemon* to certain asteroids: Distributional and phylogenetic implications. *Canadian J. Zool.* 66:2484.

Elliott, J. K., D. M. Ross, C. Pathirana, S. Miao, R. J. Andersen, P. Singer, W. C. M. C. Kokke, and W. A. Ayer. 1989. Induction of swimming in *Stomphia* (Anthozoa: Actiniaria) by imbricatine, a metabolite of the asteroid *Dermasterias imbricata. Biol. Bull.* 176:73.

Lawn, I. D., and D. M. Ross. 1982. The behavioural physiology of the swimming sea anemone *Boloceroides mcmurrichi. Proc. Royal Soc. London* 216B:315.

Mariscal, R. N., E. J. Conklin, and C. H. Bigger. 1977. The ptychocyst, a major new category of cnida used in tube construction by a cerianthid anemone. *Biol. Bull.* 152:392.

Robson, E. A., 1961. Some observations on the swimming behavior of the anemone *Stomphia coccinea. J. Exp. Biol.* 38:343.

Ross, D. M., and L. Sutton. 1967. Swimming sea anemones of Puget Sound: Swimming of *Actinostola* new species in response to *Stomphia coccinea. Science* 155:1419.

3. To what extent are the nutritional and respiratory requirements of hermatypic corals and other anthozoans met by resident zooxanthellae?

Battey, J. F., and J. S. Patton. 1987. Glycerol translocation in *Condylactis gigantea. Marine Biol.* 95:37.

Davies, P. S. 1991. Effect of daylight variations on the energy budgets of shallow-water corals. *Marine Biol.* 108:137.

Dunn, K. W. 1988. The effect of host feeding on the contribution of endosymbiotic algae to the growth of green hydra. *Biol. Bull.* 175:193.

Fabricius, K. E., Y. Benayahu, and A. Genin. 1995. Herbivory in asymbiotic soft corals. *Science* 268:90.

Falkowski, P. G., Z. Dubinsky, L. Muscatine, and J. W. Porter. 1984. Light and the bioenergetics of a symbiotic coral. *BioScience* 34:705.

Kevin, K. M., and R. C. L. Hudson. 1979. The role of zooxanthellae in the hermatypic coral *Plesiastrea urvellei* (Milne Edwards and Haime) from cold waters. *J. Exp. Marine Biol. Ecol.* 36:157.

Kinzie, R. A. III, and G. S. Chee. 1979. The effect of different zooxanthellae on the growth of experimentally reinfected hosts. *Biol. Bull.* 156:315.

Meyer, J. L., E. T. Schultz, and G. S. Helfman. 1983. Fish schools: An asset to corals. *Science* 220:1047.

Muscatine, L., and J. W. Porter. 1977. Reef corals: Mutualistic symbiosis adapted to nutrient-poor environments. *BioScience* 27:454.

Wethey, D. C., and J. W. Porter. 1976. Sun and shade differences in productivity of reef corals. *Nature (London)* 262:281.

4. Investigate the behavioral, morphological, or biochemical modifications of anthozoans that increase the photosynthetic contributions of their symbiotic zooxanthellae.

Dykens, J. A., and J. M. Schick. 1984. Photobiology of the symbiotic sea anemone, *Anthopleura elegantissima:* Defenses against photodynamic effects, and seasonal photoacclimatization. *Biol. Bull.* 167:383.

Lasker, H. R. 1979. Light-dependent activity patterns among reef corals: *Montastrea cavernosa. Biol. Bull.* 156:196.

Lewis, J. B. 1984. Photosynthetic production by the coral reef anemone, *Lebrunia coralligens* Wilson, and behavioral correlates of two nutritional strategies. *Biol. Bull.* 167:601.

Pearse, V. B. 1974. Modification of sea anemone behavior by symbiotic zooxanthellae: Expansion and contraction. *Biol. Bull.* 147:641.

Redalje, R. 1977. Light adaptation strategies of hermatypic corals. *Pacific Sci.* 30:212.

Schlicter, D., H. W. Fricke, and W. Weber. 1986. Light harvesting by wavelength transformation in a symbiotic coral of the Red Sea twilight zone. *Marine Biol.* 91:403.

Sebens, K. P., and K. DeRiemer. 1977. Diel cycles of expansion and contraction in coral reef anthozoans. *Marine Biol.* 43:247.

Sutton, D. C., and O. Hoegh-Guldberg. 1990. Host-zooxanthella interactions in four temperate marine invertebrate symbioses: Assessment of effect of host extracts on symbionts. *Biol. Bull.* 178:175.

Vareschi, E., and H. Fricke. 1986. Light responses of a scleractinian coral (*Plerogyra sinuosa*). *Marine Biol.* 90:395.

5.   Notwithstanding the obvious nutritional contributions of photosynthesizing symbionts to the cnidarian host, the algal cells may themselves exploit their situation within the host in obtaining nutrients for themselves. When the cnidarian captures and digests particulate food from the surrounding seawater, the resulting nutrients become potentially accessible to the intracellular algae. To what extent do zooxanthellae and zoochlorellae compete with their host for nutrients obtained by the host's feeding activities?

Blanquet, R. S., D. Emanuel, and T. A. Murphy. 1988. Suppression of exogenous alanine uptake in isolated zooxanthellae by cnidarian host homogenate fractions: Species and symbiosis specificity. *J. Exp. Marine Biol. Ecol.* 117:1.

Cook, C. B., C. F. D'Elia, and G. Muller-Parker. 1988. Host feeding and nutrient sufficiency for zooxanthellae in the sea anemone *Aiptasia pallida*. *Marine Biol.* 98:253.

McAuley, P. J. 1987. Quantitative estimation of movement of an amino acid from host to *Chlorella* symbionts in green hydra. *Biol. Bull.* 173:504.

McDermott, A. M., and R. S. Blanquet. 1991. Glucose and glycerol uptake by isolated zooxanthellae from *Cassiopea xamachana*: Transport mechanisms and regulation by host homogenate fractions. *Marine Biol.* 108:129.

Rees, T. A. V. 1991. Are symbiotic algae nutrient deficient? *Proc. Royal Soc. London B* 243:227.

Steen, R. G. 1986. Evidence for heterotrophy by zooxanthellae in symbiosis with *Aiptasia pulchella*. *Biol. Bull.* 170:267.

Thorington, G., and L. Margulis. 1981. *Hydra viridis*: Transfer of metabolites between *Hydra* and symbiotic algae. *Biol. Bull.* 160:175.

6.   The nematocyst is a morphologically and functionally complex structure. Through what process are nematocysts formed by the cnidoblasts?

Skaer, R. J. 1973. The secretion and development of nematocysts in a siphonophore. *J. Cell Sci.* 13:371.

7.   What is the relative importance of chemical versus physical stimuli in triggering nematocyst discharge?

Conklin, E. J., and R. N. Mariscal. 1976. Increase in nematocyst and spirocyst discharge in a sea anemone in response to mechanical stimulation. In *Coelenterate Ecology and Behavior*, edited by G. O. Mackie. New York: Plenum, 549–58.

Grosvenor, W., and G. Kass-Simon. 1987. Feeding behavior in *Hydra*. I. Effects of *Artemia* homogenate on nematocyst discharge. *Biol. Bull.* 173:527.

Thorington, G. U., and D. A. Hessinger. 1990. Control of cnida discharge. III. Spirocysts are regulated by three classes of chemoreceptors. *Biol. Bull.* 178:74.

Watson, G. M., and D. A. Hessinger. 1989. Cnidocyte mechanoreceptors are tuned to the movements of swimming prey by chemoreceptors. *Science* 243:1589.

Watson, G. M., and D. A. Hessinger. 1994. Antagonistic frequency tuning of hair bundles by different chemoreceptors regulates nematocyst discharge. *J. Exp. Biol.* 187:57.

8.   To what extent is nematocyst discharge under direct nervous control?

Aerne, B. L., R. P. Stidwill, and P. Tardent. 1991. Nematocyst discharge in *Hydra* does not require the presence of nerve cells. *J. Exp. Zool.* 258:137.

Lubbock, R. 1979. Chemical recognition and nematocyte excitation in a sea anemone. *J. Exp. Biol.* 83:283.

Mire-Thibodeaux, P., and G. M. Watson. 1993. Direct monitoring of intracellular calcium ions in sea anemone tentacles suggests regulation of nematocyst discharge by remote, rare epidermal cells. *Biol. Bull.* 185:335.

Pantin, C. F. A. 1942. The excitation of nematocysts. *J. Exp. Biol.* 19:294.

Ross, D. M., and L. Sutton. 1964. Inhibition of the swimming response by food and of nematocyst discharge during swimming in the sea anemone *Stomphia coccinea*. *J. Exp. Biol.* 41:751.

Ruch, R. J., and C. B. Cook. 1984. Nematocyst inactivation during feeding in *Hydra littoralis*. *J. Exp. Biol.* 111:31.

Sandberg, D. M., P. Kanciruk, and R. N. Mariscal. 1971. Inhibition of nematocyst discharge correlated with feeding in a sea anemone, *Calliactis tricolor* (Leseur). *Nature (London)* 232:263.

Smith, S., J. Oshida, and H. Bode. 1974. Inhibition of nematocyst discharge in *Hydra* fed to repletion. *Biol. Bull.* 147:186.

9.   Sea anemones, corals, and hydrozoans must compete for space with each other, as well as with members of many other animal (and plant or algal groups). What are the roles of acrorhagi and catch tentacles in mediating the competition for space among sea anemones and corals? To what extent does the ability to distinguish self from non-self contribute to competitive ability? Through what mechanisms are such distinctions made?

Ayre, D. J. 1982. Inter-genotype aggression in the solitary sea anemone *Actinia tenebrosa*. *Marine Biol.* 68:199.

Bigger, C. H. 1980. Interspecific and intraspecific acrorhagial aggressive behavior among sea anemones: A recognition of self and not-self. *Biol. Bull.* 159:117.

Bigger, C. H. 1982. The cellular basis of the aggressive acrorhagial response of sea anemones. *J. Morphol.* 173:259.

Chadwick, N. E. 1987. Interspecific aggressive behavior of the corallimorpharian *Corynactis californica* (Cnidaria: Anthozoa): Effects on sympatric corals and sea anemones. *Biol. Bull.* 173:110.

Chornesky, E. A. 1983. Induced development of sweeper tentacles on the reef coral *Agaricia agaricites:* A response to direct competition. *Biol. Bull.* 165:569.

Francis, L. 1973. Intraspecific aggression and its effect on the distribution of *Anthopleura elegantissima* and some related anemones. *Biol. Bull.* 144:73.

Fukui, Y. 1986. Catch tentacles in the sea anemone *Haliplanella luciae*. Role as organs of social behavior. *Marine Biol.* 91:245.

Kaplan, S. W. 1983. Intrasexual aggression in *Metridium senile*. *Biol. Bull.* 165:416.

Lange R. G., M. H. Dick, and W. A. Müller. 1992. Specificity and early ontogeny of historecognition in the hydroid *Hydractinia. J. Exp. Zool.* 262:307

Miles, J. S. 1991. Inducible agonistic structures in the tropical corallimorpharian, *Discosoma sanctithomae. Biol. Bull.* 180:406.

Salter-Cid, L., and C. H. Bigger. 1991. Alloimmunity in the gorgonian coral *Swiftia exserta. Biol. Bull.* 181:127.

Sauer, K. P., M. Muller, and M. Weber. 1986. Alloimmune memory for glycoprotein recognition molecules in sea anemones competing for space. *Marine Biol.* 92:73.

Sebens, K. P., and J. S. Miles. 1988. Sweeper tentacles in a gorgonian octocoral: Their function in competition for space. *Biol. Bull.* 175:378.

Shenk, M. A., and L. W. Buss. 1991. Ontogenetic changes in fusibility in the colonial hydroid *Hydractinia symbiolongicarpus. J. Exp. Zool.* 257:80.

10. Some form of asexual reproduction is encountered among all three classes of cnidarians. What are the adaptive benefits of asexual versus sexual reproduction?

Grassle, J. F., and J. M. Schick, eds. 1979. Ecology of asexual reproduction in animals. *Amer. Zool.* 19:667.

11. How are swimming and other behaviors mediated by the cnidarian nervous system?

Leonard. J. L. 1982. Transient rhythms in the swimming activity of *Sarsia tubulosa* (Hydrozoa). *J. Exp. Biol.* 96:181.

Lerner, J., S. A. Meleon, I. Waldron, and R. M. Factor. 1971. Neural redundancy and regularity of swimming beats in scyphozoan medusae. *J. Exp. Biol.* 55:177.

Mackie, G. O. 1990. Giant axons and control of jetting in the squid *Loligo* and the jellyfish *Aglantha. Canadian J. Zool.* 68:799.

Sawyer, S. J., H. B. Dowse, and J. M. Shick. 1994. Neurophysiological correlates of the behavioral response to light in the sea anemone *Anthopleura elegantissima. Biol. Bull.* 186:195.

12. To what extent do octocorals use chemical defenses against predation?

Fenical, W., and J. R. Pawlik. 1991. Defensive properties of secondary metabolites from the Caribbean gorgonian coral *Erythropodium caribaeorum. Marine Ecol. Progr. Ser.* 75:1.

Harvell C. D., W. Fenical, and C. H. Greene. 1988. Chemical and structural defenses of Caribbean gorgonians (*Pseudopterogorgia* spp.) I. Development of an *in situ* feeding assay. *Marine Ecol. Progr. Ser.* 49:287.

La Barre, S. C., J. C. Coll, and P. W. Sammarco. 1986. Defensive strategies of soft corals (Coelenterata: Octocorallia) of the Great Barrier Reef. II. The relationship between toxicity and feeding deterrence. *Biol. Bull.* 171:565.

Mackie, A. M. 1987. Preliminary studies on the chemical defenses of the British octocorals *Alcyonium digitatum* and *Pennatula phosphorea. Comp. Biochem. Physiol.* 86A:629.

Sammarco, P. W., S. La Barre, and J. C. Coll. 1987. Defensive strategies of soft corals (Coelenterata: Octocorallia) of the Great Barrier Reef. III. The relationship between ichthyotoxicity and morphology. *Oecologia* (Berlin) 74:93.

13. What factors influence growth rates and growth forms of hermatypic corals?

Barnes, D. J. 1973. Growth in colonial scleractinians. *Bull. Marine Sci.* 23:280.

Buddemeier, R. W., and D. G. Fautin. 1993. Coral bleaching as an adaptive mechanism. *BioScience* 43:320.

Bunkley-Williams, L., and E. H. Williams Jr. 1990. Global assault on coral reefs. *Nat. Hist.* (April):46.

Dunstan, P. 1975. Growth and form in the reef-building coral *Montastrea annularis. Marine Biol.* 33:101.

Goreau, T. F., N. I. Goreau, and T. J. Goreau. 1979. Corals and coral reefs. *Sci. Amer.* 241:124.

Newell, N. D. 1972. The evolution of reefs. *Sci. Amer.* 226:54.

Taxonomic Detail

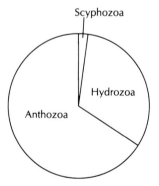

Phylum Cnidaria (= Coelenterata)

This phylum contains approximately 10,000 species, distributed among three classes.

Class Scyphozoa.

The true jellyfish (about 200 species, all marine). Four orders.

Order Stauromedusae.

Haliclystus. Unlike other scyphozoans, the medusae in this order are always sessile and develop directly from the scyphistoma stage without strobilation. Thus, only one juvenile arises from each scyphistoma; each planula larva can, however, bud off other, genetically identical planulae, so that numerous, genetically identical juveniles can still be produced from each fertilized egg. The planulae are nonciliated, creeping creatures, unlike those of other cnidarians. Adults attach to firm substrates, such as macroalgae and stones, using a central adhesive disc. Some species can move from place to place, but none can swim. Two families.

Order Coronatae.
Stephanoscyphus. This order contains over 24 mostly large, deep-water species. Members of the genus *Stephanoscyphus* are unique among scyphozoans in that the scyphistoma secretes a chitinous perisarc about itself. The medusae may reproduce sexually within the perisarc, never leaving the tube. The members of at least one species produce ephyrae that are said to develop directly into planulae, which then settle and attach to a substrate. Seven families.

Order Semaeostomeae.
Aurelia—the moon jelly. *Cyanea*—the lion's mane jellyfish, which produces the largest of all known medusae; individuals can reach 2 m in diameter and bear tentacles up to 30 m long. Contact with the oral arms or the tentacles extending from the circumference of the swimming bell brings a painful sting to unwary swimmers along the east coast of the United States. Certain species of fish and crustaceans often live in close association with these jellyfish, as commensals, external parasites, or predators. *Chrysaora*—the sea nettle. Sea nettles are bothersome to summer swimmers along the Atlantic coast of North America. Three families.

Order Rhizostomeae.
Cassiopea, Rhizostoma, Stomolophus. Instead of a central mouth, there are many small mouths opening into a complex canal system formed by the fusion of oral arms over the original mouth opening. Jellyfish in the genus *Cassiopea* grow to 30 cm or so in diameter, and all harbor symbiotic unicellular algae (zooxanthellae), which give the animals a distinctive greenish brown color. Often called the upside-down jellyfish, members of this genus typically lie upside down, pulsating on the substrate, presumably farming their zooxanthellae by maximizing their exposure to sunlight. Ten families.

Class Cubozoa.
One order.

Order Cubomedusae.
Carybdea, Tripedalia—the sea wasps. The scyphozoans contained in this order are mostly small with a highly transparent swimming bell that is essentially square in transverse section. Each animal has four tentacles; a 2.5-cm individual can have tentacles up to 30 cm long. Sea wasps swim actively, up to 6 m per minute, and are common in all tropical and subtropical oceans. Their sting is exceedingly painful and can be fatal to humans. *Chironex fleckeri,* although only the size of a thimble, is one of the most deadly of all marine animals; it is found only in parts of the Pacific Ocean. Two families.

Class Hydrozoa.
The medusae often have unknown polyp stages and vice versa, so the taxonomy is likely to be reorganized considerably once more life cycles have been worked out. Seven orders, about 2,700 species.

Order Hydroida.
This group contains most of the species in the class Hydrozoa. Three suborders.

Suborder Anthomedusae.
Hydranths lack a chitinous covering. Free-living medusae, if produced, are always bell shaped. *Bougainvillia, Sarsia, Eudendrium, Pennaria, Hydractinia, Podocoryne, Stylactis, Tubularia, Gonionemus, Hydra* (freshwater species with no medusa stage in the life cycle), *Cordylophora* (a medusa-producing hydroid from rivers and brackish waters). This suborder contains most of the species in the class. *Hydractinia* forms pinkish, bushy, highly polymorphic colonies covering marine gastropod shells inhabited by hermit crabs. The medusae produced by at least some species of *Gonionemus* have adhesive pads on the marginal tentacles; they use these pads to adhere to algae and sea grasses between bouts of swimming. Thirty-two families.

Suborder Leptomedusae.
Aequorea, Campanularia, Obelia, Sertularia. Hydranths possess a chitinous covering. Medusae, when present, are always flat and never bell shaped. All species are marine.

Suborder Limnomedusae.
Craspedacusta (the common and widespread freshwater jellyfish), *Pochella.* Most species occur only in freshwater, with some groups restricted to lakes in Africa and India. A few species in the group are marine, with highly restricted distributions. Species of *Pochella,* for example, are exclusively marine and live only on the rim of tubes built by certain sedentary polychaete annelids (sabellid polychaetes).

Order Milleporina.
Millepora—fire coral. The few species in this order are all colonial and restricted to coral reefs, and all host symbiotic zooxanthellae. The nematocysts are highly irritating to humans. All milleporine species secrete substantial calcium carbonate skeletons, making significant contributions to reef mass. The

members of this and the next order, the Stylasterina, are sometimes combined to form a single order, the Hydrocorallina.

Order Stylasterina.

The hydrocorals. *Allopora Stylaster.* These widespread, colonial, marine hydrozoans secrete substantial calcium carbonate skeletons, although they, like the milleporines previously discussed, are not true corals. The species in this group are found in both warm and cold waters, and at all depths. The members of this order are sometimes combined with the members of the preceding order to form a single order, the Hydrocorallina.

Order Trachylina.

Liriope. Most species are marine, with most restricted to warm open-ocean waters. Planulae derived from the sexual adventures of the medusa stage develop further into highly specialized, tentacled, swimming **actinula** larvae, which eventually develop directly into medusae. There is typically no attached polyp stage in the life cycle. Three suborders.

Suborder Narcomedusae.

In this group, the actinulae bud off additional actinulae asexually before developing to the medusa stage. Although most species are marine, one species is an internal parasite of freshwater fish in the Soviet Union, making it the only known parasitic adult cnidarian.

Order Siphonophora.

Physalia—the Portuguese man-of-war. All species form marine, polymorphic colonies that spend their lives swimming or floating in the water. The nematocysts inject highly toxic, paralyzing secretions, making the species in this group dangerous to humans, especially small children.

Order Chondrophora.

Porpita—the blue button; *Velella*—by-the-wind sailor. All species are marine, and individuals consist of a single, large polyp with a gas-filled float that maintains the animal at the ocean surface. Some species also possess a "sail," allowing the animal to move by wind power. *Velella* is often found associated with the gastropod snail *Janthina janthina*, which eats the "sailor" as they travel together. In all members of the order, the floating polyps bud off small, free-swimming medusae, which represent the sexual stage of the life cycle.

Order Actinulida.

Halammohydra. Individuals of all species in this group are marine, solitary (i.e., never forming colonies), and very small, typically less than 1.5 mm long. All members live interstitially, in the spaces between sand grains. There is no free-living medusa stage in the life cycle, and the polyp somewhat resembles the actinula stage of the trachylinids. The entire body is ciliated, permitting individuals to swim from place to place. Fewer than 10 species have been described.

Class Anthozoa.

All species are marine. About two-thirds of all described cnidarian species are contained in this class. Fifteen orders.

Subclass Alcyonaria (= Octocorallia).

Eight orders.

Order Stolonifera.

Tubipora—organ-pipe coral. These are found only on coral reefs of the tropical Indo-Pacific.

Order Gorgonacea.

Gorgonia—sea fans; *Plexaura, Plexaurella; Briareum*—deadman's fingers; *Leptogorgia*—sea whips; *Pseudopterogorgia*—sea feathers. Eighteen families. Most species are tropical, and all form colonies. Most colonies are supported by a firm, central, proteinaceous skeleton of gorgonin.

Order Alcyonacea.

The soft corals. *Alcyonium, Sinularia.* These colonial anthozoans lack a rigid skeleton (calcareous or proteinaceous) but do have calcareous spicules (sclerites) embedded in the tissues. Species occur in all oceans, especially in the tropics.

Order Pennatulacea.

Ptilosarcus—sea pens; *Renilla*—sea pansies. All species in this group are marine and adapted for life on soft substrates. A single, long axial polyp extends the length of the colony, with additional, much smaller polyps typically occurring on side branches (except in *Renilla*). The base of the axial polyp anchors the colony into the substrate. Sea pens stand erect. Sea pansies, which lie flat on the substrate, typically exhibit strong bioluminescence in the dark.

Subclass Zoantharia (= Hexacorallia).

Seven orders.

Order Actiniaria.

The sea anemones. *Bunodactis, Aiptasia, Anthopleura, Actinia, Diadumene, Calliactis, Bunodosoma, Edwardsia, Haliplanella, Metridium, Sagartia, Stomphia.* This large group contains about 800 species, distributed among 41 families. There are no colonial species in this order. Some members of one genus, *Stomphia*, can swim, although awkwardly, if attacked by predators. After detaching from the substrate, the anemone swims short distances by violently contracting the longitudinal body wall muscles on one side and then the other, bending the body from side to side.

Order Corallimorpharia.

Corynactis. Species in this order occur worldwide in shallow and deep water, in the tropics, in the temperate zone, and even at the poles. Members of all species are solitary, rather than colonial, and lack rigid skeletal elements.

Order Scleractinia.

The true (stony, or hard) corals. The hermatypic (reef-building) species all form symbiotic relationships with unicellular algae (zooxanthellae or zoochlorellae). Many species commonly propagate asexually from branches that break off during storms. *Acropora*—staghorn and elkhorn corals, the most important reef-building species; *Astrangia; Agaricia; Fungia*—mushroom corals, forming the largest known solitary coral polyps, up to 1 m in diameter; *Porites*—another major reef builder; *Stylophora; Siderastraea; Pocillopora; Oculina; Montastraea; Diploria*—brain coral.

Order Zoanthinaria (= Zoanthidea)

Zoanthus. Polyps may be solitary or colonial, but they never secrete a solid skeleton. Three families.

Order Ceriantharia.

Cerianthus. The species in this group are all solitary, living in tubes buried vertically in soft sediments; only the animal's oral disc is visible to the snorkeler. The musculature of the body column is largely ectodermal, permitting rapid withdrawal of the animal into its tube when attacked. The tubes are made by specialized nematocysts called **ptychocysts,** unique to members of this order. Unlike other anthozoans, the oral disc of cerianthids bears two distinct whorls of tentacles. Three families.

Order Ptychodactiaria.

These anemonelike anthozoans are all restricted to cold waters in and near the Arctic and Antarctic oceans.

Order Antipatharia.

Antipathes. The black or thorny corals. Most members of this order are deep-water tropical species. They are commonly exploited for jewelry making. The skeletal support system is composed of chitinous fibrils and protein that can be bent and shaped when heated. This axial skeleton was once thought to have medicinal properties (*anti* = L: against; *pathes* = L: disease).

Some General References About the Cnidarians

Harrison, F. W., and J. A., eds. 1991. *Microscopic Anatomy of Invertebrates, Vol. 2. Placazoa, Porifera, Cnidaria, and Ctenophora.* New York: Wiley-Liss.

Hessinger, D. A., and H. M. Lenhoff, eds. 1988. *The Biology of Nematocysts.* San Diego, Calif.: Academic Press.

Hyman, L. 1940. *The Invertebrates, Vol. 1. Protozoa Through Ctenophora.* New York: McGraw-Hill.

Morris, S. C., et al. eds. 1985. *The Origins and Relationships of Lower Invertebrates. Systematics Association, Special Vol. 28.* Oxford: Clarendon Press.

Muscatine, L., and H. M. Lenhoff, eds. 1974. *Coelenterate Biology.* New York: Academic Press.

Parker, S. P., ed. 1982. *Classification and Synopsis of Living Organisms,* vol. 1. New York: McGraw-Hill, 669–706.

Shick, J. M. 1991. *A Functional Biology of Sea Anemones.* New York: Chapman & Hall.

Thorpe, J. H., and A. P. Covich. 1991. *Ecology and Classification of North American Freshwater Invertebrates.* New York: Academic Press.

6

The Ctenophores

Introduction and General Characteristics

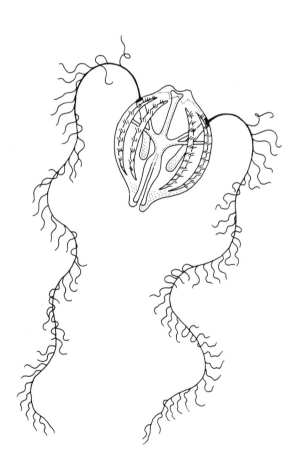

Phylum Cteno · phora
(G: comb bearing)
teen'-ō-for'-ah (or, teen-ah'-for-ah)

All ctenophores are predators, and most are nearly transparent. The 100 or so described species are exclusively marine, and most are planktonic; that is, most species are weak swimmers, carried about by ocean currents. Only one parasitic species is known.

The body architecture of ctenophores is somewhat reminiscent of cnidarian medusae. Like the body of a medusa, the ctenophore body consists of an outer epidermis, an inner gastrodermis, and a thick, gelatinous middle mesoglea layer. Both groups show a basic radial symmetry, with oral and aboral surfaces.

The digestive systems are similar in both groups as well. The mouth leads into a pharynx (also called a **stomodeum**), which serves as a site of extracellular digestion, and thence through a stomach into a series of **gastrovascular canals,** where digestion is completed intracellularly. A functional excretory system has not been documented for either group, nor are any specialized respiratory organs found. The nervous system of ctenophores takes the form of a subepidermal nerve network, as in many cnidarians. Moreover, at least one ctenophore species has a planula larval stage in its life history. One species of ctenophore is known to use nematocysts in prey capture, but these nematocysts are obtained through the ingestion of cnidarian medusae; the nematocysts are not manufactured by the ctenophore. In addition, ctenophore tentacles are solid, as in most cnidarians, rather than hollow. Lastly, ctenophores seem to be diploblastic, forming no distinct mesodermal tissue

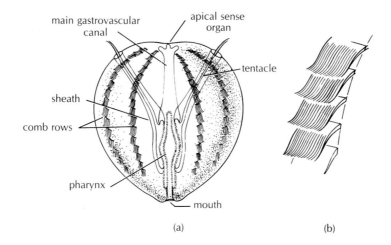

Figure 6.1

(*a*) External anatomy of a ctenophore, *Pleurobrachia* sp., showing the comb rows and several other anatomical features. The tentacles have been withdrawn into their sheaths. (*b*) Detail of a comb row, showing four ctenes. Each ctene is made up of thousands of long, compound cilia.   (a) *After Hyman.* (b) *After Hardy.*

layer during embryogenesis, as in the Cnidaria. There is some disagreement on this point: Muscle-forming cells do segregate early in ctenophore development and some workers regard these as true mesodermal cells; if they are, ctenophores could differ from cnidarians in qualifying for triploblastic status.

Ctenophores have left no fossil record, so that our best chance of unraveling their evolutionary origins probably lies in the molecular realm. Although there may be an evolutionary relationship between the Ctenophora and Cnidaria based upon the morphological similarities just discussed, ctenophores are clearly not cnidarians. I have already mentioned an embryological difference in the formation of the musculature. In addition, polymorphism, an almost diagnostic characteristic of hydrozoans and scyphozoans, is never encountered among the Ctenophora, and no ctenophores are colonial. And whereas all cnidarians have monociliated cells (one cilium per cell), ctenophore cells are always multiciliated, each bearing two or more cilia. Moreover, the type of musculature differs among the members of the two groups, as do the swimming mechanism, the system of maintaining balance, the mechanism and mode of food capture, the means of eliminating solid wastes, the nature of sexuality, and several aspects of embryonic development. Each of these characteristics will be considered in sequence.

The muscles of ctenophores develop from amoeboid cells found within the mesoglea. Thus, the resulting muscle fibers actually reside in the mesoglea layers. In contrast, the musculature of cnidarians is found within the gastrodermis and, to a lesser extent, within the epidermis. Moreover, ctenophores have genuine smooth muscle tissue, and in fact they lack the myoepithelial cells that are so characteristic of cnidarian musculature. The first known giant (up to 6 cm long) smooth muscle fibers have recently been isolated from two ctenophore species, *Mnemiopsis leydii* and *Beroe* sp.; muscle preparations from these species should provide excellent material for general studies of smooth muscle biology.[1]

In most coastal ctenophore species, the musculature plays little or no direct role in locomotion. Instead, swimming is accomplished by the activity of many bands of partially fused, remarkably long cilia. Each band is called a **ctene** because of its resemblance to a comb (*ctene* = G: a comb). This explains the name of the phylum ("comb bearer") and the common reference to its members as "comb jellies;" ctenophores are unique among animals in having their cilia fused into ctenes. The ctenes are typically organized into eight distinct rows, which are equally spaced about the body. These **comb rows,** or **costae,** extend from the oral to the aboral surface of the animal (Fig. 6.1) and are strikingly iridescent. The power stroke of the cilia comprising each ctene is toward the aboral surface, so that the typical ctenophore swims mouth first. In summary, whereas cnidarian medusae swim by means of jet propulsion, ctenophore locomotion depends largely upon the coordinated activities of the partially fused cilia in the various comb rows.[2] Some open-ocean ctenophore species use ctene activity for feeding but use vigorous muscular activity in escaping from predators.[3]

The intensity of activity in the different comb rows is under the control of a single **apical sense organ,** located at the aboral end of the ctenophore (Figs. 6.1*a* and 6.2). The **statolith,** a single sphere of calcium carbonate

---

1. See *Topics for Further Discussion and Investigation,* no. 3, at the end of the chapter.
2. See *Topics for Further Discussion and Investigation,* no. 1.
3. See *Topics for Further Discussion and Investigation,*no. 4.

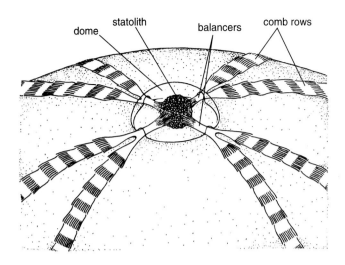

dome   statolith   balancers   comb rows

Figure 6.2
Detail of an apical (aboral) sense organ and its transparent covering. Tilting of the animal causes the statolith to press against particular balancers, stimulating activity in the associated comb rows and thus restoring the animal's orientation.

(CaCO₃), sits atop four tufts of fused cilia called **balancers,** or **springs.** Each balancer may consist of several hundred cilia. A ciliated groove radiates from each balancer and bifurcates to service two adjacent comb rows. Experiments have shown these grooves to be agents of nerve impulse conduction from the apical sense organ to the ctenes of the comb rows. If the animal becomes tilted, the statolith presses against one of the balancers more than the others, causing the cilia of the comb rows associated with that balancer to increase their beat frequency until a satisfactory body orientation is restored. If the apical sense organ is surgically removed, the ctenophore continues to swim. However, the surgery obliterates any coordination of ciliary beating in different comb rows, attesting to the synchronizing role of the apical sense organ.

The epidermal nerve net also seems to play a role in coordinating the activities of the ctenes. For example, mechanical stimulation of the oral end of the animal results in a sudden reversal of the ciliary beat in all comb rows. This response is observed even if the apical sense organ has been surgically removed.

Similarly to medusae, many ctenophore species capture their prey using tentacles (Research Focus Box 6.1). Unlike the tentacles of cnidarians, ctenophore tentacles can be completely retracted into proximal pits or sheaths. Moreover, with one exception, the tentacle and/or the general epidermis of ctenophores is studded not with nematocysts but rather with quite different structures called **colloblasts** (Fig. 6.3). Each colloblast consists of a bulbous, sticky head connected to a long,

straight filament and a spiral, contractile filament. Prey organisms become stuck to the tentacles, which are then retracted. In many species, the body rotates to bring the mouth in contact with the tentacles once the food items are brought within range. The tentacles of some species may be extended more than 100 times the length of the body and retracted in seconds.

In other species, the food-catching function of the tentacles is much reduced. Instead, the body surface area is increased by lateral compression, and major areas of the body are coated with a sticky mucus and with colloblast cells. The body itself thus becomes the major organ of food collection, and the small tentacles found in some of these species merely aid in transporting food to the mouth. The foods primarily captured by most ctenophores are small crustaceans (phylum Arthropoda) and the larval stages of various fish and shellfish species, including some of commercial importance, such as oysters. Some ctenophores are carnivorous on other ctenophores or on gelatinous animals in other groups (especially the Cnidaria and the Urochordata).

The digestive systems of ctenophores and cnidarians differ in one interesting respect. As discussed earlier, the digestive tract of a medusa has but one opening, which serves as both mouth and anus. In ctenophores, on the other hand, four **digestive canals** lead from the roof of the stomach to the animal's aboral surface (Fig. 6.4). Although two of the digestive canals terminate as blind sacs, the other two canals open to the outside. Undigested wastes are discharged through these **anal pores.**

Most ctenophore species described to date are simultaneous hermaphrodites; that is, a single individual has both male and female gonads. In contrast, cnidarian medusae, and many anemone species, are dioecious, with one sex per individual.

Ctenophore development differs in several respects from that of cnidarians. In particular, ctenophore cleavage is highly determinate; cell fates are fixed at the first cell division. Cell fates of cnidarian embryos become fixed later in development. Moreover, the mechanism of **gastrulation** in ctenophores (i.e., the formation of distinct inner and outer germ layers) is quite unlike that found in most cnidarians. Among ctenophores, gastrulation is achieved either by **epiboly,** a process in which a sheet of micromeres spreads over what were the adjacent macromeres (Fig. 6.5a), or by **invagination,** in which groups of cells push into the blastocoelic space (Fig. 6.5b). Although gastrulation by invagination occurs among cnidarians, epiboly does not, and most cnidarians gastrulate by a process of **delamination** (Fig. 6.5c). In delamination, the cells of the blastula divide with the cleavage plane approximately parallel to the surface of the embryo. Thus, the cells essentially divide into the blastocoel, forming an inner and outer cell layer, between which the mesoglea is later secreted. Gastrulation in still

Research Focus Box 6.1

Ctenophore Feeding Biology

From C. H. Greene, M. R. Landry, and B. C. Monger, "Foraging Behavior and Prey Selection by the Ambush Entangling Predator Pleurobrachia bachei" in *Ecology*, 67:1493–1501, 1986. Copyright © 1986 Ecological Society of America. Reprinted by permission.

Ctenophore population densities can become quite high, particularly in coastal waters. Individual species can reach concentrations of several hundred individuals per m³. As carnivores, ctenophores are therefore likely to have substantial impact on the abundance and population dynamics of other zooplankton. In particular, ctenophores may severely deplete the numbers of copepods (Focus Fig. 6.1), which also serve as major food sources for the larvae and adults of many commercially important fish species. Quantifying the importance of ctenophores in regulating the sizes and dynamics of copepod populations requires detailed knowledge of ctenophore feeding biology. Do ctenophores preferentially select certain prey species or certain life-history stages of particular prey, or do they feed nonselectively on whatever they happen to catch? In addition, at what rates do they feed?

The study by Greene et al. (1986) documents the ability of the small (at most a few centimeters in diameter), coastal ctenophore *Pleurobrachia bachei* to feed selectively on adults of the following copepod species: *Acartia clausi, Calanus pacificus,* and *Pseudocalanus* sp. The authors also examined feeding on different larval and juvenile stages of *C. pacificus.*

In the basic experiment, the researchers set up a number of 3.8 l (liter) jars of seawater and added one ctenophore and either 25 or 50 copepod prey to each. The ctenophores were allowed to feed in the dark for 12 hours. Like other tentaculate ctenophore species, *P. bachei* drifts passively in the water for long periods, trailing its long fishing tentacles behind. When prey are ensnared by colloblasts, the ctenophore draws the successful tentacle across the mouth and ingests the prey. At the end of 12 hours, the number of copepods remaining in each container was determined. There was no decline in copepod density in control jars (which contained copepods but no ctenophores), so that the disappearance of copepods in experimental treatments must have reflected predation by the ctenophores.

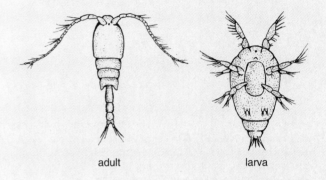

Focus Figure 6.1
Adult and larval stages of a typical planktonic copepod (not to scale). Copepods are crustaceans (phylum Arthropoda). Adults are typically less than 1 mm long, and larvae are much smaller.

For comparative purposes, the data on rates of copepod disappearance were expressed in terms of the volume of water each ctenophore must have processed each hour to have captured and eaten the recorded number of copepods ingested. For example, if the jar initially contained 215 copepods in 3.8 l of seawater and the ctenophore ate 50 copepods over the next 12 hours, the average concentration of copepods in the jar during that time was 50 copepods per liter ([215 + 165]/2 copepods ÷ 3.8 l). To have eaten 50 copepods, the ctenophore must have filtered 1 l of seawater in the 12-hour period.

The results of the feeding studies clearly demonstrate that, at least in the laboratory, this ctenophore species feeds selectively (Focus Fig. 6.2). Of particular interest, feeding rates on the various larval and juvenile stages of *C. pacificus* varied unpredictably from stage to stage. One might have expected feeding rates to be consistently higher for older—larger and faster-swimming—prey, but the data for *C. pacificus* do not

other cnidarians is by **ingression,** in which certain cells become detached from their neighbors and simply move into the blastocoel, creating a second layer of cells.

Ctenophore embryos, in contrast to cnidarian embryos, rarely develop into ciliated planula larvae. Rather, the ctenophore embryo usually develops directly into a miniature ctenophore called a **cydippid.** The cydippid is approximately spherical in shape; is endowed with eight comb rows, a fully formed apical sense organ, and a pharynx; and usually bears a pair of branched tentacles.

In many species, the cydippid closely resembles the adult. In other species, the cydippid undergoes a gradual, but substantial, alteration in morphology to attain the adult form.

Certain ctenophore and cnidarian characteristics are compared in Table 6.1.

Nearly all ctenophores are **bioluminescent.** Unlike iridescence, in which colors are generated by the diffraction of incident light, bioluminescence results from a chemical reaction in which much of the excess energy is

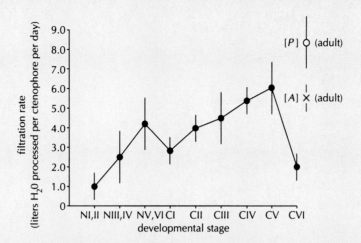

Focus Figure 6.2
Laboratory feeding rates of the ctenophore *Pleurobrachia bachei* on copepods of different developmental stages (*Calanus pacificus*: filled-in circles) or of different species (*Acartia clausi* adults: X; *Pseudocalanus* sp. adults: open circle). Developmental stages NI–NVI are larvae of *C. pacificus,* while stages CI–CVI are juveniles of *C. pacificus.*

show such a continuous increase in feeding rate (Focus Fig. 6.2). Moreover, feeding rates on adults of different copepod species also differed markedly. Ctenophores fed at the highest rates on *Pseudocalanus* sp., processing seawater at rates up to 8.4 l daily (Focus Fig. 6.2).

Additional experiments demonstrated that the feeding-rate differences shown when ctenophores were feeding on only one prey species at a time were maintained when the ctenophores were fed mixtures of the prey species. This clearly demonstrates that the members of this ctenophore species can feed selectively in the laboratory, ingesting more of one prey type than another even when both prey types are present at equal concentrations.

Does this mean that ctenophores actively *select* certain prey over others? An alternative explanation is that some prey are simply less likely to be captured and ingested; that is, differential predation might reflect differences in prey behavior rather than deliberate choices made by the predator. In particular, differences in prey susceptibility could reflect (1) specific avoidance behaviors exhibited by the copepods; (2) differential ability to escape from the tentacles after capture; or (3) different swimming speeds for copepods of different species and different developmental stages (the faster one moves, the sooner one is likely to run into trouble). The authors went on to examine these possibilities. How do you suppose they designed this part of their study?

The laboratory experiments just described suggest that ctenophore predation does not have equal impact on all copepod species or on all life-history stages of any one copepod species. In the field, of course, ctenophores would be free to extend their tentacles to lengths not permitted in small glass jars, and copepods would have greater opportunities to avoid capture. Can you think of a way to test for differential predation by ctenophores under more natural field conditions?

Figure from C. H. Greene, M. R. Landry, and B. C. Monger, "Foraging Behavior and Prey Selection by the Ambush Entangling Predator Pleurobrachia bachei" in Ecology, 67:1493-1501, 1986. Copyright © 1986 Ecological Society of America. Reprinted by permission.

given off as light rather than as heat. Although the particular form of the reaction taking place in ctenophores seems to be peculiar to the members of this phylum, the phenomenon of bioluminescence is not. Indeed, at least some species from most major animal phyla display some form of bioluminescence. The functional significance of bioluminescence is often unclear. Mate location and species recognition, the luring of prey, and the startling of would-be predators are possibilities that may apply to the bioluminescing members of some phyla.

Some species may use bioluminescence to avoid detection by visual predators, producing light of ambient intensity. This would break up the silhouette of the animal when observed by potential predators from below, helping the lighted form blend into the surroundings. The adaptive significance of bioluminescence among ctenophores has not been examined, although the possibility of mate recognition can probably be ruled out by the lack of distinct photoreceptors. Regardless of its significance, the bioluminescence, together with the

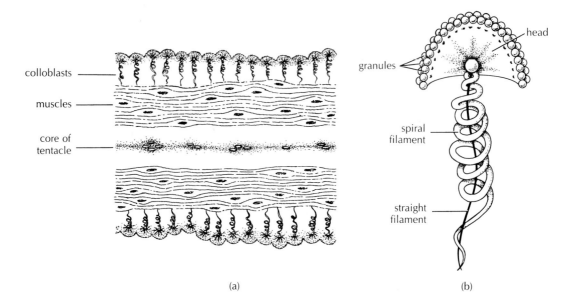

(a)   (b)

Figure 6.3

(*a*) Longitudinal section through a tentacle. (*b*) A single colloblast. The spiral filament is contractile. The head contains granules that, when discharged, produce a sticky secretion, that traps prey.

(a) From Hyman; after Hertwig. (b) From Hyman; after Komai.

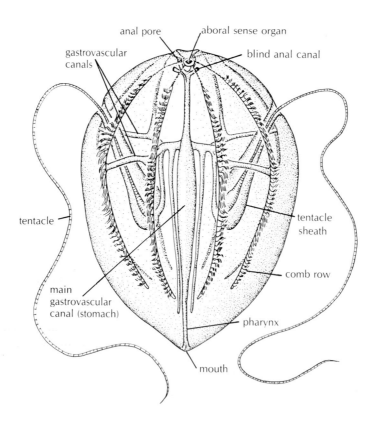

Figure 6.4

Diagrammatic illustration of the ctenophore *Pleurobrachia* sp., the sea gooseberry. Note that the main gastrovascular canal terminates as four branches near the aboral (apical) sense organ. Two of these branches end blindly, but the other two connect to the outside by means of small anal pores. All other gastrovascular canals end blindly.   *After Bayer and Owre; after Hardy.*

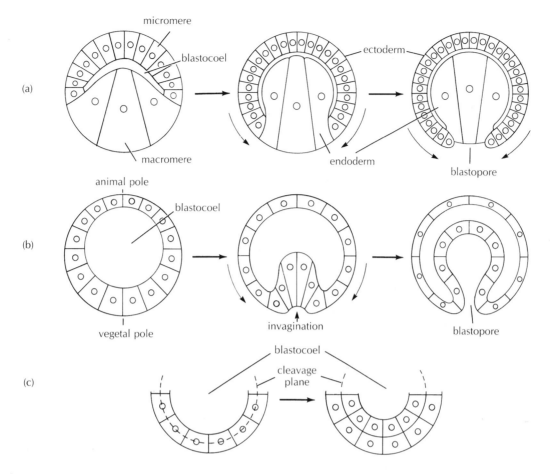

Figure 6.5
Patterns of gastrulation in ctenophores and cnidarians.
(*a*) Gastrulation by epiboly, in which the smaller cells
(micromeres) grow over larger cells (macromeres).
(*b*) Gastrulation by invagination, in which a group of adjacent cells
indents into the blastocoel. (*c*) Gastrulation by delamination, in

which the second cell layer is formed by mitotic division. In
ctenophores, gastrulation is by epiboly or by invagination. In
marked contrast, cnidarians typically gastrulate by delamination
or ingression (in which cells on the outside of the embryo
migrate inward).

iridescence of the comb rows, the delicacy of the form, and the grace of movement, make living ctenophores among the most glorious of animals to observe.

Ctenophore Diversity

Most ctenophore species live in the open sea far from coastlines, so their biology is largely unstudied. The recent switch from collecting samples with nets towed from fast-moving vessels to *in situ* observations and careful collection by divers and by submersibles has brought to light numerous species never before known, species much too fragile to have withstood the trauma of what were formerly considered standard collection techniques. As many as one dozen new species were recently captured during a single two-week expedition. At present, ctenophore species are divided into two classes, primarily on the basis of whether they possess conspicuous tentacles as adults and/or as cydippids. Representative adults are shown in Figure 6.6.

Class Tentaculata

Much of tentaculate evolution seems to be a story of modifying the mechanism of prey capture.[4] Some tentaculate species closely resemble the cydippid larvae already described, except, of course, that functional gonads are present. Long, retractable tentacles are well developed throughout life, and food is captured exclusively by these few tentacles and their side branches. These ctenophores comprise the order Cydippida (Figs. 6.6b,c and 6.7a). In other species, the body is somewhat compressed laterally, only four of the comb rows are fully developed, and the tentacles are generally much reduced in length. Large **oral lobes,** covered with mucus and colloblasts, constitute the primary food collection surfaces. These are the lobate ctenophores (order Lobata, Figs. 6.6d,e and 6.7b,f). Muscular activity of the two oral lobes aids locomotion in some species. Two pairs of ciliated, paddle- or tentacle-like structures called **auricles** (Fig. 6.7b), unique to lobate ctenophores, apparently assist in prey capture.

---

4. See *Topics for Further Discussion and Investigation*, no. 2.

Table 6.1 Similarities and Differences Between the Two Phyla
of Diploblastic Gelatinous Animals

Characteristic	Ctenophores	Cnidarians
Cleavage	Determinate	Indeterminate
Gastrulation	Epiboly or invagination	Delamination, ingression, or invagination
Common developmental stage	Cydippid	Planula
Digestive system	Gastrovascular canals	Gastrovascular canals
Nematocysts	None (unless "borrowed")	Present
Colloblasts	Present	None
Sexuality	Typically hermaphroditic	Typically dioecious
Musculature	Within mesoglea	Within gastrodermis
Ciliation	Multiciliated cells	Monociliated cells

(a)

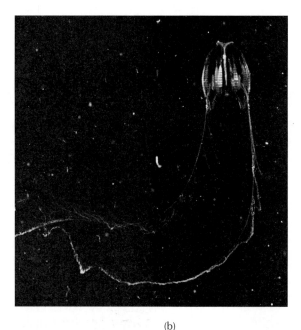

(b)

(c)

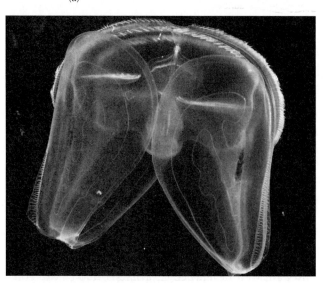

(d)

(e)

Figure 6.6

Five ctenophores photographed in the open ocean. (*a*) An open
ocean lobate ctenophore, *Eurhamphea vexilligera*. (*b*) The cydippid
ctenophore Callanira sp. (*c*) Another cydippid ctenophore,
Euplokamis dunlapae (body length approximately 1.4 cm).
(*d*) *Ocyropsis maculata*, a lobate ctenophore. (*e*) A coastal lobate
ctenophore, *Mnemiopsis macrydi*. (a,b) *Courtesy of G. R. Harbison and
M. Jones. (c) Courtesy of Claudia E. Mills. From Mackie, G.O., et al. 1992. Biol. Bull.
182:248–56. (d,e) By Anne Rudloe, Courtesy of Gulf Specimen Marine Laboratories, Inc.*

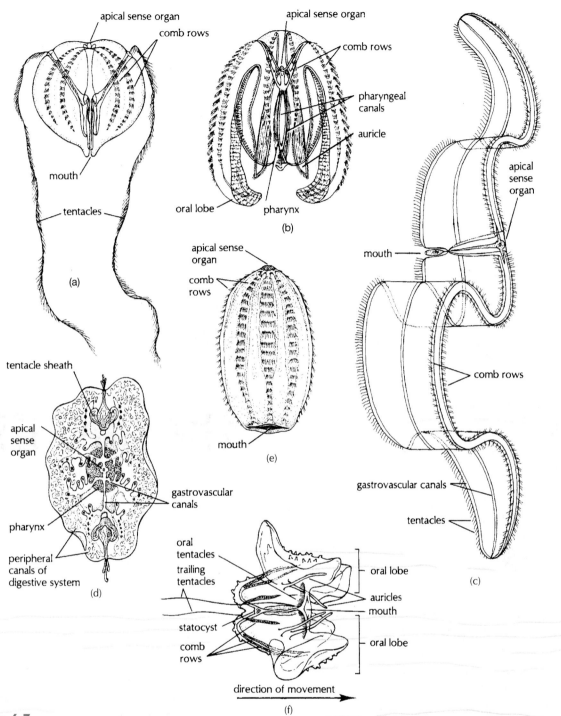

Figure 6.7

Ctenophore diversity. (*a*) The cydippid *Pleurobrachia* sp., approximately twice natural size. (*b*) A lobate ctenophore, *Mnemiopsis leidyi*. (*c*) A member of the order Cestida, *Cestum veneris*, commonly known as Venus's girdle. These animals attain lengths of about 1.5 m. (*d*) A member of the Platyctenida, *Coeloplana mesnili*, viewed from above. The animal is about 6 cm in length and spends its life in association with certain corals in the Indo-Pacific. (*e*) A member of the class Nuda, *Beroe* sp. The members of this class lack tentacles throughout life. (*f*) *Leucothea* sp., an open-ocean lobate ctenophore. The highly fragile lobes break into pieces at the slightest physical disturbance, so that these animals can be studied only by diving in their midst.

(a,b) *After Hyman.* (c) *From Hyman; after Mayer.* (d) *From Hyman, after Dawydoff.*
(e) *After Hardy.* (f) *From Matsumoto and Hamner,* Marine Biology 97–551-58, 1988.

In another group, the cestids, the body is so compressed laterally that it now forms a long ribbon, with the mouth and apical sense organ on opposite sides at its midpoint (Fig. 6.7c). Swimming is accomplished through a combination of ctene activity and sinuous, muscular movements of the body; only four of the comb rows are well developed in adults. Despite the great surface area of the body, prey are captured not by the ribbon itself, but by the numerous short tentacles extending along the extensive oral edge of the ctenophore.

Thus, selection for increased food-collection capabilities seems to have favored a redistribution of colloblasts, together with an increased body surface area, in one group of tentaculate ctenophores—the Lobata—and an increased number of feeding tentacles in another group—the Cestida. Both adaptations have involved lateral compression of the body in comparison with the cydippid form, which is believed to represent the more primitive condition. In the fourth and final major order of tentaculate ctenophores, the Platyctenida, the body is compressed in a different plane (Fig. 6.7d). Specifically, the oral and aboral surfaces have moved toward each other, so the body forms a flattened plate (*platy* = G: flat). The bottom of the plate is formed largely by the pharynx, which is extensively and permanently everted. Some species simply float in the water. Others spend their time creeping slowly over solid substrates, apparently using both pharyngeal cilia and muscular contractions for locomotion. The only non-planktonic ctenophores are members of this order. Comb rows may be present, although these are often reduced or absent in the adults of many species. Some locomotion may also be accomplished by muscular flapping of the lateral lobes of the body. The adults generally bear two long tentacles.

Class Nuda

The members of the relatively small class Nuda are contained in a single order, the Beroida. No individuals in this group have tentacles (even in the cydippid stage of development) or oral lobes. All eight comb rows, however, are well developed (Fig. 6.7e). Prey, including other ctenophores, are captured and engulfed by muscular lips surrounding the mouth. The mouth can be widened to accommodate prey substantially larger than the predator. **Macrocilia,** consisting of thousands of 9 + 2 axonemes enclosed by a single membrane, are located just inside the mouth; these macrocilia are used as teeth, to chop especially large prey into bite-sized pieces.

Other Features of Ctenophore Biology: Reproduction

Although all ctenophores appear to have substantial regenerative powers, asexual reproduction is rare within the Ctenophora. A few species (all members of the Platyctenida) are known to reproduce asexually, through fragmentation and the subsequent development of missing body parts by each fragment.

All but a few ctenophore species described to date (members of two lobate genera) are sequential or simultaneous hermaphrodites. The gonads are located on the walls of some or all of the gastrovascular canals, so that gametes are liberated into the digestive tract and are commonly discharged through the mouth. Eggs are fertilized externally (i.e., in the surrounding seawater), except in some platyctene species that fertilize their eggs internally. All ctenophore species pass through a cydippid stage of development. Indeed, some species described as cydippid ctenophores may well turn out to be juvenile stages belonging to other orders.

Taxonomic Summary

Phylum Ctenophora
   Class Tentaculata
      Order Cydippida
      Order Lobata
      Order Cestida
      Order Platyctenida
   Class Nuda
      Order Beroida

Topics for Further Discussion and Investigation

1. What role does mechanical interaction between adjacent ctenes play in coordinating the beating of cilia in a comb row?

Tamm, S. L. 1973. Mechanisms of ciliary coordination in ctenophores. *J. Exp. Biol.* 59:231.

Tamm, S. L. 1983. Motility and mechanosensitivity of macrocilia in the ctenophore *Beroë. Nature* 305:430.

2. Compare and contrast the feeding biology of lobate, cydippid, and cestid ctenophores.

Hamner, W. M., S. W. Strand, G. I. Matsumoto, and P. P. Hamner. 1987. Ethological observations on foraging behavior of the ctenophore *Leucothea* sp. in the open sea. *Limnol. Oceanogr.* 32:645.

Harbison, G. R., L. P. Madin, and N. R. Swanberg. 1978. On the natural history and distribution of oceanic ctenophores. *Deep-Sea Res.* 25:233.

Madin, L. P. 1988. Feeding behavior of tentaculate predators: *In situ* observations and a conceptual model. *Bull. Marine Sci.* 43:413.

Main, R. J. 1928. Observations of the feeding mechanism of the ctenophore, *Mnemiopsis leidyi. Biol. Bull.* 55:69.

Matsumoto, G. I., and G. R. Harbison. 1993. In situ observations of foraging, feeding, and escape behavior in three orders of oceanic ctenophores: Lobata, Cestida, and Beroida. *Marine Biol.* 117:279

Reeve, M. R., M. A. Walter, and T. Ikeda. 1978. Laboratory studies of ingestion and food utilization in lobate and tentaculate ctenophores. *Limnol. Oceanogr.* 23:740.

Swanberg, N. 1974. The feeding behavior of *Bero/au/e ovata. Marine Biol.* 24:69.

3. The feeding behavior of ctenophores commonly involves a surprisingly complex series of rapid and apparently coordinated muscle contractions. How do ctenophores achieve such coordination without a central nervous system?

Bilbaut, A., M.-L. Hernandez-Nicaise, C. A. Leech, and R. W. Meech. 1988. Membrane currents that govern smooth muscle contraction in a ctenophore. *Nature* 331:533.

4. Discuss the roles of cilia and musculature in the escape responses exhibited by planktonic ctenophores.

Mackie, G. O., C. E. Mills, and C. L. Singla. 1992. Giant axons and escape swimming in *Euplokamis dunlapae* (Ctenophora: Cydippida). *Biol. Bull.* 182:248

Matsumoto, G. I., and G. R. Harbison. 1993. In situ observations of foraging, feeding, and escape behavior in three orders of oceanic ctenophores: Lobata, Cestida, and Beroida. *Marine Biol.* 117:279

Taxonomic Detail

Phylum Ctenophora

The approximately 100 described species are distributed among seven orders and 19 families.

Class Tentaculata.
Four orders.

Order Cydippida.
Five families.

Family Pleurobrachiidae. *Pleurobrachia.* This family is represented over a very wide geographic range, from polar to tropical waters and from coastal to open-ocean habitats.

Family Euplokidae. *Euplokamis.*

Family Mertensidae. *Mertensia, Callianira.*

Order Platyctenida.
These peculiar, flattened ctenophores are restricted to shallow waters in the tropics and at the poles. Unlike other ctenophores, they reproduce asexually—by breaking off pieces of the body—as well as sexually. Fertilization is commonly internal, and the developing animal always passes through a perfectly normal cydippid stage. Four families.

Family Ctenoplanidae. *Ctenoplana.* Most individuals are smaller than about 2 cm, and all are found only in tropical waters. Although capable of swimming, they spend most of their time crawling over various benthic substrates.

Family Coeloplanidae. *Coeloplana, Vallicula.* These animals reach lengths of about 6 cm and occur in shallow water throughout the world. Some species are active swimmers, while others drift in the water or creep along the bottom. The comb rows are completely lost during development so that the animal looks superficially more like a flatworm than a ctenophore. Some species are commensal with various other benthic organisms.

Order Lobata.
All members are planktonic and all have aural lobes and auricles. Six families.

Family Bolinopsidae. *Bolinopsis, Mnemiopsis.* All species live in coastal waters, with representatives in all oceans from the tropics to the poles. The tentacles are long when animals are in the cydippid stage but short and inconspicuous in adults. Individuals may reach about 15 cm in height. *Mnemiopsis leidyi* (Fig 6.7b) was accidently introduced into the Black Sea in the early 1980's, probably transported in cargo ship ballast water from the coasts of North or South America, or from the Caribbean. It has subsequently caused the near collapse of the Black Sea anchovy fishery, by competing for food with the fish, eating their eggs and larvae, and clogging fishing nets. The species is currently spreading into the Mediterranean.

Family Ocyropsidae. *Ocyropsis.* These common, tropical, open-ocean ctenophores swim using a combination of ctene activity and vigorous flapping of the oral lobes. Early juveniles have short tentacles, but these are reduced or completely lost by adulthood. All members of the genus are dioecious.

Family Leucotheidae. *Leucothea.* The oral lobes of these open-ocean ctenophores are especially large and fragile; even moderate water turbulence will cause the lobes to break apart. The auricles are extremely long and thin. Representatives are present in all oceans, from tropical to temperate regions.

Order Cestida.

Cestum, Velamen. All individuals are pelagic. Although most are smaller than about 15 cm, members of the genus *Cestum* may grow to exceed 2 m. Development includes a normal, tentacled cydippid, which gradually elongates to adult form. Although the animals are propelled by the ctene rows when feeding, they can also wriggle like eels to escape from potential predators. The members of this order are found worldwide, with especially high concentrations in the tropics. One family (Cestidae).

Order Ganeshida.

Ganesha. This order contains two species of pelagic ctenophores whose members possess branched tentacles but lack both auricles and oral lobes. Some workers think these ctenophores may be the developmental stages of lobate ctenophores.

Order Thalassocalycida.

Thalassocalyce. This order contains a single species of pelagic, very fragile ctenophore whose members have oral lobes and tentacles but lack auricles and tentacular sheaths. The mouth and pharynx are borne on a central stalk. Individuals reach about 25 cm and have been collected from as deep as 1,000 m.

Class Nuda.

One order.

Order Beroida.

Beroe. These ctenophores are all pelagic and lack tentacles. They feed by engulfing their prey with a large, highly distensible mouth or by biting large prey into pieces with sharp-edged macrociliary bundles. These animals feed on other gelatinous animals, including salps (see Chapter 24) and other ctenophores. They occur in all the world's oceans, from the poles to the tropics, and reach lengths of about 30 cm. One family (Beroidae).

Some General References About the Ctenophores

Harbison, G. R., L. P. Madin, and N. R. Swanberg. 1978. On the natural history and distribution of oceanic ctenophores. *Deep-Sea Res.* 25:233–56.

Harrison, F. W., and J. A. Westfall, eds. 1991. *Microscopic Anatomy of Invertebrates,* vol. 2. New York: Wiley-Liss.

Hyman, L. 1940. *The Invertebrates, Vol. 1. Protozoa Through Ctenophora.* New York: McGraw-Hill.

Morris, S. C., et al., eds. 1985. *The Origins and Relationships of Lower Invertebrates. Systematics Association, Special Vol. 28,* Oxford: Clarendon Press, 78–100.

Parker, S. P., ed. 1982 *Classification and Synopsis of Living Organisms,* vol. 1, New York: McGraw-Hill, 707–15.

7

The Platyhelminthes

Introduction and General Characteristics

Phylum Platy • helminthes
(G: flat worm)
plat´-ē-hel-min´-thēs

The platyhelminths are a group of some 18,500 species, including one class of mostly free-living individuals (the turbellarians) and two classes of exclusively parasitic individuals (the trematodes and cestodes) that are believed to have evolved from free-living turbellarian ancestors. More than 80% of all platyhelminth species are parasites. All flatworms are acoelomate, triploblastic, and bilaterally symmetrical. Indeed, the free-living forms are usually considered the most primitive bilateral animals, and the first group to have evolved a true mesoderm; all coelomate animals may ultimately have evolved from flatworm-like ancestors.

On the other hand, there is also compelling evidence and logical argument suggesting that the flatworm acoelomate condition is a secondary phenomenon; that is, that the flatworms descended from coelomate ancestors. It has been suggested that the coelomic space of some flatworm ancestor filled in as an adaptive accompaniment to the evolution of reduced body size, for example, or that perhaps the acoelomates arose from a coelomate developmental stage when an embryo became sexually mature before developing a coelomic space. Both are reasonable possibilities. The fossil record is essentially nonexistent for flatworms and thus of no help in resolving the issue. Present molecular data support the primitive nature of the platyhelminth acoelomate condition, but the debate is far from over.[1]

---

1. See *Topics for Further Discussion and Investigation*, no. 9, at the end of this chapter.

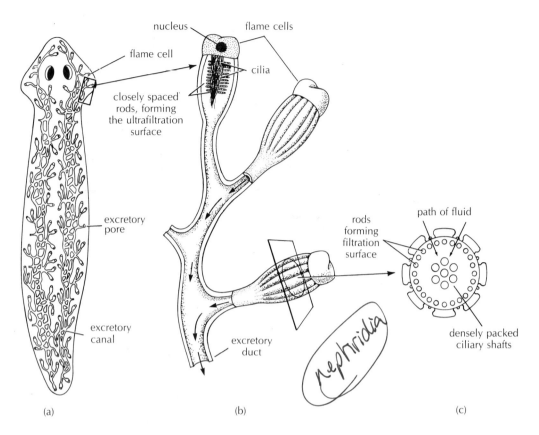

Figure 7.1

(a) The elaborately branching excretory system of a freshwater free-living flatworm. Arrows indicate direction of fluid flow.
(b) Detail of several flame cells emptying into a common collecting duct. (c) Cross section of a flame cell at the level of the ciliary bundle.   (a) *From Schmidt-Nielson, Animal Physiology: Adaptation and Environment, 3d ed. Copyright © 1983 Cambridge University Press, New York. Reprinted by permission.*

Although flatworms are acoelomate, their development is, in other respects, protostome-like: Cleavage is spiral and, at least in some species, determinate,[2] and the mouth forms from the blastopore. Most species have a conspicuous anterior brain, which is connected to at least one pair of longitudinal nerve cords. In the most advanced species, only a single pair of nerve cords is present, and these cords are always located ventrally. The mesodermal layer of the embryo develops into a loose collection of cells known as **parenchyma** tissue. This tissue occupies the entire space between the outer body wall and the endoderm of the gut. Like the cnidarians, platyhelminths have no anus; A single opening to the digestive system serves as both mouth and anus.

Perhaps the most conspicuous unifying feature of platyhelminths is that they are flat. They have no specialized respiratory organs and no specialized circulatory system, although a very few species possess hemoglobin. Gas exchange is accomplished by simple diffusion across the body surface. The rate at which such exchange can occur (milliliters of oxygen transported from the surrounding medium into the tissues per unit of time) depends upon

several factors: the oxygen concentration gradient across the body wall, the body wall's permeability to gas, the thickness of the body wall, and the total exposed surface area across which diffusion can occur. By being flat, the flatworms achieve a high surface area relative to their enclosed volume, and a sufficient amount of gas exchange can occur to support an active lifestyle, despite the lack of gills and an internal circulatory system.

Wastes probably move out of flatworms mostly by diffusing across the general body surface. Again, being flat is helpful in this regard. In addition, most platyhelminths contain a series of specialized organs called **protonephridia** (G: first kidney). The typical protonephridium consists of a group of cilia projecting into a fine-meshed cup (Fig. 7.1). The beating of the cilia within the cup has been likened to the flickering of a flame; thus, the common name for this type of cell is **flame cell.** Other protonephridia take the form of **solenocytes,** in which a single flagellum is found within the cup. In both cases, the mesh cup is attached to a long, convoluted tubule that connects to the outside of the animal through a single excretory pore. The inner workings of the protonephridium are not fully understood. Apparently, the beating of the flagella or cilia creates a negative pressure, drawing fluid through the

2. See *Topics for Further Discussion and Investigation,* no. 6.

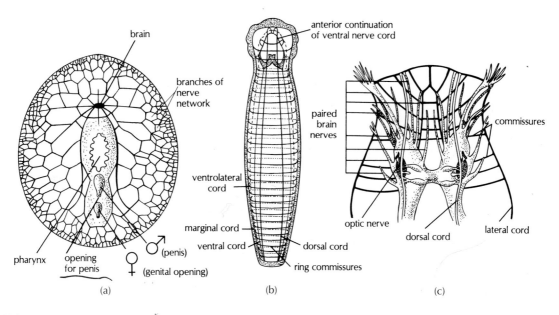

Figure 7.2
Turbellarian nervous system. (a) Nerve net system in *Planocera*.
(b) Longitudinal nerve cord system in *Bothrioplana*. (c) Detail of the
brain of *Crenobia*.   *(a) After Lang. (b) After Reisinger; after Micoletzky. (c) From Bayer and Owre, The Free-Living Lower Invertebrates. Reprinted by permission of the authors.*

mesh cup and into the protonephridial tubule. The liquid entering the protonephridium is thus ultrafiltered: Large molecules (e.g., proteins) are excluded by their inability to pass through the openings in the cup wall. Measurements made on the liquid entering and leaving protonephridia indicate that the chemical composition of the fluid changes as it moves along the tubule to the excretory pore: Ions may be selectively absorbed or secreted, and the water content may be altered as well. Thus, protonephridia likely play an important role in regulating ionic and water balance in flatworms, in addition to their possible role in eliminating metabolic wastes, such as ammonia.

The vast majority of flatworm species, in all three classes, are **simultaneous hermaphrodites;** that is, each individual can, at any one time, function as both a female and a male. As a consequence, sperm exchange and egg fertilization can occur when any individual encounters another individual of the same species. A single individual generally cannot fertilize itself, although exceptions do exist, as discussed shortly.

Class Turbellaria

Only about 16% of all flatworm species belong to the class Turbellaria (ter-bel-air´-¯e-ah). The high surface-area-to-volume ratio of turbellarians makes them especially prone to dehydration in air, so most turbellarians live in aquatic environments. About 150 species are commensal or parasitic with other invertebrates, but most of

the 3,000 turbellarian species are free-living. Most species are marine; a number of species are found in freshwater; and a few species are considered terrestrial, although these are restricted to very humid areas. Individuals are typically less than 1 cm long, regardless of habitat, although members of some terrestrial and marine species are considerably longer.

The nervous system consists of a coelenterate-style, diffuse nerve net in the most primitive turbellarian species. Increasing compactness of the system is associated with advancement in the class, culminating in the possession of a distinct brain and from one to three (rarely four) pairs of longitudinal nerve cords (Fig. 7.2). Such an advanced nervous system also characterizes the parasitic members of the phylum. Turbellarians typically bear one or more pairs of eyes anteriorly, along with a variety of cells that sense chemicals (such as potential foods), pressure changes (such as those produced by water currents), and mechanical stimuli. Individuals in about 10% of the species also have statocysts, which provide feedback regarding body orientation.

Most aquatic turbellarian species are **benthic;** that is, they live in or on the ocean, lake, pond, or river bottom. The outer surface of the body is ciliated, often more so on the ventral surface than on the dorsal surface. Most species move at least partly by secreting mucus from the ventral surface and beating the ventral cilia within this viscous mucus. As a consequence of being flat, increased size is accompanied by a substantial increase in the amount of surface area in contact with the substrate over which the animal is moving. Thus, an increased number

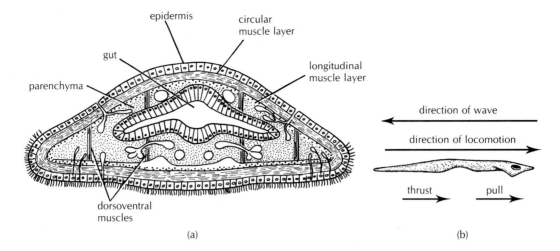

Figure 7.3

(a) Diagrammatic illustration of a turbellarian in cross section, showing the arrangement of muscle layers. (b) Locomotion by pedal waves in a turbellarian flatworm. A single wave is shown traversing the body's ventral surface. A small portion of the body is both thrust and pulled forward as the wave of dorsoventral contraction lifts that region of the body away from the substrate. The thrusting and pulling forces are generated by the regions of the body that are anchored against the substrate, on either side of the contraction. In most species, many waves of contraction travel down the body simultaneously.

of cilia in contact with the substrate compensates for the increased weight of a larger animal, and the ability to move need not suffer as the animal grows. In contrast to the monociliated condition of cnidarians and sponges, each flatworm epidermal cell is multiciliated, bearing several to many cilia.

The locomotion of many individuals involves subtle waves of muscular contraction along the animal's ventral surface. These **pedal waves** are unidirectional, moving from the anterior of the worm posteriorly. As a wave of contraction moves down the length of the body, small portions of the ventral surface are pulled up and away from the substrate. Circular muscles contract just in advance of the wave, squeezing and thrusting the body forward, and longitudinal muscles contract just behind the wave, pulling the body in the direction of locomotion (Fig. 7.3b). The magnitude of the muscle contractions involved in generating pedal waves is quite small, and a number of waves are generally in progress down the body simultaneously. Thus, the progress of the animal along the substrate is very graceful and nearly indistinguishable from that powered entirely by the action of cilia.

The musculature of the body wall includes fibers running longitudinally, circumferentially, dorsoventrally, and diagonally (Fig. 7.3a). All of this musculature is brought into play in the locomotory movements of some turbellarians. This movement, called **looping**, is quite pronounced. The individual attaches at the anterior end, pulls the posterior forward by contracting longitudinal muscles, attaches at the posterior end, releases the anterior end, and then thrusts the body forward by contracting circular muscles (Fig. 7.4).

For effective looping locomotion, the flatworm must be able to adhere locally to the substrate, to prevent slid-

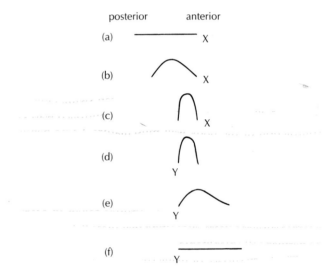

Figure 7.4

Diagrammatic illustration of looping by turbellarians. Once an attachment to the substrate is made at point X (a), contraction of the longitudinal musculature brings the rear of the animal anteriorly (b,c). The anterior attachment is released, and a new attachment is made at the posterior end (d) at point Y, allowing the head of the animal to be thrust forward (e,f).

ing backward while the pulling and pushing forces are being generated. On the other hand, attachments to the substrate must be only temporary if the animal is to progress forward. Flatworms typically possess a large number of paired secretory cells (**duo-glands**) located on the ventral surface and opening to the exterior. One cell of each pair seems to produce a viscous glue, while the other cell presumably secretes a chemical that breaks this attachment to the substrate (Fig. 7.5).

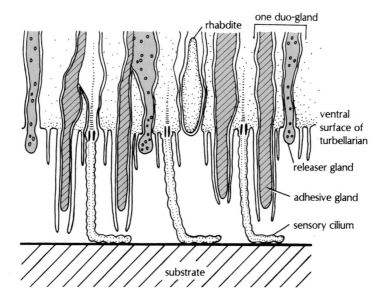

Figure 7.5

Diagrammatic illustration of the ventral surface of a free-living turbellarian flatworm, showing the arrangement of the duo-gland system. Adhesive glands produce a chemical that attaches part of the animal to a substrate; the releaser glands secrete a chemical that dissolves the attachment as appropriate. Rod–like rhabdites, as illustrated, are encountered in the epidermis of most turbellarians; a defensive function has been suggested.   *Modified from Tyler. 1976. Zoomorphology. 84:1.*

A number of turbellarian species can swim, either by means of ciliary activity or by vigorous, controlled waves of contraction of the body wall musculature. Some benthic species probably swim only when environmental conditions deteriorate, while a number of shallow-water species appear to swim routinely, remaining benthic only during low tide. Some small (< 1 mm) acoel flatworm species are routinely found in plankton samples taken in warm, oceanic surface waters, suggesting that these flatworms may be permanently planktonic.

The body surface of many species bears numerous aggregations of small, cylindrical **rhabdites** and related **rhabdoids** (Fig. 7.5). These structures are uniquely turbellarian, although their function is uncertain; they release a thick mucus that coats the animal's body, possibly in response to attempted predation or to desiccation.

The turbellarian digestive system is fairly simple, although the details vary considerably among species. Indeed, differences in the structure of the mouthparts and the gut, along with differences in reproductive morphology, are key elements in dividing the class into its 12 constituent orders.

Some species bear a simple mouth opening on the ventral surface and have no well-formed gut cavity; these are the acoel flatworms (order Acoela; *a* = G: without; *coel* = G: a cavity). In these flatworms, food is essentially thrust into a densely packed mass of specialized digestive cells (Fig. 7.6a). Acoel flatworms are exclusively marine and bear a superficial morphological resemblance to the planula lar-vae of cnidarians. In fact, many zoologists suspect that both cnidarians and flatworms evolved from a planula-like ancestor, in which case the acoel turbellarians would resemble the most primitive triploblastic metazoans, from which all other metazoans are presumably derived. Support for this attractive scenario is waning, however, as some recent studies suggest that acoels are not primitive at all, but instead have descended from more elaborate, gut-bearing (and possibly even coelomate) ancestors.

Turbellarian evolution seems to have been largely a tale of increasing complexity of the digestive system and of the means of acquiring food, and of an increasing specialization of the reproductive system. The guts of flatworms beyond the level of acoel development may be straight, three-branched, or multibranched (Fig. 7.6). The mouth of more advanced species is often borne at the end of a **protrusible pharynx** (Fig. 7.7b); other species possess a separate proboscis that spears food items and then transfers them to an adjacent mouth opening. Most turbellarian species are active carnivores, although some species ingest detritus and algae,[3] and a number harbor algal symbionts. Digestion of food is initially extracellular through secretion of enzymes. Particles are later phagocytized, and digestion is completed intracellularly.

---

3. See *Topics for Further Discussion and Investigation*, no. 1.

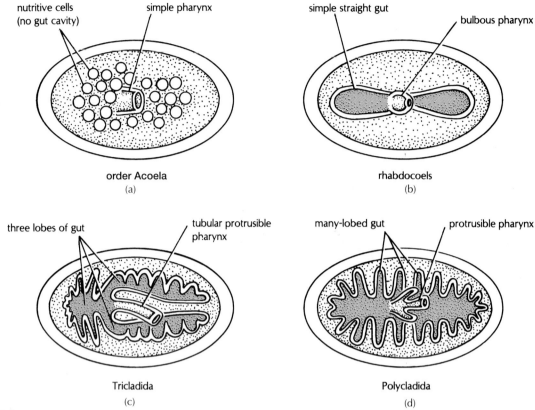

Figure 7.6
Morphological diversity of the turbellarian digestive system, ranging from (a) no well-defined gut; (b) a gut that is unbranched; (c) a three-branched gut; (d) a multibranched gut. *(a-d) From W. D. Russell-Hunter, A Life of Invertebrates. Copyright © 1979 by Macmillan Publishing company, New. Reprinted by permission of the author.*

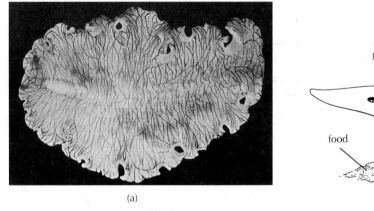

(a)

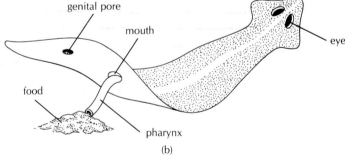

(b)

Figure 7.7
(a) A marine polyclad flatworm, *Pseudoceros crozieri*. This individual was nearly 4 cm long. (b) Typical flatworm (*Dugesia*) with pharynx extended for feeding. *(a) Courtesy of Ann Rudloe, Gulf Specimen Laboratories, Inc.*

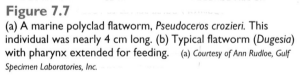

The most structurally and physiologically sophisticated system found among turbellarians is the reproductive system. Both male and female reproductive organs are found within a single individual, and the male system is particularly complex (Fig. 7.8). During the mating of triclad flatworms, each individual typically inserts its penis into the female opening of the other member of the pair, so sperm transfer is reciprocal (Fig. 7.9a). In most other turbellarian groups, copulation may occur by hypodermic impregnation, in which the stylets of the penis

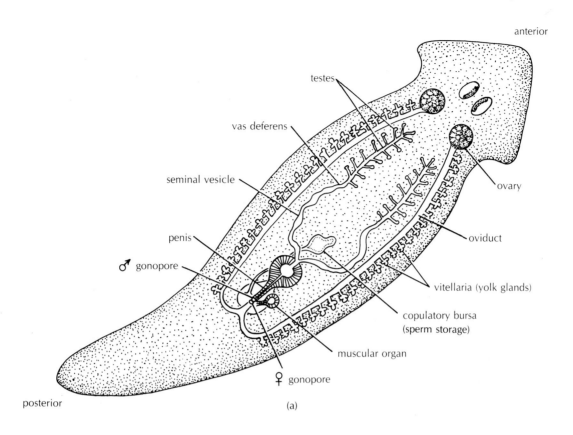

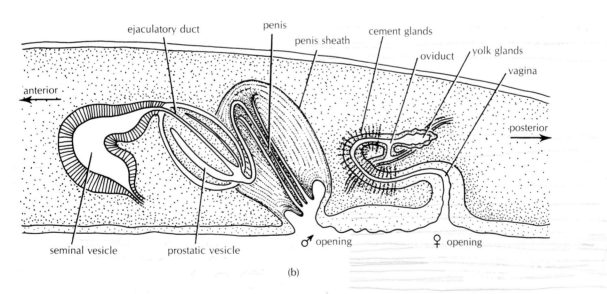

Figure 7.8

(a) Triclad turbellarian reproductive system. Note the presence of both male and female reproductive organs in a single individual. The male reproductive system consists of the testes, prostate gland, and the ducts (vas deferens and seminal vesicle) that conduct sperm from the testes to the ejaculatory duct of the muscular penis. Sperm accumulate in the seminal vesicle prior to copulation. The female reproductive tract includes the ovaries, the oviducts, and the gonopore, through which fertilized eggs are released. As eggs move down the oviduct toward the gonopore,

they become surrounded by nutritive yolk cells manufactured by the vitellaria. The copulatory bursa serves to receive and store sperm contributed by the partner. Some species possess an additional sperm storage organ, the seminal receptacle. After copulation, cement glands coat the emerging ova with a sticky substance that glues them to a variety of solid substrates, including macroalgae and other vegetation. (b) Detail of copulatory apparatus and associated structures.

(a,b) *After Steinmann.*

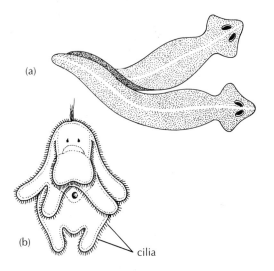

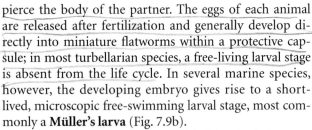

Figure 7.9

(a) Copulating flatworms. The penis of each individual is inserted into the female opening of the mate. (b) Müller's larva of a marine flatworm.   *(a) From Hyman, The Invertebrates, Vol. III. Copyright © 1951 McGraw-Hill Book Company, New York. Reprinted by permission. (b) From E. E. Ruppert, "A Review of Metamorphosis of Turbellarian Larvae" in Settlement and Metamorphosis of Marine Invertebrate Larvae, Chia and Rice, eds. Copyright © 1978 by Elsevier Science Publishing Co., Inc., New York. Redrawn by permission of the author.*

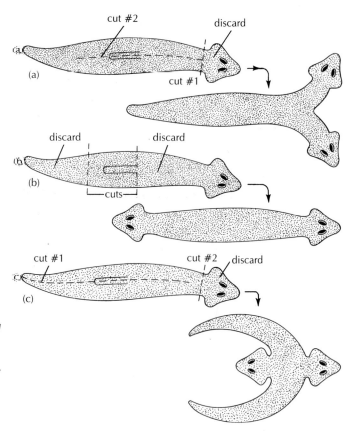

Figure 7.10

Three experiments illustrating the considerable powers of regeneration possessed by free-living flatworms. Cuts are made using a clean razorblade.   *Sherman/Sherman, The Invertebrates: Function and Form, 2e. © 1976, pp. 111, 112, 236, 172, 169, 45, 9, 15. Reprinted by permission of Prentice Hall, Upper Saddle River, New Jersey.*

pierce the body of the partner. The eggs of each animal are released after fertilization and generally develop directly into miniature flatworms within a protective capsule; in most turbellarian species, a free-living larval stage is absent from the life cycle. In several marine species, however, the developing embryo gives rise to a short-lived, microscopic free-swimming larval stage, most commonly a **Müller's larva** (Fig. 7.9b).

Turbellarians possess remarkable regenerative powers that go far beyond the ability to repair wounds,[4] as illustrated in Figure 7.10. In addition, many species reproduce routinely by asexual fission, an ability closely linked to their great regenerative capabilities. Developmental biologists have studied these phenomena intently; certain freshwater triclads in particular have long served as models for studying the processes controlling cellular differentiation.

Class Cestoda

Members of the class Cestoda (ses-to´-dah)—most of which are members of the subclass Eucestoda and commonly known as tapeworms—are all internal parasites; that is, they are **endoparasitic** (*endo* = G: within). They are primarily parasites of vertebrates, inhabiting various regions of the host digestive tract. As many as 135 million people are estimated to have tapeworm infections worldwide.

Cestodes are strikingly different from turbellarians, the differences between the two reflecting an extremely high degree of cestode specialization for an endoparasitic existence. Instead of the ciliated epidermis characterizing the Turbellaria, a nonciliated **tegument** covers the cestodes. The tegument contains numerous nuclei, but these are not separated by cell membranes; that is, the tegument is **syncytial**. The outer surface is outfolded into numerous cytoplasmic projections, vastly increasing the amount of exposed surface area across which nutrients can be taken up from the host's gut. Indeed, the cestode must receive all of its nutrients in this manner, as it has no mouth or digestive tract of its own at any point in its life cycle.

Although lacking a mouth, cestodes do have an anterior end, which in most species takes the form of a **scolex**. The scolex is studded with hooks and/or suckers and is used to maintain position within the host's gut. The relatively few cestode species that lack a scolex are mostly placed within a second subclass, the Cestodaria (Fig. 7.11a).

---

4. See *Topics for Further Discussion and Investigation*, no. 5.

The essence of tapeworm existence, however, really lies just posterior to the scolex, in a region known as the **neck.** A seemingly endless series of sections called **proglottids** bud from the neck of most cestodes at a rate of several per day (Fig. 7.11b). Only members of the small subclass Cestodaria (Fig. 7.11a) do not produce proglottids.

Each proglottid is involved primarily with the process of sexual reproduction. Not only is each tapeworm a simultaneous hermaphrodite, but so is each proglottid; that is, each proglottid contains both male and female reproductive systems (Fig. 7.12). In some species, each proglottid may contain numerous ovaries and as many as 1,000 distinct testes. The tapeworms might best be thought of as franchisers of eggs and sperm. And like the best franchising operations, tapeworms daily turn out a large amount of product, qualitatively similar from proglottid to proglottid.

Each proglottid contains perhaps 50,000 eggs. The eggs are generally fertilized by sperm from a neighboring cestode, but they can be fertilized by sperm from the same individual—or, in fact, from the same proglottid. Since proglottids are rarely more than 3 mm to 5 mm long and the total length of a cestode may exceed 10 m to 12 m, perhaps 2,000 to 4,000 proglottids are produced per individual in many species, and an incredible number of fertilized eggs (tens of thousands, or even hundreds of thousands) are produced per day per cestode.

In some species, the posteriormost proglottids break off periodically. In other species, mature proglottids burst open, releasing the fertilized eggs into the host's gut. Either way, the eggs leave the host's body along with the feces. Generally, the fertilized eggs cannot take up residence in the **definitive** (final) **host** immediately; they must first enter an **intermediate host** or, in some species, a series of intermediate hosts. Different cestode species require different intermediate hosts, which include both vertebrates and invertebrates.

When a fertilized cestode egg is ingested by the appropriate intermediate host, an **oncosphere** larva commonly hatches out. Each oncosphere has muscles, flame cells, and most significantly, three pairs of hooks with which it attaches to the wall of the host's digestive tract (*oncus* = G: a barb or hook). The oncosphere then **lyses** (dissolves) its way through the intestinal wall, taking up residence as an encysted form in the coelomic space or in specific organs and tissues of the host. An oncosphere typically produces a single resting individual called a **cysticercus,** or **bladder worm.** In some species, however, the resting stage divides asexually many times within the intermediate host to produce a large, sometimes deadly, **hydatid cyst.** Such

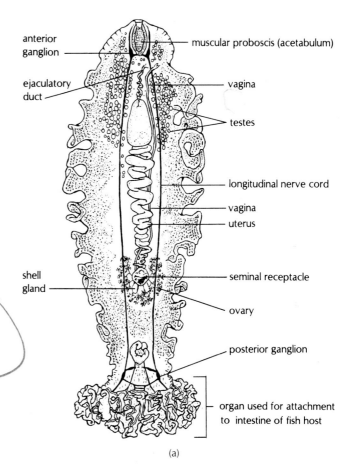

(a)

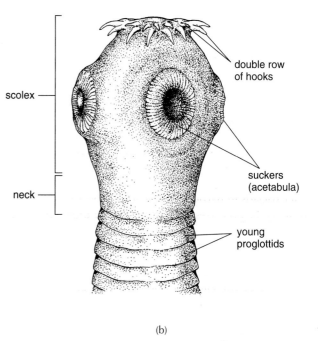

(b)

Figure 7.11

(a) *Gyrocotyle fimbriata*, a member of the small subclass Cestodaria. (b) Anterior end of the common tapeworm *Taenia* sp., magnified 12×. *(a) From T. Cheng (after Lynch, 1945), General Parasitology, 2d ed. Copyright © 1986 Academic Press, Inc. Reprinted by permission.*

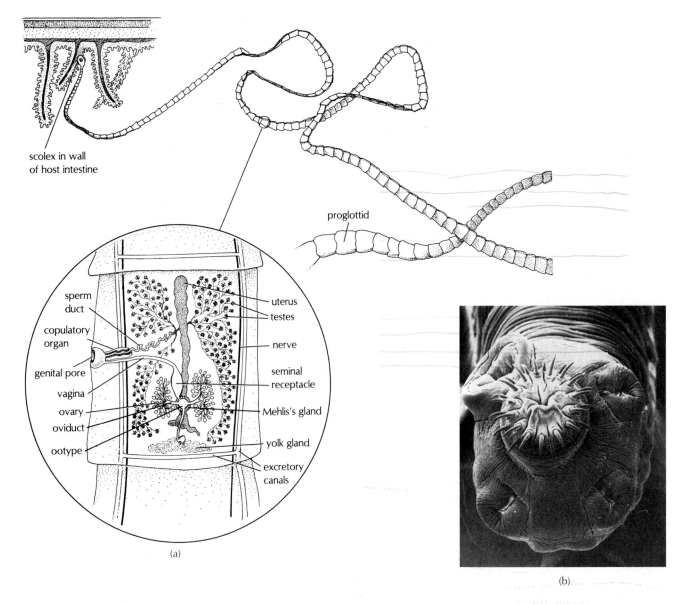

scolex in wall
of host intestine

proglottid

sperm
duct

copulatory
organ

genital pore

vagina

ovary

oviduct

ootype

uterus

testes

nerve

seminal
receptacle

Mehlis's gland

yolk gland

excretory
canals

(a)

(b)

Figure 7.12

(a) *Taenia solium*, the pork tapeworm, attached to the intestinal wall of its host. Note that the entire proglottid is dedicated to the task of reproduction. Eggs leave the ovary and are fertilized either in transit to or within the ootype (the region where the oviducts join) by sperm feeding in from the seminal receptacle. Fertilized eggs are then enclosed individually in protective capsules, possibly made from secretions of a gland (Mehlis's gland) surrounding the ootype. (b) Scanning electron micrograph of the anterior end of *Taenia hydatigena*, magnified 170 ×. (b) From D. W. Featherston, International Journal for Parasitology 5:615, fig. 1, © 1975 Pergamon Press.

asexual replication is especially common among taeniids (family Taeniidae), but is rarely encountered among other cestode species.

Further development is arrested until the intermediate host is eaten by a different host, which may be the final host or another intermediate host. Thus, the complete sequence from egg to adult is achieved only if the fertilized cestode eggs reach the appropriate intermediate hosts, which in turn must be ingested by the appropriate final, or definitive, host (Fig. 7.13). Among the vertebrates, fishes, cows, pigs, dogs, and sometimes birds may serve as intermediate

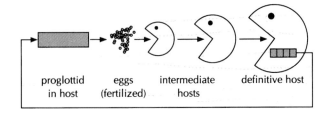

proglottid
in host

eggs
(fertilized)

intermediate
hosts

definitive host

Figure 7.13

Diagrammatic illustration of the typical cestode life cycle. The parasite cannot attain sexual maturity until the definitive host is located.

(a)

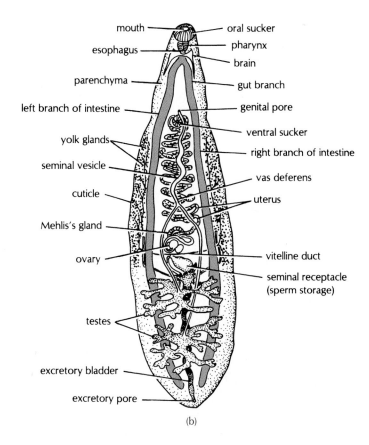

(b)

Figure 7.14

(a) Scanning electron micrograph showing external morphology of the trematode *Zygocotyle lunata,* taken from a mouse. The oral sucker is at the top of the photograph. (b) The Chinese liver fluke, *Opisthorchis sinensis,* seen in ventral view. Note the large percentage of the body devoted to reproduction. Mehlis's gland, also called the shell gland, is a conspicuous feature of the female reproductive tract; its function in trematodes is uncertain.

(a) *Courtesy of S. W. B. Irwin, University of Ulster, N. Ireland.* (b) *From Brown, 1950. Selected Invertebrate Types.* John Wiley & Sons.

hosts. Humans often serve as acceptable final and intermediate hosts. (Think twice before eating undercooked beef, pork, or fish and before letting a dog lick your face!) Among the invertebrates, most intermediate hosts are arthropods.

Class Trematoda

A parasitic existence poses a number of problems, none of them trivial. The successful parasite must

1. reproduce within the definitive host
2. get fertilized eggs or embryos out of the host
3. contact and recognize a new, appropriate host[5];
4. obtain entrance into the host
5. locate the appropriate environment within the host
6. maintain position within the host
7. withstand what is often a rather anaerobic (oxygen-poor) environment
8. avoid digestion or attack by the host's immune system[6]; and
9. avoid killing the host, at least until reproduction has been achieved.

These problems are faced by all parasites, but perhaps the most remarkable adaptations to numbers 3 and 4 are encountered among the trematodes (class Trematoda [trem-ah-tō′-dah]), or "flukes," most of which reach adulthood only as parasites in or on vertebrates.

The outer body layer of trematodes, like that of the cestodes, is an unciliated syncytial tegument. In other respects, the trematode body more closely resembles that of a turbellarian. Trematodes have a mouth opening and a blind-ended digestive tract (usually bilobed), and the body is never segmented (Figs. 7.14 and 7.15). The animal ingests the host's tissues and blood through its mouth.

Schistosomiasis, an often deadly disease prominent in many regions of the world, results from an infection by trematodes known as "blood flukes" (Fig. 7.15b). More than 200 million persons are estimated to presently suffer from schistosomiasis in 76 countries, making it the

5. See *Topics for Further Discussion and Investigation,* no. 2.
6. See *Topics for Further Discussion and Investigation,* no. 3.

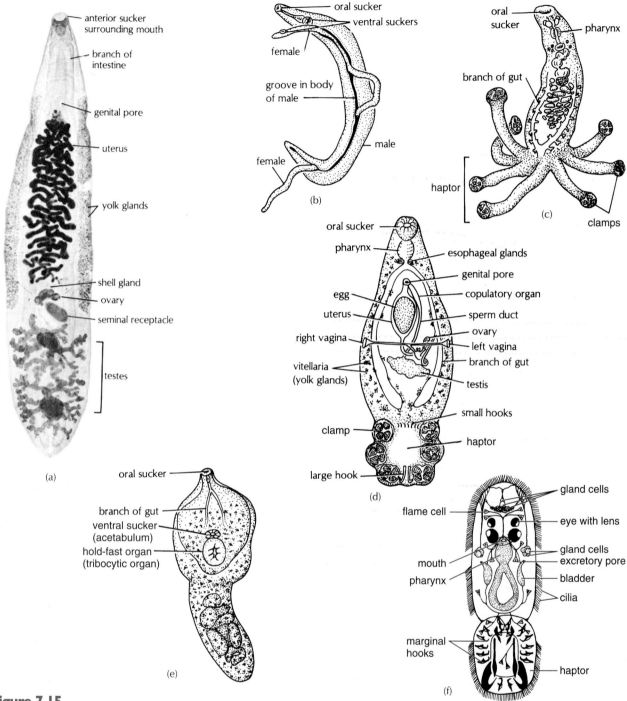

Figure 7.15

Trematode diversity. (a) The Chinese liver fluke *Opisthorchis sinensis* (a digenetic trematode). Note the two-branched gut and the proportion of the body devoted to reproduction. (b) The blood fluke *Schistosoma haematobium*, a dioecious digenetic species. During copulation, the female lies in a specialized groove in the body of the male. This and several other related species cause in humans the disease known as "schistosomiasis." "Swimmer's itch" is caused by members of the same family. (c) *Choricotyle louisianensis*, a monogenetic trematode, taken from the gills of a fish. Note the complex, multipiece haptor. (d) *Polystomoidella oblongum*, another monogenetic trematode, taken from the urinary bladder of a turtle. Note the single-unit

haptor of *P. oblongum,* studded with suckers for securing attachment to the host tissues. (e) The digenetic trematode *Neodiplostomum paraspathula,* seen in ventral view. The tribocytic organ releases a variety of digestive enzymes that solubilize host tissue at the point of attachment. (f) Oncomiracidium larva of the monogenean *Entobdella soleae* seen in ventral view. (Larval length is about 240 μm.)   *(a) Supplied by Carolina Biological Supply Company. (b) From Hyman, after Looss. (d) From Noble and Noble; after Cable. (e) From Noble and Noble; after Noble. (f) From Frederick W. Harrison, John O. Corliss, and Jane A. Westfall, Eds., Microscopic Anatomy of Intvertebrates, Vol. 3. Copyright © 1992 John Wiley & Sons, Inc., New York. Reprinted by permission.*

second most prevalent disease in the world, next to malaria (which is caused by a protozoan—see Chapter 3). Schistosomiasis kills about 800,000 people yearly, and infections in livestock cause hundreds of millions of dollars in economic damage.

The flukes are divided into two major groups, the **monogenetic trematodes** and the **digenetic trematodes** (also called "monogeneans" and "digeneans," respectively),—and one very small group, the aspidogastreans, discussed briefly at the end of this chapter. The monogenetic trematode is characterized by a highly specialized posterior attachment organ called the **haptor** (Fig. 7.15c,d). In most other respects, monogeneans and digeneans are anatomically similar as adults. But they differ greatly in the relative complexity of their life cycles and in the morphology of their larval stages.

The Monogeneans

Monogenetic flukes (Fig. 7.15c,d) are usually parasites on the skin or gills of fishes; that is, most are **ectoparasitic** on fishes (*ecto* = G: outside). This is the fluke's sole host, to which it attaches by means of suckers and hooks located at its anterior and posterior ends. There are no intermediate hosts, so the life cycle generally involves the following stages:

(1) Sexual maturity reached → (2) Egg production →
    in or on fish

→ (3) Larval stage → (4) Attachment
    (oncomiracidium)   to fish

Most monogenean species show a very high level of host specificity. The taxonomic status of the monogeneans is currently in question, as it has been for some time. Some feel that the members of this group are more closely allied with the cestodes than with digenetic trematodes, largely on the basis of the similarity between the oncosphere of cestodes and the oncomiracidium larva of monogeneans. Others feel that the monogeneans are sufficiently dissimilar from either cestodes or other trematodes to merit categorization as a separate class, the Monogenea. Molecular data supporting one or the other position should be forthcoming. Either way, the life cycles of the monogeneans and digeneans have little in common, as discussed next.

The Digeneans

As their name implies (*di* = G: two; *gena* = G: birth), digenetic flukes always require at least one intermediate host before reaching the final host. Unlike the passive process found among the cestodes, host location among digenetic trematodes is generally an active process, mediated by highly specialized, free-living larval stages.

Among the digenetic trematodes, each fertilized egg generally gives rise to a single, free-living, ciliated **miracidium** larva (Fig. 7.16). The gutless miracidium is then either eaten by, or locates and bores into, an intermediate host, which is almost always a mollusc and most commonly a snail. Miracidia produced by a given trematode species must generally enter a particular species of molluscan host to develop further. Indeed, molecular biologists may eventually eradicate schistosomiasis by inserting into the appropriate snail species genes that will prevent schistosome miracidia from penetrating or developing within those hosts.

Host penetration is accomplished by secretions from several glands located anteriorly in the miracidium (Fig. 7.16a). The miracidium then dedifferentiates into a **mother sporocyst** stage, in which most of the structures of the miracidium are lost (including the external cilia), except for the protonephridia. The mother sporocyst lives in the molluscan blood circulatory system (the hemocoel). As no mouth is present at this stage of development, the mother sporocyst grows within the host through the uptake of dissolved nutrients from the surrounding body fluids. Within each sporocyst are numerous balls of cells. Each of these **germ balls** develops either into another sporocyst (the daughter sporocysts) or into another larval stage, the **redia**. The rediae or daughter sporocysts migrate to their host's digestive gland or gonad. The rediae are active feeders, possessing a mouth and a functional, blind-ended gut (Fig. 7.16a), whereas the daughter sporocysts do not feed. Within each redia larva, or within each daughter sporocyst, are numerous germ balls, each of which develops into yet another anatomically distinct larval stage, the **cercaria.** The cercariae leave the "mother" by means of a birth canal if escaping from redia or by lysing through the body wall if escaping from sporocysts, and then lyse their way out of the intermediate host or, in some species, wait within the intermediate host until that host is eaten by a predator.

Free-swimming cercaria larvae are nonciliated, but they can swim actively by means of a muscular tail (Fig. 17.16a,b). Cercariae usually possess at least one sucker anteriorly and, in many species, may have a ventral sucker as well (Fig. 7.16 a,b). The cercaria larvae of some species transform into adults once they encounter the next host, attach to it, and enzymatically penetrate the host tissues. In other cases, the cercaria encysts on submerged vegetation likely to be eaten by the definitive host. In most species, however, the next stop is another intermediate host. In any event, penetration by the cercaria larva into the next host is accompanied by detachment of the cercarial tail, either before or after the cercaria enters the host's body (Fig. 7.17).

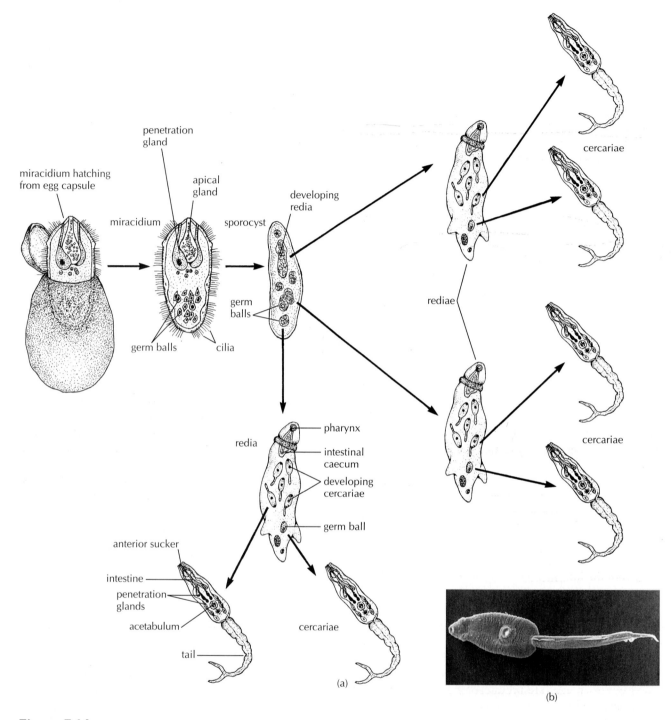

(a)

(b)

Figure 7.16

(a) Replication of larval stages in the trematode life cycle. Large numbers of cercariae are contained within each redia. Many rediae are, in turn, contained within each sporocyst. All of the larvae derived from a single fertilized egg are produced asexually and are therefore genetically identical to each other. The miracidium and cercaria stages are commonly free-swimming. The sporocyst and redia stages occur within an intermediate host, usually a snail. Note the germ balls in the miracidium and sporocyst stages; these will develop into redia larvae. Similarly, the germ balls within the rediae are future cercariae. (b) Scanning electron micrograph of the cercaria stage of *Echinostoma trivolvis*, a digenetic trematode, seen in ventral view. (Total length is about 0.5 mm) The tail is unforked in this and many other species. Note the prominent oral and ventral suckers.   (a) *Modified after Noble and Noble; after Cheng, T. C. 1986. General Parasitology, 2d ed. Academic Press. (b) Courtesy of B. Fried and M. A. Haseeb. From Fried and Haseeb 1991. Microscopic Anatomy of Invertebrates, vol. 3. F. N. Harrison and B. J. Bogitsh, eds. New York: Wiley-Liss, 141–209.*

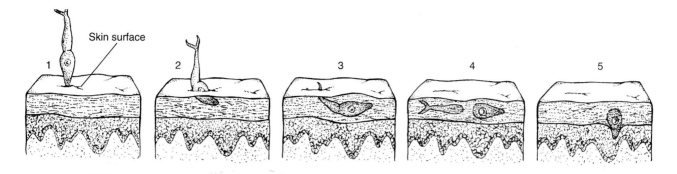

Figure 7.17

Sequential stages in penetration of a *Schistosoma mansoni* cercaria into a vertebrate host. Note loss of tail in step 4. *Modified from Ginctsinskaya, T. A. 1988. Trematodes, Their Life Cycles, Biology and Evolution. New Delhi: Amerind, Publ. Co., Pvt., Ltd., 59, Fig. 96.*

Digeneans typically show little specificity for the second intermediate host, which includes both vertebrate and invertebrate species representing most animal phyla. Even a single trematode species can often make use of a wide range of second intermediate hosts. One species in the genus *Echinostoma*, for example, can use either certain freshwater snails or the larval stage of certain amphibians as second intermediate hosts. Yet the miracidia of that species can successfully infect only snails of one species or of several closely related species.

Once in the second intermediate host, the cercaria transforms, in many trematode species, into an encysted waiting stage, the **metacercaria,** in which most of the specifically larval organs degenerate. The adult trematode develops only when the second intermediate host is ingested by an appropriate definitive host, which may take some time; metacercariae may survive for many months in the intermediate host, with no loss of infectivity. In some cases, one or more additional intermediate hosts may be involved in the life cycle before the final host is reached. Once eaten by the definitive host, the juvenile trematode migrates through the digestive tract to take up residence in the proper, species-specific location, where adulthood is reached. Once the trematode reaches sexual maturity, eggs are produced at a frightening rate; an individual adult schistosome releases more than 3,000 fertilized eggs daily.

What I have just described is a generalized digenean life cycle; details vary considerably among species (Figs. 7.18 and 7.19).

Note that in contrast to the ectoparasitic lifestyle of monogenean species, digenetic trematodes, like the cestodes, are exclusively *endoparasitic*, living within host tissue (*endo* = G: within). The digenean adult has a mouth and a blind-ended gut, usually two-branched, and often has both an anterior and a ventral sucker for maintaining position within the host (Figs. 7.14a, and 7.15e). Most species are hermaphroditic, although some have separate, anatomically distinct sexes; these species are dioecious (Research Focus Box 7.1). Most of the body is taken up by the reproductive system.

The life cycle of digenetic trematodes is, to say the least, highly complex: Several hosts are required for completion of the life cycle; the hosts often occupy very different habitats, complicating the movement from one host to the next; and both the intermediate larval stages and the adult stage require specific host species (Fig. 7.18), although specificity for the first intermediate host is always far greater than that for either the second intermediate or definitive host. The free-living larval stages are nonfeeding and can generally remain alive and infective for only a few hours, during which time the next host animal must be located (Fig. 7.20). Clearly, the probability that any given fertilized egg will reach adulthood is very small. Therein lies the adaptive significance of producing tremendous numbers of dispersal stages. Cestodes achieve a high rate of offspring production mainly by franchising egg and sperm manufacture and by adding on new franchisees (i.e., proglottids) daily. Among the digenetic trematodes, a high rate of offspring production is accomplished through the geometric multiplication of larval stages.[7] Each female generally has only a single ovary, and each fertilized egg develops into only a single miracidium larva. Yet, within each miracidium are the germ balls of the redia stage (or of an additional sporocyst stage), within which are the germ balls of the cercaria stage. A single miracidium generally gives rise to, on average, tens of thousands of cercariae—hundreds of thousands for

---

7. See Topics for Further Discussion and Investigation, no. 7.

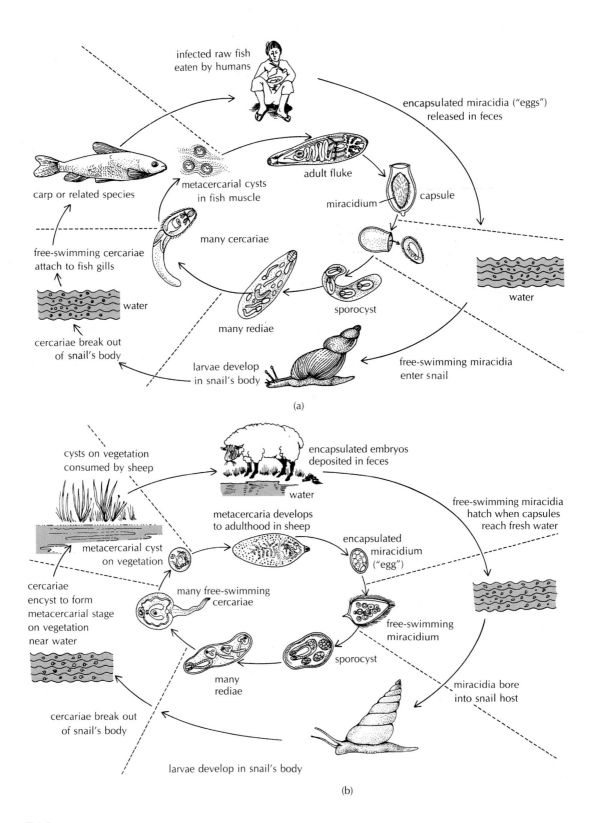

Figure 7.18

Fluke life cycles. The species are all digenetic trematodes.
(a) Chinese human liver fluke *Opisthorchis sinensis*. There are two intermediate hosts (a snail, usually in the genus *Bithynia* and a fish, of the family Cyprinidae). This fluke can reach sexual maturity only in humans, who thus serve as the definitive hosts. Encapsulated miracidia pass out of the definitive host's body along with feces. Hatching of the miracidium occurs only following ingestion by the snail. The miracidium becomes a single sporocyst, which produces (asexually) many rediae, which, in turn, each produce (asexually) many cercariae. These leave the snail and swim freely in the surrounding water. Upon encountering an appropriate fish host, the cercariae bore in and encyst within the fish muscle. The life cycle

reaches completion only when humans eat raw or undercooked fish. Adulthood is reached within the human bile duct. (b) The sheep liver fluke *Fasciola hepatica*. This parasite has a snail as the sole intermediate host. Infected sheep deposit fluke eggs along with feces. Free-swimming miracidia hatch from the eggs if the eggs reach freshwater. The miracidia bore into an appropriate snail host; sporocyst, redia, and cercaria stages follow within the snail. The cercariae leave the snail and encyst to form a metacercarial resting stage on emergent vegetation. Adulthood is reached if the encysted stage is ingested by grazing sheep. (c) *Schistosoma mansoni*, one of the species causing schistosomiasis in humans. Adults are only about 1 cm long. Members of this genus are atypical trematodes in

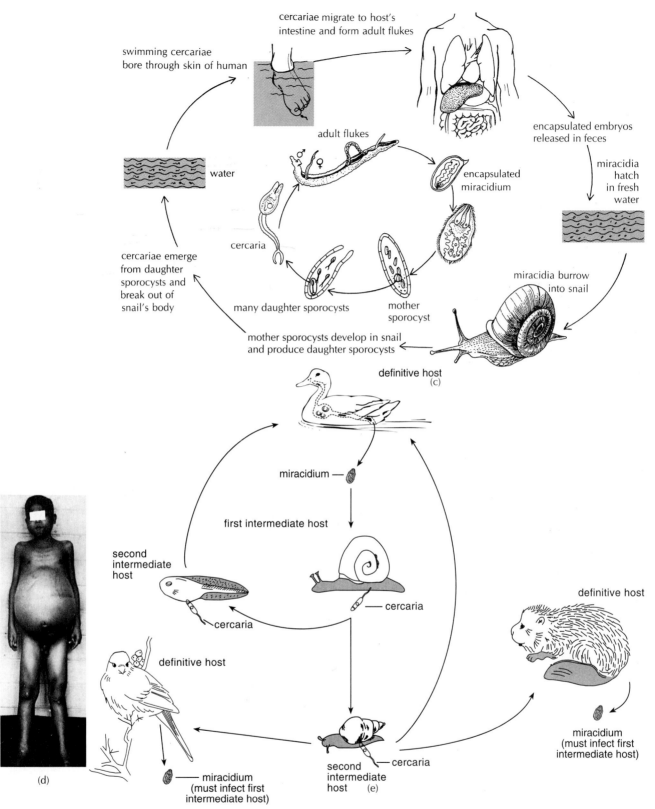

cercariae migrate to host's
intestine and form adult flukes

swimming cercariae
bore through skin of human

encapsulated embryos
released in feces

miracidia
hatch
in fresh
water

water

encapsulated
miracidium

adult flukes

cercariae emerge
from daughter
sporocysts and
break out of
snail's body

cercaria

miracidia burrow
into snail

many daughter sporocysts

mother
sporocyst

mother sporocysts develop in snail
and produce daughter sporocysts

definitive host
(c)

miracidium

first intermediate host

second
intermediate
host

cercaria

definitive host

cercaria

definitive host

miracidium
(must infect first
intermediate host)

second
intermediate
host   (e)

cercaria

miracidium
(must infect first
intermediate host)

(d)

that the adults are dioecious, with the adult female nestling within a specialized groove in the body of the larger male. Fertilized eggs (perhaps 300 per day per female) leave the human host with fecal material, and miracidia hatch if the feces contact freshwater. The miracidia must locate and bore into the tissues of the single intermediate host, a snail in the genus *Biomphalaria*. Each miracidium then develops into a single sporocyst, which then produces many daughter sporocysts. Many cercaria larvae emerge from each daughter sporocyst and exit the snail. If the free-swimming cercaria contacts a human, it bores through the skin and migrates into the circulatory system, within which it travels through the heart, to the lungs, and thence to the kidneys, where it feeds and grows. The parasite eventually reaches maturity within the blood vessels of the host intestine, where it may live for years. Pathology results mainly from inflammation caused by eggs becoming entrapped in host tissues. (d) This boy has the distended belly typical of infection by *Schistosoma japonicum*. (e) Generalized life cycle of a 37–collar-spined trematode *Echinostoma* spp. Adults live in intestines and bile ducts of many vertebrate hosts, especially aquatic birds and mammals. The adults bear 37 spines on a collar surrounding the oral sucker.   (d) *Courtesy of the Center for Disease Control, Department of Health and Human Services Atlanta, GA.* (e) *From Huffman and Fried, "Echinostoma and Echinostomiasis", in Advances in Parasitology, 29:224, 1990. Copyright © Academic Press Inc., Orlando Florida. Reprinted by permission.*

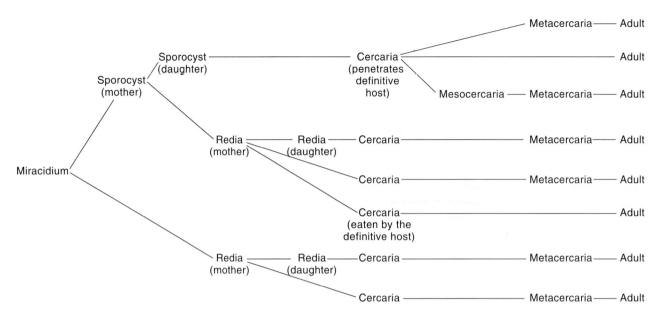

Figure 7.19
Variation in digenetic trematode life cycles. Note that mother sporocysts produce either daughter sporocysts or rediae but never both in the same species.

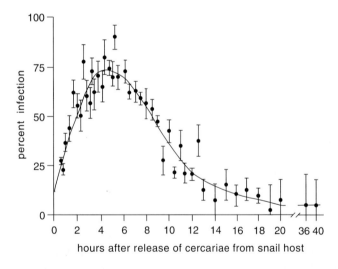

Figure 7.20
Functional lifespan for cercaria larvae of the digenetic trematode *Plagiorchis elegans*. To determine infective capability, cercariae of different ages were exposed to a suitable second intermediate host, larvae of the mosquito *Aedes aegypti*, for 15 minutes at 20°C. The number of metacercariae in each host was counted 24 hours later. Each point shows the mean infectivity to 10 mosquito larvae, with five cercariae tested per replicate. Error bars show one standard error about the mean. Ability of cercariae to infect the mosquito host declined abruptly about six hours after cercariae emerged from the first intermediate gastropod host. *From C.A. Lowenberger and M.E. Rau, in Parasitology, 109. Copyright © 1994 Cambridge University Press, New York. Reprinted by permission.*

some schistosome species—all of which are genetically identical to each other and to the original fertilized egg. A moderately infected intermediate host may release thousands of cercaria larvae per day for many years.

Several remarkable behavioral adaptations that have arisen among trematodes increase the likelihood that a given larval stage will indeed make it to the next required host. Some schistosome species, for example, emerge as cercariae from their aquatic snail hosts at only certain times of the day, times when the target vertebrate host is most likely to be in the water. But perhaps one of the most wonderful adaptations for facilitating parasite transmission is shown by a liver fluke found in sheep, cattle, pigs, and several other terrestrial vertebrates. The species is *Dicrocoelium dendriticum,* and we will go through its story in some detail, to review the events of the trematode life cycle.

The adult liver flukes are hermaphroditic and deposit their eggs into the bile ducts of their host. The eggs, which are less than 50 μm (micrometers) in diameter, leave the host with the feces. The eggs will not hatch until they are eaten by one particular species, or possibly one of several species, of land snail. Miracidia emanating from the fertilized eggs lyse their way through the gut wall of the snail host and migrate to a specific region of the snail's digestive system, where they gradually

Research Focus Box 7.1

Control of Schistosomiasis

Eveland, L. K., and M. A. Haseeb. 1989. *Schistosoma mansoni:* Onset of chemoattraction in developing worms. *Experientia* 45:309–10.

Schistosomiasis is one of the major diseases in the world today, affecting an estimated 200 million people (Fig. 7.18c, d). Many biologists are therefore seeking ways to eliminate it or at least to control its spread. Since the pathological effects of schistosome infection in humans are caused by the eggs, the parasites should be eliminated before egg laying begins. One widely attempted approach is to develop vaccines that attack schistosome-specific surface proteins on juveniles. Another approach is to control populations of the intermediate host, a freshwater snail. A third approach is to interfere with the efficacy of the schistosome digestive system, preventing growth and sexual maturation by interfering with key digestive enzymes. A fourth approach, and the focus of the paper by Eveland and Haseeb (1989), is to take advantage of the dioecious sexuality of schistosomes and somehow interfere with the essential process of mate location; eggs cannot be fertilized and released until a female worm comes to lie in the specialized groove of the male (Fig. 7.15b). Previous research has indicated that males and females locate each other by chemical means. Eveland and Haseeb's research defines the age at which worms first begin to attract each other.

Eveland and Haseeb conducted their study using a strain of *Schistosoma mansoni*, one of several species in the genus causing schistosomiasis. Cercaria larvae were obtained from the freshwater snail *Biomphalaria glabrata* (Gastropoda: Pulmonata) and applied to the shaved skin of several mice. From 20 to 28 days after the mice were infected, the researchers killed the mice and recovered the juvenile parasites to test for chemoattractive capacity.

For each bioassay, one male and one female schistosome were pipetted 6 mm to 15 mm apart into a glass container, and their subsequent movements toward or away from each other were recorded on videotape. The distance between the two worms was measured every 3 minutes for 30 minutes, and the percent change in amount of physical separation (the "migration index") was calculated for each time interval. At least two pairs, and as many as four pairs, of worms were tested in each of the age groups examined.

Clearly, worms became chemically attractive to each other sometime between 21 and 23 days after infecting the host (Focus Fig. 7.1); before that time, males and females

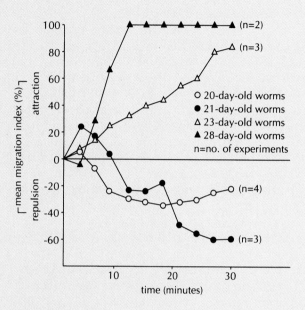

Focus Figure 7.1

Influence of schistosome age on chemoattractive capacity in 30-minute bioassays. Day zero is the day on which cercaria larvae of *Schistosoma mansoni* were allowed to infect a mouse host. The data indicate mean percent changes in the distance between male and female worms; each experiment was repeated two to four times. A value of zero indicates no net movement of worms toward or away from each other.

actually seemed to repel each other. Identifying the factors responsible for the repulsion could be as useful as determining the factors responsible for the later attraction. In both cases, the data suggest the feasibility of interfering with schistosome reproductive success by preventing worms from locating mates. This would not only aid the host, by preventing egg laying and the associated inflammatory responses, but would also reduce the spread of the disease to other individuals, by reducing the output of infective eggs into the environment.

transform to sporocysts. Each mother sporocyst produces a large number of daughter sporocysts, each of which, in turn, gives birth to large numbers of cercariae; this species has no redia stage. Each cercaria is about 600 µm long. The cercariae migrate to the "lung" of the snail, where groups of about 500 larvae become engulfed in mucus. These mucous balls are expelled from the snail to the outside. The outer surface of each slime ball dries to form a water-resistant outer coat, while the cercariae persist in the watery environment within.

From here, the tale becomes even less believable. The slime balls are routinely collected and ingested by several species of ant. Within the ant, the cercariae encyst in various tissues, including the brain, to form the metacercarial stage, which somehow succeeds in altering the ant's behavior. Each evening, the ants are now obliged to crawl upward on a blade of grass and to bite down firmly upon its tip. They apparently remain in this position, unable to open their jaws, until the air temperature rises during the following day. Meanwhile, they are especially vulnerable to grazing by local herbivores, which tend to feed primarily during the evening and early morning. Only if the ants are ingested by an appropriate host can the life cycle be completed. Adult flukes of this species can reside in the bile ducts and gallbladders of many mammals, including horses, sheep, cattle, dogs, rabbits, pigs, and humans.

Exactly how the trematode metaceraria is able to influence ant behavior is not known, but it is easy to imagine how this ability would be selected for once it appeared. Those metacercariae possessing this capability would have a greater chance of infecting the definitive host and leaving offspring. Some proportion of those offspring would also possess the capacity to alter the behavior of the ant intermediate host, since the capability is genetically determined, and gradually the trait would become a species-specific characteristic.

The life cycle of this particular trematode is perhaps more remarkable than most. Equally remarkable is the patience and persistence of the people who have pieced this and other trematode life cycles together.[8]

The Evolution of Digenean Life Cycles

What could account for the evolution of the complex, seemingly roundabout digenean life cycle? Let's summarize the key features: (1) The life cycle typically includes five distinct and unique developmental stages (miracidium, sporocyst, redia, cercaria, and metacercaria) and typically two or more intermediate hosts. (2) The first intermediate host almost always must be a mollusc: Of the approximately 10,000 known trematode species all but two require a gastropod or bivalved mollusc as the first intermediate host. (3) Species specificity for the first intermediate host is extremely high. (4) Specificity for subsequent intermediate hosts and for the definitive host is far less. (5) The digenean life cycle includes asexual replication of developmental stages, something not found in any turbellarian species, and that asexual replication occurs entirely within the molluscan intermediate host. (6) Finally, most species can reach adulthood only in a vertebrate host, usually after a period of encystment in the environment or in a previous host.

The evolutionary origin of such a life cycle presents a most intriguing puzzle. It is easy to imagine that the present parasitic relationship began as a benign, commensal one between a free-living flatworm (probably a rhabdocoel) and another animal. But with what other animal was the flatworm initially associated? Which was the original host: the mollusc or the vertebrate? If the relationship developed before the evolution of vertebrates (about 500 million years ago), the answer to that question is obvious; but how long ago *did* the first association develop? And was the original parasite the larval stage of the flatworm or the adult stage? In what order were the three standard hosts most likely acquired? Was asexual replication a part of the ancestral life cycle before the group became parasitic, or did it evolve later, as an adaptation to parasitic life? Would that the answers were as readily available as the questions.

Although there has been considerable speculation and much lively debate on these issues, no definitive answers are possible. But it is always a wonderful intellectual challenge to try to come up with hypotheses that others have difficulty refuting![9] Certainly the evolution of trematode life cycles poses such a challenge. One particularly attractive scenario* is that originally the flatworm larva parasitized a molluscan host. This would explain the almost complete present specificity for a molluscan first intermediate host and the high degree to which that host tolerates the parasite, assuming that, over evolutionary time, parasites and their hosts tend to become better adapted to each other. It is easy to imagine that the adult flatworm remained free-living for a time. Eventually, asexual replication within the molluscan host evolved, and the free-living cercaria developed the ability to encyst on vegetation, perhaps as an adaptation to environmental stresses; cercariae of some trematode species still encyst on vegetation rather than within a host. The ingestion of those free-living cysts by some vertebrate could have then led to the adult stage becoming parasitic in what is now the definitive vertebrate host; if so, the second intermediate host would have become incorporated into the life cycle later, probably as a vehicle for increasing throughput of the parasite to the vertebrate. This could account for the particularly wide range of acceptable second intermediate hosts even within individual

---

8. See *Topics for Further Discussion and Investigation*, no. 4.

9. See *Topics for Further Discussion and Investigation*, no. 10.

*Based on Ginetsinskaya, T. A. 1988. *Trematodes, Their Life Cycles, Biology and Evolution.* New Delhi: Amerind Publishing Company.

trematode species: Essentially all animals routinely eaten by the final hosts can serve as acceptable intermediate hosts. Again, this is not the only possible pathway to the present trematode life cycle, but it seems a very reasonable one.[9]

The Aspidogastreans

As promised, we will now consider one last, small group of trematodes. The aspidogastrean (as-pid´-o-gas-tray´-un) trematodes, a mere 32 species, bear similarities to monogeneans and digeneans, but they fit neatly into neither group. Most species have a simple life cycle involving a single host, as in the Monogenea. That host is always a mollusc (freshwater mussels or gastropods). However, aspidogastreans never have the posterior attachment organ (opisthaptor) characterizing monogenean species. Instead, the entire ventral surface of the body is modified to form a powerful attachment sucker (Fig. 7.21). Also unlike monogeneans, some species require an intermediate host to complete the life cycle; these species reach adulthood in fishes or turtles, using the mollusc as an intermediate host. In this respect, they resemble the digenetic trematodes, but unlike the digenetic life cycle, the developmental stages never exhibit asexual replication within the intermediate molluscan host.

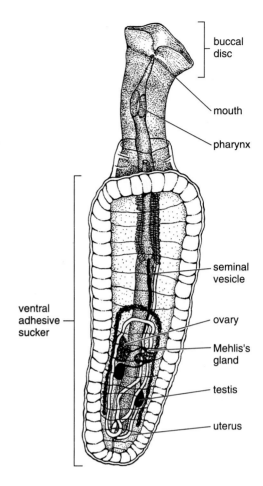

Figure 7.21
Young juvenile of the aspidogastrean *Cotylogaster occidentalis* taken from the gill of a freshwater snail or bivalve. Note the single large ventral sucker divided into a series of compartments. *From B. Fried and M. A. Haseeb.. In Frederick W. Harrison, John O. Corliss, and Jane A. Westfall, Eds., Microscopic Anatomy of Invertebrates, Vol. 3. Copyright © 1992 John Wiley & Sons, Inc., New York. Reprinted by permission.*

Taxonomic Summary

Phylum Platyhelminthes—the flatworms
    Class Turbellaria—the free-living flatworms
    Class Cestoda
        Subclass Cestodaria
        Subclass Eucestoda—the tapeworms
    Class Trematoda—the flukes
        Subclass Monogenea (sometimes considered a separate class)
        Subclass Digenea
        Subclass Aspidogastrea

Topics for Further Discussion and Investigation

1. Discuss the feeding biology of turbellarian flatworms.

Calow, P., and D. A. Read. 1981. Transepidermal uptake of the amino acid leucine by freshwater triclads. *Comp. Biochem. Physiol.* 69A:443.

Dumont, H. J., and I. Carels. 1987. Flatworm predator (*Mesostoma* cf. *lingua*) releases a toxin to catch planktonic prey (*Daphnia magna*). *Limnol. Oceanogr.* 32:699.

Jennings, J. B. 1989. Epidermal uptake of nutrients in an unusual turbellarian parasitic in the starfish *Coscinasterias calamaria* in Tasmanian waters. *Biol. Bull.* 176:327.

Jennings, J. B., and J. I. Phillips. 1978. Feeding and digestion in three entosymbiotic graffillid rhabdocoels from bivalve and gastropod molluscs. *Biol. Bull.* 155:542.

2. How do trematode larvae locate and penetrate their future hosts?

Blankespoor, H. D., and H. van der Schalie. 1976. Attachment and penetration of miracidia observed by scanning electron microscopy. *Science* 191:291.

Curtis, L. A. 1987. Vertical distribution of an estuarine snail altered by a parasite. *Science* 235:1509.

Curtis, L. A. 1993. Parasite transmission in the intertidal zone: vertical migrations, infective stages, and snail trails. *J. Exp. Marine Biol. Ecol.* 173:197.

Feiler, W., and W. Haas. 1988. Host-finding in *Trichobilharzia ocellata* cercariae: Swimming and attachment to the host. *Parasitology* 96:493.

Fried, B., and B. W. King. 1989. Attraction of *Echinostoma revolutum* cercariae to *Biomphalaria glabrata* dialysate. *J. Parasitol.* 75:55.

Hass, W., M. Gui, B. Haberl, and M. Ströbel. 1991. Miracidia of *Schistosoma japonicum*: Approach and attachment to the snail host. *J. Parasitol.* 77:509

Mason, P. R. 1977. Stimulation of the activity of *Schistosoma mansoni* miracidia by snail-conditioned water. *J. Parasitol.* 75:325.

Roberts, T. M., S. Ward, and E. Chernin. 1979. Behavioral responses of *Schistosoma mansoni* miracidia in concentration gradients of snail-conditioned water. *J. Parasitol.* 65:41.

Salafsky, B., Y-S. Wang, A. C. Fusco, and J. Antonacci. 1984. The role of essential fatty acids and prostaglandins in cercarial penetration (*Schistosoma mansoni*). *J. Parasitol.* 70:656.

Wilson, R. A. N., and J. Dennison. 1970. Short chain fatty acids as stimulants of turning activity by miracidia of *Fasciola hepatica*. *Comp. Biochem. Physiol.* 32:511.

3. How do flatworm parasites evade the host immune system?

Clegg, J. A., S. R. Smithers, and R. J. Terry. 1971. Acquisition of human antigens by *Schistosoma mansoni* during cultivation "in vitro." *Nature* 232:653.

Damian, R. T. 1967. Common antigens between adult *Schistosoma mansoni* and the laboratory mouse. *J. Parasitol.* 53:60.

Dineen, J. K. 1963. Immunological aspects of parasitism. *Nature* 197:268.

Hopkins, C. A., and H. E. Stallard. 1974. Immunity to intestinal tapeworms: The rejection of *Hymenolepis citelli* by mice. *Parasitology* 69:63.

Smithers, S. R., R. J. Terry, and D. J. Hockley. 1969. Host antigens in schistosomiasis. *Proc. Royal Soc. London* 171B:483.

4. How do researchers go about documenting the complex life histories of trematode flatworms?

Reversat, J., R. Leducq, R. Marin, and F. Renaud. 1991. A new methodology for studying parasite specificity and life cycles of trematodes. *Int. J. Parasitol.* 21:467.

Shinn, G. L. 1985. Infection of new hosts by *Anoplodium hymanae*, a turbellarian flatworm (Neorhabdocoela, Umagillidae) inhabiting the coelom of the sea cucumber *Stichopus californicus*. *Biol. Bull.* 169:199.

Stunkard, H. W. 1941. Specificity and host-relations in the trematode genus *Zoogonus*. *Biol. Bull.* 81:205.

Stunkard, H. W. 1964. Studies on the trematode genus *Renicola*: Observations on the life-history, specificity, and systematic position. *Biol. Bull.* 126:467.

Stunkard, H. W. 1980. The morphology, life-history, and taxonomic relations of *Lepocreadium aveolatum* (Linton, 1900) Stunkard, 1969 (Trematoda: Digenea). *Biol. Bull.* 158:154.

5. Flatworms are capable of considerable regeneration of lost body parts, including the head. Investigate regeneration in turbellarian flatworms.

Baguna, J., E. Salo, and C. Auladell. 1989. Regeneration and pattern formation in planarians. III. Evidence that neoblasts are totipotent stem cells and the source of blastema cells. *Development* 107:77.

Best, J. B., A. B. Goodman, and A. Pigon. 1969. Fissioning in planarians: Control by brain. *Science* 164:565.

Goldsmith, E. D. 1940. Regeneration and accessory growth in planarians. II. Initiation of the development of regenerative and accessory growths. *Physiol. Zool.* 13:43.

Mead, R. W. 1985. Proportioning and regeneration in fissioned and unfissioned individuals of the planarian *Dugesia tigrina*. *J. Exp. Zool.* 235:45.

Nentwig, M. R. 1978. Comparative morphological studies of head development after decapitation and after fission in the planarian *Dugesia dorotocephala*. *Trans. Amer. Microsc. Soc.* 97:297.

6. How can one determine whether cleavage is determinate (= mosaic) or indeterminate (= regulative) in spirally cleaving embryos?

Boyer, B. C. 1990. The role of the first quartet micromeres in the development of the polyclad *Hoploplana inquilina. Biol. Bull.* 177:338.

7. Discuss the similarities and differences in the life cycles of digenean flatworms and members of the cnidarian class Scyphozoa. Or, compare the life cycle of digenetic trematodes with that of malarial parasites in the protozoan phylum Apicomplexa (p. 57).

8. Like the flatworms, many members of the phylum Cnidaria lack specialized respiratory structures and blood circulatory systems. How are cnidarians able to meet their gas exchange requirements without being flat?

9. The acoelomate condition of flatworms is of considerable interest to those attempting to work out the phylogenetic interrelationships among the animal phyla. If we can safely assume that the flatworm body plan preceded that of the coelomates and pseudocoelomates, then it is very likely that arthropods, molluscs, annelids, and other major metazoan groups evolved from flatworm ancestors. If we cannot, the multicellular ancestors of coelomates must lie elsewhere. Debate the issue of whether the acoelomate condition is primitive or advanced.

Ax, P. 1985. The position of the Gnathostomulida and Platyhelminthes in the phylogenetic system of the Bilateria. S. C. Morris et al., eds. *The Origins and Relationships of Lower Invertebrates.* Oxford: Clarendon Press, 168–80.

Hyman, L. H. 1951. *The Invertebrates: Platyhelminthes and Rhynchocoela,* vol. 2. New York: McGraw-Hill, 52–219.

Rieger, R. M. 1985. The phylogenetic status of the acoelomate organization with the Bilateria: A histological perspective. S. C. Morris et al., eds. *The Origins and Relationships of Lower Invertebrates.* Oxford: Clarendon Press, 101–22.

Smith, J. P. S. III, and S. Tyler. 1985. The acoel turbellarians: Kingpins of metazoan evolution or a specialized offshoot? S. C. Morris et al., eds. *The Origins and Relationships of Lower Invertebrates.* Oxford: Clarendon Press, 123–42.

10. The order in which the various hosts were acquired in the evolution of digenetic trematode life cycles is very controversial. Here are the three most plausible scenarios:

Scenario	Stage I	Stage II	Stage III
A	molluscan host	acquire 2d intermediate host	acquire vertebrate host
B	molluscan host	acquire vertebrate host	acquire 2d intermediate host
C	vertebrate host	acquire molluscan host	acquire 2d intermediate host

Suggest one sort of evidence that would support each scenario and one sort of evidence that would argue against each scenario.

11. What does a trematode cercaria have in common with a long-staying house guest?

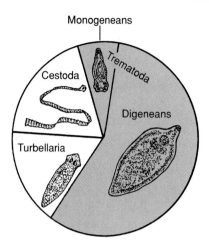

Phylum Platyhelminthes

The approximately 18,500 species are distributed among three classes. Members of the Trematoda (monogeneans and digeneans) are shown in blue.

Class Turbellaria.

The approximately 3,000 species in this class are distributed among 12 orders and about 120 families.

Order Acoela.

This order contains about 200 flatworm species, all of which are marine. These worms have no permanent digestive cavity, and they have a statocyst housing a single statolith. There are no distinct gonads. Most individuals are smaller than about 1 mm, and none are larger than several millimeters. Most species occur in or on marine sediments, but a few species (e.g., *Ectocotyla paguris* and *Avagina* spp.) are commensal with other invertebrates, particularly echinoderms, and a few species are planktonic, spending their lives swimming in the water. Some benthic species within the genus *Convoluta* harbor algal symbionts within the parenchyma and display a rhythmic "sunning" behavior that promotes algal photosynthesis. At least a few other acoel species thrive in sediments low in oxygen, living on the energy gained from oxidizing hydrogen sulfide gas. Swimming acoels can be common components of warm, ocean surface plankton; these flatworms always house algal symbionts. Surprisingly, at least one acoel species shows indeterminate cleavage rather than the determinate pattern characteristic of protostomes. Fifteen families.

Order Rhabdocoela.

This group originally contained all flatworms with linear guts, but its members are now distributed among the following three new orders: Catenulida, Macrostomida, and Neorhabdocoela. Catenulids and macrostomids are thought to be closely related to the acoel flatworms, while neorhabdocoels are believed to be considerably more advanced. All turbellarian species with straight guts are still referred to collectively as rhabdocoels, but the term no longer has formal taxonomic status.

Order Catenulida.

Catenula, Stenostomum (freshwater flatworms), *Rectonectes* (marine, interstitial flatworms). Most of the 70 species occur only in freshwater lakes and ponds, although a few species are marine. Some species can use their cilia to swim short distances. Members of the freshwater species reproduce primarily by asexual means: Organ systems are replicated, and the worms then divide into two individuals. The worms in this order are extremely fragile.

Order Macrostomida.

Microstomum, Macrostomum. Most species are interstitial, living between sand grains either in freshwater or marine sediments. Most species show numerous adhesive glands and rhabdites, but none have statocysts. Members of the genus *Microstomum* are somehow able to sequester the nematocysts (defensive structures) of the freshwater hydras on which they feed and use them for their own defense. Three families, containing nearly 200 species.

Order Nemertodermatida.

Nemertoderma. These free-living flatworms resemble the acoels in many respects (including a 9 + 2 microtubular array in the sperm flagellum; most other flatworms show a 9 + 1 arrangement), but they possess a permanent gut cavity and a statocyst with two statoliths rather than one. One family, containing only five species.

Order Lecithoepitheliata.

Prorhynchus, Geocentrophora. The freshwater and terrestrial species contained in this small order are common worldwide. The order also includes a few deep-water, marine species. Two families.

Order Polycladida.

This large group of flatworms contains no parasitic species, and all but one species (which occurs only in freshwater) are exclusively marine. Polyclads exhibit a gut that is highly branched (*poly* = G: many; *clad* = G: branch). Individuals are typically several centimeters long; most marine flatworms visible with the unaided eye belong to this order. Some can swim short distances by undulating the lateral margins of the body. Twenty-nine families.

Suborder Acotylea.

Seventeen families, many with no more than 12 species.

Family Stylochidae.   *Stylochus*—the oyster leech. A number of species prey on juvenile and adult oysters, making them commercially important shellfish pests. They burrow through the oyster's shell enzymatically and then destroy the adductor muscles holding the oyster's two shell valves closed. The oyster's tissues are then easy prey. Other species are commensal with other invertebrates, including hermit crabs and ascidians (sea squirts—phylum Chordata).

Family Discocelidae.   *Adenoplana, Coronadena.* A widespread group in temperate and tropical marine environments.

Family Hoploplanidae.   *Hoploplana.* At least one species is symbiotic in the mantle cavities of marine snails, although most species are free-living carnivores.

Family Planoceridae.   A number of species in this family are transparent and pelagic throughout their lives.

Suborder Cotylea.

Twelve families, many with fewer than six species. Some species are interstitial.

Family Pseudoceridae.   *Pseudoceros.* This large family of more than 200 described species includes numerous, brightly colored, tropical flatworms. The margins of the body are always folded anteriorly to form a pair of pronounced tentacles (*pseudo* = G: false; *cera* = G: a horn). All species are marine.

Order Prolecithophora (= Holocoela).

Plagiostomum. This group contains both marine and freshwater species, most of which are free-living. A few species are external parasites on certain crabs (decapod crustaceans). The sperm of all species are unusual in lacking flagella. Nine families.

Order Proseriata.

Most species are marine; they are divided among eight families.

Family Monocelididae. *Monocelis.* This group contains about 60 species, several of which live in freshwater. Most are carnivorous, but some parasitize or live commensally with crabs.

Family Coelogynoporidae. This is a widely distributed group of marine interstitial flatworms.

Family Bothrioplanidae. This family contains only one species, *Bothrioplana semperi,* but it lives in restricted freshwater habitats (e.g., ephemeral ponds, groundwater, and drainage ditches) all over the world. It reproduces by parthenogenesis and has a generation time of only about three weeks.

Order Tricladida.
All individuals exhibit a gut with three distinct branches (*tri* = G: three; *clad* = G: branch). The freshwater members of this order are known collectively as "planarians." Seven families.

Family Bdellouridae. *Bdelloura.* Most species are free-living, but the most famous members of this family are commensal on the book gills of horseshoe crabs; they attach to the host, using sticky secretions and a specialized, posterior sucker. Females deposit their eggs on the host's gills.

Family Procerodidae. *Procerodes.* The members of this family live in a great variety of marine and freshwater habitats. Some species are free-living, while others are commensal with chitons, skates, or Japanese horseshoe crabs. Freshwater species occur in Tahiti and in Mexican caves.

Family Planariidae. *Planaria, Polycelis.* This familiar group of some 80 freshwater species occurs only in the Northern Hemisphere.

Family Dugesiidae. *Dugesia.* These flatworms are common in streams, lakes, and ponds and are frequently used in studies of regeneration. They differ from individuals in the family Planariidae, particularly in possessing eyes with multicellular retinas and pigment cups. Over 100 species have been described, many of which reproduce primarily (or exclusively) by asexual fission.

Family Bipaliidae. *Bipalium.* A sizable group (about 100 species) of terrestrial planarians. Individuals may attain lengths exceeding 0.5 m. At least one species is especially common in greenhouses worldwide.

Family Geoplanidae. *Geoplana.* This is another large group of terrestrial triclads, including about 200 species.

Family Dendrocoelidae. This group includes both marine and freshwater species, including many blind cave dwellers. About 150 species are known, all from the Northern Hemisphere.

Order Neorhabdocoela.
Four suborders, containing 34 families. All members bear straight guts and a single pair of ventral nerve cords. The members of the different suborders differ largely with respect to morphology of the male genitalia. Some species can use their external cilia to swim.

Suborder Dalyellioida.
All of the parasitic groups (monogeneans, digeneans, aspidogastreans, and cestodes) are believed to have had their origins among the members of this suborder. Eight families.

Family Dalyelliidae. *Castrella, Dalyellia, Microdalyellia.* The 145 species in this family are widely distributed geographically and are particularly abundant in freshwater lakes and ponds. The group also contains a few marine species and one species commensal with certain small crustaceans (amphipods). Individuals in several species harbor symbiotic zoochlorellae.

Family Provorticidae. *Provortex.* Most of the approximately 50 species are marine or freshwater, but one (*Archivortex*) is terrestrial, and another (*Oekiocolax*) parasitizes other marine turbellarians.

Family Graffillidae. *Graffilla, Paravortex.* All seven species are marine and widespread. Several species are exclusively parasitic in bivalves or gastropods, while the others are free-living. *Paravortex* spp. are among the few hemoglobin-containing flatworms.

Family Umagillidae. *Anoplodium.* All approximately 35 species are parasitic, mostly in the guts or coelomic cavities of echinoderms.

Family Acholadidae. *Acholades.* The family contains a single species, parasitic in the tube feet of a particular sea-star species from Tasmania. These turbellarians lack a mouth or digestive tract, apparently subsisting on dissolved organic nutrients obtained through the enzymatic digestion of adjacent sea-star tissue.

Family Hypoblepharinidae. *Hypoblepharina.* All four species seem to be commensal on amphipod crustaceans in the antarctic. These flatworms lack eyes.

Family Fecampiidae. *Fecampia, Kronborgia.* All species (about 12) are parasitic in the hemocoel of certain marine crustaceans. These flatworms lack mouths, digestive tracts, and eyes as adults, and sometimes in the larval stage as well. Adults leave the host to deposit eggs. Members of the genus *Kronborgia,* reaching lengths of over 39 cm, castrate the host and then kill it upon exiting. Of small consolation to the host, the worm itself dies soon after egg laying.

Suborder Typhloplanoida.

Promesostoma, Typhloplana. This group includes about 175 species, mostly free-living in marine, freshwater, or terrestrial habitats. One species is ectoparasitic, living only between the parapodia of one particular polychaete annelid species (*Nephtys scolopendroides*). Members of a number of marine species appear to be very active and powerful swimmers. Some species in the genus *Typhloplana* harbor symbiotic green algae (zoochlorellae). Nine families, most of which contain fewer than 12 species each.

Suborder Kalyptorhynchia.

Most of the approximately 140 species live in marine sediments, although a few live in freshwater. Fifteen families, most of which contain fewer than six species each.

Suborder Temnocephalida.

Temnocephala. All are freshwater commensals, living especially on the gills of decapod crustaceans. Members of a few species live with freshwater snails or turtles. Unlike that of most other turbellarians, the body is not externally ciliated. Most of the nearly 40 species occur only in South America, Australia, or New Zealand. These small (less than 2 mm long) flatworms typically crawl in leech-like fashion, using anterior tentacles and a large, posterior, adhesive disc. They feed mostly on other smaller invertebrates rather than on host tissue. The anterior tentacles apparently function also in prey capture and defense. Two families.

Class Cestoda.

The approximately 3,500 species are divided among nearly 60 families. Two subclasses.

Subclass Cestodaria. *Amphilina, Gyrocotyle.*

These gutless flatworms are endoparasites of fish and, to a lesser extent, turtles. The life cycles are poorly known, probably because none of the species has any economic or medical importance. Unlike true tapeworms in the subclass Eucestoda, cestodarians lack a scolex and show no external segmentation, and the first larval stage has 10 hooks instead of 6. No individual houses more than a single set of male and female reproductive organs. Individuals of some species may exceed 35 cm in length. Three families.

Subclass Eucestoda. The true tapeworms are

distributed among 12 orders, largely based on differences in scolex morphology. Fifty-six families.

Order Caryophyllidea.

Archegetes, Caryophyllaeides. The species in this small group are unusual in that the body is unsegmented and bears only a single set of male and female organs. Where known, the life cycles involve freshwater oligochaete worms as intermediate hosts. Adults parasitize freshwater fish and oligochaetes. One family.

Order Spathebothriidea.

Diplocotyle, Spathebothrium. These cestodes are unusual in lacking any attachment organs, including a scolex. Moreover, they show no external segmentation, although the reproductive organs are arranged internally in a linear series. Adults parasitize marine teleosts. One family.

Order Trypanorhyncha.

Grillotia, Lacistorhyncus. All species in this order parasitize the stomachs or spiral valves of sharks and other elasmobranchs. Sixteen families.

Order Pseudophyllidea.

These widespread flatworms mostly parasitize marine and freshwater fishes, but some species infect toads, birds, amphibians, or humans, sometimes with severe consequences. Ten families.

Family Diphyllobothriidae. *Diphyllobothrium, Polygonoporus, Spirometra.* This family of some 60 species contains the largest of all tapeworms: *Polygonoporus giganticus,* a parasite of whales, reaches 30 m in length. *Diphyllobothrium latum* grows to 15 m in human hosts and contains many thousands of proglottids. Humans become infected by ingesting undercooked fish; abdominal pain and weight loss characterize the infection.

Order Tetraphyllidea.

These tapeworms parasitize sharks and other elasmobranchs. Their life cycles are poorly explored. The order contains one of the few known dioecious tapeworms (genus *Dioecotaenia*). Four families.

Order Cyclophyllidea.

This important and well-studied order includes most of the tapeworm species infecting people, pets, and livestock. All species have four suckers on the scolex. Fourteen families, containing several hundred species.

Family Triplotaeniidae. All species parasitize Australian marsupials exclusively.

Family Tetrabothriidae. The approximately 60 species in this family all parasitize marine birds and mammals, including whales and seals.

Family Dioecocestidae.   This small group of bird parasites includes most of the few known dioecious tapeworm species. (See also order Tetraphyllidea.)

Family Taeniidae.   *Taenia, Taeniarhynchus, Echinococcus.* This family of approximately 100 species is of great economic and medical importance. The group includes *Taenia solium,* which uses a pig as its intermediate host; humans eating infected, undercooked pork can develop severe neurological problems, including blindness and paralysis, which can lead to death. Adults of *T. solium* can reach 7 m in length. A more common human parasite is the beef tapeworm, *Taeniarhynchus (= Taenia) saginata,* currently infecting some 61 million to 77 million people worldwide. As the common name implies, infection is transmitted through consumption of infected, undercooked beef. Adults typically possess about 1,000 proglottids, but they rarely exceed lengths of 60 cm (although individuals 3 m to 5 m long have been reported). The pathology associated with infection is far less severe than that associated with infection by the pork tapeworm, *T. solium.* Members of the related genus *Echinococcus* reach adulthood in dogs and are rarely larger than about 3 mm, possessing no more than five proglottids. However, in ruminants and humans, which serve as intermediate hosts, the larval stages multiply asexually to form fluid-filled, orange- or even grapefruit-sized **hydatid cysts,** each of which may house many thousands of scolices. The cysts must be removed surgically. Humans typically become infected from dogs, which often carry the eggs on their tongues.

Family Dilepididae.   This is a major group of tapeworms, particularly common in birds and mammals, with arthropods as intermediate hosts. Members of *Dipylidium caninum* parasitize cats and dogs worldwide and infect humans (especially children) who ingest the intermediate hosts (fleas and lice). Infection causes no major medical difficulties.

Family Hymenolepididae.   This group includes hundreds of species parasitizing birds and mammals, including humans. *Hymenolepis diminuta* is the major cestode model for medical studies. *Vampirolepis nana* infects humans but causes few medical problems; this species is unique among cestodes in needing no intermediate host.

Class Trematoda.

The approximately 10,000 species are distributed among more than 200 families. All members of this class are parasitic. Three subclasses.

Subclass Monogenea.

This group is sometimes considered a separate class within the Platyhelminthes. All individuals of all species are simultaneous hermaphrodites, and there are no intermediate hosts in the life cycle. Forty-four families, containing about 1,100 species.

Suborder Monopisthocotylea.

Gyrodactylus, Dactylogyrus, Protogyrodactylus, Trivitellina, Capsala. The species in this suborder bear a haptor (posterior attachment organ) with a single mass of hooks, suckers, or other attachment devices. Most parasitize marine or freshwater fishes (skin, gills, urinary tracts, respiratory tracts, or rectal cavities, in particular), but some species parasitize certain marine copepods (which are themselves parasites of fish), and others parasitize amphibians. Most species feed on host epidermal tissue and mucus. Fifteen families.

Suborder Polyopisthocotylea.

Diplozoon, Discocotyle, Microcotyle, Polystoma. All members are characterized by having a haptor (posterior attachment organ) subdivided into at least two distinct parts. Most species parasitize the gills or skin of marine or freshwater fishes, but some parasitize the urinary tracts of amphibians or reptiles. Individuals in most species feed on blood. Twenty-nine families.

Subclass Digenea.

This subclass contains about 90% of all trematode species. Five orders, 160 families.

Order Strigeidida.

Strigea, Cotylurus, Alaria. The species in this group parasitize such diverse habitats as the digestive tracts of birds, crocodiles, turtles, snakes, and mammals; the mouth or esophagus of reptiles and birds; the scales of freshwater or estuarine fishes; the rectum of birds and mammals; and the circulatory systems of fishes, birds, and mammals. Some species require four successive hosts to complete the life cycle. Thirteen families, containing over 1,350 species.

Family Schistosomatidae.   Bilharziella, Schistosoma, Trichobilharzia—blood flukes. Adults are all parasitic in the blood vessels of mammals and birds. The life cycle requires only two hosts—a snail and a vertebrate—so there is no metacercaria stage. Neither is there any redia stage in the life history; cercaria larvae are produced directly from the sporocysts. Several species are responsible for the widespread and debilitating disease called schistosomiasis. Also, the cercaria larvae of some marine and some freshwater species penetrate the skin of humans despite their inability to survive within that host; this produces a short-lived (typically several weeks in duration) but frustrating dermatitis appropriately called "swimmer's itch." This family contains some of the only dioecious trematode species.

Order Azygiida.

Azygia, Hemiurus. Species (over 500) in this group commonly parasitize the coelomic cavity or various parts of the digestive tracts of marine and freshwater fishes. Some species in this order parasitize the respiratory tract or Eustachian tubes of amphibians. A variety of invertebrates and vertebrates, including barnacles, copepods, cnidarians, and fishes, serve as intermediate hosts. Twenty-six families, many of which contain only one or a few species. About 80% of all species are contained in just two families, the Hemiuridae and Didymozoidae.

Order Echinostomida.

Echinostoma, Fasciola. Members of this large order of some 1,360 species commonly parasitize the cloaca and intestines of birds and mammals (including dogs, cats, herbivores, and marine mammals); the respiratory systems of birds and other vertebrates; the intestines of fishes; and the lungs of turtles. People can become infected, but medical problems are minor. Twenty-eight families.

Order Plagiorchiida.

Plagiorchis, Paragonimus, Dicrocoelium, Nanophyetus. The life cycles often require three hosts. Adults parasitize a variety of hosts and habitats, including the gallbladder, kidney, liver, ovaries, rectum, and lungs of virtually all vertebrates (including fishes, snakes, amphibians, birds, and mammals); the intestines of marine and freshwater fishes; and the coelom and bladder of teleosts, elasmobranchs, amphibians, and turtles. Individuals of some species may reach 12 m in length. *Nanophyetus salmincola* is pathogenic to salmonid fishes, which serve as intermediate hosts for the metacercariae. This fluke also carries the agent of rickettsial disease, which usually kills dogs that eat raw, infected salmon. The lancet liver fluke, *Dicrocoelium dendriticum,* is unusual among trematodes in that its life cycle is entirely terrestrial; a land snail and an ant serve as intermediate hosts. Fifty-three families, containing many hundreds of species.

Order Opisthorchiida.

This order contains about 700 species parasitizing a variety of organs in marine and freshwater fishes, reptiles, birds, and mammals. The life cycle always requires three hosts for completion. Seven families.

Family Opisthorchiidae. *Opisthorchis, Clonorchis sinensis*—the Chinese liver fluke. Adults live in the bile ducts of a human host, causing serious medical problems. Twenty million people currently may be infected with species of *Opisthorchis* and *Clonorchis.* The family includes over 140 species.

Family Heterophyidae. *Cryptocotyle, Heterophyes, Haplorchis.* These worms are common intestinal parasites of birds and mammals, including dogs, cats, and humans. Infection in humans can cause intense intestinal pain; also, the eggs may collect in the heart, causing cardiac arrest.

Family Nasitrematidae. *Nasitrema.* The 10 species in this family all parasitize the nasal cavities of whales.

Subclass Aspidogastrea (= Aspidobothrea).

Aspidogaster, Stichocotyle. Most of the 32 species in this subclass are internal parasites of freshwater mussels, gastropods, fishes, or turtles. The entire ventral surface of the body is modified to form a powerful attachment sucker or series of suckers. Most species require an intermediate host to complete the life cycle; in this respect, they resemble the digenetic trematodes, but unlike the digenetic life cycle, the developmental stages never exhibit asexual replication within the intermediate host. Three families.

Two Possible Flatworm Relatives: Gnathostomulids and Mesozoans

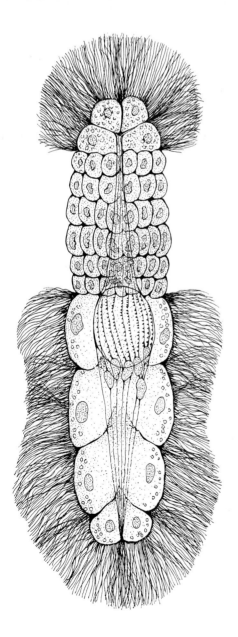

Both of the groups discussed in this chapter—gnathosto-mulids and mesozoans—contain small, exclusively marine, acoelomate worms. All gnathostomulids are free-living, while all mesozoans are parasitic. The members of both groups have uncertain relationships to other phyla (and to each other), but they bear enough similarities to flatworms to suggest some affinity, albeit slight.

Phylum Gnathostomulida

Phylum Gnatho • stomulida
(G: jaw mouth)
nath-ō-stom-ū-lē´-dah

The gnathostomulids (Fig. 8.1) are discussed first because they seem more morphologically similar to the turbellarian flatworms discussed in Chapter 7 than do the mesozoans. Gnathostomulids were first discovered only about 40 years ago. Members of the approximately 80 species described so far are all very small, soft-bodied, mostly interstitial worms living in the spaces between sand grains. Most individuals are less than 1 mm long, but worm concentrations as high as 6,000 individuals per kilogram of sand have been reported, with the greatest worm aggregations occurring in deeper sediment layers high in hydrogen sulfide. Like turbellarian flatworms, gnathostomulids are acoelomate; are externally ciliated (but with only one cilium per cell, rather than the several per cell characterizing flatworms and most other spirally cleaving animals); are triploblastic; and have a mouth but no functional anus. Moreover, they lack any specialized circulatory or respiratory systems, and their excretory organs are simple protonephridia. The gnathosto-mulids' most complex structures are the mouthparts,

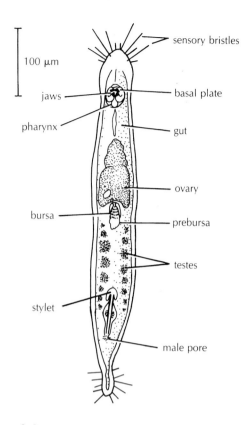

Figure 8.1

A gnathostomulid, *Gnathostomulida jenneri*. Most individuals are less than 1 mm long.   *From Sterrer, 1972. Syst, Zool. 21:151.*

consisting of a pair of hardened jaws and an associated basal plate; the jaws are used to scrape bacteria and fungi from sand grains and other substrates.

Most species are hermaphroditic, as are flatworms, and eggs are fertilized internally and then released. As in the flatworms, cleavage is spiral, but no free-swimming larvae are known; all gnathostomulid zygotes develop directly into small worms.

The resemblances between gnathostomulids and flatworms may be purely superficial. Gnathostomulids are almost alone among spirally cleaving animals in having a monociliated epidermis, with a single cilium per cell. Another argument against a close affinity between the Gnathostomulida and the Platyhelminthes are the pronounced differences in sperm morphology. Gnathostomulid sperm have the 9+2 arrangement of microtubule doublets in the sperm tail that is standard among animal sperm. In contrast, although the sperm of some flatworm species also show a 9+2 arrangement, those of most species show a 9+1 arrangement, with only a single microtubule in the center of the ring of nine outer doublets, and others show a 9+0 arrangement, or they have no tail at all. In most other respects, the morphological similarities between the two groups are impressive. In addition, few other groups show any affinity with

gnathostomulids at all, although several features (particularly the monociliated epidermis and peculiarities of jaw morphology and ultrastructure) suggest possible affinities with some aschelminths (Chapters 11 and 13) or with the planula larva of cnidarians (Chapter 5).

The Mesozoan Phyla: Orthonectida and Rhombozoa

Phylum Meso • zoa
(G: middle animal)
mē-sō-zō´-ah

The phylogenetic relationship between mesozoans and other multicellular animals is very uncertain. As the name "mesozoa" implies, these animals are somewhere between the unicellular protists and the triploblastic flatworms in their level of organization. They are definitely multicellular, but they are neither diploblastic nor triploblastic animals. Their cells lie in only one or two layers, and these bear no clear relationship to the tissue layers of other metazoans. The extent to which their simple morphology is original and the extent to which it instead reflects extreme degeneration as an adaptation to parasitism is hotly debated. The 85 mesozoan species are now distributed between two distinct phyla: the Orthonectida and the Rhombozoa (most of which are contained in a single class, the Dicyemida).

Like the cestodes and trematodes, all mesozoans parasitize other animals (but exclusively marine invertebrates in this case), and like the body of turbellarian flatworms, the outer body surface is ciliated in at least some parts of the life cycle (Fig. 8.2), with each cell bearing two or more cilia. Like the cestodes, mesozoans lack any trace of a mouth or digestive system, and they are also without any specialized circulatory, respiratory, or nervous systems. In most other respects, however, their peculiar biology is unique. Probably the most remarkable aspect of their biology is that the reproductive cells and embryos develop *within* other cells; that is, they develop intracellularly. Such intracellular development is otherwise unknown among animals, and suggests that mesozoans represent a branch of multicellular evolution separate from that leading to, or from, any other metazoans.

Although orthonectids and rhombozoans were long united in a single phylum, the Mesozoa, the catalog of differences between them exceeds the short list of similarities, suggesting that they have no close ancestor in common. While orthonectids and rhombozoans may both represent degenerate parasitic flatworms, it seems more likely that they represent independent evolutionary lines. In consequence, they are now generally considered to represent separate phyla.

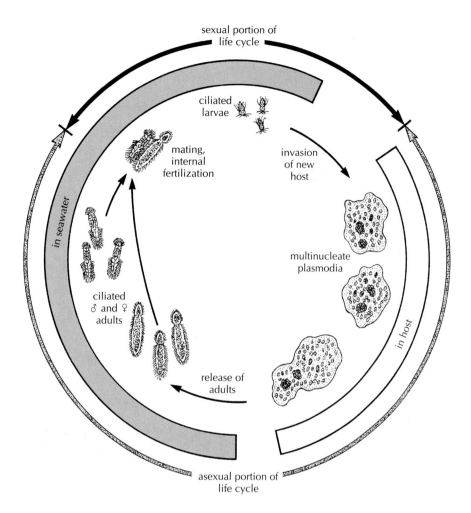

Figure 8.2
The orthonectid life cycle.

Phylum Orthonectida

The orthonectids parasitize a variety of marine invertebrates, including bivalved molluscs such as clams and oysters, polychaete annelids, echinoderms, and turbellarian flatworms. Individuals rarely exceed lengths of 300 μm, and most show dioecious sexuality, with a single individual possessing either male or female gonads but not both. Once in the host, juveniles take the form of asymmetrical, multinucleate but syncytial, amoeba-like individuals called **plasmodia** (Fig. 8.2). These live mostly in the host gonad, typically rendering the host impotent by castration. The plasmodia are juveniles in the sense that they are incapable of sexual reproduction, but they do not "mature" in the normal sense of the word. Rather, within itself, each plasmodium produces cellular, ciliated sexual forms, which might as well be called "adults." A single plasmodium typically produces both male and female adults. These small adults—rarely longer than about 150 μm—leave the plasmodium and then leave the invertebrate host, to swim freely in the sea. When two individuals of opposite sex meet, they juxtapose their genital pores, and the male releases sperm into the female opening, fertilizing her eggs. The eggs develop into two-layered, ciliated larval forms that leave through the female's genital opening and seek new hosts to infect. Upon entering the genital duct of a new invertebrate host, the orthonectid larva loses its outer layer of ciliated cells. The inner cells then move throughout the host gonad, and each becomes a plasmodium through mitotic division. Only about 20 orthonectid species have been described.

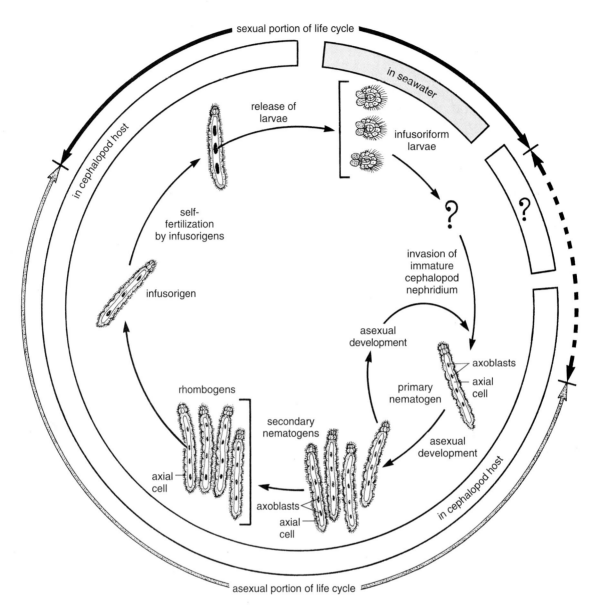

Figure 8.3
The rhombozoan life cycle.

Phylum Rhombozoa

The phylum Rhombozoa contains about 65 species. The rhombozoan life cycle shows little resemblance to that of orthonectids or to other metazoans. The associated terminology is correspondingly cumbersome, but unavoidably so.

Rhombozoans all parasitize the nephridia and associated structures of cephalopod molluscs, particularly the cuttlefish and octopuses. The first stage found in the cephalopod host is called a **stem nematogen.** Like adult orthonectids, rhombozoan nematogens are vermiform in shape. Some individuals are 7.5 mm long, but even the largest animals have bodies composed of no more than about 30 cells. The 20 to 30 externally ciliated cells of the

outer body layer surround one (or several) elongated **axial cell(s),** which forms a central shaft in the animal (Fig. 8.3). This central shaft is the animal's reproductive center, but it is not a gonad in the normal sense of the term; rather, reproduction is a function of other cells— up to about 100 in some species—that develop *within* this axial cell. Each intracellular **axoblast cell** (also called an **agamete**) develops into more asexually reproducing nematogens, which in turn develop more axoblasts that develop into more nematogens (Fig. 8.3). Once each nematogen attains a species-specific number of cells early in its development, all further growth results from increases in cell size rather than by mitotic replication.

This asexual production of nematogens continues through several, or many, generations of nematogens until the kidney of the cephalopod host is filled; sexual maturation of the host may also play a role in ending nematogen production. At that point, either existing nematogens develop into a sexual stage, called the **rhombogen,** or the last generation of nematogens produces rhombogens from their axoblast cells. The rhombogen is morphologically identical to the nematogen and has its own set of axoblast cells internally, within its axial cell(s) (Fig. 8.3). These axoblasts, not the rhombogen itself, differentiate into nonciliated, sexual individuals called **infusorigens.** These are essentially the adult rhombozoans, but they never emerge from the rhombogen that produced them. To some extent, the situation is analogous to a human becoming sexually mature within the mother's uterus. To improve the analogy, the "child" would have to have arisen asexually, as would have the mother. Moreover, infusorigens are simultaneous hermaphrodites, producing both sperm and eggs by meiosis within a single individual. Surprisingly, there is no cross-fertilization; gametes produced by one infusorigen fertilize only other gametes produced by that same individual. (When it comes to rhombozoans, it does *not* "take two to tango.") Self-fertilization is also common among cestodes, as discussed in Chapter 7.

Following fertilization, the zygotes differentiate to form ciliated **infusoriform larvae** that escape from the rhombogen and then from the cephalopod host, emerging into the sea in host urine and then sinking to the bottom. The larvae do not seem capable of infecting a new cephalopod host directly and may require an intermediate host, although nothing is known about this part of the life cycle. The nephridia of another juvenile cephalopod are eventually invaded by something—we don't know by what or from where—that becomes a nematogen and goes on to produce more nematogens asexually.

Rhombozoans cause little detriment to their hosts, despite their extensive asexual proliferation in the host's kidney. About 65 species have been described.

Taxonomic Summary

Phylum Gnathostomulida
The Mesozoans:
Phylum Orthonectida
Phylum Rhombozoa—dicyemids and others

Topic for Further Discussion and Investigation

Compare and contrast the life cycle of a rhombozoan with that of a digenean flatworm (Chapter 7).

Taxonomic Detail

Phylum Gnathostomulida
The 80 or so species in this phylum are distributed among only two orders, based largely on differences in reproductive anatomy and sperm morphology. All species are marine.

Order Filospermoidea.
Haplognathia. Two families.

Order Bursovaginoidea.
Agnathiella, Gnathostomula. This order contains about three-fourths of all gnathostomulid species. The reproductive anatomy is more complex for species in this order, and males have a stylet-bearing penis, which injects sperm through the body wall of females. Many members of this order can reverse the direction of ciliary beating, thereby swimming backward. Eight families, some of which contain a single species.

Phylum Orthonectida.
Rhopalura, Stoecharthrum. All species in this small class (about 20 species) are endoparasitic in various invertebrates. Two families.

Phylum Rhombozoa.
This group contains about 65 species, all of which parasitize the nephridial system of cephalopods. Two orders.

Class Dicyemida.
Dicyema, Pseudicyema. This order contains all but two of known rhombozoan species. One family.

Class Heterocyemida.
Conocyema, Microcyema. The nematogens lack cilia and have a syncytial outer cell layer. The life cycles for the two species in this order are incompletely described.

Some General References About the Gnathostomulids and Mesozoans

Cheng, T. C. 1986. *Parasitology,* 2d ed. New York: Academic Press.

Higgins, R. P., and H. Thiel, eds. 1988. *Introduction to the Study of Meiofauna.* Washington, D.C.: Smithsonian Institution Press.

Lammert, V. 1991. Gnathostomulida. Harrison, F. W., and E. E. Ruppert, eds. *Microscopic Anatomy of Invertebrates, Vol. 4. Aschelminthes.* New York: Wiley-Liss.

Lapan, E. A., and H. Morowitz. 1972. The Mesozoa. *Sci. Amer.* 227:94–101.

Morris, S. C., et al., eds. 1985. *The Origins and Relationships of Lower Invertebrates. Systematics Association, Special Vol. 28.* Oxford: Clarendon Press.

Parker, S. P., ed. 1982. *Classification and Synopsis of Living Organisms,* vol. 1, New York: McGraw-Hill, 880–929.

Schmidt, G. D., and L. S. Roberts. 1989. *Foundations of Parasitology,* 4th ed. St. Louis: C. V. Mosby and Company.

Stunkard, H. W. 1954. The life history and systematic relations of the Mesozoa. *Q. Rev. Biol.* 29:230–44.

9

Introduction to the Hydrostatic Skeleton

The word *skeleton* invariably conjures up an image of the articulated bones hanging in the corner of the high-school biology classroom, or perhaps in the corner of the general practitioner's office. However, jointed bones are only one form of skeletal system. Nearly all multicellular animals, even invertebrates, require a skeleton for movement. The only exceptions to this rule are those small, aquatic metazoans that may move exclusively by cilia. A functional definition of the word *skeleton* is

> A solid or fluid system permitting muscles to be stretched back to their original length following a contraction. Such a system may or may not have protective and supportive functions as well.

A skeletal system is essential simply because a muscle is capable of only two of the three activities required for repeated movements: A muscle can shorten or relax, but it cannot actively extend itself. To bend your arm at the elbow, one set of muscles, the biceps, must contract. This contraction of the biceps not only causes your arm to bend at the elbow, but also serves to stretch another muscle in your arm, the triceps (Fig. 9.1). The triceps can now contract, making it possible for you to reextend your arm. Reextension of your arm, in turn, serves to stretch the biceps. The bones in your arm have functioned in these movements as the vehicle through which the triceps and biceps take turns stretching each other back to precontraction length; that is, the muscles **antagonize** each other, making controlled, repeatable movement possible. In vertebrates, the mutual antagonism of muscles is mediated through a solid skeleton. A rigid skeletal system is essential in a terrestrial environment, in part because the skeleton must also serve to support the body in a nonsupportive medium (Chapter 1). Aquatic organisms are supported by the medium in which they live, so a rigid skeletal system is not required.

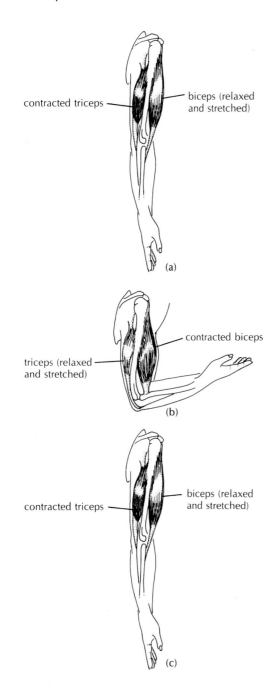

(a)

(b)

(c)

Figure 9.1
Antagonistic interaction between the biceps and triceps in the human arm. Contraction of the biceps (a) results not only in movement of the arm (b), but also in stretching of the opposing muscle, the triceps. Contraction of the triceps then returns the arm to its initial position (c) and stretches the biceps. In vertebrates, muscle pairs antagonize each other through a rigid skeleton, which is internal and jointed.

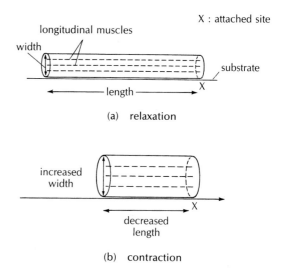

(a)   relaxation

(b)   contraction

Figure 9.2
(a,b) Shape changes possible in a worm-like organism equipped with only longitudinal muscles. Because the fluid volume is constant, a change in the width of the hypothetical animal must be compensated for by a change in length, brought about by the increase in internal hydrostatic pressure during muscle contraction. Similarly, a shortening of the worm is accompanied by an increase in width, as seen in (b).

Invertebrates commonly make use of a **hydrostatic skeleton,** a system in which fluid serves as the vehicle through which sets of muscles interact. A functional hydrostatic skeleton requires

1. the presence of a cavity housing an incompressible fluid that transmits pressure changes uniformly in all directions;
2. that this cavity be surrounded by a flexible outer body membrane, permitting deformations of the outer body wall to take place;
3. that the volume of fluid in the cavity remain constant; and
4. that the animal be capable of forming temporary attachments to the substrate, if progressive locomotion is to occur on or within a substrate.

Let us assume that these four attributes are met in the hypothetical organism shown in Figure 9.2. This cylindrical being is equipped with **longitudinal muscles** only. If this animal attaches at point X (as shown in Fig. 9.2a) and then contracts its musculature, the increase in internal hydrostatic pressure will deform the outer body wall, resulting in a shorter, fatter animal (Fig. 9.2b). This

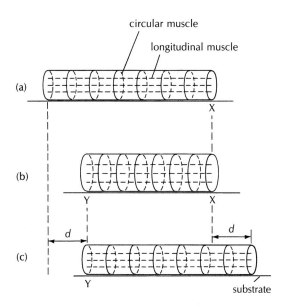

circular muscle

longitudinal muscle

(a)

(b)

(c)

d

d

X

Y

X

Y

substrate

Figure 9.3

Locomotion in a hypothetical worm possessing both circular and longitudinal muscles and a continuous, fluid-filled internal body cavity of constant volume. (a) The animal attaches to the substrate at point *X*, relaxes its circular muscles, and contracts its longitudinal muscles. The contraction increases the hydrostatic pressure within the body cavity because the volume of the cavity cannot be decreased and the fluid within the cavity is incompressible. This pressure is relieved by permitting the circular muscles to stretch. (b) The animal releases its anterior attachment, forms an attachment at point *Y*, and relaxes its longitudinal muscles. (c) The circular muscles have contracted, causing another increase in pressure, which, in turn, is relieved as the longitudinal muscles are extended. The animal has now advanced by distance *d*, and regained its initial body shape.

animal can regain its initial shape only if it is surrounded by a stiff, elastic covering that will spring back to its original shape upon relaxation of the longitudinal musculature. Such a stiff covering could be difficult to deform in the first place and is not commonly encountered among unjointed invertebrates.

Instead, we add a second set of muscles (**circular muscles**) to our hypothetical animal. Forward locomotion then results from the series of contractions illustrated in Figure 9.3. In (a), the circular muscles are contracted, and the longitudinal muscles are stretched. The longitudinal muscles now contract while the circular

muscles relax, producing the shorter, wider animal of Figure 9.3b. In (b), the animal releases its anterior attachment to the substrate and forms a new temporary attachment posteriorly. The circular muscles then contract while the longitudinal muscles relax, thrusting the animal's anterior end forward. In Figure 9.3c, we see that the animal has advanced by a distance *d* and has regained its initial shape, ready to repeat the cycle of muscular contractions.

From this discussion, it is obvious that one addition must be made to the previous list of requirements for a functional hydrostatic skeleton:

5. the presence of a deformable but elastic covering or the presence of at least two pairs of musculature oriented in different directions.

Clearly, the skeletal system in our hypothetical organism is a fluid. Temporary increases in internal pressure are caused by the contraction of one set of muscles, and this temporary pressure increase elongates another set of muscles. I emphasize that the internal pressure increase is temporary; elongation of the opposing set of muscles relieves the pressure. The fluid thus makes possible the mutual antagonism of the two sets of muscles, resulting in repeatable locomotory movements. This simple, hydrostatic skeleton plays some role in the movements made by representatives of nearly every invertebrate phylum.

Topics for Further Discussion and Investigation

1. Which features of a hydrostatic skeleton do sponges (Chapter 4) possess? Which features are absent from sponges?

2. The parenchyma of turbellarians (Chapter 7) functions as a hydrostatic skeleton. What sequence of muscle contractions and relaxations is likely to be involved in mediating (*a*) locomotion via pedal waves and (*b*) locomotion via looping?

10

The Nemertines

Introduction and General Characteristics

Phylum Rhyncho • coela (rink´-ō-sēl´-ah)
(G: snout cavity)
(= phylum Nemertea; nem-er´-tē-ah)
(= phylum Nemertinea; nem-er-tē´-nē-ah)

The nemertines (also called "nemerteans") are a small group of elongated, unsegmented, soft-bodied worms. Most of the approximately 900 described species in this phylum are marine. Nemertines are common inhabitants of marine shallow-water environments, crawling over solid substrates, burrowing into sediment, or lurking under stones, rocks, or mats of algae. Some other species (mostly in a single genus) are found in freshwater, and the members of several more small genera are terrestrial. The terrestrial species are most common in tropical, moist environments. The freshwater and terrestrial species most likely descended from marine ancestors. Many nemertine species live commensally with invertebrates from other phyla, particularly the Arthropoda and Mollusca, but only a few parasitic nemertine species are known.

In several respects, the nemertines superficially resemble the free-living flatworms (phylum Platyhelminthes, class Turbellaria). Like the flatworms, nemertines are ciliated externally and secrete a mucus through which they progress. Small nemertines can move over substrates exclusively by means of ciliary beating. Larger individuals often generate waves of muscle contraction for locomotion over hard substrates or through soft substrates. A few species can swim, by generating relatively violent waves of muscular contraction.

As with the turbellarians, most nemertines are flattened dorsoventrally and possess circular, longitudinal, and dorsoventral muscles. In at least some nemertine species, the area lying between the outer body wall and the

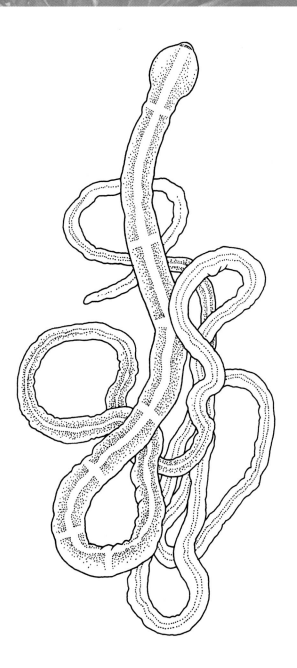

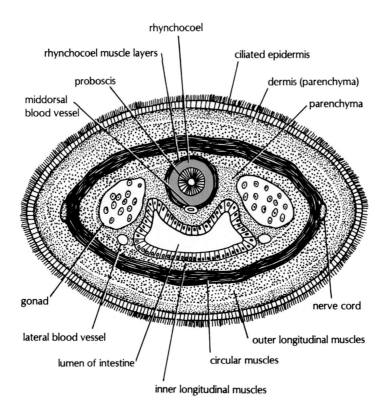

Figure 10.1
Diagrammatic cross section through the body of a nemertine worm. The functions of the proboscis and rhynchocoel are discussed later in the chapter.   *After Harmer and Shipley.*

gut is solid to loosely packed, mesodermally derived tissue referred to as **parenchyma** (Fig. 10.1). The nervous system of nemertines is also reminiscent of the turbellarian plan: The cerebral ganglia form a ring anteriorly and give rise to a ladder-like arrangement of longitudinal nerves with lateral connectives. Nemertines are equipped with chemoreceptors and mechanoreceptors, located in specialized pits and grooves on the body surface, and sensory bristles. Most species also possess pigmented photoreceptors ("eyes"), and a few species possess balance organs (**statocysts**). The number of eyes varies widely among species, from zero to more than 80 per animal. As in the flatworms, nemertines generally have protonephridial excretory systems, and digestion is, as in the Platyhelminthes, largely intracellular. The body of most species is of considerable length, typically from several to 20 cm but in some cases as long as several meters. "Ribbon worm," the common name for members of this phylum, is especially appropriate for such elongated nemertines.

Nemertines differ from the flatworms in a number of respects, important enough to warrant the placement of these two groups in separate phyla. Indeed, recent molecular evidence argues persuasively that flatworms and nemertines have no recent ancestor in common and that nemertines are in fact far more closely related to certain coelomates than to any flatworms (Research Focus Box 10.1). Flatworms and nemertines differ conspicuously with respect to their systems for gas exchange, food capture, and digestion.

Because the bodies of nemertines are flat and permeable, a fair amount of gas exchange likely occurs by diffusion across the general body surface. In addition, and in contrast to the platyhelminths, nemertines possess a true circulatory system. The blood circulates throughout the body through well-defined vessels, many of which are contractile (Fig. 10.2). Recent ultrastructural studies suggest that these vessels are highly modified coelomic spaces.[1] There is no true heart, and the blood vessels lack one-way valves, so the blood does not circulate unidirectionally. Instead, it ebbs and flows erratically, propelled largely by muscular contractions associated with routine movements of the animal. A few species have hemoglobin in their blood, but most species lack any such oxygen-carrying blood pigment.

Unlike flatworms, ribbon worms have a one-way digestive tract, with a mouth anteriorly and a separate anus

---

1. See *Topics for Further Discussion and Investigation*, no. 1, at the end of the chapter.

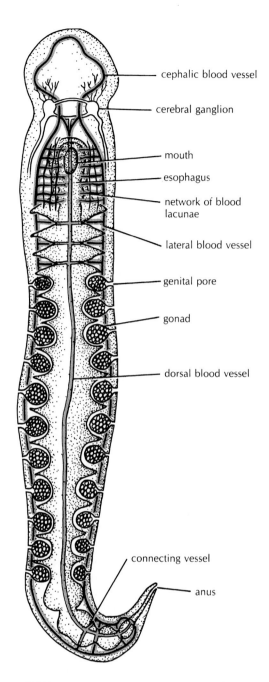

Figure 10.2
Schematic illustration of *Cerebratulus,* a common genus of ribbon worm. Note the well-developed circulatory system. *After Bayer and Owre; after Joubin.*

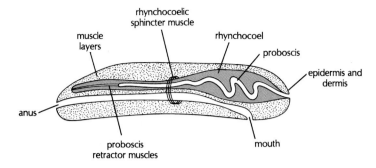

Figure 10.3
Diagrammatic longitudinal section through a nemertine, showing the tubular gut. *Modified from Turberville and Ruppert, 1983, Zoomorphol. 103:103.*

posteriorly (Fig. 10.3). The unidirectional flow of food material through this tubular gut makes an orderly digestive process possible. Digestive enzymes may be secreted in sequence as the food is moved through the gut by ciliary activity, and the animal can continue to eat without disrupting the digestive process. As in the Turbellaria, most nemertines are meat eaters. Their food

preferences are often highly specific. Nemertines seem to have a predilection for small annelids and crustaceans, and a given nemertine species often shows a preference for a particular species of prey.[2]

Also unlike flatworms, the ribbon worms possess a hollow, muscular **proboscis.** This structure is not homologous with the turbellarian pharynx, which is essentially a protrusible extension of the gut. In contrast, the nemertine proboscis is not directly associated with the digestive tract (Figs. 10.3 and 10.4b,c). Rather, the proboscis floats in a separate, fluid-filled, tubular cavity called the **rhynchocoel** (Fig. 10.4a). The rhynchocoel forms during development as a split in the mesodermal tissue, and in this sense it is coelomic.

The proboscis is generally everted through a distinct **proboscis pore,** although in some species it is discharged through the mouth opening. The proboscis can be shot out explosively from the animal, and in one major group of species, it is armed with a piercing **stylet** (see Figs. 10.4a, 10.5, and 10.6d). A potent paralytic toxin found in the proboscis is often discharged into wounds made by the stylet; nemertines are carnivores to be reckoned with, even for rapidly moving prey.

It is difficult to imagine a pharynx being shot out of the mouth of a flatworm. The turbellarian pharynx is protruded when the animal contracts the circular musculature of the tube and relaxes the longitudinal musculature, but this is not a rapid process. What, then, is the functional difference between the operation of the flatworm pharynx and that of the nemertine proboscis that permits the nemertine proboscis to be extended with such rapidity?

The secret of proboscis functioning lies in the fluid-filled rhynchocoel. Although the nemertine proboscis is highly muscular, it is discharged not by its own muscular activity, but by the musculature of the tissue surrounding the rhynchocoel. When this musculature contracts, the pressure within the rhynchocoel rises. Even a small

---

2. See *Topics for Further Discussion and Investigation,* no. 2.

Research Focus Box 10.1

Assessing Phylogenetic Relationships

Turbeville, J. M., K. G. Field, and R. A. Raff. 1992. Phylogenetic position of phylum Nemertini, inferred from 18s rRNA sequences: Molecular data as a test of morphological character homology. *Molec. Biol. Evol.* 9:235–49.

For many decades, the nemertine worms have been closely allied with the platyhelminths as triploblastic acoelomates, both derived from a common ancestor. Two key characteristics that appear only among nemertines—a fluid-filled rhynchocoel and a blood circulatory system—were believed to have evolved in nemertines after they diverged from their flatworm ancestors. Studies made by the senior author in the 1980s, however, confirmed suggestions made by German scientists many years earlier: (a) the rhynchocoel forms by a splitting of mesodermal tissue, as in the formation of coelomic cavities, and (b) the "blood vessels" have a continuous epithelial lining, musculature, and several other characteristics suggesting they are modified coelomic spaces.

But many biologists were not convinced that nemertines were true coelomates. At issue is the definition of coelom: Is it any cavity arising as a split in mesodermal tissue? Or must it specifically fill the space between the gut and body wall musculature? If we accept the latter, more restrictive definition, the nemertine rhynchocoel and blood vessels could be analogues, but not homologues, of the cavities found in annelids and other noncontroversial coelomates, with the structural resemblances among the cavities resulting from convergent evolution.

Without the rapid advances in molecular technology made during the 1980s there would be no way to resolve the controversy. In this paper, James Turbeville and colleagues tested the hypothesis that nemertines are indeed closet coelomates, by comparing certain slow-evolving nucleotide sequences of 18s ribosomal RNA (rRNA) taken from the nemertine worm *Cerebratulus lacteus* with similar sequences documented previously from nine other animals representing six other phyla. Three of the animals were flatworms (including one parasitic fluke species), two were annelids (including the common earthworm), two were other protostomous coelomates (one chiton, one peanut worm—a sipunculan, see p. 322), and the last two were deuterostomous coelomates: a common sea star and a cephalochordate (a member of our own phylum, the Chordata, p. 497).

The researchers extracted all of the RNA from unfertilized nemertine eggs and then used six oligonucleotide primers to bind six particular regions of 18s rRNA. Using the enzyme reverse transcriptase, they then made DNA fragments complementary to the six primed regions of rRNA. The linear sequence of over 1,000 nucleotide bases obtained from the six complementary DNA fragments (cDNA) was then compared, base by base, with those from the other nine organisms. A greater degree of difference between sequences implies a longer period of independent evolution between the groups compared; the greater the similarities, the closer the evolutionary relationship between the groups being compared.

But how does one decide how close is close enough to implicate a closely shared evolutionary origin? How much difference does it take to establish that two groups are *not* closely related? And when differences in sequence patterns are detected, how does one tell which is the ancestral (plesiomorphic) situation and which is the more advanced (apomorphic) situation? In other words, how does one determine which groups are the ancestors and which are the descendants of other groups?

The answers to these questions are still controversial. To construct phylogenetic trees from their data, Turbeville et al. (1992) used three different computer techniques, including two techniques that provide a statistical indication of how likely it is that any given conclusion (i.e., phylogenetic tree) is in fact the correct conclusion for that particular data set. The analyses require comparison to comparable data from what is called an "outgroup," a phylum believed to be the sister group to all the other phyla being compared; this sister group is believed to share a common ancestor with all of the other phyla being studied. The outgroup provides a means of assessing the direction of evolutionary change in the gene sequences, with outgroup characteristics

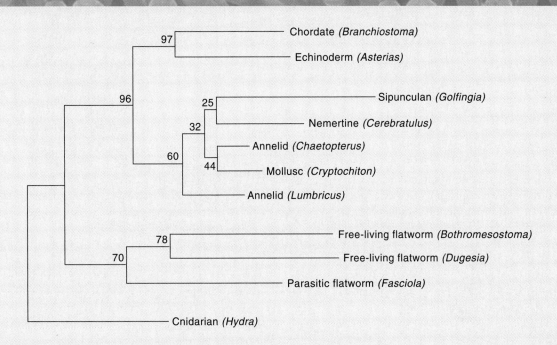

Focus Figure 10.1

One representative result of 18S rRNA nucleotide base sequence analysis. This particular analysis produces the phylogenetic tree requiring the smallest number of nucleotide base substitutions (874 to produce this particular tree). The number on each branch point was obtained by a statistical technique termed "bootstrapping," representing the degree of confidence in that particular branching pattern on a scale of 0–100.

believed to be ancestral (plesiomorphic). For this study, the researchers chose a cnidarian (*Hydra*) for the outgroup comparison.

The results were very clear and very similar for all three analyses. The phylogenetic tree resulting from one of the three analyses is shown in Focus Figure 10.1. It shows the three flatworm species clustering as a single cohesive acoelomate unit, the two deuterostome species clustering as a separate cohesive unit, and the nemertine species nestling comfortably with the protostomous coelomates. There is considerable uncertainty about exactly who is *most* closely related to the nemertine star of the show, but the grouping of nemertines with the protostomous coelomates is all but certain. Do these results prove that nemertines are in fact coelomates? No, they do not, unless the traditional definition of "coelomic cavity" is altered. But the available data

argue clearly against the nemertines having descended directly from any flatworm ancestors, and they place the nemertines clearly on the path of coelomate evolution.

How much do you think the results might differ if more representatives of each group, including more nemertine species, were included in the study? How much do you think the results might differ if different RNA sequences were examined? Based on the ideas discussed in Chapter 2, what other factors can you think of that might alter the outcomes of these sorts of analyses?

Figure from J. M. Turbeville, et al., "Phylogenetic position of phylum Nemertini, inferred from 18s rRNA sequences: Molecular data as a test of morphological character homology", in Molecular Biology and Evolution, *9:235–49, 1992. Copyright © 1992 University of Chicago Press, Chicago, Illinois. Reprinted by permission.*

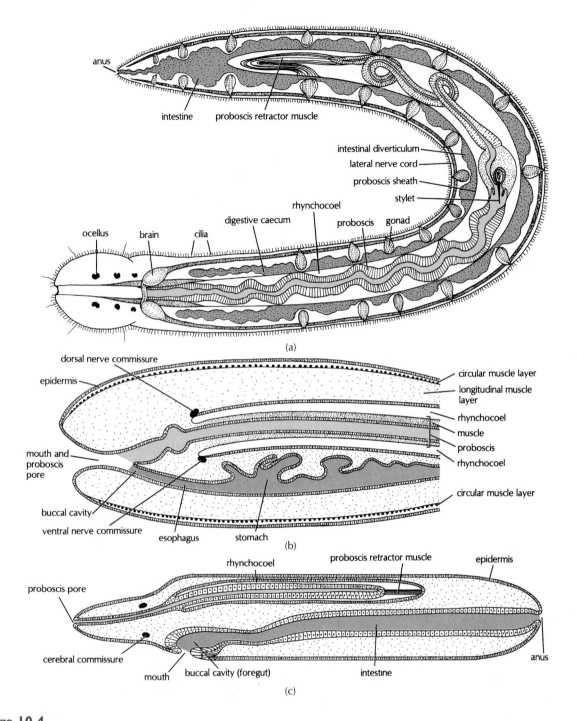

Figure 10.4

(a) Diagrammatic illustration of *Prostoma graecense*, a freshwater nemertine species. The intestine connects with various bulges (diverticulae) and sacs (caecae). (b) Diagrammatic illustration of *Nemertopsis* sp., anterior end in longitudinal section. Note that this species has a single opening for both the proboscis and the mouth. The nerve commissures connect to form a ring around the anterior portion of the rhynchocoel. (c) In *Cephalothrix bioculata*, the proboscis is ejected through an opening separate from the mouth. (a) *From Pennak, Fresh-Water Invertebrates of the United States, 2d ed. Copyright © 1978 John Wiley & Sons, New York. Reprinted by permission of John Wiley & Sons, Inc. (b) After Bayre and Owre; after Burger. (c) From J.B. Jennings and Ray Gibson, 1969. Biological Bulletin, 136:405. Reprinted by permission.*

amount of contraction creates a substantial increase in pressure, since the rhynchocoel has a constant volume and the fluid within it is essentially incompressible; the fluid can neither escape nor be compressed. By default, hydro-static pressure increases. The elevated pressure is relieved by relaxing the sphincter muscles surrounding the anterior end of the rhynchocoel, allowing the proboscis to shoot out of the cavity with great speed and force (Fig. 10.5).

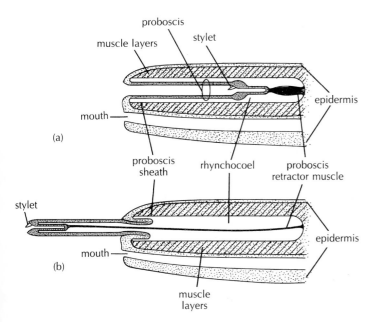

(a)

(b)

Figure 10.5

Diagrammatic illustration of the proboscis in (a) retracted and (b) extended positions. The digestive system has been omitted for clarity. (a, b) *After Harmer and Shipley; after Alexander.*

The workings of a proboscis can be visualized by picturing a rubber glove with one of the fingers pushed inside. If the wrist of the glove is tied off so that the air cannot escape, the glove becomes a constant-volume, fluid-filled cavity—a rhynchocoel, of sorts. The inverted finger is the proboscis in this rhynchocoel. If the glove is squeezed, the "proboscis" will be everted, turning inside out as it goes. If a hole is poked in the glove and the experiment repeated, the importance of the rhynchocoel's constant volume will be evident.

For those nemertine species possessing a stylet on the proboscis, prey may be harpooned directly. More commonly, the everted proboscis is first wound tightly around the prey, and the struggling animal is then stabbed repeatedly with the barb (Fig. 10.6). Species lacking a stylet capture prey by simply coiling the prehensile proboscis around the victim. A sticky mucus is generally secreted by the proboscis to help hold the prey.[3]

A **proboscis retractor muscle** runs from the tip of the everted proboscis to the inner wall of the rhynchocoel (Figs. 10.4a,c and 10.5). Contraction of this muscle draws the proboscis back inside the rhynchocoel.

3. See *Topics for Further Discussion and Investigation,* no. 2.

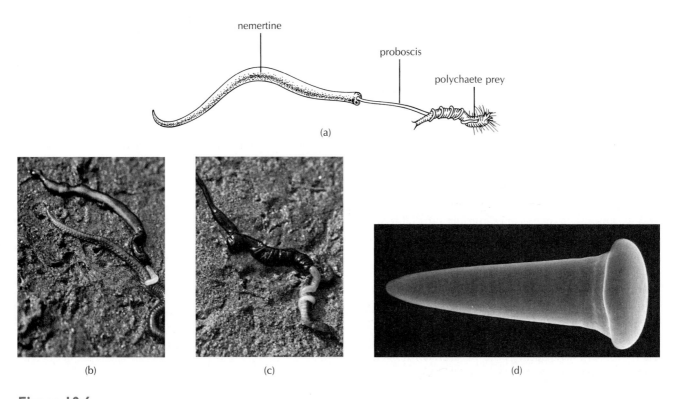

(a)

(b)

(c)

(d)

Figure 10.6

(a) Diagrammatic illustration of a nemertine (*Paranemertes peregrina*) subduing its prey. (b,c) *P. peregrina* attacking a polychaete worm (*Platynereis bicanaliculata*). (d) Scanning electron micrograph of the stylet of *Zygonemertes virescens;* the stylet is approximately 175 μm long. (a) *After MacGinitie and MacGinitie.* (b, c) *Courtesy of S.A. Stricker. From Stricker and Cloney. 1983. J. Morphol. 177:89.* (d) *Courtesy of S.A. Stricker. From Stricker, 1983. J. Morphol. 175:153.*

For those species lacking stylets, captured food is usually transferred to the mouth as the proboscis is retracted. Nemertines with stylets first paralyze the prey and then retract the proboscis, completely losing contact with the stricken animal. The nemertine then moves to the prey to eat. In some species, prey are ingested whole, particularly if they are **vermiform** (worm shaped). In other species, the prey are typically too large or awkwardly shaped to be engulfed; in such cases, the nemertine inserts its foregut into the prey and, by peristaltic waves of muscular activity, pumps the juices of the victim into the gut of the victor.

The nemertine proboscis does more than make the capture of fast-moving prey possible. It also enables the animal to burrow into soft substrates, something the flatworms cannot do. Contraction of the body wall musculature elevates the internal pressure of the rhynchocoel. The proboscis is then everted into the sediment and allowed to widen at its tip, through a relaxation of the associated circular muscles. This distal dilation anchors the end of the proboscis in the sediment. The body of the worm can then be pulled downward, into the sediment, upon contraction of the proboscis retractor muscles and the longitudinal muscles of the body wall. Contracting the longitudinal muscles also forces additional fluid into the proboscis, under considerable pressure, causing additional dilation and anchoring. We shall encounter similar burrowing mechanisms in several other animal phyla, phyla in which the coelom or blood sinuses subserve the same function as the rhynchocoel of nemertines.

Here, then, we see the principles of the hydrostatic skeleton (Chapter 9) in action. Advances in locomotion and food capture are achieved by exploiting the properties of a constant-volume, fluid-filled cavity. Such cavities have been independently evolved at least four different times in metazoans: by enterocoely, by schizocoely, by the persistence of a blastocoel, and by the formation of a rhynchocoel. The selection pressures supporting the evolution of secondary body cavities obviously have been strong. Among the nemertines, the adaptive value would clearly seem to lie in mechanical advantages gained and in new lifestyles thereby made possible.

Other Features of Nemertine Biology

Classification

Nemertines are divided into two major classes: the Anopla and the Enopla. Members of the Anopla lack stylets; that is, the proboscis is unarmed. In addition, the mouth of anoplan nemertines is posterior to the brain. In contrast, the proboscis may be armed with a stylet among the Enopla—with all of the armed species belonging to a single order (the Hoplonemertea)—and the mouth is anterior to the brain.

Reproduction and Development

Most nemertine species have separate sexes; that is, they are **dioecious.** The few hermaphroditic species that have been described are **protandric;** that is, a single individual may be both male and female but not simultaneously (as it ages, each male becomes a female). Following fertilization, cleavage is usually spiral and determinate, as in the Platyhelminthes. In some groups of nemertines, a distinctive, microscopic pilidium larva is formed (Fig. 10.7a,b). This is a ciliated, swimming, feeding individual that resembles a football helmet with earflaps. Only a small portion of the pilidium larva develops into the juvenile ribbon worm, which eventually abandons the pilidium. What remains of the larva apparently swims off in some species and eventually starves to death, since the juvenile takes the mouth with it when it leaves (Fig. 10.7c). Alternatively, the worm may ingest the larval tissues following metamorphosis.

Many nemertines also reproduce asexually, by fragmenting into two or more pieces; each piece then regenerates any missing parts.

Taxonomic Summary

Phylum Rhynchocoela (= Nemertea)—the ribbon worms
    Class Anopla—the unarmed nemertines (proboscis lacks a stylet)
    Class Enopla—includes the armed nemertines (proboscis may be equipped with a stylet)

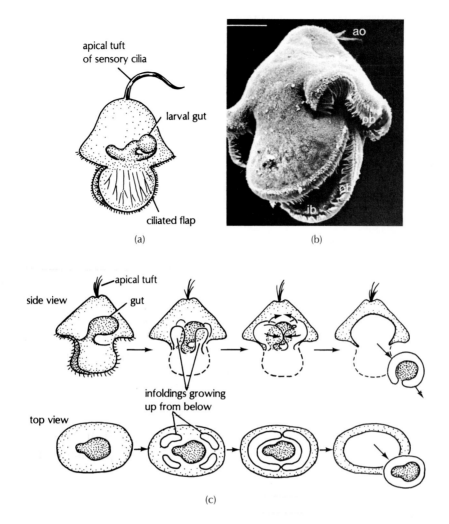

apical tuft
of sensory cilia

larval gut

ciliated flap

(a)

ao

(b)

side view   apical tuft

gut

infoldings growing
up from below

top view

(c)

Figure 10.7

(a) The pilidium larva of a nemertine worm. (b) Scanning electron micrograph of an unidentified pilidium larva, about 400 μm long. ("ib" and "ob" = inner and outer ciliary bands, respectively; "ao" = apical tuft.) (c) Stages in the metamorphosis of the pilidium larva. Infoldings form around the mouth, growing upward and toward each other. When these infoldings completely surround the gut and fuse together, the central mass detaches from the rest of the larva to take up residence on the sea bottom as a small nemertine. The other portion of the larva continues swimming for a time but in the absence of a digestive system eventually starves to death.   (a) *After Harmer and Shipley.* (b) *Courtesy of Claus Nielsen. From Nielsen. 1987. Acta Zoologica (Stockholm) 68:205–62. Permission of Royal Swedish Academy of Sciences.* (c) *From Hardy, The Open Sea: Its Natural History. Copyright © 1965 Houghton Mifflin Company, Boston, Massachusetts.*

Topics for Further Discussion and Investigation

1.   Recent morphological studies indicate that the nemertine rhynchocoel and blood circulatory system arise during embryonic development from splits in mesodermal tissue. What is the evidence on which this suggestion is based, and to what extent does it argue a coelomate ancestry for nemertines?

Turbeville, J. M. 1986. An ultrastructural analysis of coelomogenesis in the hoplonemertine *Prosorhochmus americanus* and the polychaete *Magelona* sp. *J. Morphol.* 187:51.

Turbeville, J. M., and E. E. Ruppert. 1985. Comparative ultrastructure and the evolution of nemertines. *Amer. Zool.* 25:53.

Willmer, P. 1990. *Invertebrate relationships.* New York: Cambridge University Press, 204–6.

2.   Compare and contrast the different methods of feeding encountered among the Rhynchocoela.

Fisher, F. M., Jr., and J. A. Oaks. 1978. Evidence for a nonintestinal nutritional mechanism in the rhynchocoelan, *Lineus ruber. Biol. Bull.* 154:213.

Jennings, J. B., and R. Gibson. 1969. Observations on the nutrition of seven species of rhynchocoelan worms. *Biol. Bull.* 136:405.

McDermott, J. J. 1976. Observations on the food and feeding behavior of estuarine nemertean worms belonging to the order Hoplonemertea. *Biol. Bull.* 150:157.

Roe, P. 1976. Life history and predator-prey interactions of the nemertean *Paranemertes peregrina* Coe. *Biol. Bull.* 150:80.

Stricker, S. A., and R. A. Cloney. 1981. The stylet apparatus of the nemertean *Paranemertes peregrina*: Its ultrastructure and role in prey capture. *Zoomorphology* 97:205.

3. Discuss the limitations placed on shape changes in nemertines imposed by the structure of the body wall.

Clark, R. B., and J. B. Cowey. 1958. Factors controlling the change of shape of certain nemertean and turbellarian worms. *J. Exp. Biol.* 35:731.

Turbeville, J. M., and E. E. Ruppert. 1983. Epidermal muscles and peristaltic burrowing in *Carinoma tremaphoros* (Nemertini): Correlates of effective burrowing without segmentation. *Zoomorphology* 103:103.

4. How is the functioning of a nemertine proboscis like the Wild West?

Taxonomic Detail

Phylum Rhynchocoela (= Nemertea)

The approximately 900 species are distributed among two classes.

Class Anopla.

In these nemertines, the mouth and proboscis openings are separate, and the proboscis is usually unarmed. Two orders, nine families.

Order Palaeonemertea (= Palaeonemertini).

All species are marine bottom-dwellers, living in sand or mud. None have free-living larvae. Four families.

Family Tubulanidae. *Tubulanus.* These nemertines may be striped or banded, and they are often brightly colored. The largest individuals are about 2 m long. Most species are restricted to shallow water, although some live as deep as 2,500 m.

Order Heteronemertea.

Most species are marine, although some are found in brackish or freshwaters. Many of the marine species have a free-swimming larval stage in the life history. Five families.

Family Baseodiscidae. *Baseodiscus.* Individuals are up to 2 m long and brightly colored, with conspicuous stripes or bands. Members are widely distributed in shallow waters of the Atlantic, Pacific, and Indian oceans and the Mediterranean Sea.

Family Lineidae. *Cerebratulus, Micrura, Lineus.* Most species are marine and free-living, although a few occur only in freshwater, and one (*Uchidana parasita*) lives only in the mantle cavity of marine bivalve molluscs. Members of the genus *Cerebratulus* can swim short distances by undulating. The family includes the largest known nemertines (*Lineus longissimus*), which can exceed 30 m in length.

Class Enopla.

Two orders, 29 families. The proboscis of these nemertines typically exits through the mouth, rather than through a separate opening, and it is armed with one or more stylets in many species. All terrestrial nemertine species reside in this class.

Order Hoplonemertea (= Hoplonemertini).

Each member of this order possesses an armed, stylet-bearing proboscis. Studies of nervous and sensory systems have convinced some biologists that hoplonemertines gave rise to the vertebrates. Two suborders, 28 families.

Suborder Monostilifera.

All species bear a proboscis armed with a single, central, piercing stylet and up to 12 accessory pouches housing replacement stylets. Members of this suborder occur in both marine and terrestrial habitats, and many feed suctorially on various arthropods, pumping out the body contents of their prey. Seven provisional families. The systematics of this group are particularly uncertain at present.

Family Carcinonemertidae. *Carcinonemertes.* The proboscis is not eversible. All species live on decapod crustaceans (crabs and spiny lobsters) as parasites or ectosymbionts. Members of the genus *Carcinonemertes* can apparently reach adulthood only on female hosts. Individuals normally live on the gill of the host crab, but they migrate to the host's abdomen to feed on developing embryos when the host is reproductively active. Members of at least some species can produce haploid larvae by parthenogenesis, although it is not yet clear if these larvae are viable or if they contribute to future population growth.

Family Cratenemertidae. *Nipponnemertes.* The members of some species in this family

can swim, and individuals in at least one species are completely planktonic, spending their entire lives in the water.

Family Ototyphlonemertidae.
Ototyphlonemertes. All individuals are small and live in the spaces between sand grains in marine shallow-water and intertidal habitats.

Family Prosorhochmidae. *Geonemertes, Gononemertes, Prosorhochmus.*
Most species are marine and free-living, although members of the genus *Gononemertes* live only in association with sea squirts (ascidian urochordates), and members of several other genera live exclusively in terrestrial habitats (e.g., *Geonemertes:* Indo-Pacific and West Indies; *Leptonemertes:* European greenhouses, Azores and other N. Atlantic islands; *Argonemertes:* Australian species with up to 120 eyes!).

Family Tetrastemmatidae. *Tetrastemma, Prostoma.*
Most members live in marine or brackish water, but some species (in the genus *Prostoma*) live only in freshwater. Species in this family are among the smallest nemertines, some attaining adult lengths of only 1 mm to 3 mm. A number of species live interstitially, in the spaces between sand grains.

Suborder Polystilifera.
These nemertines are all marine and exhibit a proboscis armed with many small stylets. Twenty families.

Tribe Reptantia.
All species are bottom-dwellers, some living at depths of several hundred meters.

Tribe Pelagica. *Nectonemertes, Pelagonemertes, Planktonemertes.*
All species live free in the water, either swimming or floating, at depths of hundreds or even thousands of meters. About 10 families.

Order Bdellonemertea.
Malacobdella. These strange nemertines are leech-like, with a posterior ventral sucker. All are commensal, living in the mantle cavities of certain marine bivalves and feeding on small food particles in suspension. The proboscis is unarmed. One family.

Some General References About the Nemertines

Harrison, F. W., and B. J. Bogitsh, eds. 1991. *Microscopic Anatomy of Invertebrates, Vol. 3: Platyhelminthes and Nemertinea.* New York: Wiley-Liss, 285–328.

Hyman, L. H. 1951. *The Invertebrates, Vol. 2. Platyhelminthes and Rhynchocoela: The Acoelomate Bilateria.* New York: McGraw-Hill.

Parker, S. P., ed. 1982. *Classification and Synopsis of Living Organisms,* vol. 1. New York: McGraw-Hill, 823–46.

Roe, P., and J. L. Norenburg, eds. 1985. Comparative biology of nemertines. *Amer. Zool.* 25:3–151.

Thorpe, J. H., and A. P. Covich. 1991. *Ecology and Classification of North American Freshwater Invertebrates.* New York: Academic Press.

II

The Nematodes

Introduction and General Characteristics

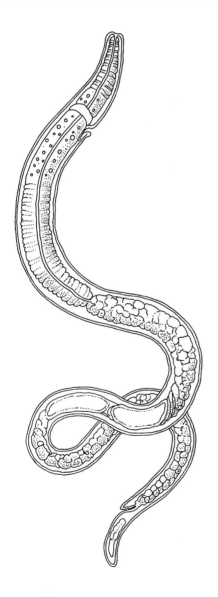

Phylum Nematoda
(G: thread)
nem-ah-tō´-dah

The nematodes are ubiquitous, unsegmented, pseudocoelomate worms. They are probably the most abundant multicellular animals alive today; nematode concentrations of one million individuals per square meter are typically encountered in shallow-water sediments in both fresh- and salt water, and concentrations exceeding four million per square meter have been reported from some marine habitats. Other nematode species are found free-living on land, and still others are major parasites of vertebrates, invertebrates, and plants. Fewer than 20,000 species have so far been described, but authorities estimate that 4 to 50 times this number of nematodes (mostly free-living species) may be awaiting description. The difficulty in describing and recognizing nematode species lies in the generally small size of these animals and in the extreme uniformity of both internal and external anatomy and morphology. Often, species determinations must be based upon biochemical attributes or morphological details not readily visible, such as the size and placement of microscopic sensory structures or the morphological details of the supposed excretory system.

In the literature, the Nematoda is sometimes listed as a class within another phylum, the Aschelminthes ("cavity worms," in reference to their possession of a pseudocoel). The major groups contained within the Aschelminthes (Nematoda, Rotifera, Gastrotricha, Kinorhyncha, Nematomorpha, Acanthocephala, and Gnathostomulida) are now generally considered more distantly related to each other than was previously thought, so each class has been elevated to phylum status.

Figure 11.1
The free-living nematode *Caenorhabditis elegans*. Two individuals are shown mating. *Courtesy of Anne M. Villeneuve.*

The relationships between the various aschelminth phyla, however, are still far from clear.[1] Here, as in most other textbooks, the term "aschelminth" refers to most of the pseudocoelomate invertebrates as a group, but the term has uncertain taxonomic validity.

Body Coverings and Body Cavities

The typical nematode is 1 mm to 2 mm long; shows no external segmentation; is pointed at both ends; and is covered by a thick, multilayered **cuticle** (a noncellular covering) of collagen secreted by the underlying epidermis (Fig. 11.1).

As pseudocoelomates, nematodes have an internal body cavity lying between the outer body wall musculature and the gut, and the cavity is not lined with mesodermally derived tissue; the cavity is derived directly from the embryonic blastocoel (pp. 9–10). The body organs, therefore, are not enveloped by **peritoneum** (the mesodermal lining of a coelomate body cavity), but instead lie freely in the pseudocoel. Pseudocoel fluid serves as the circulatory medium, and it sometimes contains hemoglobin; nematodes lack a closed circulatory system of discrete blood vessels.

The epidermis of nematodes is often **syncytial;** that is, nuclei are not separated from each other by complete cell membranes. The cuticle is composed of a highly complex network of fibers that are virtually inelastic (Fig. 11.2a). The trellis-like arrangement of these fibers permits bending, stretching, and shortening of the cuticle (Fig. 11.2b). The cuticle is permeable to water and to gases, so gas exchange can take place across the entire body surface. On the other hand, since the cuticle offers little protection against dehydration, all free-living nematodes must live in water or at least in a film of water. The cuticle is also selectively permeable to certain ions and organic compounds, regulating the movement of these substances between the internal and external environment.

The cuticle is shed (molted) and resecreted four times during development from the juvenile to the reproductively mature adult. Within a species, the cuticle of each developmental stage may be structurally and biochemically distinct, providing one of several nematode models for studies of gene expression and its control during development. Unlike the situation in arthropods (Chapter 18), nematodes continue to increase in size between molts and even after the final molt. But unlike most other animals, nematodes grow mainly by increasing the size of individual cells rather than by increasing the number of cells; this phenomenon is referred to as eutelic growth, or **eutely.**

In some parasitic nematode species, the first two juvenile stages are free-living. In many of these parasitic species, the animal may become enclosed within two envelopes before completing the second molt. The outer envelope is termed the **sheath.** Escape from the sheath (**exsheathment**), the subsequent two molts, and development to adulthood often do not occur until the encapsulated form is eaten by a suitable host.[2] Once in the host, much of the cuticle may be lost and replaced by a microvillar surface, presumably facilitating uptake of soluble nutrients from the host.

Musculature, Internal Pressure, and Locomotion

The body wall of nematodes contains no circular muscles. This is most unusual for animals of **vermiform** (worm-like) shape, and it places great limitations on their locomotory potential in that, for example, peristaltic waves of contraction cannot be generated.

Another obstacle to graceful locomotion is that the body is quite turgid in most species, due to substantial hydrostatic pressure within the pseudocoel (Fig. 11.3a). Internal hydrostatic pressures as high as 225 mm Hg (millimeters of mercury) have been measured inside some nematode species. (For comparison, atmospheric pressure is approximately 760 mm Hg.) The internal pressure within nematodes averages 70 mm Hg to 100 mm Hg, perhaps 10 times higher than pressures reported for most other invertebrates. At least two factors contribute to generating and maintaining such high internal pressures. First, the cuticle cannot expand to relieve pressure. Second, the musculature is always in a partially contracted state, trying to compress an incompressible fluid. The high internal pressure gives the nematode a very circular cross section. Thus, the common name for a member of this phylum is "roundworm."

1. R. M. Kristensen. 1983. Loricifera, a new phylum with Aschelminthes characters from the meiobenthos. *Z. Zool. Syst. Evol.-forsch.* 21:163.

2. See *Topics for Further Discussion and Investigation*, no. 2, at the end of the chapter.

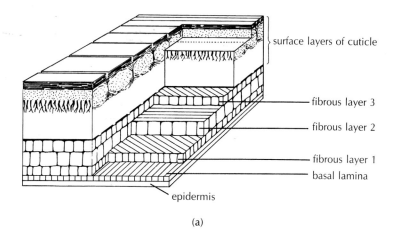

(a)

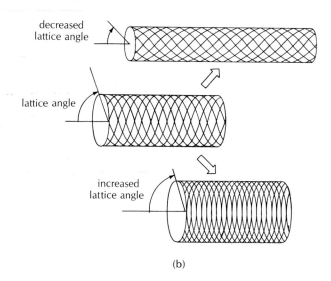

(b)

Figure 11.2

(a) Diagrammatic illustration of the multilayered nematode cuticle. Note the three layers of crossed fibers. (b) Although the collagenous fibers of the cuticle are individually inelastic, changes in the angle at which the fibers cross each other permit the animals to change shape. (a) *From A. F. Bird and K. Deutsch, "The Structure of the Cuticle of Ascaris Lumbridides vas suie" in* Parasitology, *47:319-328. Copyright © Cambridge University Press, New York. Reprinted by permission. (b) From Wells,* Lower Animals. *Copyright © 1968 McGraw-Hill, Inc., New York. Reprinted by permission.*

The rigid cuticle, high internal pressure, and lack of circular muscles preclude generation of pedal waves or peristaltic waves for locomotion. Nematodes completely lack locomotory cilia, so ciliary movement is also an impossibility. Instead, nematodes generally must move by undulating the body into sinusoidal waves, through alternating contractions of the longitudinal muscles on each side of the body[3] (Fig. 11.3b). Contracting one set of muscles causes bending and stretches out muscles elsewhere in the body. Thus, muscles antagonize each other by means of pressure changes transmitted through the fluid skeleton of the pseudocoel, according to the basic principles of the hydrostatic skeleton (Chapter 9). Reextension of contracted muscles is also brought about by the stiff cuticle surrounding the animal and the high pressure within the animal, both acting to spring the body back to a linear configuration when the muscles relax. Clearly, nematode design is not well suited to a free-swimming existence. Instead, most free-living nematodes are found, often in incredibly high numbers, living in soil, in aquatic sediments, in fruits, on surface films, and in other similar situations where either the substrate or the surface tension of a fluid at the air-water interface can provide resistance against which a thrust can be generated.

Muscle contraction is under the control of a simple nervous system consisting of an anterior brain (nerve ring plus associated ganglia) and four or more major longitudinal nerve cords (ventral, dorsal, and at least one pair of lateral nerve cords) (Fig. 11.4a). Strangely, the nerve cords seem not to send out processes innervating

3. See *Topics for Further Discussion and Investigation*, no. 1.

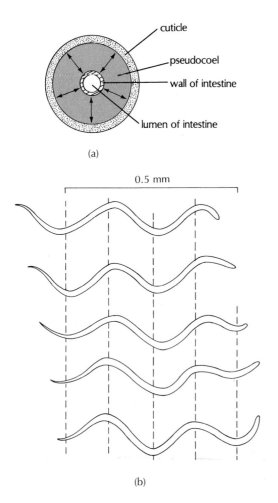

(a)

0.5 mm

(b)

Figure 11.3

(a) Diagrammatic representation of a nematode in cross section. Arrows represent the high pressure within the pseudocoel, acting to maintain the rounded body shape and to collapse the gut.
(b) Nematode locomotion on a solid surface. By contracting the muscles on each side of its body alternately, the animal forms a series of sinusoidal waves that thrust against the substrate surface, propelling the animal forward. In this figure, successive silhouettes of the nematode have been displaced below each other for greater clarity. The time elapsed between stages is approximately 0.33 seconds.   *(b) From Gray and Lissman, in Journal of Experimental Biology, 41:35, 1964. Copyright © 1964 Company of Biologists, Colchester, England. Reprinted by permission.*

the muscles. Rather, noncontractile extensions of the muscle fibers hook up to the nerve cords (Fig. 11.4b). While not unique to nematodes—a similar arrangement has been found in some flatworm and echinoderm species, for example—this pattern of innervation is certainly unusual, and it characterizes all nematode species examined to date.

Organ Systems and Behavior

The high-pressure pseudocoel poses potential digestive as well as locomotory difficulties. Nematodes have a linear digestive system, with a mouth (**stoma**) at the anterior end leading, in sequence, through a muscular pharynx,

intestine, and rectum and thence out of the body through an anus located near the body's posterior end (Fig. 11.5a). The difficulty in processing food lies in preventing the tubular, nonreinforced digestive tract from collapsing in response to the high pressure of the surrounding pseudocoel (Fig. 11.3a). The inner surface of the nematode gut is usually not lined by cilia. Indeed, ciliary beating probably would not be very effective in countering the high positive pressure exerted on the gut by the fluid in the pseudocoel. Instead, the gut is kept open by the activity of a highly muscular pharynx, which pumps in fluid at rates of up to four pulses per second in some species (Fig. 11.5b). At the animal's posterior end, the high pressure in the pseudocoel keeps the terminal end of the digestive tract tightly closed. A dilator muscle at the anus must open to allow discharge of wastes.

In addition to its role in keeping the gut lumen open against the high pressure of the pseudocoel, the pharynx (and associated glands) also adds lubricants and digestive enzymes to the food. Digestion is mainly extracellular, and nutrients are absorbed by the very thin wall (only one cell thick) of the intestine. Wastes are voided from the anus at intervals, about once every one or two minutes.

The respiratory and excretory systems are easily described: There are no specialized organs for gas exchange, no specialized circulatory system, and no nephridia. Metabolic waste products are apparently discharged along with other materials leaving the gut, or they diffuse across the body wall. A glandular system, the **renette**, or a modification of this system is present in most nematodes. The renette system varies considerably in complexity among species. It is often referred to as an excretory system, but its actual function has never been convincingly documented.

Despite their small size and limited locomotory abilities, nematodes are capable of fairly sophisticated behaviors. Various species are known to respond to temperature, light, mechanical stimulation, and a variety of chemical cues, including chemical cues produced by other individuals of the same species (**pheromones**). The general body surface may be light-sensitive in some species, possibly reflecting direct sensitivity of underlying nerve fibers. Many species have simple, pigmented light receptors (**ocelli**) as well. The major chemosensory organs, called **amphids** (Fig. 11.6), are anteriorly located pits lined with highly modified, nonmotile cilia (**sensillae**). Similar structures, called **phasmids,** are located at the posterior ends of some nematodes, and these are also thought to be chemosensory. The anterior and posterior ends of the body often have **cephalic** and **caudal papillae,** respectively, which also contain modified cilia. These structures are arranged around the mouth or anus, and are believed to be sensitive to mechanical stimulation. Many species possess external setae at various locations on the body; these are also thought to be **mechanoreceptors.**

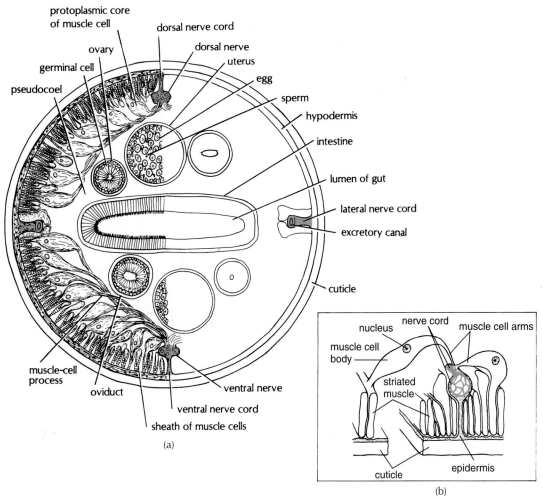

Figure 11.4
(a) *Ascaris lumbricoides* in cross section. The processes sent out from the muscle fibers to the nerve cords are clearly illustrated. For clarity, muscle cells are shown only in half of the figure.
(b) Detail of nematode muscle cell from *Ascaris suum*. Note the long, noncontractile extension of the muscle cell leading from its position in the body wall to one of the major nerve cords.

(a) *From Frank Brown,* Selected Invertebrate types. *Copyright © John Wiley & Sons, Inc., New York. Reprinted by permission of Mrs. Frank Brown.* (b) *From J.T. Debell, in* Quarterly Reveiw of Biology, *1965, 40:233–251. Copyright © 1965 The University of Chicago Press. Reprinted by permission.*

Reproduction and Development

Most nematode species are **dioecious** (G: two houses); that is, they have separate sexes. Individuals copulate, so fertilization is internal. Males possess one to two copulatory spicules posteriorly that are inserted into the female gonopore during sperm transfer (Fig. 11.5a, male). Cleavage is determinate (mosaic), so cell fates are permanently set at the first cleavage, as in protostomes. Certain other aspects of reproduction and development, while not unique to nematodes, are highly unusual. For example, the sperm of nematodes are amoeboid, rather than flagellated. A second unusual characteristic is the phenomenon of **chromosome diminution,** which has been observed in at least a dozen parasitic nematode species. Following fertilization of the egg, the zygote undergoes a normal first cleavage to the two-cell stage. Prior to the second cleavage, however, the chromosomes of one of the cells fragment, and much of the chromatin from the ends of the original chromosomes is destroyed. The pieces of the chromosomes remaining after fragmentation and disintegration replicate, and are distributed in normal fashion to the two daughter cells of the subsequent division (Fig. 11.7). The other two daughter cells (**germ cells,** or **stem cells**) of the four-cell embryo retain the full chromosome complement. Chromosome diminution occurs several more times during the next few cleavages. By the 64-cell stage, only two cells retain the complete genetic information present at fertilization. These two cells give rise to the gonad and produce the gametes for the next generation. The remaining cells, which retain as little as about 20% of the original genome, produce all of the somatic (i.e., nongamete producing) tissues (Research Focus Box 11.1). Loss of genetic information by chromosome diminution has also been reported to occur in some insects, crustaceans, and ciliated protozoans.

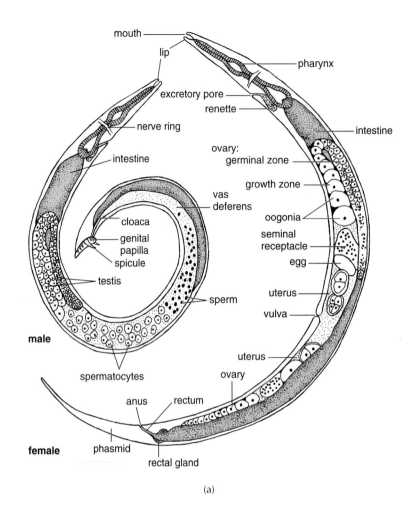

(a)

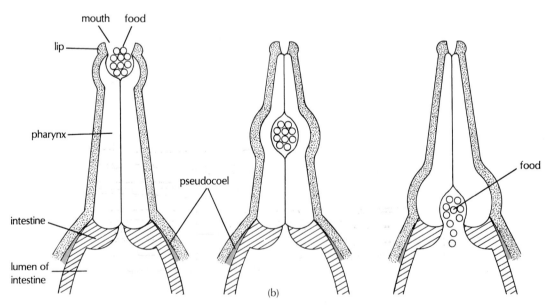

(b)

Figure 11.5

(a) Internal anatomy of a free-living nematode, *Rhabditis* sp.
(b) Sequence of movements of the pharynx musculature during
swallowing.   (a) *From Cleveland P. Hickman, Biology of the Invertebrates, 2d ed.*
Copyright © 1973 The C. V. Mosby Company, St. Louis, Missouri. Reprinted by permission.
(b) *After Sherman and Sherman; after Clark.*

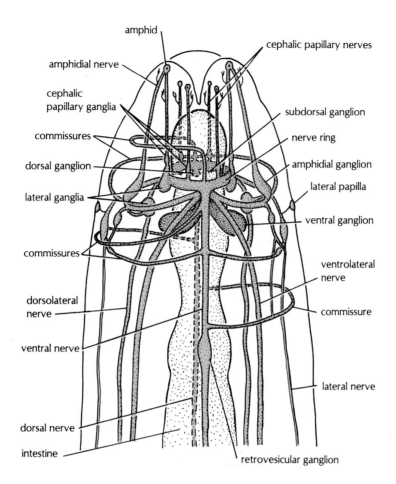

Figure 11.6

Nervous system of a nematode. Note in particular the brain ganglia; lateral, dorsal, and ventral nerve cords; and innervation of the amphids. *Modified from Brown, 1950. Selected Invertebrate Types. New York: John Wiley & Sons. After Hyman; after Chitwood and Chitwood.*

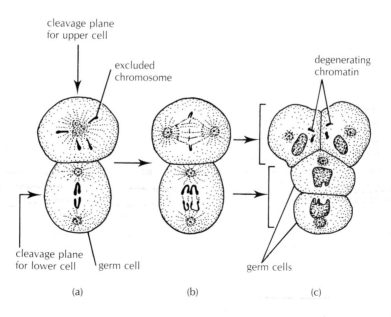

(a)   (b)   (c)

Figure 11.7

(a–c) Illustration of chromosome diminution in the parasitic nematode *Ascaris megalocephala*. In the second cleavage, only the chromosomes of the germ cell (= stem cell) undergo normal mitosis. Some of the chromosomal material in the other cell is excluded from participation in mitosis (a), and at the four-cell stage (c) is seen to lie outside the nucleus. This material will later degenerate. *Figure from Biology of Developing Systems by Philip Grant, copyright © 1978 by Saunders College Publishing, reproduced by permission of the publisher.*

Research Focus Box 11.1

Nematode Development

Bennett K. L., and S. Ward. 1986. Neither a germ line-specific nor several somatically expressed genes are lost or rearranged during embryonic chromatin diminution in the nematode *Ascaris lumbricoides* var. *suum*. *Devel. Biol.* 118:141–47.

Perhaps a dozen parasitic nematode species are reported to undergo chromosome diminution during their embryonic development. In *Ascaris lumbricoides,* the somatic cells lose about 30% of their DNA by weight (dropping from 0.63 pg [picograms, 10^{-12} grams] of DNA per cell to 0.46 pg of DNA per cell), whereas the germ-line cells retain the full 0.63 pg of DNA per cell throughout their lives. What is the nature of the lost chromatin, and why is it lost from some cells but not others? More than 100 years ago, Theodor Boveri suggested that the somatic cells might eliminate genetic material needed only for germ-cell development—a very reasonable case of the somatic cells ridding themselves of unnecessary genetic baggage. Bennett and Ward (1986) attempted to determine whether DNA unique to the germ line is, in fact, completely eliminated from somatic cells as predicted. To do this, the researchers had to find a specific protein produced only by germ-line cells and then determine whether the genes coding for that protein were indeed absent from somatic cells.

Bennett and Ward chose to work with the Major Sperm Protein, a protein detected only in the sperm of both parasitic and free-living nematodes. The Major Sperm Proteins of different nematode species must be very similar, since antibodies prepared against the Major Sperm Protein from the free-living species *Caenorhabditis elegans* also react with the protein from *A. lumbricoides.*

The researchers obtained adult worms from a pig-slaughtering company and used standard procedures to isolate DNA from the sperm. This DNA is known to contain the gene coding for the Major Sperm Protein since the protein is synthesized by developing sperm. Bennett and Ward also isolated DNA from the oocytes; the first two larval stages, which lack significant gonadal tissue; and adult intestines, obtained by careful dissection. The trick then was to determine how germ-line DNA differed from that contained in the cells of the somatic tissues.

The DNA isolated from the various tissues was first chemically fragmented. The fragments from each tissue were then separated by size, using gel electrophoresis, and next denatured with a strong base, to separate the double-stranded DNA into separate strands. The DNA was then transferred from the gel to a nitrocellulose filter, by laying the filter atop the gel and allowing the DNA to blot onto the filter. The task then was to locate the particular DNA fragment coding for the Major Sperm Protein in testicular tissue and to see whether that same gene was present in the other tissues; if Boveri's hypothesis about gene elimination is correct, that gene would be absent from all samples except those prepared from sperm.

To locate the gene coding for the Major Sperm Protein, Bennett and Ward used purified messenger RNA that, when translated, produces that protein; they used this messenger RNA to synthesize a complementary DNA strand that was labeled with a radioactive tag. This synthesized DNA fragment will code for the Major Sperm Protein; making it radiocactive enables researchers to recognize it again later. The radioactive DNA probe was then spread over the nitrocellulose filter, allowed to sit for a time, and then rinsed off; tissue DNA complementary to that radioactive DNA probe binds to it and will show up as a distinct band on X-ray film. If the X ray shows such a band, the tissue DNA must contain the gene coding for the Major Sperm Protein; if no such band appears on the X ray, the gene for the Major Sperm Protein must be absent from the sampled tissue. Focus Figure 11.1 clearly indicates that the Major Sperm Protein gene was present in all samples tested: oocyte, sperm, larvae, and adult intestine.

This work clearly demonstrates that the DNA coding for the Major Sperm Protein is present in all cells of the nematode, even after chromosome diminution. Moreover, the Major Sperm Protein genes in the different tissues fragmented into units virtually identical in molecular weight

Yet a third peculiar feature of nematode development is constancy of cell numbers (**eutely**). Once organogenesis is completed, mitosis ceases in all of the somatic cells. Thus, except for the cells of the germ line, further growth is due not to increases in cell numbers, but to increases in cell size. This characteristic is shared by a number of other aschelminth species.

Finally, nematodes are among the few groups of invertebrates that lack a morphologically distinctive free-swimming larval stage. The young animals emerge from their sturdy egg coverings as miniature adults. However, free-living species can enter a state of developmental arrest, called the **dauer larva,** at the second molt. The dauer "larva" (a juvenile stage, really) is essentially

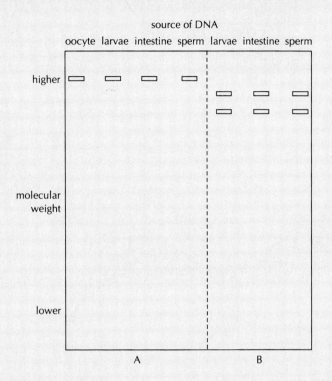

source of DNA

oocyte larvae intestine sperm larvae intestine sperm

higher

molecular
weight

lower

A   B

Focus Figure 11.1

Drawing made from X-ray film showing that the gene for nematode Major Sperm Protein is found in all nematode tissues examined. After the DNA fragments were transferred from an electrophoretic gel to a nitrocellulose filter, the filter was coated with radiolabeled DNA fragments coding specifically for the Major Sperm Protein. This radioactive DNA binds only to its complement on the nitrocellulose filter, and the unbound fragments are then washed off. If there are no complementary sequences on the filter, all the radioactive fragments wash off. Although the filter contains fragments of many different genes, only those coding for the Major Sperm Protein can become radioactive. The radioactive bands are located by placing X-ray film over the filter for about one week and then developing the film. The left and right sides (A and B) of the X ray show results when the DNA from each tissue is fragmented using two different enzymes.

(Focus Fig. 11.1), suggesting that these genes are apparently not altered in any substantial way from one tissue to the next; DNA molecules differing markedly in sequence would have been cut by enzymes into fragments of different sizes. Thus, whatever function is served by chromosome diminution, the process does *not* seem to rid somatic cells of all DNA that functions only in gonadal tissues.

However, as Bennett and Ward note, perhaps *other* germ-cell specific genes *are* eliminated by chromosome diminution; the gene for the Major Sperm Protein may have been an unrepresentative choice. Bennett and Ward also acknowledge that the Major Sperm Protein might have

some other, unknown function in somatic tissues; they have been unable to detect the Major Sperm Protein, or the messenger RNA coding for that protein, in somatic tissues, but it may nevertheless be present, in concentrations too low to measure. Obviously if the gene functions in somatic cells, it is not surprising to find it there. Until the sensitivity of available techniques improves, the functional significance of chromosome diminution must remain uncertain.

Figure from Karen L Bennett and Samuel Ward, "Neither a Germ Line-specific Nor Several Somatically Expressed Genes are Lost or Rearranged During Embryonic Chromatin Diminution in the Nematode Ascaris Lumbricoides var. suum" in Developmental Biology, 118:141–147, 1986. Copyright © 1986 Academic Press, Inc. Reprinted by permission of the author.

inactive, has a very low metabolic rate, and cannot feed. It may, nevertheless, live for many months until environmental conditions improve, at which time normal development resumes and the animal completes the final two molts to adulthood. The genetic basis for entry into and departure from the dauer state has been much studied over the past decade.

Parasitic Nematodes

Many nematodes show modifications of the previous description as adaptations to a parasitic existence. In fact, much of the research on nematode biology has been driven by the need to control the potentially devastating impact of these parasitic species (Fig. 11.8). Nematodes are

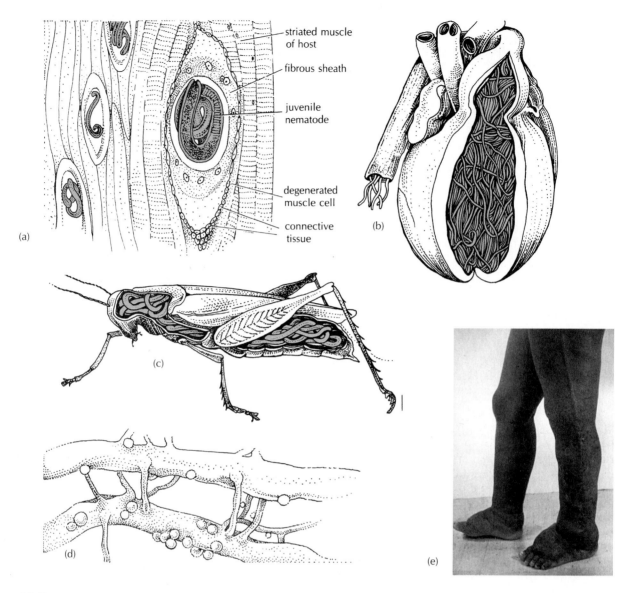

Figure 11.8

Examples of parasitism by nematodes and its effects on hosts. (a) *Trichinella spiralis* (responsible for trichinosis), shown encysted in vertebrate striated muscle. (b) Heart of a dog infested with *Dirofilaria immitis* (heartworm). (c) Cutaway view of a grasshopper infected with *Agamermis decaudata*. (d) Gall formation by nematodes (*Heterodera rostochiensis*) parasitic in plant roots. (e) Human male with elephantiasis of the legs.

(a) *Modified from Brown, 1950. Selected Invertebrate Types. New York: John Wiley & Sons and from Harmer and Shipley, The Cambridge Natural History, Macmillan.* (b) *From Harmer and Shipley.* (c) *From Hyman; after Christie.* (d) *From T. Cheng, General Parasitology, 2d ed. Copyright © 1986 Academic Press, Inc. Reprinted by permission.* (e) *Courtesy of the Centers for Disease Control, Department of Health and Human Services, Alanta, GA.*

parasitic in humans, cats, dogs, and domestic animals of economic importance, such as cows and sheep. Considerable veterinary and clinical research activity is now focused on how the various parasitic species suppress or otherwise outfox the host's immune response, and on why some individuals appear less susceptible to infection than others. Nematodes are also parasitic on the roots, stems, leaves, and flowers of plants, including species of great economic importance, such as soybeans, potatoes, oats, tobacco, onions, and sugar beets. Some of these parasitic nematodes attain

great length, although they may be extremely thin. The largest nematode so far described is 9 m long and resides in the placenta of female sperm whales.

Hookworms and pinworms are two groups of nematodes well known to many humans (Fig. 11.9). The damage done by parasitic nematodes is generally indirect, resulting from competition with the host for nutrients. One hookworm, for example, may imbibe more than 0.6 ml (milliliters) of blood per day. Someone with a respectable infection of 100 hookworms would then be

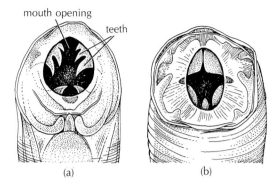

mouth opening

teeth

(a)   (b)

Figure 11.9
Hookworm mouth regions. (a) *Ancylostoma duodenale.*
(b) *Necator americanus.*   *From Chandler,* Introduction to Parasitology, *8th ed.*
Copyright © 1949 John Wiley & Sons, New York. After Looss. Reprinted by permission.

losing perhaps 60 ml of blood daily. Infections of 1,000 hookworms per host are not uncommon. Other nematode species do most of their damage by becoming so densely packed in their preferred tissues that flow of nutrients or fluids is blocked. Some ascarids (members of the genus *Ascaris*), for example, may block the host's intestine or bile duct. Other nematode species (e.g., *Wuchereria bancrofti* and *Onchocerca volvulus*) plug the lymphatic system, which sometimes results in a substantial buildup of fluid and subsequent dense growth of connective tissue in various regions of the body (Fig. 11.8e). The scrotum of one afflicted human male reportedly grew to a weight of about 18 kg (kilograms) before its removal by surgery. Less extreme forms of the disease (elephantiasis) presently afflict nearly 100 million people; elephantiasis is said to be one of the world's fastest-spreading diseases.

Many nematodes that are parasitic in animals make extensive migrations during their development within the host—from intestine, to liver, to heart, to lungs, to esophagus, and back to intestine, for instance. Various organs, including the intestinal wall, lungs, and eyes, can be damaged from these migrations. The larvae of *Onchocerca* spp. often migrate to the victims' eyes, causing blindness (so-called river blindness, or onchocerciasis); indeed, *Onchocerca* infection is one of the major causes of blindness in the world. About 40 million people are presently affected, mostly in western Africa.

Plant parasites also tend to do their damage indirectly, by (1) causing wounds, which are then susceptible to bacterial or fungal infection; (2) injecting plant viruses; or (3) damaging root, leaf, or stem transport systems.

The life cycles of parasitic animals are generally more complex than those of their free-living relatives,

and the nematodes are no exception. Fecundities are enormous (a single female *Ascaris* releases some 200,000 fertilized eggs per day—that's 73 million per year!), and one or more intermediate hosts are often obligate in the life cycle; the life histories of some important nematode parasites are outlined in the *Taxonomic Detail* section at the end of this chapter. Here I will describe just a few examples.

The so-called American hookworm (*Necator americanus*), infecting some 900 million people worldwide, has a rather simple life cycle with only a single host: humans. Once in the human host, the parasite undergoes an extensive internal migration before ending up in the host's intestine, where it matures. Fertilized eggs pass out in the host's feces and soon become infective to another human host (Fig. 11.10a). Another globally distributed nematode parasite of humans is pinworm, infecting in temperate regions alone some 500 million people, particularly children. Again, the life cycle involves no intermediate host, and the juvenile exhibits an impressive migration within the host before reaching its final location, this time in the host's large intestine and colon (Fig. 11.10b). Improving human sanitary practices is an essential step in limiting infection by both parasites.

The guinea worm, *Dracunculus medinensis*, presents us with a more complex and far more colorful example of a parasitic nematode life cycle; the generic name suggests that this nematode is up to no good. The female, only about 1 mm wide but generally 1 m or more long, lives just beneath the skin in humans, releasing a highly irritating secretion. When the skin comes into contact with water (as when the host bathes), the nematode protrudes its posterior end through the sore on the host's skin and ejects a considerable number of young from its uterus—up to about 1.5×10^7 offspring per day! These juveniles cannot reinfect humans directly but must first be ingested by a species of microscopic aquatic crustacean (phylum Arthropoda). Humans become infected by drinking water containing these crustaceans. The parasite, liberated from the intermediate host during digestion, migrates through the intestinal wall of the primary host and into the host connective tissue. Male parasites, which are relatively small, die soon after inseminating the females.

Once females reach adulthood beneath the skin, the cure is charmingly simple. Commonly, an incision is made and the worm is rolled out on a stick, very slowly (only a few centimeters daily), to prevent breakage (Fig. 11.11).

Dracunculus medinensis is found only in Africa, South America, and Western Asia. The World Health Organization hopes soon to eliminate the problem worldwide, something it has previously accomplished for only

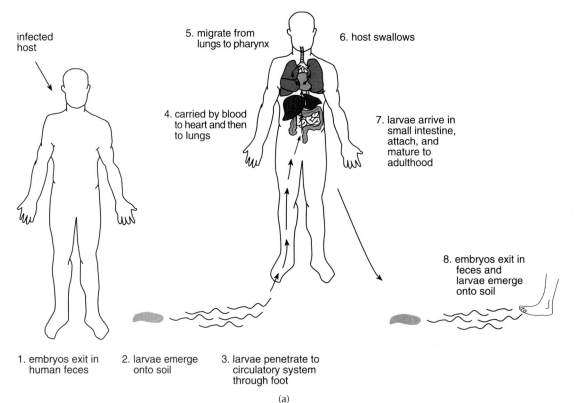

infected host

5. migrate from lungs to pharynx

6. host swallows

4. carried by blood to heart and then to lungs

7. larvae arrive in small intestine, attach, and mature to adulthood

8. embryos exit in feces and larvae emerge onto soil

1. embryos exit in human feces

2. larvae emerge onto soil

3. larvae penetrate to circulatory system through foot

(a)

Figure 11.10

(a) Life cycle of the hookworm *Necator americanus*. The parasite matures in the host's small intestine and begins releasing fertilized eggs, which pass out with the host's feces. After hatching out of the egg case and molting twice, the parasite becomes infective to humans, typically lysing its way in through the foot. Adulthood is reached only after undergoing an extensive migration within the host's body. (b) Life cycle of the pinworm, *Enterobius vermicularis*.

one other disease, smallpox. The goal is far more feasible for the guinea worm than for most other parasites, since the guinea worm appears to have no alternate hosts. Thus, its elimination is mainly a question of keeping infected persons away from public water supplies and teaching people in target areas to boil water before drinking it.

Wuchereria, Loa loa, Brugia, Onchocerca, and other **filarial nematodes** (so named because they produce a characteristic infective stage called a **microfilaria**) also exhibit complex life cycles involving an arthropod intermediate host and extensive migration within the definitive host (Fig. 11.12).

Beneficial Nematodes

It is unfortunate that parasitic species have given nematodes such a bad name. Species that require insects as intermediate hosts are increasingly being used to decrease population sizes of various insects, including mosquitoes and several major agricultural pests; the parasite often does mortal harm to its insect vector, particularly if infection densities are high. Besides, most nematode species are free-living, innocuous detritivores, not parasites. These species likely play valuable ecological roles in the cycling of nutrients and energy in a variety of environments. Free-living nematodes may also have nascent commercial value as potential food sources in the aquaculture of some edible animals, such as penaeid shrimp and certain fish, and may soon be exploited as sensitive monitors of environmental contamination. In addition, much of what is currently known about aging, inheritance, and the factors controlling gene expression and programmed cell death during development comes from the study of nematodes, particularly the free-living species *Caenorhabditis elegans* (Fig. 11.1). This animal has a

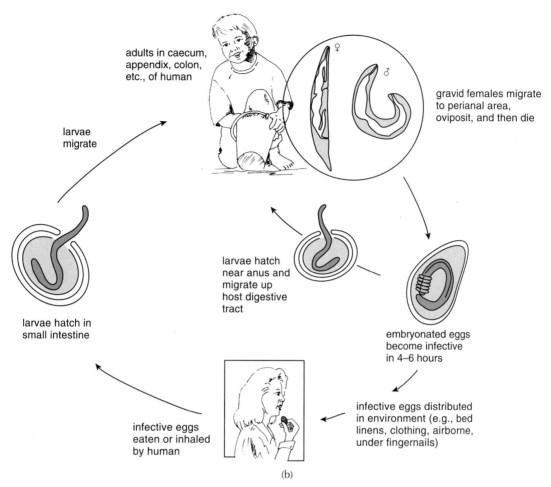

adults in caecum, appendix, colon, etc., of human

larvae migrate

gravid females migrate to perianal area, oviposit, and then die

larvae hatch near anus and migrate up host digestive tract

larvae hatch in small intestine

embryonated eggs become infective in 4–6 hours

infective eggs distributed in environment (e.g., bed linens, clothing, airborne, under fingernails)

infective eggs eaten or inhaled by human

(b)

Figure 11.10 (continued)
Mature females deposit their fertilized eggs at the host's anus and then die, releasing even more eggs. The embryos are transmitted to new hosts on the child's hands or through the air. Embryos remaining in the host's anal region emerge from their capsules after the second molt and migrate up the child's intestinal tract to reinfect the same host.   (b) *Figure from* Human Parasitology *by Burton J. Bogitsch and Thomas C. Cheng, copyright © 1990 by Saunders College Publishing, reproduced by permission of the publisher.*

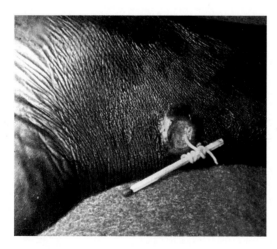

Figure 11.11
Using a matchstick to wind *Dracunculus medinensis* out of an infected human leg.   *Courtesy of the Centers for Disease Control, Department of Health and Human Services, Atlanta, GA.*

small genome with only six chromosomes and some 100 million base pairs; biologists expect to sequence the entire genome by the year 2000. The eventual role of every cell in development is now known, and the interconnections of all 302 neurons have been completely described, making these nematodes especially suitable for studying the molecular basis of differentiation, behavior, and learning. Genetic and developmental studies are further facilitated by the extremely short generation time in the laboratory—only about three days at room temperature—and by the ease with which the worms can be reared at high densities, up to about 10,000 individuals on a single petri dish. Molecular biologists have recently succeeded in implanting foreign genes into the genome of *C. elegans;* these genes are not only expressed in their foreign environment,

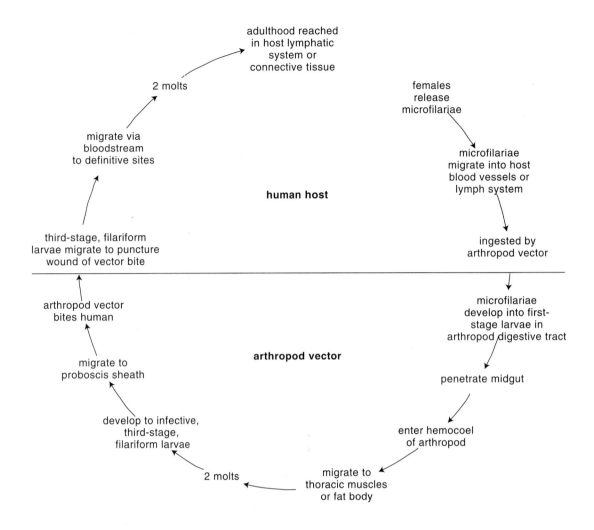

Figure 11.12

Generalized life cycle of *Wuchereria, Onchocerca, Loa loa,* and other filarial nematodes. *Figure from* Human Parasitology *by Burton J.*

Bogitsch and Thomas C. Cheng, copyright © 1990 by Saunders College Publishing, reproduced by permission of the publisher.

but are transmitted to succeeding generations, opening the door to important studies in genetic manipulation and gene therapy. At least 60 research laboratories worldwide are currently studying the development of *C. elegans.*

Topics for Further Discussion and Investigation

1. Discuss nematode locomotion.

Clark, R. B. 1964. Nematode locomotion. *Dynamics in Metazoan Evolution.* New York: Oxford University Press, 78–83.

Gray, J., and H. W. Lissman. 1964. The locomotion of nematodes. *J. Exp. Biol.* 41:135.

Harris, J. E., and H. D. Crofton. 1957. Structure and function in the nematodes: Internal pressure and cuticular structure in *Ascaris. J. Exp. Biol.* 34:116.

Wallace, H. R. 1959. The movement of eelworms in water films. *Ann. Appl. Biol.* 47:350.

Wallace, H. R., and C. C. Doncaster. 1964. A comparative study of the movement of some microphagous, plant-parasitic and animal-parasitic nematodes. *Parasitology* 54:313.

2. Investigate the factors that induce the hatching or exsheathment of parasitic nematodes.

Barrett, J. 1982. Metabolic responses to anabiosis in the fourth stage juveniles of *Ditylenchus dipsaci* (Nematoda). *Proc. Royal Soc. London Ser. B* 216:159.

Clarke, A. J., and A. H. Sheperd. 1966. Picrolonic acid as a hatching agent for the potato cyst nematode *Heterodera rostochiensis* Woll. *Nature (London)* 211:546.

Lackie, A. M. 1975. The activation of infective stages of endoparasites of vertebrates. *Biol. Rev.* 50:285.

Ozerol, N. H., and P. H. Silverman. 1972. Enzymatic studies on the exsheathment of *Haemonclius contortus* infective larvae: The role of leucine aminopeptidase. *Comp. Biochem. Physiol.* 42B:109.

Rogers, W. P., and R. I. Sommerville. 1957. Physiology of exsheathment in nematodes and its relation to parasitism. *Nature (London)* 179:619.

Wilson, P. A. G. 1958. The effect of weak electrolyte solutions on hatching rate of *Trichostrongylus retortaeformis* (Zeder) and its interpretation in terms of a proposed hatching mechanism of Strongylid eggs. *J. Exp. Biol.* 35:584.

3. Investigate the distribution and abundance of the free-living nematodes found in shallow-water marine environments.

Bell, S. S., M. C. Watzin, and B. C. Coull. 1978. Biogenic structure and its effect on the spatial heterogeneity of the meiofauna in a salt marsh. *J. Exp. Marine Biol. Ecol.* 35:99.

Hopper, B. E., J. W. Fell, and R. C. Cefalu. 1973. Effect of temperature on life cycles of nematodes associated with the mangrove (*Rhizophora mangle*) detrital system. *Marine Biol.* 23:293.

Tietjen, J. H. 1977. Population distribution and structure of the free-living nematodes of Long Island Sound. *Marine Biol.* 43:123.

Warwick, R. M., and R. Price. 1979. Ecological and metabolic studies on free-living nematodes from an estuarine mud flat. *Est. Coast. Marine Sci.* 9:257.

Taxonomic Detail

Phylum Nematoda (= Nemata)

This phylum has two classes, containing about 185 families of pseudocoelomate worms. About 16,000 species have been described to date, about 40% of them parasitic.

Class Adenophorea.
This class contains 87 families, distributed among 12 orders, with both free-living and parasitic species; most species are free-living. Moreover, most species possess caudal glands (for attaching to substrates) and epidermal glands that may secrete a lubricating covering at the body surface (*adenophorea* = G: gland-bearing). No members have phasmids, and the excretory systems lack collecting tubules. Most marine species are placed in this class.

Subclass Enoplia.
Seven orders, containing about 3,000 described species.

Order Dorylaimida.
This is the largest order in the class and the only one containing parasites of plants. (Other plant parasites are found in the class Secernentea.) Over 20 families.

Family Tricodoridae.   *Trichodoris.* An important parasite of plant roots, and one of the few known nematode vectors for plant viruses.

Order Trichocephalida.
All species parasitize vertebrates. Three families.

Family Trichuridae.   *Trichuris*—whipworm. This family includes many parasites of mammals. *Trichuris trichuris* is one of the most common gastrointestinal worm parasites of humans, infecting the colons of perhaps 500 million people worldwide. One female parasite can produce over 45,000 eggs per day. The mammalian host becomes infected by ingesting embryos in contaminated water or food.

Family Trichinellidae.   *Trichinella spiralis.* A parasite widespread in mammals and responsible for trichinosis in humans eating undercooked pork or some other meats. Approximately 40 million people are presently infected around the world.

Order Mermithida.
The adult stage in this order is typically free-living. However, the larval stages all parasitize invertebrates, especially insects, which can be killed by heavy parasite infestations. For this reason, considerable interest has been shown in using members of this order as biological control agents for target insects, including blackflies and mosquitoes, in which the larvae develop.

Subclass Chromadoria.
Most species in this subclass are free-living, and none are parasitic; a few species are symbiotic with fish (on the gills). Members are distributed among 35 families contained within five orders. Species occur in terrestrial, marine, and freshwater habitats.

Class Secernentea.
The members of this class are nearly all terrestrial. About 50% of all species parasitize plants or animals, both vertebrate and invertebrate. None are parasitic throughout the life cycle; the free-living stages commonly feed in dung and depend on insects for transport from one dung pile to the next. Unlike members of the Adenophorea, no species in this class exhibit epidermal or caudal glands, but all possess plasmids. Eight orders containing nearly 100 families and over 5,000 described species. Three subclasses.

Subclass Rhabditia.
Nearly 60 families distributed among three orders.

Order Rhabditida.
These nematodes include free-living species (all terrestrial) and parasitic species (all aquatic). Twenty-two families.

Family Panagrolaimidae   *Turbatrix aceti*—vinegar eels. These free-living nematodes thrive in high-acidity liquid environments, such as vinegar, and feed on bacteria.

Family Rhabditidae. *Caenorhabditis elegans.* These small, free-living nematodes, less than 1 mm long when full-grown, have become very important animal models for studying the molecular basis of aging and development. Unlike most nematode species, members of *C. elegans* are simultaneous hermaphrodites, rather than dioecious. Individuals live about three weeks, and they can reproduce when only three to four days old.

Family Strongyloididae. *Strongyloides stercoralis.* The complete life cycle is typically nonparasitic in soil or dung. Under unfavorable environmental conditions, however, the nematode larvae metamorphose into nonfeeding, infective filariform larvae. These may then penetrate the skin of a suitable mammalian host, traveling in the bloodstream to the heart and thence to the lungs. Juveniles then burst out of the lung capillaries, causing considerable damage in the process, and migrate up the host's throat. A cough and a swallow sends the parasite to its final resting place in the small intestine, where it dedicates its life to producing eggs by parthenogenesis. The embryos and larvae leave with the host feces and may then develop into free-living adults or infective filariform larvae, depending on environmental conditions. Humans get infected by contacting contaminated soil or feces or through contact with parasitized cats and dogs. Some 50 to 60 million people are currently infected, mostly in tropical and subtropical countries.

Order Ascaridida.

All members are parasitic in vertebrates. Chromosome diminution is known only from members of this order, occurring in at least 12 species. Twenty-three families.

Family Ascarididae. *Ascaris, Parascaris.* All species are intestinal parasites of mammals. Ascarids are the most common human parasites. There are estimated to be more than 1,000 million human infestations of *A. lumbricoides* around the world (affecting more than 25% of the earth's population), with over 75% of the cases occurring in Asia. The incidence of human infestation is perhaps 10 times higher in Europe than in the United States, but about three million cases are presently known in North America. The worms reside in the host's small intestine, feeding mostly on chyme and less frequently on blood sucked from the mucosal lining of the digestive tract. Their reproductive capacity is impressive: One female can produce over 25 million offspring in her short lifetime, at rates of up to several hundred thousand per day. Adult ascarids are among the largest of parasitic nematodes,

reaching lengths of 30 cm to 40 cm. Fertilized eggs leave the host in feces and become infective after a time in soil. If the infective larval stage is swallowed, it hatches in the host's intestine and begins a long, damaging journey through the host, only to return finally to the intestine, where it reaches maturity. The migration through the host is similar to that previously described for *Strongyloides* spp., including passage through the heart and lungs, rupture from the alveolar capillaries, and an upward crawl to the throat, followed by the inevitable cough and swallow.

Family Anisakidae. *Anisakis.* All species are obligate parasites of aquatic vertebrates in both marine and freshwater habitats. Fish, including such popular types as sardines, salmon, flounder, and mackerel, typically act as intermediate hosts. The final host, usually a mammal and sometimes a human, contracts anisakiasis by eating raw or under-cooked fish or squid. The parasite causes acute abdominal pain in the host. Infection can be avoided by cooking, salting, or freezing contaminated fish.

Family Oxyuridae. This family contains parasites of vertebrates and invertebrates. *Enterobius vermicularis*—pinworm. Gravid females, residing in the host's large intestine, migrate to the anus nightly and there deposit their fertilized eggs, creating a very itchy situation. When the sleeper scratches, the embryos collect on the fingers and under the fingernails, or become airborne. Other humans can then become infected (or reinfected) by breathing in or ingesting the fertilized eggs. Larvae hatch in the digestive tract and migrate directly to the large intestine, where they become adults. At least 300 million people are currently infected with *E. vermicularis* worldwide.

Order Strongylida.

Fourteen families distributed among five superfamilies.

Superfamily Ancylostomatoidea. Hookworms. As adults, all members parasitize the mammalian intestinal tract, feasting on host blood; the first two larval stages are generally free-living on feces. More than 20% of the world's human population presently may be infected with various hookworm species. Three families.

Family Uncinariidae. *Necator americanus*—the North American hookworm. Although the species occurs commonly in Asia, Africa, Central and South America, and the Caribbean, there are presently about one million human infestations in North America, primarily in southern states.

Worldwide, the species infects some 900 million people. Infestations with fewer than about 100 worms cause little trouble and may escape attention altogether. Infestations of more than 500 worms are highly pathological, due to extreme blood loss: Hosts are anemic and physically weak; periodically suffer severe abdominal pain, fever, and dizziness; and often experience a strong craving to eat dirt or wood. Fertilized eggs pass out with the host feces and hatch into the soil as long-lived, infective larvae. These larvae infect the next host (or reinfect a host) by penetrating the skin, often between the toes of the foot. The developing worm then undergoes an extensive, damaging journey through the host as described earlier for *Strongyloides* spp. and *Ascaris lumbricoides.* Once the host coughs and swallows the infective agent, the worm chomps down on the mucosa of the small intestine and begins its blood-sucking business.

Family Ancylostomatidae. *Ancylostoma duodenale*—the Asian hookworm, the foreign equivalent of *Necator americanus.* Species are common parasites of dogs, cats, and humans in Europe, China, Africa, and India. Some 900 million people probably harbor this parasite. Each worm typically ingests blood at twice the rate of its North American counterpart, so equivalent levels of pathology occur with smaller infestations.

Subclass Spiruria.

Twenty-one families distributed among two orders.

Order Spirurida.

All members parasitize annelids or vertebrates. Arthropods, usually insects, generally serve as obligate intermediate hosts.

Family Filariidae. *Wuchereria, Brugia, Onchocerca, Loa.* Members of this family—the filarial parasites—parasitize all vertebrates but fish, and they cause a number of infamous human disorders, including elephantiasis (a consequence of infection by *Wuchereria bancrofti* or *Onchocerca volvulus*), loa loa (a consequence of infection by a nematode of the same name, *Loa loa*), and river blindness (a common consequence of infection by *Onchocerca volvulus*). More than 200 million people (and equal or even greater numbers of domesticated and wild animals) presently may be infected with filarial nematodes, particularly in India, the Philippines, South America, and the Caribbean. The family also includes *Dirofilaria immitis,* the agent of canine (and feline) heartworm; infection can be transmitted from pet to human through physical contact. All filarial species are parasitic throughout their lives. The first juvenile stage, the colorless and transparent **microfilaria** (which gives the group its common name), circulates in the host's blood, while the adults live a more sedentary existence in host tissues and lymphatic glands.

Wuchereria bancrofti requires a mosquito as an intermediate host, as do many other members of this family (including dog heartworm). In at least some strains of this parasite, infective microfilariae circulate in the host bloodstream mainly at night; mosquitoes feeding while the host sleeps ingest the microfilariae with their blood meal. The parasite then develops to the third juvenile stage within the mosquito's thoracic muscles and soon migrates to the mouth, where it accesses the definitive host the next time the mosquito feeds. If the human host houses more than about 5,000 microfilariae per milliliter of blood, the indulging mosquitoes will likely die.

Onchocerca volvulus, the agent of river blindness and one cause of elephantiasis, requires blackflies (*Simulium* spp.) as intermediate hosts. Most of the serious damage to humans results from activities of the small microfilariae, rather than from the 60-cm-long, but merely annoying, adults. Lesions in the eye, resulting from the buildup of dead juveniles, eventually cause blindness. Nearly 40 million people currently may be blind or going blind from onchoceriasis. As with *Wuchereria bancrofti,* heavy infestations may kill the insect vectors, primarily by damaging the midgut epithelium and Malpighian tubules.

Loa loa, and other members of the genus, are restricted to African primates, including people. Over 13 million people currently may be infected. The worms are small and slender, perhaps 40 mm to 50 mm long and less than 0.5 mm wide. Rather than residing in any permanent location, adults usually migrate about the host's body subcutaneously; they are sometimes observed migrating under the conjunctiva of the eye. The adults secrete toxins that may cause large, egg-sized swellings on the host's body. The densities of microfilariae, living as always in blood, are highest in the daytime; transfer to a new host is therefore mediated by a daytime blood-sucking insect, the biting tabanid flies.

Order Camallanida.

Five families.

Family Dracunculidae. *Dracunculus medinensis*—the guinea worm. This large, debilitating parasite probably infects over 30 million people worldwide, particularly in Africa and India. Adult females reach lengths of some 100 cm (about 3 ft) in their hosts, some 250 times larger than the largest males. Adults secrete substances that cause considerable itching and burning in the host. (See p. 185 for a complete life-cycle description.)

Subclass Diplogasteria.

Ditylenchus, Heterodera. This group contains a variety of insect and plant parasites, and it is thus of considerable economic importance in agriculture and pest management. About 20 families.

Family Heteroderidae. *Heterodera.* This family contains some of the most important agricultural parasites in the world. The larvae invade underground roots, and the parasites spend their lives there, destroying root tissue.

Some General References About the Nematodes

Bird, A. F., and J. Bird. 1991. *The Structure of Nematodes,* 2d ed. New York: Academic Press.

Cheng, T. C. 1986. *Parasitology,* 2d ed. New York: Academic Press.

Desowitz, R. S. 1981. *New Guinea Tapeworms and Jewish Grandmothers: Tales of Parasites and People.* New York: W. W. Norton.

Harrison, F. W., and E. E. Ruppert. 1991. *Microscopic Anatomy of Invertebrates, Vol. 4. Aschelminthes.* New York: Wiley-Liss.

Heip, C., M. Vine, and G. Vranken. 1985. The ecology of marine nematodes. *Oceanogr. Marine Biol. Ann. Rev.* 23:399–489.

Hope, W. D., ed. 1994. *Nematodes: Structure, Development, Classification, and Phylogeny.* Washington, D.C.: Smithsonian Institution Press.

Maggenti, A. 1981. *General Nematology.* New York: Springer-Verlag.

Parker, S. P., ed. 1982. *Classification and Synopsis of Living Organisms,* vol. 1. New York: McGraw-Hill, 880–929.

Schmidt, G. D., and L. S. Roberts. 1989. *Foundations of Parasitology,* 4th ed. St. Louis: C. V. Mosby and Company.

Thorpe, J. H., and A. P. Covich. 1991. *Ecology and Classification of North American Freshwater Invertebrates.* New York: Academic Press.

12

Three Phyla of Uncertain Affiliation: Nematomorpha, Acanthocephala, and Priapulida

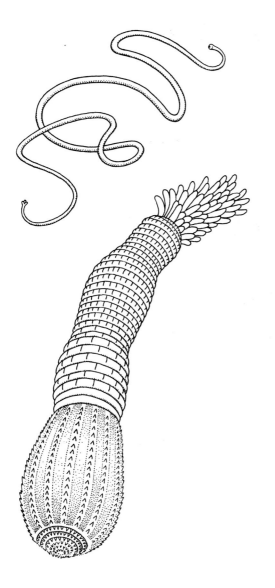

The three phyla unceremoniously lumped together in this chapter—Nematomorpha, Acanthocephala, and Priapulida—bear at least a superficial resemblance to nematodes. All species of all three phyla are worm shaped (vermiform), and like many nematode species, members of two of the phyla (Nematomorpha and Acanthocephala) are parasitic for at least part of their lives. At least two of the phyla, and possibly all three, share pseudocoelomate status with the nematodes. None of the three groups is of substantial medical, veterinary, or agricultural importance, however, and none of their members have been exploited as model systems for the study of any basic biological phenomena; their biology is therefore not nearly as well studied as that of many nematode species. Still, the animals live fascinating lives and present interesting evolutionary puzzles.

Phylum Nematomorpha

Phylum Nemato · morpha
(G: thread body)
nem-at-ō-mor´-fah

The least phylogenetically troublesome of the three phyla are the nematomorphs, also called "horsehair worms." About 250 species have been described worldwide. The adults are free-living, aquatic pseudocoelomates with a decidedly nematode external appearance: circular in cross section, long (commonly 0.5 m to 1.0 m in length), thin (rarely more than 1 mm wide), lacking body segmentation, and enclosed in an external cuticle (Fig. 12.1a). Like most nematodes, nematomorphs

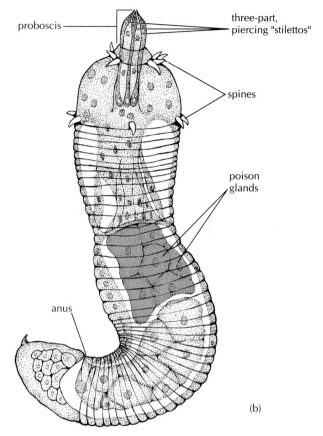

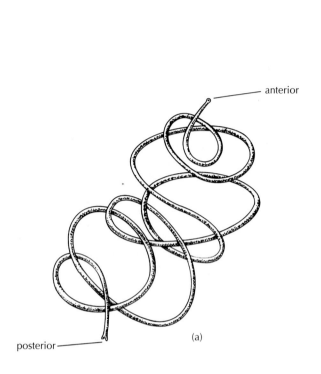

Figure 12.1
(a) A male horsehair worm, from the phylum Nematomorpha.
(b) The infective larva of the nematomorph *Gordius aquaticus*.
(a) *After Noble and Noble.* (b) *From Gosta Jagersten,* Evolution of the Metazoan Life.

are dioecious, and the eggs are fertilized internally. Also like nematodes, nematomorphs molt their collagenous external cuticle as they grow, possess only longitudinal body wall muscles, and lack locomotory cilia, as well as specialized circulatory and respiratory systems. They differ from nematodes primarily in lacking any excretory system and in the morphological details of their nervous and reproductive systems. Also unlike nematodes, they do not show constancy of cell numbers. Moreover, horsehair worms have nonfunctional digestive tracts that are degenerate at both the anterior and posterior ends. The adults do not feed at all, but typically live rather inactive lives, metabolizing nutrients acquired as juveniles. Females are especially inactive, devoting most of their energy to egg production: A female nematomorph commonly deposits more than one million fertilized eggs in her lifetime.

Juvenile nematomorphs also lack a functional digestive system; they live as internal parasites in arthropod hosts, exploiting dissolved nutrients in the host tissues and fluids. The parasite enters the arthropod host as a small larva, through ingestion or penetration. The larvae possess a spined, reversible proboscis (Fig. 12.1b), similar to that found in the next phylum to be discussed, the Acanthocephala. Once in their unlucky host, the nematomorphs develop gradually to full adult size (up to 1 m long) and shape; they ultimately emerge into a suitable aquatic environment by bursting out, killing the host miserably in the process. In the days before horses gave way to cars, the sudden appearance of these large but slender worms into the external world suggested that they were hairs from horses' tails somehow come to life. Their actual life history is, perhaps, almost as startling.

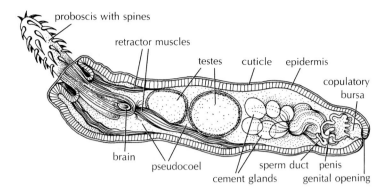

Figure 12.2
A male acanthocephalan (*Acanthocephalus* sp.). The copulatory bursa is eversible and holds the female during copulation, using secretions from the cement glands.   *From Hyman; after Yamaguti.*

Phylum Acanthocephala

Phylum Acantho · cephala
(G: spine head)
ah-can-thō-sef´-ah-lah

The phylum Acanthocephala includes fewer than 1,200 species, all of which are gut parasites in vertebrates. They are especially common in the small intestines of various fishes, particularly freshwater species, but also infect birds (including chickens and turkeys), mammals, and to a lesser extent, reptiles and amphibians. A single host may harbor hundreds of individual acanthocephalans. Like the nematodes, acanthocephalans are typically cylindrical and generally small (a few millimeters to a few centimeters long), show a constant number of cells (**eutely),** lack any specialized respiratory structures, and exhibit a large pseudocoel. In addition, like most nematodes, acanthocephalans are dioecious, and like many parasitic nematodes, arthropods are often required as intermediate hosts to complete the life cycle.

Despite the similarities, any evolutionary relationship between acanthocephalans and nematodes is likely a distant one. Acanthocephalans differ from nematodes in a number of major respects. First, most acanthocephalan species lack a specialized excretory system, although a number of species have a system of protonephridia, reminiscent of those found in flatworms; no nematode has such an excretory system. Second, acanthocephalans possess circular muscles, unlike any nematode. Third, whereas the nematode body is uniform and unsegmented, that of acanthocephalans is clearly divided into three parts: proboscis,

neck, and trunk (Fig. 12.2). In all but a few species, the proboscis can be retracted into a specialized pouch. The proboscis always bears numerous biochemically hardened hooks and spines (which give the phylum its name—spinehead), the sizes and arrangements of which are important in species identification. Calling someone an acanthocephalan makes a splendid insult—satisfying, but more likely to cause puzzlement than physical retaliation.

Intestinal parasites invariably require some means of attaching to the host's gut wall to avoid being swept downstream to the anus. Acanthocephalans use their hooked proboscis to remain in place. As with most other parasites, the nongonadal organ systems are greatly reduced; the capacious pseudocoel of the acanthocephalan trunk contains mostly gonad and associated glands (Fig. 12.2). Following copulation and internal fertilization, the fertilized eggs develop within the female's pseudocoel to an advanced, differentiated state—the **acanthor** stage (Fig. 12.3). An individual female may release several hundred thousand acanthors daily, which soon exit the host with host fecal matter, encased in protective shells.

The acanthor can develop further only within certain invertebrate hosts; it cannot infect another vertebrate host directly. If the shelled embryo is eaten by the proper intermediate host—usually particular insects for terrestrial life cycles and particular crustaceans for aquatic life cycles—the acanthor emerges from its shell, bores through the gut tissue and into the blood sinus (hemocoel) of the arthropod, and develops to a more advanced state. Adulthood is reached only if the intermediate host is eaten by a suitable vertebrate, but this need not occur directly: The acanthocephalan can typically pass through one or more **transport hosts** before eventually being taken into the final, definitive host. For many

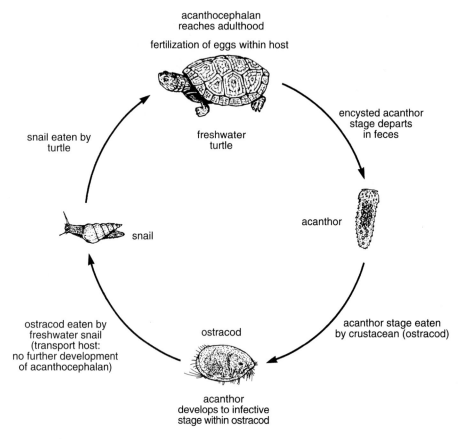

acanthocephalan
reaches adulthood

fertilization of eggs within host

encysted acanthor
stage departs
in feces

snail eaten by
turtle

freshwater
turtle

acanthor

snail

ostracod eaten by
freshwater snail
(transport host:
no further development
of acanthocephalan)

ostracod

acanthor stage eaten
by crustacean (ostracod)

acanthor
develops to infective
stage within ostracod

Figure 12.3
Life cycle of an acanthocephalan, *Neoechinorhynchus emydis*. The
transport host is essential for completing the life cycle in this
species.

acanthocephalan species, these transport hosts, in which the
parasite does not develop further, are essential links in the
life history; the definitive host often feeds not on arthropods
directly, but only on small fish, snails, or other transport
hosts that themselves feed on arthropods (Fig. 12.3).

Throughout the life cycle, acanthocephalans must
subsist on solubilized nutrients absorbed from the various hosts; acanthocephalans display no trace of a digestive tract at any stage of development. This is a common
correlate of parasitic existence, along with the general reduction of sensory and many other organ systems.

The pronounced adaptations shown by acanthocephalans for the endoparasitic lifestyle leave few clues about their
evolutionary history. The protonephridia found in a few
acanthocephalan species suggest at least a distant relationship
to the free-living flatworms (phylum Platyhelminthes). Recent fossil evidence, however, suggests a much closer relationship to a small phylum of free-living worms called "priapulids" (discussed next in this chapter); several species of
fossilized priapulids recovered from the Burgess shale
(formed about 530 million years ago) show apparent acanthocephalan morphologies.[1] This suggests that acanthocephalans have descended from free-living, marine mud
dwellers, possibly developing their present association with

arthropods as an adaptation to intense predation by arthropods long ago: Parasitize or perish. Their present life histories
would have evolved later, after the evolutionary appearance of
vertebrates. Some biologists believe that acanthocephalans
may be more closely related to another group of pseudocoelomates, the rotifers, discussed in the next chapter.

Phylum Priapulida

Phylum Priapulida
(G: the penis)
prē-ap-´ū-lē´-dah

The phylum Priapulida contains only about 15 described
species, most of which live in muddy sediments. Like
most nematodes and many acanthocephalans, most priapulids are only a few millimeters long (although some
species reach 20 cm in length) and possess an extensive
body cavity. Like nematodes and nematomorphs, priapulids secrete an external cuticle and molt it periodically;
this cuticle is chitinous, as in arthropods. Like that of

1. Morris, S. C., and D. W. T. Crompton. 1982. The origins and evolution
of the Acanthocephala. *Biol. Rev.* 57:85–113.

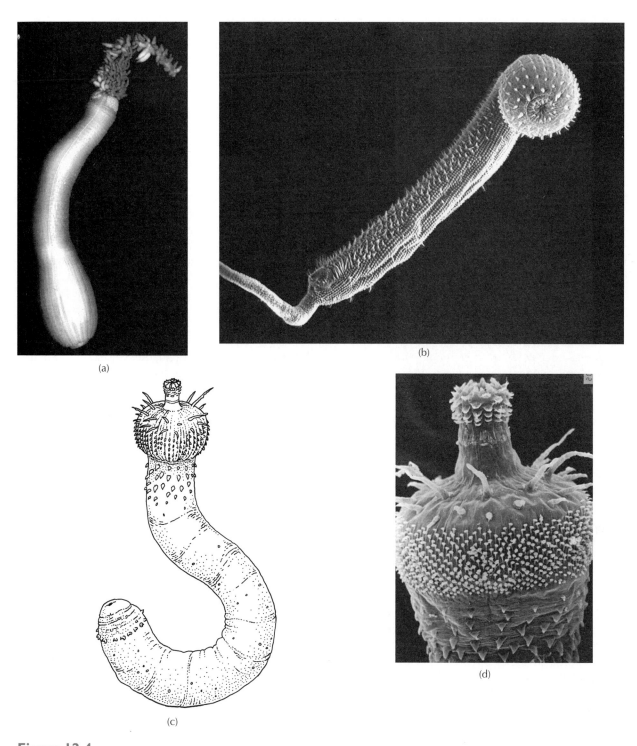

Figure 12.4

(a) *Priapulus caudatus.* The animal is 4 cm long, excluding the tail. The longitudinal nerve cord is visible on the ventral surface of the body. (b) Scanning electron micrograph of the interstitial priapulid *Tubiluchus corallicola* with introvert everted. The body is about 2 mm long, excluding the tail. (c) Diagram of the external morphology of *Meiopriapulus fijiensis,* an interstitial priapulid species. (d) Scanning electron micrograph of the anterior end of *M. fijiensis.* (a,b) *Courtesy of C. Bradford Calloway. From Calloway, 1975. (Springer-Verlag) Marine Biology, Vol. 31, pp. 161-74. "Morphology of the Introvert and Associated Structures of the Priapulid Tubiluchus corallicola from Bermuda." (Fig. 1, p. 163, in Marine Biology.) (c) From M. P. Morse, in Transactions of the American Microscopic society 100: 239, 1981. Courtesy of Dr. M. P. Morse. (d) Courtesy of Dr. M. P. Morse.*

Taxonomic Summary

Phylum Nematomorpha—horsehair worms
Phylum Acanthocephala
Phylum Priapulida

both acanthocephalans and nematomorphs, the priapulid body lacks internal septa and is circular in cross section, cylindrical, and not annulated. Adults of all three groups are dioecious.

Unlike the other groups discussed in this chapter, however, priapulids are all marine and free-living throughout their life cycles, and they typically show external fertilization rather than copulation and internal fertilization. The fluid of the body cavity serves as a hydrostatic skeleton for locomotion within the sediment and also acts as the circulatory medium. It contains cells bearing the unusual blood pigment hemerythrin, an iron-based oxygen-binding protein also found in such unrelated animals as polychaete worms and brachiopods.

Like the body of most acanthocephalans, the priapulid body bears an eversible anterior end (Fig. 12.4). However, unlike any acanthocephalan, priapulids possess a complete digestive tract, which may float freely in the nonseptate body cavity. Thus, the eversible "proboscis" of priapulids bears a centrally placed mouth at its end and is more properly termed an **introvert;** all priapulids are introverted. Most of the larger priapulids and at least some of the others seem to be active predators, while many of the smaller priapulids apparently feast exclusively on detritus. The excretory system is conspicuous and includes protonephridia. There are no free-swimming larvae in the life history; developmental stages resemble adult priapulids (except for having a cuticular covering about the abdomen) and live in sediment. Individuals of many species develop a pronounced tail as they approach adulthood; the adult tail may be single or multiple (Fig. 12.4a,b).

Priapulids occur in both warm- and cold-water habitats from the intertidal to abyssal depths, but they are seldom found in high densities. The larger priapulids seem to do best in habitats intolerable to most other animals, such as anoxic muds and high-salinity pools. The smaller species tend to be found within thriving interstitial communities.

The place of priapulids in the evolutionary scheme is very uncertain, despite their free-living lifestyle. Until the early part of the 20th century, priapulids were grouped with two other types of unsegmented worms, the Echiura and the Sipuncula (see Chapter 16), to form a single phylum, the Gephyrea. However, unlike the members of these other two groups, and unlike protostomes

in general, priapulids show a radial cleavage pattern during early development. One of the more frustrating aspects of current knowledge of this group is not knowing whether the body cavity is a true coelom or whether it is a pseudocoel. The necessary, careful embryological studies have simply never been done. Priapulids may be related most closely to acanthocephalans, as discussed earlier. Other biologists would affiliate them most closely with rotifers and kinorhynchs (see Chapter 13) or perhaps the loriciferans (see Chapter 13). Both of these hypotheses assume the priapulids to be pseudocoelomate. If the body cavity turns out to be a true coelom rather than a pseudocoel, the presumed affiliations between priapulids and other invertebrate groups would be altered considerably.

Topics for Further Discussion and Investigation

1. Until the early part of the 20th century, priapulids were grouped with two other types of marine worms, the Sipuncula and the Echiura (Chapter 15), in a single phylum, the Gephyrea. In what respects do the priapulids resemble sipunculans and echiurans? In what respects are they similar to the pseudocoelomates?

2. Nematomorphs are unusual parasites in that the adults are nonfeeding and free-living, while the larvae are all parasitic in a variety of arthropod hosts. What characteristics do nematomorphs share with the parasitic nematodes? What characteristics set the Nematomorpha apart from the Nematoda?

3. In what respects does the life cycle of acanthocephalans resemble that of parasitic flatworms (phylum Platyhelminthes, Chapter 4)?

Nicholas, W. L. 1973. The biology of the Acanthocephala. *Adv. Parasitol.* 11:671.

Van Cleve, J., II. 1941. Relationships of the Acanthocephala. *Amer. Nat.* 75:31.

4. Investigate the influence of acanthocephalan infection on the crustacean host's tolerance of environmental pollutants.

Brown, A. F., and D. Pascoe. 1989. Parasitism and host sensitivity to cadmium: An acanthocephalan infection of the freshwater amphipod *Gammarus pulex. J. App. Ecol.* 26:473.

Taxonomic Detail

Phylum Nematomorpha

The approximately 240 species in this phylum are divided among two classes.

Class Nectonematoida.
Nectonema. This small class of only about 20 species contains the marine horsehair worms, which parasitize crabs and other decapod crustaceans.

Class Gordioida.
Gordius. This class contains most nematomorph species. All live in freshwater or semiterrestrial habitats, and all are endoparasites of various insects, including beetles, grasshoppers, and cockroaches. Four families.

Phylum Acanthocephala

The approximately 1,150 described species in this phylum are distributed among three classes.

Class Archiacanthocephala.
The only acanthocephalans with specialized excretory systems are contained in this class. All species parasitize birds and mammals, with insects, centipedes, and millipedes serving as intermediate hosts. The class includes *Macracanthorhynchus hirudinaceus,* the largest acanthocephalan and the only one of economic significance; adults parasitize swine. Females are three to four times larger than the males, and they reach lengths of about 70 cm. The larvae develop within developing scarabid beetles that ingest swine feces, and the swine become infected by eating the beetle grubs or adults. Four families.

Class Eoacanthocephala.
Neoechinorhynchus (= Neorhynchus). Most species parasitize fish. Crustaceans serve most often as intermediate hosts. Three families.

Class Palaeacanthocephala.
This class contains most acanthocephalan species. These acanthocephalans parasitize a wide range of vertebrate hosts: Definitive hosts include fish (including bass), amphibians, reptiles, birds, and mammals. Crustaceans usually serve as intermediate hosts. Some species parasitize turkeys and chickens, using terrestrial isopods as obligate intermediate hosts. Some species can alter host behavior in ways that increase the likelihood of capture by the definitive host. Thirteen families.

Phylum Priapulida

Maccabeus, Meiopriapulis, Priapulus, Tubiluchus. The entire phylum contains only 15 described species, divided among three families (the Priapulidae, Tubiluchidae, and Chaetostephanidae); a few more species are awaiting formal description. All species are marine. Many priapulids are only a few millimeters long when fully grown, but the largest individuals, found in the genus *Priapulus,* reach lengths of up to 20 cm. Priapulids are found at all depths, from intertidal to abyssal, and in both warm and cold waters.

Some General References About the Nematomorphs, Acanthocephalans, and Priapulids

Harrison, F. W., and E. E. Ruppert, eds. 1991. *Microscopic Anatomy of Invertebrates, Vol. 4: Aschelminthes.* New York: Wiley-Liss.

Morris, S. C., and D. W. T. Crompton. 1982. The origins and evolution of the Acanthocephala. *Biol. Rev.* 57:85–115.

Morris, S. C., et al., eds. 1985. *The Origins and Relationships of Lower Invertebrates. Systematics Association, Special Vol. 28.* Oxford: Clarendon Press.

Parker, S. P., ed. 1982. *Classification and Synopsis of Living Organisms,* vol. 1. New York: McGraw-Hill, 931–44.

Roe, P., and J. L. Norenburg, eds. 1985. Comparative biology of nemertines. *Amer. Zool.* 25:3–151.

Thorpe, J. H., and A. P. Covich, eds. 1991. *Ecology and Classification of North American Freshwater Invertebrates. Nematomorpha.* New York: Academic Press, 273–80.

13

The Rotifers and Three Related Phyla

Introduction and General Characteristics

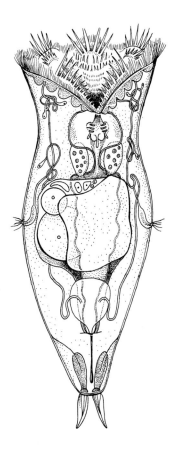

I call it a Water Animal, because its Appearance as a living Creature is only in that Element. I give it also for Distinction Sake the Name of Wheeler, Wheel Insect, or Animal; from its being furnished with a Pair of Instruments, which in Figure and Motion appear much to resemble Wheels. It can, however, continue many Months out of Water, and dry as Dust; in which Condition its Shape is globular, its Bigness exceeds not a Grain of Sand, and no Signs of Life appear. Notwithstanding, being put into Water, in the Space of Half an Hour a languid Motion begins, the Globule turns itself about, lengthens by slow Degrees, becomes in the Form of a lively Maggot, and most commonly in a few Minutes afterwards puts out its Wheels, and swims vigorously through the Water in Search of Food.

So wrote a Mr. Baker, in a letter addressed to the President of the Royal Society in London, in 1744, on the subject of rotifers. Rotifers were so named because of the two ciliated anterior lobes present in many species. This ciliated surface is called the **corona** (Figs. 13.1 and 13.2). The cilia of the corona do not beat in synchrony. Instead, each cilium is at a slightly earlier stage in the beat cycle than the preceding cilium in the sequence; that is, the cilia beat **metachronally.** A wave of ciliary beating therefore appears to pass around the periphery of the ciliated lobes, giving the impression of rotation. The degree and pattern of coronal ciliation vary considerably among species.

Phylum Roti · fera
(G: wheel bearer)
rō-tih′-fer-ah

Approximately 2,000 rotifer species have been described, mostly (about 95%) from freshwater environments, including lakes, ponds, and the surface films of mosses and other semiterrestrial vegetation. Some freshwater rotifers live **interstitially,** in the spaces between the sand grains of freshwater beaches. In fact, rotifers are generally considered one of the most characteristic groups of freshwater animals. Typically, 40 to 500 individuals are found per liter of lake or pond water, with densities as high as 5,000 individuals per liter having been recorded on several occasions. About 5% of rotifer species are found in shallow-water marine environments, and some of these marine species are interstitial.

Like all aschelminths, rotifers are pseudocoelomate. As in several other aschelminth phyla, mitotic divisions cease early in development, so further increases in body size are due to an increase in cell size rather than in cell number; that is, like nematodes, rotifers are **eutelic,** and the different organs and tissues are characterized by species-specific numbers of nuclei. Probably as a consequence of eutely, rotifers have poor regenerative powers. The epidermis of rotifers, like that of nematodes, is syncytial; that is, cell membranes between nuclei are incomplete. The synctial epidermis produces a nonchitinous, intracellular "cuticle" that is never molted. Rotifers possess both smooth and striated muscle fibers (Fig. 13.2), the latter being used for the rapid movement of spines and other appendages. Specialized respiratory and blood circulatory systems are lacking.

Most species are free-living and short-lived. Typically, the life span of a rotifer is between one and two weeks, although individuals of a few species can survive for up to about five weeks. Some free-living species spend their adult lives permanently attached to a substrate, while others are capable of moving from place to place. Some rotifers are parasitic, but their hosts are always invertebrates, especially arthropods and annelids (Fig. 13.3). Because parasitic rotifers have minimal impact on humans, their life histories have never commanded the attention accorded those of most other parasitic animals. The free-living species have never been much in the public eye either. Most are only 50 μm to 500 μm (micrometers) long, which is remarkably small for an adult metazoan. The largest individuals never exceed a length of 3 mm. Interest in rotifer biology seems to be increasing somewhat, as several free-living rotifer species have been found to be good food sources for the rearing of some commercially important fish and crustaceans. Rotifers may also play important roles in determining aquatic community structure and in mediating the flow of energy through freshwater ecosystems. Quite recently, rotifers have been utilized as models for research on the process of **senescence** (aging).

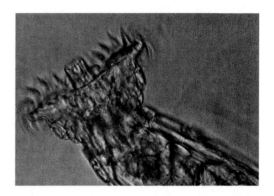

Figure 13.1
Light micrograph of rotifer corona, showing pattern of ciliation.
Photo copyright © 1975 by T. E. Adams.

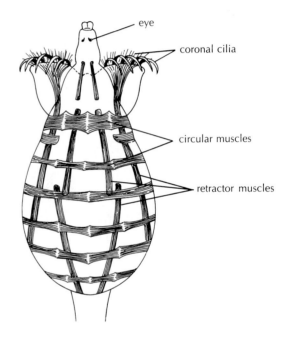

Figure 13.2
Illustration of rotifer musculature (*Rotatoria* sp.). The musculature of a given individual rotifer may be smooth, striated, or a combination of both. *From Hyman, The Invertebrates, Vol. III. Copyright © 1951 McGraw-Hill Book Company, New York. Reprinted by permission.*

Nonparasitic (free-living) rotifers feed on a variety of items. Most species are omnivores, selectively ingesting algae of appropriate size and chemical composition, small free-living animals (**zooplankton**), and detritus of appropriate size. Some other species feed on the intracellular juices of algae, and a number of species are carnivores, preying on a variety of animals smaller than themselves, including other rotifers.

Certain elements of the rotifer feeding and digestive system—notably the mastax and the trophi—are unique. The **mastax** is a prominent, muscular swelling located between the pharynx and the esophagus (Fig. 13.4). In some parasitic species, the mastax is modified for attachment to the host. Within the mastax of all species are found a number of rigid structures, the **trophi,** which are

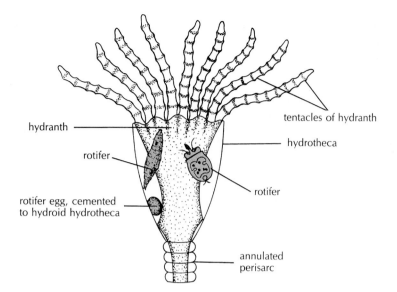

labels for top figure:
- tentacles of hydranth
- hydranth
- hydrotheca
- rotifer
- rotifer
- rotifer egg, cemented to hydroid hydrotheca
- annulated perisarc

Figure 13.3

Cnidarian hydranth with rotifer parasites (*Proales gonothyraea*) in hydrotheca. The rotifer feeds on the cnidarian tissues and lays its eggs in the hydranth. Other species within this genus are parasites in or on protozoans, oligochaetes (Annelida), gastropod embryos (Mollusca), crustaceans (Arthropoda), and filamentous algae. Several other genera are common internal parasites of oligochaetes, leeches (Annelida), and slugs (Mollusca). *From Hyman; after Remane.*

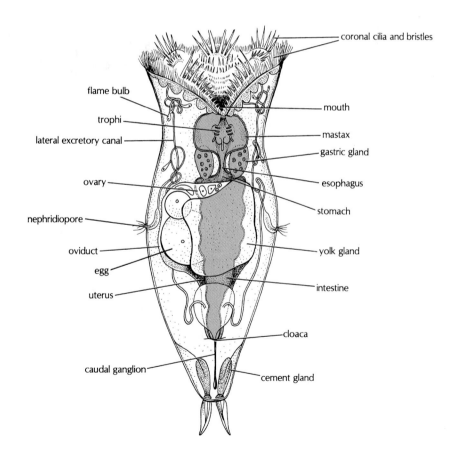

labels for bottom figure:
- coronal cilia and bristles
- flame bulb
- trophi
- lateral excretory canal
- ovary
- nephridiopore
- oviduct
- egg
- uterus
- caudal ganglion
- mouth
- mastax
- gastric gland
- esophagus
- stomach
- yolk gland
- intestine
- cloaca
- cement gland

Figure 13.4

Epiphanes senta, a common free-living rotifer from freshwater. Note the placement of the mouth and feeding apparatus. The trophi/mastax complex is used to grind ingested food into smaller particles. In some species, the trophi can be protruded from the mouth for prey capture. *From Frank Brown,* Selected Invertebrate Types. *Copyright © John Wiley & Sons, Inc., New York. Reprinted by permission of Mrs. Frank Brown.*

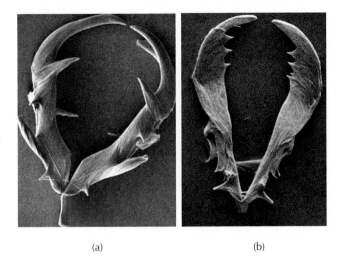

(a)   (b)

Figure 13.5
Scanning electron micrographs of rotifer trophi. (a,b) *Courtesy of George Salt. From Salt et al. 1978. Trans. Amer. Microsc. Sci. 97:469.*

often used to grind food following ingestion (Fig. 13.5). In some species, the trophi are used to suck food in through the mouth; in other species, the trophi can be protruded from the mouth to grab and/or pierce prey. The structure and sculpturing of trophi vary considerably in different species, in accordance with how the trophi are used. They thus are important tools in species identification. Trophi ultrastructure is almost unique to rotifers; only the jaws of gnathostomulids (Chapter 8) are structurally similar to those of rotifers, suggesting a close evolutionary relationship between the two groups.

Free-living rotifers are propelled primarily by the coronal cilia. In some species, particularly well-developed appendages also contribute to locomotion (Fig. 13.6). Some species are entirely planktonic; that is, the rotifers swim throughout their lives. Members of many other species stop swimming periodically and form temporary attachments to a solid substrate by secreting a cementing substance from a pair of pedal glands. The pedal glands open to the outside through pores in the **toes** of the foot. Rotifer feet possess up to four toes, depending upon the species (Fig. 13.7). Once the rotifer is attached to the substrate, the coronal cilia generate water currents for respiration and food collection. In a number of planktonic and sedentary rotifers, the coronal cilia form two parallel bands, with a ciliated food groove lying between. The cilia of the two bands beat toward each other, sweeping particles into the food groove (Fig. 13.8). Cilia in the food groove then conduct these captured particles to the mouth for ingestion.

Many free-living species can move upon solid substrates between bouts of swimming. The foot and toes play a role in such locomotion by forming temporary attachments to the substrate. Once the attachment is made, the body can be elongated through the contraction of circular muscles, which are distributed about the body in discrete bands. Contracting the circular musculature elongates the body and stretches out the relaxed longitu-

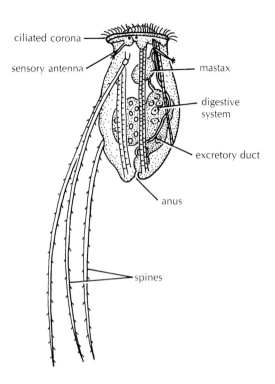

Figure 13.6
Filinia longiseta, a pelagic rotifer commonly found in lakes. Note the lack of a foot, the lack of toes, and the presence of conspicuous spines for swimming. *From Hyman; after Weber.*

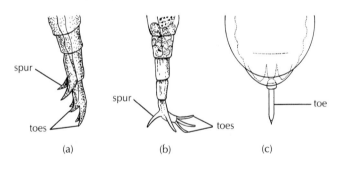

(a)   (b)   (c)

Figure 13.7
Rotifers may have zero, one, two, three, or four toes on the foot, depending on the species, and may also bear a pair of nonsecretory protuberances called spurs. (a) *Philodina roseola,* with four toes and two spurs. (b) *Rotaria* sp., with three toes and two spurs. (c) *Monostyla* sp., with a single toe and no spurs. *From Hyman, The Invertebrates, Vol. III. Copyright © 1951 McGraw-Hill Book Company, New York. Reprinted by permission.*

dinal musculature. The pseudocoel of these rotifers thus functions as a hydrostatic skeleton, permitting the mutual antagonism of the circular and longitudinal musculature through the generation of temporary increases in hydrostatic pressure.

As the rotifer's body elongates, the corona is withdrawn, turning the animal's anterior end into a suction-generating **proboscis** (Fig. 13.9). The proboscis can be applied to the substrate, forming a new attachment site. The posterior attachment is then broken, and the body becomes shorter and fatter through contraction of the longitudinal musculature and stretching of the relaxed circular

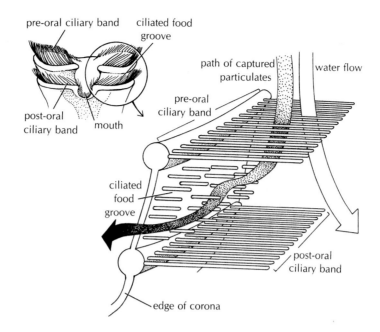

Figure 13.8
Double-banded ciliary system of suspension-feeding rotifers. As water is swept between the cilia of the pre-oral ciliary band, suspended food particles are captured and conducted to the mouth by the cilia of the food groove.   *Modified from Strathmann, et al., 1972. Biological Bulletin, 142:505–519.*

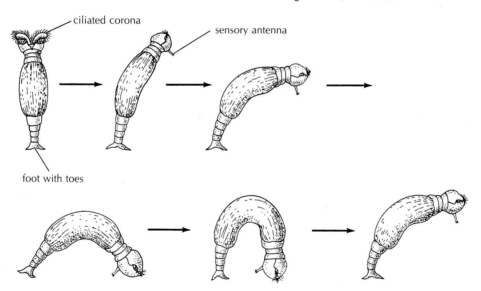

Figure 13.9
Looping locomotion of *Philodina roseola*. Contraction of circular muscles elongates the body, which is then bent toward the substrate by the differential contraction of longitudinal muscles. Once the anterior end of the animal grips the substrate, the foot releases its attachment and the rear of the body may be pulled forward. The body may then reattach posteriorly and extend forward again. Note that the corona is withdrawn during this sequence of movements.   *After Harmer and Shipley.*

musculature. Finally, the anterior attachment is released while the foot regains its grip on the substrate, and the body is again elongated. The animal can thus progress by **looping,** somewhat as described previously for free-living flatworms (phylum Platyhelminthes, class Turbellaria—Chapter 7) and as yet to be described for leeches (phylum Annelida, p. 302). Some rotifers exhibit extraordinary shape changes during locomotion, which are often facili-

tated by the foot and trunk of the body being divided into a number of sections that can be telescoped into each other, much like a collapsible drinking cup.

Other nonparasitic rotifers are **sessile** as adults; that is, they are incapable of locomotion. Pedal gland secretions attach these animals to a substrate permanently. Members of many species secrete protective tubes, often incorporating debris, sand grains, or even fecal pellets into their walls

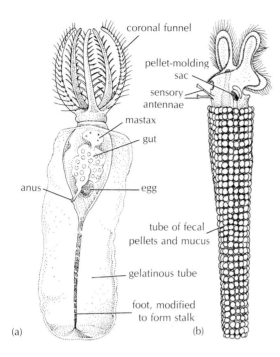

Figure 13.10
Two sedentary rotifer species: (a) *Stephanoceros fimbriatus.*
(b) *Floscularia ringens*, showing region of pellet formation.   (a) *After
Jurczyk; and after Jägersten. (b) After Pennak.*

(Fig. 13.10). All species that are sessile as adults have free-swimming young resembling the young of planktonic species. Thus, even sedentary species are capable of locomotion—at least for a short time between adult generations.

Rotifer reproduction is unusual in a number of respects. As among nematodes and other aschelminths, reproduction by fission or fragmentation is unknown, again a likely consequence of eutely. Where asexual reproduction does occur among rotifers, it is by **parthenogenesis,** the development of unfertilized eggs. The details of sexual and asexual reproduction among rotifers are best discussed on a class-by-class basis. There are three classes in the phylum.

Class Seisonidea

The small class Seisonidea (sī-sun-id′-ē-ah) is composed exclusively of ectoparasites of marine crustaceans. As might be expected, the corona of these species is often greatly reduced in size (Fig 13.11a). Reproduction seems to be exclusively sexual, and all individuals have separate sexes; that is, the members of this class are always **dioecious.** Fertilization is internal, either through true copulation or by means of hypodermic impregnation of the female by the male. In the latter case, sperm are injected by the male into the pseudocoel of the female, from which they find their way into the ovary.

Class Bdelloidea

Members of the class Bdelloidea (del-oy′-dē-ah) are all free-living and mobile; that is, this class of rotifers contains

no tube-dwelling, sessile species and no parasitic species. Most members are omnivorous suspension feeders, and the corona is correspondingly well developed and bilobed (Fig. 13.11b,c). Reproduction among bdelloid rotifers appears to be exclusively by parthenogenesis, since males have never been discovered. Thus, all known rotifers in this class are female.

Bdelloid rotifers commonly inhabit environments that periodically expose the animals to physiologically stressful conditions, such as freezing, dehydration, or high temperatures. Species living in polar lakes and ponds, in temporary ponds, and among emergent mosses and lichens are typically capable of entering a state of extremely low metabolism, or **cryptobiosis.** Some species secrete a gelatinous covering during the early stages of drying. This covering then hardens to form a **cyst** (Fig. 13.12). Like the nematodes already discussed in Chapter 11 and the tardigrades to be discussed in Chapter 19, rotifers in the cryptobiotic state can withstand environmental extremes, including extensive desiccation, for prolonged time periods. Indeed, some bdelloid rotifers have been successfully rehydrated following 50 years or more of desiccation! How cryptobiosis permits rotifers to withstand what would otherwise be lethal environmental conditions is not yet known.

Class Monogononta

Most rotifers belong to the class Monogononta (mah-nō-gon-on′-tah). Monogonont rotifers may be free-swimming or sessile. In fact, this class contains the only free-living sessile rotifers. The sessile rotifers are generally found attached to macroscopic plants, to filamentous algae, or to the tubes of other sessile rotifers of the same or of different species. Some sessile species use the corona to collect food particles as described previously, or with a modification of that theme. In other species, the corona is poorly ciliated or is nonciliated, but it may bear long spines surrounding a funnel-like anterior end. The spines can be moved to entrap small metazoans that come too close, and the captured prey are then forced into the mouth for ingestion. Thus, many members of this class are carnivores. Ciliation of the **buccal field** surrounding the mouth presumably aids ingestion.

Many sessile species live in protective tubes (Figs. 13.10 and 13.11f). In some instances, these are merely gelatinous secretions from specialized glands opening externally. In other species, particles are collected by the corona, coated with mucus, and cemented in place, thereby elongating the tube as the animal grows in size. In some species, fecal pellets are incorporated into the tube. In both free-swimming and sessile monogonont species, the cuticle is commonly thick and rigid, forming a protective **lorica.** Lorica production is encountered only among members of this class. Within a species, the shape of the lorica may be modified by environmental factors (Research Focus Box 13.1).

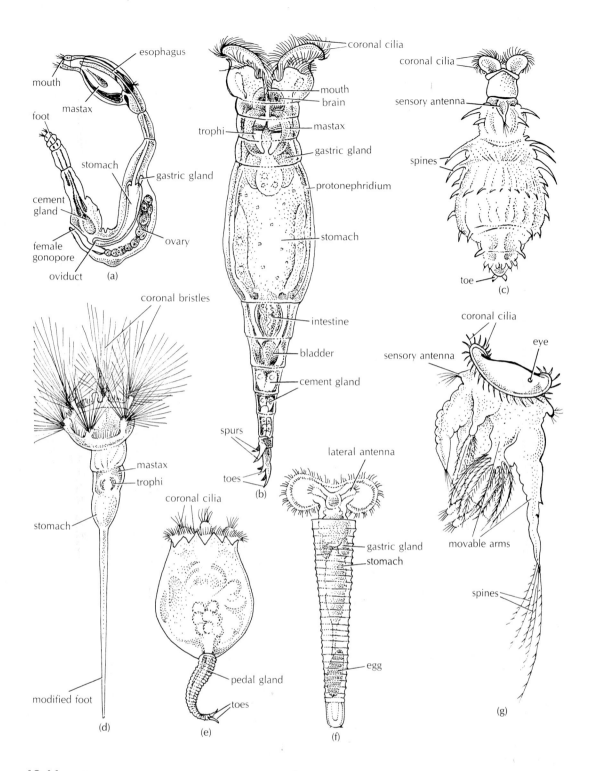

Figure 13.11

Representatives of the three rotifer classes. (a) Seisonidea (*Seison* sp.). All species are ectoparasitic on marine crustaceans and are dioecious. The illustrated individual is a female. Note the reduced ciliation associated with a parasitic lifestyle. (b) Bdelloidea (*Philodina roseola*). All species are free-living and motile suspension-feeders, and no males have ever been described; all reproduction is parthenogenetic. (c) Bdelloidea (*Macrotrachela multispinosus*). (d) Monogononta (*Collotheca* sp.). Note the highly modified corona, with seven distinct lobes. The animal is sessile and secretes a gelatinous tube (not illustrated). (e) Monogononta (*Brachionus rubens*). This species is free-swimming and is commonly used as a food source for rearing larval fish. (f) Monogononta (*Limnius* sp., a sessile tube dweller). (g) Monogononta (*Pedalia mira,* a free-swimming species). Members of this class show both sexual and asexual reproduction; the switch to sexual reproduction is triggered by alterations in the physical environment. *(a) From Meglitsch; after Plate. (b) From Hyman; after Hickernell. (c) From Hyman; after Murray. (d) After Hyman. (e) From Pennak; after Halbach. (f) From Meglitsch; after Edmondson. (g) From Hyman; after Hudson and Gosse.*

Research Focus Box 13.1

Adaptive Value of Rotifer Morphology

Gilbert, J. J., and R. S. Stemberger. 1984. *Asplanchna*-induced polymorphism in the rotifer *Keratella slacki*. *Limnol. Oceanogr.* 29:1309–16.

Certain aquatic invertebrates from several phyla exhibit **cyclomorphosis**—morphological changes in the subsequent generation induced by changing environmental factors, such as temperature, food level, predation pressure, or dissolved organic substances. The freshwater, suspension-feeding rotifer *Brachionus calyciflorus*, for example, exhibits such a change in response to some (uncharacterized) chemical released by a predaceous relative, rotifers in the genus *Asplanchna*. The chemical does not alter existing adult morphology, but it does affect the development of early-stage embryos. In the presence of *Asplanchna* spp., the offspring of *B. calyciflorus* develop much larger spines than usual, which probably protects them from predation by *Asplanchna*.

Do other suspension-feeding rotifers show similar responses to *Asplanchna*? Do the morphological changes actually reduce predation? And can the same morphological changes be induced by other predators, or are they specific responses to *Asplanchna*? These were the issues addressed by Gilbert and Stemberger (1984). They chose to work with the herbivorous rotifer *Keratella slacki* because it co-occurs with the predatory *Asplanchna* spp. and exhibits a wide range of body morphology in the field; indeed, field-collected individuals have longer spine lengths only when *Asplanchna* is present, suggesting a causal relationship worth examining. The questions are, is *Asplanchna* directly responsible for the morphological variation, and how does the variation benefit *K. slacki*?

Changes in body morphology of a species in the field over time can be explained in one of three ways: (1) individuals with certain morphologies survive better at some times than others; (2) individuals with certain genetically determined morphologies leave more offspring at some times than at others; or (3) the phenotype of most individuals is altered under some conditions without any underlying genetic change. These possibilities are readily evaluated for *K. slacki* (and many other rotifers) because it reproduces parthenogenetically and has a short generation time; a large, genetically identical population can be established rapidly from an individual female. In the experiments described, all individuals of *K. slacki* were clones derived from one individual.

To conduct their studies, Gilbert and Stemberger maintained high-density laboratory cultures of the predatory rotifer *Asplanchna girodi* and of a small, planktonic, predatory crustacean, *Tropocyclops prasinus*. Before each experiment, they passed the culture water through a glass-fiber filter to remove solid materials. Several adult females of *K. slacki* were placed into each of a number of small petri dishes in 5 ml of the *Asplanchna* filtrate, and they were fed on a small unicellular alga, *Cryptomonas* sp. As a control, additional populations of *K. slacki* were cultured with the same unicellular algae in filtrates of the protozoan *Paramecium aurelia*, which was used as a food source in culturing the *Asplanchna*. Other rotifer populations were all cultured in filtrates of the predatory crustacean. Under laboratory conditions, the rotifer populations increased rapidly. The researchers changed the food and water every few days for up to nine days, and then they preserved all rotifers for measurement. The body and spine lengths of about 50 individuals from each treatment were measured, using a compound microscope equipped with a measuring scale (ocular micrometer).

In the presence of the predatory rotifer filtrate, the average body size in the population of *K. slacki* increased by about 15%, the right posterior spine length increased by an average of about 130%, and the lengths of the anterior spines increased by about 30%. Some individuals also developed conspicuous posterior spines on the left side (Focus Fig. 13.1). In contrast, filtrate from the predatory crustaceans or from the protozoan controls had no effect on the rotifer's body size or morphology (Focus Fig. 13.2). These data clearly indicate that *K. slacki* is not simply responding to higher levels of some general waste product, such as ammonia, since the crustaceans and protozoans were cultured at comparably high densities and excretory rates probably do not differ dramatically among these animals. Also, the response seems to be triggered by a specific water-soluble substance produced by *Asplanchna*, rather than by any physical contact, and it clearly reflects a developmental polymorphism rather than any genetic change in the population. As mentioned earlier, all the offspring of *K. slacki* in these experiments were produced parthenogenetically.

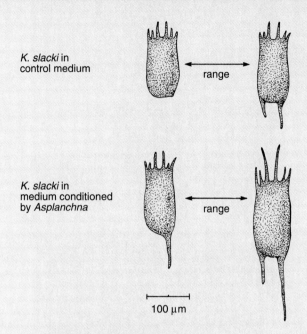

K. slacki in
control medium

range

K. slacki in
medium conditioned
by Asplanchna

range

100 μm

Focus Figure 13.1
Most common body types encountered in control cultures of
Keratella slacki and those reared in *Asplanchna*-conditioned
water. Animals in the experimental treatments have somewhat
larger bodies and substantially longer spines.

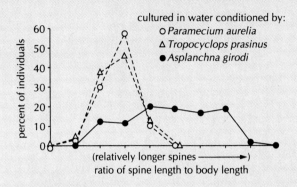

cultured in water conditioned by:
○ *Paramecium aurelia*
△ *Tropocyclops prasinus*
● *Asplanchna girodi*

percent of individuals

(relatively longer spines ———➤)
ratio of spine length to body length

Focus Figure 13.2
The effect of water conditioned by a protozoan and two
predators on the length of the right posterior spine in
Keratella slacki.

But is the response adaptively beneficial? To examine
this, Gilbert and Stemberger placed an equal number of
short- and long-spined *K. slacki* into each of several dishes
of water, and then they added some *Asplanchna* predators to
each dish. About six hours later, they counted the number
of short- and long-spined *K. slacki* remaining in each dish.
If short-spined individuals are more vulnerable to predation, fewer of these individuals should be left at the end of

such an experiment. Indeed, this was the result obtained in
each of the three separate studies conducted (Focus Table
13.1). Using a dissecting microscope to directly observe
predator-prey interactions, the researchers confirmed these
results: The predator-induced, large-spined morphs were
captured and eaten significantly less often by *Asplanchna,*
even though they were attacked just as often as the short-
spined individuals.

As usual, the results of one scientific study trigger additional questions, leading to further studies. In particular, if
having spines is such an advantage to *K. slacki,* why does it
develop spines only in the presence of predaceous rotifers?
Why are long spines and larger bodies not genetically programmed for all individuals? Gilbert and Stemberger have
gone on to address this and related issues in subsequent papers. Can you envision how they might have designed their
experiments?

Focus Table 13.1   The Influence of Body Size and Morphology on the Susceptibility of the
Herbivorous Rotifer *Keratella slacki* to Predation by *Asplanchna girodi*

Experiment	Percent Eaten Short-Spined	Long-Spined	Original Number of Each Type	Differences Statistically Significant?
1	40	10	10	No
2	62	0	8	Yes
3	50	20	18–20	Yes
All three combined	50	12	36–38	Yes

Note: The predaceous rotifers were allowed to feed for about six hours; the extent of predation on short- and long-spined rotifers was then assessed by the disappearance of individuals from containers and confirmed by finding remains of prey in predator guts or on the bottoms of the
dishes. A statistically significant result is one that would occur by chance alone in fewer than five trials out of 100.

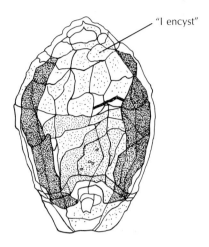

Figure 13.12
Philodina roseola, encysted. The adult is illustrated in Figure 13.11(b).   *From Hyman; after Hickernell.*

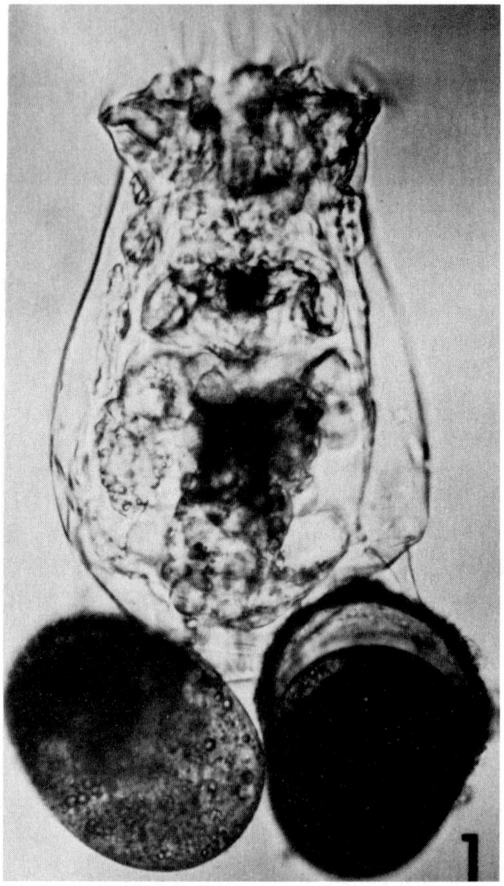

The reproductive pattern of monogonont rotifers is unique. Typically, monogononts reproduce by means of parthenogenesis. Females generally produce diploid eggs by mitosis, usually one at a time, and each egg develops into another diploid female—all in the absence of males. Such females are termed **amictic,** referring to the production of eggs without the "mixing in" of genes from any other individual. Young rotifers emerge from amictic eggs soon after the eggs are released; thus, amictic eggs are also known as **subitaneous** eggs (*subit* = G: sudden). Through the production of subitaneous eggs, the population size of a given rotifer species can double in as little as 15 hours. Typically, each female produces 4 to 40 amictic eggs per lifetime, and 20 to 40 or more generations of genetically identical amictic females can occur yearly.

Under certain conditions, however, a different type of monogonont female is produced.[1] These **mictic** females produce their eggs through meiosis, so the eggs are haploid (Fig. 13.13). In the absence of fertilization, mictic eggs develop into haploid males. The males are usually smaller and morphologically dissimilar to the females of the same species, but they are always fast swimmers. The males are nonfeeding, lacking both mouth and anus. Their job must be done quickly, since they usually cannot survive for more than a few days. Typically, the males are ready to fertilize eggs within an hour of hatching! The fertilized eggs form **resting eggs,** also known as **winter eggs** (Fig. 13.14). These resting eggs are highly resistant to a variety of physical and chemical stresses, permitting the developing embryos to withstand unfavorable conditions. During such periods, the embryos are in a state of developmental arrest—a state of **diapause** or **cryptobiosis.** Hatching occurs after

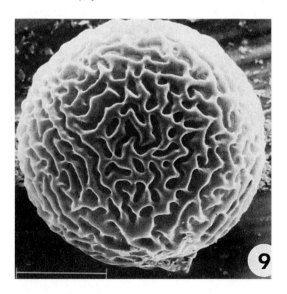

Figure 13.13
Photograph of female *Brachionus calyciflorus* with mictic resting eggs.
Courtesy of John J. Gilbert and Allen Press. From Wurdak, Gilbert, and Jagels, 1978. Trans. Amer. Microsc. Soc. 97:49.

Figure 13.14
Scanning electron micrograph of resting egg of *Asplanchna intermedia.   Courtesy of John J. Gilbert and Allen Press. From Gilbert and Wurdak, 1978. Trans. Amer. Microsc. Soc. 97:330.*

---

1. See *Topics for Further Discussion and Investigation,* no. 1, at the end of the chapter.

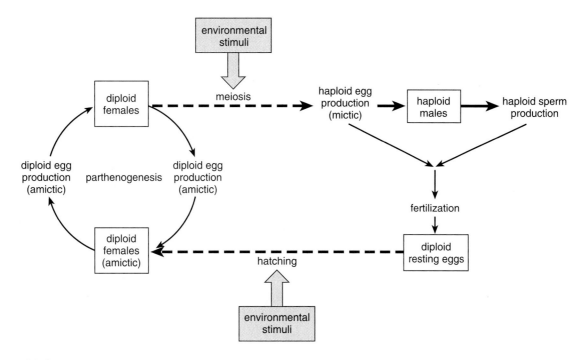

Figure 13.15

The life cycle of monogonont rotifers. In the absence of specific stimuli, reproduction is exclusively asexual, by means of parthenogenesis. Environmental factors that may influence reproduction include changes in food quality and quantity, changes in photoperiod, and changes in temperature.

conditions improve. Resting eggs always give rise to amictic females, which go on to reproduce by means of parthenogenesis (Fig. 13.15). Generally, only one or two mictic generations occur per year.[2]

Recently, researchers have discovered that a small percentage (< 0.5%) of females in a population may be characterized as neither mictic nor amictic. Instead, these females are **amphoteric;** a single female produces some haploid eggs by meiosis (which become males if unfertilized) and also produces some diploid eggs by mitosis.

It should be clear by now that most free-living rotifers encountered in the field, whether the rotifers are mobile or sessile, are females.

Other Features of Rotifer Biology

Digestive System

Most rotifer species have a tubular digestive system with an anterior mouth and a posterior anus (Figs. 13.4 and 13.11). Cilia lining the inner surface of the gut move food through the digestive system. Digestion is largely extracellular and takes place in the stomach. **Gastric glands,** or **gastric caeca,** associated with the stomach, contribute digestive enzymes. Undigested wastes pass through a short intestine and discharge into a **cloaca,** which, by definition, also receives the terminal ducts of the excretory and repro-

ductive systems. The anus opens dorsally, near the junction of the trunk and the foot. Some bdelloid rotifers have no pronounced stomach cavity and no anus. Instead, the "stomach" is a continuous syncytial mass through which food is circulated. In such species, digestion is primarily intracellular. Except for members of the Seisonidea, male rotifers lack a functional digestive system.

Nervous and Sensory Systems

The rotifer brain consists of a bilobed mass of ganglia lying dorsal to the mastax (Fig. 13.16). Nerves extend throughout the body, connecting the brain to the musculature and organ systems and to a variety of sensory receptors. Sensory bristles and, usually, three antennae (two lateral and one median-dorsal) serve as chemoreceptors and mechanoreceptors. A pigmented, cuplike photoreceptor often lies directly on the brain. Additional photoreceptors may occur in the corona.

Excretion and Water Balance

Excretion is at least partly accomplished by diffusion across the general body surface. In addition, all rotifers contain a pair of protonephridia, resembling those of flatworms. Flagella activity is presumed to create a negative pressure

2. See *Topics for Further Discussion and Investigation,* no. 2.

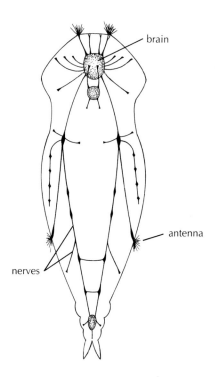

brain

antenna

nerves

Figure 13.16

Rotifer nervous system.   *From Pennak, Fresh-Water Invertebrates of the United States, 2d ed. Copyright © 1978 John Wiley & Sons, New York. Reprinted by permission of John Wiley & Sons, Inc.*

within each nephridium, drawing fluid in from the pseudocoel. As many as 50 flame bulbs may be associated with each protonephridium. The two collecting tubules, one from each nephridium, lead to a common **bladder** (Fig. 13.11b). Bladder contractions, up to six per minute, expel the fluid into the **cloaca,** which also receives the products of the digestive and reproductive systems. The tissues and body fluids of freshwater rotifers contain a significantly higher concentration of dissolved materials than is found in the surrounding medium, so water continually diffuses into the animal across the permeable body surface. The major role of the protonephridia would thus seem to be the maintenance of water balance and body volume, rather than the removal of excretion products, at least in most species. The osmotic gradient that exists across the body wall of freshwater rotifers, and the accompanying diffusional influx of water, may aid the rotifer in maintaining body turgor.

Three Phyla of Possible Rotifer Relatives

Phylum Gastrotricha

Phylum Gastro · tricha
(G: belly hair)
gas-tro-trik´-ah

Gastrotrichs are about the size of rotifers—rarely as large as 1 mm—and are common members of benthic (bottom-dwelling) communities, both freshwater and marine, sometimes occurring in concentrations of up to 100,000 individuals per m². About 200 gastrotrich species live in freshwater, and another 200 or so species are marine. Gastrotrichs have long been phylogenetically allied with rotifers and other aschelminths, but the association has become uneasy of late. Present histological studies suggest that gastrotrichs are acoelomate; a blastocoel forms during development, but it does not persist into adulthood. The pseudocoelomate status of gastrotrichs is now questionable at best. But how best to realign the gastrotrichs with other phyla is not yet clear. Perhaps the definition of an aschelminth will have to change instead.

Most gastrotrichs live interstitially in the spaces between sediment particles. The head bears a number of sensory bristles, and the body has numerous spines (Fig. 13.17). Despite their superficial resemblance to rotifers, gastrotrichs are clearly beasts of a different phylum: In particular, they lack the characteristic rotifer corona and mastax. Many biologists now feel that gastrotrichs are far more closely related to nematodes than to rotifers. In any event, all gastrotrichs possess both circular and longitudinal body wall muscles. The ventral surface is amply ciliated (as the name "gastrotrich" implies), so gastrotrichs can glide over a substrate and even swim short distances. Epidermal cells are monociliated in some species, with a single cilium per cell, and multiciliated in others. The pattern of ciliation is an important taxonomic characteristic. Gastrotrichs also can form temporary attachments to solid surfaces, as can rotifers. More like the free-living flatworms, however, they possess a double-gland system in which one gland secretes the glue and the other secretes a de-adhesive to release the attachment. They cling so tightly to particles that zoologists must first anesthetize them with magnesium chloride ($MgCl_2$) to dislodge them for enumeration and further study.

There are no known parasitic species or carnivorous species; all gastrotrichs eat detritus, bacteria, diatoms, or protozoans. All have a linear digestive system, with an anterior mouth and a posterior anus.

Like rotifers and other aschelminths, gastrotrichs lack any specialized respiratory or circulatory systems. Despite their small size, however, they do possess a discrete, protonephridial excretory system, which differs in morphological detail from those of both flatworms and rotifers; protonephridia are especially common in the freshwater species. In freshwater species, the protonephridia most likely function in maintaining osmotic concentration and body volume, in addition to their presumed function in removing soluble wastes.

Like nematodes and rotifers, gastrotrichs demonstrate **eutely,** with all adults of any given species having the same number of cells; cell numbers increase only in early development. Similar to that of nematodes but

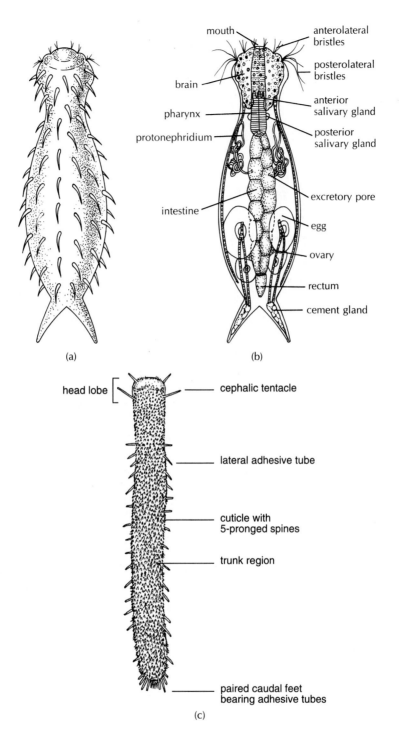

(a)

mouth

anterolateral
bristles

posterolateral
bristles

brain

pharynx

anterior
salivary gland

posterior
salivary gland

protonephridium

excretory pore

intestine

egg

ovary

rectum

cement gland

(b)

head lobe

cephalic tentacle

lateral adhesive tube

cuticle with
5-pronged spines

trunk region

paired caudal feet
bearing adhesive tubes

(c)

Figure 13.17

Typical gastrotrich (*Chaetonotus* sp.). (a) External appearance.
(b) Internal anatomy. (c) *Tetranchyroderma* sp., a marine
gastrotrich. The animal forms temporary attachments to sand
grains using its adhesive tubes. The specimen illustrated was

about 500 μm long.   *(a) After Brunson; after Zelinka. (b) From Brown; after
Remane. (c) From L. Margulis and K. Schwartz, Five Kingdoms, 2d ed. Copyright © 1988
W. H. Freeman and Company, New York. Reprinted by permission of the authors.*

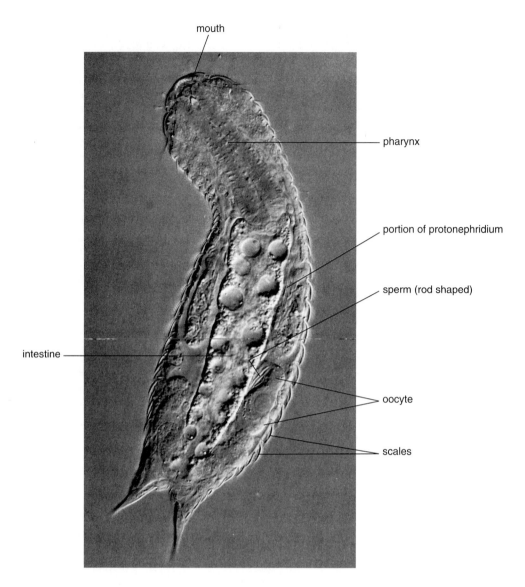

mouth

pharynx

portion of protonephridium

sperm (rod shaped)

intestine

oocyte

scales

Figure 13.18
Photograph of *Lepidodermella squamata*, a freshwater gastrotrich.
Courtesy of Mitchell J. Weiss.

unlike that of rotifers, the gastrotrich body is covered with an external cuticle; the rotifer cuticle is typically intracellular. Also unlike rotifers, gastrotrichs—particularly the marine species—are mostly hermaphroditic (Fig. 13.18); no other aschelminth can so boast. Freshwater gastrotrichs commonly reproduce by parthenogenesis. In sexual reproduction, fertilization is always internal, and the fertilized eggs display determinate cleavage, as in rotifers and other aschelminths. However, the cleavage pattern is bilateral and radial rather than spiral. There are no free-living larvae in the life cycle.

Phylum Kinorhyncha (= Echinoderida)

Phylum Kino · rhyncha (= Echino · derida)
(G: movable snout; G: hedgehog hide)
kī-nō-rink´-ah   ē-kī-nō-deer´-id-ah

Kinorhynchs are bona fide pseudocoelomates, and the approximately 80 described species are exclusively marine. They look somewhat like gastrotrichs (Fig. 13.19) but lack external cilia. Instead of gliding or swimming,

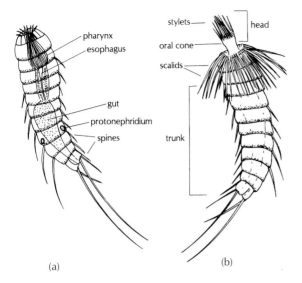

Figure 13.19

Typical kinorhynch (*Echinoderella* sp.), with head withdrawn (a) and protruded (b). (a,b) *After Hyman.*

they crawl through the mud in which they live, using the active, eversible head to thrust forward; a series of recurved spines along the length of the body prevents the animal from backsliding as the posterior part of the body is then pulled along.

The body is covered by an external, chitinous cuticle. The cuticle, musculature, and nervous elements are distinctly segmented. Such segmentation is unique among aschelminths. The first segment is the head, which displays several rings of curved spines called **scalids** and an additional, inner ring of piercing stylets surrounding the mouth opening (Fig. 13.19b). Since the head bears the mouth and is eversible—giving rise to the most widely used phylum name (Kinorhyncha, meaning "movable snout")—it is properly termed an **introvert.** The second segment, called the "neck," typically bears numerous cuticular plates that seal the opening once the head has been retracted. In some other kinorhynch species, both the head and neck are retracted, and the opening is sealed with plates on the third segment. The trunk, always consisting of 11 segments, bears numerous spines and adhesive tubes; these are important to the animal in locomotion and to the systematist in species identification. Like other free-living aschelminths, kinorhynchs are small; few are as long as 1 mm. As is so common for such small organisms, kinorhynchs lack specialized respiratory or circulatory systems. The excretory system consists of one pair of solenocytic protonephridia.

Kinorhynchs can occur at impressively high densities: Aggregations of about two million individuals per square meter of sediment have been reported. Kinorhynchs mostly eat bacteria and all have linear digestive tracts. All species are dioecious, with separate males and females, but developmental details are poorly known. Juveniles molt the cuticle periodically as they grow, passing through six stages before reaching adulthood. There are no free-living larvae in the life cycle.

Phylum Loricifera

Phylum Lorici · fera
(L: armor bearing)
lor-ih-sif´-er-ah

One would think that all the animal phyla would have been discovered years ago. Even the kinorhynchs, for example, despite their small size and specialized habitat, have been well known for about 150 years. Yet, in 1983, a new phylum, the Loricifera, was erected to hold a number of newly discovered little animals, *Nanaloricus mysticus* (Fig. 13.20). Nearly 100 more loriciferan species are now known, although most have not yet been formally described. All known loriciferans are marine. They are only about 200 μm to 300 μm long, and all live interstitially in subtidal sediments, clinging tightly to the surrounding sand grains. Loriciferans were discovered by accident, when newly collected sediments were rinsed with tap water instead of the usual seawater; the osmotic shock apparently caused the animals to release their grips on the sand grains, bringing them to zoologists' attention for the first time.

Like the kinorhynchs just discussed, loriciferans begin with an anterior introvert surrounded by recurved spines, called **scalids;** like that of the kinorhynchs, the introvert is retractable and bears the mouth at its end (by definition). The mouth is encircled by piercing stylets. The neck is made of several segments, rather than the single segment of kinorhynchs, and bears numerous plates that likely protect the anterior end when it is withdrawn, as with kinorhynchs.

The posterior half of the body is covered by six overlapping plates comprising an external cuticle, or **lorica,** which is molted as the juveniles grow. Many rotifer species also exhibit a distinct lorica, making them superficially resemble loriciferans. However, like kinorhynchs (and unlike gastrotrichs and rotifers), loriciferans lack

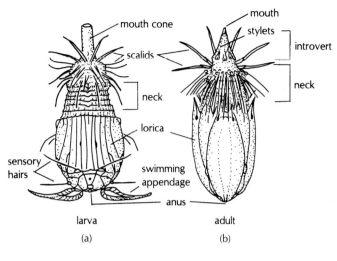

Figure 13.20
Nanaloricus mysticus, the first known member of the recently erected phylum Loricifera. (a) Higgins larva. (b) Adult. *From Kristensen, 1983. Zeit. Zool. Syst. Evol.-forsch. 21:163.*

external cilia. The animals are dioecious and fertilization is probably internal, but no details of their sex lives or development are yet known. Until the appropriate histological and embryological studies are done, the nature of the conspicuous body cavity cannot be assessed: Is it a pseudocoel, or is it coelomic? The preadult stage, called a **Higgins larva,** differs morphologically from the adult mainly in having a pair of posterior, unjointed swimming appendages and several pairs of locomotory spines on the ventral surface of the lorica (Fig. 13.20a). The Higgins larva shares certain morphological characteristics with the juveniles of priapulids, nematomorphs, kinorhynchs, and rotifers, but the phylogenetic affiliations between loriciferans and any of these other groups are presently unclear. Certainly, the loriciferans, priapulids, kinorhynchs, and nematomorphs seem closely related, and several researchers now advocate grouping all of these animals within a single phylum. The discovery of so many shared characteristics may eventually bind at least some of the various aschelminth groups more closely together again.

Taxonomic Summary

Phylum Rotifera
  Class Seisonidea
  Class Bdelloidea
  Class Monogononta
Phylum Gastrotricha
Phylum Kinorhyncha
Phylum Loricifera

Topics for Further Discussion and Investigation

1.   Investigate the environmental factors that appear to trigger the production of mictic females in monogonont rotifers.

Birky, C. W., Jr. 1964. Studies on the physiology and genetics of the rotifer, *Asplanchna.* I. Methods and physiology. *J. Exp. Zool.* 155:273.

Gilbert, J. J. 1963. Mictic-female production in the rotifer *Brachionus calyciflorus. J. Exp. Zool.* 153:113.

Gilbert, J. J., and J. R. Litton Jr. 1978. Sexual reproduction in the rotifer *Asplanchna girodi:* Effects of tocopherol and population density. *J. Exp. Zool.* 204:113.

Lubzens, E., and G. Minkoff. 1988. Influence of the age of algae fed to rotifers (*Brachionus plicatilis* O. F. Müller) on the expression of mixis in their progenies. *Oecologia (Berl.)* 75:430.

Pourriot, R., and P. Clement. 1975. Influence de la durée de l'éclairement quotidien sur le taux de femelles mictiques chez *Notommata copeus* Ehr. (rotifère). [Influence of photoperiod on the mictic-female production rate in *Notommata copeus* Ehr. (rotifer).] *Oecologia (Berl.)* 22:67.

Snell, T. W., and E. M. Boyer. 1988. Thresholds for mictic female production in the rotifer *Brachionus plicatilis* (Muller). *J. Exp. Marine Biol. Ecol.* 124:73.

2.   Discuss the physical and biological factors influencing the dynamics of natural rotifer populations.

Edmondson, W. T. 1945. Ecological studies of sessile Rotatoria. II. Dynamics of populations and social structures. *Ecol. Monog.* 15:141.

Gilbert, J. J. 1989. The effect of *Daphnia* interference on a natural rotifer and ciliate community: Short-term bottle experiments. *Limnol. Oceanogr.* 34:606.

Gilbert, J. J., and C. E. Williamson. 1978. Predator-prey behavior and its effect on rotifer survival in associations of *Mesocyclops edax, Asplanchna girodi, Polyarthra vulgaris,* and *Keratella cochlearis. Oecologia (Berl.)* 37:13.

King, C. E. 1972. Adaptation of rotifers to seasonal variation. *Ecology* 53:408.

Ricci, C., L. Vaghi, and M. L. Manzini. 1987. Desiccation of rotifers (*Macrotrachela quadricornifera*): Survival and reproduction. *Ecology* 68:1488.

Snell, T. W., and M. A. Nacionales. 1990. Sex pheromone communication in *Brachionus plicatilis* (Rotifera). *Comp. Biochem. Physiol.* 97A:211–16.

Wallace, R. L., and W. T. Edmondson. 1986. Mechanism and adaptive significance of substrate selection by a sessile rotifer. *Ecology* 67:314.

Taxonomic Detail

Phylum Rotifera

The approximately 2,000 species are divided among three classes.

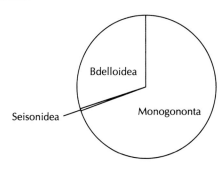

Class Monogononta.
This class contains nearly 70% of all rotifer species. Three orders.

Order Ploima.
Most rotifer species are contained in this order. Under most conditions, the populations consist of amictic females reproducing parthenogenetically, but mictic eggs are produced under certain environmental conditions. Most species live in freshwater, but a few species are marine. Fourteen families.

Family Brachionidae.   *Brachionus, Keratella, Epiphanes, Lepadella.* Individuals are mostly planktonic, although the family also includes some benthic (bottom-dwelling) species, such as *Lepadella.*

Family Asplanchnidae.   *Asplanchna.* All species are planktonic predators, often preying on each other. Individuals get as long as about 2 mm.

Family Synchaetidae.   *Synchaeta, Polyarthra.* These rotifers are common in both marine and freshwater habitats. Most of the truly planktonic marine rotifers are found in this family.

Family Proalidae.   *Proales.* A number of species are internal parasites of animals (living in such bizarre hosts as snail eggs or certain protozoans) or plants, while others live in symbiotic association with such diverse animals as colonial protozoans, hydrozoans, crustaceans, or insect larvae.

Order Flosculariaceae.
Four families.

Family Flosculariidae.   *Floscularia.* This group includes some common benthic rotifers that build tubes on freshwater plants, using pellets formed from detritus and carefully cemented into place. Some species, including *Sinantherina,* form colonies in a common gelatinous mass, which may be attached to substrates or free-floating in the water.

Order Collothecaceae.
One family.

Family Collothecidae.   *Collotheca, Stephanoceros.* Most species in this family are sessile and live attached to solid freshwater substrates, encased in gelatinous secretions. The anus empties anteriorly, on the neck, as an adaptation to living as a tube dweller.

Class Bdelloidea.
Members of this group occupy a wide range of habitats, including damp soil, wet moss, hot springs, and Antarctic lakes, and a few species live in the sea. Bdelloid rotifers are extremely numerous at the bottoms of freshwater lakes and ponds worldwide. The life cycles seem entirely amictic; reproduction is exclusively by parthenogenesis, and males are unknown. Individuals can tolerate a remarkable range of temperatures and long periods of desiccation. Four families.

Family Philodinidae.   *Philodina, Rotaria, Zelinkiella.* Most species are free-living in freshwater and moist terrestrial environments, but a few species (e.g., *Embata*) are parasitic on the gills of certain crustaceans (phylum Arthropoda). *Zelinkiella* lives only in skin depressions on the bodies of sea cucumbers (Echinodermata: Holothuroidea).

Class Seisonidea.
One family, containing only two species.

Family Seisonidae.   *Seison.* These rotifers are all marine symbionts. Instead of a foot with toes, the animals' posterior end terminates in a leech-like adhesive disc. *Seison* lives exclusively on the gills of a European crustacean. The animals are typically 2 mm to 3 mm long (large, for rotifers), but they have a highly reduced corona. Males and females are equally abundant and equally well developed.

Phylum Gastrotricha

The approximately 430 species are divided among two orders.

Order Chaetonotida.
Chaetonotus, Lepidodermella. Although most species live in freshwater (including freshwater bogs), representatives occur in all aquatic habitats. Individuals are always small, with members of most species less than 300 μm long. The order includes nearly 65% of all gastrotrich species, many of which reproduce mostly by parthenogenesis. Seven families.

Order Macrodasyida.
Turbanella, Tetranchyroderma, Macrodasys, Dactylopodola. Members of this order occur in marine, brackish, and estuarine habitats; there are no freshwater species. This order contains the largest gastrotrichs, some of which reach lengths of about 3.5 mm. Most species are either simultaneous hermaphrodites or sequential hermaphrodites. Six families.

Phylum Kinorhyncha (= Echinoderida)

All approximately 80 species are marine. They are distributed among only two orders.

Order Cyclorhagida.
Echinoderes. This order contains a highly diverse group of kinorhynchs. Some species are intertidal, while others are found only at depths of several thousand meters. Many species are free-living, while some appear to be commensal with such other invertebrates as sponges, bryozoans, or hydrozoans. The order contains slightly more than 60% of all kinorhynch species, and all are small, certainly less than about 500 μm long. Four families.

Order Homalorhagida.
Pycnophyes, Kinorhynchus. This group contains the largest of the kinorhynchs, with some individuals growing to nearly 1 mm in length. Individuals commonly live at depths of several thousand meters. Two families.

Phylum Loricifera

Nanaloricus. All known loriciferans, about 100 species, live interstitially in marine sediments. Most species have yet to be formally described.

Some General References About Rotifers, Kinorhynchs, Gastrotrichs, and Loriciferans

Harrison, F. W., and E. E. Ruppert, eds. 1991. *Microscopic Anatomy of Invertebrates, Volume 4: Aschelminthes.* New York: Wiley-Liss.

Higgins, R. P., and H. Thiel, eds. 1988. *Introduction to the Study of Meiofauna.* Washington, D.C.: Smithsonian Institution Press.

Hyman, L. H. 1951. *The Invertebrates, Volume 3. Acanthocephala, Aschelminthes, and Entoprocta.* New York: McGraw-Hill.

Morris, S. C., et al., eds. 1985. *The Origins and Relationships of Lower Invertebrates. Systematics Association, Special Volume 28.* Oxford: Clarendon Press.

Parker, S. P., ed. 1982. *Classification and Synopsis of Living Organisms,* vol. 1. New York: McGraw-Hill, 857–77.

Thorpe, J. H., and A. P. Covich, eds. 1991. *Ecology and Classification of North American Freshwater Invertebrates.* New York: Academic Press, 173–87 (gastrotrichs and rotifers).

Willmer, P. 1990. *Invertebrate Relationships.* New York: Cambridge University Press.

14

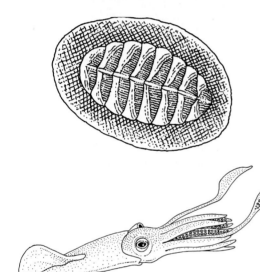

The Molluscs

Introduction and General Characteristics

> *Phylum Mollusca*
> (L: soft)
> mō -lusk´-ah

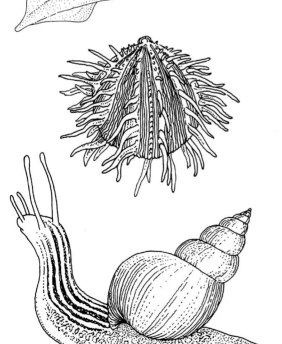

The phylum Mollusca is one of the largest of all animal phyla, including between 50,000 and 110,000 living species, depending upon who is doing the counting. These species are distributed among some extremely dissimilar-looking organisms, making the molluscan body plan probably the most malleable in the animal kingdom. Remarkably, clams, snails, and octopuses are all members of the phylum Mollusca!

There really is no "typical" mollusc. Most, but not all, molluscs have shells consisting primarily of calcium carbonate set in a protein matrix. Organic material may comprise about 35% of the shell's dry weight in some gastropod species and up to about 70% of the dry weight in bivalves. The shells of most molluscs (including all gastropods and bivalves) have a thin, outer organic layer (the **periostracum**); a thin, innermost calcareous layer (the **nacreous layer**); and a thick, calcareous middle layer (the **prismatic layer**) (Fig. 14.1). Both the organic and inorganic components of the shell are secreted by specialized tissue known as the **mantle.** If a grain of sand, parasite, or other foreign particle becomes trapped between the mantle and the shell's inner surface, a pearl may form over a period of years. Natural pearl formation is a fairly rare event: Perhaps only one oyster in 1,000 is likely to harbor a valuable pearl naturally. Humans increase the frequency of pearl production by surgically implanting pieces of shell (usually from freshwater bivalves) or plastic spheres between the shell and mantle of mature oysters, and then keeping the oysters alive and

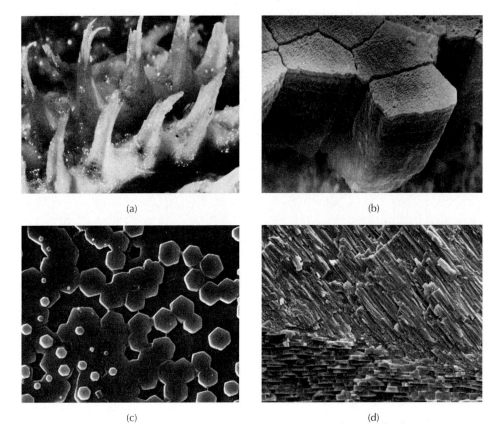

(a)

(b)

(c)

(d)

Figure 14.1

Molluscan shell structure. (a) Periostracal spines from the shell of the gastropod *Trichotropis cancellata*. (b) Tall calcite prisms from the prismatic layer of an oyster, *Crassostrea virginica* (Gmelin). The shell has been fractured and treated briefly with Chlorox to dissolve away the proteinaceous matrix (conchiolin) that normally occurs between individual prisms. (c) Individual calcareous tablets from the nacreous layer of a mussel shell, *Geukensia demissa* (Dillwyn). Each nacreous tablet is about 7 μm from side to side. (d) Fractured shell section of a deep-sea mussel, illustrating the outer prismatic layer (top) and calcareous tablets of the underlying nacreous layer.   (a) *From D. J. Bottjer, Third North American Paleontological Convention, Proceedings, 1982, Vol. 1, pp. 51–56. (b) Courtesy of M. R. Carriker, from M. R. Carriker et al., 1980. Proc. Nat. Shellf. Assoc. 70:139. (c) From Dr. R. A. Lutz, "Anaerobiosis and a Theory of Growth Line Formation." Science 198:1222, fig. 1, 12-23-77. Copyright 1977 by the AAAS. (d) Courtesy of R. A. Lutz.*

protected for five to seven years. Note that cultured pearls *form* in a perfectly normal manner; humans intervene only in getting the process started.

Although the mantle is a major molluscan characteristic, its role varies substantially in different molluscan groups. Similarly, most molluscs have a **foot,** but it is also highly modified for a variety of functions in different groups.

The "average" mollusc has a characteristic cavity lying between the mantle and the viscera. This **mantle cavity** usually houses the comb-like molluscan gills, known as **ctenidia** (*ctenidi* = G: comb), and also generally serves as the exit site for the excretory, digestive, and reproductive systems. The ctenidium, when present, may have a purely respiratory function or may function in the collection and sorting of food particles as well. A chemoreceptor/tactile receptor known as the **osphradium** (*osphra* = G: a smell) is generally located adjacent to the ctenidium (Figs. 14.2 and 14.13).

The molluscan coelom is very small, being restricted largely to the area surrounding the heart and gonads. Some zoologists have suggested that this cavity is in fact not homologous with the larger body cavities of annelids, sea urchins, and other noncontroversial coelomates, and that molluscs evolved directly from acoelomate flatworm ancestors; by this line of reasoning, molluscs are in fact acoelomate! Current molecular data, however, support an alternative view—that molluscs have descended instead from some coelomate ancestor and that the body cavity experienced a substantial reduction in size in the course of subsequent evolution. Either way, the molluscan "coelom" is small, and has no locomotory role. On the other hand, blood sinuses comprising a **hemocoel** ("blood cavity") are well developed. This hemocoel serves as a hydrostatic skeleton in the locomotion of some molluscs.

Many molluscs possess a feeding structure known as the **radula.** The radula consists of a firm ribbon,

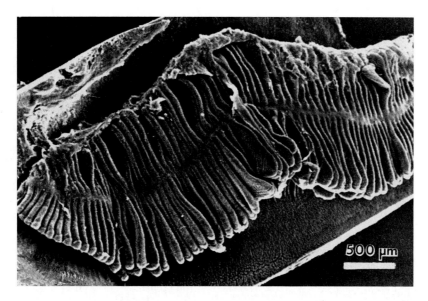

Figure 14.2

The osphradium (scanning electron micrograph) from the mantle cavity of the southern oyster drill, *Thais haemastoma canaliculata* (Mollusca: Gastropoda). The leaflets of the osphradium, which is 4 mm to 5 mm long in this species, hang down into the incurrent water stream within the mantle cavity; water thus contacts the osphradium before contacting the gill. *Courtesy of D. W. Garton and R. A. Roller. From Garton et al., 1984. Biological Bulletin 167: 310–21. Permission of Biological Bulletin.*

composed of chitin and protein, along which are found rows of sharp, chitinous teeth (Fig. 14.3). The ribbon is produced from a **radular sac** and is underlain by a supportive cartilage-like structure called the **odontophore** (literally, G: tooth bearer). The odontophore-radular assembly, together with its complex musculature, is known as the **buccal mass** (*bucca* = L: cheek). For feeding, the buccal mass is protracted so that the odontophore extends just beyond the mouth. The radular ribbon is then moved forward over the leading edge of the supporting odontophore and then pulled back. As each row of teeth passes back over the edge of the odontophore, the teeth automatically stand upright and rotate laterally, rasping food particles from the substrate and bringing them into the mouth as the radula is withdrawn. As old teeth are worn down or broken off at the anterior end of the radular ribbon, new teeth are continually being formed and added onto the ribbon's posterior end in the radular sac.

As might be inferred from the cautious wording in the preceding paragraphs, generalizations about molluscs are difficult to make.

Extant molluscs are distributed among seven classes. Six of these seven classes are represented by fossils formed some 450 million years ago, along with one additional class of molluscs, the Rostroconchia, whose clam-like members went extinct some 225 million years ago (Fig. 14.4). All told, there are at least 35,000 molluscan species known only as fossils. Only one class of molluscs, the Aplacophora, has left no fossil record.

Class Polyplacophora

Class Poly · placo · phora
(G: many plate bearing)
pol-ē-plah-koff´-or-ah

The 800 species in the class Polyplacophora are known as "chitons" (not to be confused with chitin, a polysaccharide!). Chitons are typically 3–10 cm (centimeters) long and generally found close to shore, particularly in the intertidal zone; they are restricted to living on hard substrate, especially rocks. A chiton's most distinctive external feature is its shell, which occurs as a series of eight overlapping and articulating plates covering the dorsal surface (Fig. 14.5a). These plates are partially or largely embedded in the mantle tissue that secretes them. Because the shell is multisectioned, the body can bend to conform to a wide variety of underlying substrate shapes. A chiton's thick lateral mantle is called the **girdle.** In most species, the girdle bears numerous calcareous spicules, secreted independently of the shell plates.

The mantle cavity of chitons takes the form of two lateral grooves on either side of the body (Fig. 14.5b). Up to about 80 **bipectinate** (L: double-combed, i.e., two-branched) ctenidia hang down from the roof of each groove, dividing each elongated mantle cavity into incurrent and excurrent chambers. Water is drawn into the

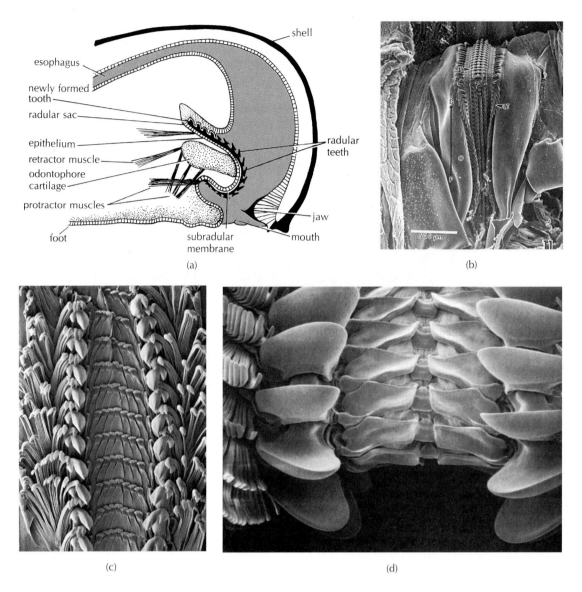

(a)

(b)

(c)

(d)

Figure 14.3

(a) Longitudinal section through the anterior of a gastropod, showing the relationship between the odontophore, radular ribbon, and mouth. (b) The radula and odontophore (scanning electron micrograph) projecting from the mouth of a marine snail, the southern oyster drill *Thais haemastoma canaliculata*.

O=odontophore; RT and LT=radular teeth. (c,d) Scanning electron micrographs of radular teeth of two different snail species: (c) *Montfortula rugosa*; (d) *Nerita undata*.   (a) *Modified after Runham and various sources.* (b) *Courtesy of Roller et al., 1984. American Malacological Bulletin 2:63–73.* (c,d) *Courtesy of C. S. Hickman, from Hickman, 1981. Veliger 23:189.*

incurrent chamber by the action of the gill cilia. The flow of water is anterior to posterior, so waste products are discharged posteriorly in the excurrent stream. The flow of blood through the individual gill lamellae is opposite in direction to the flow of water, facilitating gas exchange as discussed later (p. 236).

The foot extends along the animal's entire ventral surface and is completely covered by the overlying shell and girdle. Locomotion is accomplished by waves of muscular activity called "pedal waves," as in gastropods (p. 232). When disturbed, the chiton can press the girdle tightly against the substrate. By then lifting up the central portion of the foot (and the inner margin of the mantle tissue as well, if required), while retaining a tight

seal against the substrate along the entire outer margin of the foot (and girdle), the chiton is able to generate a suction that holds the animal tightly against the substrate. Mucus secretion along the girdle helps to maintain the grip. This ability to cling tightly to the substrate is a particularly effective adaptation for life in areas of heavy wave action.

The chiton nervous system is simple and ladder-like (Fig. 14.5f). Ganglia are lacking in many species, and in others they are only poorly developed. Sensory systems are also reduced: Adult chitons lack statocysts, tentacles, and eyes on the head. **Aesthetes**—abundant organs derived from mantle tissue and extending through holes in the shell plates—are thought to function, at least in some

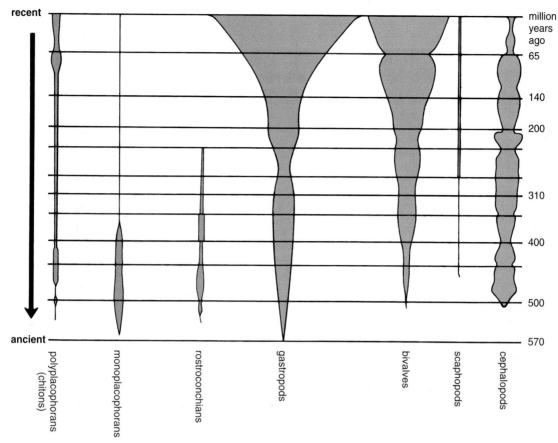

Figure 14.4
Molluscan species diversity as represented in the fossil record. The width of each bar is proportional to the number of species in each class. Members of the Rostroconchia superficially resembled the clams and related bivalves, except that they had hingeless, gaping shells. Bivalved molluscs and scaphopods may be descendants of rostroconchian ancestors.   *From R.S. Boardman, et al., Eds., Fossil Invertebrates. Copyright © 1987 Blackwell Scientific Publications, Osney Mead, England. Reprinted by permission.*

species, as general light receptors. Recent ultrastructural studies, however, suggest that aesthetes may function primarily in secreting periostracum, replacing material that is naturally abraded away in the highly turbulent environment that most chitons live in.

The chiton mouth and anus are at opposite ends of the body, and the digestive tract is linear (Fig. 14.5d). Food particles, usually algae, are typically scraped from the substrate by a radula/odontophore complex, although a few species are carnivores. Even though most chitons ingest algae, a crystalline style is not a component of the digestive system; instead, a pair of pharyngeal glands, often called **sugar glands,** release amylase-containing secretions into the stomach (Fig. 14.5d).

Chitons have a fossil record extending back some 500 million years (Fig. 14.4). The evolutionary relationships between the Polyplacophora and other molluscan classes are unclear, although there is no reason to suspect that any other molluscs evolved directly from chiton ancestors. Chitons probably diverged from the main line of molluscan evolution early on, which is also likely to be the case for the worm-like aplacophoran molluscs to be described next.

Class Aplacophora

Class A · placo · phora
(G: not shell bearing)
ā-plak-off´-or-ah

Aplacophorans are worm-shaped (**vermiform**) molluscs (Fig. 14.6) found in all oceans, mostly in deep water. Most species are quite small—usually only a few millimeters and rarely more than a few centimeters long. Like the cephalopods, scaphopods, monoplacophorans, and chitons, aplacophorans are entirely marine. The body is unsegmented and bears numerous calcareous spines or scales embedded in an outer cuticle. The spines or scales are secreted by individual cells in the underlying epidermis; there is no true shell. At least some species possess a style sac (often complete with style and gastric shield), small posterior mantle cavity with ctenidia, and a radula, although in most species the radula is apparently used for grasping rather than rasping. Aplacophorans have no conspicuous foot, although the members of one

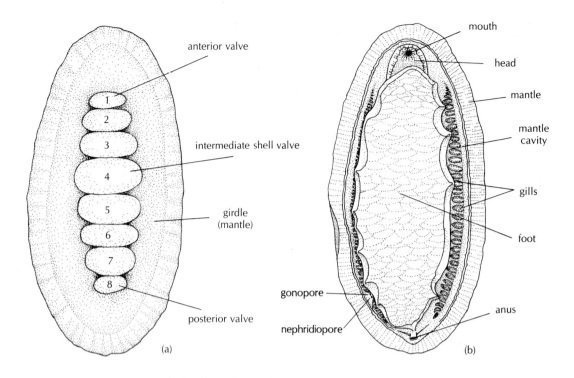

Figure 14.5

The polyplacophoran *Katharina tunicata*. (a) Dorsal view.
(b) Ventral view. All chitons are dorsoventrally flattened, with the
foot forming a large suction cup for clinging to firm substrates.
The shell is composed of eight articulating plates, permitting the
entire animal to curl and therefore conform to the topography of
the underlying surface. (c) *Tonicella lineata* firmly attached to rock
substrate in the intertidal zone off the coast of California. Note
the eight shell plates and conspicuous girdle. (d) Internal anatomy
of a polyplacophoran. (e) Same animal viewed dorsally, after the
shell plates were removed and the mantle tissue was dissected
away. (f) Nervous system of a typical chiton. *(a,b) Beck/Braithwaite,
Invertebrate Zoology, Laboratory Workbook, 3/e, © 1968. Reprinted by permission
of Prentice Hall, Upper Saddle River, New Jersey. (c) Courtesy of T. M. Niesen. (d,e) From
Sherman and Sherman, 1976. (f) Modified from Gardiner, M. S. 1972. Biology of the
Invertebrates. New York: McGraw–Hill. After Grasse.*

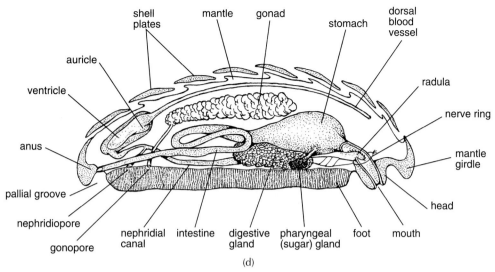

(d)

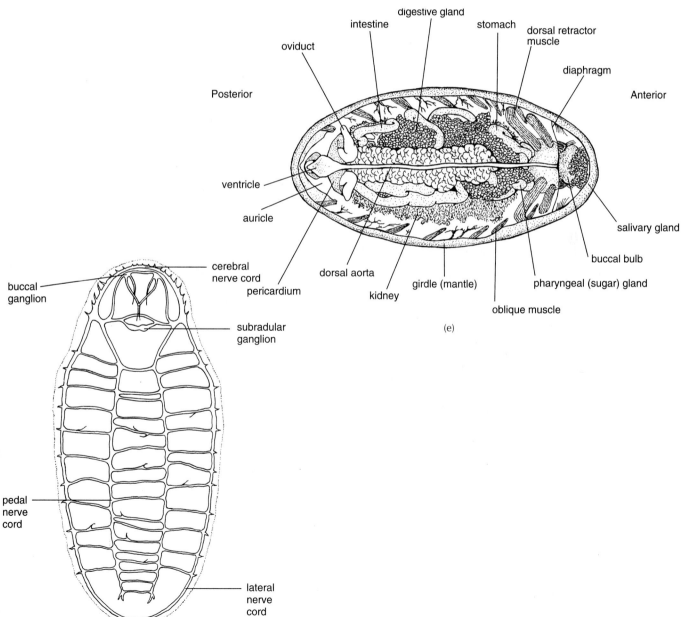

(e)

(f)

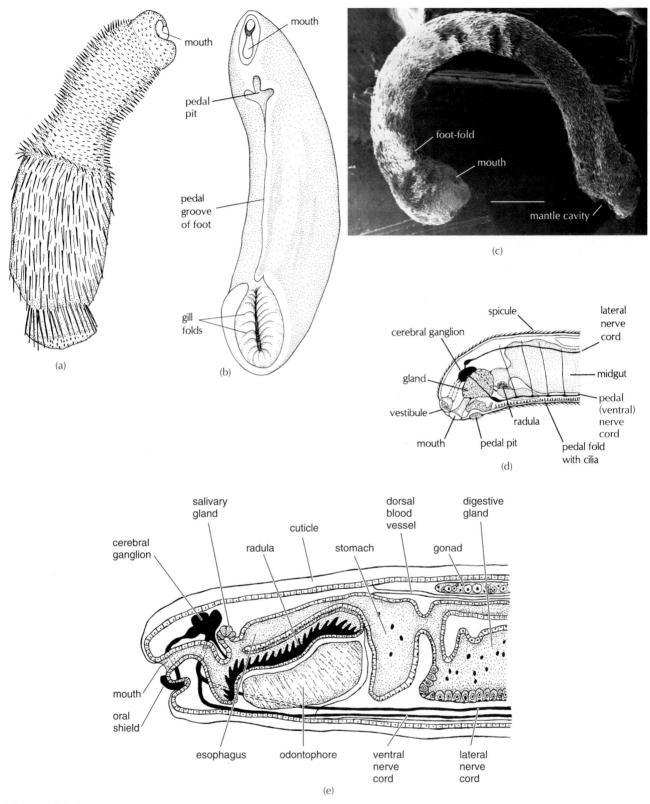

Figure 14.6

Members of the Aplacophora: (a) *Falcidens sp.* (b) *Neomenia carinata.* (c) Scanning electron micrograph of *Lyratoherpia* sp. (Scale bar = 0.5 mm.) (d) Anterior portion of an aplacophoran, showing the arrangement of nerve cords. (e) Internal anatomy, anterior end of *Limifossor talpoideus.* Note the well-developed radula and supporting odontophore. (a) *Courtesy of Dr. A. H. Scheltema.* (b) *From Hyman.* (c) *From Scheltema. A. H. et al. 1994. Microscopic Anatomy of*

Invertebrates, Vol. 5, New York: Wiley-Liss, 13–54. Photograph kindly provided by A. H. Scheltema. (d) *From R. S. Boardman, Fossil Invertebrates. Copyright © 1987 Blackwell Scientific Ltd., Oxford, England. Reprinted by permission.* (e) *From Harrison, Microscopic Anatomy of Invertebrates, Vol. 5. Copyright © 1994. John Wiley & Sons, Inc., New York, New York. Reprinted by permission.*

group (the Neomeniomorpha, or solenogastres) possess a nonmuscular, ciliated ridge located in a groove on the body's ventral surface and believed to be homologous with the foot of other molluscs. Solenogastres use the "foot" cilia to glide over the sediment along a mucous trail that they secrete as they go.

The Aplacophora are the only class of molluscs to have left no fossil record. About 300 extant aplacophoran species have been described to date, with many of the species known from only one or two specimens. Many aplacophorans burrow into or meander about on mud; other species live on cnidarians—primarily soft corals—on which they prey. A few species are interstitial, living in the spaces between sand grains and feeding on interstitial hydrozoans (phylum Cnidaria). The evolutionary relationships to other members of the phylum remain particularly obscure, in large part because aplacophoran biology has been so little studied.

For many years, aplacophorans were considered close relatives of the chitons and were placed with them in a single taxonomic group, the Amphineura. In the 1970s and 1980s, studies of aplacophoran spicule formation, radula structure, and anatomy argued against a close relationship with the chitons, but arguments in favor of such a relationship are now again being made. In particular, the aplacophoran nervous system consists of paired cerebral ganglia, giving rise to four linear, ganglionated nerve cords—two lateral, two ventral—interconnected in a ladder-like arrangement as in chitons (Fig. 14.6d). In addition, members of both groups form calcareous spicules in apparently identical fashion, through extracellular secretions from single cells. Molecular data that might help resolve the uncertainties have not yet been reported. In any event, it seems likely that aplacophorans are a primitive offshoot, diverging early from the main line of molluscan evolution before the advent of shell formation.

Class Monoplacophora

Class Mono · placo · phora
(G: one shell bearing)
mon-ō-plak-off´-or-ah

Prior to 1952, the class Monoplacophora was known only from the fossil record. Actually, some representatives were collected in the 1890s but were categorized as gastropods and ignored. Since their rediscovery and recognition as a distinct class of molluscs in 1952, about a dozen living species have been described, all marine and all collected from very deep water. A sin-

gle, unhinged shell is present (Fig. 14.7), as in many limpet-like gastropods. The shell of adult monoplacophorans is flattened rather than spirally wound, although the larval shell is spiral. The monoplacophoran foot is flattened, as in gastropods and polyplacophorans.

The mantle cavity takes the form of two lateral grooves, as in polyplacophorans; five or six pairs of gills hang down within the mantle grooves. Whether these gills are homologous with the typical molluscan ctenidium is, however, uncertain. In addition to the gills, the pedal retractor muscles, auricles and ventricles of the heart, gonads, and nephridia occur in multiple copies. Both a radula and a crystalline style are present, and the gut is linear, with the mouth being anterior and the anus posterior. As in the polyplacophorans and aplacophorans, the nervous system includes both lateral and pedal nerve cords (Fig. 14.7b).

Despite the small number of living species, monoplacophorans are well deserving of additional study: All of the molluscan groups remaining to be discussed—the scaphopods, gastropods, bivalves, and cephalopods—may well have evolved either from monoplacophorans directly or from an ancestor shared with the Monoplacophora.

Class Gastropoda

Class Gastro · poda
(G: stomach foot)
gas-trop´-ō-dah

The Gastropoda is the largest molluscan class, comprising 40,000 to 75,000 living species of snails and slugs distributed among marine, freshwater, and terrestrial environments. About 75% to 80% of all living molluscs are gastropods. Gastropods occupy very diverse habitats, including rivers, lakes, trees, deserts, the marine intertidal zone, the plankton, and the deep sea, and they exhibit a striking diversity of lifestyles, including suspension-feeding, carnivorous, herbivorous, deposit-feeding, and ectoparasitic species. The typical snail consists of a **visceral mass** (i.e., all of the internal organs) sitting atop a muscular **foot** (Fig. 14.8). The visceral mass is commonly protected by a univalved shell that is typically coiled, probably as an adaptation for efficient packaging of the visceral mass. Shell morphology differs considerably among species.[1]

The animal is attached to the inside of its shell by a **columellar muscle** (Fig. 14.8), which extends from within the animal's foot to the central axis of the shell;

---

1. See *Topics for Further Discussion and Investigation*, no. 1, at the end of the chapter.

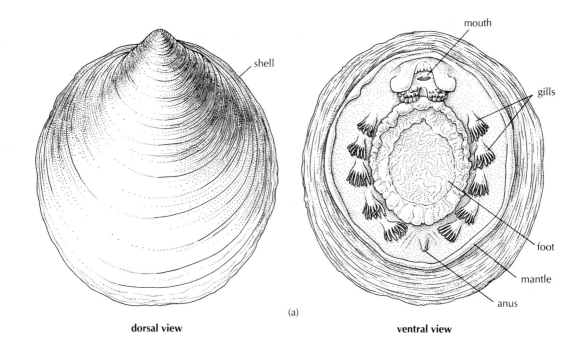

dorsal view | ventral view

(a)

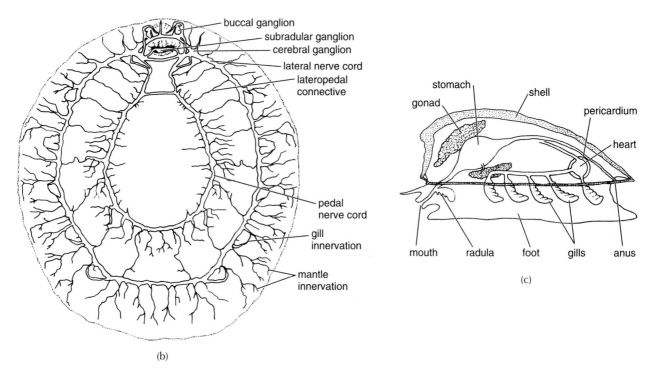

(b)

(c)

Figure 14.7

The monoplacophoran *Neopilina galatheae.* This animal was known only as a fossil until 1952, when it was dredged off the Pacific coast of Mexico from 5,000 m depth. (a) Dorsal and ventral views. The adult shell is about 3 cm in length. (b) Nervous system, dorsal view. (c) Internal anatomy, lateral view. (a) *After Lemche.* (b–c) *From E. N. K. Clarkson,* Invertebrate Palaeontology & Evolution, *2d ed. Copyright © 1986 Chapman & Hall, Div. International Thomson Publishing Services Ltd., U.K. Reprinted by permission.*

this central axis is known as the **columella** (Fig. 14.9b). The columellar muscle is important in most major body movements: protraction from the shell, retraction into the shell, twisting, raising the shell above the substrate, and lowering it back down.

The shell is typically carried so that it leans to the left side of the body. The shell axis is thus oblique to the long axis of the body, balancing the animal's center of mass over the foot. The shells of most gastropod species coil clockwise, to the right (Figs. 14.9 and 14.10a). That is, the shells are "right-handed" or **"dextral"** (*dextro* = L: the right-hand side). Probably as a consequence of space limitations within the coiled shell, the ctenidium, osphradium, kidney (nephridium), and heart auricle on

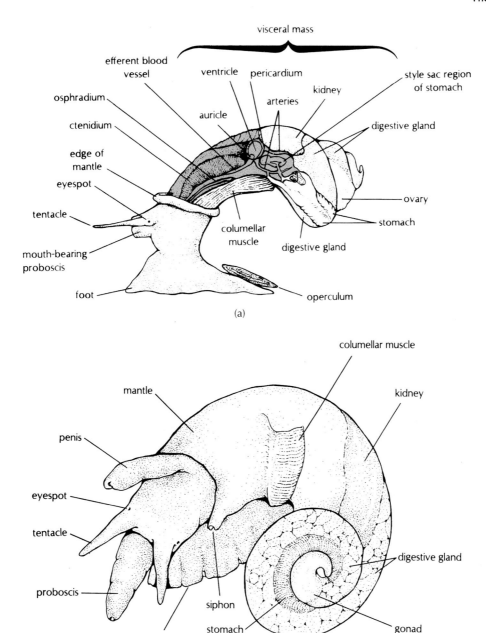

Figure 14.8

(a) Internal anatomy and gross morphology of the female periwinkle, *Littorina littorea*, removed from its shell. The animal occupies the entire inside of the shell, coiling around the columella as it grows. Blue structures are visible only by dissection. (b) The marine whelk Busycon sp., removed from its shell. (a) *From Fretter and Graham, A Functional Anatomy of Invertebrates. Copyright © 1976 Academic Press. Reprinted by permission of Harcourt Brace & Company Ltd., London.*

the right side of the body tend to be reduced or absent; only the primitive gastropods (many members of the order Archaeogastropoda) still exhibit paired structures (Fig. 14.11c,d). A relatively few snail species form shells that coil counterclockwise, to the left (Fig. 14.10b), and show a corresponding reduction or absence of the ctenidium, osphradium, kidney, and heart auricle on the left side of the body. Such species are said to be "left-handed," or **sinistral** in their coiling (*sinister* = L: the left-hand side). Note that it is possible for a dextrally coiled snail to produce a shell that appears to be left-

handed (Fig. 14.10c,d); some gastropods exhibit such a pattern, particularly in the larval stage.

Many gastropod species possess, in addition to their shells, fairly elaborate behavioral or chemical defenses against predators. These adaptations commonly take one of the following forms: (1) the gastropod senses the presence of potential predators, either chemically or by touch, and initiates appropriate escape, avoidance, or deterrent behavior; (2) the gastropod chemically senses the presence of injured individuals of its own species (i.e., conspecific individuals) and initiates appropriate escape

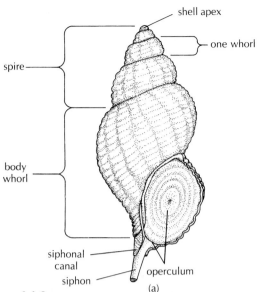

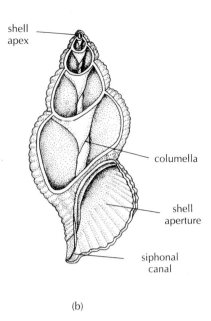

(a)

(b)

Figure 14.9

(a) Major external features of a gastropod shell. The siphon is a fold of mantle tissue through which water enters the mantle cavity. The operculum is a rigid, proteinaceous shield attached to the foot; when the animal fully withdraws into its shell, the operculum seals the aperture, protecting the animal from predators and physical stresses. (b) Longitudinal section through a typical gastropod shell. As the snail grows, its body coils around the columella.   (b) *After Hyman; after Dakin*

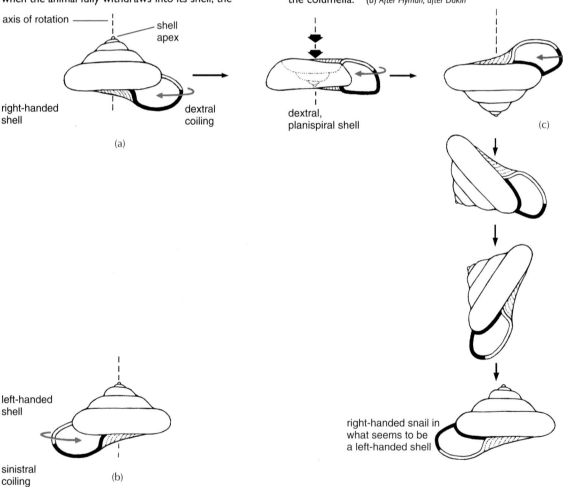

Figure 14.10

Major patterns of shell coiling among gastropods. Blue arrows indicate direction of shell coiling. Members of most species coil to the right and live in right-handed shells (a), although members of some species coil to the left and live in left-handed shells (b). As an intriguing complication, a right-handed snail can produce what appears to be a left-handed shell, by coiling up the central axis rather than down the central axis (c); only anatomical observations can confirm the true direction of body coiling. To visualize getting from (a) to (c), imagine pushing downward on the apex of the shell in (a), producing a planispiral shell, with all shell whorls in one plane, as an intermediate step. Continue pushing downward until you get to (c); note that the direction of shell coiling has not been altered in the process.   *From R. S. Boardman, et al., Eds.* Fossil Invertebrates. *Copyright © Blackwell Scientific Publications, Osney Mead, England. Reprinted by permission.*

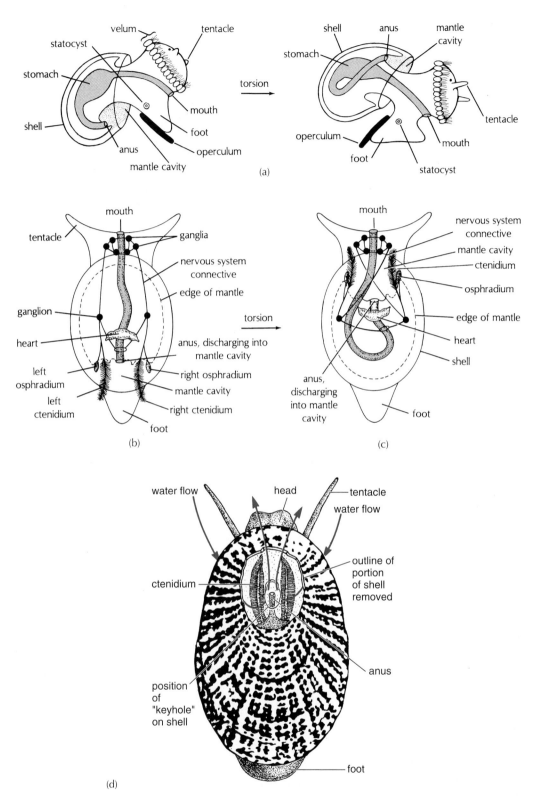

Figure 14.11

(a) Torsion in the free-swimming larva of a primitive prosobranch gastropod, Patella sp. Note that the mantle cavity is moved, along the right side of the animal, from the posterior to the anterior of the larva. Following torsion, the head and foot can be fully retracted into the mantle cavity and the aperture tightly sealed by the rigid operculum. (b,c) The consequences of torsion to the adult gastropod. (b) The untorted state of a hypothetical ancestral gastropod-like mollusc. (c) The rearrangement of internal anatomy following the evolution of torsion. Note that the primitive gill has leaflets extending from both sides of the central axis. As will be discussed later, this is termed a bipectinate gill. (d) Pathway of water circulation through the mantle cavity of a primitive gastropod—a keyhole limpet, order Archaeogastropoda—with paired gills. Water enters on both sides of the head and leaves through a circular opening (the "keyhole") in the shell.   (b,c) *From Hyman,* The Invertebrates, *Vol. III. Copyright © 1951 McGraw-Hill Book Company, New York. Reprinted by permission.*

behavior; or (3) the gastropod accumulates noxious organic compounds in its tissues, thereby becoming distasteful to potential predators.[2]

Of great importance in the evolutionary history and present-day biology of gastropods is the phenomenon of **torsion,** an anticlockwise twisting of most of the body (the **visceral mass**) through 180° during early development. As a consequence of torsion, the nervous and digestive systems become obviously twisted, and the mantle cavity moves from the rear of the animal to become positioned over the head (Fig. 14.11). This dramatic rearrangement of the internal and external anatomy is brought about largely through the particular orientation and asymmetric development of the retractor muscles attaching the head-foot of the developing embryo or larva to its shell. Because the retractor muscles on the body's right side develop before those on the left, the mantle cavity and associated organs get pulled along the animal's right side toward the front.

Torsion may occur within a few hours or even minutes in some species. It has no direct relationship to shell coiling: The two are separate and independent processes.

The adaptive significance of torsion has been the subject of considerable speculation.[3] The controversy focuses largely on whether torsion benefits the larva or the adult, or both, and part of the difficulty in interpreting the "why" of torsion is that movement of the mantle cavity anteriorly would seem to be a mixed blessing at best. Certainly, through torsion, the ctenidia and osphradia come to be located at the front of the animal, in the direction of locomotion; but torsion also shifts the anus so that it discharges over the head, creating a potentially serious (and seemingly distasteful) sanitation problem. Moreover, the story of subsequent gastropod evolution clearly involves compensating for the results of embryonic or larval torsion; that is, the more advanced gastropods, which are contained in the subclasses Opisthobranchia and Pulmonata, exhibit a marked reduction in the degree to which torsion occurs during development. In some species, an apparent **detorsion** occurs subsequent to torsion. The selective pressures responsible for the evolution of torsion, as with all "why did it happen?" evolutionary questions, are ultimately unknowable. Yet the phenomenon is too dramatic to be ignored (Research Focus Box 14.1). Torsion occurs in no other class of mollusc.

Small gastropod species may move largely through the action of cilia located on the ventral surface of the foot, but most species move by means of **pedal waves** of muscle contraction. Unlike peristaltic waves, pedal waves generally do not involve circular muscles or muscular contractions of great magnitude, and they are restricted to the animal's ventral surface.

The musculature of the foot is predominantly vertical (dorsoventral) and transverse (Fig. 14.12a). At the start of a pedal wave, the dorsoventral musculature contracts at the anterior portion of the foot. Apparently, the transverse muscles do not relax, which means that the foot cannot widen; instead, the foot is squeezed forward. A wave of contraction of the dorsoventral musculature then moves posteriorly, allowing the rest of the foot to catch up with the anterior (Fig. 14.12a). The edges of the foot are temporarily sealed against the substrate with mucus secreted by glands on the foot's ventral surface, so a small negative pressure (suction) is generated in the space between the substrate and the raised portion of the foot. The dorsoventral muscles are reextended when they relax, at least in part, by this negative external pressure; this small space thus acts as a hydrostatic skeleton, even though it is external to the body, allowing the musculature at the forward edge of the wave to antagonize that at the trailing edge. In this case, the hydrostatic skeleton operates through a temporary pressure decrease, essentially sucking the raised portion of the foot downward and extending the associated muscles, rather than through a temporary pressure increase.

The previous description applies to **retrograde waves** (*retro* = L: back; *grad* = L: step); the wave of muscular contraction travels in the direction opposite that in which the snail is moving. Pedal waves may also be **direct** (i.e., moving in the same direction as the animal) (Fig. 14.12b). The mechanical details of pedal wave formation may differ substantially among species.[4]

Gastropods play an important, although indirect, role in transmitting several major human diseases, with many species serving as obligate intermediate hosts in the life cycles of parasitic flatworms (phylum Platyhelminthes, class Trematoda; Chapter 7). Indeed, much research on the control of these flatworm parasites has focused on regulation of the snail populations.

Subclass Prosobranchia

Subclass Proso · branchia
(G: anterior gill)
prō-sō-brank´-ē-ah

Members of the Prosobranchia, the largest of the three gastropod subclasses, are mostly marine, although a small percentage live in freshwater or terrestrial environments. At least 20,000 species have been described. Prosobranchs are generally free-living and mobile, although some species have evolved sessile or even parasitic lifestyles.

---

2. See *Topics for Further Discussion and Investigation,* no. 5.
3. See *Topics for Further Discussion and Investigation,* no. 3.

---

4. See *Topics for Further Discussion and Investigation,* no. 16.

Research Focus Box 14.1

Gastropod Torsion

Pennington, J. T., and F. S. Chia. 1985. Gastropod torsion: A test of Garstang's hypothesis. *Biol. Bull.* 169:391–96.

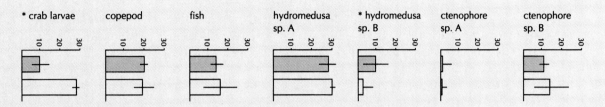

Focus Figure 14.1

Predation by seven predators on pretorted (blue bars) and torted abalone larvae (*Haliotis kamtschatkana*). Data are mean number of larvae eaten (mean of five replicates, 50 larvae per replicate) within a 15-hour test period. Error bars show the standard deviation about the mean. A * indicates that predation on pretorted larvae was significantly different from predation on torted larvae for that predator ($P < 0.05$). Crab larvae = 5 species, mixed; copepod = *Epilabidocera longipedata*; fish = *Oncorhynchus gorbuscha*; hydromedusa sp. A = *Phialidium gregarium*; hydromedusa sp. B = *Aequorea victoria*; ctenophore sp. A = *Pleurobrachia bachei*; ctenophore sp. B = *Bolinopsis infundibulum*.

In the late 1920s, the English zoologist Walter Garstang hypothesized that torsion arose as an adaptation for larval life rather than for adult life. Before torsion, he reasoned, the head and velum, the uniquely larval organ used for food collection and swimming, were more vulnerable to predation; because the untorted shell opened posteriorly, the head and velum were withdrawn last. Torsion brought the mantle cavity anteriorly, providing a space in which to rapidly retract these anterior structures when danger threatened. According to Garstang's hypothesis, torted individuals should therefore have survived better than their untorted counterparts, and the genes for larval torsion would quickly have become permanent residents in the life history.

Although Garstang's ideas have been much discussed and widely approved over the past 60 years, they were never directly tested until Pennington and Chia (1985) exposed abalone larvae (*Haliotis kamtschatkana*) to seven different potential planktonic predators: crab larvae (the final larval stage—the megalopa), copepods, two ctenophore species, two species of hydromedusae, and one species of fish. Both pretorted and torted abalone larvae were tested, with 50 larvae put into each of five containers for each predator studied. Larvae that disappeared from test containers over 15-hour test periods were assumed to have been eaten; control containers, each with 50 abalone larvae but without predators, were used to adjust for any losses due to the counting and handling errors inevitably associated with working with very small aquatic animals—these larvae are only a few hundred μm (micrometers) long. If Garstang's larval adaptation hypothesis is correct, there should be less predation on the abalone larvae after they undergo torsion.

The results did not support Garstang's hypothesis. Five of the seven predators ate as many pretorted as torted larvae during the 15-hour experiment (Focus Fig. 14.1). Only the hydromedusae of *Aequorea victoria* showed a statistically significant preference for pretorted larvae. The crab larvae (species not determined) actually ate more torted than pretorted abalone larvae. On balance, there was no indication that torsion protected abalone larvae from predation.

In fairness to Garstang's hypothesis, it should be noted that Pennington and Chia did not test fully torted larvae; their "torted" larvae had only completed the first 90° of torsion. Even though each such larva could retract fully into its shell when provoked and could completely seal off the shell aperture with the operculum, fully (180°) torted larvae might be less vulnerable to attack. Also, larvae of other gastropod species might be better protected by torsion than are those of the abalone. In addition, the sorts of predators that may have brought about selection for torsion in ancestral gastropods perhaps no longer exist, or at least were not tested in this study. We cannot conclude that Garstang's ideas about the importance of predation on larvae in selecting for torsion are wrong; the limited experimental evidence now available can only suggest that torsion today does not effectively protect abalone larvae against most planktonic predators.

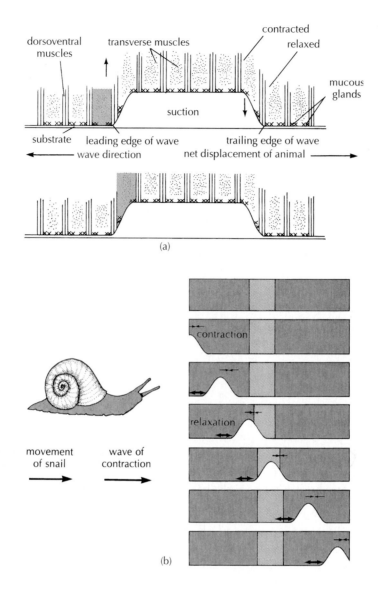

Figure 14.12

(a) Diagrammatic longitudinal section through the foot of a moving gastropod. A pedal wave is traversing the foot from anterior to posterior, while the animal advances in the opposite direction; this is called a retrograde pedal wave. Localized contraction of dorsoventral muscles raises a small portion of the foot away from the substrate, producing a small but measurable suction. This suction helps stretch the dorsoventral muscles when they relax, pulling the area at the rear of the wave back against the substrate. Note the forward progression of the blue region of the foot. (b) Direct waves in the locomotion of a terrestrial snail (subclass Pulmonata). Direct waves move in the direction of locomotion, from posterior to anterior. Only the portion of the foot lifted away from the substrate moves forward (note the forward progression of the blue region of the foot). Many waves travel down the foot simultaneously, as shown in the snail at the left.   (a) *From Jones and Trueman, in* Journal of Experimental Biology, *52:201, 1970. Copyright © 1970 Company of Biologists, Colchester, England. Reprinted by permission.* (b) *From Lissman, in* Journal of Experimental Biology, *21:58, 1945. Copyright © 1945 Company of Biologists, Colchester, England. Reprinted by permission.*

Free-living prosobranchs may be herbivores, deposit feeders, omnivores, suspension feeders, or carnivores, depending on the species. A number of carnivorous species—notably the warm-water cone snails (*Conus* spp.)—produce potent venoms (Fig. 14.14b) that are injected into the fish, molluscan, or annelid prey through a harpoon-like, hollow radular tooth. The various toxins exert their effects by binding to very specific classes of cell surface receptors and ion channels; they are being used widely by neurobiologists to study receptor and ion channel function in both invertebrates and vertebrates.

An entirely different nutritional mode seems likely for a still-undescribed gastropod recently discovered in the deep sea. These snails contain symbiotic bacteria living in the gill tissue; the bacteria may provide the snail with nutrients (p. 250).

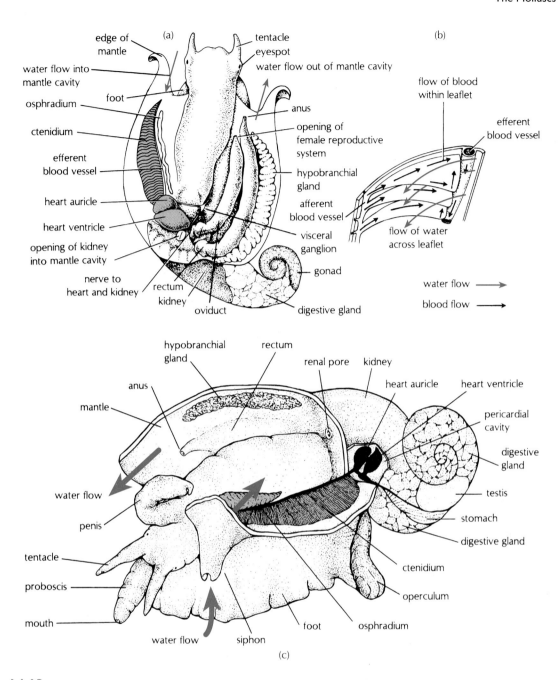

Figure 14.13

(a) The periwinkle *Littorina littorea,* removed from its shell and with its mantle cut middorsally to reveal the arrangement of organs within the mantle cavity. Note the major blood vessel (efferent blood vessel) that conducts blood from the gill to the auricle of the heart. From here, the blood moves to the ventricle and then to the tissues. The hypobranchial gland secretes mucus for binding particles carried by the ctenidial filaments. (b) Detail of ctenidium, showing direction of blood flow within individual gill sheets (black arrows) and water flow across the gill sheets (blue arrows). (c) The whelk *Busycon* sp., with mantle cavity exposed and the top of the pericardium dissected away. (a,b) *From Fretter and Graham, 1976. A Functional Anatomy of Invertebrates. Academic Press Inc. Copyright © Academic Press. Reprinted by permission of Harcourt Brace & Company Ltd., London.* (c) *Modified from Brown, 1950. Selected Invertebrate Types.*

Prosobranchs are the most primitive of gastropods; that is, the other two gastropod subclasses most likely evolved from prosobranch-like ancestors. Most species possess a well-developed shell, mantle cavity, osphradium, and radula, and the foot of most prosobranch gastropods bears a rigid disc of protein (sometimes strengthened with calcium carbonate) called the **operculum.** When the foot is withdrawn into the shell, the operculum may completely seal the shell aperture, thus protecting the snail from such physical stresses as dehydration and from predators (Figs. 14.9a, 14.14a).

The typical prosobranch gill is a **ctenidium,** consisting of a series of flattened, triangular sheets (filaments) lying one adjacent to the next (Fig. 14.13). Deoxygenated blood enters an afferent blood vessel from the animal's open system of blood sinuses (the **hemocoel**). Once distributed to the individual sheets of the ctenidium, the blood moves through the sheet, where it becomes oxygenated, and then on to the auricle of the heart through an efferent blood vessel. From the auricle, the blood is pumped into the single associated ventricle and is then distributed to the tissues through a single aorta leading to the blood sinuses of the hemocoel. Primitive prosobranch species (order Archaeogastropoda) possess a pair of auricles, a pair of efferent blood vessels, and a pair of ctenidia (Fig.14.11 d).

Water is drawn into the mantle cavity and across the gill sheets by the movements of gill cilia. In many prosobranch species, a portion of the mantle is drawn out into a cylindrical extension called the **siphon** (Figs. 14.14a and 14.13c); water is drawn through this siphon, by the action of the gill cilia, into the mantle cavity and across the **osphradium** (a chemical and tactile receptor organ—Figs. 14.2 and 14.13a,c). The snail moves the muscular siphon back and forth, sampling the water from different directions. In burrowing species, the siphon is extended through the substrate to the water above. The gastropod siphon is especially well developed in carnivores and scavengers—which often hunt their prey by chemical sensing—and generally reduced or absent in suspension feeders or deposit feeders.

Water flow across the molluscan gill is typically unidirectional. In most gastropod species, water enters the mantle cavity at the left side of the head, passes over or between the gill filaments, and exits at the right side of the head (Fig. 14.13). In primitive archaeogastropod species with paired gills, water is necessarily drawn instead into the mantle cavity at both sides of the head; unidirectional water flow in such species is made possible by the presence of conspicuous lateral or posterior slits or circular openings in the shell, as in keyhole limpets (Fig. 14.11d) and abalone (Fig. 14.16c). In all species, the direction of water movement across the gill is counter to the direction of blood flow. This **countercurrent exchange system** increases the efficiency of gas exchange between the ctenidium and the water, as discussed shortly. In the alternative situation, in which water and blood move in the same direction (Fig. 14.15b), the magnitude of the oxygen concentration gradient between the water and the blood would con-

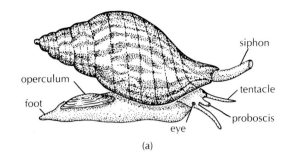

(a)

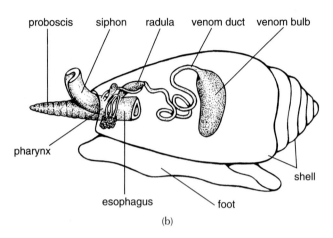

(b)

Figure 14.14
(a) Prosobranch gastropod (*Fasciolaria tulipa*) showing operculum and extended siphon. (b) Schematic diagram illustrating the venom apparatus in the carnivorous prosobranch gastropod *Conus* sp.   (a) *After Niesen.* (b) *From F. E. Russell, in* Advances in Marine Biology, *21:59, 1984. Copyright © Academic Press. Reprinted by permission of Harcourt Brace & Company Limited, London, England.*

tinually decrease along the surface of the gill sheet as the water lost oxygen to the adjacent blood; the rate of gas exchange between water and blood would decrease correspondingly.

In the countercurrent situation (Fig. 14.15a,c), however, a concentration gradient exists at all points along the gill. Between points 1 and 2 in both (b) and (c) of Figure 14.15, oxygen is diffusing from water to blood. However, at point 2 in Figure 14.15b, the water has already given up much of its oxygen to the adjacent blood and is nearly at the same oxygen concentration as the blood next to it. Essentially, no further gas exchange between the two fluids can occur as the blood continues to move across the gill past point 2; in terms of oxygen exchange, the gill surface beyond point 2 is, in effect, wasted.

This is not the case with countercurrent exchange. At point 1 in Figure 14.15c, even though the blood has

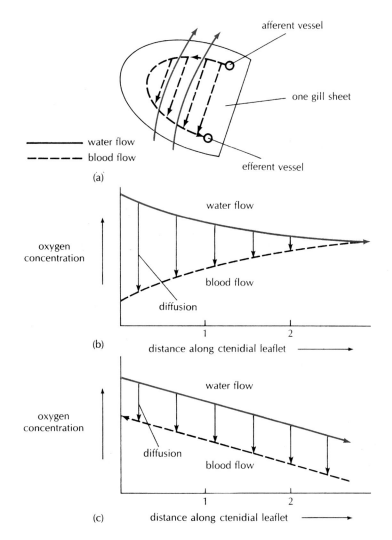

Figure 14.15

Countercurrent exchange in the molluscan gill. (a) The direction of water flow across the surface of each ctenidial leaflet is opposite to the direction of blood flow through the gill capillaries. (See Fig. 14.13b.) The afferent vessel carries deoxygenated blood to the ctenidial leaflets from the tissues. Oxygenated blood leaves the gill through the efferent vessel, carrying the blood to the heart and thence to the tissues. (b,c) Illustration of the principle of countercurrent exchange.

Arrow length signifies magnitude of the concentration gradient for oxygen between water and blood. In (b), water and blood flow in the same direction, so the oxygen concentration in the two fluids quickly comes to equilibrium. In (c), the countercurrent situation, equilibrium is never attained; diffusion of oxygen occurs along the entire gill-sheet surface, permitting a greater percentage of the oxygen in the water to be taken up by the blood.

acquired oxygen from the water during its travel from point 2, it continues to come into contact with water of even higher oxygen content as it travels past point 1, farther along the gill, so diffusion of gas from water to blood continues. In other words, as blood moves from point 2 to point 1 and beyond in (c), it is always adjacent to water whose oxygen content is higher than that of the blood, so diffusion between blood and water occurs along the gill's entire length.

Much of prosobranch evolution is a story of changes in gill numbers (from two in the more primitive species, contained within the order Archaeogastropoda, to one in the more advanced species, found within the Mesogastropoda and Neogastropoda) and changes in the orientation of gill filaments extending from the ctenidial axis. In the primitive **bipectinate** condition, gill filaments extend from both sides of the ctenidial axis, while in the relatively advanced **monopectinate** condition (Figs. 14.13 and 14.15a), the filaments project from only one side (the "downstream" side) of the supporting axis. Regardless of the number of gills and the placement of gill filaments in different prosobranch species, however, the principle of countercurrent exchange applies to all. Indeed, the countercurrent principle applies to all ctenidia in the phylum Mollusca, except those of cephalopods.

Considerable anatomical and functional diversity exists within the Prosobranchia (Fig. 14.16). The members of one group, the **heteropods,** show especially striking modifications of the basic prosobranch body plan and lifestyle. The heteropods are planktonic, voracious carnivores, whose shell is reduced or absent and whose foot is a thin, undulating paddle that propels the animal through the water (Fig. 14.16a,b). Except for the viscera, the body is nearly transparent, an excellent adaptation for inconspicuous water travel. As illustrated in Figure 14.16, heteropods swim upside down.

Subclass Opisthobranchia

Subclass Opistho · branchia
(G: posterior gill)
ōpis-tō-brank´-ē-ah

Members of the subclass Opisthobranchia, which includes the sea hares, sea slugs, and bubble shells, are almost all marine. Fewer than 2,000 species have been described. The characteristics that distinguish this group from the prosobranchs are (1) a trend toward reduction or loss of the shell, (2) reduction or loss of the operculum, (3) limited torsion during embryogenesis, (4) reduction or loss of the mantle cavity, and (5) reduction or loss of the ctenidia. Most species that have lost the ctenidia have evolved other respiratory structures that are unrelated to the ancestral gill. For example, in many sea slugs (the nudibranchs—order Nudibranchia), gas exchange occurs across brightly colored dorsal projections called **cerata** (Fig. 14.17b,c), which also contain extensions of the digestive system. In at least one species, the cerata exhibit rhythmic muscular contractions, apparently serving to move blood through the hemocoelic sinuses for gas exchange.

Shell reduction or loss potentially increases vulnerability to predators, and it is reasonable to expect that pressures selecting for alternate means of defense have been quite strong. In particular, opisthobranch cerata often house unfired defensive organelles (**nematocysts**) usurped from cnidarian prey; these nematocysts can then function in defense of the nudibranch. Instead of cerata, many other nudibranchs possess feathery gills arising from the dorsal surface (Fig. 14.17a). Some nudibranch species (including the sea hares, *Aplysia* spp.) produce chemical defenses against predation.

The opisthobranch head typically bears in addition to a pair of tentacles adjacent to the mouth as in prosobranchs, a second pair of tentacles located dorsally, called **rhinophores.** The rhinophores are believed to be chemosensory, making them analogous to the osphradium of prosobranch gastropods and to the osphradium of those opisthobranchs bearing a mantle cavity.

Opisthobranchs show various degrees of departure from the ancestral, prosobranch-like condition. Adult sea slugs, for example, have no mantle cavity, ctenidia, osphradium, shell, or operculum, and some species show no evidence of torsion as adults. Sea slug larvae, on the other hand, have a pronounced mantle cavity, shell, and operculum, indicating a clear affinity with the prosobranchs. Another common opisthobranch group, the sea hares (order Aplysiacea) have a mantle cavity (with gill and osphradium) as adults, and most adults also have a shell. However, the mantle cavity is very small and on the animal's right side, and the shell is reduced and internal. Clearly, the sea hares are closer to the ancestral, prosobranch-like condition than are the sea slugs. Even so, there is little evidence of torsion in the adult sea hare.

The most primitive of the opisthobranchs are the bubble shells (order Cephalaspidea), which include the greatest number of opisthobranch species. Although shell reduction, internalization, or loss is exhibited by most members of this order, a few species possess a conspicuous, external, spirally coiled shell, with operculum. Moreover, the mantle cavity (containing a gill and osphradium) is well developed, and the snails often show little sign of detorsion: In many species, the mantle cavity is still located anteriorly, and the nervous system still shows the fully twisted, "streptoneurous" condition (see Fig. 14.47c).

Although locomotion is generally by means of cilia and pedal waves along the ventral surface of the foot, some opisthobranchs, such as the sea hares, can swim in short spurts by flapping lateral folds of the foot called **parapodia.** In other members of this subclass, the entire foot is drawn out into two thin lobes, also called **parapodia,** which are used for swimming. These animals are known as pteropods ("wing-footed") (Fig. 14.18). Pteropods may or may not have shells, depending on the species, but all are permanent members of the plankton. Pteropods typically have no specialized respiratory organs, so gas exchange is accomplished across the general body surface.

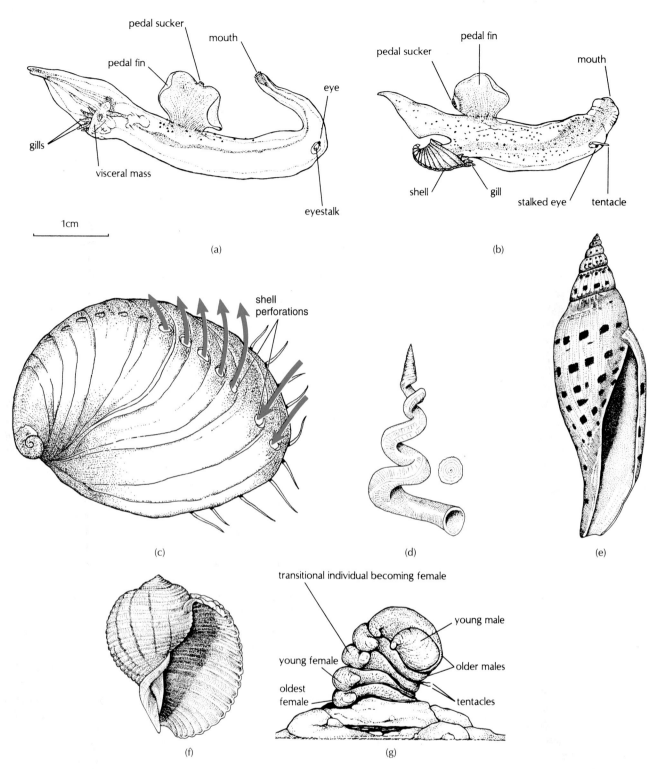

Figure 14.16

Prosobranch diversity. (a) *Pterotrachea hippocampus,* a Hawaiian heteropod. The shell is lost at metamorphosis from the larval stage. The visceral mass contains the digestive tract, heart, nephridia, and much of the reproductive system. All heteropods are planktonic. (b) *Carinaria lamarcki,* a shelled heteropod. Note the small size of the shell relative to the rest of the animal. (c) The abalone, *Haliotis* sp. Gill cilia draw water into the mantle cavity at the anterior end of the body. Water leaves through a series of posterior shell perforations. The openings of the digestive and excretory systems are located beneath one of these openings. (d) A worm-shell snail, *Vermicularia* sp. This is a sessile gastropod that lives attached to solid substrates, including rocks and other shells; the foot is reduced appropriately. As a juvenile, the animal produces a typical, spirally coiled shell, as observed near the apex of the adult shell. As the animal grows, however, the coiling

becomes very loose, so that the whorls become disconnected. This corkscrew-shaped shell resembles that secreted by some species of sessile, polychaete worm. (e) A volute shell, *Scaphella* sp. All volutes are marine carnivores, feeding on other invertebrates. (f) The giant tun shell, *Tonna galea.* Note the conspicuous shell ridges, the low spire, and the large body whorl. The adult shell lacks an operculum. (g) The slipper shell snail, *Crepidula fornicata.* Adults live in stacks, as shown, with the large foot becoming little more than a strong suction device. Each individual eventually transforms from male to female, so that the oldest snails (at the base of the stack) are females. Females are typically 3 cm to 4 cm long.   (a) *From R. Seapy, 1985, in* Malacologia, *26:125–135. Reprinted by permission.* (c) *After Hyman.* (d) *From Hyman,* The Invertebrates, Vol. III. *Copyright © 1951 McGraw-Hill Book Company, New York. Reprinted by permission.* (e) *After Hardy.* (f) *From Pimentel, after Abbott.* (g) *Modified from the drawings of W. R. Coe and others.*

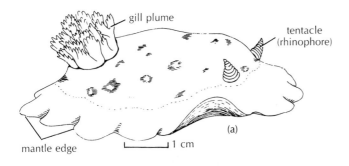

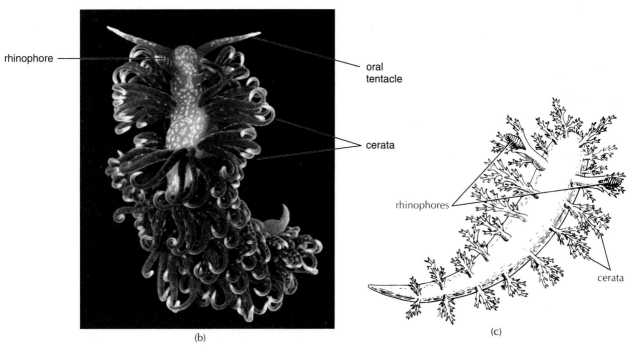

Figure 14.17

(a) A dorid nudibranch, *Dialula sandiegensis*. Note the gills arranged as a plume around the anus. (b) An eolid nudibranch, *Spurilla neapolitana*. The conspicuous dorsal projections are cerata, which serve for gas exchange and also contain outfoldings of the digestive system. (c) A dendronotid nudibranch, Dendronotus arborescens. Note the elaborate branching of the cerata. This and related species are capable of simple swimming by flexing the body from side to side. *(a) From Bayard H. McConnaughey and Robert Zottoli, Introduction to Marine Biology, 4th ed. Copyright © 1983 The C. V. Mosby Company, St. Louis, Missouri. Reprinted by permission. (b) Copyright L. S. Eyster. (c) After Kingsley.*

Subclass Pulmonata

Subclass Pulmo · nata
(L: lung)
pul´-mō-not´-ah

In contrast to the prosobranch and opisthobranch gastropods, few of the 17,000 pulmonate species are marine, and those few species occur only intertidally and in estuaries. Most pulmonate species are found in terrestrial or freshwater environments; slugs and "escargot" (Fig. 14.19) are terrestrial members of this subclass.[5] A coiled shell is present in most pulmonate species, but the shell is reduced, internalized, or completely lost in others (the slugs) (Fig. 14.20c,d). Only a few species have an operculum on the foot. Most pulmonates possess a long radula, in keeping with their generally herbivorous diet, and the head commonly bears two pairs of tentacles. Torsion is limited to about 90°, so the nervous system is not so greatly twisted, and the mantle cavity opens on the right side of the body, as in many opisthobranchs.

A major feature distinguishing the pulmonates from members of the other two gastropod subclasses is that the mantle cavity is highly vascularized and functions as a lung (Fig. 14.19b). Downward movement of the floor of

---

5. See *Topics for Further Discussion and Investigation*, no. 10.

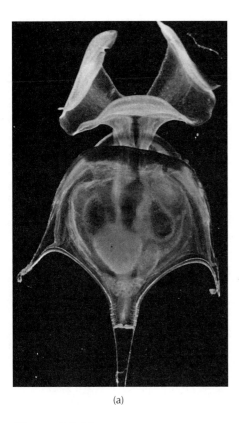
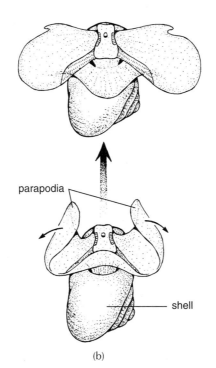
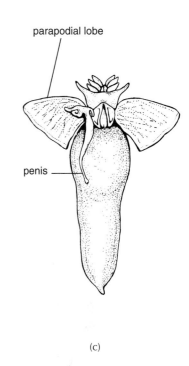

(a)   (b)   (c)

Figure 14.18

Pteropods. (a) *Cavolina* sp., with parapodia exposed. (b) *Spiratella* sp., swimming. The animal is shown executing its power stroke, in which the parapodia are swept downward forcefully, generating forward thrust. The muscular parapodia are elaborations of the ancestral foot. (c) A naked pteropod, *Clione limacina*. The animal loses its shell when the larva metamorphoses.   (a) *Courtesy of R. W. Gilmer.* (b) *After Morton* (c) *From Hyman,* The Invertebrates, *Vol. III. Copyright © 1951 McGraw-Hill Book Company, New York. Reprinted by permission.*

the mantle cavity increases the cavity's volume so that air, or in some cases water, is drawn into the mantle cavity for respiration. The fluid is then expelled by decreasing the volume of the mantle cavity. Flow of the air or water into and out of the lung occurs through a single small opening called the **pneumostome** (*pneumo* = G: lung; *stoma* = G: mouth) (Figs. 14.19 a and 14.20 a,c). Although ctenidia are lacking in pulmonates, a gill has been secondarily evolved in some species. This gill takes the form of folds of mantle tissue near the pneumostome.

Class Bivalvia (= Pelecypoda)

> *Class Bi · valvia (= Pelecy · poda)*
> (L: two valved[G: hatchet foot])
> bī-val′-vē-ah (pel-iss-ih-pō′-dah)

The class Bivalvia contains over 7,000 contemporary species, including clams, scallops, mussels, and oysters. Largely on the basis of gill structure, the bivalves may be divided into two major subclasses—the Protobranchia and the Lamellibranchia—and one very small subclass, the Septibranchia.[6] Major bivalve characteristics include (1) a hinged shell, the two valves (left and right sides) of which are joined together by a springy ligament that springs the shell valves apart when the adductor muscles relax; (2) lateral compression of the body and foot; (3) virtual absence of a head and associated sensory structures; (4) a spacious mantle cavity, relative to that found in other molluscan classes; (5) a sedentary lifestyle; and (6) the absence of a radula/odontophore complex. Bivalves are primarily marine, but about 10% to 15% of all species occur in freshwater. No bivalves are terrestrial.

The hinged portion of a bivalved shell is dorsal (Fig. 14.21). The shell valves, then, are on the animal's left

---

6. A more complex classification system, based largely upon morphological characteristics of the hinge area, is also widely used, as in the *Taxonomic Detail* section of this chapter.

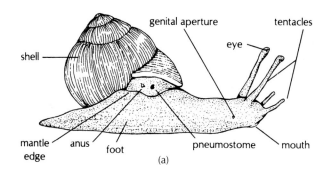

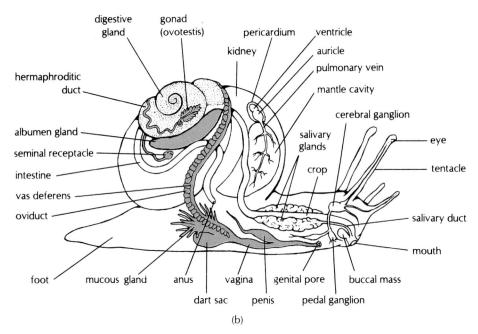

Figure 14.19
(a) External and (b) internal anatomy of the terrestrial pulmonate *Helix* sp. This is the animal eaten as "escargot." (a) *Sherman/Sherman, The Invertebrates: Function and Form, 2/e, © 1976, pp. 111, 112, 236, 172, 169, 45, 9,*

15. Reprinted by permission of Prentice Hall, Upper Saddle River, New Jersey. (b) Reprinted by permission of Prentice Hall, Inc. from Invertebrate Zoology, *3d ed by Joseph Engemann and Robert Hegner. Copyright © 1981 by Joseph Engemann and Robert Hegner.*

and right sides. The shell opens ventrally. A conspicuous bulge in the shell is frequently seen on the dorsal surface, adjacent to the hinge. This bulge, termed the **umbo,** is comprised of the earliest shell material deposited by the animal. Distinct **growth lines** typically run parallel to the shell's outer margins, as illustrated in Figure 14.22. The foot projects ventrally and anteriorly, in the direction of movement, and the siphons, when present, project posteriorly.

Subclass Protobranchia

> *Subclass Proto · branchia*
> (G: first gill)
> prō-tō-brank´-ē-ah

Members of the Protobranchia retain what biologists regard as a morphologically primitive state of bivalve organization. The group is entirely marine, and all species live in soft substrates (Fig. 14.23a). One pair of gills is present in the mantle cavity, with one gill on either side of the foot (Fig. 14.23b,c). Each gill consists of two parts, called **demibranchs** (Middle English: half-gills), extending from opposite sides of a central gill axis (Fig. 14.23b,c). Thus, the gills of protobranch bivalves are always **bipectinate** (L: double-combed). The many units making up each demibranch may be flat, rounded sheets (Fig. 14.23b), as in the typical gastropod ctenidium, or they may be finger-like structures (Fig. 14.23c) that have a more circular cross section; in both cases, each unit is called a **filament.** The gill filaments hang down into the mantle cavity, dividing it into an **incurrent** (ventral) chamber and an **excurrent** (dorsal) chamber. Water enters the mantle cavity ventrally, generally passes between

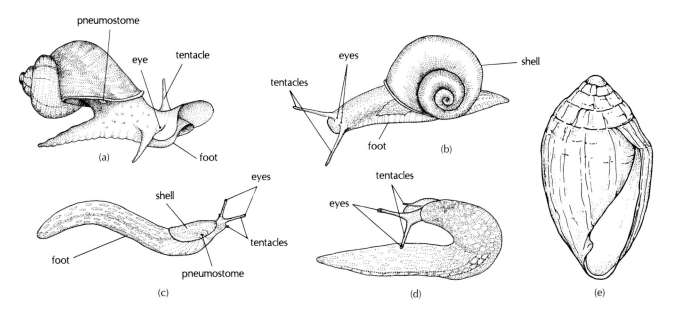

Figure 14.20

Pulmonate diversity. (a) *Lymnaea* sp., with conspicuous pneumostome. Members of this genus are common inhabitants of freshwater lakes and ponds throughout the world, and serve as intermediate hosts for a variety of trematodes and cestodes (Platyhelminthes). (b) *Helisoma* sp., another freshwater pulmonate. Note the planospiral shell, with all the whorls lying in a single plane. (c) A terrestrial slug, *Arion fuscus,* with a reduced, external shell. (d) *Limax flavus,* a terrestrial slug with a small internal shell (not shown). (e) Shell of *Melampus bidentatus,* a common salt-marsh snail. This is one of the few pulmonates with a free-living, marine larval stage in the life cycle.   (a,b) *From Pennak, Fresh-Water Invertebrates of the United States, 2d ed. Copyright © 1978 John Wiley & Sons, New York. Reprinted by permission of John Wiley & Sons, Inc. (c,d) After Kingsley.*

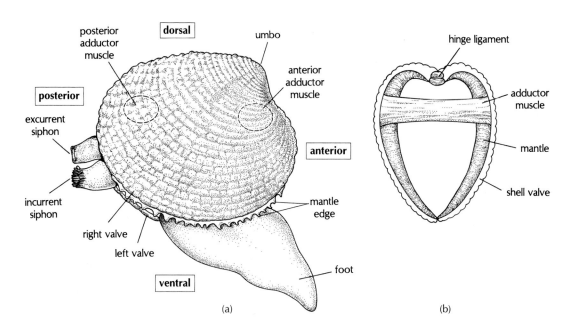

Figure 14.21

A bivalve, indicating the orientation of the body within the shell valves. Note that the hinge is located dorsally and the valves open ventrally. The siphons protrude posteriorly. (a) Lateral view. (b) Cross section through shell; siphons would face out of page.

Figure 14.22
Growth lines in the shell of the bivalve *Arctica islandica*. The age of
the shell can be determined by counting growth lines, since the
patterns of shell growth are seasonal. The most recent growth
lines are near the outer margin of the shell valve. This individual
was estimated to be 149 years old.   *Courtesy of Douglas S. Jones.*

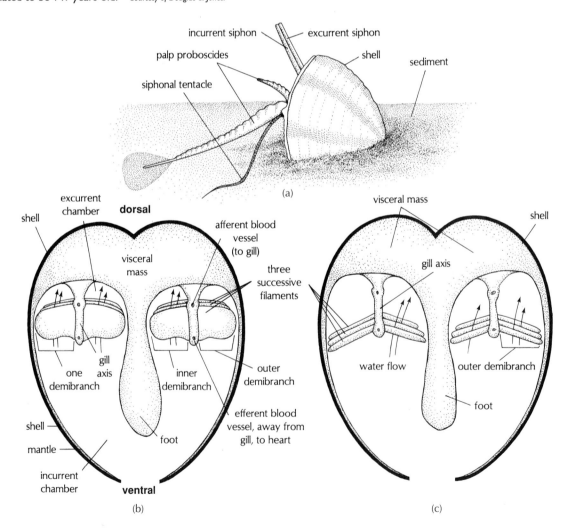

Figure 14.23
(a) A protobranch bivalve, *Yoldia* limatula, with its foot and the
lower part of its shell buried in the sediment. (b,c) Two types of
protobranch gill, with the more primitive gill type on the left. The
anterior end of the animal projects into the page. Arrows
indicate the path of water flow through the mantle cavity. The
gills function primarily in gas exchange, not food collection.
(a) *After Meglitsch.* (b,c) *After Russell-Hunter.*

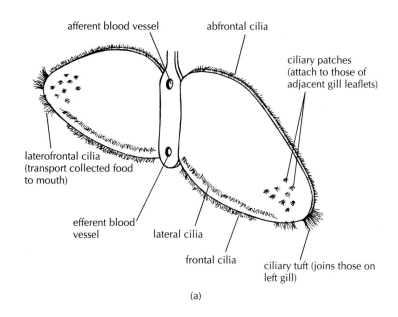

afferent blood vessel

abfrontal cilia

ciliary patches (attach to those of adjacent gill leaflets)

laterofrontal cilia (transport collected food to mouth)

efferent blood vessel

lateral cilia

frontal cilia

ciliary tuft (joins those on left gill)

(a)

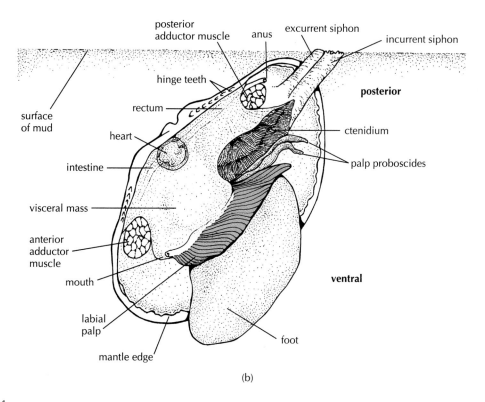

posterior adductor muscle

anus

excurrent siphon

incurrent siphon

hinge teeth

posterior

surface of mud

rectum

heart

ctenidium

intestine

palp proboscides

visceral mass

anterior adductor muscle

mouth

labial palp

mantle edge

ventral

foot

(b)

Figure 14.24
(a) Detail of right gill of the protobranch bivalve *Nucula* sp., showing ciliation on a single filament. (b) Internal anatomy of the protobranch bivalve *Yoldia eightsi* (Courthouy). Note the location of the mouth at the base of the labial palps. (a) *After Orton, 1912. J. Marine Biol. Assoc., U.K. 9:444* (b) *From J. Davenport, in* Proceedings of the Royal Society, 232:431–442, 1988. Copyright © 1988 The Royal Society, London.

gill filaments, and then exits dorsally (Fig. 14.23b,c). The protobranch gill functions primarily in gas exchange (as in most prosobranch gastropods), although it may also play some role in food collection by filtering unicellular algae from the water.

Gill cilia are confined to the area near the edges of the gill filaments, the **lateral cilia** always playing the primary role in driving water through the mantle cavity (Fig. 14.24a). The **frontal cilia** function primarily in cleaning the gill of sediment and debris.

Each gill filament is attached to adjacent filaments by circular patches of stiff, nonmotile cilia on the anterior and posterior surfaces of each sheet. The cilia of each sheet actually interlock with those of adjacent sheets, stabilizing the gill structure; ionic interactions between the ciliary surfaces apparently further strengthen the attachment between adjacent sheets. The ciliated discs thus fasten together adjacent filaments; that is, they form **interfilamental junctions**—stabilizing junctions between individual gill filaments (Fig. 14.24a). The ubiquitous fastening system known as Velcro®* works on a similar principle.

Most of the food collection in protobranch bivalves is accomplished not by the gills, but rather by **palp proboscides**—long, thin, muscular extensions of the tissue surrounding the mouth (Fig. 14.24b). The palp proboscides protrude between the shell valves and probe the surrounding mud substrate, entangling particles in mucus. The sediment-laden mucus is then transported into the mantle cavity by cilia along the ventral surface of the proboscides. Attached to the palp proboscides at their bases, within the mantle cavity, are flattened structures called **labial palps** (Fig. 14.24b); these sort the particles by ciliary action, transporting light particles to the mouth for ingestion and transferring heavy particles (which are less likely to be nutritional) to the margins of the labial palps, where these particles are ejected into the mantle cavity and expelled. This rejected material is termed **pseudofeces**, since it is material that has never been ingested. Protobranch bivalves' mode of feeding clearly restricts them to soft sediments. This type of feeding, in which sediment is taken in and the organic fraction is digested, is called **deposit feeding**, and is quite common among invertebrates. An individual protobranch may process more than 1.5 kg (kilograms) of sediment each year.

Although protobranchs are present in shallow water, they are much more common in deep water. Protobranch bivalves may account for about three-fourths of all bivalves in sediments sampled from depths of 1,000 m (meters) or more.

Subclass Lamellibranchia

Subclass Lamelli · branchia
(G: plate gill)
lam-el-ih-brank´-ē-ah

Most bivalves are lamellibranchs. Although the majority of lamellibranchs are marine, all freshwater bivalve species are also members of this subclass. Most freshwater species are contained within a single family, the Unionidae.

In bivalves of all subclasses, water typically enters and exits the mantle cavity posteriorly; the water generally enters through an **incurrent siphon,** passes dorsally between adjacent gill filaments, and then exits through a more dorsally located **excurrent siphon** (Figs. 14.25b,c and 14.30c). The siphons, where present, are tubular extensions of mantle tissue that often can be protruded far beyond the posterior shell margins, permitting the rest of the animal to live safely, deep within the surrounding substrate (Fig. 14.25c). As in the prosobranch gastropods and the protobranch bivalves, water currents are generated by the action of gill cilia—not by muscular activity of the siphons—and the gill is the primary gas exchange surface. But lamellibranch gills also play another important role.

The lamellibranch ctenidium is often much larger than that of protobranchs and is variously modified to provide an enormous surface area for collecting, sorting, and transporting suspended particles. Discussing the morphology of these ctenidia requires some unavoidably confusing terminology; the reader is urged to go through the next few pages slowly and with frequent reference to the appropriate illustrations. The discussion is limited to the basic terminology essential for laboratory observation and any discussion of bivalve evolution.

The individual filaments of the lamellibranch gill are thin and greatly elongated; they typically bend to form a V shape (Fig. 14.26). A ciliated ventral food groove lies between the two limbs of each filament, at the base of each V (Figs. 14.26 and 14.27c), and the cilia within this groove pass food particles from one filament to the next, toward the mouth.

The two arms of each V are named in reference to the central axis from which each filament hangs. A **descending limb** descends from the central axis, and an **ascending limb** bends upward from the bottom of the descending limb (Fig. 14.27b). Two V-shaped filaments usually unite to form a W on each side of the foot (Figs. 14.26 and 14.27b). Adjacent Ws in a gill are always attached to each other, sometimes by **interfilamental** (i.e., "between filaments") **ciliary junctions** (Fig. 14.27b), as in the protobranch gill. These lamellibranch gills—consisting of individual filaments linked together solely by ciliary disc junctions—are termed **filibranch gills.** In the most highly modified bivalve gills, called **eulamellibranch gills,** the junctions between adjacent filaments are made of tissue rather than cilia; these **interfilamental tissue junctions** completely and firmly attach adjacent filaments together (Fig. 14.27d). In both filibranch and eulamellibranch gills, the series of ascending and descending limbs form continuous sheets of tissue, or **lamellae.** Water passes through the gill lamellae—between adjacent gill filaments—before leaving the mantle cavity.

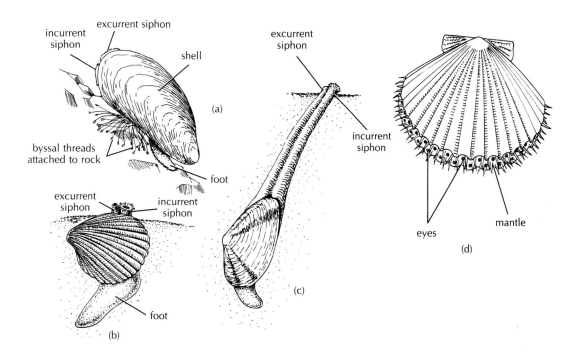

Figure 14.25

Lamellibranch diversity. (a) A blue mussel *Mytilus edulis* attached to a rock by proteinaceous threads. These byssal threads are secreted by a gland located at the base of the foot. (b) A cockle, *Cardium* sp., in its normal feeding position. (c) The soft-shell clam, *Mya arenaria*, in its normal feeding position. The fused siphons may extend more than 30 cm from the shell, permitting this clam to live well below the surface of the sediment. (d) A scallop, *Pecten* sp. Numerous light receptors are present along the edge of the mantle. The scallop can swim by forcefully contracting the well-developed posterior adductor muscle, ejecting water through openings near the shell hinge. Scallops lack siphons and have lost the anterior adductor muscle. It is the large, well-developed posterior adductor muscle that is served as "scallops" in restaurants.   (a–d) *After Niesen.*

As with protobranch gills, lamellibranch gills are bipectinate, with one demibranch ("half-gill") extending from each side of the central gill axis; each demibranch typically consists of an ascending and a descending lamella, formed from the ascending and descending limbs of the numerous V-shaped filaments. The demibranch nearest the foot is termed the **inner demibranch;** that on the side nearest the mantle is termed the **outer demibranch** (Figs. 14.26 and 14.27b).

To summarize, on the left side of the foot and visceral mass is one gill. That gill consists of one inner and one outer demibranch, which are united dorsally at the gill axis. Each demibranch consists of many filaments; the filaments typically consist of descending and ascending limbs, so each demibranch consists of one descending and one ascending lamella. On the right side of the foot and visceral mass is another gill, the mirror image of the gill on the left side.

Commonly, the ascending and descending limbs of a single filament are connected by crosspieces of tissue, forming **interlamellar** (i.e., "between lamellae") **junctions** (Fig. 14.27b). In eulamellibranch gills, the interlamellar junctions between the ascending and descending limbs of each filament may be so extensive as to form a complete, solid sheet of tissue across the space between the two limbs. As shown in Figure 14.27d, the interlamellar and interfilamental junctions in such gills essentially turn pairs of adjacent filaments into lidless, rectangular boxes. Water must now pass between gill filaments, from the incurrent chamber of the mantle cavity to the excurrent chamber, through minute holes, called **ostia**, located in the sides of each box (Fig. 14.27d,e). In addition, the ascending lamellae of the inner demibranchs (e.g., the two demibranchs closest to the foot) often attach, at their tips, to the bivalve's foot, and the ascending lamellae of the two outer demibranchs often attach, at their tips, to the mantle (Figs. 14.27d and 14.28b). With the tips of the ascending lamellae firmly attached to other tissues of the bivalve, *all* water must pass between adjacent gill filaments before exiting the mantle cavity; this must increase the gill's filtering effectiveness considerably, since no particles can now escape the clutches of the food-gathering cilia.

The surfaces of lamellibranch gill filaments that face the incurrent chamber of the mantle cavity show complex patterns of ciliation (Fig. 14.27c,e). **Lateral cilia** along the sides of each filament create the water currents responsible for moving water into, and out of, the mantle

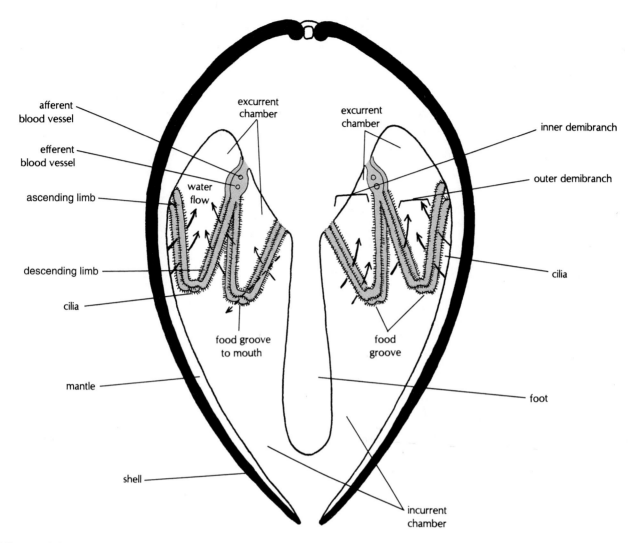

Figure 14.26

A simple lamellibranch gill. Additional gill filaments project in a row out of the page on each side of the foot. *From W. D. Russell-* *Hunter, A Life of Invertebrates. Copyright © 1979 by Macmillan Publishing Company, New York. Reprinted by permission of the author.*

cavity, as in protobranch bivalves; remember, even though water enters the mantle cavity through an incurrent siphon, the water is pulled in through the actions of these lateral gill cilia, not through any direct muscular action of the siphon itself. Even so, the rate of water flow through the bivalve mantle cavity can be impressive indeed: An American oyster moves 30 l (liters) to 40 l of water past its gills each hour at 24°C.

Details of particle capture and transport by bivalve gills are still being worked out.[7] The traditional model has food particles being captured by compound **laterofrontal** cilia, which then pass the captured particles to the nearby frontal cilia. More recent work indicates that particles are captured instead largely by hydrodymanic forces, without any physical contact between the particles

and the laterofrontal cilia; in addition, some particles are intercepted directly by the gill filaments. In both cases, **frontal cilia** then move the captured food particles to specialized food grooves located at the ventral and dorsal margins of each demibranch (Figs. 14.26, 14.27c,e, and 14.29). Members of some bivalve species transport captured food particles primarily in the ventral food grooves, members of some other species transport captured food particles primarily in the dorsal food grooves, and the remaining bivalves transport food particles in both the ventral and dorsal food grooves (Fig. 14.29).

Periodically, the lamellibranch bivalve closes its valves forcefully, expelling water and unwanted particles (pseudofeces) from the mantle cavity through the incurrent siphon.

---

7. See *Topics for Further Discussion and Investigation*, no. 18.

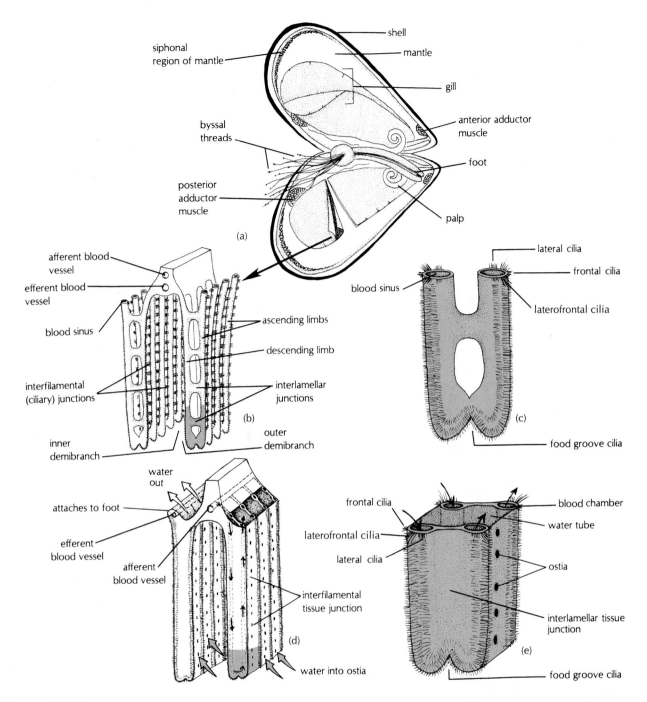

Figure 14.27

(a) The blue mussel, *Mytilus edulis,* with its shell opened to reveal the gills. (b) Filibranch gill of the blue mussel, *M. edulis.* In these relatively simple lamellibranch gills, adjacent filaments are locked together only by sporadic patches of adjoining cilia (interfilamental ciliary junctions). The ascending and descending lamellae of each demibranch are supported by thin crosspieces (interlamellar junctions). (c) Transverse section through one filibranch gill filament, showing pattern of ciliation. Lateral cilia generate water currents for feeding and gas exchange. Food particles are trapped by hydrodynamic forces or by laterofrontal cilia, which then transfer the food to frontal cilia. Frontal cilia move food particles along the gill filament to the food groove. If you remove one shell valve from a bivalve and lay the animal on its remaining valve, the frontal cilia project directly upward at you. (d) Diagrammatic representation of a eulamellibranch gill, such as that found in the hard-shell clam *Mercenaria mercenaria.* The interfilamental junctions between adjacent filaments are complete sheets of perforated tissue, rather than sporadic patches of interlocking cilia. The intrafilamental junction between the inner and outer lamellae of a single demibranch is now a solid sheet of tissue rather than a series of thin, separate bars. (e) Detail of one demibranch, showing basic structure and pattern of ciliation. All water enters through ostia in the interfilamental junctions and must flow past the blood vessels along the water tubes before exiting.

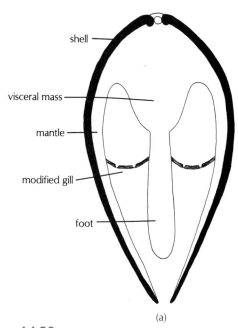

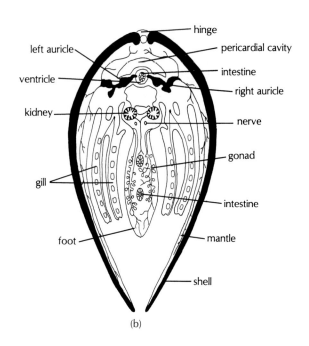

Figure 14.28

(a) Cross section through a septibranch bivalve, showing the highly modified gill. Compare with cross section through a freshwater eulamellibranch, the clam Anodonta sp., shown in (b). In (b), note that the heart is wrapped around the intestine, as in all bivalves.

(a) *From M. S. Gardiner,* Biology of the Invertebrates. *Copyright © 1972 McGraw-Hill, Inc., New York. Reprinted by permission.* (b) *Beck/Braithwaite,* Invertebrate Zoology, Laboratory Workbook, *3/e, © 1968. Reprinted by permission of Prentice Hall, Upper Saddle River, New Jersey.*

After food particles are ingested at the mouth and passed down the esophagus, entangled in strings of mucus, they are drawn into the stomach, stirred, and, in part, digested by the action of a rotating translucent rod known as the **crystalline style.** The crystalline style is composed of structural protein and several digestive enzymes. One end of the rod lies in a **style sac,** a pouch of the intestine lined by cilia (Fig. 14.30b). The activity of these cilia causes the rod to rotate. The end of the style protruding into the stomach abrades against a chitinous **gastric shield** as the rod rotates, breaking food into smaller pieces. The abrasion also causes the rod to slowly degrade at the end, releasing digestive enzymes into the stomach. Additions to the crystalline style are made in the style sac. A morphologically and functionally similar crystalline style apparatus is also found in the stomachs of monoplacophorans and some suspension-feeding gastropod species. In contrast, protobranch bivalves possess a style sac containing mucus and digestive enzymes but do not have a crystalline style. As in most other molluscs, the bivalve stomach connects with larger **digestive glands (digestive diverticula),** which serve as the major sites of digestion and absorption.[8]

The nutrition of some lamellibranch bivalves differs considerably from what has just been described as the norm. In the early 1980s, bacteria were found living symbiotically in the gill tissue of a number of shallow- and deep-water bivalve species.[9] The bacteria-rich gills are unusually thick and typically more than three times heavier than the gills of bivalve species lacking the bacterial symbionts. The bacteria within the gills generate ATP and reducing power (NADH or NADPH) by oxidizing highly reduced chemical substrates (sulfides—HS^-—or methane) and use the energy so entrapped to fix carbon dioxide (CO_2) into organic compounds that then become available for the nutrition of the host bivalve. The bacteria, then, like plants, are autotrophs. In fact, to fix CO_2 into carbohydrates, they use the same enzymatic pathways of the Calvin cycle that plants use in photosynthesis, but the bacteria are **chemoautotrophs** rather than solar autotrophs, using chemical energy rather than light energy to fix carbon from CO_2. Bacterial chemosynthesis undoubtedly plays a role in the nutrition of the host, although the exact contribution has yet to be formally or fully quantified for any species. Many bivalves hosting gill bacteria have a greatly reduced digestive system, and in some species the digestive system has been lost completely; symbiotic bacteria surely contribute substantially to host nutrition in such animals.

The symbiosis between bivalves and sulfur-oxidizing bacteria is accompanied by certain morphological (Research Focus Box 14.2) and physiological modifications on the part of the host, modifications that still are being actively studied. The relationship with bacterial symbionts typically correlates with a reduction of the labial palps and digestive system, the loss of the outer gill

8. See *Topics for Further Discussion and Investigation,* no. 9.

9. See *Topics for Further Discussion and Investigation,* no. 14.

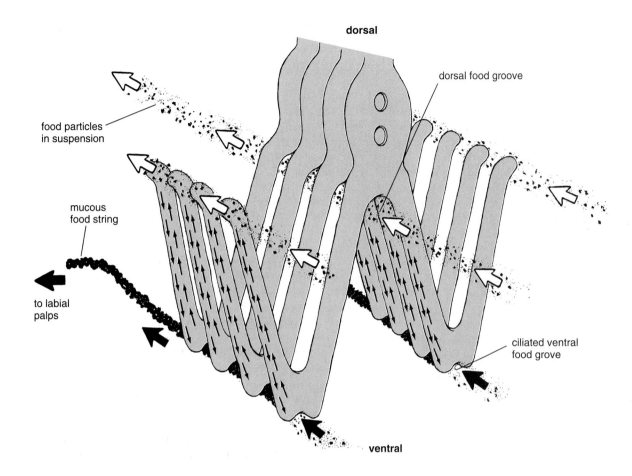

dorsal

dorsal food groove

food particles
in suspension

mucous
food string

to labial
palps

ciliated ventral
food grove

ventral

Figure 14.29

Capture and transport of food particles by lamellibranch bivalve gills. Particles entering the ventral food grooves are incorporated into a mucous string and transported by cilia from gill filament to gill filament toward the labial palps (Figs. 14.27a and 14.30a,c), where some or even most sorting of particles occurs before ingestion as in protobranchs. Particle transport by mucociliary mechanisms is indicated by solid arrows. Particles entering the dorsal food grooves, on the other hand, are captured by hydrodynamic mechanisms and carried to the labial palps in suspension (hollow arrows).   *From J. E. Ward, et al., in Limnol. Oceanogr, 38:265–272, 1993. Copyright © 1993 American Society of Limnology & Oceanography. Reprinted by permission.*

demibranchs, and further modifications in gill morphology presumably associated with decreasing dependence on particulate food.

Physiological modifications are equally intriguing. HS⁻ is normally toxic to living tissues even at extremely low concentrations, disabling cytochrome-*c* oxidase, a key enzyme in the electron transport chain involved in ATP synthesis. Allowing symbiotic bacteria access to the sulfide needed for bacterial ATP production is thus a potentially risky proposition for the host. Some species have specialized blood proteins that bind HS⁻, preventing free sulfide from contacting sulfide-sensitive enzymes of the electron transport system. This adaptation has the added benefit of protecting the sulfide from spontaneous oxidation while in transit to the bacteria; such spontaneous oxidation of sulfide would greatly

reduce the sulfide's value as an energy source. Other species may detoxify HS⁻ by partially oxidizing it to thiosulfate, a nontoxic substance that can then be further processed by the bacteria. Finally, some of the host tissues may rely mostly on anaerobic metabolic pathways for ATP generation, reducing their dependency on the cytochrome-*c* oxidase system and thus decreasing their susceptibility to sulfide poisoning.

Bivalves exhibit a modest variety of lifestyles. Some lamellibranchs, such as mussels, live attached to hard substrates by means of proteinaceous secretions called **byssal threads.** A proteinaceous liquid is secreted by a byssal gland at the base of the foot, within the mantle cavity, and is quickly transported to the substrate along a groove in the foot. Shortly after contacting seawater, the secretion solidifies to form many thin—but very sturdy—

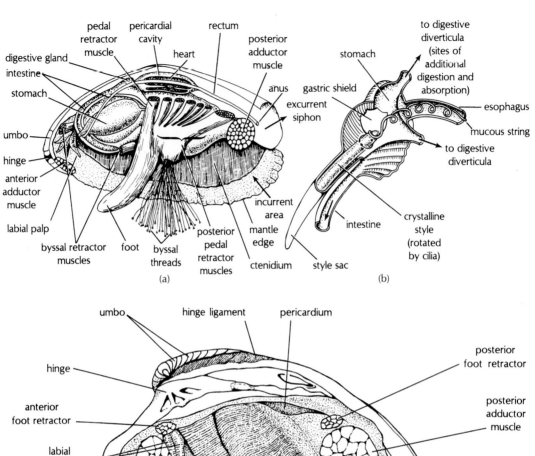

(a)

(b)

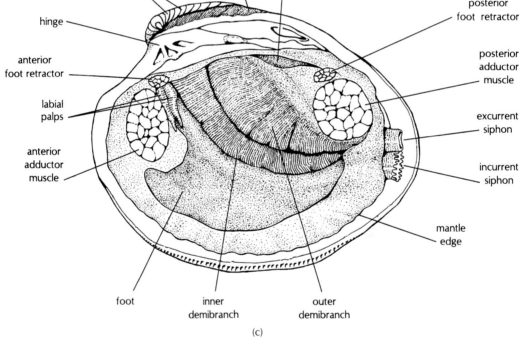

(c)

Figure 14.30

(a) Internal anatomy of a lamellibranch bivalve, the marine mussel *Mytilus edulis*. The left gill has been removed to reveal underlying anatomy. (b) The digestive system of a suspension-feeding lamellibranch bivalve, such as the blue mussel or soft-shell clam. The crystalline style aids digestion; rotated by cilia in the style sac, the style releases digestive enzymes as the tip of the rod is abraded against the gastric shield. A crystalline style apparatus is also commonly encountered among prosobranch gastropods. (c) Internal anatomy of the hard-shell clam, *Mercenaria mercenaria*.

The animal is lying on its right shell valve, with the left valve removed. Only the left gill is illustrated. The right gill lies beneath the foot in this drawing, and thus cannot be seen. (d) Same animal as in (c), but with the left gill removed and internal organs exposed by dissection. Note that the viscera extend well into the foot. (e) Internal structure of the American oyster, *Crassostrea virginica*, with the right shell valve removed. Adult oysters have no foot and lack an anterior adductor muscle.   (a) *After Turner.*
(b) *After Morton.* (c) *After Kellogg.* (e) *Modified from Kellogg and other sources.*

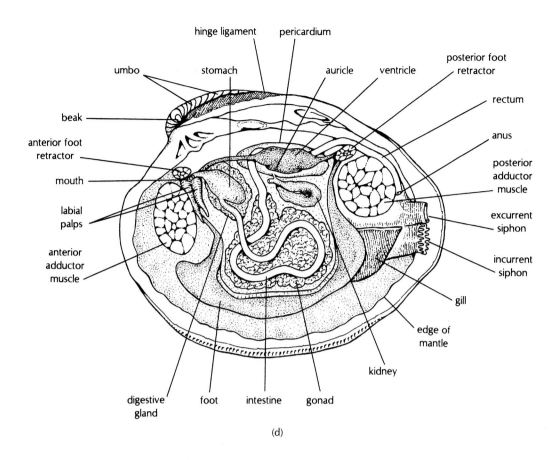

(d)

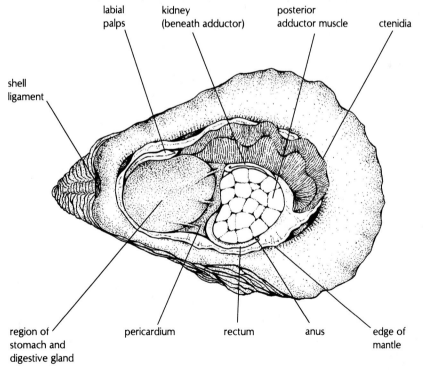

(e)

Research Focus Box 14.2

Bacterial Symbiosis

Distel, D. L., and H. Felbeck. 1987. Endosymbiosis in the lucinid clams *Lucinoma aequizonata*, *Lucinoma annulata*, and *Lucina floridana*: A reexamination of the functional morphology of the gills as bacteria-bearing organs. *Marine Biol.* 96:79–86.

The gill structure of typical bivalves—scallops, chowder clams, soft-shell clams, oysters, and mussels—has been well understood since the early 1900s. Members of the bivalve family Lucinidae, however, have unusually large gills that recently have been shown to harbor large numbers of symbiotic bacteria; the bacteria fix dissolved carbon dioxide (CO_2) into carbohydrates, using energy released through the oxidation of sulfides dissolved in seawater. The study by Distel and Felbeck (1987) asks, "How do lucinid gills differ from those of bivalves lacking endosymbiotic bacteria, and to what extent are the differences adaptively advantageous for the symbiosis?"

Distel and Felbeck collected representatives of two lucinid species (*Lucinoma aequizonata* and *L. annulata*) off the coast of California and one species (*Lucina floridana*) off the coast of Florida, and then they dissected out the gills. How are the bacteria distributed within the gill tissue? Study of the gill ultrastructure required that the gill tissue be killed, stabilized, and prepared for electron microscopy with appropriate chemicals. Even when chemically stabilized ("fixed"), the tissue is still too soft to be sliced ("sectioned") without being torn, crushed, or deformed. To stiffen the tissue, the researchers removed water by dehydrating the tissue in alcohol and then replaced the water with a plastic epoxy-like solution. Once the plastic hardened, the tissue samples were cut into sections using a microtome and glass or diamond knives. Thick sections (1–4 µm) were stained for light microscopy, and thin sections (less than 0.1 µm) were stained for transmission electron microscopy (TEM) to make certain cell components more visible.

Focus Figure 14.2b shows a chunk of tissue cut from a gill of one of the clams; these species have only one demibranch—the inner one—on each side of the foot/visceral mass system. When a gill is sectioned in the plane shown and stained appropriately, the light microscope reveals a normal gill structure at the surface: thin, parallel filaments bearing lateral, laterofrontal, and frontal cilia as in other bivalve gills (Fig. 14.27c,e). Deeper into the gill tissue, however, the filaments are unusual in being greatly swollen with large bacteria-containing cells called **bacteriocytes** (Focus Fig. 14.2c,d), each of which contains dozens of bacteria enclosed in membrane-bound vesicles. Cells in the bacteriocyte zone also house numerous dark-staining granules, termed "pigment granules."

After studying hundreds of adjacent sections of gill tissue, Distel and Felbeck constructed a three-dimensional model of the gill (Focus Fig. 14.2e). As illustrated, the bacteria-containing cells are packed into stacks forming hollow cylinders; with this arrangement, each bacteriocyte has equal access to sulfide in the water flowing through the gill. Water flows into the gill tissue through these cylinders and moves past, but does not directly contact, the bacteriocytes; the bacteriocytes are separated from direct contact with seawater by a thin layer of epithelial cells. These epithelial cells bear numerous microvilli, greatly increasing the absorptive surface area available for sulfide uptake and for exchange of O_2 and CO_2. After passing through the hollow cylinders of the bacteriocytes and epithelial cells, water flows out of the gill into the space between adjacent lamellae of the demibranch, and then it exits the mantle cavity through the excurrent siphon.

Although Distel and Felbeck successfully documented the structure of the gill and the location of the bacteria within the gill, their work raises a number of interesting questions about the functioning of this symbiotic relationship between bivalve and bacterium:

1. Because the sulfide-rich water comes first into contact with the gill itself, there is no opportunity for other tissues to detoxify the sulfide before the water reaches the gill. What protects the outer layer of epithelial cells from sulfide poisoning? For that matter, how are other tissues in direct contact with the seawater, such as the mantle, protected from sulfide poisoning?
2. Why are the bacteriocytes not in direct contact with the seawater? What is the function of the thin layer of epithelial cells covering the bacteriocytes?
3. What is the function of the prominent pigment granules present in the bacteriocyte layer of the gill? Might they be involved in eliminating waste products or in digesting bacterial cells?
4. How does the host obtain nutrients from the bacteria? Are the bacteria digested by the bivalve? Or do the bacteria remain intact, releasing carbohydrates and other organic molecules to the host?

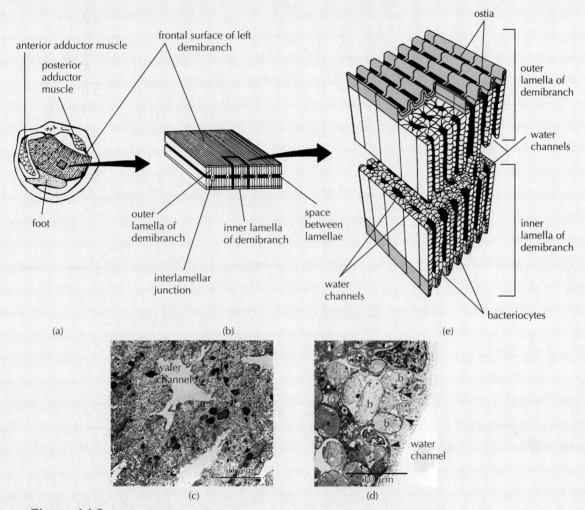

(a)   (b)   (e)

(c)   (d)

Focus Figure 14.2

(a) The deep-water bivalve *Lucinoma aequizonata* lying on its right side, with the left shell valve removed to expose the unusually thick gill. (b) Enlargement of square section shown in (a). The section has been rotated about 90° to show portions of the two lamellae composing the left demibranch. The shaded region is morphologically identical to the gills of other lamellibranch bivalves. (c) Transmission electron micrograph of sagittal section through the bacteria-laden portion of the gill lying between the two thin layers shaded blue in (b). Bacteria (labeled b) occur intracellularly in clusters around narrow water channels that extend perpendicular to the frontal surface of the gill. The bacteria are seen clearly at the higher magnification of (d). (e) Schematic reconstruction of gill structure based on electron micrographs, showing thick columns of bacteriocyte cells surrounding water channels.

5. Where do the symbiotic bacteria come from? Are they continually taken up from the external environment? Distel and Felbeck consider this unlikely, since their electron microscope studies showed no evidence of endocytotic activity in the gill tissue; the bacteria seem to be self-sustaining within the gill. But how does an individual gill first become "infected"? Are the bacteria passed directly from parent to fertilized egg? Do the bacteria first appear in the free-swimming larvae, or do they not appear until after metamorphosis?

It will take years, and the work of many people, to resolve even some of these questions. How might each be addressed?

Figures from "D. L. Distel and H. Felbeck, "Endosymbiosis in the Lucinid clams Lucinoma aequizonata, Lucinoma annulata and Lucina floridana: A reexamination of the Functional Morphology of the Gills as Bacteria-bearing Organs" in Marine Biology, *96:79–86, 1987. Copyright © 1987 Springer-Verlag, Heidelberg, Germany. Reprinted by permission.*

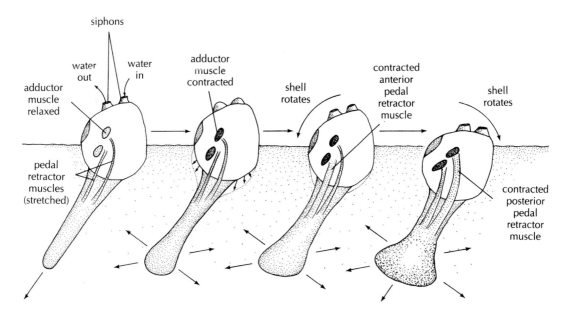

Figure 14.31

Burrowing by bivalves. The foot serves both to penetrate into the substrate and to anchor the animal as the body is pulled downward. See text for discussion.   *From Jones and Trueman, in* Patella *Locomotion, 52:201, 1970. Copyright © 1970 Company of Biologists, Colchester, England. Reprinted by permission.*

anchoring threads (Figs. 14.25a and 14.27a). Some other bivalves (e.g., oysters, Fig. 14.30e) cement one shell valve permanently to a substrate. Biochemists have scrutinized these processes for many years; a sturdy glue that hardens quickly under water could have a bright commercial future as, for instance, a new dental adhesive. Most bivalve species, however, do not form such attachments.

A few lamellibranch species (some scallops, Fig. 14.25d) live unanchored atop the substrate and are capable of short bursts of "swimming," achieved by repeatedly clapping the shell valves together; rapid expulsion of water from the mantle cavity quickly displaces the animal, permitting escape from potential predators.

Most bivalves, however, avoid predators by living within a substrate—sediment, wood, or even rock—in burrows of their own making. Burrowing into a soft substrate, such as sand or mud, is accomplished by the muscular foot (Fig. 14.31). The foot is initially extended into the substratum by hydraulic or hydrostatic means, through contractions of the appropriate musculature in the foot. The **adductor muscles,** which attach the animal to the shell in all bivalve species, are relaxed at this time. This relaxation of the adductor muscles permits the release of energy that was stored in the compressed hinge ligament, so that the shell valves now move apart and press tightly against the surrounding substrate. This laterally directed force of the shell valves provides anchorage for the shell as the foot extends down into the

substratum. Thus, downward extension of the foot need not eject the animal from the burrow. The adductor muscles then contract, drawing the shell valves toward each other. This action releases the shell anchor. Contracting the adductor muscles also pumps blood into the foot, which then dilates at the tip to form another anchor.

Abruptly closing the shell valves, in addition to swelling the foot, forces water out of the mantle cavity, blowing away and loosening some of the sediment adjacent to the opening between the shell valves. With the foot firmly anchored in the substrate, the shell can now be pulled downward by contracting the **pedal retractor muscles** that extend from the foot to the shell. The pedal retractor muscles do not contract simultaneously, but rather sequentially, so the bivalve shell "rocks" forward and then backward as it progresses, slicing into the substrate.

Many marine bivalve species burrow, as newly metamorphosed juveniles, into wood, shell, coral, and other hard substrates. The shell plays a major role in burrow formation by such species, actually carving into the substrate. As the animal grows, the body remains hidden within, and protected by, the substrate in which the burrow has been formed, with only the siphons protruding to the outside; the siphons of all burrowing species tend to be long, permitting the rest of the animal to live deep within the substrate. The damage done by wood-boring bivalves can be considerable, as any marina owner or boat owner knows all too well. Wood-boring bivalves

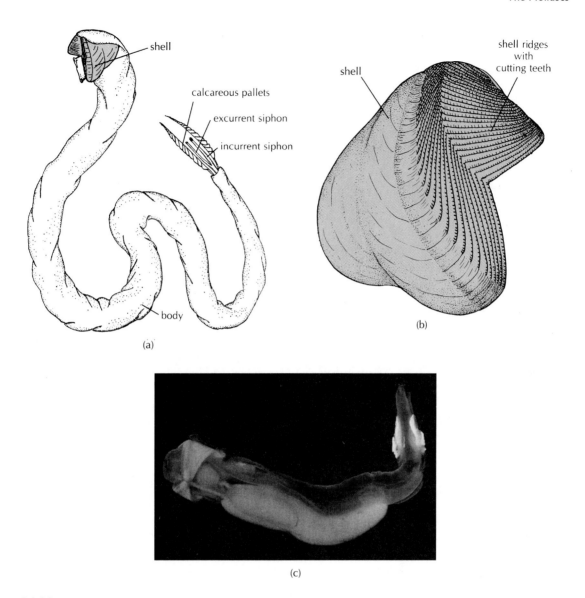

Figure 14.32

(a) External morphology of a shipworm removed from its wood burrow. The shell is used to excavate the burrow. The pallets seal off the opening of the burrow when the animal withdraws. The pallets are thus analogous to the operculum of gastropods.

(b) Detail of the shell, showing the cutting teeth. (c) Photograph of the shipworm *Teredora malleolus*. (a,b) *From Ruth D. Turner, in* A Survey and Illustrated Catalogue of the Teredinidae. *Copyright © 1966 Dr. Ruth D. Turner. Reprinted by Permission.* (c) *Courtesy of C. Bradford Calloway.*

have been found in all marine habitats, including the deep sea at several thousand meters.[10]

A major group of wood-boring lamellibranch bivalves are misleadingly referred to as "shipworms" (Figs. 14.32 and 14.33). Filtering food particles from suspension in typical lamellibranch fashion may actually play a minor role in the feeding biology of these bivalves, since shipworms appear to meet many of their nutritional requirements with the aid of symbiotic bacteria living in high concentrations within a specialized portion of the shipworm stomach, called the **wood-storing caecum.** The bacteria (possibly assisted by cohabitating protozoans) fix nitrogen and digest the wood eaten by the shipworm.

Subclass Septibranchia

Subclass Septi · branchia
(G: fence gill)
sep-ti-brank´-ē-ah

The septibranchs are a small group of carnivorous bivalves that feed on small planktonic animals (zooplankton) and on pieces of decomposing animal tissue. All species are marine, and most are found in very deep water. Only a few hundred species of septibranch bivalves have been described.

10. See *Topics for Further Discussion and Investigation,* no. 11.

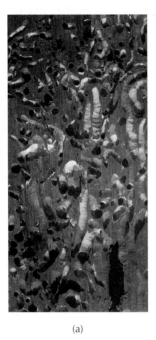 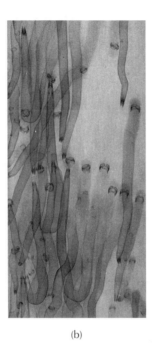

(a)   (b)

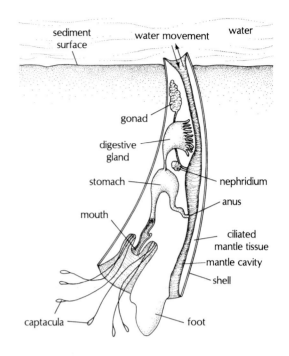

Figure 14.33
(a) A piece of wood riddled with wood-boring bivalves. (b) X-ray photo of a piece of wood containing shipworms.   *Courtesy of R. D. Turner, Museum of Comparative Zoology.*

Figure 14.34
Dentalium sp., a scaphopod, in its normal feeding orientation. Food particles are captured by the captacula.   *After Borradaile; after Naef.*

The septibranch ctenidium is highly modified, lacking filaments and forming a muscular septum (Fig. 14.28a). The septum divides the mantle cavity into ventral and dorsal chambers, as in other bivalves, but it is perforated by a series of ciliated openings. One-way valves regulate water flow through these openings in some species. The septum can be moved forcefully upward within the mantle cavity, drawing water in through the incurrent siphon and simultaneously expelling water through the excurrent siphon. Septibranchs thus feed as organic vacuum cleaners, commonly sucking in small crustaceans and annelids. The stomach is lined with hardened chitin, serving to grind up ingested food. Labial palps, although present, are quite small and do not have any sorting function. A very reduced style sac is found but never a crystalline style. Septibranchs are now commonly included in a new subclass, the Anomalodesmata.

Class Scaphopoda

Class Scapho · poda
(G: spade foot)
skaf-ō-pō´-dah

Of the seven molluscan classes, six were well represented in the Cambrian fauna of some 550 million years ago. The morphological diversity of these fossils makes it clear that molluscan radiation had been going on for a considerable time. In comparison, the scaphopods are a young group, first appearing in the fossil record in the middle Ordovician—a mere 450 million years ago (Fig.14.4). Present research suggests that the scaphopods had the same ancestors as the bivalves: The now extinct class of molluscs known as "rostroconchs" as mentioned earlier (p. 221).

The 300 to 400 species in the class Scaphopoda are all marine and live sedentary lives in sand or mud substrates, mostly in deep water. Scaphopods possess these "typical" molluscan features: foot, mantle tissue, mantle cavity, radula, and shell. The scaphopod shell is never spirally wound, but rather grows linearly as a hollow, curved tube; hence, the common names "tooth shell" and "tusk shell." The shell has an opening at each end (Fig. 14.34). Water enters at the narrower end, which protrudes above the substrate. Inflow of water is due to the action of ciliated cells restricted to ridges of mantle tissue. Periodically, water is expelled through this same opening by a sudden contraction of the foot musculature.

Unlike many other molluscs, scaphopods possess no ctenidia. A heart and circulatory system are also lacking; instead, the blood circulates through the various large sinuses of the hemocoel as a consequence of the foot's rhythmic movements.

Burrowing into soft substrate is accomplished by the foot, essentially as described previously for the bivalves. Scaphopods capture small food particles, including foraminiferans (a group of shelled protozoans), from the

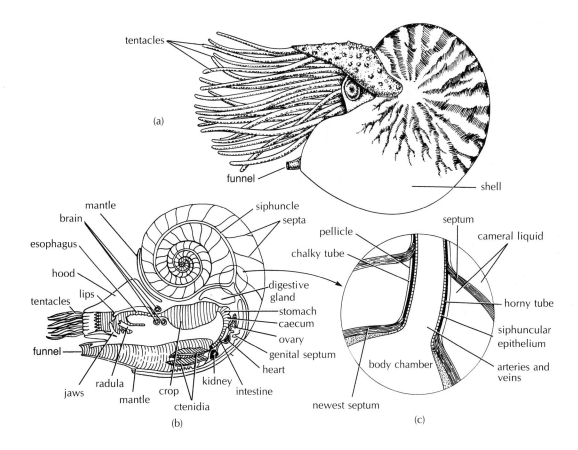

Figure 14.35

(a) *Nautilus* sp., in its normal swimming position. (b) Longitudinal section through the shell of the same individual, showing the septa isolating the different shell compartments, the siphuncle, and the internal anatomy. The entire body of the animal is found in the outermost chamber of the shell. Of the other shell chambers, all but the newest are filled with gas. (c) Detail of the newest and second most recent shell chambers. (b) *After Engemann and Hegner; after Borradaile and Potts.* (c) *Adapted with permission from "The Buoyancy of the Chambered Nautilus" by Peter Ward, Lewis Greenwald, and Olive E. Greenwald, in* Scientific American, *October 1980.*

surrounding sediment and water using specialized, thin tentacles known as **captacula** (Fig. 14.34). A typical individual possesses 100 to 200 such tentacles. Each captaculum terminates in a ciliated, sticky bulb; for feeding, the tentacles are extended by the creeping action of the cilia until the bulb contacts food. The food is then transported to the mouth by captacular cilia or, in the case of large food particles such as foraminiferans and small bivalves, by muscular contractions of the tentacles themselves.

Class Cephalopoda

> *Class Cephalo · poda*
> (G: head foot)
> sef´-ahl-ō-pō´-dah

It is truly wondrous that such animals as the squid and octopus belong to the same phylum as the clams, snails, scaphopods, and chitons. Unlike most molluscs, cephalopods are often fast-moving, active carnivores capable of considerable behavioral complexity. Members of this class are exclusively marine. The largest cephalopods—the giant squid—weigh perhaps 1,000 kg and reach lengths of up to 18 m, including the tentacles, which themselves can be 5 m or more long. The smallest cephalopod species are less than 2 cm (centimeters) long, including the tentacles.

Ctenidia and a radula are present in all cephalopod species. A mantle cavity and foot are present as well, but they do not generally function in "typical" molluscan fashion. The head and associated sensory organs of cephalopod molluscs are extremely well developed. Cephalopods are the supreme testament to the impressive plasticity of the basic molluscan body plan.

Of the 600 or so extant cephalopod species, only the five or six species in the genus *Nautilus* possess a true external shell. The shell of *Nautilus* is spiral, but unlike that of gastropods, it is divided by **septa** into a series of compartments (Figs. 14.35 and 14.36). The living animal is found only in the largest, outermost chamber. The septa are penetrated by the **siphuncle**—a calcified tube and its enclosed strand of vascularized tissue that spirals through the shell from the visceral mass, traversing all shell chambers. Liquid may be slowly transported to and

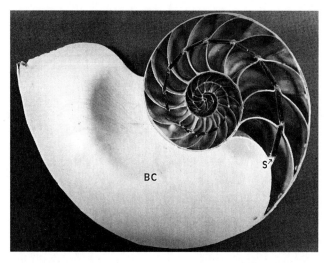

Figure 14.36
Section through the shell of *Nautilus*, showing the large body chamber (BC), siphuncle (S), and smaller chambers separated by septa. Holes have been drilled through the shell in several of the most recent chambers for experimental purposes; they are not natural. *Courtesy of L. Greenwald.*

from the shell chambers through the siphuncle, gas diffusing into or out of the chambers as the fluid volumes are altered.

This liquid transport is apparently made possible by enzymes in the siphuncular tissue; these enzymes actively concentrate solutes—probably ions—either inside or outside the siphuncular tissue, thereby establishing local osmotic gradients. If the osmotic concentration is higher within the siphuncular tissue, water will diffuse along the osmotic gradient from the liquid in the chamber—termed the **cameral fluid**—into the siphuncle and thence into the blood, to be discharged by the kidney. Gas probably diffuses from the blood—remember, the siphuncle is well vascularized—into the shell chamber as cameral liquid is removed.

As the gas content of each chamber is changed by this mechanism, the buoyancy of the shell, and therefore of the animal, also changes. In this way, the nautilus can maintain neutral buoyancy as it grows, exactly compensating for increased weight of shell and tissue through

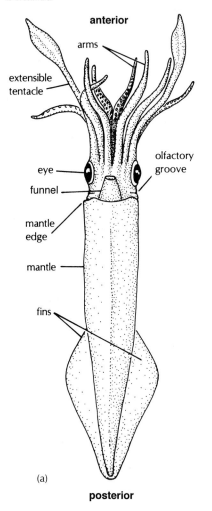

Figure 14.37
(a) External anatomy of the squid *Loligo* sp. (b) The internal anatomy of a male squid (ventral surface facing up). Digestion is initiated in the stomach and completed in the large, associated caecum. (c) Dissected ripe female, showing major reproductive organs. The nidamental glands secrete a gelatinous coating around each egg.   (c) *Beck/Braithwaite, Invertebrate Zoology, Laboratory Workbook, 3/e, © 1968. Reprinted by permission of Prentice Hall, Upper Saddle River, New Jersey.*

appropriate discharge of cameral fluid. Cameral fluid flux also may be involved in buoyancy regulation during the chambered nautilus's vertical migrations, in which the nautilus may ascend and then descend hundreds of meters each day.

The nautilus locomotes by jet propulsion, expelling water from the mantle cavity through a flexible, hollow tube called the **siphon** or **funnel** (Fig. 14.35a). The expulsion of water is brought about by contraction of the head retractor muscles. The muscular funnel, which is derived from the foot of the "typical" mollusc, may be turned in various ways to move the animal in different directions.

Other cephalopods also move by jet propulsion, but they expel a much greater volume of water and do so more forcefully. In contrast to the nautiloids, other cephalopods lack an external shell, and the mantle can thus play a major role in movement (Fig. 14.37a). The mantle tissue is thick and replete with both circular and radial musculature. Contracting the radial musculature

of the mantle tissue while the circular muscle fibers are relaxed causes the volume of the mantle cavity to increase. Water then enters the mantle cavity along the anterior mantle margin, often through one-way valves. (If you think of a squid or octopus as wearing a jacket [the mantle], the mantle cavity lies between the jacket and the body, and water enters the mantle cavity around the neck.) This influx of water causes the animal to be drawn forward slightly. When the radial muscles relax and the circular muscles then contract, the margins of the mantle tissue form a tight seal against the neck, and a large volume of water is forcefully expelled entirely through the flexible, hollow funnel. Since thrust = mass of expelled fluid × velocity, the expulsion of so much water through the funnel at great velocity enables these cephalopods to move at high speeds; for brief periods at least, squid probably can reach speeds of 5 m/sec (meters per second) to 10 m/sec. In such rapidly moving squid, the mantle cavity refills partly by radial muscle contractions, partly by elastic recoil of the compressed mantle, and partly (or largely) by pressure gradients set up along the sides of the body as the squid moves through the water.[11] Jet propulsion is used most commonly in escaping from predators. For more leisurely locomotion, many species use the arms (as in *Octopus*) or muscular lateral fins (as in squid and cuttlefish).

The mantle's active role in locomotion is incompatible with having an external shell. Most modern cephalopods that retain a shell (the cuttlefish and squid) bear it internally (Fig. 14.38a,b). Cuttlefish, for example, have an internal chambered shell that is involved in buoyancy regulation, as in *Nautilus*.[12] The squid shell is also internal, but it is little more than a thin, stiff, proteinaceous sheet, called the **pen** (Fig. 14.37b). This nonchambered "shell" plays no role in buoyancy regulation; instead, squid compensate for their own body weight by accumulating high concentrations of ammonium ions in the coelomic fluid. In *Octopus*, no vestiges of a shell remain. The so-called paper nautilus (*Argonauta*) also lacks a true shell; it is much more closely related to the octopus than to *Nautilus*, despite its common name. When reproductively active, however, females produce a chamberless, spiral, fragile "shell," using glands on the arm—not the mantle tissue; the female paper nautilus lives in this shell while brooding her young.

Thus, of all living cephalopods, only the chambered nautilus forms a true external shell. Loss of the shell has clearly been a major theme in cephalopod evolution over the last several hundred million years; over 7,000 species of shelled cephalopods are known from fossils (Fig. 14.38c), some of which have shells up to 4.5 m in diameter. As external shells became reduced or lost in evolution, cephalopods must have become increasingly

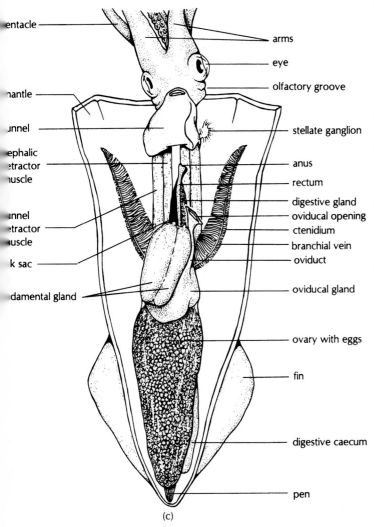

tentacle —
arms —
eye —
olfactory groove —
mantle —
funnel —
stellate ganglion —
cephalic retractor muscle —
anus —
rectum —
digestive gland —
funnel retractor muscle —
oviducal opening —
ctenidium —
branchial vein —
oviduct —
ink sac —
oviducal gland —
nidamental gland —
ovary with eggs —
fin —
digestive caecum —
pen —

(c)

11. See *Topics for Further Discussion and Investigation*, no. 12.
12. See *Topics for Further Discussion and Investigation*, no. 2.

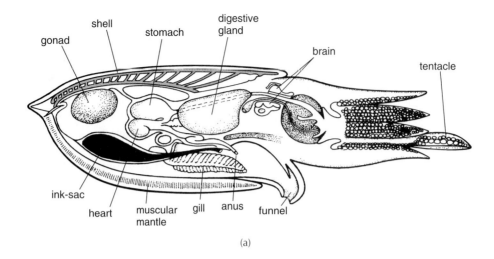

(a)

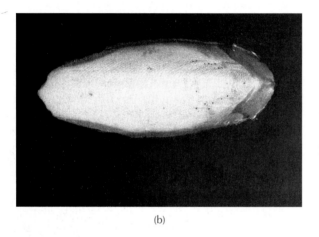

(b)

(c)

Figure 14.38

(a) Internal anatomy of the cuttlefish *Sepia* sp., showing location of internal shell. (b) The internal shell of a cuttlefish. (c) Fossilized ammonite (about 5 cm in diameter), with the coiled shell typical of most cephalopods for several hundred million years. Shells of some species exceeded 4.5 m in diameter. Most shelled cephalopods were driven to extinction at the end of the Mesozoic, quite likely through competition with an increasingly successful evolutionary upstart: the bony fishes. (a) *Boardman et al. 1987. Fossil Invertebrates. Palo Alto, Calif: Blackwell Scientific Publications.* (b) *Courtesy of Kerry M. Zimmerman.* (c) *Courtesy of H. Klinger.*

vulnerable to predation. Such vulnerability probably led to selection for other means of protection. For example, the skin of most cephalopods contains several layers of tiny colored cells called **chromatophores,** which overlay reflective cells called **iridocytes** (Fig. 14.39). Expansions and contractions of the chromatophores are mediated by muscle elements in the skin and are under direct nervous control. Thus, the coloration of cephalopod skin may change extremely quickly. Defensive, camouflaging, and courtship-related changes of color have been described.[13] Not surprisingly, the chambered *Nautilus* lacks chromatophores.

Many cephalopod species, particularly midwater and deep-water squid, possess numerous light organs, called **photophores.** These are distributed on the body in species-specific patterns, particularly on the ventral sur-

face (Fig. 14.40). Light is produced in the photophores by biochemical reactions that parallel to a remarkable degree those demonstrated by many arthropods (including fireflies) and fishes. Although such **bioluminescence—** the biochemical production of light with minimal heat—may play a role in attracting or recognizing mates and in luring potential prey, it almost certainly also protects against predation. At night, or in very deep water where it is perpetually dark, bright coordinated flashes of light may startle would-be predators. At intermediate depths, the photophores probably act instead to make the squid less visible when viewed from below.[13] Imagine looking up at an opaque object silhouetted against its surroundings during the day. Now imagine that object breaking up its silhouette by producing light of ambient intensity at many points along its ventral surface. Note that to achieve camouflage in this manner the squid

13. See *Topics for Further Discussion and Investigation,* no. 5.

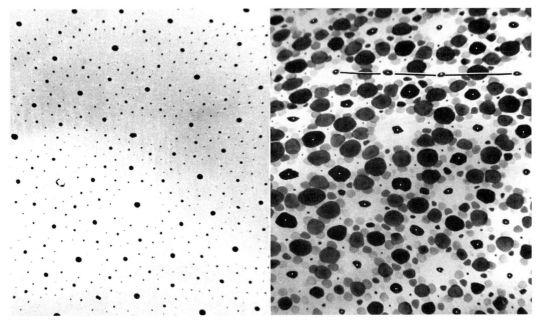

Figure 14.39
Chromatophores photographed from a living squid. The chromatophores are contracted at the left and expanded at the right. *Courtesy of R. Hanlon. 1982. Malacologia 23:89.*

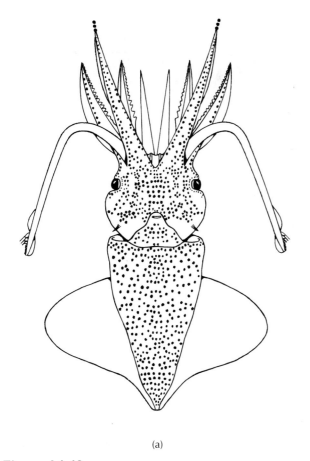

(a)

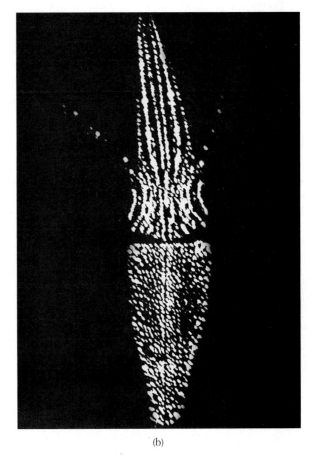

(b)

Figure 14.40
(a) Ventral surface of the squid *Abraliopsis* sp., showing its numerous photophores. During the day, the squid is found mainly 500 m to 600 m below the ocean surface; it migrates into shallower water at night. (b) Photograph of bioluminescence, showing pattern of photophores on the squid Abralia veranyi.
(a) From R. E. Young, in Science 191:1046–1048, 1976. Copyright © 1976 AAAS, Washington DC. Reprinted by permission. (b) Courtesy of E. A. Widders. From Herring, P. J., et al. 1992. Marine Biol. 112:293–98.

must be able to assess ambient light intensity and adjust the intensity of its own light production to match that intensity.

In addition, most cephalopods other than the chambered *Nautilus* have an **ink sac** associated with the digestive system (Fig. 14.37b,c). The dark-pigmented fluid secreted by the ink sac may be discharged deliberately through the anus, forming a cloud that presumably confuses potential predators and that also may act as a mild narcotic. The chambered nautilus lacks an ink sac; its absence is consistent with the notion that the ink sac has been selected for in other species by the increased predation pressure that must have accompanied reduction and loss of the external shell.

The cephalopod mantle cavity generally contains a pair of ctenidia (Fig. 14.37b,c), but unlike the situation in other molluscs, blood and water both flow in the same direction rather than in countercurrent fashion. Moreover, the ctenidia of cephalopods are not ciliated. Water circulation is maintained by the continual emptying and refilling of the mantle cavity, accomplished through contraction of the mantle musculature.

Cephalopods are also unique among molluscs in having a completely closed circulatory system, in which blood flows entirely through a system of arteries, veins, and capillaries. The blood sinuses found in other molluscs are not present in cephalopods. In addition to a single **systemic heart,** which receives oxygenated blood from the gills and sends it back to the tissues, an accessory **(branchial)** heart is associated with each gill (Fig. 14.37b). The two branchial hearts increase the blood pressure, helping to push blood through the gill capillaries. Concentrations of oxygen-binding blood pigments (hemocyanin, p. 269) are also unusually high in cephalopod blood. The cephalopod circulatory system is thus more efficient than that of other molluscs, supporting a far more active lifestyle.

In addition to forming the funnel, derivatives of the molluscan foot form the cephalopods' muscular **arms** and extensible **tentacles** (Fig. 14.41). The mouth thus lies in the center of the "foot." The total number of arms and tentacles is usually either 8 or 10, depending on the species, although nautiloids possess nearly 40. The tentacles of most cephalopods—all except those of the nautiloids—have small suction cups that are used for clinging to the substrate or to objects, including potential prey. The arms are also studded with receptors sensitive to touch and taste; all cephalopods tested to date have well-developed chemical sensory capabilities.

The degree of **cephalization** (concentration of sensory and nervous tissue at the anterior end of an animal) found among the Cephalopoda exceeds that found in any other invertebrate. Cephalopods possess a large,

complex, highly differentiated brain (Fig. 14.42a). The brain of the common octopus, *Octopus vulgaris,* has over 10 distinct lobes. Studies of cephalopod behavior indicate a clear capability for memory and learning (Research Focus Box 14.3). In the basic experiments, an individual—usually an octopus—is trained, through appropriate rewards (food) and discouragements (mild electric shocks), to attack an object of a certain shape, color, or texture or to attack prey only if the particular object is or is not present. Once training has been completed, researchers test the individual at intervals to determine how long the animal remembers what it has learned. By changing an object's shape, texture, orientation, color, size, or weight, they can discover what differences the cephalopod can perceive: Can it tell the difference, for example, between a red circle and an otherwise identical white circle, or between a horizontal rectangle and a vertical rectangle, or between a heavy sphere and an identically shaped, but lighter, sphere, or between a rough sphere and a smooth one, or between a scallop shell and a clam or mussel shell?

The results of such experiments clearly indicate that the octopus perceives shape, color intensity, and texture but, surprisingly, that it cannot distinguish between objects differing only in weight. Even more surprising, cephalopods show no behavioral response to sound except at very low frequencies. While direct electrical recordings clearly indicate that the statocysts are sensitive to sound stimuli, the information apparently is not suitably processed in the brain. Cephalopod deafness may be an adaptive response to millions of years of predation by toothed whales and dolphins, which may stun their prey with sound.[14] By tuning most sounds out, cephalopods might thus be immune to such a diabolical hunting mechanism; deaf cephalopods would have an edge over hearing cephalopods and so would be selected for over many generations, if we assume deafness to be a genetically determined trait.

Each cephalopod has two eyes. In *Nautilus* spp., the eyes are simple and function on the pinhole camera principle; there is no lens (Fig. 14.42c). The eyes form an image, but visual acuity is not great. In contrast, all other cephalopods have image-forming eyes that are incredibly similar to those of mammals (Fig. 14.42b). The eyes of these cephalopods and those of mammals are among the most beautiful examples of convergent evolution encountered among animals; the two groups of animals have independently evolved eyes that are amazingly similar in structure. Like the mammalian eye, the cephalopod eye possesses a cornea, lens, iris, diaphragm, and retina; both eyes are focusable and image-forming, although the process of image formation in mammals and cephalopods differs in detail. In most mammals, light is

---

14. See *Topics for Further Discussion and Investigation,* no. 17.

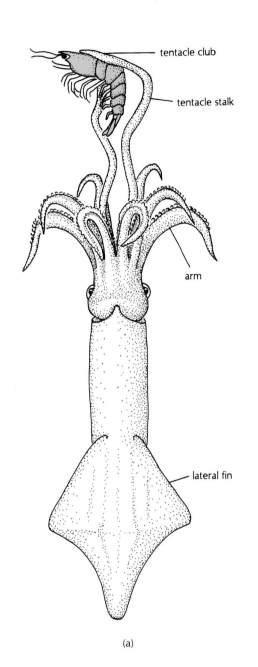

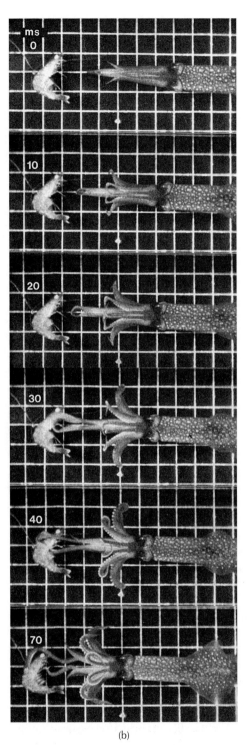

(a)

(b)

Figure 14.41

(a) The squid *Loligo pealei*, capturing prey with its suckered, extensible tentacles. (b) In this sequence, the crustacean prey was dropped in front of the squid at T_0 (time zero). The events shown occurred within 70 milliseconds.   (a) *From Kier, in* Journal of Morphology, *172:179, 182. Copyright © 1982 Alan R. Liss, Inc., New York. Reprinted by permission of John Wiley & Sons, Inc., New York.* (b) *Courtesy of W. M. Kier.*

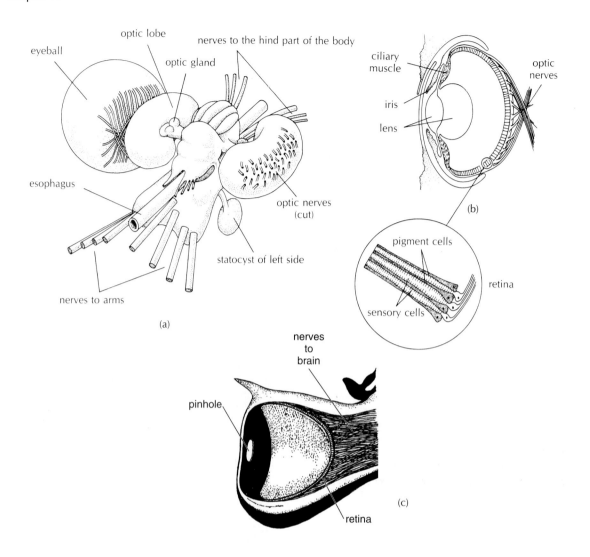

Figure 14.42

(a) A cephalopod brain. Cephalopods possess the most complex brains found among invertebrates. (b) The eye of *Octopus*, demonstrating convergent evolution with the eye of vertebrates. Unlike the vertebrate eye, cephalopod eyes are focused by moving the lens back and forth, rather than by altering the shape of the lens. (c) Vertical section through the pinhole camera eye of the chambered *Nautilus*. *(a,b) From Wells, Lower Animals. Copyright © 1968 McGraw-Hill, Inc., New York. Reprinted by permission. (c) From Muntz, W. R. A. 1991. American Malacological Bulletin. 9:69–74.*

focused by altering the shape of the lens. In contrast, light is focused within the cephalopod eye by moving the lens toward or away from the retina.

Clearly, the story of cephalopod evolution over the past several hundred million years has been one of developing adaptations for an active, carnivorous lifestyle: reduction and elimination of the shell; evolving alternate means of protection from predators; replacement of ciliary activity with muscular activity for locomotion and respiration; modification of the foot, mantle, and blood circulatory system; and extensive development of the head, sensory organs, brain, and nervous system. What might account for this dramatic evolution away from the standard molluscan body plan and lifestyle? What pressures could possibly have selected against a protective device as effective as a sturdy, external shell? Here is one attractive scenario: By the early Paleozoic, some 500 million years ago, the cephalopods, with their buoyant, chambered shells, were undoubtedly the most mobile and successful of all marine predators. With the subsequent diversification and success of bony fishes must have come intense competitive pressures between fish and cephalopod. With the intriguing exception of the chambered nautilus, only those cephalopods that had or evolved a more fish-like morphology and lifestyle apparently met those pressures and persist today.[15]

---

15. See *Topics for Further Discussion and Investigation*, no. 19.

Research Focus Box 14.3

Cephalopod Behavior

Fiorito, F., and P. Scotto. 1992. Observational learning in *Octopus vulgaris*. *Science* 256:545–47.

How we learn and remember are two of the great mysteries of human biology. Biologists have long studied aspects of behavior in other animals, hoping to discover general principles of brain function that may apply equally to us, and to determine just how unique is our ability to learn, to remember, and to teach. Many such studies have been conducted on *Octopus vulgaris*, a species common in European coastal waters. Fiorito and Scotto (1992) are the first to ask whether an octopus can learn from other octopuses.

They performed their study with *Octopus vulgaris* collected from the Bay of Naples, Italy. First they trained each of 44 octopuses to grab either a red plastic ball or a white plastic ball, by mildly shocking the animal if it chose incorrectly and rewarding it with food if it chose correctly. After about 17 to 21 trials, the octopuses consistently would select balls of the color they were trained to prefer, even without additional reward or punishment. *Octopus vulgaris* is color-blind, so the animals must have been perceiving differences in brightness rather than in color *per se*. Then the researchers let each octopus demonstrate its acquired color preference to a freshly caught test octopus that had received no prior training. Each demonstrator octopus consistently chose either a red or a white ball four times (Focus Fig. 14.3a,c), while a test octopus looked on. Then each of the 44 test octopuses were allowed to choose between a red ball and a white ball, again without reward or punishment. Would the observer octopuses adopt the preferences of the demonstrators?

The results were truly remarkable. The 30 test octopuses exposed only to octopuses demonstrating a preference for red balls mostly chose red balls themselves (Focus Fig. 14.3b), even though they received no reward or punishment for choosing one ball over the other. Similarly, test octopuses exposed only to octopuses demonstrating a preference for white balls mostly chose white balls themselves (Focus Fig. 14.3d). Moreover, the preferences shown by observer octopuses persisted when the animals were tested again five days later. These results are all the more impressive in that the octopuses showed a clear preference for red balls for some reason; so choosing a white ball over a red ball went against their apparent preference for red.

It seems clear that individuals of *Octopus vulgaris* can learn from other individuals of the same species. In fact, they

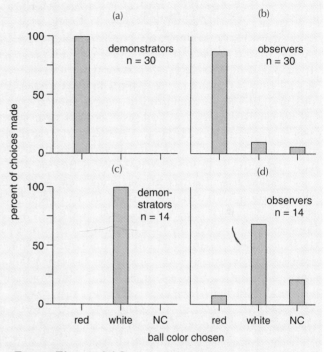

Focus Figure 14.3
The impact of training and observation on choices made by *Octopus vulgaris*. Demonstrators were trained to prefer either a red ball (a) or a white ball (c). Observers were then allowed to watch the choices made by demonstrators but received no training themselves. NC = no choice made. n= number of individuals participating in each experiment.

seem to learn more quickly by observing what other octopuses do than through human training; it took the octopuses at least 17 trials to learn by shock/reward training but only four trials to learn by example! We fully expect to find this sort of learning by copying among chimpanzees ("monkey see, monkey do") and other vertebrates, but to find it also among molluscs is indeed intriguing. One wonders about the extent to which octopuses can learn more complex behaviors by example, and about the things octopuses might actually be learning from each other in the field.

Other Features of Molluscan Biology

Reproduction and Development

Although some gastropods are parthenogenetic, molluscs usually reproduce sexually. Most species are **dioecious** (i.e., sexes are separate), but there are many exceptions to this generalization. Some prosobranch gastropods, opisthobranch gastropods, and lamellibranch bivalves, for example, are **protandric hermaphrodites;** that is, the sex of a single individual changes from male to female with age. All pulmonates, and those opisthobranchs that are not protandric hermaphrodites, are **simultaneous hermaphrodites,** with a single individual producing both eggs and sperm simultaneously; the gonad of such an individual is termed an **ovotestis.** Simultaneous hermaphrodites often have reciprocal copulation, resulting in a mutual exchange of sperm.

Generally, the molluscan genital ducts are associated with a portion, or a modified portion, of the excretory system. Fertilization of eggs is exclusively external in the Scaphopoda and probably in the Monoplacophora as well. External fertilization is quite common among the bivalves, chitons, and aplacophorans; less so among the gastropods; and nonexistent among the cephalopods. Terrestrial and freshwater molluscs (i.e., some gastropods and bivalves) fertilize only internally, as adaptations to stressful conditions that would otherwise be imposed upon the gametes and embryos by the environment (Chapter 1). Cephalopods show a particularly distinctive set of adaptations for achieving internal fertilization. In particular, one arm of the male is modified as a copulatory organ (Fig. 14.43). This sometimes highly modified arm, called a **hectocotylus** (*hecto* = G: one hundred; *cotylo* = G: sucker), transfers packets of sperm (**spermatophores,** Chapter 25) to the female; in the paper nautilus, the hectocotylus detaches from the male and stays behind in the female's mantle cavity until sperm transfer is completed.

Not surprisingly, free-living larval stages are associated with the development of most species that fertilize their eggs externally in the surrounding seawater. The embryo passes through a conspicuous **trochophore** stage (Fig. 14.44a,b), resembling that of polychaete annelids (Fig. 15.18a,b). Indeed, this similarity of larval stages has long been thought to indicate a close evolutionary relationship between annelids and molluscs; however, many biologists now believe that trochophore larvae may have instead arisen in the two phyla independently, by convergence. In any event, the prototroch of the trochophore

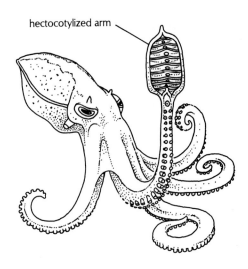

hectocotylized arm

Figure 14.43
Octopus lentus male with hectocotylized arm. The arm is turned up to show where the spermatophores will be carried. *After Huxley; after Verrill.*

stage of gastropods, bivalves, and scaphopods gradually becomes outfolded into a distinctive, ciliated organ known as the **velum.** A larva with a velum is called a **veliger** (Figs. 14.44d,e,g,h and 25.13e–f). The velum may be used for locomotion, food (phytoplankton) collection, and gas exchange and is lost upon metamorphosis to adult form. Veligers may spend hours, days, weeks, or months swimming in the plankton before metamorphosing. Among the shelled molluscs, metamorphosis often marks an abrupt change in shell morphology or ornamentation; in many species, the transition between larva and juvenile is thus clearly indicated on the shell (Fig. 14.45b). Among unshelled opisthobranchs, the larval shell is lost at metamorphosis.

Even when fertilization is internal, a free-swimming dispersive larva often occurs at some point during development. In the more advanced species, particularly among the gastropods, the free-living larval stage is often suppressed. Development to the juvenile stage may take place entirely within a jelly mass, an egg capsule (Figs. 14.44f and 25.12d), or a specialized brood chamber of the female. Most pulmonates develop to miniature snails within well-protected, calcified egg coverings, although some species retain a free-living veliger; obviously, none of the truly terrestrial gastropods have free-living larvae. Only among the cephalopods is a free-living, morphologically and ecologically distinct larval stage absent from the life cycle of all species. Cephalopod eggs develop within gelatinous masses, often attentively protected and ventilated by the mother until the young emerge as fully formed miniatures of the adult.

Circulation, Blood Pigments, and Gas Exchange

All molluscs have a blood circulatory system. In cephalopods, the system is completely closed; all blood flow occurs through arteries, veins, and capillaries. In all other molluscs, the circulatory system is largely an open one, blood moving through a series of large sinuses comprising the **hemocoel** (Fig. 14.46). In gastropods, the turgor of the tentacles and foot depends upon the amount of blood in the sinuses of these tissues. In most molluscs, including the apparently primitive monoplacophorans, the blood is pumped by a heart. However, in scaphopods, which lack a heart, muscular contractions of the foot have primary responsibility for moving the blood through the large sinuses.

The bloods of many mollusc species lack a specialized oxygen transport pigment. Where one is found, it is either hemoglobin, as in some bivalves and pulmonates, or hemocyanin, as in most pulmonates and all prosobranchs and cephalopods. **Hemocyanin** is a pigment structurally similar to hemoglobin, except that the heme group contains copper rather than iron.

Gas exchange may take place across gills housed within a mantle cavity (e.g., in cephalopods, bivalves, chitons, and prosobranch gastropods), across external gills (e.g., in some opisthobranch gastropods), or across other vascularized tissues (e.g., the cerata of opisthobranch gastropods and the tissue lining the mantle cavity of pulmonates and scaphopods).

Nervous System

The degree of development of the molluscan nervous system corresponds to the activity level of its possessor. Molluscan ganglia range from being nonexistent in many chitons to being extremely well developed in the cephalopods. Among the cephalopods, the ganglia form a true brain (Fig. 14.42a). In other molluscs, the major ganglia are generally paired and are connected by nerve fibers, forming a ring through which the esophagus passes (Fig. 14.47a). Major nerve cords may run to the mantle, foot, gills and osphradium, viscera, and radula. The ganglia associated with innervation of these tissues are named, respectively, the pleural, pedal, parietal, visceral, and buccal ganglia. Details vary among members of the different classes.

In a number of molluscs, certain neurons are specialized for very rapid impulse conduction. These unusually wide **giant fibers** are especially prominent among cephalopods, connecting the cerebral ganglia with the musculature of the mantle (Fig. 14.48). Giant fibers are extremely important in synchronizing contractions of the mantle musculature for jet propulsion, making rapid escape responses possible. The fibers also have been important to humans; present knowledge of how nerve impulses are generated and transmitted comes largely from studies of these cephalopod giant fibers.

Digestive System

Most molluscs have a complete digestive system with a separate mouth and anus. The mouth leads into a short esophagus, which, in turn, leads to a stomach. Associated with the stomach are one or more **digestive glands** or **digestive caeca**. Digestive enzymes are secreted into the lumen of these glands. Additional extracellular digestion takes place in the stomach. In cephalopods, digestion is entirely extracellular. In most other molluscs, the terminal stages of digestion are completed intracellularly, within the tissue of the digestive glands. The absorbed nutrients enter the blood circulatory system for distribution throughout the body or are stored in the digestive glands for later use. Undigested wastes pass through an intestine and out through the anus.

Other aspects of food collection and processing have already been discussed where appropriate for each group.

Excretory System

Molluscan urine is typically formed as the coelomic fluid passes through one or more pairs of metanephridia (often called "kidneys"). As in annelids, coelomic fluid generally enters the nephridium through a nephrostome. Recall that the molluscan coelom is little more than a small cavity (**pericardium**) surrounding the heart (*peri* = G: around; *cardio* = G: heart) (Fig. 14.46). Thus, the coelomic fluid appears to be largely a filtrate of blood, containing small waste molecules, such as ammonia, that are forced across the heart wall with the filtrate. Additional wastes are actively secreted into the coelomic fluid by glands lining the pericardium. As in annelids, the primary urine is further modified by selective resorption and secretion as it travels through the metanephridial tubules. The penultimate urine is then discharged into the mantle cavity through a **nephridiopore** and carried away by water currents.

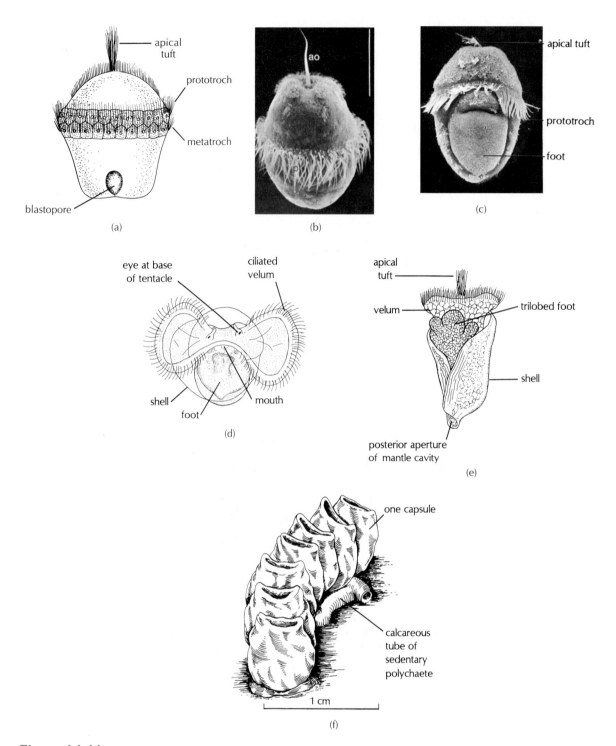

(a)

(b)

(c)

(d)

(e)

(f)

Figure 14.44

Molluscan development. (a) The trochophore larva of *Patella* sp., a primitive prosobranch gastropod. (b) Early trochophore larva of the chiton *Lepidochitona dentiens*. The larva is about 150 μm wide. ao = apical organ; pt = prototroch. (c) Advanced larva (seven days old) of the chiton *Lepidochitona hartweiqii*; note the pronounced developing foot. (d) A gastropod veliger larva in frontal view. (e) A scaphopod veliger. (f) Egg capsules of the marine prosobranch gastropod *Conus abbreviatus*. Each capsule is about 1 cm high and contains numerous embryos. (g) Larva of the American oyster, *Crassostrea virginica*, in lateral view. (h) *Veliger* of the gastropod *Nassarius reticulatus*, showing internal anatomy in lateral view. Arrows in velar groove show movement of captured food particles toward mouth.

(a) *From Hyman.* (b) *Courtesy of C. Nielsen. From Nielsen, 1987. Acta Zoologica (Stockholm) 68: 205–62. Permission of Royal Swedish Academy of Sciences.* (c) *Courtesy of D. J. Eernisse. From D. J. Eernisse, 1988. Biological Bulletin 174:287–302. Permission of Biological Bulletin.* (d) *Courtesy Dr. Rudolph S. Scheltema.* (e) *From Giese and Pearse, Reproduction of Marine Invertebrates, Vol. 5 (after Lucaze-Duthiers, 1856). Copyright © 1979 Academic Press, Inc. Reprinted by permission.* (f) *Reprinted from Kohn, in Pacific Science, 15:163–179, 1961, by permission of University of Hawaii Press, Honolulu, Hawaii.* (g) *From H. F. Prytherch, in Ecological Monographs, 4:56. Copyright © 1934 Ecological Society of America. Reprinted by permission.* (h) *from Fretter and Montgomery, in Journal of the Marine Biological Association of the United Kingdom, 48:504, 1968. Copyright © 1968 Cambridge University Press, New York. Reprinted by permission.*

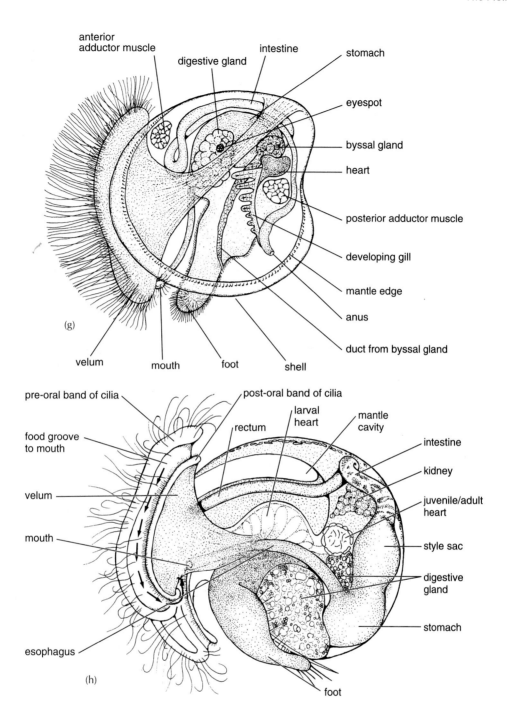

(g)

anterior adductor muscle
digestive gland
intestine
stomach
eyespot
byssal gland
heart
posterior adductor muscle
developing gill
mantle edge
anus
duct from byssal gland
velum
mouth
foot
shell

(h)

pre-oral band of cilia
post-oral band of cilia
food groove to mouth
rectum
larval heart
mantle cavity
intestine
kidney
juvenile/adult heart
style sac
digestive gland
stomach
velum
mouth
esophagus
foot

(a)

(b)

Figure 14.45
(a) The adult shell (scanning electron micrograph) of the prosobranch gastropod *Cyclostremiscus beauii* from Florida. The shell, 8 mm in diameter, is viewed from above; the aperture opens on the right. (b) High magnification of the shell apex, showing the transition between larval (protoconch) and juvenile shell (arrow); the larva metamorphosed when the shell was about 450 μm in diameter. At about 9 o'clock, there is another conspicuous line, which probably indicates an unsuccessful predatory attack on the larva; the larva survived, repaired its shell, and continued to grow.   (a,b) *Courtesy of R. Bieler and P. M. Mikkelson. From Bieler and Mikkelson. 1988. The Nautilus 102:1–29.*

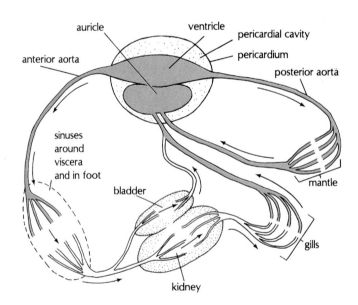

Figure 14.46
The pattern of blood circulation in the freshwater lamellibranch Anodonta sp. The shaded areas indicate the route of oxygenated blood. *From Cleveland P. Hickman, Biology of the Invertebrates, 2d ed. Copyright © 1973 The C. V. Mosby Company, St. Louis, Missouri. Reprinted by permission.*

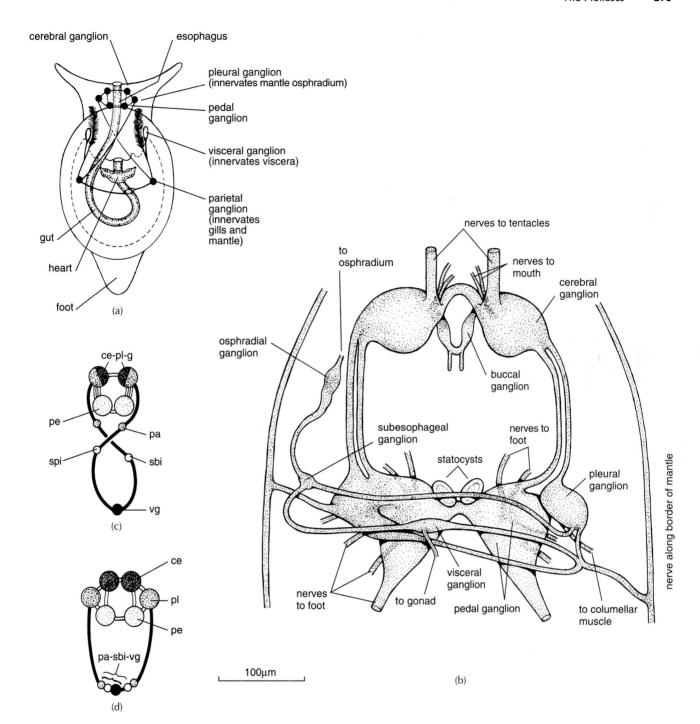

Figure 14.47

(a) Generalized nervous system of a prosobranch gastropod, demonstrating the effects of torsion. (b) Anterior portion of nervous system of the deep-sea limpet (class Gastropoda) *Cocculinella minutissima*. (c) Nervous system of a primitive opisthobranch gastropod, the bubble-shell snail *Acteon*. The members of this genus show no sign of detorsion; the nervous system still has the fully twisted, "streptoneurous" configuration (ce-pl-g=fused cerebral and plural ganglia; pa=parietal ganglion; pe=pedal ganglion; sbi=subintestinal ganglion; spi=supraintestinal ganglion; vg=visceral ganglion). (d) The nervous system of an advanced opisthobranch gastropod, the sea hare *Aplysia* sp. There is no longer any sign of torsion; the animal displays the "detorted" condition (pa-sbi-vg=parietal-subintestinal-visceral ganglia; ce=cerebral ganglion; pe=pedal ganglion; pl=pleural ganglion). (a) *From Hyman,* The Invertebrates, Vol. III. *Copyright © 1951 McGraw-Hill Book Company, New York. Reprinted by permission.* (b) *From Haszprunar, in* Journal of Molluscan Studies, 54:1-20, 1988. *Copyright © 1988 Malacological Society of London. Reprinted by permission.* (c,d) *From Louise Schmekel, in* The Mollusca, Vol. 10, 1985. *Copyright © 1985 Academic Press, Inc. Reprinted by permission.*

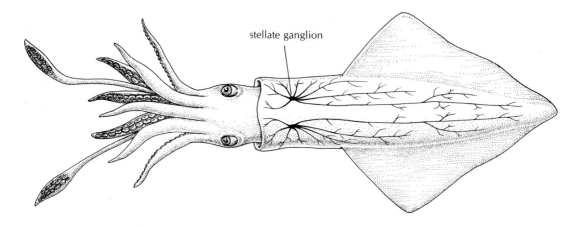

stellate ganglion

Figure 14.48

The nervous system of a cephalopod (squid), showing the system of giant fibers. Individual axons may be as large as 1 mm in diameter. *From Richard D. Keynes, "The Nerve Impulse and the Squid", illustrated by Bunji Tagawa in Scientific American, December 1958. Reprinted by permission.*

Taxonomic Summary

Phylum Mollusca
    Class Polyplacophora—chitons
    Class Aplacophora
    Class Monoplacophora
    Class Gastropoda—snails and slugs
      Subclass Prosobranchia
      Subclass Opisthobranchia
      Subclass Pulmonata
    Class Bivalvia (= Pelecypoda)—clams, mussels, oysters, and shipworms
      Subclass Protobranchia
      Subclass Lamellibranchia
      Subclass Septibranchia
    Class Scaphopoda—tooth shells, tusk shells
    Class Cephalopoda—squid, octopus, cuttlefish, and chambered nautilus

Topics for Further Discussion and Investigation

1. What is the functional significance of differences in shell morphology among gastropods and bivalves?

Appleton, R. D., and A. R. Palmer. 1988. Water-borne stimuli released by predatory crabs and damaged prey induce more predator-resistant shells in a marine gastropod. *Proc. Nat. Acad. Sci.* 85:4387.

Bertness, M. D., and C. Cunningham. 1981. Crab shell-crushing predation and gastropod architectural defense. *J. Exp. Marine Biol. Ecol.* 50:213.

Bottjer, D. J. 1981. Periostracum of the gastropod *Fusitriton oregonensis:* Natural inhibitor of boring and encrusting organisms. *Bull. Marine Sci.* 31:916.

Bottjer, D. J., and J. G. Carter. 1980. Functional and phylogenetic significance of projecting periostracal structures in the Bivalvia (Mollusca). *J. Paleontol.* 54:200.

Carefoot, T. H. and D. A. Donovan, 1995. Functional significance of varices in the muricid gastropod *Ceratostoma foliatum. Biol. Bull.* 189:59.

Conover, M. R. 1979. Effect of gastropod shell characteristics and hermit crabs on shell epifauna. *J. Exp. Marine Biol. Ecol.* 40:81.

Currey, J. D., and J. D. Taylor. 1974. The mechanical behavior of some mollusc hard tissues. *J. Zool., London* 173:395.

Johannesson, B. 1986. Shell morphology of *Littorina saxatilis* Olivi: The relative importance of physical factors and predation. *J. Exp. Marine Biol. Ecol.* 102:183.

Palmer, A. R. 1977. Function of shell sculpture in marine gastropods: Hydrodynamic destabilization in *Ceratostoma foliatum*. *Science* 197:1293.

Palmer, A. R. 1979. Fish predation and the evolution of gastropod shell sculpture: Experimental and geographical evidence. *Evolution* 33:697.

Perry, D. M. 1985. Function of the shell spine in the predaceous rocky intertidal snail *Acanthina spirata* (Prosobranchia: Muricacea). *Marine Biol.* 88:51.

Schmitt, R. J. 1982. Consequences of dissimilar defenses against predation in a subtidal marine community. *Ecology* 63:1588.

Stanley, S. M. 1969. Bivalve mollusk burrowing aided by discordant shell ornamentation. *Science* 166:634.

Thomas, M. L. H., and J. H. Himmelman. 1988. Influence of predation on shell morphology of *Buccinum undatum* L. on Atlantic coast of Canada. *J. Exp. Marine Biol. Ecol.* 115:221.

Vermeij, G. J. 1979. Shell architecture and causes of death of Micronesian reef snails. *Evolution* 33:686.

Vermeij, G. J., and A. P. Covich. 1978. Coevolution of freshwater gastropods and their predators. *Amer. Nat.* 112:833.

Vermeij, G. J., and J. D. Currey. 1980. Geographical variation in the strength of thaidid snail shells. *Biol. Bull.* 158:383.

2. A number of cephalopod species retain shells as adults, either internally or externally. What role do these shells play in regulating buoyancy?

Clarke, M. R., E. J. Denton, and J. B. Gilpin-Brown. 1979. On the use of ammonium for buoyancy in squids. *J. Marine Biol. Assoc. U.K.* 59:259.

Denton, E. J. 1974. On buoyancy and the lives of modern and fossil cephalopods. *Proc. Royal Soc. London B* 185:273.

Denton, E. J., and J. B. Gilpin-Brown. 1961. The buoyancy of the cuttlefish *Sepia officinalis* (L.). *J. Marine Biol. Assoc. U.K.* 41:319.

Greenwald, L., C. B. Cook, and P. D. Ward. 1982. The structure of the chambered nautilus siphuncle: The siphuncular epithelium. *J. Morphol.* 172:5.

Greenwald, L., P. D. Ward, and O. E. Greenwald. 1980. Cameral liquid transport and buoyancy in chambered nautilus (*Nautilus macromphalus*). *Nature* 286:55.

Ward, P. D., and L. Greenwald. 1981. Chamber refilling in *Nautilus*. *J. Marine Biol. Assoc. U.K.* 62:469.

3. Discuss the adaptive significance of gastropod torsion.

Crofts, D. R. 1955. Muscle morphogenesis in primitive gastropods and its relation to torsion. *Proc. Zool. Soc. London* 125:711.

Fretter, V. 1969. Aspects of metamorphosis in prosobranch gastropods. *Proc. Malac. Soc. London* 38:375.

Ghiselin, M. T. 1966. The adaptive significance of gastropod torsion. *Evolution* 20:337.

Goodhart, C. B. 1987. Garstang's hypothesis and gastropod torsion. *J. Moll. Stud.* 53:33.

Hardy, A. C. 1959. Pelagic larval forms. *The Open Sea: Its Natural History.* Boston: Houghton Mifflin, 187–88.

Kriegstein, A. R. 1977. Stages in the post-hatching development of *Aplysia californica*. *J. Exp. Zool.* 199:275.

Pennington, J. T., and F.–S. Chia. 1985. Gastropod torsion: A test of Garstang's hypothesis. *Biol. Bull.* 169:391–96.

Thompson, T. E. 1967. Adaptive significance of gastropod torsion. *Malacologia* 5:423.

Underwood, A. J. 1972. Spawning, larval development and settlement behavior of *Gibbula cineraria* (Gastropoda: Prosobranchia) with a reappraisal of torsion in gastropods. *Marine Biol.* 17:341.

4. A number of molluscs demonstrate a symbiotic relationship with photosynthetic protists that is reminiscent of that between zooxanthellae and cnidarians. How are the photosynthetic symbionts acquired by the molluscs, and what benefits are obtained by the hosts?

Fitt, W. K., and R. K. Trench. 1981. Spawning, development, and acquisition of zooxanthellae by *Tridacna squamosa* (Mollusca, Bivalvia). *Biol. Bull.* 161:213.

Gallop, A., J. Bartrop, and D. C. Smith. 1980. The biology of chloroplast acquisition by *Elysia ciridis*. *Proc. Royal Soc. London B* 207:335.

Klumpp, D. W., B. L. Bayne, and A. J. S. Hawkins. 1992. Nutrition of the giant clam *Tricdacna gigas* (L.). I. Contribution of filter feeding and photosynthates to respiration and growth. *J. Exp. Marine Biol. Ecol.* 155:105.

Masuda, K., S. Miyachi, and T. Maruyama. 1994. Sensitivity of zooxanthellae and non-symbiotic microalgae to stimulation of photosynthate excretion by giant clam tissue homogenate. *Marine Biol.* 118:687.

Rees, T. A. V., W. K. Fitt, B. Baillie, and D. Yellowlees. 1993. A method for temporal measurement of hemolymph composition in the giant clam symbiosis and its application to glucose and glycerol levels during a diel cycle. *Limnol. Oceanogr.* 38:213.

Trench, R. K., D. S. Wethey, and J. W. Porter. 1981. Observations on the symbiosis with zooxanthellae among the Tridacnidae (Mollusca: Bivalvia). *Biol. Bull.* 161:180.

5. Investigate the behavioral and chemical defenses of gastropods, bivalves, or cephalopods against predation.

Alexander, J. E., Jr., and A. P. Covich. 1991. Predator avoidance by the freshwater snail *Physella virgata* in response to the crayfish *Procambarus simulans*. *Oecologia* 87:435.

Bullock, T. H. 1953. Predator recognition and escape responses of some intertidal gastropods in the presence of starfish. *Behavior* 5:130.

Cimino, G., S. De Rosa, S. De Stefano, and G. Sodano. 1982. The chemical defense of four Mediterranean nudibranchs. *Comp. Biochem. Physiol.* 73B:471.

Crowl, T. A., and A. P. Covich. 1990. Predator life-history shifts in a freshwater snail. *Science* 247:949.

Di Marzo, V., A. Marin, R. R. Vardaro, L. De Petrocellis, G. Villani, and G. Cimino. 1993. Historical and biochemical bases of defense mechanisms in four species of Polybranchioidea ascoglossan molluscs. *Marine Biol.* 117:367.

Fainzilber, M., I. Napchi, D. Gordon, and D. Zlotkin. 1994. Marine warning via peptide toxin. *Nature* 369:192.

Feder, H. M. 1963. Gastropod defensive responses and their effectiveness in reducing predation by starfishes. *Ecology* 44:505.

Feifarek, B. P. 1987. Spines and epibionts as antipredator defenses in the thorny oyster *Spondylus americanus* Hermann. *J. Exp. Marine Biol. Ecol.* 105:39.

Ferguson, G. P., J. B. Messenger. 1991. A countershading reflex in cephalopods. *Proc. Royal Soc. London B* 243:63.

Fishlyn, D. A., and D. W. Phillips. 1980. Chemical camouflaging and behavioral defenses against a predatory seastar by three species of gastropods from the surf grass *Phyllospadix* community. *Biol. Bull.* 158:34.

Garrity, S. D., and S. C. Levings. 1983. Homing to scars as a defense against predators in the pulmonate limpet *Siphonaria gigas* (Gastropoda). *Marine Biol.* 72:319.

Gillette, R., M. Saeki, and R. -C. Huang. 1991. Defense mechanisms in notaspid snails: Acid humor and evasiveness. *J. Exp. Biol.* 156:335.

Gilly, W. F., B. Hopkins, and G. O. Mackie. 1991. Development of giant motor axons and neural control of escape responses in squid embryos and hatchlings. *Biol. Bull.* 180:209.

Greenwood, P. G., and R. N. Mariscal. 1984. Immature nematocyst incorporation by the aeolid nudibranch *Spurilla neapolitana*. *Marine Biol.* 80:35.

Hanlon, R. T., M. J. Smale, and W. H. H. Sauer. 1994. An ethogram of body patterning behavior in the squid *Loligo vulgaris reynaudii* on spawning grounds in South Africa. *Biol. Bull.* 187:363.

Manuel, J. L., and M. J. Dadswell. 1991. Swimming behavior of juvenile giant scallop, *Placopecten magellanicus,* in relation to size and temperature. *Canadian J. Zool.* 69:2250.

Margolin, A. S. 1964. A running response of *Acmaea* to seastars. *Ecology* 45:191.

Marko, P. B., and A. R. Palmer. 1991. Responses of a rocky shore gastropod to the effluents of predatory and non-predatory crabs: avoidance and attraction. *Biol. Bull.* 181:363.

Packard, A., and G. D. Sanders. 1971. Body patterns of *Octopus vulgaris* and maturation of the response to disturbance. *Anim. Behav.* 19:780.

Parsons, S. W., and D. L. Macmillan. 1979. The escape responses of abalone (Mollusca, Prosobranchia, Haliotidae) to predatory gastropods. *Marine Behav. Physiol.* 6:65.

Phillips, D. W. 1975. Distance chemoreception-triggered avoidance behavior of the limpets *Acmaea (Collisella) limatula* and *Acmaea (Notoacmaea) scutum* to the predatory starfish *Pisaster ochraceus*. *J. Exp. Zool.* 191:199.

Prior, D. J., A. M. Schneiderman, and S. I. Greene. 1979. Size-dependent variation in the evasive behaviour of the bivalve mollusc *Spisula solidissima*. *J. Exp. Biol.* 78:59.

Rice, S. H. 1985. An anti-predator chemical defense of the marine pulmonate gastropod *Trimusculus reticulatus* (Sowerby). *J. Exp. Marine Biol. Ecol.* 93:83.

Young, R. E., C. F. E. Roper, and J. F. Walters. 1979. Eyes and extraocular photoreceptors in midwater cephalopods and fishes: Their role in detecting downwelling light for counterillumination. *Marine Biol.* 51:371.

6. Gastropods continue to have pronounced impact on their communities even after death. Investigate the ways in which empty gastropod shells influence the populations of other invertebrates living in the same locality.

Conover, M. R. 1979. Effect of gastropod shell characteristics and hermit crabs on shell epifauna. *J. Exp. Marine Biol. Ecol.* 40:81.

McLean, R. 1983. Gastropod shells: A dynamic resource that helps shape benthic community structure. *J. Exp. Marine Biol. Ecol.* 69:151.

Spight, T. M. 1977. Availability and use of shells by intertidal hermit crabs. *Biol. Bull.* 152:120.

Vance, R. 1972. Competition and mechanism of coexistence in three sympatric species of intertidal hermit crabs. *Ecology* 53:1062.

Wilber, T. P., Jr., and W. F. Herrnkind. 1984. Predaceous gastropods regulate new-shell supply to salt-marsh hermit crabs. *Marine Biol.* 79:145.

7. Do molluscs show an immune response?

Cheng, T. C., K. H. Howland, and J. T. Sullivan. 1983. Enhanced reduction of T4D and T7 coliphage titres from *Biomphalaria glabrata* (Mollusca) hemolymph induced by previous homologous challenge. *Biol. Bull.* 164: 418.

Noël, D., R. Pipe, R. Elston, E. Bachère, and E. Mialhe. 1994. Antigenic characterization of hemocyte subpopulations in the mussel *Mytilus edulis* by means of monoclonal antibodies. *Marine Biol.* 119:549.

Tripp, M. R. 1960. Mechanisms of removal of injected microorganisms from the American oyster, *Crassostrea virginica* (Gmelin). *Biol. Bull.* 119: 273.

8. Investigate adaptations for carnivorous behavior or parasitism in prosobranch gastropods.

Carr, W. E. S. 1967. Chemoreception in the mud snail, *Nassarius obsoletus*. I. Properties of stimulating substances extracted from shrimp. *Biol. Bull.* 133:90.

Carriker, M. R., D. van Zandt, and T. J. Grant. 1978. Penetration of molluscan and nonmolluscan minerals by the boring gastropod *Urosalpinx cinerea*. *Biol. Bull.* 155:511.

Hermans, C. O., and R. A. Satterlie. 1992. Fast-strike feeding behavior in a pteropod mollusk, *Clione limacina* Phipps. *Biol. Bull.* 182:1.

O'Sullivan, J. B., R. R. McConnaughey, and M. E. Huber. 1987. A blood-sucking snail: The Cooper's Nutmeg, *Cancellaria cooperi* Gabb, parasitizes the California electric ray, *Torpedo californica* Ayres. *Biol. Bull.* 172:362.

Perry, D. M. 1985. Function of the shell spine in the predaceous rocky intertidal snail *Acanthina spirata* (Prosobranchia: Muricacea). *Marine Biol.* 88:51.

Rittschoff, D., L. G. Williams, B. Brown, and M. R. Carriker. 1983. Chemical attraction of newly hatched oyster drills. *Biol. Bull.* 164:493.

Smith, T. B. 1984. Ultrastructure and function of the proboscis in *Melanella alba* (Gastropoda: Eulimidae). *J. Marine Biol. Assoc. U.K.* 64:503.

9. Investigate the cycles of digestive activity encountered among bivalved molluscs and prosobranch gastropods.

Curtis, L. A. 1980. Daily cycling of the crystalline style in the omnivorous, deposit-feeding estuarine snail *Ilyanassa obsoleta*. *Marine Biol.* 59:137.

Hawkins, A. J. S., B. L. Bayne, and K. R. Clarke. 1983. Coordinated rhythms of digestion, absorption and excretion in *Mytilus edulis* (Bivalvia: Mollusca). *Marine Biol.* 74:41.

Morton, J. E. 1956. The tidal rhythm and the action of the digestive system of the lamellibranch *Lasaea rubra*. *J. Marine Biol. Assoc. U.K.* 35:563.

Palmer, R. E. 1979. Histological and histochemical study of digestion in the bivalve *Arctica islandica* L. *Biol. Bull.* 156:115.

Robinson, W. E., and R. W. Langton. 1980. Digestion in a subtidal population of *Mercenaria mercenaria* (Bivalvia). *Marine Biol.* 58:173.

Yonge, C. M. 1926. Structure and physiology of the organs of feeding and digestion in *Ostrea edulis*. *J. Marine Biol. Assoc. U.K.* 14:295.

10. Investigate the adaptations of gastropods for terrestrial existence.

Boss, K. J. 1974. Oblomovism in the mollusca. *Trans. Amer. Microsc. Soc.* 93:460.

Sloan, W. C. 1964. The accumulation of nitrogenous compounds in terrestrial and aquatic eggs of prosobranch snails. *Biol. Bull.* 126:302.

Verderber, G. W., S. B. Cook, and C. B. Cook. 1983. The role of the home scar in reducing water loss during aerial exposure of the pulmonate limpet, *Siphonaria alternata* (Say). *Veliger* 25:235.

Wells, G. P. 1944. The water relations of snails and slugs. III. Factors determining activity in *Helix pomatia* L. *J. Exp. Biol.* 20:79.

Welsford, I. G., P. A. Banta, and D. J. Prior. 1990. Size-dependent responses to dehydration in the terrestrial slug, *Limax maximus* L: Locomotor activity and huddling behavior. *J. Exp. Zool.* 253:229.

11. Discuss the ecological significance of wood-boring bivalves in the deep sea.

Turner, R. D. 1973. Wood-boring bivalves, opportunistic species in the deep sea. *Science* 180:1377.

12. Discuss the evidence indicating that radial muscle contractions alone do not completely account for refilling the cephalopod mantle cavity during jet propulsion.

Gosline, J. M., and R. E. Shadwick. 1983. The role of elastic energy storage in swimming: An analysis of mantle elasticity in escape jetting in the squid, *Loligo opalescens. Canadian J. Zool.* 61:1421.

Vogel, S. 1987. Flow-assisted mantle cavity refilling in jetting squid. *Biol. Bull.* 172:61.

Ward, D. V. 1972. Locomotor function of the squid mantle. *J. Zool., London* 167:437.

13. How humans learn and remember things is still one of the major unsolved biological mysteries. Much of what is currently understood about learning and memory and the biological basis of behavior comes from the study of invertebrates, including the gastropod molluscs. Describe the types of experiments that have been performed on gastropods and the extent to which each has increased our understanding of how animals learn.

Barnes, D. M. 1986. From genes to cognition. *Science* 231:1066.

Colebrook, E., and K. Lukowiak. 1988. Learning by the *Aplysia* model system: Lack of correlation between gill and gill motor neurone responses. *J. Exp. Biol.* 135:411.

Kandel, E. R., and J. H. Schwartz. 1982. Molecular biology of learning: Modulation of transmitter release. *Science* 218:433.

Lin, S. S., and I. B. Levitan. 1987. Concanavalin A alters synaptic specificity between cultured *Aplysia* neurons. *Science* 237:648.

Moffett, S., and K. Snyder. 1985. Behavioral recovery associated with the central nervous system regeneration in the snail *Melampus. J. Neurobiol.* 16:193.

14. Bacteria have been found in the gills of various marine bivalve species living in sediments rich in decomposing organic matter. What evidence suggests that these bacteria live in a symbiotic relationship with the bivalves, and what is the nature of that relationship?

Anderson, A. E., J. J. Childress, and J. A. Favuzzi. 1987. Net uptake of CO_2 driven by sulphide and thiosulphate oxidation in the bacterial symbiont-containing clam *Solemya reidi. J. Exp. Biol.* 133:1.

Cavanaugh, C. M. 1983. Symbiotic chemoautotrophic bacteria in marine invertebrates from sulfide-rich habitats. *Nature (London)* 302:58.

Cavanaugh, C. M., R. R. Levering, J. S. Maki, R. Mitchell, and M. E. Lidstrom. 1987. Symbiosis of methylotrophic bacteria and deep-sea mussels. *Nature (London)* 325:346.

Childress, J. J., C. R. Fisher, J. M. Brooks, M. C. Kennicutt II, R. R. Bidigare, and A. E. Anderson. 1986. A methanotrophic marine molluscan (Bivalvia: Mytilidae) symbiosis: Mussels fueled by gas. *Science* 233:1306.

Kochevar, R. E., J. J. Childress, C. R. Fisher, and E. Minnich. 1992. The methane mussel: Roles of symbiont and host in the metabolic utilization of methane. *Marine Biol.* 112:389.

Powell, M. A., and G. N. Somero. 1986. Adaptations to sulfide by hydrothermal vent animals: Sites and mechanisms of detoxification and metabolism. *Biol. Bull.* 171:274.

15. Discuss how molluscs have been used to detect and assess the impact of environmental pollutants.

Amiard, J. C., C. Amiard-Triquet, B. Berthet, and C. Métayer. 1986. Contribution to the ecotoxicological study of cadmium, lead, copper and zinc in the mussel *Mytilus edulis. Marine Biol.* 90:425.

Axiak, V., and J. J. George. 1987. Effects of exposure to petroleum hydrocarbons on the gill functions and ciliary activities of a marine bivalve. *Marine Biol.* 94:241.

Berger, B., and R. Dallinger. 1989. Accumulation of cadmium and copper by the terrestrial snail *Arianta arbustorum* L.: Kinetics and budgets. *Oecologia* 79:60.

Byrne, C. J., and J. A. Calder. 1977. Effect of the water-soluble fractions of crude, refined and waste oils on the embryonic and larval stages of the quahog clam *Mercenaria* sp. *Marine Biol.* 40:225.

Gilfillan, E. S., D. Mayo, S. Hanson, D. Donovan, and L. C. Jiang. 1976. Reduction in carbon flux in *Mya arenaria* caused by a spill of no. 6 fuel oil. *Marine Biol.* 37:115.

Stickle, W. B., S. D. Rice, and A. Moles. 1984. Bioenergetics and survival of the marine snail *Thais lima* during long-term oil exposure. *Marine Biol.* 80:281.

16. Compare the nature and role of hydrostatic skeletons in the locomotion of different gastropod species.

Bernard, F. R. 1968. The aquiferous system of *Polinices lewisi* (Gastropoda, Prosobranchiata). *J. Fish Res. Bd. Canada* 25:541.

Dale, B. 1973. Blood pressure and its hydraulic functions in *Helix pomatia* L. *J. Exp. Biol.* 59:477.

Jones, H. D. 1973. The mechanism of locomotion in *Agriolimax reticulatus* (Mollusca; Gastropoda). *J. Zool., London* 171:489.

Kier, W. M. 1988. The arrangement and function of molluscan muscle. In *The Mollusca, Vol. 11: Form and Function.* Trueman, E. R., and M. R. Clarke, eds. New York: Academic Press, 211–52.

Trueman, E. R., and A. C. Brown. 1976. Locomotion, pedal retraction and extension, and the hydraulic systems of *Bullia* (Gastropoda: Nassaridae). *J. Zool., London* 178:365.

Voltzow, J. 1986. Changes in pedal intramuscular pressure corresponding to behavior and locomotion in the marine gastropods *Busycon contrarium* and *Haliotis kamtschatkana*. *Canadian J. Zool.* 64:2288.

17. Discuss the hypothesis that deafness in cephalopods is an adaptive response to predation.

Hanlon, R. T., and B.-U. Budelmann. 1987. Why cephalopods are probably not "deaf". *Amer. Nat.* 129:312–317.

Moynihan, M. 1985. Why are cephalopods deaf? *Amer. Nat.* 125:465.

Packard, A., H. E. Karlsen, and O. Sand. 1990. Low frequency hearing in cephalopods. *J. Comp. Physiol.* A166:501.

18. How has the application of new technology altered our understanding of how lamellibranch bivalves collect food?

Nielsen, N. F., P. S. Larsen, H. U. Riisgård, and C. B. Jørgensen. 1993. Fluid motion and particle retention in the gill of *Mytilus edulis*: Video recordings and numerical modelling. *Marine Biol.* 116:61.

Ward, J. E., B. A. MacDonald, R. J. Thompson, and P. G. Beninger. 1993. Mechanisms of suspension feeding in bivalves: Resolution of current controversies by means of endoscopy. *Limnol. Oceanogr.* 38:265.

Ward, J. E., R. I. E. Newell, R. J. Thompson, and B. A. MacDonald. 1994. *In vivo* studies of suspension-feeding processes in the Eastern oyster, *Crassostrea virginica* (Gmelin). *Biol. Bull.* 186:221.

19. Evaluate the hypothesis that shell reduction and loss among cephalopods was driven by competition with bony fishes. What additional evidence could be sought that might argue for or against the hypothesis?

Packard, A. 1972. Cephalopods and fish: The limits of convergence. *Biol. Rev.* 47:241.

Taxonomic Detail

Phylum Mollusca[15]

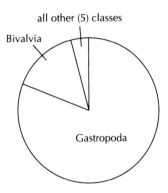

all other (5) classes

Bivalvia

Gastropoda

Class Polyplacophora.
The approximately 500 species are divided among 13 families.

> **Family Lepidopleuridae.** *Lepidopleurus.* Unlike most chitons, which occur in shallow water or intertidally, most species in this family are collected only in deep water, down to 7,000 m depth.
>
> **Family Ischnochitonidae.** *Ischnochiton, Lepidochitona, Tonicella.* These are mostly shallow-water species living on stones and in oyster beds. This family contains up to 40% of all chiton species.
>
> **Family Chaetopleuridae.** *Chaetopleura.*
>
> **Family Mopaliidae.** *Mopalia*—the hairy, or mossy, chiton; *Katharina.* The girdle of *Mopalia* spp. is studded with conspicuous spines (the "hairy girdle" syndrome). Members of one genus, *Placiphorella*, are carnivores, preying on polychates and crustaceans.
>
> **Family Chitonidae.** *Chiton, Acanthopleura.* About 20% of all chiton species are members of this family.
>
> **Family Acanthochitonidae.** *Cryptochiton*—The largest of the chitons, up to 40 cm long. Shell valves are completely covered by the extensive

girdle. *Cryptoplax*—a strange, worm-like chiton with a highly extensible body; it lives in rock crevices in the Indo-Pacific. About 20% of all chiton species are placed in this very diverse family.

Class Aplacophora.

Approximately 280 species.

Subclass Neomeniomorpha (= Solenogastres).

Members all possess a slender foot, seated in a narrow groove that runs along the ventral surface from just posterior to the mouth to the posterior mantle cavity. Many species lack a radula, and the rest possess a highly simplified radula. None have true ctenidia. All species are hermaphroditic. Members of this subclass either crawl over the sediment or live among cnidarians, on which they feed. All are marine. The 21 families contain about 70% of all aplacophoran species.
Family Neomeniidae. *Neomenia*.

Subclass Chaetodermomorpha (= Caudofoveata).

Members possess neither a foot nor a ventral groove. All species possess a radula and a mantle cavity housing one pair of bipectinate ctenidia. All species are marine and burrow through muddy substrate. Three families.

Family Chaetodermatidae. *Chaetoderma, Falcidens*.

Class Monoplacophora.

The 10 species are all in one family (Neopilinidae). *Neopilina, Vema*.

Class Gastropoda.

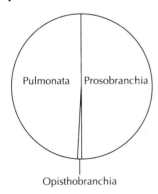

This class contains at least 40,000 described species.

Subclass Prosobranchia.

This subclass contains at least 20,000 species, distributed among over 140 families.

Order Archaeogastropoda.

These snails possess many primitive prosobranch characteristics, with many species bearing a pair of hypobranchial glands, a pair of osphradia, a pair of auricles, a pair of kidneys, and a pair of ctenidia (always bipectinate). Ganglia are paired, rarely fused, and typically separated by long commissures. The radula has many teeth. Most archaeogastropod species are marine herbivores, including about 50 species of abalone; a few terrestrial species and a few freshwater species are also known. Archaeogastropods typically fertilize externally and are unique among gastropods in commonly developing as free-swimming trochophore larvae. Approximately 5,000 species, distributed among 26 families.

Superfamily Pleurotomariacea.   Four families.

Family Haliotidae. *Haliotis*—the abalone. Shallow-water herbivorous snails—about 50 species—living on solid substrates, to which they can cling very tightly. The tissue is commonly eaten by humans, and the broad shells are commonly used by humans as ornaments.

Family Neomphalidae. *Neomphalus*. A deep-sea archaeogastropod recently collected at depths of 2,400 m in the Galapagos Rift of the Pacific Ocean. Considered a living fossil because of its many primitive characteristics.

Superfamily Fissurellacea.   One family (Fissurellidae). *Fissurella, Emarginula, Emarginella, Diodora*—keyhole limpets. So-named because the shell bears a conspicuous hole at the top, through which water exits after passing over the gills. All marine.

Superfamily Patellacea.   Five families.

Families Acmaeidae, Patellidae. *Acmaea, Patella*. True limpets, all marine. Acmaeids have lost the right ctenidium, but patellids retain both ctenidia.

Superfamily Cocculinacea.   Two families.

Family Cocculinidae. *Cocculina*. Deep-sea limpets, collected from depths of over 9,000 m. Most species eat wood, but some species feed on such restricted substrates as cephalopod beaks, polychaete tubes, and the bones of dead fish.

Superfamily Trochacea.   Eight families.

Family Trochidae. *Gibbula, Margarites, Calliostoma, Cittarium, Tegula, Monodonta*—top shells. All species are marine, and most are herbivores.

Family Turbinidae. *Astraea*—star shells. All species are marine, and most are herbivores. The operculum is calcified.

Superfamily Neritacea. Six families.

Family Neritidae. *Nerita.* Among the most advanced of the archaeogastropods, reflected by the loss of the right kidney and the right ctenidium and by unusually complex reproductive anatomy and biology. Most species are marine, but some occupy estuarine, freshwater, or terrestrial habitats.

Family Hydrocenidae. *Hydrocena.* Terrestrial archaeogastropod with mantle cavity forming a lung; no ctenidia.

Family Helicinidae. *Helicina, Alcadia.* Terrestrial snails living on the ground or, more rarely, in trees.

Order Mesogastropoda (= Taenioglossa).[16]

The right ctenidium has been lost, and the remaining (left) ctenidium is always monopectinate; some terrestrial species have lost both ctenidia. Ganglia tend to be fused. The head typically bears a protrusible proboscis and one pair of cephalic tentacles, with a simple eye at the base of each. The right kidney is lost or modified for reproduction. The heart has a single auricle. The left hypobranchial gland is lost. The radula usually has seven teeth in each row (the "taenioglossate" condition). All species have internal fertilization, and most produce free-living veliger larvae. In some species, the mantle edge forms an incurrent siphon. Most mesogastropods are free-living and marine, but some species are terrestrial, freshwater, or parasitic. Approximately 10,000 species, distributed among 95 families.

Superfamily Viviparacea. Three families.

Family Viviparidae. *Viviparus.* Freshwater suspension feeders. Right tentacle of male modified for copulation.

Family Bithyniidae. *Bithynia.* Freshwater suspension feeders that serve as intermediate hosts in the life cycle of the trematode (fluke) intestinal parasite *Opisthorchis tenuicollis.*

Superfamily Littorinacea. Five families.

Family Lacunidae. *Lacuna.* Common, small, cold-water marine snails.

Family Littorinidae. *Littorina, Tectarius*—the true periwinkles. All are marine herbivores, typically smaller than about 2.5 cm. Most live intertidally; some species are found only above high tide.

Family Pomatiasidae. *Pomatias.* Terrestrial snails living among fallen leaves and moss. Possess a unique form of bipedal, walking locomotion, in which the two sides of the foot move forward in alternation.

Superfamily Rissoacea. Twenty-five families; highly diverse.

Family Hydrobiidae. *Hydrobia.* Small (approximately 6 mm high), mostly freshwater prosobranchs; some brackish species and a few terrestrial species. Most hatch as small juvenile snails, but one species (*H. ulvae*) has a free-living veliger larva.

Family Caecidae. *Caecum.* Very small (typically a few millimeters long) marine species with tubular shells.

Superfamily Cerithiacea. Fifteen families.

Family Cerithiidae. *Cerithium, Bittium, Litiopa, Batillaria.* All are marine, mostly in shallow water, and most eat detritus.

Family Turritellidae. *Turritella, Vermicularia.* Suspension-feeding marine species.

Family Vermetidae. *Vermetus, Serpulorbis, Petaloconchus, Dendropoma.* Suspension-feeding, immobile marine snails living in loosely coiled or uncoiled shells resembling those made by many sedentary polychaete worms. The shells are permanently cemented to rocks.

Superfamily Heterogastropoda. Four families. A polyphyletic assemblage containing gastropods not easily categorized as either prosobranchs or opisthobranchs. May or may not have ctenidia; the radula departs from the taenioglossate (seven teeth per row) condition. Most species are carnivores.

Family Architectonicidae. *Architectonica, Philippia*—sundial shells. Warm-water marine species with conical or disc-shaped shells, living as deep as 2,000 m. Typically feed on cnidarians.

---

16. The mesogastropods and neogastropods are now commonly grouped in one of the following formal taxonomic categories: Caenogastropoda (*caeno* = G: recent), Monotocardia (G: single heart), or Pectinobranchia (G: comb gills). In this system, the terms "Mesogastropoda" and "Neogastropoda" have no formal taxonomic significance, but they still are widely used for grouping prosobranchs into different levels of general organization.

Family Pyramidellidae. *Odostomia, Pyramidella.* All are marine predators or ectoparasites on other invertebrates, including other molluscs; they pierce their host with a sharp stylet and then pump in body fluids through the proboscis. Some workers consider these snails to be members of the subclass Opisthobranchia.

Superfamily Epitoniacea. Three families.

Family Epitoniidae. *Epitonium*—wentletraps. Typically ectoparasitic on anthozoans.

Family Janthinidae. *Janthina, Recluzia*—violet snails. All are pelagic, drifting attached to a float composed of secreted mucus and air. All are carnivores, feeding on planktonic hydrozoans (siphonophores), and all species are sequential hermaphrodites, with each individual being first male and later female.

Superfamily Eulimacea. Six families.

Family Eulimidae. *Eulima, Stilifer.* Marine snails, mostly ectoparasitic on echinoderms; body fluids are pumped in through a long proboscis.

Family Entoconchidae. *Entoconcha, Thyonicola.* All species are internal parasites of sea cucumbers (holothurians). Adult females are typically shell-less and vermiform, and they may attain lengths of up to 1.3 m. Males are microscopic (they are "dwarf" males) and usually embedded in tissues of the female as little more than testicular sacs.

Superfamily Strombacea. Three families.

Family Strombidae. *Strombus*—conchs. Marine. Economically important as a food source in some tropical areas.

Superfamily Calyptraeacea. Four families.

Family Calyptraeidae. *Calyptraea*—cup-and-saucer shells; *Crepidula*—slipper shells. Limpet-like marine herbivores. Many species are suspension feeders, collecting food on the ciliated gill. Some species live in communal stacks, with females at the bottom and males at the top. All species are sequential hermaphrodites.

Superfamily Cypraeacea. Six families.

Family Cypraeidae. *Cypraea*—the cowries. All marine, mostly in shallow, tropical seas.

The shells are very smooth and glossy, usually about 4.0 cm to 7.5 cm long. Juveniles and adults lack an operculum.

Family Ovulidae. *Cyphoma*—flamingo tongue snails; *Ovula, Simnia.* Mostly tropical marine carnivores, feeding on colonial cnidarians.

Superfamily Heteropoda. Three families. All are pelagic carnivores.

Family Atlantidae. *Atlanta.* Heteropods with small, thin, fragile, coiled shells. Animal can pull completely into shell. Head has large eyes and proboscis. Apparently feeds preferentially on other planktonic gastropods (pteropods).

Family Carinariidae. *Carinaria.* Heteropods with thin, flattened shell too small to contain entire animal. Proboscis extremely extensible. Body up to 0.5 m long.

Family Pterotracheidae. *Pterotrachea.* Heteropods lacking shells, mantle, mantle cavity, and proboscis. Body up to 20 cm long, transparent and cylindrical.

Superfamily Naticacea. One family (Naticidae). *Natica, Polinices, Lunatia*—moon snails. Marine, usually sand- or mud-dwelling predators on bivalves and other gastropods; they bore circular holes through prey shells using glandular secretions and radular activity and then insert a highly protrusible proboscis for feeding. Naticids have an extensive, water-filled cavity in the foot; the water must be expelled from pores at the rear of the foot before the foot can be fully retracted into the shell.

Superfamily Tonnacea. Five families.

Family Tonnidae. *Tonna*—tun shells. Marine, typically sand-dwelling carnivores living to depths greater than 5,000 m. Prey on a variety of invertebrates and fish by injecting sulfuric acid and paralytic secretions.

Family Cymatiidae. *Cymatium, Fusitriton, Charonia*—the tritons. Marine, primarily warm-water snails. Prey on molluscs and echinoderms, incapacitating prey with sulfuric acid.

Family Bursidae. *Bursa*—frog shells. Mostly shallow-water marine carnivores feeding on various worms, which they paralyze with an acidic secretion and then ingest whole through a greatly expandable proboscis and esophagus.

Order Neogastropoda.[16]

These are the most highly evolved gastropods. All species are marine, and most are carnivores. Like the mesogastropods, the neogastropods have a single, monopectinate ctenidium, a single kidney, and a heart with one auricle. The radula, however, has no more than three teeth per row (the "stenoglossate" condition), and the osphradium is especially well developed and bipectinate. Fertilization is always internal. Many species develop to the juvenile stage without having a free-living veliger. The mantle edge always forms an incurrent siphon. Approximately 5,000 species, distributed among 21 families.

Superfamily Muricacea.   Seventeen families.

Family Muricidae.   *Urosalpinx*—the oyster drill; *Thais, Nucella*—dogwinkles; *Concholepas*—the economically important, highly edible South American "loco"; *Drupa, Murex, Ocenebra.* Shallow-water marine predators that bore holes through shells of barnacles, gastropods, and bivalves, using a specialized acid-secreting gland on the foot, in conjunction with radular rasping. Typically lacking free-living veliger larvae in the life cycle; most emerge as juveniles from egg capsules cemented to hard substrates. Phoenicians and Romans dyed their ceremonial robes purple using a substance secreted by many species of *Murex*. In fact, that's how the Phoenicians got their name (*phoenix* = G: reddish purple).

Family Buccinidae.   *Buccinum, Neptunea, Colus*—the whelks. All marine. A very large family of mostly carnivorous species.

Family Columbellidae.   *Columbella, Anachis*—dove shells. All marine. Mostly small snails, less than 0.5 cm high. Carnivorous and herbivorous species.

Family Nassariidae.   *Nassarius, Ilyanassa*—mudsnails, basket shells. All are marine and include carnivores, scavengers, and herbivores. Live mostly on mud or sand in marine habitats; some brackish, some freshwater species.

Family Melongenidae.   *Busycon*—whelks; *Melongena*—crown conchs. Large marine snails, up to 60 cm long. Mostly carnivores and scavengers in shallow water; most species are tropical.

Family Fasciolariidae.   *Fasciolaria*—tulip shells; *Pleuroplaca.* Large marine snails, up to 60 cm long.

Superfamily Conacea.   Three families.

Family Conidae.   *Conus*—cone shells. Marine carnivores, about 500 species, mostly tropical on coral reefs. Highly toxic venom is produced by poison glands and injected into prey through the sharp lateral teeth of the radula.

Subclass Opisthobranchia.

Approximately 2,000 species, mostly marine, are distributed among over 120 families.

Order Cephalaspidea.

Bulla, Haminoea, Hydatina, Retusa—the bubble shells. All members have shells, which may be either external, as in most prosobranchs, or hidden internally. As with the prosobranchs, the nervous system is fully torted (i.e., displays streptoneury). Cephalaspids are typically marine carnivores and ingest prey whole, crushing them using hard, calcareous plates in the gizzard. Thirty-one families.

Order Runcinoidea.

Two families.

Family Runcinidae.   Long, vermiform snails, less than 0.8 cm long; some species show a reduced shell, and none have an operculum. All are marine hermaphrodites, with ctenidia, and feed on filamentous algae in tidal pools.

Order Acochlidioidea.

Seven families.

Family Acochlidiidae.   *Acochlidium, Microhedyle.* Mostly marine, interstitial, vermiform snails living in the spaces between sand grains. No mantle cavity, ctenidia, or shell. A few species live in freshwater in Indonesia, Palau, and the West Indies. Most species only 2 mm to 5 mm long. All individuals are hermaphroditic, and the penis is in some species armed with a sharp stylet that injects sperm into a mate by hypodermic impregnation.

Order Sacoglossa (= Ascoglossa).

The sacoglossans, or ascoglossans. Seven families.

Family Elysiidae.   *Elysia.* Members of this family lose the shell at metamorphosis and lack cerata. The foot usually has lateral folds, called "parapodia," which may be folded over the dorsal surface. The mantle cavity, associated ctenidium, and osphradium have been lost. The anus opens on the animal's right side. All are hermaphroditic herbivores, and most adults are less than 1 cm long. At least some species contain unicellular algae living symbiotically in the tissues of the snail, giving the animal a distinctly green color.

Family Juliidae. *Berthelinia, Julia.* These opisthobranchs form a two-valved shell, superficially resembling that of the bivalved molluscs; however, the remains of a coiled larval shell, like those of other gastropods, can be found on the left shell valve. Mantle cavity, osphradium, and gill are present; cerata are lacking. Species are typically green due to symbiotic algae in the tissues, smaller than 1 cm, and tropical, feeding on a particular genus of macroalgae (*Caulerpa*).

Order Anaspidea (= Aplysiacea).
Five families.

Family Aplysiidae. *Aplysia*—the sea hares. Large, herbivorous animals, up to 75 cm long and weighing up to 16 kg. Rhinophores resemble rabbit ears. A thin, internal shell is covered by the mantle. A mantle cavity (containing one ctenidium) is present, but it has shifted to the animal's right side as a result of detorsion. All species are simultaneous hermaphrodites. Large, lateral projections called "parapodia" extend from the foot; by flapping these, the animal can swim.

Order Notaspidea.
Three families.

Family Pleurobranchidae. *Pleurobranchus, Berthella, Pleurobranchaea.* The shell has been lost in many species; if present, the shell is internal. Individuals retain a gill. Highly acidic secretions are produced by glands opening into the pharynx and by glands distributed on the mantle. Members of this family are predators, especially on sponges and ascidians. Some species can swim.

Order Thecosomata.
The shelled pteropods. All are marine. All use a modified foot (the parapodia) to swim, and many secrete an external, mucous web for collecting food. Five families.

Family Limacinidae (= Spiratellidae). *Limacina.* Pteropods living in small (less than 1 cm), sinistrally coiled shells; mantle cavity, operculum, and osphradium are present. All species are protandric hermaphrodites.

Family Cavoliniidae (= Cuvieriidae). *Cavolinia, Clio.* Pteropods living in uncoiled shells up to 5 cm long. Shells may be bottle shaped, bulbous and shield-like, or conical. The operculum has been lost, but the mantle cavity, osphradium, and ctenidium are retained. After death, empty shells often form an important component of temperate and warm-water sediments.

Family Cymbulidae. *Gleba, Corolla, Cymbulia.* The typical gastropod spiral shell is discarded at metamorphosis, and the adult then secretes a transparent, internal, gelatinous "shell." All species are hermaphroditic.

Order Gymnosomata.
The unshelled pteropods. All are marine. Seven families.

Family Clionidae. *Clione.* Unshelled pteropods, also lacking mantle cavity and gills; an osphradium is present. Typically less than 4 cm long. All are hermaphroditic. *Clione limacina* exhibits highly specialized feeding appendages to capture the shelled pteropods on which it exclusively feeds.

Order Nudibranchia.
The nudibranchs, with 60 families containing 40% to 50% of all opisthobranch species. The shells are discarded at metamorphosis, and adults lack a mantle cavity and ctenidia. Cerata typically serve as secondary gills. All are marine hermaphrodites.

Suborder Doridoidea.
Adults lack a shell, ctenidia, and mantle cavity. All members have evolved secondary gills, arranged in a circular tuft around the anus; true cerata, containing extensions of the digestive system, are lacking. All are marine hermaphrodites. Most are carnivores, feeding most commonly on sponges. Twenty-seven families.

Family Polyceratidae. *Polycera.*

Family Hexabranchidae. *Hexabranchus.* A widespread tropical group containing snails up to 30 cm long and weighing 350 g (grams). The members of this order can swim as well as crawl.

Family Rostangidae. *Rostanga.* Small snails, less than 0.2 cm long. The gills are pinnate or bipinnate.

Family Dorididae. *Doris, Austrodoris.*

Family Archidorididae. *Archidoris.* Plump, particularly soft-bodied snails, usually less than 10 cm long.

Family Discodorididae (= Diaululidae). *Discodoris, Diaulula.*

Suborder Dendronotoidea.
All species are marine and hermaphroditic, and most have true cerata. The cerata are highly branched in many species. Ten families.

Family Tritoniidae (= Duvaucellidae). *Tritonia.* Most are carnivores, feeding on soft corals (alcyonarians).

Family Dendronotidae. *Dendronotus.* These snails typically feed on hydroids.

Family Tethyidae. *Melibe, Tethys.* Up to 30 cm long. The cerata are unusually wide and flattened, and the head bears a conspicuous oral hood, fringed with tentacles. The hood may be as wide as 15 cm when fully expanded and is used to capture active prey, including crustaceans and fish.

Family Dotoidae. *Doto, Tenellia.*

Suborder Arminoidea.
Another group of shell-less marine hermaphrodites. Nine families.

Family Arminidae. *Armina.* Smallish slugs, typically less than 5 cm long. Rather than cerata, secondary gills are found under the surface of the unusually thick mantle. These shallow-water carnivores feed at night, often on bioluminescent sea pansies (colonial anthozoans); they themselves may become bioluminescent as a consequence of their feeding activities.

Suborder Aeolidoidea.
Slugs with numerous cerata on the dorsal surface. Most species feed on cnidarians in shallow waters, appropriating their prey's nematocysts for their own defense. Fourteen families.

Family Coryphellidae. *Coryphella.*

Family Pseudovermidae. *Pseudovermis.* Small slugs, less than 0.6 cm long, adapted for interstitial life in the spaces between sand grains. Cephalic tentacles and rhinophores are absent, and the cerata are very short and positioned laterally.

Family Tergipedidae (= Cuthonidae). *Cuthona, Tenellia, Cratena, Phestilla.* Individuals are less than 2.5 cm long. Some species feed on hydroids.

Family Glaucidae. *Glaucus, Hermissenda, Phidiana.* All are carnivores, feeding on cnidarians, annelids, and molluscs. Species in— and closely related to—the genus *Glaucus* are pelagic, feeding on siphonophores; by usurping their prey's nematocysts, these slugs themselves become dangerous, even to humans. The penis is often armed with spines.

Family Aeolidiidae. *Aeolidia, Spurilla.*

Subclass Pulmonata.
Members of most species possess a shell, which in slugs is often enclosed within the mantle and therefore not readily seen. The mantle cavity lacks a ctenidium, but sometimes it houses secondarily evolved gills. All species are hermaphroditic, and most are oviparous; some species protect their fertilized eggs with calcareous shells. Approximately 70 families containing about 17,000 species.

Order Archaeopulmonata.
Melampus, Ovatella. Shallow-water, or intertidal, typically marine or estuarine pulmonates (a few species are terrestrial), with spirally coiled external shells; an operculum and osphradium are lacking. All are hermaphroditic. A few species release free-living planktonic veliger larvae closely resembling those produced by prosobranch and opisthobranch gastropods. The penis often bears a sharp stylet. Two families.

Order Basommatophora.
These are smallish snails (usually less than 10 cm), with eyes usually located at the base of the tentacles. Fifteen families, containing about 1,000 species.

Family Siphonariidae. *Siphonaria.* The shell is cap shaped, with an irregular bulge on the right side, and the mantle cavity houses a secondary gill. Some species release free-living veliger larvae. All species are marine. Many are tropical, living in the high intertidal zone.

Family Amphibolidae. *Amphibola.* Snails in this family are unique among pulmonates in having an operculum as adults. Although gills of any sort are lacking, amphibolids do possess an osphradium in the mantle cavity. Species are estuarine and produce a distinct veliger, but the veliger stage never leaves the egg capsule in which it develops.

Family Lymnaeidae. *Lymnaea.* Pond snails, with a spiral, external shell. All species live in freshwater; they commonly serve as intermediate hosts in the life histories of various parasitic trematode flatworm species.

Family Physidae. *Physa.* These freshwater hermaphrodites can self-fertilize. The shell is always sinistral.

Family Planorbidae. *Biomphalaria, Planorbis, Helisoma*—ram's horn snails. All are freshwater snails, often with a planispiral shell, which coils sinistrally. Members of this family are extremely important as intermediate hosts in the transmission of schistosomiasis, a devastating disease caused by trematode flatworms in the genus *Schistosoma.* The blood of these snails is often rich in hemoglobin, allowing them to inhabitat environments—including polluted areas—with very low dissolved oxygen concentrations. Populations are difficult to control, in part because these snails can self-fertilize.

Order Stylommatophora.

Individuals of most species have spiral shells that, in slugs, may be completely enveloped by the mantle. An operculum is never present. All individuals possess two pairs of tentacles, with the upper ones each bearing an eye at the tip. Over 50 families, with about 15,000 species.

Family Achatinellidae.   *Achatinella.* These are tree-dwelling terrestrial snails found primarily on Pacific islands, including Hawaii.

Family Pupillidae.   *Pupilla, Orcula.* Small, terrestrial snails, usually less than 1 cm long. This large group contains nearly 500 species.

Family Clausiliidae.   *Clausila, Vestia*—door snails. Exclusively terrestrial, ovoviviparous snails distributed among more than 200 genera in Asia, South America, Europe, and Asia Minor. All members possess a unique, morphologically and functionally complex mechanism (the "clausilium") for sealing the aperture after withdrawing the body into the shell.

Family Succineidae.   The amber snails. All are terrestrial, with very thin, fragile shells smaller than 2 cm.

Family Athoracophoridae.   Slug-like terrestrial snails, with the shell reduced to a mass of calcified pieces embedded in the integument. The unusual respiratory system resembles the tracheal system of insects. The snails usually live in trees and bushes, mostly in Australia and New Zealand.

Family Achatinidae.   *Achatina.* All are terrestrial, with shells as large as 23 cm; these are the largest of all terrestrial pulmonates. The African species estivate during the dry season, forming a particularly strong epiphragm. *A. fulica* is a major agricultural pest and is commonly eaten by humans in some parts of the world; the shells are used as utensils and as decorations.

Family Streptaxidae.   A large group of tropical, terrestrial snails, with perhaps 500 species. They live in leaf litter and under logs and can withstand long periods of drought by estivating. All are carnivorous on other snails and on annelids.

Family Limacidae.   *Deroceras.* Slug-like terrestrial snails, with the shell reduced to a flat plate and mostly enveloped by the mantle. These are major agricultural pests, particularly in Europe and Africa. Individuals can self-fertilize.

Family Helicidae.   *Helix, Cepaea.* These terrestrial, distinctly edible snails ("escargot") have an external, spiral shell. They overwinter by burrowing in soil, covering the burrow with a substantive epiphragm. A specialized dart sac is associated with the vagina of each individual. During mating, each partner injects a calcareous dart into the flesh of its mate, as happens in members of several other pulmonate groups and in some opisthobranchs.

Order Systellommatophora.

These slugs have no trace of a shell, internally or externally. The pulmonary cavity is always located posteriorly. The head bears two pairs of tentacles, with the upper pair bearing the eyes as in basommatophorans. Most species are terrestrial, although a few are amphibious, living partly in air and partly in the sea. Some species are herbivores; others feed on other pulmonates. Three families.

Family Rhodopidae.   *Rhodope.* A bizarre, very small (less than 0.4 cm long), shell-less, vermiform snail. These snails lack tentacles, mantle cavity, gills of any kind, and heart; there is not even any trace of a pericardium. Calcareous spicules are embedded in the flesh. Members of this family are exclusively interstitial, living in the spaces between sand grains in both the Atlantic Ocean and Mediterranean Sea. Formerly, these species were classified as nudibranchs, but there is growing support for relating them to pulmonates.

Class Bivalvia.[17]

Approximately 7,650 species are distributed among over 90 families.

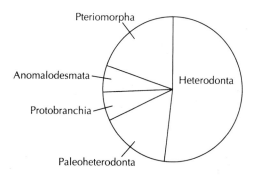

---

17. In this classification scheme, members of the Lamellibranchia are now distributed among the following four subclasses: Pteriomorphia, Paleoheterodonta, Heterodonta, and Anomalodesmata. The septibranch bivalves are now treated as two families (Poromyidae and Cuspidariidae) within the subclass Anomalodesmata.

Subclass Protobranchia (= Paleotaxodonta = Cryptodonta).

Seven families, all marine, with approximately 500 species.

> **Family Nuculidae.** *Nucula*—nut shells. These are deposit feeders living in sandy sediments, with shells typically only 2 cm to 3 cm long. These bivalves lack siphons, and water enters the mantle cavity anteriorly.
>
> **Family Nuculanidae.** *Yoldia.* These small protobranchs (less than 7 cm long) live mostly in deep-water sediments. They possess both incurrent and excurrent siphons posteriorly.
>
> **Family Solemyidae.** *Solemya.* Solemyids burrow in sand and mud and occur over an extremely wide depth range. The digestive system is substantially reduced, or even absent, in some species; nutrients may be obtained through activities of symbiotic sulfur bacteria living in the gills. These clams lack siphons, and water enters the mantle cavity anteriorly.

Subclass Pteriomorphia.

Twenty-four families with approximately 1,500 species.

> **Family Mytilidae.** *Modiolus, Mytilus, Lithophaga*—mussels. Most species live attached by byssal threads to solid substrates in marine and estuarine habitats; a few species live in freshwater. A few of the marine species (*Lithophaga* spp.) bore into calcareous substrate (including coral) or live commensally with sea squirts (ascidians). *Mytilus edulis* has become a major pollution bioindicator species.
>
> **Family Pinnidae.** *Pinna*—pen shells. The shells are thin and fragile, and may be as long as 1 m. The posterior adductor muscles are much larger than the anterior adductor muscles. The animals live in shallow tropical seas, partly buried in sediment and attached to the underlying solid substrate by silky byssal threads.
>
> **Family Ostreidae.** *Ostrea, Crassostrea*—oysters, of considerable commercial importance. These bivalves lie on the left valve, which may be firmly cemented to the substrate. Adults lack a foot and do not secrete byssal threads. There is no anterior adductor muscle, and the shells lack a nacreous layer. Individuals alter their sex every few years throughout life. Each female may produce more than one million eggs yearly.
>
> **Family Pectinidae.** *Chlamys, Pecten, Aequipecten, Argopecten, Placopecten*—scallops. Many species can swim by snapping the two shell valves closed, but a few species cannot swim and lie on the substrate or live attached to firm substrate by byssal threads. The anterior adductor muscle is lacking. The posterior adductor muscle, however, is large and is the only part of the scallop that humans eat.
>
> **Family Anomiidae.** *Anomia*—jingle shells. The shells are round or oval and decidedly shiny. Jingle shells live attached to solid substrates; a chitinous, often calcareous plug of byssal material reaches the substrate through a hole in the right shell valve. There is no anterior adductor muscle, and the posterior adductor muscle is much reduced.

Subclass Paleoheterodonta.

Eight families, with approximately 1,200 species.

> **Family Unionidae.** *Lampsilis, Ligumia, Medionidus, Villosa, Unio, Anodonta* (now *Pyganodon*). All members of this family live in freshwater. Females brood embryos in the gills and release glochidia, which continue their development parasitically on fish. The adults are free-living, with a particularly well-developed periostracum and two strong adductor muscles.

Subclass Heterodonta.

Forty-two families, with approximately 4,000 species. The adult foot usually lacks a byssal gland.

> **Family Lucinidae.** *Lucina, Lucinoma.* These clams are common in sulfide-rich sediments and build (with the foot) fairly complex tunnels that extend deep into the sediment below the animal. The long, worm-like foot of these suspension-feeding bivalves typically builds an anterior feeding tube and lines it with mucus. The position of this inhalant tube is changed frequently. All lucinids studied so far have chemoautotrophic bacteria living symbiotically in the gills; the bacteria form carbohydrates from CO_2, using energy obtained in sulfide oxidation. The gills are unusually thick and have only the inner demibranch.
>
> **Family Thyasiridae.** *Thyasira.* Like their close relatives, the lucinids, these bivalves commonly have enlarged gills harboring symbiotic chemoautotrophic bacteria, and build a mucus-lined anterior feeding tube with the foot. Unlike that of the lucinids, the feeding tube is fixed in position and the gills are complete, bearing both the inner and outer demibranchs. Populations as dense as 4,000 clams/m^2 have been described in the North Atlantic Ocean, in sediments rich in organic matter. The foot

constructs an extensive network of tunnels that extend deep into the sediment below the clams; these tunnels are more complex than those of the lucinids.

Family Lasaeidae. *Lasaea, Montacuta.* A group of small (typically less than 2 cm), actively crawling bivalves that are commonly parasitic or commensal on other marine invertebrates, especially annelids, echinoderms, and crustaceans. *Lasaea* spp. are abundant worldwide, living in rock crevices and other inconspicuous habitats. Individuals in this family are hermaphroditic and typically brood embryos in the mantle cavity before releasing offspring as free-living veliger larvae.

Family Galatheavalvidae. *Galatheavalva holothuriae.* A bizarre marine bivalve that, although it has a normal foot, byssus, anterior adductor muscle, and posterior adductor muscle, has an internal shell embedded in the mantle and lives exclusively inside one species of deep-sea sea cucumber (holothurian).

Family Carditidae. *Cardita*—little heart shells. These shallow-water suspension feeders attach to firm substrates by byssal threads. The blood contains hemoglobin. The sexes are separate, and females brood embryos in the gill.

Family Cardiidae. *Cardium, Laevicardium*—cockles. This group contains about 200 species of shallow-water suspension feeders, generally living in sandy substrates. The foot is highly muscular and used for burrowing, jumping, and even swimming, although the swimming is hardly graceful. The periostracum is poorly developed.

Family Tridacnidae. *Tridacna, Hippopus*—giant clams. Individuals can weigh up to 180 kg. Most species have a very small foot and live attached to substrates by a huge byssus. The mantle is replete with unicellular algae (zooxanthellae) living symbiotically in the tissue. All six species live in the tropical Indo-Pacific, in shallow water.

Family Mactridae. *Mactra, Spisula*—the surf clams; *Mulinia, Rangia.* Most species are marine, although a few live only in freshwater. Marine species burrow in shallow-water sediments, using a large foot that lacks a byssus. Several species are commercially important foods.

Family Cultellidae. *Ensis*—razor clams. Rapidly burrowing suspension feeders in marine and estuarine habitats.

Family Tellinidae. *Tellina, Macoma.* These are all marine and primarily deposit feeders, living in sand or mud. The outer demibranch is very small and lacks all or most of the ascending lamellae.

Family Donacidae. *Donax.* Burrowing suspension feeders, all marine. At least some species migrate up and down sandy beaches by leaping out of the sand and being carried by the incoming or outgoing tide.

Family Arcticidae. *Arctica islandica.* These clams live in fairly deep water off the New England coast. They are of some commercial importance as chowder clams.

Family Corbiculidae. *Corbicula.* Burrowing suspension feeders found in estuaries and in freshwater. About 100 species, one of which was introduced to the United States from Asia probably in the 1930s and is now a major economic nuisance throughout much of the United States. Populations can exceed 1,000 individuals per m^2. Most species release free-swimming veliger larvae. The introduced species, *C. fluminea,* can self-fertilize, broods embryos in the gill, and releases small juvenile bivalves that can be carried considerable distances downstream by currents.

Family Dreissenidae. *Dreissena*—zebra mussels. These small (< 5 cm) European and African bivalves recently have invaded the eastern United States, probably carried here in ship ballast water in 1986. They thrive in fresh- and salty water, and they are rapidly becoming major economic pests; they compete for suspended food with commercially important fish and native bivalve species and clog the water pipes of power plants, boats, and industrial cooling systems. Adults attain densities exceeding 30,000 individuals per m^2, reproduce prolifically (fertilizing eggs externally and releasing free-swimming veliger larvae), and form thick mats of byssal threads that make them extremely difficult to dislodge from clogged pipes. No other North American freshwater bivalves form byssal threads as adults.

Family Pisidiidae. *Pisidium*—fingernail clams. Freshwater suspension feeders. The shells are usually smaller than about 0.5 cm. Some species live out of the water in moist leaf litter along the shores of ponds and streams, so they are, in a sense, terrestrial.

Family Veneridae. *Mercenaria* (= *Venus*)—quahogs (pronounced kwō-hogs), or "hard-shell clams"; *Gemma, Tapes.* A large group of some 500 suspension-feeding species, all of which are marine. The northern quahog, *Mercenaria mercenaria,* is used in chowders and eaten on the half-shell as "cherry-stone" clams.

Family Petricolidae. *Petricola, Mysia.* These clams typically bore into a variety of substrates, such as mud, chalk, and coral. All are marine.

Family Myidae. *Mya*—soft-shell clams, steamers. Most species are burrowing suspension feeders. The siphons are fused and covered by a periostracal sheath.

Family Hiatellidae. *Panopea*—the geoduck (pronounced "gooey-duck"). Geoducks are large clams found along the U.S. Pacific coast. Shells grow to lengths of 20 cm, with siphon lengths exceeding 75 cm, permitting the animals to live far beneath the sediment's surface.

Family Pholadidae. *Martesia, Xylophaga, Zirphaea*—the piddocks. These marine bivalves have three adductor muscles—anterior, posterior, and ventral—instead of the usual pair, and they bore into hard substrates including shales, shells, and wood. The siphons protrude outside the substrate, for suspension feeding. These animals do considerable damage to wooden boats, docks, and pilings.

Family Teredinidae. *Teredo, Bankia*—the shipworms. These bivalves have three adductor muscles—anterior, posterior, and ventral—instead of the usual two, and they mostly bore into wood. They line their burrows with calcium deposits. Their shell is quite small—typically about 0.4 cm in an individual that is 6 cm to 7 cm long—and most of the animal protrudes posteriorly from the shell as a long "worm." The animal lives hidden inside the wood except for the siphons, which protrude for feeding. Most species are marine or estuarine. Some are dioecious; others are hermaphroditic. Young are brooded in the gills of some species. These bivalves cause considerable economic damage to wooden ships and pilings, which they enter during larval metamorphosis.

Subclass Anomalodesmata.
This class includes all "septibranch" species. Twelve families with approximately 450 species.

Family Pandoridae. *Pandora.* A small group (about 25 species) of marine, burrowing suspension feeders living in shallow-water sediments. The inner demibranchs of the ctenidia are fully developed, but the outer demibranchs are greatly reduced. All species are hermaphroditic.

Family Poromyidae. *Poromya.* These species are carnivores—especially on annelids—and most live in the deep sea. The ctenidia are substantially modified. All species are marine and hermaphroditic.

Family Cuspidariidae. *Cuspidaria.* These animals are highly modified carnivores, living in deep-sea sediments. The foot is small but byssate; the labial palps may be missing entirely; the mantle cavity is divided into two chambers by a muscular, perforate septum; true ctenidia are absent. All species are marine, and individuals eat mostly crustaceans and annelids.

Class Scaphopoda.

The approximately 350 species are divided among eight families.

Family Dentaliidae. *Dentalium.* The shells are up to 15 cm long, which is as large as any scaphopod gets. Species in this group are very widespread, with some living in shallow water and others at great depths, in all the major oceans: Atlantic, Pacific, Indian, and Arctic.

Class Cephalopoda.

The approximately 600 species are divided among 44 families. All species are marine, dioecious, and carnivorous.

Subclass Nautiloidea.
Nautiloids are the only cephalopods with true external shells secreted by the mantle. Although this group once contained thousands of species distributed among many families, only six species have escaped extinction, and these species are so closely related as to be placed in a single family.

Family Nautilidae. *Nautilus*—the chambered nautilus. The external, calcareous shell is coiled and divided internally by transverse septa; the animal lives only in the outermost chamber. Shells of adults can reach about 27 cm across. Nautiloids have simple lens-less eyes, 80 to 90 tentacles, two pairs of ctenidia, and two pairs of osphradia; none have an ink sac. All six species occur in the Indo-Pacific, typically living at depths between 50 and 500 m to 600 m.

Subclass Coleoidea (= Dibranchiata).
Most species possess internal shells that are completely surrounded by mantle tissue. Four orders, 43 families.

Order Sepioidea.
All members bear eight arms and two tentacles. Five families.

Family Spirulidae. *Spirula*—the ram's horn. The internal shell is spiralled and calcareous, and functions in buoyancy regulation. These animals are pelagic in deep water (200–600 m), and they have an ink sac and, posteriorly, a specialized external bioluminescent organ; they lack a radula.

Family Idiosepiidae. *Idiosepius.* These cephalopods never exceed lengths of about 1.5 cm, and they lack even an internal shell.

Family Sepiidae. *Sepia*—the cuttlefish. The lightweight, calcareous shell ("cuttlebone") is internal and functions in buoyancy regulation. The tentacles can be completely retracted into a special pocket.

Order Teuthoidea (= Decapoda).
The squids. Shells are always internal and uncalcified. The head is surrounded by eight arms and two tentacles, and it bears advanced eyes with lenses. All members have a well-developed radula. Squid are commercially important, some 200 million metric tons of certain species being caught for food each year. Twenty-five families.

Family Loliginidae. *Loligo*—the common Atlantic squid; *Sepioteuthis, Lolliguncula. Loligo* squid attain lengths of about 50 cm and tend to live in large aggregations.

Family Ommastrephidae. *Illex, Todarodes*—arrow squids. These squid support enormous N. Atlantic and Japanese fisheries, respectively.

Family Lycoteuthidae. *Lycoteuthis.* These are small (less than 10 cm long), deep-water squid, found as deep as 3,000 m. They possess bioluminescent organs in species-specific patterns. Particularly patriotic species emit red, blue, and white light in different regions of the body.

Family Architeuthidae. *Architeuthis*—giant squids. These are the largest of all invertebrates, attaining lengths of 20 m, including the tentacles, and weights exceeding 1 ton. The eyes are up to 20 cm in diameter, making them the largest animal eyes on record. These squid have no bioluminescent organs and live at depths of 500 m to 1000 m. They are heavily preyed on by sperm whales.

Order Vampyromorpha.
One family (Vampyroteuthidae). *Vampyroteuthis*—vampire squids. These deep-water (300–3,000 m), dark-bodied squids have eight normal arms plus a pair of highly modified, thin, elongated, tendril-like arms. Conspicuous sheets of tissue occur between arms, forming a vampire-like cloak. Species bear well-developed bioluminescent organs, a radula, and large, red eyes. The shell is internal, uncalcified, and nearly transparent.

Order Octopoda.
These cephalopods possess eight arms and no tentacles. Twelve families, containing about 200 species.

Family Cirroteuthidae. *Cirrothauma.* These octopods occur as deep as 4,000 m, and at least one species is unique among cephalopods in apparently being blind. They lack a radula and ink sac, and some species have a gelatinous appearance, looking more like large jellyfish than cephalopods.

Family Octopodidae. *Octopus*—the octopuses. These shallow-water cephalopods are up to 3 m long and 2.5 kg in weight. They are bottom dwellers, not swimmers, and individuals tend to live by themselves in small, protective caves. The well-developed brain is surrounded by a cartilaginous skull. Octopuses have complex behavior; they can learn and remember.

Family Argonautidae. *Argonauta*—the paper nautilus. These octopods are exclusively pelagic. The females are up to 30 cm long and secrete a large, but paper-thin, chamberless shell in which the female lives and broods her developing embryos; the shell is produced by specialized glands on several of the tentacles, and is not homologous with other molluscan shells. The male is only about 1.5 cm long and shell-less. During mating, the disproportionately long hectocotyl arm of the male detaches completely after entering the female's mantle cavity.

Family Ocythoidae. *Ocythoe.* These Atlantic, Pacific, and Mediterranean pelagic octopods are sexually dimorphic, with dwarf males (about 3–4 cm long) living in abandoned salp tests; females are up to 30 cm long and free-living. Females of *Ocythoe tuberculata* are unique among cephalopods in having a fish-like swim bladder for buoyancy regulation.

Some General References About the Molluscs

Boardman, R. S., A. H. Cheetham, and A. J. Rowell, eds. 1987. *Fossil Invertebrates.* Palo Alto, Calif: Blackwell Scientific, 270–435.

Fretter, V., and A. Graham. 1994. *British Prosobrach Molluscs,* 2d ed. London: The Ray Society.

Harrison, F. W., ed. 1992. *Microscopic Anatomy of Invertebrates, Vols. 5 and 6: Molluscs.* New York: Wiley-Liss.

Hyman, L. H. 1967. *The Invertebrates, Vol. 6. Mollusca I.* New York: McGraw-Hill.

Kat, P. W. 1984. Parasitism and the Unionacea (Bivalvia). *Biol. Rev.* 59:189–207.

Lalli, C. M., and R. W. Gilmer. 1989. *Pelagic snails: The Biology of Holoplanktonic Gastropod Mollusks.* Stanford, Calif.: Stanford Univ. Press.

Morton, J. E., 1979. *Molluscs,* 5th ed. London: Hutchinson Univ. Press.

Parker, S. P., ed. 1982. *Classification and Synopsis of Living Organisms,* vol. 2. New York: McGraw-Hill, 946–1166.

Thorpe, J. H., and A. P. Covich, eds. 1991. *Ecology and Classification of North American Freshwater Invertebrates.* New York: Academic Press, 285–399.

Vermeij, G. 1993. *A Natural History of Shells.* Princeton N.J.: Princeton Univ. Press.

Ward, P. D. 1987. *The Natural History of Nautilus.* Boston: Allen & Unwin.

Wilbur, K. M., ed. 1983–1988. *The Mollusca.* (Vol. 1: Metabolic Biochemistry; Vol. 2: Environmental Biochemistry and Physiology; Vol. 3: Development; Vols. 4 and 5: Physiology; Vol. 6: Ecology; Vol. 7: Reproduction; Vols. 8 and 9: Neurobiology and Behavior; Vol. 10: Evolution; Vol. 11: Form and Function; Vol. 12: Paleontology and Neontology of Cephalopods.) New York: Academic Press.

15

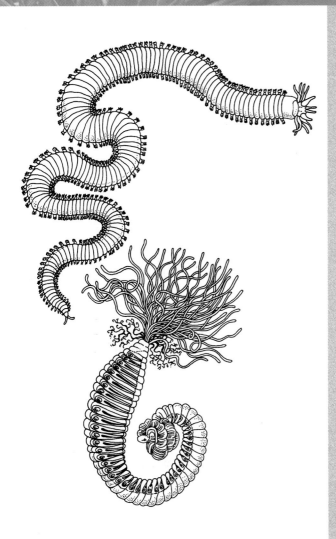

Introduction and General Characteristics

> *Phylum Annelida*
> (annulus = L: ring)
> an-el-ē´-dah

The nearly 12,000 species comprising the phylum Annelida are **vermiform** (worm-like), as are the members of a number of other invertebrate phyla; that is, these animals are soft-bodied, reasonably circular in cross section, and longer than they are wide. Unlike those of most other vermiform animals, however, the bodies of annelids consist of a series of repeating segments. This serial repetition of segments and organ systems (skin, musculature, nervous system, circulatory system, reproductive system, and excretory system) is known as **metamerism,** or metameric segmentation (Fig. 15.1).

The annelid outer body wall is generally flexible and can play an active role in locomotion. Moreover, the thin body wall can serve as a general surface for gas exchange, provided that it is kept moist. Even when the epidermis secretes a protective cuticle, the cuticle remains permeable to both water and gases. For this reason, annelids are restricted to moist environments.

The individual annelid segments are generally separated from each other to a large degree by **septa,** which are thin sheets of mesodermally derived tissue (**peritoneum**) that essentially isolate the coelomic fluid in one segment from that in adjacent segments (Fig. 15.1). This allows for localized deformation of the outer body wall, brought about by contractions of circular and longitudinal musculature within a single segment. Thus, muscle contractions in any one segment do not alter the hydrostatic pressure in other parts of the animal.

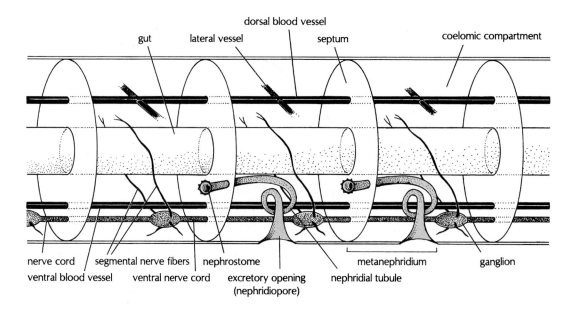

gut   lateral vessel   dorsal blood vessel   septum   coelomic compartment

nerve cord
ventral blood vessel

segmental nerve fibers
ventral nerve cord

nephrostome
excretory opening
(nephridiopore)

metanephridium
nephridial tubule

ganglion

Figure 15.1

Schematic illustration of metameric organization in annelids. The annelid body consists of a linear series of segments, separated from each other by transverse, mesodermally derived septa. Much of the internal anatomy, including excretory, nervous, coelomic, and muscle systems, is segmentally arranged. Body wall musculature has been omitted for clarity. Note the ventral nerve cord; the vertebrate nerve cord is located dorsally.

Although some wastes are excreted across the general body surface, excretion generally occurs by means of structures called **nephridia** ("little kidneys"). Most annelid segments contain two nephridia, each of which is open at both ends. This type of nephridium is called a **metanephridium.**

Coelomic fluid is drawn into the nephridium at the **nephrostome** (kidney-mouth; *stoma* = G: mouth) by the action of cilia (Fig. 15.2). As the fluid passes through the convoluted tubule of the nephridium, some substances (including salts, amino acids, and water) may be selectively resorbed, and other substances (including metabolic waste products) may be actively secreted into the lumen of the tubule. The final urine emerging from the nephridiopore is thus quite different in chemical composition from the primary urine entering at the nephrostome. In addition to providing an outlet for metabolic waste products, nephridia may be used to regulate the water content of the coelomic fluid.[1]

In many annelid species, ducts leading from the gonadal tissue merge with the nephridial tubule. Thus, the nephridium generally plays a role in discharging gametes as well as urine.

Annelids are distributed among three classes: the Polychaeta, the Oligochaeta, and the Hirudinea. The evolutionary relationships among the groups—particularly between the Polychaeta and Oligochaeta—have long been controversial.[2]

Class Polychaeta

Class Poly · chaeta
(G: many setae)
pahl´-ē-kē´-tah

Approximately 70% of all annelid species are placed in the class Polychaeta. Nearly all polychaetes live in salt water. Polychaetes generally possess at least one pair of eyes and at least one pair of sensory appendages (**tentacles**) on the anteriormost part of the body (the **prostomium**). Generally, the body wall is extended laterally into a series of thin, flattened outgrowths called **parapodia** (Figs. 15.3 and 15.4). Parapodial morphology differs significantly among species and therefore plays an important role in polychaete identification. These outfoldings increase the animal's exposed surface area and, because parapodia are highly vascularized (Fig. 15.3b), they function in gas exchange between the worm and its environment. The parapodia also have a locomotory function in many species, being stiffened by the presence of chitinous support rods called **acicula** (Fig. 15.3a). In addition, siliceous, chitinous, or, more rarely, calcareous bristles called **setae** protrude from each parapodium; setal morphology also differs substantially among polychaete species. The body is covered by a series

1. See *Topics for Further Discussion and Investigation*, no. 4, at the end of the chapter.

2. See *Topics for Further Discussion and Investigation*, no. 13.

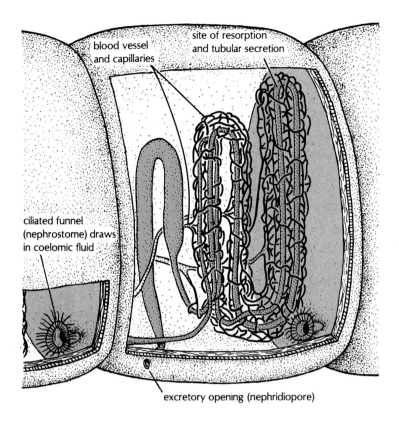

blood vessel
and capillaries

site of resorption
and tubular secretion

ciliated funnel
(nephrostome) draws
in coelomic fluid

excretory opening (nephridiopore)

Figure 15.2

Diagrammatic representation of a typical metanephridium. The chemical composition of the primary urine drawn through the nephrostome is altered by selective resorption and secretion as fluid moves through the nephridial tubules. The final urine is discharged through the nephridiopore.

of overlapping protective plates (**elytra**) in some species (Figs. 15.3a and 15.4b).

The septa present between most segments in polychaete worms enable the hydrostatic skeletal system to function independently in each segment. Localized contractions in one part of the body result in localized deformations without interfering with the musculature in the worm's other segments. Setae form temporary attachment sites and prevent backsliding during locomotion on or within the substrate or burrow. The septa between anterior adjacent segments are often absent or incomplete (**perforate**) in active, burrowing forms, enabling a greater volume of coelomic fluid to be utilized. This, in turn, permits a greater change of shape to be associated with extending and anchoring the penetration organ (termed the **proboscis**) into the sediment.[3]

A cross section of a polychaete worm reveals a secreted, nonliving cuticle and a layer of circular muscles underlying the epidermis (Fig. 15.3a). Beneath the layer of circular muscles lies a layer of longitudinal muscle fibers. Oblique muscles are often found as well, serving to maintain body turgor and to operate the parapodia, which can be used as oars for locomotion through the water (i.e., swimming), over surfaces, or within burrows.

Many polychaetes form burrows in the sediment, largely by the mutual antagonism of the longitudinal and circular muscles through the hydrostatic skeleton of each coelomic compartment. In some burrowing polychaetes, the proboscis is everted into the sediment (Fig. 15.5a), penetrating and pushing aside the sand. The segments just behind the proboscis flange markedly as the circular muscles in those segments contract (Fig. 15.5b), preventing the worm from backsliding. Protrusion of setae also deters backsliding. The longitudinal muscles then contract in the anterior segments, while the circular muscles in these segments relax. This dilates the worm's anterior segments (Fig. 15.5c). The anterior end is thus firmly anchored in the substrate, and the more posterior segments can then be drawn forward by contracting the longitudinal muscles. The worm is then ready to repeat the cycle. Note that burrow formation is accomplished by exploiting the properties of a compartmentalized hydrostatic skeleton (Chapter 9).

Polychaetes may be divided into two general groups. One group includes the generally active, mobile species (i.e., the **errant** species, which are sometimes placed in a separate subclass, the Errantia). Circular muscles typically play a minor role in the locomotion of these species; instead, the polychaetes are levered forward by the action of

---

3. See *Topics for Further Discussion and Investigation*, no. 2.

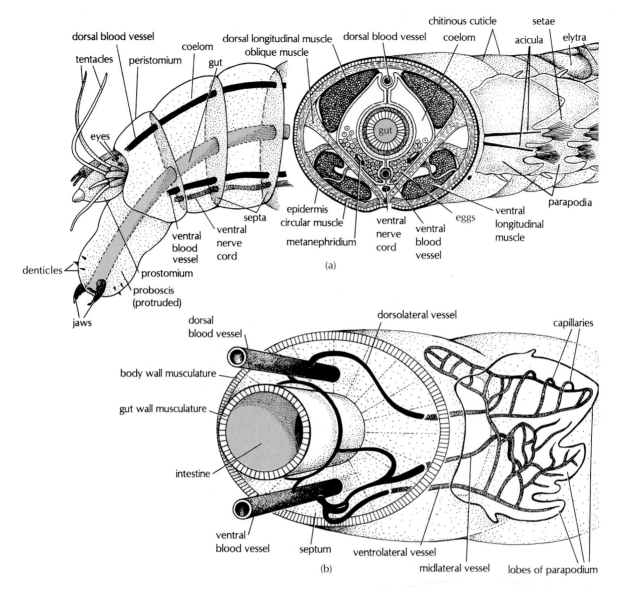

Figure 15.3
(a) Hypothetical polychaete worm, a composite of several species, showing typical major features. (b) Detail of a parapodium, showing the high degree of vascularization.

parapodia, which are operated in a complex pattern as oars. The acicula play an important role as stiffening elements during this activity, preventing the thin parapodial tissue from collapsing as thrust is applied against the substrate. Errant polychaetes typically augment parapodial movements with carefully coordinated contractions of the longitudinal body wall musculature, generating sinusoidal waves of activity; here, the muscles on one side of the body antagonize those on the opposite side, resulting in rapid, eel-like movement over the substrate (Fig. 15.6) or, in some species, swimming.

Not all errant polychaetes are surface dwellers: Members of some "errant" species form simple or complex burrows within substrates, but these polychaetes nevertheless have well-developed parapodia, complete with acicula and setae, and a well-developed head equipped with

a toothed or jawed protrusible pharynx. Most errant species are carnivorous, but the group also includes suspension feeders, detritus feeders, and omnivores.

In contrast to these errant polychaetes, other species (sometimes placed in a separate subclass, the Sedentaria) typically spend their entire lives in simple burrows in the sediment or in rigid, protective tubes.[4] These tubes vary in construction from simple organic secretions mixed with sand grains or mud to tubes composed of calcium carbonate and/or complex mixtures of proteins and polysaccharides. Other species actively bore into calcareous substrates and live in the resulting burrows. The parapodia tend to be greatly reduced, highly modified, or absent among the sedentary polychaetes (Fig. 15.7). Not surprisingly, acicula are absent as well, confirming their locomotory function in errant species. Most sedentary

4. See *Topics for Further Discussion and Investigation*, no. 1.

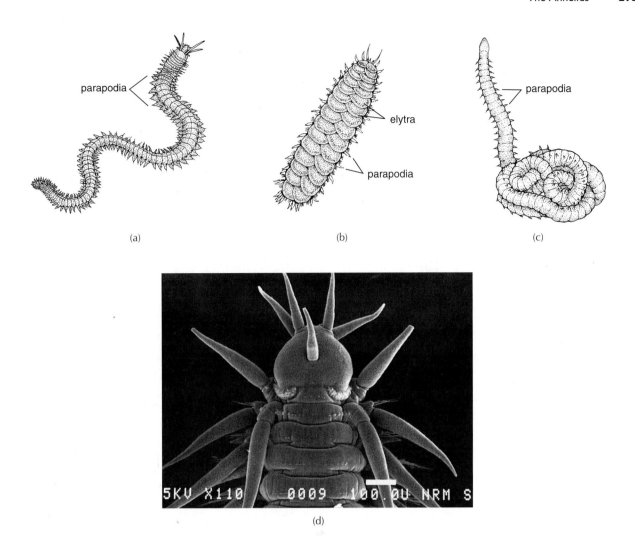

Figure 15.4

Representative errant polychaetes. (a) *Nereis virens.* (b) *Harmothoe imbricata.* (c) *Arabella iricolor.* (d) Anterior end of the phyllodocid polychaete *Sige fusigera.*   (a,b) *Modified from McConnaughey and Zottoli,* *Introduction to Marine Biology, 4th ed., 1983.* (c) *Modified from Ruppert and Fox. Seashore Animals of the Southeast United States. South Carolina Press, Columbia, S.C., 1988.* (d) *Courtesy of Fredrik Pleijel. From Pleijel, F. 1990. Zool. J. Linnean Soc. 98: 161–84.*

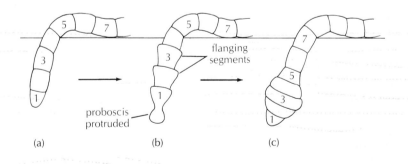

Figure 15.5

Sequence of movements involved in burrowing by the marine lugworm *Arenicola marina.* (a) The worm has already burrowed partway into the sediment. (b) The proboscis is protruded anteriorly. Much of the sediment is ingested; the rest is forced aside anteriorly. Appropriate muscle contractions cause adjacent segments to flange, preventing the worm from being pushed backward, out of the burrow, by the forward thrust of the proboscis. (c) Sequential contraction of the longitudinal musculature of each segment draws the worm downward. The increased width of the anterior segments, reflecting relaxation of circular muscles, anchors the worm as the more posterior segments are drawn forward.   *After Trueman.*

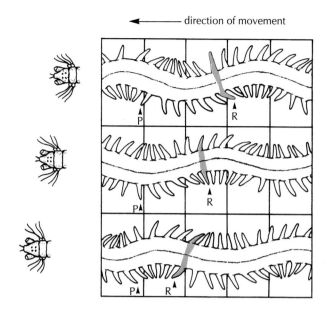

← direction of movement

Figure 15.6

The errant polychaete *Nereis virens,* crawling rapidly over the substrate. *P* indicates a parapodium executing its power stroke. Its tip is applied to the substrate, with setae protruded for added stability; note that the tip remains stationary as the body is levered forward. *R* indicates a parapodium executing its recovery stroke. Its setae are retracted as the parapodium is lifted up and swung forward above the substrate; the parapodium is then lowered to contact the substrate, ready to perform a power stroke. Simultaneously, waves of longitudinal muscle contraction pass posteriorly; the longitudinal muscles on each side of the body contract out of phase, producing a rapid wriggling motion.

Source: J. Gray, 1939, "Studies in Animal Locomotion". VIII. Nereis diversicolor in Journal of Experimental Biology, 16:9-17, Plate I.

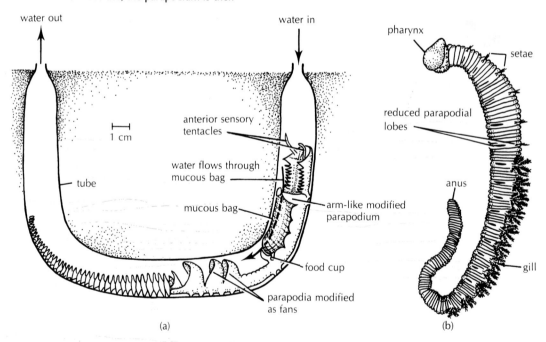

(a)   (b)

Figure 15.7

Representative sedentary polychaetes, showing a variety of morphological adaptations for a nonmotile existence. Most animals are drawn removed from their tubes or burrows. (a) *Chaetopterus variopedatus.* Feeding currents are generated by rhythmic movements of fanlike parapodia. One pair of anterior parapodia is modified for holding open a mucous bag, which filters out food particles as water moves through the tube. The bag is held posteriorly by a ciliated food cup, which continuously gathers the food-laden mucus into a ball. Periodically, the ball is released from the rest of the net and passed forward along a ciliated tract to the mouth for ingestion. (b) *Arenicola marina,* the lugworm. This worm is a deposit feeder and lives burrowed in sand. For gas exchange, water is pumped through the burrow by peristaltic waves of muscle contraction; parapodia play no role in locomotion or water flow and are greatly reduced in size in most segments. In a number of segments, a small portion of the parapodial lobes has become modified to form gills—clusters of thin, vascularized, branching filaments. (c) *Sabellaria alveolata,* in lateral view. Members of this species build tubes of sand grains and mucus and attach them to a variety of solid substrates, including the tubes of neighboring individuals. Segments have fused anteriorly to form a densely bristled surface that occludes

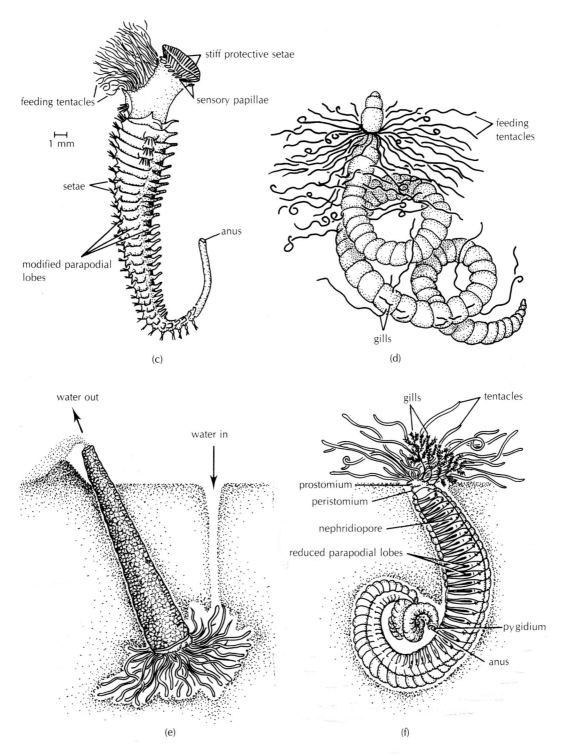

(c)

(d)

(e)

(f)

the opening of the tube when the animal pulls inside. (d) *Cirratulus cirratus.* These animals live within muddy substrates. Although the parapodia are much reduced, a portion of the parapodium has become greatly elongated in some segments, forming gills for gas exchange. (e) *Pectinaria belgica,* in its conical tube of mucus and cemented sand grains. The worm lives within this tube, head-down in the sediment. Members of the genus *Pectinaria* (=*Cistenides*) are selective deposit feeders, ingesting particles high in organic content. The undigestible fraction is expelled at the posterior, elevated end of the tube, as illustrated. The long tentacles, seen anteriorly, are involved in food collection. (f) *Amphitrite ornata,* a species living in shallow water, in burrows of sand or mud lined with mucus. The first three segments each bear a pair of highly branched gills for gas

exchange. Numerous grooved tentacles project from the prostomium. When the animal is feeding, these tentacles lie against the substrate, trapping food particles. Food is moved to the mouth by cilia lining the tentacular grooves or by shortening the tentacles themselves and bringing the food directly to the mouth. Note the absence of conspicuous parapodia; water is moved through the burrow by peristalsis of the body wall musculature.

(b) From Frank Brown, Selected Invertebrate Types. Copyright © John Wiley & Sons, Inc., New York. Reprinted by permission of Mrs. Frank Brown. (c,e) From Fretter and Graham. 1976. Functional Anatomy of Invertebrates. Academic Press. (d) From Bayard H. McConnaughey and Robert Zottoli, Introduction to Marine Biology, 4th ed. Copyright © 1983 The C. V. Mosby Company, St. Louis, Missouri; after W. C. McIntosh, 1915, British Marine Annelids, Vol. III, The Royal Society. Reprinted by permission. (f) After Brown; after Barnes.

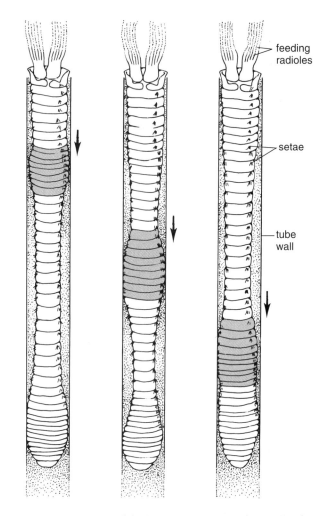

feeding
radioles

setae

tube
wall

Figure 15.8
The tube-dwelling sabellid polychaete *Eudistylia vancouveri*
irrigating its tube with a peristaltic wave of muscle contraction.
From A. Giangrande, "Irrigation of Eudistylia Tube" in Journal of the Marine Biological
Association of the U.K., 71:27-35. Copyright © 1991 Cambridge University Press, New
York. Reprinted by permission.

species also lack a protrusible proboscis. Movement of
water through the tubes of sedentary species for respiration, feeding, and/or waste removal is accomplished by
ciliary action, rhythmic movements of modified parapodia, or waves of muscle contraction passing from one
end of the worm to the other (Fig. 15.8). Many species of
sedentary polychaetes display an abundance of thread-
like or feathery appendages at the anterior end, some
serving for food capture and others being highly vascularized and serving as gills for gas exchange (Figs. 15.7
and 15.9). All polychaetes in this ecological grouping are
either suspension feeders or deposit feeders.

Species in a number of polychaete families have protonephridia similar to those found among the Platyhelminthes and several other invertebrate groups. Thus,
in these polychaete species, coelomic fluid is probably ultrafiltered as it is drawn into the nephridial tubule by
ciliary beating. In other respects, protonephridia and

(a)

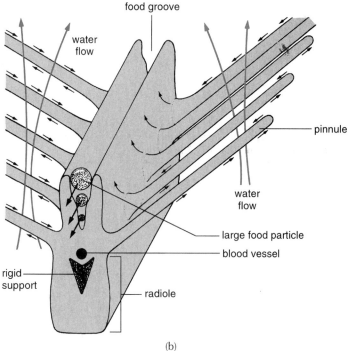

food groove

water
flow

pinnule

water
flow

large food particle

blood vessel

rigid
support

radiole

(b)

Figure 15.9
(a) The sedentary polychaete *Sabella pavonia*. The stiff ciliated
filaments (called "radioles") form a fan surrounding the mouth.
Water currents pass between the tentacles, and captured food
particles are conducted to the mouth area for sorting and
subsequent ingestion or rejection. When the worm is startled,
the tentacles are rapidly withdrawn within the protective tube.
(b) Detail of a single filament (radiole) seen in cross section. Food
particles are captured by the ciliated *pinnules* and transported to
a ciliated food groove, which runs the length of each filament to
the animal's mouth. Blue arrows show direction of water flow;
small black arrows show path taken by captured food particles
along the pinnules to the major food groove of the radiole.
Particles of different sizes are transported to the mouth in
different regions of the food groove as shown.   (a) *Courtesy of D. P.
Wilson/Eric and David Hosking.* (b) *From W. D. Russell-Hunter,* A Life of Invertebrates.
Copyright © 1979 W. D. Russell-Hunter. Reprinted by permission.

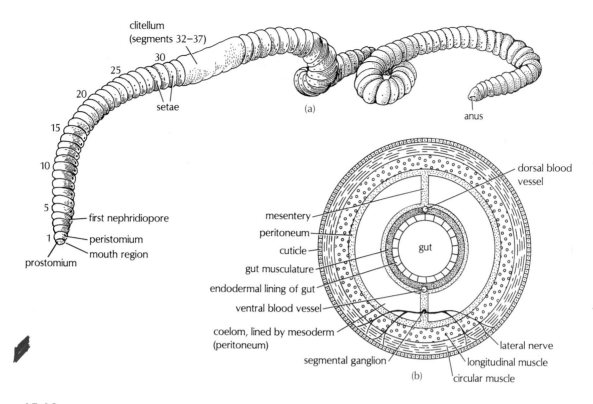

Figure 15.10

(a) External morphology of the terrestrial oligochaete *Lumbricus terrestris*. Note the absence of anterior and lateral appendages. As discussed later (p. 305–307), the clitellum is a glandular region characteristic of both oligochaetes and leeches. It produces a

mucus that aids copulation and encloses fertilized eggs in a protective cocoon. (b) Oligochaete in diagrammatic cross section, showing the arrangement of muscle layers of the body wall and gut.   (a) *After Sherman and Sherman.* (b) *After Russell-Hunter.*

metanephridia are believed to function similarly. All other polychaetes—indeed, all other annelids—have metanephridia.

Feeding habits are highly diverse among polychaetes and include predation, scavenging, suspension feeding, and deposit feeding (Research Focus Box 15.1).

Class Oligochaeta

> *Class Oligo · chaeta*
> (G: few setae)
> ō-lig´-ō-kē´-tah

Approximately 3,500 oligochaete species have been described. In contrast to the Polychaeta, only about 6.5% of oligochaete species are marine; most are found in freshwater or terrestrial habitats. The common earthworm, *Lumbricus terrestris*, is a familiar example of this class (Fig. 15.10a); this species has become widely used as a

biomonitor of pollution stress, and has gained considerable attention for its potential role in transferring pollutants to birds and other terrestrial vertebrates.[5]

Compared with most polychaete annelids, oligochaetes have a more streamlined appearance. In particular, parapodia are lacking and the anteriormost region of the body, the prostomium, lacks conspicuous sensory structures, such as eyes and tentacles. Oligochaetes do have setae, but the setae are less densely distributed along the body. In contrast to the diversity of body plans found among polychaete worms, the oligochaete body plan is relatively invariant among species. There are usually no specialized respiratory organs; gas exchange is accomplished by diffusion across a moist body wall.

As in the Polychaeta, septa divide the body coelomic cavity into a series of semi-isolated compartments. The musculature is arranged as in the polychaetes, with an inner layer of longitudinal muscles overlain by a layer of well-developed circular muscles (Fig. 15.10b). To move, oligochaetes generate a continuous series of localized contractions and relaxations of

5. See *Topics for Further Discussion and Investigation*, no. 14.

Research Focus Box 15.1

Feeding Biology of Deposit-Feeding Annelids

Grémare, A. 1988. Feeding, tube-building and particle-size selection in the terebellid polychaete *Eupolymnia nebulosa*. *Marine Biol.* 97:243–52.

The organic content of sediment is, at most, only about 5% of the sediment's dry weight, and much of that organic material may itself be indigestible plant debris. Deposit feeding would therefore seem to be a rather difficult way to make a living: A deposit feeder must process huge amounts of sediment each day to acquire enough food to meet its metabolic requirements. Previous studies have shown annelids of many species to ingest at least several times their own weight—and as much as 120 times their own weight—in sediment daily. Nevertheless, deposit-feeding annelids dominate mud bottoms in both marine and freshwater habitats. Since a mud flat may have many hundreds of worms per square meter of sediment surface, such extensive sediment reworking will markedly affect soft-bottom community dynamics. Consequently, there has been considerable interest in understanding the biological details of sediment processing.

A common adaptation to maximizing the nutritional content of ingested food is to somehow sort through sediment particles and ingest only those especially high in nutritive content. Among terebellid polychaetes (members of the marine family Terebellidae), this selection is thought to occur at the tips of the ciliated tentacles. These tube-building polychaetes employ the same physical apparatus—the tentacles and mouth parts—to capture both food particles and particles for tube construction. Do they select similarly sized particles for both activities?

Grémare (1988) addressed this question for terebellid polychaetes by analyzing the size distribution of particles in the animals' guts and in the walls of their tubes. The study was conducted using the widely distributed marine species *Eupolymnia nebulosa*. Determining whether the worms preferentially remove particles of particular sizes for feeding would involve comparing the size distributions of particles in the gut with those in the surrounding sediment. To do this, Grémare allowed his worms to feed on sediment for 12 hours and then removed ingested particles from the guts of 15 worms by careful dissection. He then compared the abundance of different-sized particles in the gut with that in the surrounding sediment and calculated the degree of selectively for particles in each of six size ranges. In this "selectivity index," a calculated value of zero for particles in a particular size range would indicate comparable abundances in gut and sediment; a value substantially below zero would indicate that the particles in a particular size range were more abundant in the sediment than in the gut and thus must have been avoided by the worms; a value substantially exceeding zero would indicate a higher incidence of the particles in the gut than in the sediment, implying preferential uptake by the worm.

As seen in Focus Figure 15.1, Grémare found that the worms preferentially ingested particles in the smallest size class (less than 10 μm in diameter). This may be adaptively beneficial insofar as smaller particles have a higher relative surface area than larger particles; per weight of particle then, smaller particles should have a larger amount of encrusting bacteria and other nutritious microorganisms.

---

the circular and longitudinal musculature. These cycles of contraction, known as **peristaltic waves,** are also commonly used among burrowing polychaetes, both during burrow formation and for driving water through the burrow for oxygenation and waste removal.

The principle of peristaltic locomotion is illustrated in Figure 15.11. In Figure 15.11a, the musculature is partially relaxed, and the worm is uniform in diameter. A portion of the body wall in each segment is in contact with the substrate. In Figure 15.11b, the circular musculature of segment 1 has contracted. Since the adjacent segments are pushing laterally against the substrate, the anteriormost region of the body (the prostomium) moves forward, pushing off against segment 2 and other,

more posterior segments. The setae are protruded in the segments behind segment 1, helping to prevent slippage. The circular muscles in segments 2, 3, and 4 then contract in succession, resulting in anterior extension of the body (and retraction of setae), as illustrated in Figure 15.11c. At about this time, a wave of contraction of the longitudinal musculature of each segment begins at the animal's anterior end (Fig. 15.11d), and this wave also passes from anterior to posterior.

As the peristaltic wave of circular muscle contraction passes along the body of the worm, the adjacent uncontracted segments serve as temporary anchors, so that the contraction of circular muscles generates a forward thrust. As the longitudinal musculature subsequently contracts, the anterior segments become fat and anchor

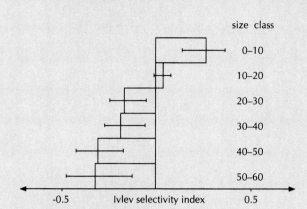

Focus Figure 15.1
The extent to which the deposit-feeding polychaete *Eupolymnia nebulosa* selects food particles of particular size (size ranges in microns). A value greater than zero indicates that particles in that size range were more common in the gut contents than in the surrounding sediment. A value below zero indicates that particles in that size range were more common in the sediment than in the worms' guts.

The selective uptake of smaller particles had been thought to be due simply to the poor adhesive properties of the mucus secreted at the tips of the tentacles: Larger particles simply do not adhere, and so they are not transported to the mouth. This hypothesis made the results of the next set of studies, on tube-building, surprising. To study particle selection during tube building, Grémare incubated a number of worms for 12 hours in an artificial sediment of tiny glass spheres of assorted sizes. After 12 hours, he used a microscope to examine the size distribution of the spheres incorporated into the worm tubes and those in the "sediment." The worms clearly avoided the smaller, food-sized particles and

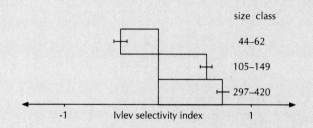

Focus Figure 15.2
The extent to which *Eupolymnia nebulosa* selectively chooses particles of particular size (size ranges in micrometers) in building its tubes.

instead built their tubes primarily using particles larger than 100 µm (Focus Fig. 15.2). Grémare did not determine whether any of the glass spheres were ingested. Can you see any value to including that additional step in the experiment?

In summary, Grémare's study suggests that individuals of E. nebulosa choose small particles for ingestion and large particles for tube building. Apparently, the worms collect and transport particles with a single mechanism. Details regarding this mechanism are unclear, although the hypothesis that particle selection is simply due to properties of the mucus seems defeated; such a passive mechanism would provide particles of similar sizes for both ingestion and tube building— unless the tentacles can secrete mucus with different properties at different times. What other possibilities migh enable the worms to select different-sized particles at different times? Try designing an experiment that could distinguish between any two of these possibilities.

Figures from A. Grémare, "Feeding, Tube-building and Particle-size Selection in the Terebellid Polychaete Eupolymnia nebulosa" in Marine Biology, 97:243–252, 1988. Copyright © 1988 Springer-Verlag, Heidelberg, Germany. Reprinted by permission.

the animal. The adjacent thin segments can then be pulled forward. Note that the muscular waves pass in one direction, but the animal moves in the opposite direction (Figs. 15.11 and 15.12). Such waves are known as **retrograde,** as opposed to **direct.**[6]

Peristaltic waves are especially effective in burrow formation and in movement within a burrow or tube. When the longitudinal musculature is contracted, thickening the worm's body, the entire surface area of the segments becomes tightly pressed against the walls of the burrow, greatly increasing the magnitude of the pushing and pulling forces that can be generated without causing the

worm to slip within the burrow. Protrusion of setae by the worm, at appropriate times in the cycles of muscular contraction and relaxation, aids in anchoring these portions of the worm and in preventing backward slippage when thrusting and pulling forces are generated (Fig. 15.12).

Class Hirudinea

Class Hirudinea
(G: leech)
here-oo-din´-ē-ah

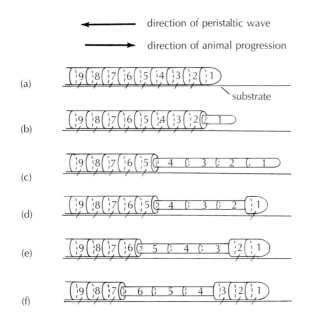

Figure 15.11
Diagrammatic illustration of peristaltic locomotion in oligochaetes. (a–c) A wave of contraction of the circular muscles begins passing down the body, resulting in forward extension of the worm. As the wave continues to travel down the worm's body, from segment to segment, a consolidation phase is initiated anteriorly; that is, a wave of longitudinal muscle contractions moves posteriorly (d–f), shortening and thickening each segment in preparation for the next extension phase.

The class Hirudinea, which includes the leeches, has approximately only 500 described species, which are believed to have evolved from oligochaete stock. Reflecting their presumed oligochaete ancestry, most leeches occupy freshwater or terrestrial habitats; only a small proportion of species is marine. Like the oligochaetes, the basic body plan of the Hirudinea varies little among species. Conspicuously absent are parapodia and head appendages, as in oligochaetes (Fig. 15.13).

Both oligochaetes and leeches are **hermaphroditic** (both male and female reproductive apparatus are contained within a single individual) and demonstrate a specialized region of the epidermis known as the **clitellum.** In both classes, the clitellum secretes a cocoon within which embryos develop. Specialized gill structures are absent in most leech species, gas exchange being accomplished by diffusion across the general body wall surface. The morphological similarities between the oligochaetes and the leeches are sufficiently great to warrant classification of both groups as orders within a single class, the Clitellata, according to some invertebrate zoologists.

In contrast to both polychaete and oligochaete annelids, most leeches lack setae. Moreover, the body is generally not separated into compartments by septa, and the continuous coelomic space is largely filled with connective tissue, or **mesenchyme** (Fig. 15.14). The remaining channels and sinuses within this tissue serve a blood

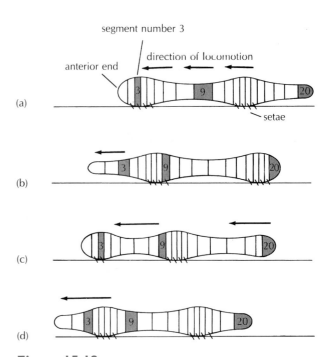

Figure 15.12
Locomotion of an earthworm, based upon movie footage. (a) The circular muscles of the first five to six anterior-most segments are relaxed, and the longitudinal muscles of these segments are fully contracted. In (b), a wave of circular muscle contraction has been initiated anteriorly; segment 3 and its immediate neighbors have become longer and thinner. Segment 3 has also moved forward, in part due to the contraction of its own circular musculature (and relaxation of its longitudinal musculature), and in part due to the forward thrust generated by the muscle contractions of segments 4 and 5. At the same time, the wave of longitudinal muscle contraction has passed posteriorly, encompassing segment 9. Note that segments 6–9 are prevented from backsliding by the protrusion of setae as the anterior segments are thrust forward. A second wave of longitudinal muscle contraction has passed from segments 12–15 to the terminal segments of the worm; consequently, segment 20 has progressed forward. In (c) and (d), waves of longitudinal and then circular muscle contraction continue to be generated anteriorly, thrusting and pulling the segments of the worm forward. *From Purves and Orians, Life: The Science of Biology, 2d ed. Copyright © 1983 Sinauer Associates, Sunderland MA. Reprinted by permission.*

transport function in most leech species; thus, the circulatory medium is the coelomic fluid. Not surprisingly, considering the lack of septa, the locomotion of leeches differs considerably from that of other annelids. Leeches do not move by generating peristaltic waves; instead, they locomote over solid substrates by using suckers as temporary anchors. These suckers are formed from groups of segments at the body's anterior and posterior ends (Figs. 15.13 and 15.15). With the posterior sucker attached to the substrate, the worm extends itself forward by contracting the circular muscles of the body wall (Fig. 15.16a–c). The anterior sucker is then applied to the substrate, the circular muscles are relaxed, and the longitudinal muscles are contracted. The leech becomes shorter and, as the circular muscles are stretched, fatter

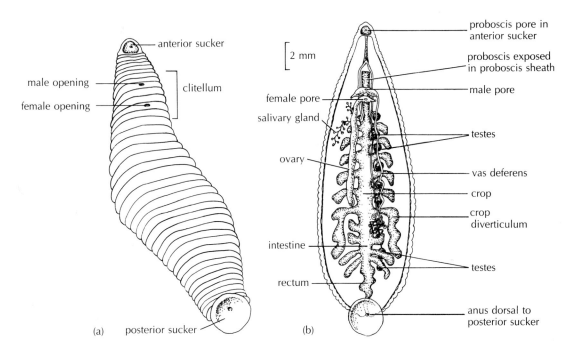

Figure 15.13
(a) External anatomy of a leech. (b) Internal anatomy of
Glossiphonia complanata.   (a) After Pimentel. (b) Mann (after Harding), Leeches.
Oxford, England: Pergamon Press, plc, 1962.

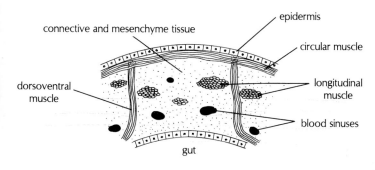

Figure 15.14
A leech in cross section.   After Mann.

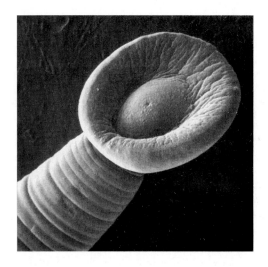

Figure 15.15
Scanning electron micrograph showing anterior end of a
freshwater leech, *Johanssonia arctica.*   Courtesy of Dr. R. A. Khan and Allen
Press, from Khan and Emerson, Trans. Amer. Microsc. Soc. 100:51, 1981.

(Fig. 15.16d–g). By further contracting the longitudinal
musculature on the ventral surface, the leech arches its
body, bringing the posterior and anterior suckers in close
proximity (Fig. 15.16h). The posterior sucker is then applied to the substrate, the anterior sucker is released, and
the cycle can be repeated. The mouth is included within
the anterior sucker.

Leeches are generally ectoparasitic, feeding either on
the blood of other invertebrates or, more commonly, on
the blood of vertebrates. Most parasitic species possess
three toothed jaws within the mouth; the leech uses these
jaws to make an incision in the host. Other species have a
protrusible proboscis, through which blood is removed
from the host.

The blood-sucking habits of leeches have been long
employed in the practice of medicine, for bloodletting

through much of the nineteenth century and, more recently, to alleviate fluid pressure following damage to vascular tissue (e.g., following a snakebite or the reattachment
of a severed finger or ear). When a finger is surgically reattached, the arteries supplying blood can be rejoined to
those in the finger, but the veins, because of their smaller
size, generally cannot. Thus, blood flows into the reattached finger but cannot escape. An application of *Hirudo
medicinalis,* the medicinal leech, helps in two ways: The

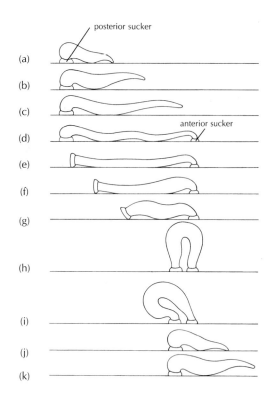

posterior sucker

(a)

(b)

(c)

anterior sucker

(d)

(e)

(f)

(g)

(h)

(i)

(j)

(k)

Figure 15.16
Leech locomotion. In (a–d), the posterior sucker is affixed to the substrate while the circular muscles of the leech contract, extending the animal forward. The longitudinal muscles are stretched in the process. In (e–g), the posterior sucker is released while the anterior sucker is applied. Contraction of the longitudinal musculature shortens the animal and reextends the circular muscles. In (h), selective contraction of the longitudinal musculature places the posterior sucker immediately behind the anterior sucker, and the entire sequence is repeated (i–k).   *After Gray, Lissman, and Pumphrey.*

leech (1) drains off the excess blood from the finger and (2) injects an anticoagulant—hirudin—that keeps the blood flowing out from the bite-wound long after the sated leech has dropped off. Eventually, the patient's venous system reestablishes itself in the reattached appendage, and the leech's services are no longer needed.

Leeches are also attracting the interest of modern drug companies as sources of such useful substances as local anesthetics (which enable leeches to bite unnoticed by the host) and certain antibiotics. Very recently, an Amazonian leech has been found to produce a protein that dissolves blood clots; the hirudin produced by *H. medicinalis* only prevents such clots from forming. If the genes coding for these substances can be isolated, commercial mass production of these substances should become possible. Finally, leeches are becoming leading research models in the study of nervous system development and function.

Less than 100 years ago, selling leeches to the medical establishment was a booming business, and perhaps it will become so again. As described by the late Reverend J. G. Wood (*Animal Creation,* 1885), the collection of leeches was a colorful enterprise:

> The Leech-gatherers take them in various ways. The simplest and most successful method is to wade into the water and pick off the leeches as fast as they settle on the bare legs. This plan, however, is by no means calculated to improve the health of the Leech-gatherer, who becomes thin, pale, and almost spectre-like, from the constant drain of blood, and seems to be a fit companion for the old worn-out horses and cattle that are occasionally driven into the Leech-ponds in order to feed these bloodthirsty annelids.

Today, leeches are reared commercially, particularly in England.

The nonparasitic leeches, about 25% of all known species, are predators upon other invertebrate animals.

Other Features of Annelid Biology

Reproduction and Development

Class Polychaeta

Reproduction is exclusively sexual in most polychaete species, and most polychaetes are **dioecious** (i.e., they have separate sexes). Gametes are produced by peritoneal tissue, rather than in distinct gonads, and are then released into the associated coelomic compartments, where they mature. At least six adjacent segments of a given individual are involved in gamete production, and in some species gametes are produced within nearly all segments.

In several families of errant polychaetes, many tube-dwelling or otherwise sedentary species undergo **epitoky,** a marked morphological transformation in preparation for reproductive activity. The result of this transformation is an **epitoke:** a sexually mature being that is highly specialized for swimming and sexual reproduction (*tokus* = G: birth). In many species, epitoky involves asexual budding (Fig. 15.17a). One or more new, reproductive individuals (epitokes) are budded, one segment at a time, from the posterior portion of the original animal (the **atoke**).[7] These epitokes subsequently detach from the atoke and swim off to commingle with other epitokes and discharge gametes. The atoke remains safely behind in its burrow or tube.

In some other epitokous polychaete species, epitoky involves the remodeling of preexisting structures, rather

---

7. See *Topics for Further Discussion and Investigation,* nos. 10, 12.

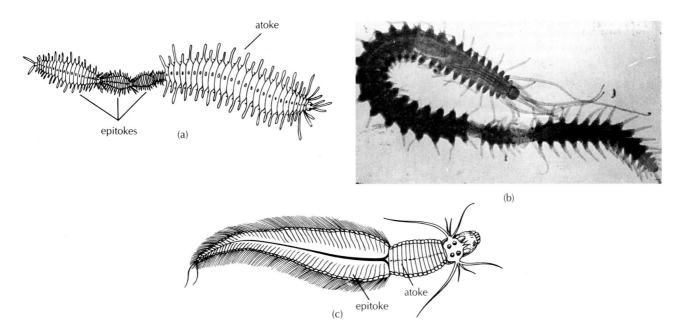

Figure 15.17

(a) Epitoke formation by asexual budding in *Autolytus*. (b) *Autolytus* sp. Female atoke with several epitokes. (c) Epitoke formation via remodeling of original structures. Note that in (a) and (b) the original worm becomes the atoke, left behind in the burrow, and does not participate directly in reproduction, while in (c) a single individual is both epitoke and atoke and participates directly in reproduction.   *(a) After Harmer and Shipley; after Malaquin. (b) Courtesy of K. L. Schiedges. (c) After Fauvel.*

than the budding off of new segments. The original worm thus becomes the epitoke, through adaptations for gamete production and storage, and for active swimming (Fig. 15.17c). In these species, the entire worm swims off to reproduce. Gamete maturation and epitoke formation have been shown to be under environmental and hormonal control in a number of polychaete species.

Fertilization among polychaetes usually occurs externally, in the surrounding water. Typically, the free-living embryo soon develops a digestive system and two rings of cilia (Fig. 15.18). These rings are the **prototroch** (G: first wheel), located around the equator of the animal, anterior to the mouth, and the **telotroch** (G: tail wheel), located posteriorly on what will become the terminal portion of the adult, the **pygidium.** The prototroch is the larva's major locomotory organ. Because of the ciliated rings, the larva is called a **trochophore,** meaning "wheel-bearer" (Fig. 15.18a,b). A third band of cilia, the **metatroch** ("between wheel"), later forms between the prototroch and the telotroch. A trochophore stage is not found among the other annelid classes, although it does occur in several other animal phyla, including the Mollusca; whether this shared trochophore stage indicates a close evolutionary relationship between groups or arose independently by convergent evolution is hotly debated.

As the polychaete trochophore larva swims and, generally, feeds in the water, body segments are repeatedly budded from its posterior region, just anterior to the pygidium. Each new segment bears a ring of locomotory cilia, permitting the larva to continue swimming despite the increasing body weight (Fig. 15.18c). The larva becomes more like the adult as it continues to develop (Fig. 15.18d,e), and eventually it metamorphoses to adult form and habitat, often after many weeks of dispersing in the plankton. Some polychaete species produce a nonfeeding, short-lived trochophore stage that subsists on stored yolk reserves and metamorphoses after swimming for only a few hours or, in some cases, only a few minutes.

Some species lack any free-living larval stage in the life history. Instead, the embryos may develop within a gelatinous egg mass anchored in the sediment or to the inside surface of the female's tube. In some species, embryos may be directly protected by the parent, developing to a larval or juvenile stage within specialized adult brooding chambers.

Some species also reproduce asexually, by fragmenting and subsequently regenerating missing parts.

Class Oligochaeta

Unlike polychaetes, all oligochaete species are hermaphroditic, and generally only a few segments of each individual produce gametes. Moreover, gametes are produced within distinct testes and ovaries, rather than from the peritoneum lining the coelomic cavity. Sperm are generally exchanged simultaneously between two mating individuals (Fig. 15.19a); the sperm are stored for later use in specialized organs called **spermathecae** (G: sperm boxes). Eventually, eggs and sperm are extruded from separate openings into a complex cocoon secreted by the

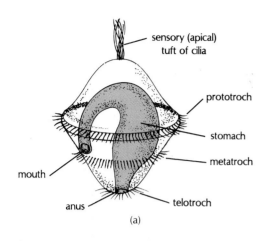

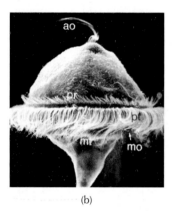

(a)

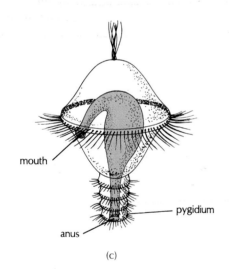

(b)

(c)

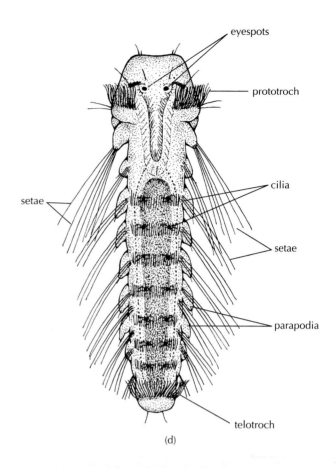

(d)

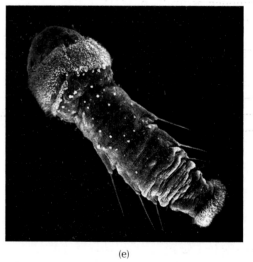

(e)

Figure 15.18

(a) Polychaete trochophore larva. The larva feeds on unicellular algae collected by the ciliary bands. (b) Scanning electron micrograph of the trochophore larva of *Serpula vermicularis*. The larva is about 110 μm wide. (ao = apical tuft; mo = mouth opening; mt = metatrochal cilia; pr = pretrochal cilia; pt = prototrochal cilia.) (c) Advanced trochophore larva, showing additional ciliated segments budded posteriorly from the initial trochophore. Parapodia will soon develop from each of these segments. New segments are added just anterior to the pygidium, which bears the anus. Thus, the youngest segments are located posteriorly, and the oldest segments of an annelid are anterior. (d) Advanced larva, about 600 μm long, of the polychaete *Polydora ciliata* in dorsal view. (e) Scanning electron micrograph of an advanced larva of the polychaete *Eupolymnia nebulosa*, viewed dorsally. The animal is about 13 days old. Note the conspicuous anterior prototroch and posterior telotroch.

(a) From Purves and Orians, Life: The Science of Biology, 2d ed. Copyright © 1983 Sinauer Associates, Sunderland MA. Reprinted by permission. (b) Courtesy of Claus Nielsen, from Nielsen, Acta Zoolog. (Stockholm) 68:205–62, 1987. Permission of the Royal Swedish Academy of Sciences. (c) From Hardy, The Open Sea: Its Natural History, 1965. Boston: Houghton Mifflin Company. (d) From D. P. Wilson, in Journal of Marine Biology Association, U. K. 15:567, 1928. (e) Courtesy of Michel Bhaud. From Bhaud and Grémare, 1988. Zoologica Scripta 17:347. Courtesy of Pergamon Press.

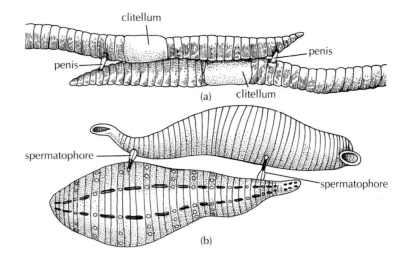

Figure 15.19

(a) Copulating oligochaetes, *Pheretima communissima*.
(b) Copulating leeches, *Glossiphonia* sp. The members of this family generally lack a penis; spermatophores are forcefully injected into the partner, and the sperm migrate to the reproductive tract for storage and subsequent fertilization of eggs. In both oligochaetes and leeches, the fertilized eggs are incubated in a gelatinous cocoon secreted by the clitellum.

(a) *From Fretter and Graham*, A Functional Anatomy of Invertebrates. *Copyright © 1976 Academic Press, Inc. Reprinted by permission. After Oishi.* (b) *After Meglitsch; after Brumpt.*

clitellum. The embryos develop in and feed upon a nutritive fluid found within the cocoon. When development is completed, miniature worms emerge from the cocoons; oligochaetes lack free-living larval stages, even in marine environments.

Although asexual reproduction does occur in some polychaete groups, it is more common among oligochaetes, particularly in freshwater species. The process of asexual reproduction involves the transverse (i.e., crosswise) division of the "adult" into a number of separate sections and the subsequent regeneration of each section into a complete individual. In addition, quite a few oligochaete species, especially those living in terrestrial environments, are **parthenogenetic;** that is, eggs may develop normally in the absence of fertilization.

Class Hirudinea

As in the Oligochaeta, adult leeches are simultaneous hermaphrodites and a free-living larval stage is absent from the life cycle. Only a few segments of each individual are directly involved in **gametogenesis** (gamete production). As with oligochaetes, fertilization occurs internally, either through copulation or, in species lacking a penis, by jabbing packets of sperm (**spermatophores**) into the partner's body. Sperm exchange is mutual between two mating individuals (Fig. 15.19b), and the fertilized eggs of each individual generally develop within external cocoons. The clitellum generally functions in producing the cocoon and a nutritive fluid, as in the Oligochaeta. A miniature leech eventually emerges from the cocoon.

In contrast to the other annelid classes, within the Hirudinea asexual reproduction is completely unknown.

Digestive System

The annelid gut is linear and unsegmented, with a mouth opening on the peristomium and an anus opening at the posterior end of the animal (pygidium). Food is moved through the gut by cilia and/or by muscular contractions. Digestion is primarily extracellular, although some species show an intracellular component as well.[8]

Class Polychaeta

The digestive tract of polychaetes is typically divided into a pharynx, esophagus, intestine, and rectum (Fig. 15.20a). In some species, evaginations of the gut form blind-ending **digestive glands** or **digestive caeca** (*caec* = L: blind), increasing the amount of surface area available for digestion and absorption.

Class Oligochaeta

Oligochaetes generally show more modification of the basic arrangement just described for polychaetes. The esophagus may be modified to form a **crop,** for food storage, and/or a **gizzard,** a highly muscular structure lined with hardened cuticle, for grinding food (Fig. 15.20b). **Calciferous glands** are associated with the esophagus; these may function in regulating blood pH by controlling the concentration of carbonate ions.

The intestine of many terrestrial oligochaete species is thrown into a ridge or fold (**typhlosole**), which increases the gut's effective surface area (Fig. 15.21).

8. See *Topics for Further Discussion and Investigation*, no. 8.

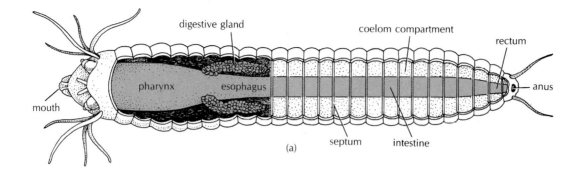

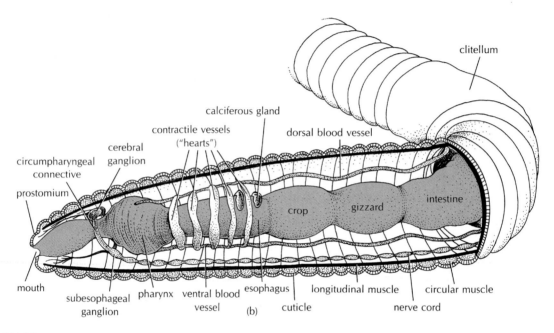

Figure 15.20
(a) Digestive system of the errant polychaete *Nereis virens*.
(b) Digestive system of an oligochaete, *Lumbricus terrestris*.
Elements of the nervous and circulatory systems are also shown.

Associated with the intestine (and dorsal blood vessel) of oligochaetes is a characteristic yellow tissue called **chloragogen.** Chloragogen cells play major roles in protein, carbohydrate, and lipid metabolism of oligochaetes.

Class Hirudinea

The mouth of a leech opens into a muscular, pumping pharynx. **Salivary glands** associated with the pharynx (Fig. 15.13b) secrete **hirudin,** an anticoagulant. A crop and digestive glands are found in some species.

Nervous System and Sense Organs

A mass of **ganglia** (aggregations of nerve tissue) forming a brain is present at the anterior end of an annelid (Fig. 15.22a). A solid ventral nerve cord (or a pair of nerve cords in the primitive condition) passes from the anterior to the posterior end of each individual. Swellings of the cord in each segment form segmental ganglia. A variety of sense organs, including touch receptors, statocysts, light receptors, vibration receptors, and chemoreceptors, are distributed along the length of the body. These receptors are either connected to the ventral nerve cord by means of segmental nerves or (as in the case of the eyes) connected directly to the brain by other nerves. The eyes of most species are not image forming, but they are sensitive to light and changes in light intensity. Lateral nerves extending from the ventral nerve cord also innervate the digestive tract and the parapodial and body wall musculature of each segment.

Generally, a few of the nerves in the ventral cord are of considerably greater diameter than the others (Fig. 15.22b). These "giant" fibers may conduct nerve impulses 20 to more than 1,000 times faster than other fibers, making possible the nearly simultaneous contraction of

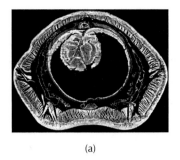

(a)

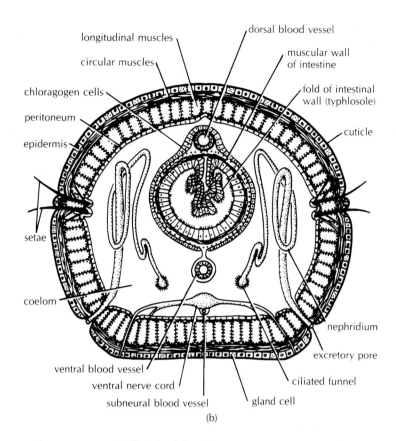

(b)

Figure 15.21

(a) *Lumbricus terrestris* in cross section. (b) Diagrammatic illustration of *L. terrestris* in cross section. A pronounced infolding of the intestinal wall is characteristic of terrestrial oligochaetes; this typhlosole increases the gut surface area available for

absorbing nutrients. Note the chloragogen tissue surrounding the intestine. This tissue, found in oligochaetes and polychaetes, plays a role in excretion, hemoglobin synthesis, and basic metabolism.

(a) Copyright L. S. Eyster. (b) After Buchsbaum and other sources.

appropriate musculature throughout the worm's body. These giant fibers permit a very rapid, coordinated response to potential predators.[9]

Circulatory System

Annelids generally have a closed circulatory system, consisting of a dorsal vessel (carrying blood anteriorly), a ventral vessel (carrying blood posteriorly), and capillaries connecting the two (Figs. 15.3b and 15.20b). There is no specialized heart. Instead, circulation is maintained by contractions of the blood vessels themselves, especially of the dorsal vessel. Valves assure a unidirectional flow of blood throughout the body. This form of blood circulatory system is much reduced or even absent among members of the Hirudinea. Among the leeches, coelomic fluid assumes all or part of the circulatory role, reaching the tissues via contractile coelomic sinuses and channels.

Oxygen-carrying blood pigments are found in the circulatory fluid of most (but not all) annelid species in each of the three classes. Several chemically and functionally distinct pigments are found among annelids. The blood of most species contains **hemoglobin,** in which iron atoms serve as the binding sites for oxygen. Hemoglobin may occur in corpuscles or may be found in solution in the blood fluid, depending upon the species examined. Hemoglobin-containing blood is especially characteristic of leeches and oligochaetes, although it is found as well in the blood of some polychaete species. In addition to hemoglobin, two other blood pigments have been encountered among the Polychaeta. **Chlorocruorin,** another iron-containing pigment, is found in solution in the blood of several polychaete species. This pigment is chemically quite similar to hemoglobin, but it has a greenish coloration. Yet a third iron-containing pigment, **hemerythrin,** is found in at least one polychaete species. Hemerythrin is structurally quite dissimilar to the other two pigments, and it is always contained within cells. More than one type of blood pigment (including several structurally and functionally different types of hemoglobin) may occur simultaneously in the blood of a single annelid.[10]

---

9. See *Topics for Further Discussion and Investigation*, no. 9.

10. See *Topics for Further Discussion and Investigation*, no. 5.

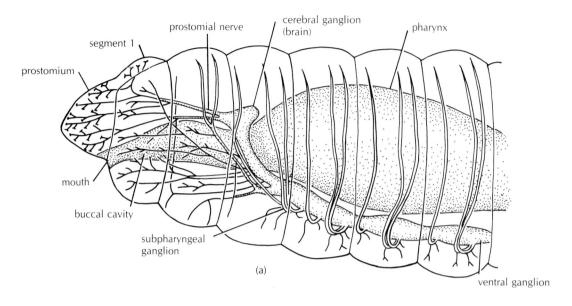

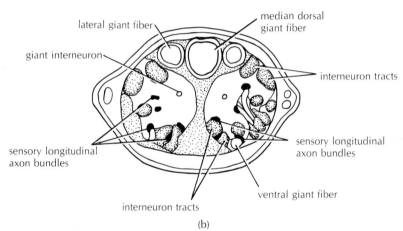

Figure 15.22

(a) Nervous system of *Lumbricus terrestris*, anterior end of animal. (b) Diagrammatic illustration of ventral nerve cord in cross section, showing a complex array of nerve fibers, including giant fibers. The giant fibers mediate rapid motor responses. The interneuron tracts are primarily involved in coordinating the activities of the various body segments. The giant fibers conduct nerve impulses up to about 1,600 times faster than the smaller nerve fibers.   (a) *From Barnes; after Hess; from Avel.* (b) *Reprinted from P. J. Mill in* Comparative Biochemistry and Physiology, *73A:641, 1982 with kind permission from Elsevier Science Ltd., The Boulevard, Langford Lane, Kidlington 0X5 1GB, UK.*

Taxonomic Summary

Phylum Annelida
  Class Polychaeta
    The Errantia
    The Sedentaria
  Class Oligochaeta
  Class Hirudinea—the leeches

Topics for Further Discussion and Investigation

1. Explore the morphological, behavioral, and physiological adaptations associated with a sedentary existence among polychaetes.

Aller, R. C., and J. Y. Yingst. 1978. Biogeochemistry of tube dwellings: A study of the sedentary polychaete *Amphitrite ornata* (Leidy). *J. Marine Res.* 36:201.

Barnes, R. D. 1965. Tube-building and feeding in chaetopterid polychaetes. *Biol. Bull.* 129:217.

Brenchley, G. A. 1976. Predation detection and avoidance: Ornamentation of tube-caps of *Diopatra* spp. (Polychaeta: Onuphidae). *Marine Biol.* 38:179.

Brown, S. C. 1975. Biomechanics of water-pumping by *Chaetopterus variopedatus* (Renier). Skeletomusculature and kinematics. *Biol. Bull.* 149:136.

Brown, S. C., and J. S. Rosen. 1978. Tube-cleaning behavior in the polychaete annelid *Chaetopterus variopedatus* (Renier). *Anim. Behav.* 26:160.

Busch, D. A., and R. E. Loveland. 1975. Tubeworm-sediment relationships in populations of *Pectinaria gouldii* (Polychaeta: Pectinariidae) from Barnegat Bay, New Jersey, USA. *Marine Biol.* 33:255.

Chughtai, I., and E. W. Knight-Jones. 1988. Burrowing into limestone by sabellid polychaetes. *Zool. Scripta* 17:231.

Dales, R. P. 1955. Feeding and digestion in terebellid polychaetes. *J. Marine Biol. Assoc. U.K.* 34:55.

Dauer, D. M. 1985. Functional morphology and feeding behavior of *Paraprionospio pinnata* (Polychaeta: Spionidae). *Marine Biol.* 85:143–51.

Flood, P. R., and A. Fiala-Médioni. 1982. Structure of the mucous feeding filter of *Chaetopterus variopedatus* (Polychaeta). *Marine Biol.* 72:27.

Hedley, R. H. 1956. Studies of serpulid tube formation. I. The secretion of the calcareous and organic components of the tube of *Pomatoceros triqueter*. *Q. J. Microsc. Sci.* 97:411.

Hoffmann, R. J., and C. P. Mangum. 1972. Passive ventilation in benthic animals? *Science* 176:1356.

MacGinitie, G. E. 1939. The method of feeding of *Chaetopterus*. *Biol. Bull.* 77:115.

Merz, R. A. 1984. Self-generated *versus* environmentally produced feeding currents: A comparison for the sabellid polychaete *Eudistylia vancouveri*. *Biol. Bull.* 167:200.

Strathmann, R. R., R. A. Cameron, and M. F. Strathmann. 1984. *Spirobranchus giganteus* (Pallas) breaks a rule for suspension feeders. *J. Exp. Marine Biol. Ecol.* 79:245.

Völkel, S., and M. K. Grieshaber. 1992. Mechanisms of sulphide tolerance in the peanut worm, *Sipunculus nudus* (Sipunculidae), and in the lugworm, *Arenicola marina* (Polychaeta). *J. Comp. Physiol. B* 162:469.

Whitlatch, R. B., and J. R. Weinberg. 1982. Factors influencing particle selection and feeding rate in the polychaete *Cistenides* (*Pectinaria*) *gouldii*. *Marine Biol.* 71:33.

Zottoli, R. A., and M. R. Carriker. 1974. Burrow morphology, tube formation, and microarchitecture of shell dissolution by the spionid polychaete *Polydora websteri*. *Marine Biol.* 27:307.

2. What mechanical advantages are gained and lost through the elimination of septa among annelids?

Chapman, G. 1958. The hydrostatic skeleton in the invertebrates. *Biol. Rev.* 33:338.

Chapman, G., and G. E. Newell. 1947. The role of the body fluid in relation to movement in soft-bodied invertebrates. I. The burrowing of *Arenicola*. *Proc. Royal Soc. London B* 134:431.

Elder, H. Y. 1973. Direct peristaltic progression and the functional significance of the dermal connective tissues during burrowing in the polychaete *Polyphysia crassa* (Oersted). *J. Exp. Biol.* 58:637.

Gray, J., H. W. Lissmann, and R. J. Pumphrey. 1938. The mechanism of locomotion in the leech (*Hirudo medicinalis* Ray). *J. Exp. Biol.* 15:408.

Trueman, E. R. 1966. The mechanism of burrowing in the polychaete worm, *Arenicola marina* (L.). *Biol. Bull.* 131:369.

3. Compare the locomotion of errant polychaetes with that of oligochaetes.

Gray, J. 1939. Studies in animal locomotion. VIII. *Nereis diversicolor*. *J. Exp. Biol.* 16:9.

Seymour, M. K. 1969. Locomotion and coelomic pressure in *Lumbricus terrestris*. *J. Exp. Biol.* 51:47.

4. What differences in nephridial structure and function might you expect to see among marine, freshwater, and terrestrial annelids?

5. A variety of blood pigments are found within the Annelida. All blood pigments must have one key feature to be functional: They must be able to combine reversibly with oxygen. The amount of oxygen potentially carried per milliliter of blood, and the conditions under which a blood will become saturated with oxygen, are determined by the abundance and properties of the particular blood pigment present. In addition, blood pigments differ in how readily they give up oxygen under any given set of environmental conditions. What are the structural and functional differences among the various annelid blood pigments, and how do these differences relate to the environments in which the different species occur?

Baldwin, E. 1964. *An Introduction to Comparative Biochemistry.* New York: Cambridge University Press, 88–106.

Kayar, S. R. 1981. Oxygen uptake in *Sabella melanostigma* (Polychaeta: Sabellidae): The role of chlorocruorin. *Comp. Biochem. Physiol.* 69A:487.

Mangum, C. P. 1985. Oxygen transport in invertebrates. *Amer. J. Physiol.* 248:R505.

Mangum, C. P., J. M. Colacino, and J. P. Grassle. 1992. Red blood cell oxygen binding in Capitellid polychaetes. *Biol. Bull.* 182:129.

Toulmond, A., F. E. I. Slitine, J. de Frescheville, and C. Jouin. 1990. Extracellular hemoglobins of hydrothermal vent annelids: Structural and functional characteristics in three alvinellid species. *Biol. Bull.* 179:366.

Wood, S. C., ed. 1980. Respiratory pigments. *Amer. Zool.* 20:3.

6.   What is the influence of deposit-feeding polychaetes on the distribution and abundance of other annelids in the community?

Wilson, W. H., Jr. 1981. Sediment-mediated interactions in a densely populated infaunal assemblage: The effects of the polychaete *Abarenicola pacifica. J. Marine Res.* 39:735.

Woodin, S. A. 1985. Effects of defecation by arenicolid polychaete adults on spionid polychaete juveniles in field experiments: Selective settlement or differential mortality? *J. Exp. Marine Biol. Ecol.* 87:119.

7.   Do annelids show an immune response? What is the evidence?

Anderson, R. S. 1980. Hemolysins and hemagglutinins in the coelomic fluid of a polychaete annelid, *Glycera dibranchiata. Biol. Bull.* 159:259.

Chain, B. M., and R. S. Anderson. 1983. Antibacterial activity of the coelomic fluid of the polychaete, *Glycera dibranchiata.* I. The kinetics of the bactericidal reaction. *Biol. Bull.* 164:28.

Cooper, E. L., and P. Roch. 1984. Earthworm leukocyte interactions during early stages of graft rejection. *J. Exp. Zool.* 232:67.

Çotuk, A., and R. P. Dales. 1984. The effect of the coelomic fluid of the earthworm *Eisenia foetida* Sav. on certain bacteria and the role of the coelomocytes in internal defence. *Comp. Biochem. Physiol.* 78A:271.

Fitzgerald, S. W., and N. A. Ratcliffe. 1989. In vivo cellular reactions and clearance of bacteria from the coelomic fluid of the marine annelid, *Arenicola marina* L. (Polychaeta). *J. Exp. Zool.* 249:293.

8.   More than 99% of all the organic matter in the ocean is in dissolved form, rather than particulate (Hedges, J. I. 1987. Organic matter in sea water. *Nature* [London] 330:205). What is the potential role of this dissolved organic material in meeting the nutritional requirements of aquatic annelids?

Ahearn, G. A., and S. J. Townsley. 1975. Transport of exogenous D-glucose by the integument of a polychaete worm (*Nereis diversicolor* Müller). *J. Exp. Biol.* 62:243.

Manahan, D. T. 1990. Adaptations by invertebrate larvae for nutrient acquisition from seawater. *Amer. Zool.* 30:147.

Stephens, G. C. 1975. Uptake of naturally occurring primary amines by marine annelids. *Biol. Bull.* 149:397.

Taylor, A. G. 1969. The direct uptake of amino acids and other small molecules from seawater by *Nereis virens* Sars. *Comp. Biochem. Physiol.* 29:243.

9.   Investigate the role of giant fibers in mediating escape responses by aquatic oligochaetes.

Drewes, C. D., and C. R. Fourtner. 1989. Hindsight and rapid escape in a freshwater oligochaete. *Biol. Bull.* 177:363.

Zoran, M. J., and C. D. Drewes. 1987. Rapid escape reflexes in aquatic oligochaetes: Variations in design and function of evolutionarily conserved giant fiber systems. *J. Comp. Physiol. (A)* 161:729.

10.   In what respects is epitoke formation similar to the strobilation of polyps in the life cycle of scyphozoans (jellyfish) (p. 88)?

11.   Where does a polychaete go when its feet hurt?

12.   Compare and contrast the phenomenon of epitoky among polychaetes with the life cycle of marine hydrozoans (p. 93).

13.   The evolutionary origins of the three annelid classes are far from certain. Some regard the primitive annelid as polychaete-like, with oligochaetes evolving subsequently through the reduction and loss of parapodia. Others place the streamlined oligochaetes, or even the leeches, in the ancestral position. But although most agree that leeches evolved from oligochaetes, it may well be that polychaetes and oligochaetes evolved independently from early coelomate, segmented ancestors. What types of evidence might persuade you to accept one of these scenarios over another?

Brinkhurst, R. O. 1982. Evolution in the Annelida. *Canadian J. Zool.* 60:1043.

Clark, R. B. 1964. *Dynamics in Metazoan Evolution.* London: Oxford Univ. Press.

Willmer, P. 1990. *Invertebrate Relationships: Patterns in Animal Evolution.* Cambridge: Cambridge Univ. Press.

14.   Ecological risk assessment is widely used to manage waste disposal, waste site cleanup, and other complex environmental problems. As deposit feeders and important components of many terrestrial food webs, earthworms may play major roles in transferring various pollutants from soil to vertebrates. What are the benefits and complications associated with using earthworms to monitor soil pollutant concentrations and to assess the potential risks associated with those concentrations?

Menzie, C. A., D. E. Burmaster, J. S. Freshman, and C. A. Callahan. 1992. Assessment of methods for estimating ecological risk in the terrestrial component: A case study at the Baird & McGuire superfund site in Holbrook, Massachusetts. *Environ. Toxicol. Chem.* 11:245.

Morgan, J. E., and A. J. Morgan. 1993. Seasonal changes in the tissue-metal (Cd, Zn, and Pb) concentrations in two ecophysiologically dissimilar earthworm species: Pollution-monitoring implications. *Environ. Poll.* 82:1.

Phylum Annelida

This phylum has approximately 12,000 species.

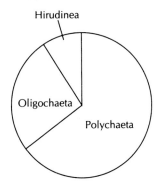

Class Polychaeta.

The approximately 8,000 species are distributed among 25 orders. Most species are marine or estuarine.

Order Phyllodocida.

Twenty-seven families.

Superfamily Phyllodocidacea.   Eight families, many of which contain small numbers of exclusively pelagic, swimming worms. Most phyllodocid species are benthic. All members are marine.

Family Phyllodocidae.   *Eulalia, Notophyllum, Phyllodoce, Eteone, Sige.* Adults are benthic, and only a few species burrow; most are active carnivores, crawling over hard substrates. There are over 300 described species. When sexually mature, the entire worm may become epitokous and swim up into the water to reproduce. Individuals are composed of up to 700 segments.

Family Tomopteridae.   *Tomopteris.* These small (fewer than 40 segments), transparent worms live their entire lives swimming in open-ocean waters. All are voracious carnivores.

Family Typhloscolecidae.   *Typhloscolex.* Most of the 15 species are free-swimming in the deep sea.

Superfamily Glyceracea.   Three families, all marine.

Family Glyceridae.   *Glycera*—bloodworms. All species use the muscular proboscis to burrow forcefully through sediment. Glycerids are notable carnivores, subduing prey with potent toxins discharged through oral fangs. Morphological alterations occurring during sexual maturation (epitoke formation) include degeneration of the proboscis, gut, and body wall musculature and elongation of the parapodia and setae. Fertilization is external, with gametes departing through the mouth. Adults die after mating. *Glycera dibranchiata* is a commercially important cold-water bait worm, and one that can inflict a nasty bite on its collector. Like some other polychaetes, glycerides possess blind-ended protonephridia rather than metanephridia.

Superfamily Nereididacea.   Six families.

Family Hesionidae.   *Hesionides, Leocrates, Podarke.* Most hesionids are surface-dwelling scavengers, although some species are interstitial, living between sand grains. Many species are commensal with other invertebrates, including echinoderms and other polychaetes. One species has recently been reported from freshwater sandy beaches, but all other species are marine.

Family Antonbruuniidae.   This family contains a single species (*Antonbruunia viridis*) restricted to the waters off Mozambique. These worms live exclusively within the mantle cavities of one bivalve species (*Lucina fosteri*), with each host housing only one male worm and one female worm.

Family Syllidae.   *Syllis, Odontosyllis, Exogone, Autolytus.* Some species are interstitial, and all are active carnivores. One species (*Calamyzas amphictenicola*) lives ectoparasitically on certain other polychaetes. Syllids show a variety of complex reproductive adaptations. In some species, the entire animal becomes an epitoke, whereas in others, epitokes bud off posteriorly as stolons. The epitokes often swarm and may do so at marvelously precise times in the lunar cycle.

Family Nereididae.   *Nereis, Platynereis.* Most species live in tubes or burrows, while some are commensal with hermit crabs or certain other polychaetes. A few species live in freshwater. The marine sandworm *Nereis virens* is an economically important bait worm along the northern New England coast. Some species in this family form extremely modified epitokes, while others do not form epitokes at all. Animals often die after spawning.

Superfamily Nephtyidacea.   Two families.

Family Nephtyidae.   *Nephtys.* Most species are shallow-water, marine carnivores that use a muscular proboscis to burrow through sediment. The group contains some interstitial species, some deep-water species, and some freshwater species. Members of this family are unlike other polychaetes and most other animals in that the striated muscle fibers commonly contain high concentrations of intracellular calcium phosphate granules; the granules possibly strengthen the muscle fibers and aid the worms in their vigorous burrowing activities. Like some other polychaetes, nephtyids possess blind-ended protonephridia rather than metanephridia.

Superfamily Aphroditacea.   All members are marine and have characteristic dorsal scales (elytra). Six families.

Family Aphroditidae.   *Aphrodita*—the sea mouse. The common name refers to the often thick, hair-like dorsal surface, made of numerous fine setae. These bottom-dwelling polychaetes spend most of their time partly buried in fine muds, using thick, tubular, highly muscular parapodia to walk over the substrate. The largest individuals are about 22 cm long, but none have more than 60 segments.

Family Polynoidae.   *Harmothoe, Hermenia, Lepidonotus.* These are mostly shallow-water marine carnivores, although some species are commensal with other invertebrates. This is a large family, containing over 600 species. The largest individuals, which live in the Antarctic, grow as long as 19 cm.

Family Pholoidae.   *Pholoe.* These widely distributed worms live in muddy sediments or crawl under rocks.

Family Sigalionidae.   *Sigalion.* Most species are active, burrowing carnivores, composed of up to 300 segments.

Family Polyodontidae.   *Polyodontes*—the sea wolf. These tube-dwelling, fierce predators can become over 0.5 m long.

Family Pisionidae.   *Pisionella.* These small, carnivorous worms are highly modified for interstitial life between sand grains.

Order Spintherida.
Spinther. The nine species in this order all live as ectoparasites on sponges.

Order Eunicida.
Nine families.

Family Onuphidae.   *Diopatra, Kinbergonuphis, Onuphis.* These polychaetes live in characteristic parchment-like tubes or, less commonly, in burrows. They are predators or scavengers and are common over a wide depth range. Some species "farm" algae growing on their tubes. The largest individuals may reach 200 cm in length.

Family Eunicidae.   *Eunice, Palola, Marphysa.* This family contains some of the largest of all polychaetes, with the members of some species exceeding 2 m in length and possessing more than 600 segments. All are predators, mostly in shallow waters, and most live in crevices and burrows in hard substrates. The reproductive behavior of palolo worms has been documented especially well; epitokes break free from the parent and swarm at very predictable times in the lunar cycle.

Family Lumbrinereidae.   *Lumbrinereis, Lumbrinerides.* These marine, free-living annelids resemble earthworms in appearance. Most feeding types, including deposit feeders and carnivores, are represented within the group.

Family Arabellidae.   *Arabella.* All species are free-living and predaceous as adults, but some are parasitic in other invertebrates as juveniles. Arabellids resemble earthworms; parapodia are inconspicuous and anterior appendages are absent, as adaptations to burrowing.

Family Histriobdellidae.   *Histriobdella.* This group of small (usually less than 2 mm long) worms contains only six known species. All are symbiotic with certain crustaceans (including lobsters and crayfish) and highly modified for this association: The anterior and posterior appendages form flattened, adhesive discs, with which the worms walk about and attach within the host's gill chamber. Setae are absent. The worms feed on bacteria and other microorganisms coating the walls of the crustacean gill chamber. The group includes both marine and freshwater species.

Order Spionida.
Five families.

Family Spionidae.   *Polydora, Scolelepis, Spio, Streblospio.* Most species burrow in sediments, but species in the genus *Polydora* burrow into calcareous substrates, such as snail and bivalve shells. Some species secrete their own tubes. The group includes both deposit-feeding and suspension-feeding species, and members of some species can alternate between the two feeding modes as conditions warrant. Over 300 species have been described, including a few freshwater species.

Order Chaetopterida.
One family (Chaetopteridae). *Chaetopterus, Phyllochaetopterus.* These sedentary polychaetes are entirely marine and live in secreted, leathery tubes. The body is always divided into three distinct tagmata. The worms capture suspended food particles with mucous nets, with most species drawing water through the tube and across the net by ciliary beating. Individuals of *Chaetopterus variopedatus* pump water through their tubes by the synchronized waving of three highly modified central notopodia. This species also differs from other family members in producing a remarkably bioluminescent mucus when disturbed.

Order Magelonida.
One family (Magelonidae). *Magelona.* These worms are all detrital feeders and are common in shallow coastal waters. Among the annelids, only worms in this family possess the blood pigment hemerythrin.

Order Psammodrilida.
Psammodrilus. One family (Psammodrilidae). Both species in this small order are marine and highly modified for interstitial life. They are small (less than about 9 mm) and narrow, and are streamlined by having very reduced parapodia. The animals are widespread in cooler waters of the Northern Hemisphere, feeding mostly on benthic diatoms.

Order Cirratulida.
Three families.

Family Cirratulidae.　*Cirratulus.* Most species form burrows in mud, collecting sediment with thin tentacles that creep actively over the substrate surface. Numerous filamentous gills project anteriorly, lying on the substrate. Species in the genus *Dodecaceria* instead burrow into bivalve mollusc shells and other calcareous substrates. Reproductive patterns among the members of this family are complex and include asexual reproduction by fragmentation, with the subsequent regeneration of missing parts.

Order Flabelligerida.
Three families.

Family Flabelligeridae.　*Flabelligera.* These deposit-feeding marine worms live mostly under stones and in shallow burrows, collecting

sediment with their palps and tentacles. One species (*F. commensalis*) lives exclusively among the spines of sea urchins, feasting on their fecal products.

Order Opheliida.
Two families.

Family Opheliidae.　*Armandia, Ophelia.* All members are burrowing deposit feeders, typically restricted to sediments of specific particle sizes. Most species can swim for short periods by undulating the entire body, and some become completely free-swimming as epitokes.

Order Capitellida.
Three families.

Family Capitellidae.　*Capitella, Mediomastus, Notomastus.* In general morphology, capitellids resemble thin earthworms. The body is thread-like, with inconspicuous parapodia and no anterior appendages. Some species are opportunistic, having very short generation times (less than 30 days) and being capable of living under environmental conditions not tolerated by most other species; as such, they are excellent indicators of pollution and other environmental disturbances. A few species occur in freshwater, but most are marine or estuarine. All capitellids are deposit feeders, ingesting sediment and digesting the organic components.

Family Maldanidae.　*Clymenella*—bamboo worms. In external appearance, these worms resemble a stick of bamboo. They live head down in the sediment, in vertical tubes of their own manufacture.

Family Arenicolidae.　*Arenicola, Abarenicola*—lugworms. Lugworms are sluggish, deposit-feeding worms common in intertidal areas of estuaries and bays. They live head down in their burrows, defecating conspicuous, coiled fecal castings on the sand at the burrow opening. They commonly deposit their eggs in enormous gelatinous cylinders that sometimes approach 2 m in length. The worms themselves may grow to about 1 m long. The family is fairly small, containing only about 30 described species, but representatives are widespread.

Order Oweniida.
One family (Oweniidae). *Owenia.* These worms live in tubes in sand or mud bottoms. Individuals of most species have ciliated tentacles for suspension feeding, while members of some species subsist instead on ingested sediment.

Order Terebellida.
Six families.

Family Amphictenidae (= Pectinariidae). *Cistena* (= *Pectinaria, Cistenides*). These worms are active deposit feeders, living head down in conical tubes open at both ends. The tubes are commonly constructed of sand grains, selected and cemented in place with remarkable precision.

Family Sabellariidae. *Sabellaria, Phragmatopoma.* These suspension-feeding worms typically live in tubes of cemented sand and shell bits, often in massive, reef-forming aggregations; these reefs can extend along hundreds of kilometers of coastline. The body is divided into four distinct tagmata.

Family Ampharetidae. *Melinna.* This group of some 230 deposit-feeding marine polychaetes is particularly common in muddy substrates in shallow water. The animals live protected within tubes, collecting sediment with thin, active, feeding tentacles.

Family Terebellidae. *Amphitrite, Eupolymnia, Polycirrus.* Most species in this family are tube dwellers, but the structure and composition of the tubes vary widely among species. All are specialized deposit feeders. Nearly 400 species have been described, some of them reaching nearly 0.33 m in length; the tentacles of such large individuals can extend more than a meter from the tube or burrow.

Order Sabellida.
The species in this order are known as feather-duster worms or fanworms. Four families.

Family Sabellidae. *Sabella, Schizobranchia, Myxicola, Eudistylia, Fabricia, Manayunkia.* The 300 fanworm species in this group are mostly tube-dwelling suspension feeders. The worm's posterior end is at the bottom of the blind-ended tube or burrow; ciliated tracts carry fecal wastes upward for expulsion. Many species produce a freestanding tube elevated above the substrate (Fig. 15.9a). A number of other species burrow into coral, rock, or other calcareous substrates, and others burrow into sediment; in most such cases, the worms secrete mucous tubes within the burrows. In addition to reproducing sexually, many species routinely reproduce asexually by fission.

Family Caobangidae. *Caobangia.* The half-dozen or so species in this small family burrow into the shells of freshwater gastropods and bivalves in Southeast Asia. As an adaptation to living in blind-ended burrows in the shells, the digestive tract has become U-shaped, with the anus opening anteriorly. The worms are hermaphroditic suspension feeders, with internal fertilization. Despite their freshwater habitat, they release a swimming larva, which eventually metamorphoses on a suitable shell.

Family Serpulidae. *Hydroides, Serpula; Spirobranchus*—the tropical "Christmas-tree worm." All species live in permanent, calcareous tubes, usually attached to rocks, shells, ship bottoms, or other solid substrates. Elaborate, ciliated filaments are projected from the tube opening for suspension feeding; these are retracted rapidly if danger threatens. Although most of the 350 serpulid species are marine, at least one (*Ficopomatus* [= *Mercierella*] *enigmaticus*) occurs in freshwater.

Family Spirorbidae. *Spirorbis.* All 170 species live in permanent calcareous tubes cemented to algae, shells, rocks, and other solid objects. The tubes are typically small (a few millimeters or less in diameter) and coiled. The short-lived larvae have been widely used in studies of habitat selection. All species are marine.

Order Protodrilida.
Protodrilus, Saccocirrus. A small group (fewer than 50 species) of common interstitial polychaetes. Many species lack parapodia, and a few live in freshwater; most are marine or estuarine. Two families.

Order Myzostomida.
Myzostomum. The 120 or so species in this order are exclusively parasitic in or on sea stars (asteroids), brittle stars (ophiuroids), and especially, sea lilies and feather stars (crinoids). Seven families.

Archiannelids.
Troglochaetus (freshwater), *Polygordius, Dinophilus, Protodrillus.* Formerly placed in a separate class of mostly marine annelids, these segmented worms are now considered to be highly modified polychaetes and are distributed among several polychaete families (including the Protodrilida) accordingly. Archiannelids are mostly small and highly modified for life between sand grains. Individuals of many species retain certain larval characteristics, including external cilary bands.

Class Oligochaeta.
The approximately 3,500 species are distributed among three orders.

Order Lumbriculida.
One family (Lumbriculidae). *Lumbriculus.* These worms live in freshwater or are semiterrestrial.

Order Haplotaxida.

Suborder Tubificina.
Six families.

Family Tubificidae.
Limnodrilus, Clitellio, Tubifex—sludge-worms. This group contains numerous marine species as well as many freshwater species; recently, some species have even been found living in the deep sea. Members of this family typically thrive in heavily polluted water, as the common name implies. Individual worms are thread-like and rarely longer than a few centimeters. They live head down in sedimentary tubes, waving their posteriors about in the water, presumably to facilitate gas exchange. Asexual reproduction by fission is common.

Family Naididae.
Branchiodrilus, Chaetogaster, Dero, Nais, Pristina. The members of this widespread family often resemble some polychaetes in possessing eyes and gills and in being able to swim. Asexual reproduction by fisson is extremely common and may be the primary or sole form of reproduction in some species. Most of the approximately 100 species are free-living, but species of *Chaetogaster* are either carnivorous or symbiotic in the mantle cavity or tissues of some freshwater gastropods and bivalves. A few species live in estuaries or in the ocean.

Suborder Lumbricina.
This large group of over 3,000 species includes the earthworms and their freshwater counterparts. About 14 families.

Family Lumbricidae.
Lumbricus. This group of some 300 species includes the most widespread and familiar earthworms.

Family Ailoscolecidae.
Komarekiona. These earthworms are especially common in North Carolina, Tennessee, and Indiana.

Family Glossoscolecidae.
Pontoscolex. These are the most commonly encountered earthworms in South America and in the Caribbean. Although found mostly in forests, one species lives primarily in coastal beach sand. Members of this family can reach 2 m in length. These species are apparently eaten by some South American forest peoples.

Family Megascolecidae.
This family of earthworms holds over 1,000 species, mostly in Asia and Australia. The family includes the largest known annelids, members of *Megascolides australis,* which may grow to lengths exceeding 3 m; the diameters are correspondingly great—about 3 cm or more. These worms, listed as one of the world's dozen most endangered animal species, construct permanent burrows as deep as 2 m underground.

Family Eudrilidae.
Eudrilus. This family of some 500 African earthworm species includes the most structurally complex of all oligochaetes.

Class Hirudinea.
The approximately 500 species are distributed between two orders.

Order Rhynchobdellae.
These species lack jaws, obtaining the host's blood through a rigid, muscular proboscis instead. All are aquatic parasites or predators in both freshwater and marine habitats. They feed on hosts as diverse as elephants and small invertebrates. Three families.

Family Glossiphoniidae.
Glossiphonia, Placobdella. This group includes parasites and predators, all restricted to freshwater. The bodies are broad and leaf shaped.

Family Piscicolidae.
Calliobdella, Myzobdella, Piscicola, Pontobdella. Most marine species are included in this family, although some freshwater species are included as well. All species are parasitic on fish or other aquatic vertebrates.

Family Ozobranchidae.
Ozobranchus. This family includes both marine and freshwater species, which are mostly parasitic on turtles and crocodiles.

Order Arhynchobdellae.
Nine families.

Family Hirudinidae.
Macrobdella; Hirudo medicinalis—the medicinal leech. The more than 80 species in this group are mostly freshwater bloodsuckers with three well-developed, toothed jaws. The largest known leeches (*Haemopis* spp.) occur in this family, reaching lengths of 45 cm.

Family Haemadipsidae.
Haemadipsa. These so-called "land leeches" are all terrestrial bloodsuckers, typically living in leaf litter or in bushes and trees. Most species are restricted to Australia and the Orient. Like hirudinids, these leeches possess toothed jaws.

Family Erpobdellidae.   *Erpobdella.* These "worm leeches" are cylindrically shaped, lack jaws, and for the most part, lack teeth. All are predators, living in terrestrial and freshwater habitats worldwide.

Class Branchiobdellida.

Cambarincola, Stephanodrilus, Xironodrilus. The 130 or so species in this group live primarily as commensals or parasites in the gill chambers of freshwater crustaceans. Morphologically, these worms are intermediate between the true leeches and the oligochaetes; they attach to their host by means of anterior and posterior suckers. A single freshwater crayfish may be infected with several hundred worms simultaneously. Many workers consider branchiobdellids to be a specialized family of the Oligochaeta.

Some General References About the Annelids

Clark, R. B. 1964. *Dynamics in Metazoan Evolution: The Origin of the Coelom and Segments.* London: Oxford Univ. Press.

Dales, R. P. 1967. *Annelids,* 2d ed. London: Hutchinson University Library.

Fauchald, K., and P. A. Jumars. 1979. The diet of worms: A study of polychaete feeding guilds. *Oceanogr. Marine Biol. Ann. Rev.* 17:193–284.

Harrison, F. W., and S. L. Gardiner, eds. 1992. *Microscopic Anatomy of Invertebrates,* vol. 7. New York: Wiley-Liss.

Parker, S. P., ed. 1982. *Classification and Synopsis of Living Organisms,* vol. 2. New York: McGraw-Hill, 1–43.

Thorpe, J. H., and A. P. Covich, eds. 1991. *Ecology and Classification of North American Freshwater Invertebrates.* New York: Academic Press, 285–399.

16

Echiurans and Sipunculans—Likely Annelid Relatives

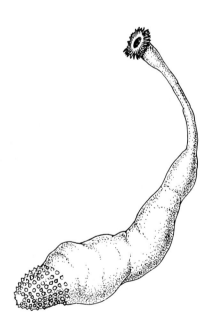

Introduction and General Characteristics

Members of both phyla considered in this chapter—the Echiura and the Sipuncula—are entirely marine. All species in both groups are **vermiform** (worm shaped), sedentary, and found primarily in shallow waters, although some species in both phyla occur at depths of 7,000 m to 10,000 m. Neither echiurans nor sipunculans secrete tubes, but both may form burrows. Most species of both phyla are deposit/detritus feeders. The body wall of both groups is annelid-like (i.e., composed, in sequence, of an outer cuticle, an epidermis, a layer of circular muscles, a layer of longitudinal muscles, and a peritoneum lining the coelom), as are the nervous system, excretory system, mode of gamete formation, and embryology. Echiurans and sipunculans depart from the annelid plan significantly, however, in their lack of metameric segmentation.

Phylum Echiura

Members of the phylum Echiura (ek-ē-your´-ah) live in sandy or muddy burrows or, more rarely, in rock crevices. About 140 echiuran species have been described, mostly from shallow water. The length of the cylindrical, sausage-like body varies considerably among species, from several millimeters to more than 8 cm. Perhaps the echiurans' most conspicuous feature is an anterior cephalic projection commonly called the **proboscis** (Fig. 16.1). This organ contains the brain and may well be homologous with the prostomium of annelids; that is, the annelid prostomium and the echiuran proboscis may have had the same evolutionary origin. The proboscis is muscular and quite mobile—in some species, it can be extended more than 25 times the length of the body proper (200 cm in an 8-cm individual).

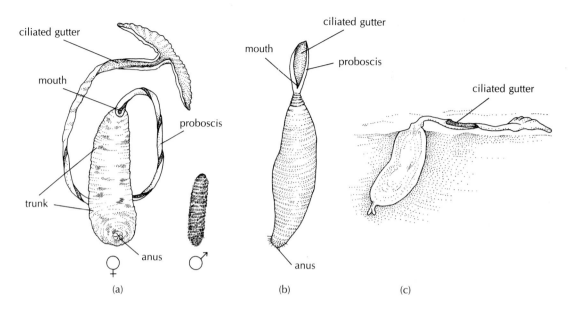

Figure 16.1

Representative echiurans. (a) *Bonellia viridis* (female). The trunk is about 2.5 cm in length, and the proboscis may exceed 1 m when fully extended. Note the ciliated gutter of the proboscis, leading to the mouth. The nondescript male has a total body length of only a few millimeters. (b) *Echiurus echiurus*, in ventral view. Males and females are similar in shape. The trunk length may reach 15 cm, while the proboscis extends about 2 cm to 4 cm. Both species live in soft sediments, in burrows. (c) Typical echiuran in feeding position. *(b) After Harmer and Shipley. After Zenkevitch; modified from Grasse, Traité de Zoologie, vol. 5.*

The proboscis serves as the organ of food collection in all echiuran species and is ciliated on the ventral surface. Although the length of the proboscis can be altered greatly by its owner, the proboscis can never be retracted into the body. The edges of the proboscis curl to form a gutter; mud and detritus, trapped by mucous secretions of the proboscis, are moved by cilia posteriorly toward the mouth, which is located at the base of the proboscis near the junction of the proboscis and the body proper (the **trunk**). The digestive tract is contained within the trunk and is very long and convoluted, with the anus opening posteriorly.

Although most echiurans are deposit feeders, all members of the genus *Urechis* (Fig. 16.2) are suspension feeders, straining out food particles as water passes through a mucous net that the animals secrete across the lumen of their burrows. Water flow through the U-shaped burrow characteristic of this genus is maintained by peristaltic waves of muscular contraction moving down the body (Fig. 16.3).[1] The adjoining sediments are often rich in hydrogen sulfide, a metabolic poison. The coelomic fluid of *U. caupo* has recently been found to contain millimolar (mM) concentrations of an iron-containing compound that oxidizes this sulfide, thereby detoxifying it.

Aside from the **papillae** (small external bumps) found in some species, the only conspicuous projections from the echiuran trunk are a single pair of large, chitinous, annelid-like setae located just posterior to the proboscis. Some echiuran species bear additional setae in the form of a ring just anterior to the anus. The setae probably aid in burrowing.

The trunk contains a single, uninterrupted, large coelomic space that extends into the proboscis (Fig. 16.4). There is no metameric segmentation of the coelomic space, nor of the ventral nerve cord that runs through it. In addition to containing the digestive system, the coelomic space houses from one to many hundreds of annelid-like metanephridia; the members of most species possess one to five pairs of nephridia. Coelomic fluid is presumably drawn into each nephrostome by ciliary action, and a final urine is excreted through nephridiopores opening to the surrounding seawater.

The one additional organ within the echiuran coelom is also believed to serve an excretory role. A pair of **anal sacs** outpocket from the rectum (Fig. 16.4). The sacs are muscular, and their surface is dotted with many thousands of ciliated funnels resembling nephrostomes. Like metanephridia, these funnels collect coelomic fluid, but there is no indication that the collected fluid is then modified by either secretion or resorption. Moreover, the fluid is discharged to the outside through the anus, rather than through a nephridiopore. Discharge is accomplished by periodic muscular contractions of the sacs; one-way valves at the base of the funnels prevent fluid from moving back into the coelom.

---

1. See *Topics for Further Discussion and Investigation*, no. 3, at the end of the chapter.

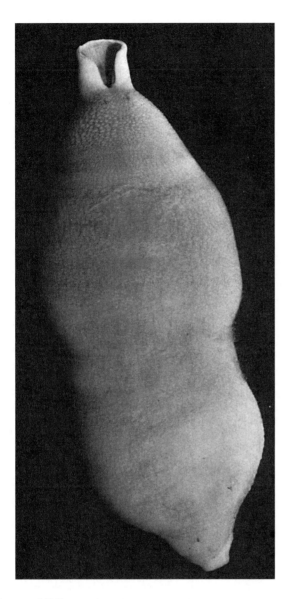

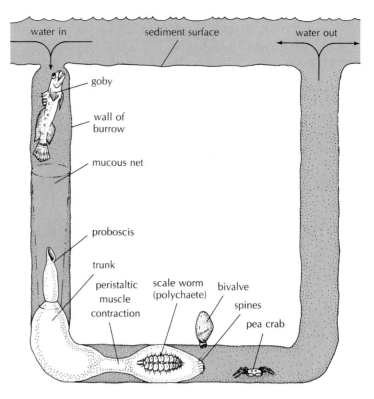

Figure 16.3
Urechis caupo, the innkeeper worm, driving water through its U-shaped burrow by waves of muscle contraction (peristaltic waves). The adults reach approximately 50 cm in length. The burrows of *U. caupo* are commonly shared with various symbionts, most notably a polychaete worm (*Hesperonoe adventor*), several species of pea crab (especially *Pinnixa franciscana* and *P. schmitti*), a small bivalve (*Cryptomya californica*), and the goby fish (*Clevelandia ios*).   *From Pimentel; after Fisher and MacGinitie.*

Figure 16.2
Photograph of *Urechis caupo* (about 8 cm long), removed from its U-shaped burrow. The conspicuous constriction of the body wall represents a peristaltic contraction. These waves of muscle contraction drive water through the burrow for food collection, gas exchange, and waste removal, and circulate the blood, which lies free in the body cavity; blood vessels are lacking in this species. To facilitate gas exchange, water is periodically taken into the thin-walled intestine through the anus and expelled through this same opening.   *Courtesy of C. Bradford Calloway and B.D. Opell.*

The adult echiuran generally possesses a closed circulatory system with heart and blood vessels. Although the circulatory system itself lacks oxygen-carrying blood pigments, hemoglobin is found in the coelomic fluid of some species. No specialized respiratory organs are found; diffusion across the general body surface apparently satisfies requirements for gas exchange in most species.[2]

Echiurans have no distinct gonads. Instead, gametes are produced by the peritoneal lining of the coelom and are released into the coelomic cavity. This is highly reminiscent of the situation encountered among many polychaetes. The gametes of echiurans leave the body by passing through the metanephridia, exiting at the nephridiopores. Gamete formation and release are by essentially identical methods among sipunculans.

Echiuran sex lives appear to be rather unexciting, with only a few exceptions (as discussed in the next paragraph). Gametes are always liberated through nephridiopores into the surrounding seawater, where fertilization occurs. Development is typically protostomous, culminating in the production of a **trochophore larva,** as among the polychaetes (Fig. 16.5). Of particular phylogenetic interest is the observation that segmented coelomic pouches form during embryogenesis. Although this condition does not persist into adulthood, even the transient appearance of metameric segmentation in echiurans is a

---

2. See *Topics for Further Discussion and Investigation,* no. 4.

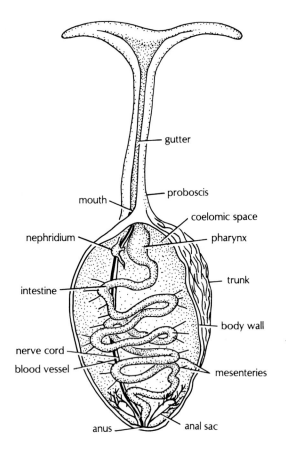

Figure 16.4
Internal anatomy of *Bonellia viridis.* Note the capacious coelomic cavity. *After Harmer and Shipley.*

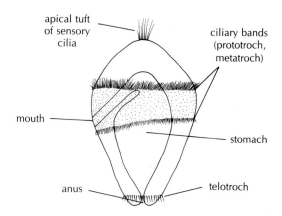

Figure 16.5
A typical echiuran trochophore larva with three conspicuous bands of cilia. *From Gosta Jägersten, Evolution of the Metazoan Life Cycle. Copyright © 1972. Academic Press. Reprinted by permission of Harcourt Brace & Company Limited, London, England.*

Phylum Sipuncula

Like the echiurans, members of the phylum Sipuncula (si-punk´-ū-lah) are all marine. These vermiform creatures are commonly found in burrows in shallow-water muddy or sandy sediments[4] and in empty mollusc shells or polychaete tubes. One very small species is known to inhabit the calcareous shells built by foraminiferans (protozoa). Some other species are found in rock crevices, and a few species can bore into the calcareous substrates of coral reefs. There are about 2.5 times as many sipunculan species as there are echiuran species. Most of the 350 sipunculan species are only a few millimeters long, although some species attain lengths in excess of 1 m.

The sipunculan body consists of a plump, unsegmented trunk and an anterior **introvert.** Unlike the echiuran proboscis, the sipunculan introvert is fully retractable into the body (*intro* = L: within; *verta* = L: turn; i.e., "turn within"), and the mouth opens at the end of the introvert (Fig. 16.6). Contracting the body wall musculature causes the introvert to be thrust out of the body (Fig. 16.6b). Contracting well-developed introvert retractor muscles draws the anterior end of the animal back inside the trunk (Fig. 16.6a). The introvert commonly bears numerous mucus-covered tentacles surrounding or adjacent to the terminal mouth. These tentacles trap particles from the surrounding water or are pressed into the substrate, trapping mud and detritus. The entire introvert may then be withdrawn into the body and the captured particulate material ingested, or the particles may be moved to the mouth by ciliary tracts on the tentacles. Thus, most sipunculans, like echiurans, are deposit feeders. During burrow formation, sipunculans may ingest mud directly.

tantalizing indication of probable annelid relationships. In contrast, in the sipunculan life history, metameric segmentation is not apparent at any stage.

All echiuran species are **dioecious** (one sex per individual), and males and females are generally similar in appearance. However, *Bonellia viridis* and a few close relatives are sexually **dimorphic** (characterized by two body forms) (Fig. 16.1a). Compared with the females, the males of *B. viridis* are much smaller, lack a proboscis and a specialized circulatory system, and possess a degenerate digestive tract. These males are not free-living; rather, they dwell within the body of the female—often in the nephridia, which take on the function of uteri. A single female commonly has 20 males living within her body at any one time, awaiting their opportunity to fertilize eggs. Perhaps the most bizarre aspect of this peculiar life history is that the sex of most embryos is not determined at fertilization. Instead, for the majority of individuals, maleness is only induced following contact of the larva with the proboscis of an adult female. In the absence of such contact, the larva generally becomes a female.[3]

---

3. See *Topics for Further Discussion and Investigation,* no. 1.

4. See *Topics for Further Discussion and Investigation,* no. 5.

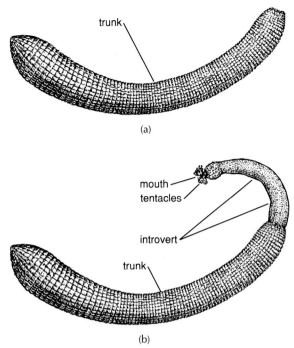

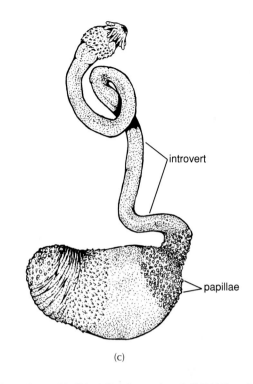

Figure 16.6
Sipunculus nudus, with its introvert withdrawn (a) and
extended (b). (c) *Phascolion* sp., a sipunculan with conspicuous
papillae. The introvert is fully extended.   (c) *From R. I. Smith,* Key to

Marine Invertebrates of the Woods Hole Region. *Copyright © 1964 Marine Biological
Laboratory, Woods Hole, MA. Reprinted by permission.*

Unlike the echiuran digestive tract, that of sipunculans is U-shaped, with the anus opening at about the
midpoint of the trunk. This would seem to be an adaptation for a sedentary existence in burrows having only a
single opening to the outside; solid wastes are discharged
near the mouth of the burrow.

Other than the tentacles borne on the introvert, the
sipunculan body lacks substantial projections; setae and
body appendages are absent.

The sipunculan body contains an enormous
coelomic cavity (Fig. 16.7b) that shows no sign of segmentation or septa at any stage in the life history. Surprisingly, the hollow tentacles of the introvert are not associated with the coelomic cavity. Instead, the tentacles
are connected to a series of sacs, called **contractile vessels** or **compensatory sacs** (Fig. 16.7b). These sacs are attached to the surface of the esophagus; they may be quite
extensive, ramifying throughout the trunk coelom and
forming a type of circulatory system in some species.
Contraction of the sac musculature drives fluid into the
tentacles, bringing about their extension. Muscular retraction of the tentacles drives fluid back into the sacs. A
similar system works in the locomotion of sea urchins
and sea stars and in operating the tentacles of sea cucumbers (phylum Echinodermata), as will be discussed
in Chapter 21.

In contrast to the echiurans, sipunculans lack a circulatory system with heart and blood vessels. However,
like the echiurans, oxygen-binding pigment is found
within cells in the coelomic fluid. In the case of sipunculans, this pigment is not hemoglobin, but rather **hemerythrin,** another iron-containing compound of quite different structure. Hemerythrin is found among only a few
other invertebrate species, including a small number of
polychaetes.

Excretion by sipunculans is accomplished largely by
metanephridia, which resemble those of echiurans and
annelids. However, most sipunculan species bear only a
single pair of these organs (Fig. 16.7b). The nephridial
system is supplemented by a peculiar system of **urns.**
These urns are clusters of cells that arise and detach from
the peritoneal lining of the coelom (Fig. 16.8). Floating
freely in the coelomic fluid, urns collect solid wastes and
eventually deposit these wastes in the body wall of the
animal or exit the animal through the nephridial system.

The reproduction and development of sipunculans
is very similar to that of echiurans. Most species are dioecious, cleavage is spiral and determinate, coelom formation is by schizocoely, and the embryo generally develops
into a free-swimming trochophore larva (Fig. 16.9a). In
some sipunculans, the trochophore develops further to
form a **pelagosphera** larva (Fig. 16.9b), which at one

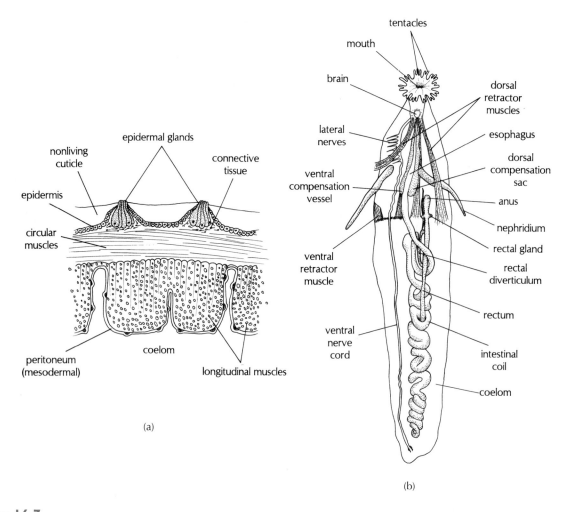

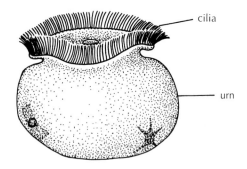

Figure 16.7
(a) Cross section through the body wall of the sipunculan *Phascolosoma gouldi,* illustrating the typical wormlike arrangement of circular and longitudinal musculature. (b) Internal anatomy of *Sipunculus nudus.* Note that the anus terminates anteriorly rather than posteriorly; that is, the gut is U-shaped. Also note the spacious coelomic cavity.   (a) *From Frank Brown,* Selected Invertebrate Types. *Copyright © John Wiley & Sons, Inc., New York. Reprinted by permission of Mrs. Frank Brown. (b) From* Invertebrate Zoology, Third Edition, *by Paul A. Meglitsch. Copyright © 1967, 1972, 1991 by Oxford University Press, Inc. Reprinted by permission.*

Figure 16.8
An urn cell recovered from the coelomic fluid of *Sipunculus nudus.*   *From Hyman; after Selensky.*

time was thought to be a free-swimming adult sipunculan and placed in the now-defunct genus *Pelagosphaera.* These larvae are typically large (several millimeters long) and long-lived: They can disperse over great distances.

Both sipunculans and echiurans bear a number of resemblances to annelids and to each other, as summarized in Table 16.1. The lack of metameric segmentation in adults, however, clearly distinguishes the Sipuncula and Echiura from the Annelida. The brief metameric condition encountered during the embryological development of echiurans suggests that the Echiura diverged from the presumed ancestral annelid stock later than did the Sipuncula.

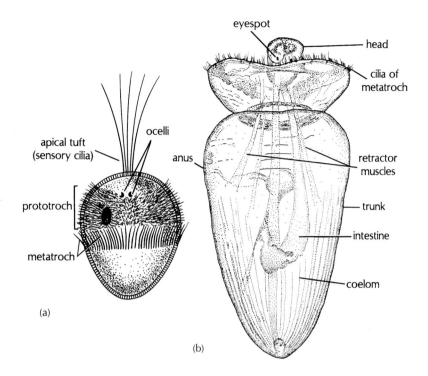

Figure 16.9

(a) Trochophore larva of *Golfingia* sp. (b) Pelagosphera larva of *Sipunculus polymyotus*. The trochophore larva is always a nonfeeding stage in sipunculans, whereas the pelagosphera larva may feed on microscopic algae in the plankton. *(a) From Hyman; after Gerould. (b) Courtesy Dr. Rudolph S. Scheltema.*

Table 16.1. Comparison of Features Encountered among Annelids, Echiurans, and Sipunculans.

Characteristic	Annelida	Echiura	Sipuncula
Body wall musculature	Outer: circular; inner: longitudinal	Outer: circular; inner: longitudinal	Outer: circular; inner: longitudinal
Excretory system	Metanephridia (many pairs)	Metanephridia (one to many pairs) plus anal sacs	Metanephridia (one pair) plus urns
Nervous system	Brain with ventral nerve cord	Brain with ventral nerve cord	Brain with ventral nerve cord
Anteriormost feature	Prostomium	Proboscis (not retractable)	Introvert (fully retractable)
Blood circulatory system	Heart plus dorsal and ventral blood vessels	Heart plus dorsal and ventral blood vessels	None
Oxygen-binding pigments	Hemoglobin, hemerythrin, chlorocruorin	Hemoglobin	Hemerythrin
Setae	Present	Present	None
Metamerism	Present throughout life	Only during development	None
Coelom formation	Schizocoely	Schizocoely	Schizocoely
Cleavage pattern	Spiral, determinate	Spiral, determinate	Spiral, determinate
Larval form	Trochophore	Trochophore	Trochophore, pelagosphera
Gamete production	May arise from peritoneum and mature in coelom; may exit through nephridiopores	Arise from peritoneum; mature in coelom; exit through nephridiopores	Arise from peritoneum; mature in coelom; exit through nephridiopores
Digestive tract	Linear (i.e., mouth and anus at opposite ends of body)	Linear	U-shaped

Other Features of Echiuran and Sipunculan Biology: Nervous and Sensory Systems

Echiurans have no specialized sensory systems, other than sensory cells studding the proboscis. Sensory cells also are found on the surface of the sipunculan introvert. In addition, some sipunculan species possess a pair of presumed chemoreceptors, called **nuchal organs,** on the introvert. They also may have several pairs of light-sensitive ocelli within the brain.

Taxonomic Summary

Phylum Echiura
Phylum Sipuncula

Topics for Further Discussion and Investigation

1. A green pigment called "bonellin" has been isolated from the body wall of the echiuran *Bonellia viridis*. Discuss the evidence implicating bonellin as the factor that causes metamorphosing larvae to become males.

Aguis, L. 1979. Larval settlement in the echiuran worm *Bonellia viridis*: Settlement on both the adult proboscis and body trunk. *Marine Biol.* 53:125.

Jaccarini, V., L. Aguis, P. J. Schembri, and M. Rizzo. 1983. Sex determination and larval sexual interaction in *Bonellia viridis* Rolando (Echiura: Bonellidae). *J. Exp. Marine Biol. Ecol.* 66:25.

2. Compare and contrast burrow formation by sipunculans (or echiurans) and polychaetes.

Trueman, E. R. 1966. The mechanism of burrowing in the polychaete worm, *Arenicola marina* (L.). *Biol. Bull.* 131:369.

Trueman, E. R., and R. L. Foster-Smith. 1976. The mechanism of burrowing of *Sipunculus nudus*. *J. Zool. (London)* 179:373.

Wilson, C. B. 1900. Our North American echiurids. A contribution to the habits and geographic range of the group. *Biol. Bull.* 1:163.

3. Compare and contrast the feeding mechanism of the echiuran *Urechis caupo* with that of the polychaete *Chaetopterus variopedatus*.

Barnes, R. D. 1965. Tube-building and feeding in chaetopterid polychaetes. *Biol. Bull.* 129:217.

MacGinitie, G. E. 1939. The method of feeding of *Chaetopterus*. *Biol. Bull.* 77:115.

MacGinitie, G. E. 1945. The size of the mesh openings in mucous feeding nets of marine animals. *Biol. Bull.* 88:107.

4. The innkeeper worm, *Urechis caupo* (Echiura), supplements gas exchange across its body surface by pumping water in and out of the cloaca. Discuss the mechanism by which this movement of water is accomplished.

Wolcott, T. G. 1981. Inhaling without ribs: The problem of suction in soft-bodied invertebrates. *Biol. Bull.* 160:189.

5. Both echiurans and sipunculans may be exposed to substantial concentrations of the metabolic poison hydrogen sulfide in their burrows. Discuss the physiological adaptations to sulfide exposure exhibited by members of both phyla.

Arp, A. J., B. M. Hansen, and D. Julian. 1992. Burrow environment and coelomic fluid characteristics of the echiuran worm *Urechis caupo* from populations at three sites in northern California. *Marine Biol.* 113:613.

Völkel, S., and M. K. Grieshaber. 1992. Mechanisms of sulfide tolerance in the peanut worm, *Sipunculus nudus* (Sipunculidae) and in the lugworm, *Arenicola marina* (Polychaeta). *J. Comp. Physiol. B* 162:469.

Taxonomic Detail

Phylum Echiura

The approximately 140 species are distributed among three orders. All echiurans are marine.

Order Echiura.
This order contains nearly 95% of all echiuran species. Two families.

Family Bonelliidae. *Bonellia.* The various members of this family occur over a remarkable depth range, from the intertidal zone to the abyss. All species show pronounced sexual dimorphism, with females typically being at least 20 times larger than males. The males live symbiotically on or in the females.

Family Echiuridae.   *Echiuris.* This is the largest echiurid family, holding nearly 60% of all species. Most species occur in shallow water and typically show little or no sexual dimorphism.

Order Xenopneusta.

Urechis. The members of this well-known genus are atypical echiurans in having an unusually short proboscis, an open circulatory system, and an enlarged cloacal region that serves in gas exchange. One family, with only four species.

Order Heteromyota.

Ikeda taenioides. This small order contains a single family, which itself contains a single species found only around Japan. The proboscis can be over 1 m long, more than three times longer than the rest of the animal. Unlike other echiurans, the members of this species have hundreds of unpaired nephridia in the coelomic cavity. The arrangement of the body wall musculature also departs from the standard echiuran plan.

Phylum Sipuncula

The approximately 350 species of sipunculans (peanut worms) are distributed among four families. All species are marine.

Family Golfingiidae.   *Golfingia, Themiste, Phascolion.* This family contains nearly half of all sipunculan species. *Golfingia minuta* is unusual in being the only known hermaphroditic sipunculan. Some species in the genus *Themiste* bore into coral and other calcareous substrates.

Various species in the genus *Phascolion* commonly inhabit structures made by other animals, including the tubes of polychaete worms and the empty shells of dead gastropods.

Family Phascolosomatidae.   *Phascolosoma.* Many of these species bore into rock.

Family Sipunculidae.   *Sipunculus.* Most individuals in this family are large and burrow in sand, ingesting sediment as they travel. Some species can swim short distances by flexing the body rapidly.

Family Aspidosiphonidae.   This group of nearly 60 described species is commonly encountered in coral reef rubble.

Some General References About the Echiurans and Sipunculans

Cuttler, E. B. 1994. *The Sipuncula: Their Systematics, Biology, and Evolution.* Ithaca, N.Y.: Cornell Univ. Press.

Harrison, F. W., and M. E. Rice, eds. 1993. *Microscopic Anatomy of Invertebrates, Vol. 12.* Onychophora, Chilopoda, and Lesser Protostomata. New York: Wiley-Liss.

Hyman, L. H. 1959. *The Invertebrates, Vol. 5. The Smaller Coelomate Groups.* New York: McGraw-Hill.

Parker, S. P., ed. 1982. *Classification and Synopsis of Living Organisms,* vol. 2. New York: McGraw-Hill, 65–66 (echiurans), 67–69 (sipunculans).

Rice, M. E., and M. Todorovic. 1975. *Proceedings of the International Symposium on the Biology of the Sipuncula and Echiura.* Washington, D.C.: National Museum of Natural History, Smithsonian Institution.

17

The Pogonophorans

Introduction and General Characteristics

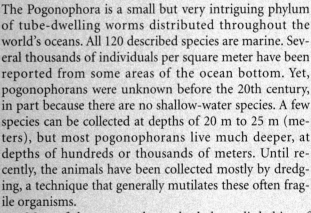

Phylum Pogono · phora
(G: beard bearer)
pō-gō-nof´-or-ah

The Pogonophora is a small but very intriguing phylum of tube-dwelling worms distributed throughout the world's oceans. All 120 described species are marine. Several thousands of individuals per square meter have been reported from some areas of the ocean bottom. Yet, pogonophorans were unknown before the 20th century, in part because there are no shallow-water species. A few species can be collected at depths of 20 m to 25 m (meters), but most pogonophorans live much deeper, at depths of hundreds or thousands of meters. Until recently, the animals have been collected mostly by dredging, a technique that generally mutilates these often fragile organisms.

Most of the pogonophoran body bears little hint of what now seems to be a close evolutionary relationship with annelids. The anteriormost region of the body typically bears a **cephalic lobe;** a "beard," consisting of one to many thousands of ciliated tentacles; and a glandular area that secretes a chitinous tube, within which the animal spends its life (Fig. 17.1). Although pogonophorans are sedentary, they are free to move up and down within their tubes, which always exceed the length of the animal's body. Each tentacle is serviced by two blood vessels, so that the tentacles form the primary gas exchange surface. The animal's anterior section contains a single small coelomic cavity that extends into each of the tentacles.

The longest part of the pogonophoran body is the **trunk** (Fig. 17.2), which contains a pair of uninterrupted coelomic cavities. The body wall of the trunk contains

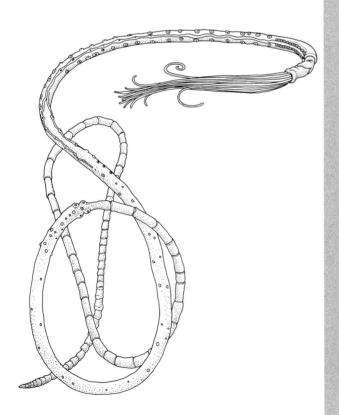

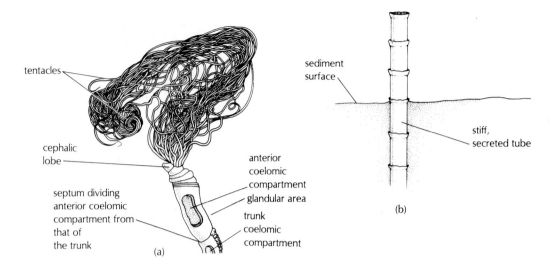

Figure 17.1
(a) Anterior end of a pogonophoran (Polybrachia). (b) A portion of the tube secreted by the same animal. Much of the tube is anchored in soft sediment. *From Hyman; after Ivanov.*

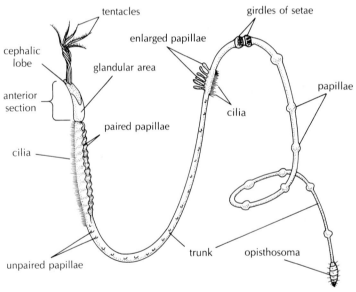

Figure 17.2
Intact pogonophoran removed from its tube. Note the three conspicuous body divisions: cephalic lobe, with tentacles; trunk; and segmented opisthosoma. *From George and Southward, in* Journal of the *Marine Biological Association, 53:403, 1973. Copyright © 1973 Marine Biological Association, U. K. Reprinted by permission.*

both circular and longitudinal muscles. Externally, the trunk is often marked by large numbers of **papillae** (small bumps), two regions of ciliation, and two conspicuous rings of setae about halfway down the body (Fig. 17.2). The trunk is unsegmented, and the body cavity of the trunk is not septate. The major organs found within this coelomic cavity are the gonads and a multilobed structure called the **trophosome** (Fig. 17.3a). The trophosome of all species examined so far contains many closely packed bacteria (Fig. 17.3b), which likely play a

major role in pogonophoran nutrition as discussed later in the chapter (*tropho* = G: nourish; *soma* = G: body).

The probable annelid affinities of the group were made clear only about 30 years ago, when the first intact specimens were collected. At this time, the third and most posterior body region, the **opisthosoma** (*opistho* = G: hind; *soma* = G: body), was discovered. The opisthosoma is conspicuously segmented, and each of the approximately 6 to 25 segments contains a coelomic compartment that is isolated from adjacent compartments

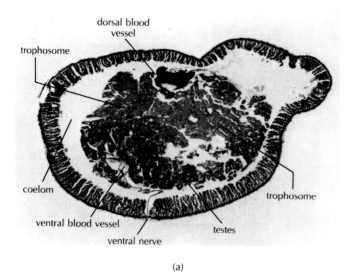

(a)   (b)

Figure 17.3

(a) Transverse section through the trunk of Riftia pachyptila Jones, showing location of the trophosome. In fact, most of the coelomic space is occupied by the bacteria-laden trophosome. (b) Scanning electron micrograph of the trophosome, showing many spherical, symbiotic, sulfide-reducing bacteria. The bacteria are globular, and 3–5 μm wide.   *(a) Courtesy of M. L. Jones, from Jones, Science 213:333, 1981. (b) From C. M. Cavanaugh, Prokaryotic cells in the hydrothermal vent tube worm riftia pachyptila Jones: possible chemoautothrophic symbionts. Science, Vol. 213, July 17, 1981, pgs. 340-41, fig. 1. Copyright 1981 by the AAAS.*

by muscular septa, as in annelids. Moreover, each segment bears annelid-like setae. The opisthosoma is used for digging in sediment by most species, or for anchoring the animal within its chitinous tube. Recent molecular analyses of 18S ribosomal RNA also indicate a close evolutionary relationship between annelids and pogonophorans.[1]

Perhaps the single most fascinating component of the biology studied to date relates to the pogonophoran digestive system: There is none. Adult pogonophorans lack a mouth, an anus, and anything resembling a digestive tract, although a complete digestive system appears for a short time in the development of some species. Biologists have long wondered how adults meet their nutritional requirements. Several possibilities exist, and some evidence supports the likelihood of each. For example, the tentacles at the animal's anterior end bear many surface microvilli and secretory cells. Food particles might be trapped by the tentacles and digested externally. The solubilized nutrients could then be absorbed across the microvillous surface of the tentacles. The evidence for this mode of feeding is strictly morphological, and it currently seems unlikely that particle feeding plays a substantial role in pogonophoran nutrition.

On the other hand, considerable experimental evidence indicates that pogonophorans, like many other soft-bodied invertebrates, can take up dissolved organic matter (DOM) from seawater, even at the low concentrations found naturally. Basically, the experiments consist of demonstrating that radioactively labeled amino acids, carbohydrates, and other organic molecules are accumulated by pogonophorans against remarkable concentration gradients. Some compounds can be taken up from concentrations as low as 1 μM (micromolar, i.e., 1×10^{-6} moles per liter). Uptake is mediated through an active, energy-requiring process. Calculations based on respiration rates and rates of DOM uptake suggest that for some, but not all, species, the uptake of dissolved organic molecules from the surrounding water may be sufficient to meet all of a pogonophoran's metabolic maintenance requirements.

Most pogonophorans—specifically the perviates, members of the class Perviata (Figs. 17.1 and 17.2)—are fairly small, perhaps 6 cm to 36 cm (centimeters) long and typically less than a millimeter wide. These species thus resemble long, thin threads. The ratio of surface area to volume in such animals is very high, increasing the likelihood that the general body surface can play a major role in taking up dissolved nutrients from seawater. The tubes of these perviate species are thin-walled and open at both ends, and some evidence indicates that the tubes are permeable to DOM. Moreover, all perviate pogonophorans are **infaunal;** that is, they live with most of the body (and tube) implanted into soft, muddy substrates, in which dissolved nutrients accumulate to high concentrations, relative to the concentrations found in seawater. DOM may well make a major contribution to the nutritional biology of these pogonophorans.

However, DOM uptake likely plays a less important nutritional role in the more recently discovered pogonophoran group known as the **vestimentiferans.** Fewer than one dozen vestimentiferan species have been described to date, all placed in a separate pogonophoran

1. Lake, J. A. 1990. Origin of the metazoa. *Proc. Nat. Acad. Sci. USA* 87:763.

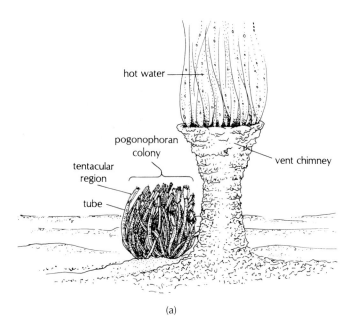

(a)   (b)

Figure 17.4

(a) Animal life thrives in the warm, sulfide-rich water near hydrothermal vents. These pogonophorans (*Riftia pachyptila*) were found at a depth of about 2,500 m by a team of scientists aboard the research submersible Alvin in 1977. The blind-ended tubes of these worms approach 3 m in length and are white, cylindrical, and flexible. (b) Dense cluster of the tube worms *Riftia pachyptila* congregated near a hydrothermal vent in the Galápagos, at a depth of over 2,500 m.   (a) *From Science at Sea: Tales of an Old Ocean by Van Andel. Copyright © 1981 by W. H. Freeman and Company. Used with permission. (b) Courtesy of H. W. Jannasch. From Jannasch,* American Society for Microbiology News *55:413–16.*

class, the Obturata or Vestimentifera. These pogonophorans are up to 2 m long and 25 mm to 40 mm wide. They were first discovered in 1977 by researchers aboard the deep-sea submersible *Alvin,* during exploration of a major ocean-spreading center near the Galápagos Islands, in the Pacific Ocean off the coast of Peru. The ocean-bottom rifts of this area are regions where new crustal material is being generated from deep within the earth. The rifts are characterized by scattered **hydrothermal vents,** small openings in the new crust that emit hot seawater rich in reduced inorganic compounds such as hydrogen sulfide and methane. The hydrothermal vent species *Riftia pachyptila* and *Ridgeia* spp. live upright in blind-ended tubes attached to solid rock, directly in the path of this outflow (Fig. 17.4), bathed in water cooled by mixing to about 10°C to 15°C. These animals live at considerable densities and form a conspicuous component of a thriving undersea community that would seem more likely in a tropical environment than in the ocean depths (Fig. 17.4b).

Similar communities have now been discovered in cold water (only about 2°C to 4°C) at a depth of nearly 3,000 m in the Gulf of Mexico and off the coast of Oregon. These so-called "seep" communities, which include large populations of approximately meter-long pogonophorans, are restricted to well-defined areas where hypersaline sulfide- and methane-containing water seeps up through the sediment.

The surface-area-to-volume ratio of these larger "vent" and "seep" vestimentiferans is considerably less than that of the smaller, thinner perviate species, making it unlikely that the larger pogonophorans could take up DOM rapidly enough to satisfy their more substantial metabolic requirements. Moreover, the thick tube walls surrounding the vestimentiferans probably act as barriers to DOM uptake across much of the body surface. Although DOM uptake has not been discounted in these species, current evidence indicates an even more intriguing mode of obtaining nutrients, one that clearly indicates why these thriving deep-water communities are uniquely associated with waters rich in reduced inorganic substrates such as hydrogen sulfide and methane.

The trophosome of these "hot-vent" and "cold-seep" pogonophorans contains vast numbers of bacteria (Fig. 17.3), at densities up to 10^{10} (10 billion) bacteria per gram (wet weight) of trophosome tissue. Recent biochemical evidence indicates that these bacteria obtain energy by oxidizing the hydrogen sulfide or methane in the surrounding seawater (Fig. 17.5). Since the seawater is outside the animal and the bacteria are inside the animal, the animal must provide the methane or sulfides to the bacteria. The sulfides and methane apparently diffuse into the pogonophoran's blood across the tentacular plume and are then delivered to the trophosome by hemoglobin transport proteins in the circulatory system. The bacteria use the energy they obtain from oxidizing these substrates in the trophosome to produce ATP, which in turn is used to synthesize organic molecules from the carbon in carbon dioxide (CO_2) as in plant photosynthesis:

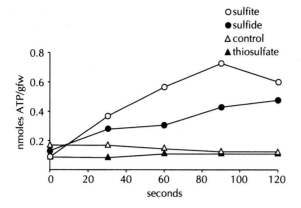

Figure 17.5

The influence of sulfides and related compounds on ATP production by homogenized, bacteria-laden trophosome tissues obtained from the hot-vent pogonophoran *Riftia pachyptila*. Each point represents the mean result from two replicates. ATP concentration was determined by adding known quantities of luciferin and firefly luciferase, which react to produce light only if ATP is present, and measuring the intensity of the light produced. The experiment was conducted at 20°C. Results are expressed as nanno-moles (10^{-9} moles) of ATP produced per gram fresh weight (gfw) of trophosome tissue. Both sulfide and sulfite additions dramatically increased ATP production above control levels within only one or two minutes. The ATP would presumably then be used by the sulfur-oxidizing bacteria to form carbohydrates. *From M. A. Powell and G. N. Somero, "Adaptations to Sulfide by Hydrothermal Vent Animals: Sites and Mechanisms of Detoxification and Metabolism" in Biological Bulletin, 171:274–290, 1986. Copyright © 1986 Biological Bulletin. Reprinted by permission.*

$$CO_2 + H_2S + O_2 \rightarrow [CH_2O]_n + H_2SO_4$$

The bacteria use these carbohydrates for their own growth and reproduction. The growing, internal bacterial population then can be digested by the host pogonophoran or can feed the host indirectly by liberating soluble carbohydrates to host tissues.

In either case, these deep-sea pogonophorans are participating in a novel food chain, one that does not depend on light as an energy source. This food chain begins with **chemosynthesis** rather than photosynthesis: The energy source for fixing carbon dioxide into organic molecules is chemical rather than light. Similar chemosynthetic bacteria have now been found in the tissues of the mud-dwelling perviate pogonophorans, as well as in the tissues of several mollusc species and at least one polychaete annelid species. A symbiotic, chemotrophic mode of nutrition may turn out to be common among animals living in water and sediments high in hydrogen sulfide or methane, regardless of depth or ambient temperature; how the animals themselves withstand exposure to such high concentrations of these toxic chemicals is still being studied (Research Focus Box 17.1).

Methane-oxidizing bacteria also are abundant and active on the exterior surface of vestimentiferan tubes, possibly serving as a major food source for vent gastropods and other grazing invertebrates.

The discovery of vent and seep pogonophorans did more than stimulate research into invertebrate nutri-

tion; it also raised questions about the systematics of the group—about the relationships within the Pogonophora and between pogonophorans and other worms. The vent and seep worms are morphologically dissimilar to other pogonophorans in possessing a conspicuous, long collar called the **vestimentum,** just posterior to the plume of tentacles (the **respiratory,** or **branchial, plume,** Fig. 17.6), and a large, solid, flat-topped anterior structure (the **obturaculum,** Fig. 17.6) that supports the respiratory plume and plugs the opening of the tube when its resident withdraws. The tentacles of these vestimentiferan species are very numerous (thousands or even several hundred thousand per individual) and fused for much or all of their length, forming flattened sheets, or lamellae, encircling the obturaculum (Fig. 17.6b). Moreover, the vestimentiferans lack trunk setae, have a far more extensive trophosome, and have opisthosomal segments with paired coelomic cavities; the perviate pogonophorans possess only one coelomic compartment per segment. Recent comparisons of 28S ribosomal RNA gene sequence data suggest that the evolutionary relationship between perviates and vestimentiferans is not a close one, but there is as yet no consensus as to whether the vestimentiferans merit status as a separate phylum. Despite their substantial anatomical differences as adults, all pogonophorans seem to develop in much the same way initially and look very much alike as juveniles. In particular, juveniles of all species examined show paired opisthosomal coelomic compartments within each segment; the mesodermal partitions between compartments are simply retained to adulthood in vestimentiferans and lost during further development in the perviate species. These observations argue in favor of retaining the perviates and the vestimentiferans in a single phylum. Further studies of coelom formation and mesoderm origination in the two groups may help resolve the controversy.

Other Aspects of Pogonophoran Biology

Reproduction and Development

Most pogonophorans are **dioecious,** with male and female gonads located in the trunks of separate individuals. Fertilization and early development have never been observed directly, so many embryological details, such as how the mesoderm originates, are uncertain. At least some of the coelomic compartments apparently form by schizocoely, linking pogonophorans with the protostomes. Cleavage seems to be a substantial modification of the protostome spiral type, however; it is neither clearly spiral nor clearly radial. The offspring of species studied to date develop for a time within the parental tube, and newly metamorphosed juveniles are sometimes

Research Focus Box 17.1

Living with Sulfides

Powell, M. A., and G. N. Somero. 1986. Adaptations to sulfide by hydrothermal vent animals: Sites and mechanisms of detoxification and metabolism. *Biol. Bull.* 171:274–90.

Much remains to be learned about the symbiotic relationship between sulfur-oxidizing bacteria and their animal hosts. For one thing, how do pogonophorans withstand exposure to sulfide? All animals rely on the sulfide-sensitive enzyme cytochrome-*c* oxidase for aerobic ATP generation. This protein, located in the inner membrane of mitochondria, transfers electrons from cytochrome *c* to oxygen, the final step in the electron transport chain; cytochrome-*c* oxidase is typically poisoned at even very low sulfide levels, levels substantially lower than those found in the water surrounding hydrothermal vents and cold-water seeps. How, then, are pogonophorans able to survive constant exposure to these high concentrations of hydrogen sulfide (H_2S)? Why are their cytochrome-*c* oxidase systems not disabled? Either (1) the pogonophoran cytochrome-*c* oxidase somehow tolerates higher sulfide concentrations than do comparable enzyme systems of animals living in sulfide-free environments, or (2) the sulfide is somehow detoxified during its transport through the pogonophoran to the bacteria living in the trophosome. Research by Powell and Somero (1986) indicates that the pogonophoran survives by detoxifying sulfide and that there are two likely detoxification sites and mechanisms.

To examine the sensitivity of pogonophoran cytochrome-*c* oxidase to sulfides, Powell and Somero collected by submarine specimens of the hot-vent species *Riftia pachyptila* from a depth of 2,600 m and then homogenized the branchial plume of tentacular tissue. This is the tissue most directly exposed to the sulfides in seawater. They then ascertained the ability of pogonophoran cytochrome-*c*oxidase

from the plume to catalyze the transfer of electrons from cytochrome *c* to oxygen in the absence of sulfides; these data served as controls for the experiment.

Different amounts of sulfide were then added to other samples of the same homogenate, and cytochrome-*c* oxidase activity was again measured by monitoring the disappearance of substrate (reduced cytochrome *c*) from the mixture. Reaction temperatures were held constant at 20°C in all experiments, so that any differences in reaction rates could not be due to temperature effects. Cytochrome-*c* oxidase activity in *Riftia pachyptila* is obviously very sensitive to sulfide (Focus Fig. 17.1).

The inhibitory effects of sulfides on cytochrome-*c* oxidase activity were dramatically counteracted by adding pogonophoran blood to the plume homogenates (Focus Fig. 17.2). The blood apparently contains high concentrations of a protein (hemoglobin, as it turns out) that binds tightly to sulfide molecules, so the sulfide is essentially inactivated during transport to the trophosome. As long as blood concentrations of *free* sulfide remain low, aerobic respiration of pogonophoran tissues is not threatened. The problem of how the sulfide becomes unbound in the trophosome was not studied.

Another way to detoxify sulfides is to oxidize them. Such oxidation by the pogonophoran would make the molecules far less useful as an energy source for the bacteria, but it would protect the pogonophoran tissues. Powell and Somero homogenized various pogonophoran tissues and used standard biochemical techniques to assess the ability of the purified homogenates to oxidize sulfur. The highest activity (per

found attached to the tubes of older individuals, so a developmental series of individuals can be obtained by carefully examining the tubes of adults. Perviate species develop as yolky, nonfeeding "larvae" within adult tubes. Detailed anatomical studies on early (less than 0.2 mm long) stages of the hot-vent pogonophorans *Ridgeia* spp. and *Riftia pachyptila* only recently have been reported. The ciliation of these individuals suggests that they develop from annelid-like trochophore larvae. But one of the most striking findings is that, at least until the juvenile is about 3 mm to 4 mm long, it bears a conspicuous, ventral appendage anteriorly, and that appendage bears a terminal mouth. The mouth leads into a ciliated gut,

which terminates in an anus; that is, the early vestimentiferan juvenile has a complete digestive tract. The gutless condition of the older juveniles and adults thus reflects a secondary degradation and loss during later development.

In 1994, field observations of natural spawning by *Riftia pachyptila* were reported for the first time. A small percentage of individuals in a vent population released eggs and sperm into the surrounding seawater, although the gametes did not appear to mix. It is not clear whether the eggs become fertilized externally, or whether females first take in sperm released by neighbors and fertilize the eggs internally before discharging them to the outside.

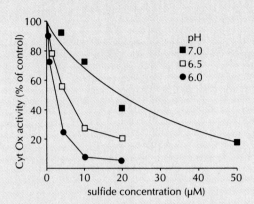

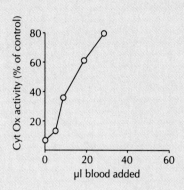

Focus Figure 17.1
The influence of sulfide on cytochrome-c oxidase activity in tissues of the hot-vent pogonophoran Riftia pachyptila. The assays were carried out at 20°C, using tentacular plume samples. Cytochrome-c oxidase activity was determined using a spectrophotometer to monitor the rate of disappearance of substrate (reduced cytochrome c) from the incubation medium. Even low sulfide levels substantially inhibited cytochrome-c oxidase activity, particularly under slightly acidic conditions.

Focus Figure 17.2
The ability of pogonophoran blood to protect mitochondrial cytochrome-c oxidase activity from sulfide poisoning in plume tissue of the vent species Riftia pachyptila. The experiment was conducted as in Focus Figure 17.1, except that different amounts of blood were added to the reaction mixture as shown. Without blood (0 µl [microliters] blood added), sulfides suppressed cytochrome-c oxidase activity dramatically, to less than 10% of its activity in the absence of sulfides. This sulfide inhibition is clearly removed by some component of pogonophoran blood, now known to be the extracellular hemoglobin itself. Adding as little as 20 µl of pogonophoran blood substantially increased the level of cytochrome-c oxidase activity despite the presence of sulfides.

gram of animal tissue) was, of course, found in the trophosome, reflecting sulfide oxidation by the dense bacterial aggregations in that tissue. However, respectable levels of sulfide oxidase activity were also found in the pogonophoran body wall, suggesting that enzymes in the outer body wall oxidize sulfide as it diffuses inward across the general body surface and thereby protect body wall tissues from sulfide poisoning. The lowest sulfide oxidase concentrations were found in plume tissue; this finding is consistent with the

plume → circulatory system → trophosome model of how the bacterial symbionts gain access to unoxidized sulfide. But how are the respiratory activities of the plume protected, since the plume tissues do not seem to detoxify sulfides (at least not by oxidation)? And do the pogonophoran body wall tissues obtain energy from sulfide oxidation? These are among the many questions that remain to be answered.

Focus figures 17.1 and 17.2 from M. A. Powell and G. N. Somero, "Adaptations to Sulfide by Hydrothermal Vent Animals: Sites and Mechanisms of Detoxification and Metabolism" in Biological Bulletin, 171:274–290, 1986. Copyright © 1986 Biological Bulletin. Reprinted by permission.

Although free-living pogonophoran larvae have never been collected in plankton samples, current thinking suggests that at least the vestimentiferans release such larvae, enabling each generation to disperse to new vents and seeps. Consistent with the idea of larval dispersal, colonization of new sites is fairly rapid: One desolate area visited by submarine in 1991 supported a thriving colony of one vestimentiferan species 11 months later and a thriving colony of a second species (*R. pachyptila*) when revisited 21 months after that. Moreover, to reach the size observed during intervening periods, the worms must have elongated their

tubes at remarkably rapid rates: at least 30 cm · yr⁻¹ (centimeters per year) for the first species and 85 cm · yr⁻¹ for the second.

Circulatory System

The pogonophoran blood circulatory system is well developed, consisting of a distinct heart anteriorly and blood vessels. Tentacles at the worm's anterior end serve as the primary gas exchange surface. The blood circulates through the trophosome, supplying chemosynthetic bacteria with sulfides and methane and with the oxygen

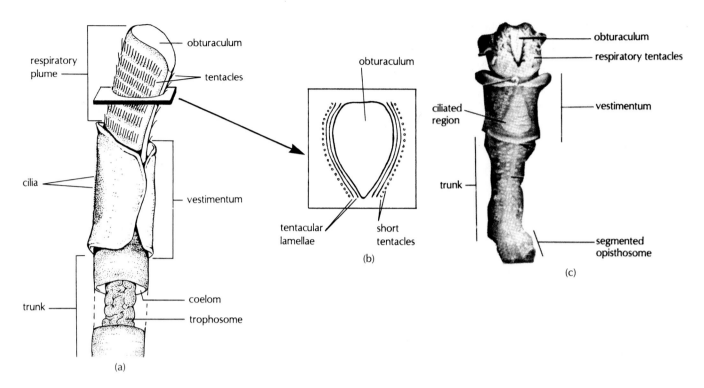

Figure 17.6

(a) Anterior portion of the hot-vent pogonophoran Riftia pachyptila. The tissue below the vestimentum has been cut open to reveal a portion of the extensive trophosome within the animal's trunk. (b) Cross section through the branchial (tentacular) plume of Riftia pachyptila, showing the lamellae formed from the partial fusion of hundreds of individual tentacles.

(c) Scanning electron micrograph of a juvenile tube worm, Ridgeia sp. This juvenile was only a few millimeters long.
(a) *Modified from Southward and Southward, Animal Energetics, Vol. 2, pp 201–28, Academic Press, 1987; and G. N. Somero, Oceanus, 27:69, 1984. (b) From Southward and Southward, 1987. (c) Courtesy of M. L. Jones. From M. L. Jones and S. L. Gardiner, Biological Bulletin 177:254–76, 1989. Permission of Biological Bulletin.*

needed for oxidizing these reduced compounds. The blood and the coelomic fluid both contain two extracellular oxygen-binding hemoglobins. These hemoglobins reversibly bind sulfides as well as oxygen.

Nervous System

Pogonophorans have a distinct brain, which is located in the cephalic lobe of perviate species and ventrally near the anterior margin of the vestimentum of other species. The nerve cord is ventral. One or more giant axons, probably involved in a rapid-withdrawal escape response, have been described in some species.

Respiration and Excretion

Pogonophorans possess no specialized respiratory structures other than the vascularized tentacles. Presumed excretory organs have been described from the anterior ends of all pogonophorans studied to date. The excretory pores open dorsally on the vestimentum of vestimentiferans.

Taxonomic Summary

Phylum Pogonophora
 Class Perviata
 Class Obturata (or Vestimentifera)—the vestimentiferans

Jones, M. L., and S. L. Gardiner. 1989. On the early development of the vestimentiferan tube worm *Ridgeia* sp. and observations on the nervous system and trophosome of *Ridgeia* sp. and *Riftia pachyptila*. *Biol. Bull.* 177:254.

Southward, E. C. 1988. Development of the gut and segmentation of newly settled stages of *Ridgeia* (Vestimentifera): Implications for relationship between Vestimentifera and Pogonophora. *J. Marine Biol. Assoc. U.K.* 68:465.

Topics for Further Discussion and Investigation

1. Discuss the evidence indicating that the uptake of dissolved organic material (DOM) from the surrounding seawater plays a role in pogonophoran nutrition.

Southward, A. J., and E. C. Southward. 1982. The role of dissolved organic matter in the nutrition of deep-sea benthos. *Amer. Zool.* 22:647.

Southward, A. J., and E. C. Southward. 1987. Pogonophora. *Animal Energetics*, vol. 2. Pandian, T. J., and F. J. Vernberg, eds. New York: Academic Press, 201–28.

2. Discuss the evidence indicating that sulfur- and methane-oxidizing bacteria play a major role in meeting the nutritional needs of pogonophorans.

Belkin, S., D. C. Nelson, and H. W. Jannasch. 1986. Symbiotic assimilation of CO_2 in two hydrothermal vent animals, the mussel *Bathymodiolus thermophilus,* and the tube worm, *Riftia pachyptila*. *Biol. Bull.* 170:110.

Cavanaugh, C. M., S. L. Gardiner, M. L. Jones, H. W. Jannasch, and J. B. Waterbury. 1981. Prokaryotic cells in the hydrothermal vent tube worm *Riftia pachyptila* Jones: Possible chemoautotrophic symbionts. *Science* 213:340.

Felbeck, H. 1981. Chemoautotrophic potential of the hydrothermal vent tube worm, *Riftia pachyptila* Jones (Vestimentifera). *Science* 213:336.

Felbeck, H. 1985. CO_2 fixation in the hydrothermal vent tube worm *Riftia pachyptila* (Jones). *Phys. Zool.* 58:272.

Southward, A. J., E. C. Southward, P. R. Dando, G. H. Rau, H. Felbeck, and H. Flügel. 1981. Bacterial symbionts and low $^{13}C/^{12}C$ ratios in tissues of Pogonophora indicate unusual nutrition and metabolism. *Nature (London)* 293:616.

Southward, A. J., E. C. Southward, P. R. Dando, R. L. Barrett, and R. Ling. 1986. Chemoautotrophic function of bacterial symbionts in small pogonophora. *J. Marine Biol. Assoc. U.K.* 66:415.

3. Evidence to date suggests that pogonophoran eggs are uncontaminated by any bacteria when they leave the oviduct. When in development, and by what mechanism, do pogonophorans acquire their endosymbiotic bacteria?

Distel, D. L., D. J. Lane, G. J. Olsen, S. J. Giovannoni, B. Pace, N. R. Pace, D. A. Stahl, and H. Felbeck. 1988. Sulfur-oxidizing bacterial endosymbionts: Analysis of phylogeny and specificity by 16S rRNA sequences. *J. Bacteriol.* 170:2506.

Taxonomic Detail

Phylum Pogonophora[2]

This phylum contains about 120 species.

Class Perviata.

Siboglinum, Oligobrachia, Polybrachia. Small, thin pogonophorans that lack a vestimentum and obturaculum. About 110 species distributed among five families.

Class Obturata (or Vestimentifera).

Riftia, Ridgeia, Lamellibrachia, Tevnia. The vestimentiferans. Large, thick-bodied pogonophorans up to 2 m long, bearing both a vestimentum and obturaculum anteriorly. About 12 species distributed among five families.

Some General References About the Pogonophorans

Gage, J. D., and P. A. Tyler. 1991. *Deep-Sea Biology: A Natural History of Organisms at the Deep-Sea Floor.* New York: Cambridge Univ. Press.

Harrison, F. W., and M. E. Rice, eds. 1993. *Microscopic Anatomy of the Invertebrates, Vol. 12: Onychophora, Chilopoda, and Lesser Protostomata.* New York: Wiley-Liss, 327–69 and 371–460.

Hyman, L. H. 1959. *The Invertebrates, Vol. 5: Smaller Coelomate Groups.* New York: McGraw-Hill.

Ivanov, A. V. 1963. *Pogonophora.* New York: Academic Press.

Morris, S. C., J. D. George, R. Gibson, and H. M. Platt, eds. 1985. *The Origins and Relationships of Lower Invertebrates. Systematics Association, Special Volume 28.* London: Oxford Univ. Press, 327–42.

---

2. Some workers suggest that the Pogonophora should be a class within a new phylum, the Brachiata, or within the phylum Annelida. Either way, this would reduce the Perviata and Obturata to subclasses within the class Pogonophora. Other workers advocate elevating the vestimentiferans to phylum status, as the phylum Vestimentifera. The phylum Pogonophora would then contain only the present members of the Perviata.

18

The Arthropods

Introduction and General Characteristics

Phylum Arthro · poda
(G: jointed foot)
ar-throp´-ō-dah

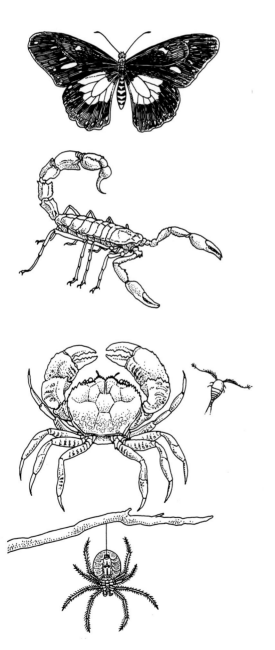

At least 75% of all animal species described to date belong to the phylum Arthropoda, making the arthropod body plan by far the best represented in the animal kingdom. Insects, spiders, scorpions, pseudoscorpions, centipedes, crabs, lobsters, brine shrimp, copepods, and barnacles are all arthropods. Like annelids, arthropods are basically metameric. In most modern members of the phylum, however, the underlying metameric, serial repetition of like segments is masked by the fusion and modification of different regions of the body for highly specialized functions. This specialization of groups of segments, known as **tagmatization,** is also seen in some polychaete annelids, but it reaches its greatest extent in the Arthropoda. Two of the major arthropod groups (Insecta and Crustacea) have three distinct tagmata: head, thorax, and abdomen.

Arthropods have one conspicuous feature in common with members of the Mollusca: Individuals of both groups generally have a hard, external, protective covering—but here the similarity ends. The external coverings in the two phyla are produced by entirely different mechanisms, differ greatly in chemical composition, have distinctly different physical properties, and perform largely different functions. Whereas the molluscan shell functions largely as protection for the soft parts within, the arthropod integument functions additionally as a locomotory skeleton.

The arthropod exoskeleton (Fig. 18.1) is secreted by epidermal cells. Its outermost layer (the **epicuticle**) is

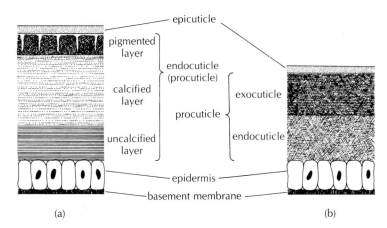

Figure 18.1

The cuticle of (a) crustaceans and (b) insects. The cuticles of both groups of animals are secreted by the underlying epidermis.

From W. D. Russell-Hunter, A Life of Invertebrates. Copyright © 1979 by Macmillan Publishing Company, New York. Reprinted by permission of the author.

generally waxy, being composed of a firm lipoprotein layer underlain by layers of lipid. The cuticle is thus water-impermeable, and the outer body surface cannot serve for gas exchange. On the other hand, the water-proof cuticle makes arthropods more resistant to water loss by dehydration than most other invertebrates. The epicuticle is quite thin, constituting perhaps only 3% of the exoskeleton's total thickness. The bulk of the exoskeleton is made up of the **procuticle,** composed largely of the polysaccharide **chitin** in association with a number of proteins.

Commercial interest in chitin has slowly increased in the past several years because chitin is strong, nonallergenic, and biodegradable. Chitin can, for example, be solubilized and then reformed into fibers, which can then be used in making fabrics and surgical sutures. It might also be used to make a biodegradable capsule that could be implanted in the body, where it would gradually release therapeutic drugs over long time periods. Since chitin can be produced as a clear film, it might someday be used to manufacture a substitute for plastic wrap. Moreover, chitin and its derivatives bind readily to numerous inorganic and organic compounds, including fats, but they are themselves nondigestible by vertebrates. As a food additive, chitin thus might reduce caloric and cholesterol uptake. Its binding abilities also make it a good candidate for removing toxic organic and inorganic compounds from drinking water and during sewage treatment.

Among arthropods, chitin functions in protection, support, and movement, providing a rigid skeletal system. The arthropod procuticle is strengthened by various hardening elements. In the Crustacea, this hardening is partly achieved through the deposition of calcium carbonate in some procuticle layers. Hardening is also accomplished by "tanning" the procuticle's protein com-

ponent. The tanning process, also called **sclerotization,** involves the formation of cross-linkages between protein chains, which contributes to the hardening of the cuticle in all arthropods. In the insects, the cuticle is hardened entirely through tanning. A similar process commonly occurs within other phyla as well: Sclerotization is involved in forming the hinges and byssal threads of bivalves; strengthening the periostracum of the molluscan shell; and hardening the molluscan radula, the setae of polychaetes, the jaws of some annelids, the egg coverings of some platyhelminths, and the trophi of rotifers.

The arthropod procuticle varies in thickness and is not hardened uniformly over the entire body, and therein lies its major functional significance. In many regions of the body, the procuticle is thin and flexible in certain directions, forming joints (Fig. 18.2). Through the presence of appropriate musculature, the arthropods thus have a jointed skeleton that functions in much the same way as does the vertebrate skeleton; pairs of muscles antagonize each other through a system of rigid levers. Some arthropod joints—such as wing joints and those involved in jumping—contain a substance called "animal rubber," or resilin, which stores energy upon compression and releases it efficiently. The development of a jointed, flexible exoskeleton is the essence of arthropod success, and as discussed later in the chapter, has opened up a lifestyle inaccessible to any other invertebrate group: flight.

The coelom can play no major role in the locomotion of an animal encased in a suit of rigid plates, and the arthropod coelom is greatly reduced accordingly. The main body cavity is instead a **hemocoel,** part of the blood circulatory system, as in the Mollusca. Spiders and other arachnids extend their legs by increasing the blood pressure within the hemocoel, but otherwise, fluids and body cavities typically have little to do with arthropod locomotion.

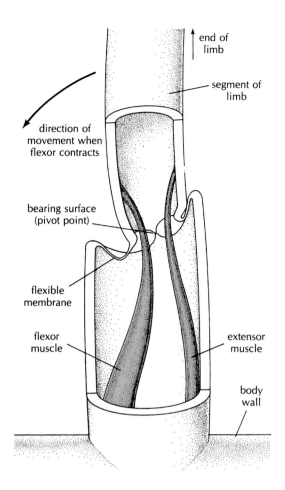

end of
limb

segment of
limb

direction of
movement when
flexor contracts

bearing surface
(pivot point)

flexible
membrane

flexor
muscle

extensor
muscle

body
wall

Figure 18.2
The articulation of a crustacean limb. The procuticle is hardened
everywhere except at the joints, as indicated. *After Russell-Hunter.*

In contrast to shell growth among molluscs, an
arthropod's outer protective covering is not added gradually at a growing edge. Instead, it is secreted over all regions of the body simultaneously. Once the hardening
process is completed, the arthropod is literally encased in
its armor, except where the armor is pierced by sensory
hairs and gland openings. Major regions of the foregut
and hindgut are also lined with cuticle. To increase in body
size, the arthropod must shed the cuticle—including that
lining the gut—grow larger, and then harden a new cuticle
to fit the larger (and, in certain instances, morphologically
altered) body. The old cuticle is partially degraded by enzymatic secretions and is split prior to its removal.

The old cuticle is split by uptake of water and air and
by increased blood pressure, which cause the body to
swell. The process of removing the existing exoskeleton is
called **ecdysis,** from the Greek word meaning "an escape"
or "a slipping out of." In practice, the new cuticle is actually secreted before the old one is shed, which may partially
explain why the arthropod does not become totally non-

functional during the molting process. High blood pressure within the hemocoel may also play a role in maintaining body form and function prior to hardening of
the new cuticle. Of course, potential collapse of the body
during molting is more of a problem in air than in water,
since air is relatively unsupportive. This may be a factor
in explaining why terrestrial arthropods are smaller than
aquatic species. The cuticle does not harden until morphological alterations, if any, and the increase in body
size have taken place. Thus, the time between ecdysis and
hardening of the new cuticle is a period of increased vulnerability to predators; most arthropods seek protective
shelter during ecdysis.

Although increases in size are discontinuous in
arthropods, growth of tissue (**biomass**) is a continuous
process. The number of cells in the epidermis, for example, increases continuously in many arthropods, the additional tissue often becoming folded or pleated until the
old cuticle is shed and the increase in body size can take
place.

The processes of ecdysis and formation of the new
exoskeleton are under both neural and hormonal control. Basically, one gland (the **Y-organ,** located in the
head of crustaceans, for example, or the **prothoracic
glands,** located in the thorax of insects) produces **ecdysteroid hormones** that stimulate molting. Among insects, ecdysteroid production is triggered by the brain's
production of another hormone, which activates the
prothoracic glands. In contrast, among crustaceans,
ecdysteroid production seems to be inhibited between
molts by a second hormone, which is produced by a
neurosecretory complex (the **X-organ**) located in the
eyestalks. When the X-organ ceases effective production
of its hormone, Y-organ activity is no longer inhibited
and ecdysone may be produced. Alternatively, X-organ
secretion may not turn the Y-organ off, but rather, it
may inhibit the action of ecdysone directly. In either
case, ecdysis cannot occur until the X-organ stops producing its inhibitory hormone (Fig. 18.3); surgical removal of crustacean eyestalks results in premature ecdysis. A number of other important arthropod functions
are known to be under neurohormonal control as well,
including regulation of the reproductive cycle, regulation of body fluid osmotic concentration, migration of
light-screening pigments in the eye, and movement of
pigment granules within **chromatophore** cells, leading
to changes in body color.

The arthropod nervous system merits special mention, since it is operationally quite different from
the nervous systems of both vertebrates and other invertebrates. In vertebrates, one muscle fiber is innervated
by a single neuron. The strength of muscle contraction
depends upon the number of fibers contracting, and the
number of fibers contracting in a given muscle

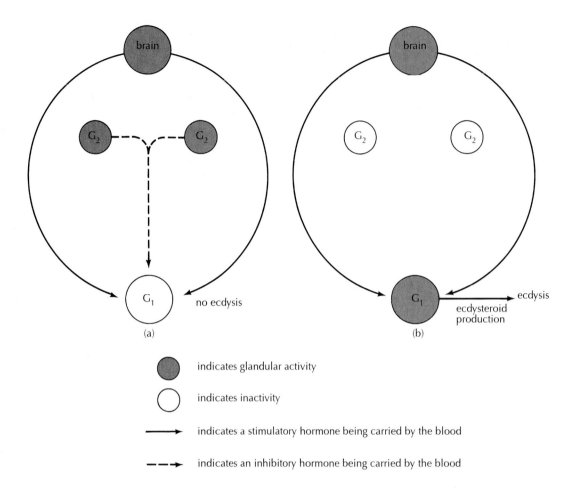

Figure 18.3

Schematic diagram of a likely mechanism regulating ecdysis in crustaceans. Glands associated with molting are indicated by G_1 and G_2. (a) The hormone produced by G_2 inhibits ecdysteroid production by G_1, and no molting occurs. (b) Gland G_2 is inactive; the brain hormone stimulates ecdysteroid production by G_1, and the animal molts.

depends upon the number of axons fired. In arthropod muscle, in contrast, the strength of contraction depends upon the rate at which nerve impulses are delivered to the fibers. Moreover, a single muscle fiber may be innervated by as many as five different types of neurons (Fig. 18.4). The type of contraction (fast but brief versus slow and sustained) depends in part upon the source of the stimulation to the muscle. In addition, some of the neurons are inhibitory; action potentials delivered to such inhibitory neurons can alter the outcome of signals delivered down other axons to the same muscle fiber. Another complication is that arthropods have several physiological and functional types of muscle fiber; that is, the rate of contraction is partly a property of the individual muscle fiber. Fine control of arthropod movement therefore depends upon the types of muscle fiber stimulated and the interaction of several types of neurons terminating on a single muscle fiber. Finally, a single arthropod neuron may innervate a large number of

muscle fibers, so a given muscle may be innervated by very few neurons (two to three, in some cases). In contrast, a given vertebrate muscle may receive innervation from hundreds of millions of neurons.

The musculature of arthropods also differs significantly from that of other invertebrate groups. Arthropod muscle is entirely striated, whereas most other invertebrates possess primarily (or entirely) smooth muscle. The functional ramifications of this difference are considerable, in that striated muscle is far more responsive than smooth muscle tissue; that is, the time required for striated muscle to complete a contraction and subsequent relaxation is generally far less than that required for smooth muscle (Table 18.1). Without striated muscle, arthropods would never have achieved flight.

The arthropod circulatory system also is of interest in that blood leaves the heart through closed vessels in most species, but it enters the heart directly from the hemocoel through perforations, called **ostia,** in the heart

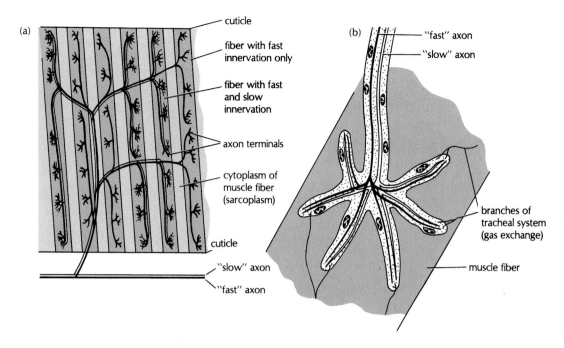

Figure 18.4
(a) Innervation of a typical arthropod muscle. Note that only some of the muscle fibers receive innervation from the "slow" axon. (b) Neuromuscular junction, in detail.   *From G. Hoyle, in*

M. Rockstein, ed., The Physiology of the Insecta, Vol. 4. Copyright © 1974 Academic Press, Inc. Reprinted by permission.

Table 18.1   Contraction Times of Invertebrate Muscle

Source	Contraction Time (Seconds)
Anthozoa	
Sphincter muscle	5.000
Circular muscle	60.000–180.000
Scyphozoa	0.500–1.000
Annelida	
Earthworm circular muscle	0.300–0.500
Bivalvia	
Anterior byssus retractor muscle	1.000
Gastropoda	
Tentacle retractor muscle	2.500
Arthropoda	
Limulus abdominal muscle	0.195
Insect flight muscle	0.025

From various sources.

wall (Fig. 18.5). The circulatory system is thus open, with the oxygenated blood moving through a series of sinuses and finally being drawn back into the heart through the ostia as the heart expands. A "heart with ostia" is one of the diagnostic features of the Arthropoda, although gas exchange is achieved by radically different means in a number of arthropod groups.

Arthropods are also unusual in lacking cilia, even in the larval stages.

In the classification scheme adopted in this text, there are 19 classes, 76 orders, and about 2,400 families contained within the phylum Arthropoda. In contrast, the phylum Mollusca contains only 7 classes, about 30 orders, and slightly more than 500 families. Among the arthropods, arachnids (mites, ticks, and spiders) alone are distributed among nearly 550 families. Even more impressively, the phylum Arthropoda contains at least 10 to 20 times the number of species found in the phylum Mollusca. Despite the large number of major and minor taxonomic groupings found within the Arthropoda, most adult arthropods show only minor deviations from the general body plan. Taxonomic distinctions among arthropods depend largely on the number, distribution, embryological origin, form, and function of appendages.

Since this book focuses on the basic vocabulary and grammar of invertebrate zoology, rather than the diversity of form and function encountered within each group, some of the arthropod classes and subclasses have

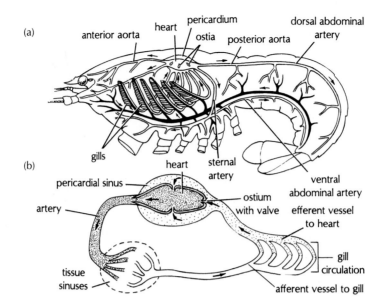

Figure 18.5

(a) Circulatory system of a typical arthropod, demonstrating the heart with ostia. The animal illustrated is a lobster.
(b) Diagrammatic illustration of the blood circulatory pattern. Oxygenated blood is transported through the unshaded vessels to hemocoelic channels, in which gas exchange between blood and tissues occurs. Deoxygenated blood (shaded vessels in [a]) collects in a series of ventral venous sinuses. From here, the blood moves to the gills, is oxygenated, and returns to the pericardial sinus surrounding the heart.   (a) *After Engemann and Hegner; after Gegenbauer.* (b) *From Cleveland P. Hickman, Biology of the Invertebrates, 2d ed. Copyright © 1973 The C. V. Mosby Company, St. Louis, Missouri. Reprinted by permission.*

been omitted from the discussion that follows. Selecting the "essential" arthropod groups is not an easy task. I have been careful to include those groups having representatives that are generally familiar to the average person and those groups having major ecological and/or evolutionary significance. The chosen groups are also those most likely to be encountered in the field and in the literature. They are sufficient to present the major principles and vocabulary of arthropod biology and architecture. The *Taxonomic Detail* section at the end of the chapter gives basic information about all of the groups omitted here.

Three groups (Arachnida, Insecta, and Crustacea) include well in excess of 95% of all arthropod species. The phylogenetic relationships among these three groups, and among these and other arthropod groups, have long been controversial; the classification scheme adopted here is only one of several that have been proposed. The nature of the relationship between the terrestrial arthropods and the other major groups (crustaceans, horseshoe crabs and spiders, and the extinct groups, including trilobites) is especially obscure. Some zoologists have argued that members of the terrestrial classes Chilopoda (centipedes), Diplopoda (millipedes), Symphyla, Pauropoda, and Insecta evolved independently of the Crustacea from entirely different ancestors,

and thus place the terrestrial arthropods within a separate phylum, the Uniramia. Recent analyses of 18S rRNA sequence data, however, along with recent reevaluations of morphological data, support the Arthropoda as a monophyletic grouping. In this edition, the Arthropoda is retained as a *bona fide* phylum, and the Uniramia is given the ranking of subphylum within the phylum Arthropoda. Be warned, however, that evidence against this arrangement is already accumulating: The centipedes, millipedes, insects, and smaller terrestrial groups may not form a natural, monophyletic grouping after all, and the Uniramia as a valid taxonomic category quickly may be approaching extinction.[1]

The origins of the Arthropod phylum itself are unclear. Over the past several decades, evidence has been mustered in support of the following: evolution of arthropods from annelid-like ancestors; evolution of annelids from arthropod-like ancestors; and independent evolution of annelids and arthropods from separate ancestors. The issue is contentious. Although arthropods have an interesting fossil record extending back some 550 million years, the fossils have to date been of little use in resolving these controversies. Some very recent analyses do not support a close link between annelids and arthropods, but I wouldn't want to predict where the matter will stand a few years from now.

---

1. See *Topics for Further Discussion and Investigation*, no. 1, at the end of the chapter.

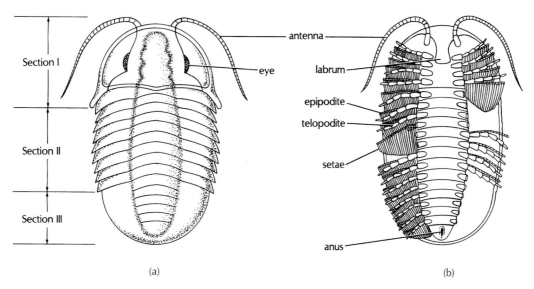

(a)   (b)

Figure 18.6

Trilobite as seen in (a) dorsal view and (b) ventral view. Note the pair of biramous appendages associated with each segment; the appendages are very uniform in structure all along the body. For a given appendage, the epipodite is the branch that bears the long filaments, the function of which remains uncertain. *Beck/Braithwaite, Invertebrate Zoology, Laboratory Workbook, 3/e, © 1968. Reprinted by permission of Prentice Hall, Upper Saddle River, New Jersey.*

Subphylum Trilobitomorpha

Class Trilobita

The class Trilobita has no living representatives, although approximately 4,000 species have been described from the fossil record. The trilobites were especially common approximately 500 million years ago, but they were extinct by about 250 million years ago, at the end of the Paleozoic era. Morphological differences apparent among the fossilized remains imply that a significant ecological diversity existed within the group at one time, varying from burrowing trilobites to walking and swimming forms.

The trilobite body was flattened dorsoventrally and divided into three sections (Fig. 18.6). Sections I and III were covered by a continuous unjointed sheet of exoskeleton (a **carapace**), so the underlying metameric segmentation was not visible when viewed from the dorsal surface. A pair of **compound eyes** was found laterally on the first body section. Each eye was composed of many elongated, light-receiving chambers (**ommatidia**); hence, the term "compound eye."

Adjacent to the mouth, on the ventral surface of one of the segments of body region I, was a chitinous lip, the **labrum**, and each body segment posterior to the mouth bore a pair of two-branched (**biramous**) appendages (Fig. 18.6b). The innermost branch was devoid of long setae and presumably functioned in walking. The seg-

ments of the outer branch bore long filaments, which may have been gill filaments or simply may have been setae used for swimming, filtering food, or digging in loose substrate. This serial repetition of identical biramous appendages along the entire body length was clearly the primitive arthropod condition. More advanced groups show increasing specialization of appendages for specific tasks. This specialization has often involved the reduction or complete loss of one of the two branches of each primitive biramous limb.

Subphylum Chelicerata

Subphylum Chelicerata
(G: claw)
chel-iss-er-ah′-tah

Chelicerates are the only group of arthropods without antennae. Indeed, the first anterior segment bears no appendages at all. The second anterior segment bears a pair of clawed appendages (**chelicerae**), adjacent to the mouth, for grabbing and shredding food. Members of the Chelicerata also lack mandibles, appendages found adjacent to the mouth in many other arthropod groups and used for chewing and grinding food during ingestion.

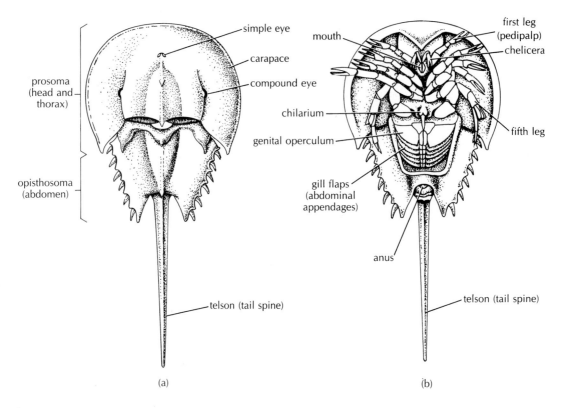

Figure 18.7
Limulus polyphemus, the horseshoe crab. (a) Dorsal view, showing major body divisions. (b) Ventral view, showing appendages. Note that the abdominal appendages are modified as flattened sheets for gas exchange.

Class Merostomata

The class Merostomata (mer-̄o-st̄o-mah´-tah) is composed primarily of extinct species. Only four species are currently living, including the so-called horseshoe crab, *Limulus polyphemus* (Fig 18.7). Despite the common name, these animals are not true crabs; the true crabs are contained in the subphylum Crustacea. Horseshoe crabs burrow through the surface layers of muddy substrate and ingest smaller animals that they come across in the process. All members of the Merostomata are marine. Curiously, living representatives are found only in the waters of eastern North America, Southeast Asia, and Indonesia. Despite the small number of horseshoe crab species and their limited geographical distributions, humans have profited considerably from studying their biology. Much of what is understood about the basic principles of vision is based on studies of horseshoe crab eyes. Moreover, components of horseshoe crab blood are routinely used to test injectable pharmaceutical solutions for contamination by bacterial endotoxins.

Characteristically, the head and thorax of merostomates are fused into a single functional unit, the **prosoma** (*pro* = G: forward; *soma* = G: body) or **cephalothorax,** and are covered with a single unjointed sheet of exoskeleton, the **carapace** (Fig. 18.7a). A pair of compound eyes is present laterally, on the dorsal surface of the prosoma. No other living chelicerate has compound eyes.

The first pair of appendages found ventrally on the prosoma are the **chelicerae** (*cheli* = G: claw) (Fig. 18.7b). These are followed by five pairs of similar appendages, the **walking legs,** all but the last of which bear claws. The first pair of walking legs (i.e., the appendages on the third anterior segment) are called **pedipalps,** but they are morphologically and functionally indistinguishable from the other walking legs. The last pair of walking legs are structurally modified for removing mud during burrowing. A pair of small, hairy appendages (**chilaria**) are found on the last segment of the prosoma; these may be involved in crushing food prior to ingestion.

The abdomen, or **opisthosoma** (*opistho* = G: behind; *soma* = G: body), bears six pairs of appendages. The first pair are modified for reproduction, and the subsequent five pairs are modified to serve as gills. The underside of each gill flap bears approximately 150 leaf-like gas exchange surfaces, called **book gills,** through which blood circulates.

Class Arachnida

Class Arachnida
(arachni = G: a spider)
ah-rak´-nid-ah

Although the earliest members of the class Arachnida were undoubtedly marine, the more than 70,000 living arachnid species so far described are primarily terrestrial (Fig. 18.8). Moreover, the relatively few living aquatic species are clearly derived from terrestrial forms. In the course of their evolution, the arachnids obviously have left the sea. This class includes many familiar but generally unpopular organisms, including spiders, mites, ticks, and scorpions. Nearly half of all arachnid species are spiders, and most of the remaining species—all but about 9,000—are mites and ticks. Spiders are major insect eaters and are increasingly used to control insect populations.

As with members of the Merostomata, the head and thorax of arachnids are fused to form a **prosoma,** which is covered by a carapace (Fig. 18.9). From zero to four pairs of eyes are found on the prosoma, with four pairs being most common. The anteriormost pair of appendages borne by the prosoma are **chelicerae,** which generally tear apart food prior to ingestion. The next pair of appendages are the **pedipalps,** which are variously modified for grabbing, killing, or reproducing and in some species may have a sensory function as well. The basal segment of each pedipalp forms a **maxilla** (endite), which, like the chelicerae, aids in the mechanical preparation of food. The pedipalps are followed by four pairs of **walking legs.** The **abdomen,** or **opisthosoma,** is generally distinct from the prosoma; in some arachnids, including the spiders, the two divisions are connected by a narrow stalk called a **pedicel,** which increases the abdomen's range of movement, facilitating the precise placement of silk threads in web building and prey capture. In a few arachnid groups, notably the ticks and mites, the prosoma and opisthosoma have fused together, and the entire dorsal surface is covered by a single carapace (Fig. 18.8c–e).

Respiration in the more primitive arachnid forms is by means of pairs of modified, internalized book gills, now known as **book lungs.** These flattened respiratory surfaces in the abdomen are connected to the outside by means of openings called **spiracles.** The spiracles of some species can be closed between "breaths," to limit water loss. In many species with small bodies, the spiracles may lead into a system of tubules known as **tracheae.** The tracheae form a system of branching tubules that ultimately terminate directly on the tissues. Gas exchange therefore occurs without use of the blood circulatory system. In some arachnid species, both book lungs and tracheae are present.

Some arachnids—the spiders (order Aranea)—bear up to four pairs of small abdominal appendages called **spinnerets** (Fig. 18.9b–d). These appendages are located ventrally and posteriorly, near the anus, and bear spigots connecting to internal abdominal glands that secrete silk proteins (Fig. 18.9c). The secretions of these silk glands are extruded through the spinnerets to produce silk, which may be used to form safety lines during climbing, egg sacs that protect developing embryos, fine threads for the aerial dispersal of newly emerged young, and webs for trapping prey, building homes, or mating.[2] Humans also have put arachnid silk to good use, notably as crosshairs in optical equipment. (The silk used for over 4,000 years to make silk clothing comes not from arachnids but from the cocoons of the silkworm *Bombyx mori* and some other lepidopterans, the order of insects including the butterflies and moths.)

An individual spider may contain seven or more different silk glands that produce different, biochemically distinct forms of silk for different uses (Fig. 18.9c). Some nonarachnids (centipedes and larval insects) also produce silk, but only one kind.

The mites and ticks are contained in a separate arachnid order, the Acari, a group containing many species of economic and medical importance despite their small physical size. Few mites exceed 1 mm in length, and some species are only 100 µm long. The largest individuals are the ticks, which can reach 5 mm to 6 mm (and considerably more after feeding). The group is tremendously diverse and includes omnivores, carnivores, herbivores, fungivores, and parasites. Like other arachnids, mites and ticks feed exclusively on fluids, which they suck in through a muscular pharynx, and some species produce silk. Acarines occupy diverse habitats, such as wood, moss, ant colonies, bird nests, bat guano, water-filled tree holes, and decomposing vertebrates. Members live in freshwater, marine, and terrestrial habitats, with some species found 30 cm to 50 cm deep in desert sand dunes. Species are parasitic, as larvae or as adults, on a tremendous variety of hosts, including birds, lizards, humans (in hair follicles), mosquitoes, bears, frogs, butterflies, scorpions, spiders, dipterans (flies), beetles, chitons, slugs, grasshoppers, sea urchins, and various crustaceans. Some species are severe agricultural pests—directly or as vectors of plant viruses—on cranberry bushes, tobacco and tea plants, fruits, vegetables, flowers, and grasses (such as wheat, oats, and corn), while others greatly lower the quality of sheep wool. Many ticks transmit to humans a variety of diseases, including Rocky Mountain spotted fever (Fig. 18.8e), Q fever, Lyme disease (Fig. 18.8d), and encephalitis. Finally, many people develop allergies to the feces or exoskeletons of mites living in household dust.

Members of other arachnid orders, including the pseudoscorpions, solfugids, and opiliones (daddy longlegs), are described briefly in the *Taxonomic Detail* at the end of this chapter.

---

2. See *Topics for Further Discussion and Investigation,* no. 5.

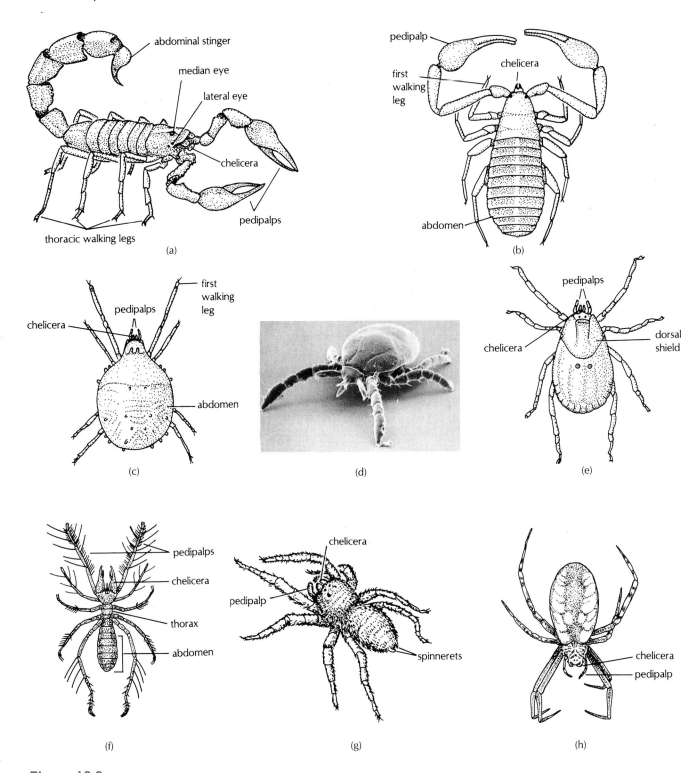

Figure 18.8

Arachnid diversity. (a) Scorpion. (b) Pseudoscorpion. (c) Red spider mite. (d) Scanning electron micrograph of the black-legged tick *Ixodes scapularis* (formerly *I. dammini*), the tick that transmits Lyme disease to humans throughout most of the United States. The vector for Lyme disease in the western U.S. is the western black-legged tick, *I. pacificus*. (e) *Dermacentor andersoni,* the tick that transmits Rocky Mountain spotted fever. (f) A solpugid,

Galeodes dastuguei. The enormous chelicerae are used to tear prey apart. (g) A jumping spider (family Salticidae). (h) A common garden spider, *Argiope* sp., which forms an orb web like those illustrated in Figure 18.9h. Most of the orb-weaving spiders are found in a single family (the Araneidae).   d) *Courtesy of A. Spielman, Dept. of Tropical Public Health, Harvard School of Public Health.*

Class Pycnogonida (= Pantopoda)

Class Pycno · gonida (= Panto · poda)
(G: thick knees [G: all leg])
pik-nō-gon´-id-ah

Zoologist Paul Meglitsch once wrote, "Pycnogonids are queer creatures, with queer habits," and that statement seems a good introduction to this group of some 1,000 chelicerate species. Pycnogonids are known as sea spiders, since all species are marine and bear conspicuously long legs; the legs are typically about 3 times the length of the body and may be nearly 16 times longer than the body in some species. Many species have bodies only a few millimeters long, while a few deep-water and antarctic pycnogonids may exceed 10 cm in body length. Most of the body is prosoma; the abdomen (opisthosoma) is reduced to a short stump (Fig. 18.10).

Unlike the true spiders, sea spiders lack specialized respiratory or excretory systems. They do, however, have a complete digestive system, with a sucking mouth that opens at the tip of an often greatly elongated proboscis. The digestive system extends well into the legs, as do the gonads.

Like arachnids, most pycnogonid species have four pairs of walking legs posterior to the pair of chelicerae and pair of palps (Fig. 18.10), although some species have six to seven pairs of walking legs. In addition, the head bears a posterior pair of **ovigers,** which are used by both sexes to groom the other legs and the trunk and by males to carry the eggs after they are fertilized.

The adults are mostly free-living, although almost comically slow moving. However, the larva, if it leaves the egg mass before completing development, apparently grows as a parasite, particularly of cnidarians (such as hydroid polyps and jellyfish). Many juvenile pycnogonids, and some adults, are parasitic or commensal in or on various marine invertebrates, including gastropods, bivalves, echinoderms, and jellyfish. Most adults are carnivores, feeding on prey that move even more slowly than they do: Bryozoans (moss animals), colonial hydrozoans, and sponges seem to be favored foods.

Subphylum Uniramia

All members of the subphylum Uniramia bear uniramous (unbranched) appendages exclusively, and the head is covered by a heavily sclerotized case called the **head capsule.**

Classes Chilopoda and Diplopoda

Subphylum Uni · ramia
(L: one branch)
ū-nē-rā´-mē-ah

The Chilopoda (chil´-ō-pō´-dah) and Diplopoda (dip´-lō-pō´-dah) contain the centipedes ("one hundred feet") and millipedes ("one thousand feet"), respectively. Because of their many legs (Fig. 18.11), the members of both groups are often referred to together as myriapods ("many feet").

Chilopods are generally fast-moving carnivores, living in soil, in humus, under logs, and occasionally, in people's homes. Although most of the 3,000 species are terrestrial, some are marine. The body is covered by a cuticle, but the cuticle is unwaxed. Moreover, respiration is accomplished by tracheae, but the spiracles cannot be closed. Thus, most centipedes are restricted to moist environments (or moist microenvironments) because of difficulty in restricting water loss. Many species conserve water by being nocturnal—that is, by avoiding the heat of the day and becoming active only at night.

The chilopod head bears a single pair of antennae, a pair of mandibles for chewing, a pair of first and second **maxillae,** and a pair of **maxillipeds.** The maxillipeds are modified for subduing prey (Chilopoda = G: jaw foot); they contain poison glands and resemble fangs. Chilopods often lack eyes; when present, the eyes generally are simple light receptors called **ocelli.**

The chilopod head is followed by 15 or more leg-bearing segments. Some species have **repugnatorial glands** on the ventral surface of each segment or on some of the legs themselves. These glands discourage predation by producing an adhesive ejaculate. A number of species produce silk from silk glands. Although most centipede species are long-legged runners, some species are adapted for burrowing through soil. In these species, the legs are reduced and thrust is generated by exploiting the properties of a hydrostatic skeleton, earthworm-style.

There are approximately 10,000 millipede species, about four times the number of known centipede species. In contrast to the chilopods, the diplopods are primarily slow-moving deposit feeders that plow through soil and decaying organic material. Some carnivorous species also exist. Pairs of segments have become fused in the millipedes, so each new segment (a diplosegment) bears two pairs of legs (Diplopoda = G: double foot) (Fig. 18.11e), as well as two pairs of spiracles

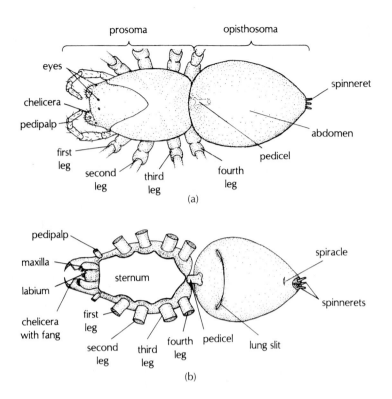

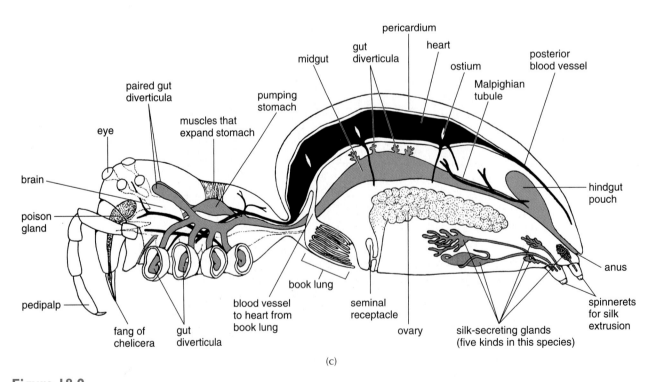

Figure 18.9

(a) Typical spider, in dorsal view. (b) Diagrammatic ventral view of spider. The legs have been removed for clarity. (c) Internal anatomy of a typical female web-spinning spider, seen in lateral view. Each silk-secreting gland secretes a different kind of silk, each specialized for a different function. (d) Ventral view showing posterior portion of spider abdomen; note the three pairs of jointed, flexible spinnerets and the platelike cribellum, which bears additional spiggots. Silk is extruded through the lumen of the hollow bristles. (e) Detail of an arachnid chelicera, consisting of a basal segment and a fang (seen here in lateral view). (f) Scanning

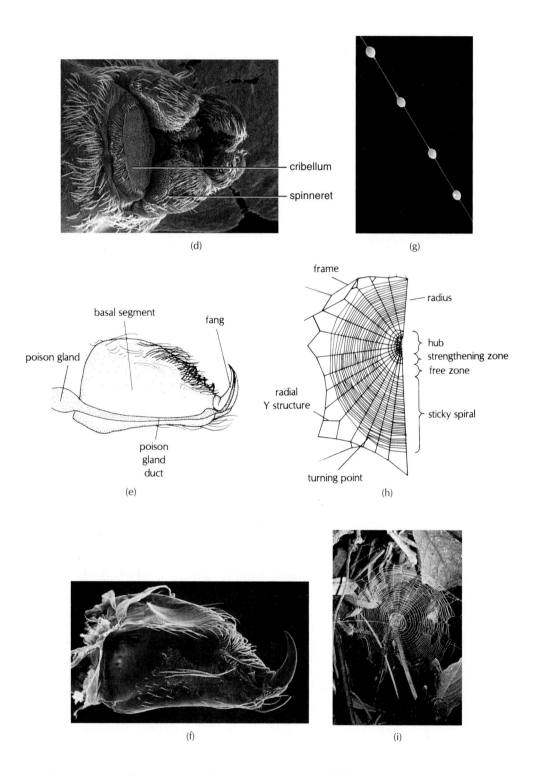

(d)

cribellum

spinneret

(g)

basal segment   fang

poison gland

poison
gland
duct

(e)

frame

radius

hub
strengthening zone
free zone

radial
Y structure

sticky spiral

turning point

(h)

(f)

(i)

electron micrograph of the chelicera of *Zosis geniculatus*. This species belongs to one of the few families lacking poison glands; the fang is used primarily for grooming rather than prey capture in this species. Prey are subdued by quickly wrapping them in silk. (g) A silk strand with adhesive droplets, from the spiral of an orb web built by *Mangora* sp. This silk can be stretched to nearly three times its resting length before it breaks. (h) Structure of an orb web. (i) Photograph of an orb web.

(a) *After Sherman and Sherman.* (b) *After the Kastons.* (c) *From L. A. Borradalle and F. A. Potts,* The Invertebrates, *2d ed. Copyright © 1935 Macmillan Publishing Company, New York. Reprinted with the permission of Cambridge University Press, New York.* (d,f,g,i) *Courtesy of Brent Opell.* (e,h) *From Foelix,* Biology of Spiders. *Copyright © 1982 Georg Thieme Verlag, Stuttgart Germany. Reprinted by permission.*

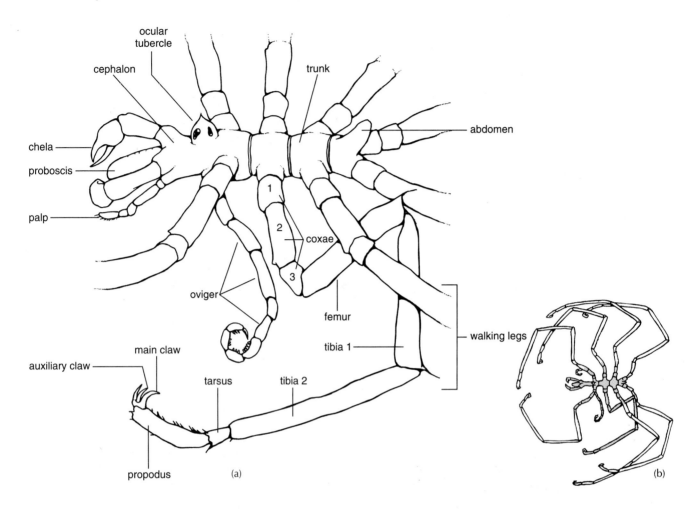

Figure 18.10

(a) Diagrammatic detailed representation of a male pycnogonid. The oviger is a leg modified for carrying an egg mass contributed by the female. (b) Entire pycnogonid in dorsal view, showing length of legs relative to body length.   (a) *From Bamber and Arnaud, 1987. Advances in Marine Biology, Vol. 24. Orlando: Academic Press; after Child, 1979.*

and ventral ganglia. In many species, the integument (body covering) is impregnated with calcium salts, as in crustaceans. The covering of millipedes is therefore more protective against abrasion and predation than that of the centipedes. As in the centipedes, however, the cuticle is not waxy. Although many diplopod species lack eyes, as many as 80 ocelli are found on the heads of some species. As with most centipedes, compound eyes are absent. The head appendages consist of a pair of uniramous (single-branched) antennae, a pair of mandibles, and a pair of maxillae. Distinct second maxillae are lacking among millipedes. Instead, the first and second maxillae on each side are fused to form a single appendage. Most species have an abundance of repugnatorial glands, which eject a variety of toxic, repellent secretions.

Both chilopods and diplopods are generally small animals, often only a few millimeters or at most 1 cm, in length, although some species of both groups have been reported to attain lengths of nearly 30 cm in the tropics. Myriapods are believed to be close relatives of the insects.

Class Insecta (= Hexapoda)

Class Insecta (= Hexa · poda)
(L: an insect [G: six-footed])

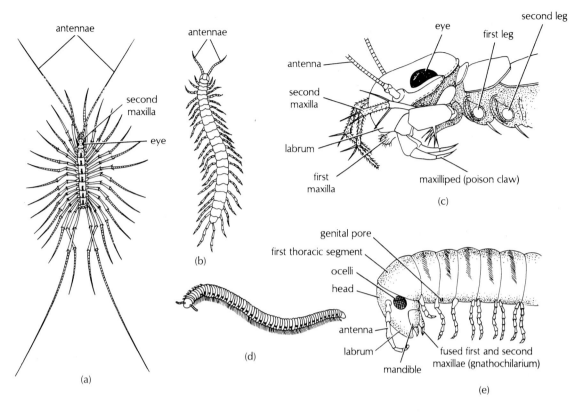

Figure 18.11

(a) A long-legged centipede, *Scutigera coleoptrata,* capable of especially rapid locomotion. Members of this species are seldom greater than 3 cm long; they are often encountered in moist areas (e.g., bathrooms) in buildings. (b) *Scolopendra* sp., a warm-climate centipede that may grow to a length of 25 cm. (c) Detail of centipede head, *Scutigera coleoptrata,* in lateral view. Note the large eye—actually a dense and organized cluster of ocelli—and the conspicuous poison claw. A poison-secreting gland is housed within the maxilliped. (d) Millipede, in dorsolateral view. (e) Detail of millipede head. Note that two pairs of legs are borne by each diplosegment of the abdomen.   (a) *After Pimentel.* (b,d) *After Huxley.* (c) *After Snodgrass.* (e) *Sherman/Sherman,* The Invertebrates: Function and Form, *2/e,* © *1976, pp. 111, 112, 236, 172, 169, 45, 9, 15. Reprinted by permission of Prentice Hall, Upper Saddle River, New Jersey.*

Insects have been reported from nearly every habitat except the deep sea. Although most species are terrestrial, many species live, either as adults or as larvae, in freshwater or in saltwater marshes (Fig. 18.12j,k). A few species (the ocean striders) live on the surface waters of the open ocean. Nearly one million species of insects have been described so far, with at least three times that number probably awaiting description. This tremendous number of species is in large part attributable to the feeding specializations, dispersal capabilities, and predator-avoidance possibilities associated with the evolution of flight. No other invertebrates, and relatively few vertebrate species, have evolved this capability. Indeed, when insects first evolved flight, they achieved access to a lifestyle previously unexploited by any other organism. Their adaptive radiation was thus unhindered by competition from other animal groups.

Insects are among the best studied of invertebrates, in large part because of their omnipresent impact on humans. Most flowering plants, including many species of agricultural importance, depend upon insects for pollination. Indeed, the close and often complex association between so many insect and angiosperm species suggests that the remarkable abundance of insect species must owe to the evolutionary proliferation of flowering plants (angiosperms). Yet this seems not to be the case. Remarkably, the fossil record shows that insect diversity did not increase dramatically after the evolution of flowering plants; if anything, insects diversified more rapidly before the evolution of angiosperms (Fig. 18.13). Similarly, most of the different insect mouthpart morphologies, including some exhibited by insects currently associated with angiosperms, are well represented in the fossil record long before terrestrial angiosperms were

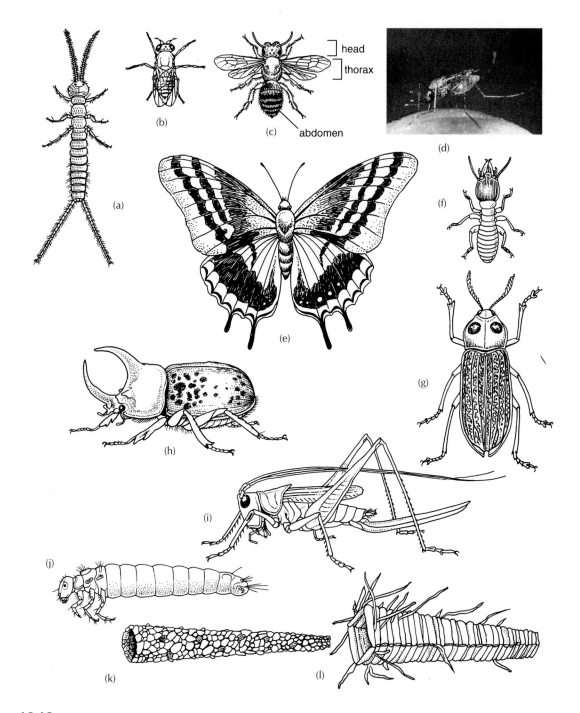

Figure 18.12

Insect diversity. (a) An apterygote, *Campodea staphylinus*. These, like the related silverfish, bristletails, and springtails, are primitive, wingless insects descended from wingless ancestors. (b) Fruit fly, *Drosophila* sp. (c) Leaf-cutter bee. (d) The mosquito *Anopheles gambiae* feeding on a mammalian host. This is one of the three main mosquito species that transmit the agent of malaria, a parasitic protozoan, throughout tropical Africa. (e) Butterfly.

(f) Termite soldier. (g,h) Beetles. (i) Grasshopper. (j) Caddisly larva. (k,l) Protective cases made by the larvae of two caddisfly species. (a) *After Huxley.* (b,i) *After Pimentel.* (d) *Courtesy of F. H. Collins. From Collins, F. H., and N. J. Besansky. 1994. Science 264:1874–75. Copyright © 1993 by the AAAS* (f) *After Romoser.* (j,k,l) *From McCafferty, Aquatic Entomology. Copyright © 1983 Jones and Bartlett Publishers. Reprinted by permission.*

widespread. The continued diversification of insects after the rise of angiosperms appears due more to an unusually low rate of extinction than to an unusually rapid rate of evolution. In any event, the lives of many insects and angiosperms are now inexorably linked, and the study of the mutual interdependence of plants and insects, and the evolution of those interactions, is a burgeoning field. Now for the bad news: The rapid rate at which many angiosperm communities are being destroyed likely will cause the extinction of many insect species around the world over the next 20 years.

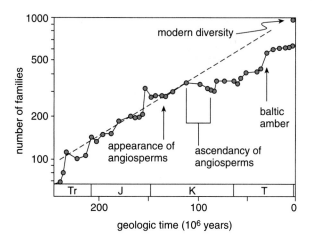

Figure 18.13

Increase in the number of insect families over geological time, from the Triassic Period to the present. Note that the Y-axis scale is logarithmic. Points along the dashed line represent an exponential increase in family-level diversification. Living insect species are now divided among nearly 1,000 families. Geological periods: Tr = Triassic; J = Jurassic; K = Cretaceous; T = Tertiary. *From Labandeira, C. C., and J. J. Sepkoski Jr. 1993. Science 261:310–14.*

Insects are also widely studied because they are significant vectors of human disease (e.g., malaria, bubonic plague, typhoid fever, yellow fever) and are major threats to agriculture, both as predators (especially as larvae) and as vectors of plant diseases. A few insect species (e.g., bees, wasps, and some beetles) produce secretions that are toxic to humans. Alternatively, other insect products (e.g., silk and honey) are of commercial importance. Finally, the life histories, social interactions,[3] and division of labor seen among some insect groups are magnificently complex and have long occupied the attention of animal behaviorists (Research Focus Box 18.1).

Please note that centipedes, millipedes, spiders, and mites are *not* insects.

The insect body is divided into three conspicuous tagmata: head, thorax, and abdomen (Fig. 18.12c). A flexible joint separates the head and thorax. Two pairs of wings are generally carried dorsally on the thorax. The wings are outfoldings of the thoracic integument and consist of two thin, chitinous sheets. In addition, the thorax generally bears three pairs of legs directed ventrally, giving rise to the alternate name applied to insects, the Hexapoda (*hexa* = G: six; *pod* = G: foot). The legs may be modified for walking, jumping, swimming, digging, or grasping, and are generally studded with a variety of sensory receptors, including receptors for taste, smell, and touch. Such receptors are also found on the mouthparts and elsewhere on the body. The main light receptors are a pair of **compound eyes,** but single-unit eyes (**ocelli**) are usually present on the head as well.

Insect Visual Systems

An **ocellus** is simply a small cup with a light-sensitive surface backed by light-absorbing pigment. Such simple photoreceptors are found in many phyla, including the Platyhelminthes, Annelida, Mollusca, and Arthropoda. The cup is often covered by a lens. As in all known visual systems, the photosensitive pigment of the ocellus is a vitamin A derivative in combination with a protein. Stimulus by light causes a chemical change in the photoreceptor pigment, generating action potentials that are then carried by nerve fibers to be interpreted elsewhere. Each insect head typically bears three ocelli. Ocelli are generally not image forming.

Compound eyes, however, *are* capable of forming images. For any eye to form an image, light must first be focused on the receptor surfaces. The animal must then be able to examine each component of the scene independently and monitor how light intensity varies across the image formed. The animal also must possess a nervous system sufficiently sophisticated to reconstruct the image detected by the sensory system. Insects satisfy all of these criteria, as do numerous other arthropods.

In the human camera-type eye, light enters through a single lens and is focused on the retina at the rear, in somewhat the same way that a motion picture image is focused on a screen. The components of the inverted image formed at the back of a human eye are sampled by myriad closely packed receptor cells in the retina—the rods and cones. Nerve impulses from the individual receptor cells are then integrated and interpreted by the brain. A compound eye works in much the same way, except that there are many lenses, the focus of each lens cannot be varied, and there are many fewer receptor cells to sample the image, which, by the way, is upright rather than inverted.

The compound eye is composed of many individual units, called **ommatidia** (*ommato* = G: eye; *ium* = G: little). The compound eyes of some insect species contain many thousands of ommatidia, each oriented in a slightly different direction from the others as a result of the eye's convex shape (Fig. 18.14). The visual field of such a multifaceted, convex eye is very wide, as anyone who has tried to surprise a fly will know. Crustaceans also have compound eyes, although it is presently uncertain whether these are homologous with those of insects or independently evolved. Some species of polychaete annelid and bivalve mollusc have independently evolved compound eyes that operate on principles remarkably similar to those demonstrated by insect and crustacean eyes.

The simplest type of compound eye is also the most common, found in insects as diverse as bees, ants, and cockroaches and in such noninsects as horseshoe crabs, true crabs, and isopods. In contrast to how human eyes work, compound eyes break up the image before it reaches the retina; one ommatidium samples only a small part of the complete image. The horror movies thus have

---

3. See *Topics for Further Discussion and Investigation,* no. 9.

Research Focus Box 18.1

Insect Behavior

Lewis, W. J., and J. II. Tumlinson. 1988. Host detection by chemically mediated associative learning in a parasitic wasp. *Nature* 331:257.

Insects show an almost overwhelming variety of lifestyles, one of which is a remarkable variation on the theme of being parasitic. A parasite, of course, is an organism that feeds on its host, generally without killing it. In contrast, perhaps 20% of all insect species are **parasitoids:** The females deposit their eggs inside the eggs, or in or on the larvae, of other insect species. The parasitoid embryo then develops at the expense of its similarly young host, which it slowly kills, eating from the inside out. The *adult* parasitoid is usually *not* parasitic and rarely even feeds on other animals at all; only the parasitoid larva feeds on another animal. But it is the adult's task to locate the ill-fated host for its offspring. Parasitoids are especially common among the true flies (order Diptera) and wasps (order Hymenoptera).

The difficulty of locating suitable hosts has exerted strong selection for behaviors that increase the likelihood of encounter. Cues used by the parasitoid to find the required host are most commonly chemical. Not surprisingly, parasitoids often use chemicals produced by the intended host species, such as mating pheromones used by the female to attract potential mates; the pheromone then brings not only males of the species but also female parasitoids that arrive in time to poison the fruit of the union. Other species use different chemical cues, such as those associated with host feces. Lewis and Tumlinson (1988) demonstrate that the parasitic wasp *Microplitis croceipes* recognizes and exploits volatile odors associated with the feces of the desired host, larvae (caterpillars) of the moth *Heliothis zea,* and that the wasps can learn to recognize a variety of such odors.

Lewis and Tumlinson found that, when a female wasp encountered feces of the preferred host caterpillar, she spent several minutes inspecting intensely with antennae and abdomen. After letting her explore the host feces, the researchers then soaked a piece of standard filter paper in a hexane extract of the feces (1.4 mg of feces per 14 µl of solvent) and suspended the filter paper at the far end of a wind tunnel. When they released the female wasps into the wind tunnel, the wasps generally flew toward, and soon alighted upon, the filter paper (Focus Table 18.1, row 2). Note that exposing the female wasp to *water* in which the feces had been soaked was just as effective in whetting her appetite for the hexane extract (Focus Table 18.1, middle two rows) and that the wasp only oriented to the hexane extract if she had had some prior experience with (i.e., had been "conditioned" on) the host feces or a water-soluble extract of those feces (Focus Table 18.1; compare, for example, rows 3 and 4).

Additional experiments clearly demonstrated that the female wasps were orienting to a hexane-soluble chemical from the feces, not to the hexane itself (Focus Fig. 18.1, first two vertical bars). Of particular interest, when wasps were first conditioned on host feces and then placed in the wind tunnel, the hexane extract of the feces was far more attractive to the wasps than were the host feces themselves (Focus Fig. 18.1, second and third vertical bars). This suggested that the wasps were *conditioned* by some nonvolatile, water-soluble substance in the feces, but that they

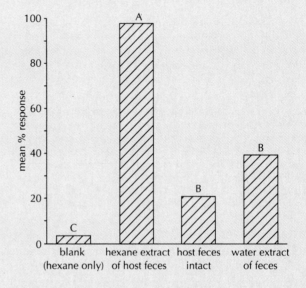

Focus Figure 18.1

Flight responses of female wasps *Microplitis croceipes* to extracts of host (*Heliothis zea*) feces suspended in a wind tunnel on filter paper. Wasps were permitted to explore water extracts of the host feces before flight responses were examined. A total of 24 wasps were tested per treatment. Bars with different letters at the top represent responses that differed significantly from each other (P<0.05).

subsequently *flew* toward some other chemical—some volatile, hexane-soluble substance (or substances) associated with the fecal pellet at the time of conditioning.

To test the possibility that the response to this volatile cue could be a learned behavior, the researchers allowed the wasps to explore water-soluble extracts of feces to which they had added common vanilla extract, a scent the wasps never would have encountered previously. Would the wasps then fly to the scent of vanilla? They did indeed, and most decisively (Focus Table 18.2). Only wasps that were simultaneously exposed to both the water extract of the feces and to the vanilla scent subsequently flew toward the smell of vanilla. Females preexposed to the water extract without vanilla *did not* fly to the source of the vanilla scent; moreover, females not preexposed to the water extract of the feces did not fly to the vanilla, whether or not they were previously exposed to the vanilla extract.

Apparently, the searching behavior has two triggers: (1) a nonvolatile, water-soluble chemical that conditions the wasp to receive a different chemical signal and (2) the second signal itself, a volatile substance present in the feces at the time of conditioning. Remarkably, the parasitoids are not necessarily fixed on any one perfume; they can learn to use different cues, perhaps those most likely to be associated with possible hosts at different times of the year or when the hosts might be feeding on different plants. Parasitoids thus have a marvelous way to track potential hosts,

Focus Table 18.1   The Effect of Conditioning to Host (Caterpillars of *Heliothis zea*) Feces, and Extracts of Those Feces, on the Tendency of the Parasitoid Wasp *Microplitis croceipes* to Fly to Hexane Extracts of the Feces

Conditioning Treatment	Mean Percentage of Wasps Alighting on a Hexane Extract of *H. zea Feces*
Hexane extract of host feces	9
Host feces (*H. zea*)	92
Water extract of host feces	97
None	5

Note: Soaking the feces in hexane solubilizes volatile compounds from the feces, while soaking in water does not; soaking in water extracts mainly nonvolatile, water-soluble substances. The table shows the mean results for four replications of the experiment, with six wasps used for each treatment in each replicate (total = 24 wasps tested per treatment).

Focus Table 18.2   The Role of Volatile Compounds and Nonvolatile Compounds in Conditioning Wasps (*M. croceipes*) to Fly to Volatile Substances

Conditioning Treatment	Mean Percentage of Wasps Flying to Vanilla Extract
No conditioning	0
Water extract of *H. zea* (host) feces	3
Water extract of host feces plus volatile vanilla scent	45

Note: Wasps were conditioned to the mostly nonvolatile (water-soluble) component of host feces (*H. zea*) with or without the presence of an artificial volatile scent (vanilla extract). Data show the mean response of the 24 wasps tested per treatment, as in Focus Table 18.1.

Focus Table 18.3   Summary of Experiments Conducted

Conditioning	Source of Volatiles	Response of Wasps
H. zea feces	Hexane extract of feces	+
Water extract of *H. zea* feces	Hexane extract of feces	+
Hexane extract of *H. zea* feces	Hexane extract of feces	−
Hexane extract of *H. zea* feces	Water extract of feces	−
Water-soluble extract of *Trichophisia ni* feces (nonhost species)	Hexane extract of *T. ni* feces or of *H. zea* feces	−

Note: A (+) response indicates that wasps flew toward, and eventually landed upon, the source of volatiles indicated. Before the flight tests, the wasps were allowed to explore caterpillar feces, or extracts of those feces, as indicated under *Conditioning.*

and without ever having to first locate the host itself. Where there are fresh feces, can their producer be far away?

Obviously, this system works well only if the wasp tracks volatiles associated with feces of the desired host. In a final experiment, the researchers exposed female wasps to water extracts of feces produced by nonhost caterpillars (*Trichophisia ni*), a species that never serves as a host for the eggs of this wasp. The wasps subsequently showed no response to hexane extracts of these feces or to hexane extracts of the feces of the proper host, *H. zea*. Clearly, the behavior is conditioned by a chemical specific to feces of the proper host. The identity of this chemical, and of the volatile chemical cues subsequently followed by the wasp, remain to be determined.

For clarity, the results of all key experiments are summarized in Focus Table 18.3.

Figure and tables reprinted by permission from W. J. Lewis and J. H. Tumlinson, "Host Detection by Chemically Mediated Associative Learning in a Parasitic Wasp" in Nature, 331:257-259, 1988. Copyright © 1988 Macmillan Journals Ltd., London.

Figure 18.14
Compound eye of a fly, containing hundreds of ommatidia. Note the eye's convex shape; no two ommatidia are oriented in precisely the same direction.   © *Thomas Eisner.*

it wrong: An insect does not see thousands of identical, complete images at once, but rather only a single, fairly coarse-grained picture at a time. Each ommatidium consists of (1) a fixed-focus lens (the cornea), which has great depth-of-field so that objects from 1 mm to several meters away are in focus at the receptor; (2) an underlying gelatinous **crystalline cone,** which serves as a lens in some insects (and in most crustaceans); (3) a series of up to eight cylindrical bodies called **retinular cells,** each containing light-sensitive pigment; (4) cylindrical cells (collars) containing **shielding pigment,** optically isolating every ommatidium from surrounding ommatidia; and (5) at the distal end, a **neural cartridge,** a cluster of neurons receiving the information carried by the retinular cells and sending action potentials to the optic ganglia for processing (Fig. 18.15a). Insect eyes are sensitive to the polarization of light, so polarized light can be used as a navigational cue during flight. Their sensitivity to ultraviolet light (UV) permits insects to see patterns (on flowers

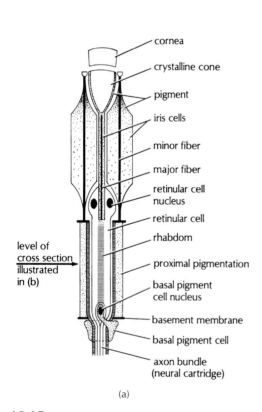

(a)

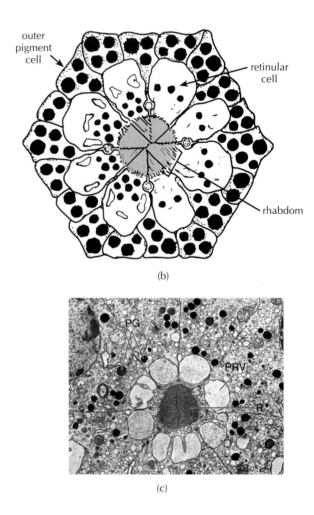

(b)

(c)

Figure 18.15
(a) Structure of a single ommatidium in a butterfly eye. The light-sensitive pigment is contained in the rhabdom. (b) Cross section through the rhabdom region of an ommatidium (at approximate level indicated in [a]), showing the relationship between the rhabdom and the reticular cells contributing to it. (c) Transmission electron micrograph of a crab ommatidium in cross section. PG = pigment granules in retinular cells; R = rhabdom; PRV = perirhabdomal vacuole.   (a,b) *Reprinted from S. L. Swihart, in Journal of Insect Physiology, 15:37, 1969 with kind permission from Elsevier Science Ltd., The Boulevard, Langford Lane, Kidlington OX5 1GB, U. K. (c) Courtesy of K. Arikawa, from Arikawa et al, 1987. Daily changes of structure, function and rhodopsin content in the compound eye of the crab Hemigrapsus santuinues. J. Comp. Physiol. A 161:161–174.*

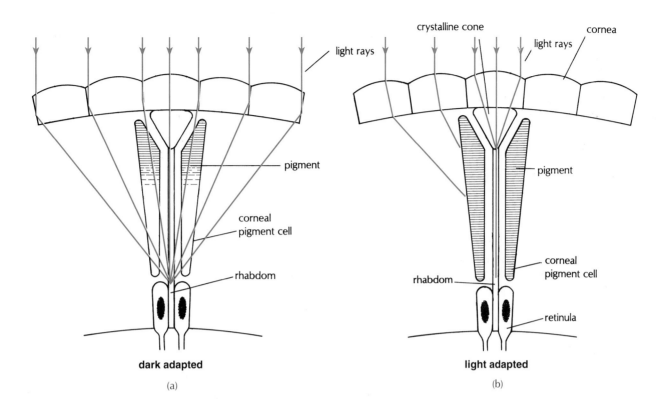

dark adapted

(a)

light adapted

(b)

Figure 18.16

A superposition eye in the (a) dark-adapted and (b) light-adapted conditions. Note that in the dark-adapted condition, light entering through the lenses of several adjacent ommatidia impinges upon a single rhabdom. Migration of pigment within the pigment collars prevents this from happening in the light-adapted eye, improving visual acuity and directional sensitivity. For better clarity, the paths of reflected and refracted light are shown separately in (a) and (b), respectively.

and other objects) that are invisible to humans. Some insects, especially those that fly in the evening and before dawn, are also sensitive to light of infrared wavelengths.

The light-sensitive pigment of the retinular cells is contained within tens of thousands of **rhabdomeres,** which are fine, microvillar outfoldings of the retinular cell walls. The rhabdomeres within one ommatidium form a discrete, ordered association (generally a central shaft) called a **rhabdom** (Fig. 18.15a–c). The rhabdom, in other words, is not really a single structure, but rather a central entity formed cooperatively by microvilli of the participating retinular cells. The rhabdom records the light intensity at the *center* of the image that falls at its tip; it does not record the entire image. The tip of one rhabdom is essentially analogous to a single rod in human eyes. The rest of the cylindrical rhabdom acts as a light guide, down which this small component of the image travels to the neural cartridge at the base. The brain then reconstructs the complete image from all the signals input from the dozens, hundreds, or many thousands of individual ommatidia. Each ommatidium is functionally isolated from its neighbors by permanent shielding pigment (Fig. 18.15a) or by reflective tracheoles, so that visual acuity is high.

This basic compound eye is called an **apposition eye,** because the lens is directly apposed to the receiving rhabdom. Because each lens is very small, each rhabdom receives only a small amount of light; apposition eyes therefore work best at fairly high light intensities. For a compound eye to work well at low light intensities, each neural cartridge must receive light from more than one ommatidium. If the screening pigment between adjacent ommatidia is lacking, many facets can combine, or superimpose, the light they receive into a single image on the retina. This type of eye is called a **superposition eye.** In such an eye, each ommatidium has a large space between the distal end of the crystalline cone and the rhabdom; without the shielding pigment in the way, light from a single point in the visual field can be received by many lenses and focused onto a single rhabdom, producing a signal of substantially greater intensity than that received through a single lens (Fig. 18.16a).

The superposition eye adapts easily, with the help of pigment-containing collars flanking each rhabdom, to different light conditions; as light intensity increases, dense pigment granules migrate down each collar, so the collar acts as an iris, blocking the light received from adjacent ommatidia (Fig. 18.16b). In bright light, then, the

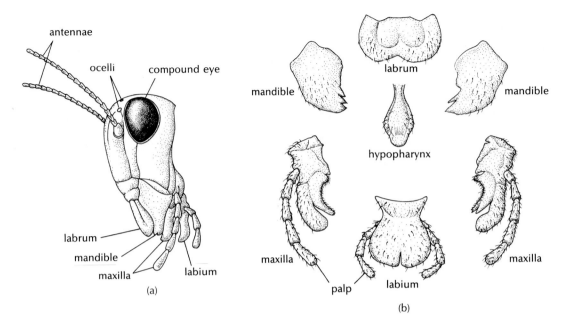

Figure 18.17
(a) Insect head, showing the various appendages and eyes.
(b) Detail of appendages, drawn in proper orientation viewed anteriorly. The morphology of the mouthparts differs widely among species and correlates with feeding biology. In butterflies and moths, for example, the maxillae are greatly elongated for taking up nectar.   (a) *After Snodgrass.* (b) *After James and Harwood.*

superposition eye functions essentially as an apposition eye. If the light intensity is reduced, the pigment granules migrate out of the way so that light rays received by the lens of one ommatidium can again cross over to adjacent ommatidia, increasing the strength of the signal received. Because pigment migration is under hormonal control, it takes some time to dark-adapt or light-adapt a superposition eye.

The sharpness of the image formed by a compound eye depends upon a number of factors: (1) the extent to which the light impinging on the rhabdomeres of a single ommatidium enters along a pathway parallel to the optic axis (i.e., the long axis) of that ommatidium (increased resolution); (2) the extent to which light from adjacent ommatidia impinges upon the receptor pigment of an ommatidium (decreased resolution); (3) the amount of difference in direction in which adjacent ommatidia are oriented (decreased angle gives increased resolution); (4) the number of ommatidia per eye (increased number gives increased resolving potential); and (5) the complexity of the information center (i.e., brain) receiving and processing the impulses sent from the ommatidia. Even the largest and most sophisticated of compound eyes must produce a rather coarse-grained image. The image that humans see is synthesized from light-intensity data collected by millions of individual rods and cones; the image seen by a compound eye results from the collaborative efforts of as few as about a half-dozen and as many as several thousands of individual neural cartridges.

Other Insect Characteristics

The insect head bears four pairs of appendages: one pair of antennae (which are always uniramous or single-branched) and three pairs of mouthparts (Fig. 18.17). In sequence, the mouthparts are the **mandibles,** the **maxillae,** and finally, a pair of **second maxillae** that have fused to form a single appendage called the **labium.** The mandibles are shielded anteriorly by a downward extension of the head called the **labrum.** The precise morphology of these mouthparts varies considerably according to the insect's feeding biology (Fig. 18.18). The abdomen lacks appendages, except for a pair of sensory **cerci** borne on the last abdominal segment. The abdomen also may house receptors that monitor the degree to which the body wall is stretched during feeding.

In keeping with a largely terrestrial lifestyle, an insect's gas exchange surfaces have been internalized. Gas exchange is generally accomplished by means of a **tracheal system** (Fig. 18.19). Although resembling the tracheal system found in more advanced arachnids, the

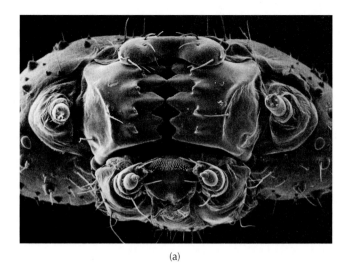

(a)

(b)

labrum

mandible

antenna

ocellus

maxilla

spinneret

labium

hypopharynx

sensory
appendage
(maxillary palp)

Figure 18.18
(a) Scanning electron micrograph of the head of an insect, the
tobacco hornworm *Manduca sexta* (caterpillar stage).

(b) Diagrammatic sketch showing major morphological features
visible in (a).   (a) *Courtesy of Nancy Milburn.*

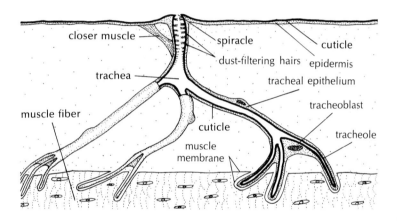

closer muscle

spiracle

cuticle

dust-filtering hairs

epidermis

trachea

tracheal epithelium

muscle fiber

cuticle

tracheoblast

muscle
membrane

tracheole

Figure 18.19
The insect tracheal system. The spiracles of terrestrial insects can
generally be closed to regulate water loss by contracting
appropriate musculature. The finest tubes of the tracheal system,

the tracheoles, develop from the tracheoblast cell.   *After Chapman;
after Meglitsch.*

insect tracheal system is thought to have been independently evolved; that is, the tracheal systems in the two groups are probably convergent, evolving independently in different ancestors. One pair of **spiracles** opening into the tracheal system is found on the thorax, with additional pairs of spiracles located on many of the abdominal segments. The spiracles of most species can be closed, deterring evaporative loss of water. The tracheae are lined by cuticle, which is shed and resecreted by the underlying epidermis each time the insect molts, and the tracheal tubules are kept from collapsing by means of chitinous rings embedded in the walls. The tracheae branch to form a network of

smaller tubules called **tracheoles,** which are less than 1 µm in diameter. These branch again and terminate directly on the insect's tissues. Thus, gas is exchanged between the tissues and the environment without the involvement of the blood circulatory system. Some insect species lack tracheae, either as adults or during development. Gas exchange in such animals must occur across general body surfaces. Such surfaces obviously cannot be waxy, and the animals are thus restricted to moist habitats.

Water conservation is another correlate of a terrestrial lifestyle.[4] Uric acid is the primary end product of protein metabolism among insects; this nontoxic

4. See *Topics for Further Discussion and Investigation,* no. 2.

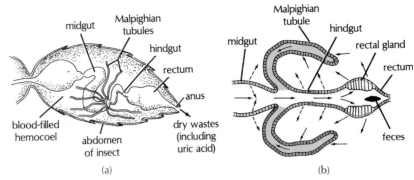

(a)   (b)

Figure 18.20

(a) Malpighian tubules in the insect abdomen. (b) Diagrammatic illustration of the relationship between the Malpighian tubules and the posterior portion of the digestive tract. Fluid moves from the hemocoel into the tubules, where it joins wastes moving toward the anus. The arrows indicate the extensive reclamation of water that occurs in the hindgut and rectum. *(a) From Purves and Orians, Life: The Science of Biology, 2d ed. Copyright © 1983 Sinauer Associates, Sunderland MA. Reprinted by permission. (b) After Wilson; after Wigglesworth.*

nitrogenous compound is excreted in nearly dry, solid form. The major excretory organs are called **Malpighian tubules** (Fig. 18.20), of which up to 250 pairs are found in the insect hemocoel. Waste products, notably a soluble derivative of uric acid, are actively transported from the blood into the distal portion of the Malpighian tubules. Increased acidity in the proximal portion of the tubules causes the uric acid to precipitate out of solution. Most of the water contained in the urine is then resorbed during its passage through the rectum.

Insect Flight

One feature that sets the insects apart from all other invertebrates, and most vertebrates as well, is the ability to fly, an innovation that first evolved some 300 to 400 million years ago. The insect wing is a lateral outfolding of the body wall. These outfoldings are very thin and lightweight, and are structurally supported by a characteristic network of veins connecting with the blood circulatory system, by tracheal tubules, and by rows of pleats radiating outward from the base of the wings to the tips.

No one knows how insect flight evolved, although various possibilities have been much debated.[5] Many flying insects use their wings to regulate body temperature, either by varying the amount of beating activity or the amount of wing exposed to sun and wind. Accordingly, some researchers have argued that the selective benefits associated with the evolution of the first insect wings likely had more to do with temperature regulation than with flight. Others believe that wings evolved from lateral outgrowths originally used to stabilize ancestral forms during jumping. Still others argue convincingly that insect wings probably evolved directly from the gills of aquatic mayfly-like developmental stages, which currently use their gills for both gas exchange and locomotion; such gills may have been used first in skimming across the water surface, as in modern stonefly nymphs, as a prelude to flapping flight. Whatever their origins, the evolution of wings opened up to insects a lifestyle that is virtually inaccessible to most other animals.

Flight requires the generation of both lift and thrust, which flying insects achieve through a combination of body shape, wing morphology, and highly complex wing behavior. Thrust is something we have an intuitive feel for; it is the force we exert in one direction that creates motion in the opposite direction (every action produces an equal and opposite reaction). The generation of lift is more mysterious. (In fact, I hesitate to discuss it, for fear you'll never want to fly again!) The secret is contained in the equation that follows, modified from the original of Daniel Bernoulli (1700–1782). When dealing with flight, the equation pertains to air moving across a solid surface, such as a wing:

$$\tfrac{1}{2}dv^2 + p + dgh = \text{a constant}$$

where d = density of the air; v^2 = the square of the air velocity relative to the wing; p = air pressure at the wing's surface; g = the gravitational constant; and h = the height of the air above the wing surface. The term dgh is related to potential energy and the $\tfrac{1}{2}dv^2$ term is related to the expression for kinetic energy. When we fly in an airplane, the principles embodied within this equation are what keep us aloft. In this situation, the air we are talking about does not change density and is always the same height above the wing, so d and dgh are constants. The equation states, then, that if the velocity of the air moving across the wing increases, the pressure above the wing surface must decrease; this must occur if the sum total of the three expressions on the left side of the equation is to remain constant. This pressure decrease above the wing produces lift; that is, the pressure below exceeds the pressure above, and the body rises. You can prove to yourself that lift is generated by differential air flow over

---

5. See *Topics for Further Discussion and Investigation*, no. 19.

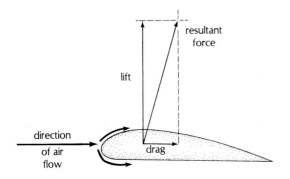

Figure 18.21
The principle of lift generation for an airfoil. Because the upper surface is convex, air moves more quickly over this surface than over the lower surface. By Bernoulli's principle, this lowers the relative pressure above the surface, creating net lift as illustrated.
From Life in Moving Fluids: The Physical Biology of Flow by Steven Vogel © 1981 by Willard Grant Press. Reprinted by permission of Wadsworth, Inc.

the upper and lower surfaces of an object by blowing along the length of a strip of paper. Try a strip about 1 in wide and 6 in to 8 in long, holding one end of the paper just below your mouth. If you blow hard enough along the length of the strip, the end of the paper will rise.

Obviously, an insect does not blow over its wings to generate lift. Neither does the airplane propeller or jet engine function by blowing air over the wings' upper surfaces. Instead, the beating of an insect wing, like the spinning of a propeller, moves the wing through the air. Wing beat frequencies vary from less than 10 to more than 1,000 beats per second in different insect species.

While flying, an insect is not simply moving its wings up and down. Instead, the wings are continually being brought either forward or backward and twisted. The net effect is that the air moving over the wing's upper surface has to travel farther (and thus faster) to reach the back of the wing than does the air moving over the lower surface, producing lift. When lift exceeds the insect's weight, the insect rises. The principle is most easily illustrated with the forward movement of a simple shape termed an **airfoil** (Fig. 18.21). Since the entire airfoil moves a given distance per unit of time, the shape of the airfoil ensures that air moves faster across the upper surface than across the lower surface, generating lift according to Bernoulli's principle. Several other factors, including the shape of the insect body, may also contribute to lift. By varying the angle at which the wing moves through the air (the "angle of attack"), the insect can alter the direction of the net forces generated, producing movement, via thrust, in the direction of choice.

Most insects have two pairs of wings, although many species have only a single pair. Beetle species, with two pairs of wings, typically use only one pair for flying, holding the hardened, protective first pair out of the way while flapping the second pair.

Insect species differ considerably with respect to wing morphology and the manner in which the wings are operated by the thoracic musculature. Much of the modification in wing structure and function encountered among different insect groups seems to reflect selection for increased energy efficiency and increased fine directional control during flight. Many insects can hover and even fly backward, a great advantage for mating and egg laying "on the wing." Other morphological modifications serve to protect the insect body or the wings themselves.

Although we will not go into the details of insect flight mechanics here,[6] the major characteristics that make flight possible in the Insecta follow:

1. abundance of striated muscle specialized for rapid, strong contractions;
2. muscle antagonism by means of a lightweight, jointed skeleton, permitting a great amount of movement to be generated from relatively short changes in muscle length;
3. small body size;
4. water-impermeable outer body covering, preventing dehydration;
5. efficient systems for gas exchange, nutrient storage, and distribution of nutrients to the musculature; and
6. highly developed nervous and sensory systems for steering, navigating, and sensing wind direction.

In many of the faster-flying insects, the flight muscle is highly specialized. In these species, the muscle fibers are capable of contracting many times following stimulation by a single nerve impulse. This type of flight is termed **asynchronous flight,** since wing beat frequency does not correspond to the frequency of nerve impulse generation. The wing beat frequencies of over 1,000 beats per second that have been recorded in mosquitoes are made possible by this unique insect invention.

Insect Social Systems

The study of insect social systems is an active and fascinating field, encompassing many important issues in behavioral ecology and evolutionary theory. Truly social insects (**eusocial** species) include many hymenopterans (all ant species, some bee species, some wasp species—all in the insect order Hymenoptera) and all termites (order Isoptera). By definition, eusocial insects form colonies composed of more or less sterile workers and one or more reproductive queens; exhibit multiple generations within a colony, so the queen is protected and cared for by her offspring; and cooperate in the care of developing embryos and larvae. A few arachnid species exhibit high degrees of social development, but insect eusociality is otherwise unique among invertebrates.

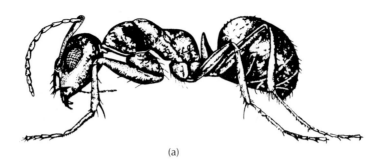

(a)

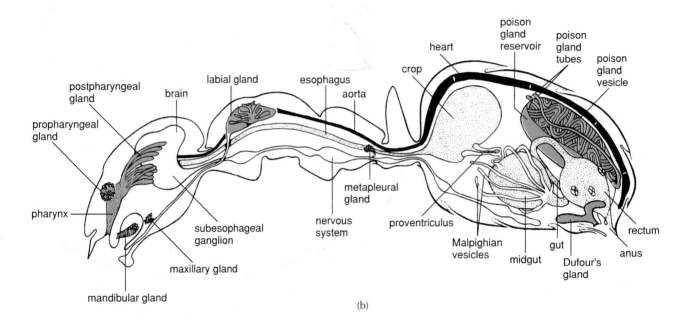

(b)

Figure 18.22

(a) A worker ant in the genus *Formica*. (b) Sagittal section through a *Formica* worker showing internal anatomy, including the numerous exocrine glands (in blue), which release their products outside the body (*exo* = G: outside). Dufour's gland secretions are mostly involved in alarm signalling and recruiting nestmates for foraging, attack, or defense; the poison gland produces formic acid for trail marking or venoms used in predation or defense; the mandibular glands produce defense and alarm pheromones; the metapleural glands secrete antibiotics, protecting the body surface and the nest itself against microbial infection. *Reprinted by permission of the publishers from* The Ants *by Bert Hölldobler and Edward O. Wilson (after Gosswald, 1985, after Otto, 1962), Cambridge, Mass: The Belknap Press of Harvard University Press, Copyright © 1990 by Bert Hölldobler and Edward O. Wilson.*

Ants evolved some 100 million years ago from nonsocial wasps, and all 9,500 described species are eusocial. Social development has surely contributed to ant success. Accounting for only about 2% of all insect species, ants make up about half of all insect biomass. A new ant colony is typically formed by a winged female shortly after she mates. The female then discards her wings, builds a nest, and produces many daughters over a number of years, all from sperm stored from matings accomplished during her one brief nuptial flight. The daughters form a worker caste, whose job it is to care for the queen, maintain the nest, care for embryos and larvae, defend the nest, and forage for food. In many ant species, some workers develop into soldiers, morphologically and behaviorally specialized for defense and aggression. Workers communicate through a complex system of mechanical and chemical cues (Fig. 18.22). Not only are all workers female, but in most colonies they are in fact all sisters, and the larvae they care for are also their sisters. Males are produced only after the colony has achieved a substantial size, which often takes several years. The queen produces males deliberately, by releasing eggs without fertilizing them.

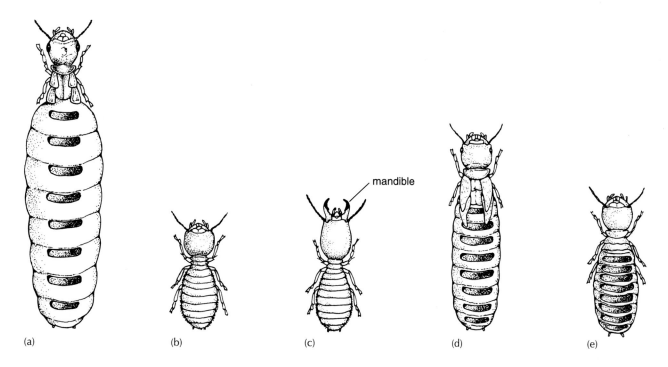

(a) (b) (c) (d) (e)

Figure 18.23
Female caste members of the termite *Ameritermes hastatus,* all drawn to the same scale. (a) Queen. (b) Worker. (c) Soldier. (d) Secondary queen. (e) Tertiary queen. Supplemental queens (d,e) replace the original queen when she dies. Note the large

mandibles of the soldier (c). *Reprinted by permission of the publishers from The Insect Societies by Edward O. Wilson (after Atkins, 1978, after Skaife, 1954), Cambridge, Mass.: The Belknap Press of Harvard University Press, Copyright © 1971 by the President and Fellows of Harvard College.*

Males are thus all haploid, developing parthenogenetically without the aid of sperm. Males are winged, and do no work; they are eventually chased out of the colony by workers. Meanwhile, perhaps by altering the quality of food provided to certain diploid embryos, hormonal titres are altered such that these particular female embryos develop into winged reproductives—future queens—which leave the colony for a short period of frenzied mating. Males die shortly after mating, while the mated females excavate new nests and initiate new colonies.

Note that the queen's female offspring obtain half their mother's alleles but all their father's alleles, since the father is haploid and therefore passes along everything he has to each daughter in turn. In consequence, the workers are more closely related to their sisters than they would be to their own daughters, if they had any; they therefore do more to propagate their own genotypes by taking care of their sisters than by having offspring of their own!

Ants exhibit a considerable diversity of lifestyles, the scope of which I can only hint at here. Members of most ant species prey on other insects (especially termites) and arachnids, but some feed largely or exclusively on excretions produced by other insects, including aphids and butterfly caterpillars. Members of some ant species (the leaf-cutter ants) actively cultivate fungal gardens within their nests for food. There are even slave-making ant species: Incapable of feeding themselves or their nestmates, the workers of such species capture larvae or

pupae from the nests of other ants; captured workers tend to the needs of the captors with apparent indifference.

Only a small proportion of bee and wasp species exhibit social development rivaling that encountered among ants; members of most bee and wasp species live as solitary individuals. Among the wasps, paper wasps, yellow jackets, and hornets show the greatest degree of social organization. Among bees, honey bees show particularly advanced social behavior. In both groups, workers are always diploid females, as in ants; they play many roles in the colony—caring for the queen and her offspring, scouting out suitable nest sites, and controlling nest temperatures, for example—and those roles typically change as workers age (and exhibit altered hormonal concentrations). As with ants, males are always haploid, do no useful work in the colony, and are eventually expelled. Queens mate for only a brief period of time, and they store enough sperm to last for life.

Termites (order Isoptera), derived from some cockroach-like ancestor some 150 to 200 million years ago, have evolved eusociality independently of the hymenopterans. Unlike hymenopteran queens, termite queens cannot control the sex of their offspring, so termite workers may be male or female. In both cases, workers are always sterile, and they are usually blind. Advanced termite species produce genuine worker, soldier, and reproductive castes (Fig. 18.23). In contrast, workers in more primitive species are made up of

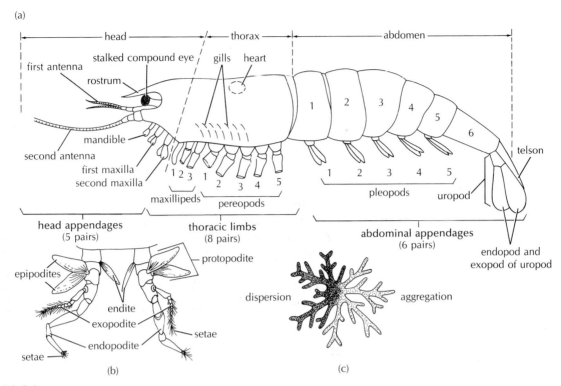

Figure 18.24

(a) General external anatomy of a crustacean, showing the head, thorax, abdomen, and associated appendages. (b) Illustration of biramous appendages. (c) Patterns of pigment dispersion in crustacean chromatophores. In the dispersed configuration, the cuticle becomes dark; pigment aggregation within the chromatophores causes the cuticle to become lighter in color. Movement of pigment within the chromatophores is under hormonal control in response to changing light intensities.

(a) *Modified after Russell-Hunter and other sources.* (c) *From Weber, 1983.* American Zoologist. *Thousand Oaks, California, American Society of Zoologists.*

developmentally arrested larval stages called **pseudergates** ("false workers"). Pseudergates can remain workers for life, metamorphose into winged reproductives and found new colonies, or metamorphose into replacement reproductives or soldiers within the same colony. Soldiers are usually blind and bear huge, heavily sclerotized mandibles (Fig. 18.23c), which they use to defend the colony against intruders. Soldiers also employ a variety of chemical defenses. The fate of workers is regulated by a complex system of pheromonal production by both queens and soldiers, which, in turn, regulates hormonal concentrations within the pseudergates.

Subphylum Crustacea

Subphylum Crusta · cea
(L: a crust)
kruss-tā´-shuh

Class Malacostraca

Class Malaco · straca
(G: soft shell)
mal´-ack-ō-strak´-ah

The class Malacostraca contains nearly 75% of all described crustacean species, including decapods, euphausiids, stomatopods, isopods, and amphipods. The most familiar malacostracans, such as the crabs, hermit crabs, shrimp, and lobsters, are decapods. The basic malacostracan body is tripartite, consisting of a **head, thorax,** and **abdomen** (Fig. 18.24a). Whereas the insect head and thorax are separated by a flexible joint, the crustacean head and thorax are almost always rigidly fused. The head and thorax may be covered by a **carapace,** extending posteriorly from the head, and therefore may function as a single unit, the **cephalothorax.** In some species, the carapace bears a prominent anterior projection called the **rostrum.** Large, stalked **compound eyes** are conspicuous, as are two pairs of head appendages, the first and second **antennae.** Often in the zoological literature, the first pair of antennae are known as the "antennules," and the second pair are simply referred to as the "antennae." In malacostracans, both pairs of antennae are primarily sensory. In other crustacean groups, the second antennae also may play roles in feeding, locomotion, and mating. In addition to the two pairs of antennae, the malacostracan head bears three pairs of smaller appendages that are involved in feeding or in generating respiratory currents. These appendages are, in sequence beginning from the mouth and moving posteriorly, the **mandibles** (which crush food) and the first and second **maxillae** (which generate water currents and manipulate food). The next eight segments of the cephalothorax are thoracic segments, commonly bearing in sequence the first, second, and third **maxillipeds** (for food manipulation) and five pairs of thoracic **walking legs,** commonly known as **pereopods.** The first three pairs of pereopods may be chelate (claw bearing), in which case they also function in feeding and in defense.

Each of the six abdominal segments bears a pair of appendages as well. The first five pairs of abdominal appendages are referred to as **pleopods;** these function primarily in generating respiratory currents and, in females, in the brooding of eggs and developing young. The last pair of abdominal appendages are the **uropods.** These flat appendages lie on either side of the telson, forming a tail (Fig. 18.24a).

Unlike those of any uniramian, malacostracan appendages are generally biramous; that is, they have two branches (bi = L: two; rami = L: branch). The portion of the limb proximal to the branch point is the protopodite (proto = G: first). The inner and outer branches are the endopodite and exopodite, respectively (Fig. 18.24b). The exopodite is often less well developed than is the endopodite. Frequently, lateral protuberances occur on the protopodites themselves. These are termed endites or epipodites, depending on whether they project inward or outward, respectively. Epipodites commonly function as gills. Some crustacean appendages no longer have both the endopodite and exopodite and are, therefore, secondarily uniramous (one branched). The first antennae and the maxillae of lobsters, for example, are biramous, whereas the second antennae and the thoracic appendages (pereopods) are uniramous. The abdominal appendages are biramous, a uniquely malacostracan characteristic.

The body surface of many malacostracan species is covered with **chromatophores** (Fig. 18.24c), which are highly branched cells containing pigment granules. The pigments come in a variety of colors, including red, black, yellow, and blue. More than one pigment may be found within a single chromatophore, and pigment distribution differs among the chromatophores of a single animal. By varying the pigment distribution in the different chromatophores over time, the animal can alter its body color considerably. The migration of pigment granules within chromatophores is under hormonal control.[7] The hormones are manufactured by the so-called X-organ located in the eyestalks and are transported a short distance to the **sinus gland** for storage. From here, the hormones are transported as needed through the bloodstream. Because chromatophore operation is under hormonal control rather than under direct nervous control, arthropod color changes never occur as rapidly as do those of cephalopods.

The description of a typical malacostracan just given applies best to members of the order Decapoda, which includes about 10,000 species and is the largest of the malacostracan orders. The term *decapod* (deca = G: ten; pod = G: foot) refers to the fact that members of the order Decapoda have only 10 thoracic legs (five pairs); the first three pairs of thoracic appendages (the **maxillipeds**) are modified for feeding. The term *decapod* is also useful in recalling that the abdomen and thorax together bear a total of 10 pairs of leg-like appendages. Decapods are probably the best-known crustaceans and include lobsters, crayfish, hermit crabs, true crabs, and shrimp (Fig. 18.25a–f).

One of the smallest malacostracan orders is the order Euphausiacea (pronounced ū-fow´-zē-ā´-sē-ah), whose members are more commonly known as "krill." Only about 85 species have been described. Yet, the euphausiids' commercial and ecological importance far exceeds their limited species diversity. The annual production of euphausiid biomass in Antarctic waters alone is estimated to at least equal the current world harvest of all other marine animals combined, some 99 million metric tons per year. Some countries, notably Japan and the Soviet Union, are now exploiting these animals for human use. An international regulatory agency has limited the amount of krill that can be captured yearly to 1.5 million tons, more than three times the amount currently harvested. The ecological impact of increased krill exploitation is uncertain: Euphausiids constitute the primary diet of seals and of many baleen whale and seabird species; some whales consume perhaps 4 tons of krill each day.

Krill have the general appearance of decapod shrimp, except that they have eight pairs of thoracic walking legs rather than five pairs (Fig. 18.25g); none of the thoracic legs are specialized as maxillipeds. Also, the thoracic appendages of euphausiids bear conspicuous, feathery gills (Fig. 18.25g). Euphausiids grow to lengths of about 6 cm. They are remarkably tolerant of starvation—individuals have survived more than 200 days without food in the laboratory—and starved euphasiids actually decrease their size at each molt, a possible adaptation to the low productivity of Antarctic winters.

As with many shrimps and copepods (to be discussed shortly), and a variety of animals sporadically distributed among most other phyla, euphausiids typically can produce bioluminescence: light produced chemically, without heat. The light-producing organs (**photophores**) of some euphausiid species are among the most complex known (Fig. 18.26). The photophores are distributed on the body in species-specific patterns and may therefore function in species and mate recognition. Photophores also may protect against predation in surface water, by breaking up the silhouette of a euphausiid when seen by a predator from below.

Stomatopods (order Stomatopoda) are another interesting group of shrimp-like malacostracans, with about

7. See *Topics for Further Discussion and Investigation*, no. 12.

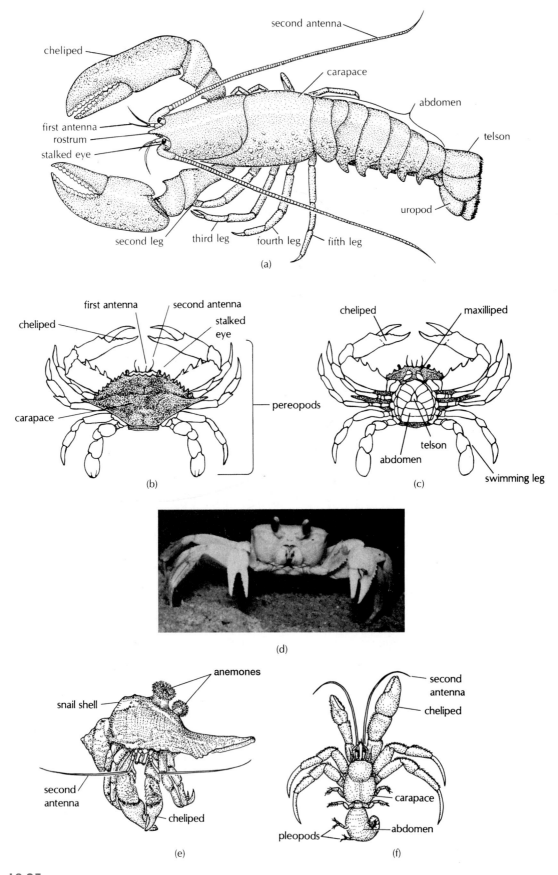

Figure 18.25
Malacostracan diversity.(a) A lobster, *Homarus americanus*.
(b) Dorsal view of the blue crab. *Callinectes sapidus*. The fifth pair
of thoracic legs have become modified into flattened paddles for
swimming. (c) Ventral view of same individual in (b), showing the
abdomen (♀). (d) Photograph of a marine crab. (e) Hermit crab
in a gastropod shell. Several dozen hermit crab species

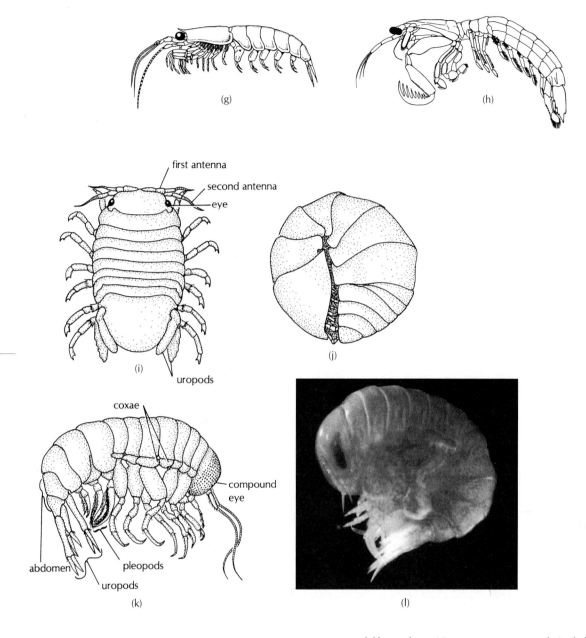

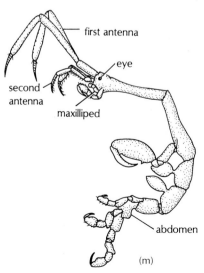

deliberately position sea anemones on their shells, gaining added protection from predators and providing the anemones with a firm, mobile substrate in return. (f) Same individual in (e), removed from the shell; note the soft abdomen, which is not tucked up under the body. (g) A euphausiid. (h) Lateral view of a stomatopod. (i) A marine isopod, *Sphaeroma quadridentatum*. Note the dorsoventral flattening of the body, the uniramous antennae, and the absence of a carapace. The eyes are not stalked. (j) Terrestrial isopod (pillbug), curled to form a ball. (k) The amphipod *Hyperia gaudichaudii* in side view. Note the large, unstalked compound eyes (only one is shown, covering nearly one entire side of the head), and the absence of a carapace. (l) A hyperiid amphipod, seen in lateral view. This individual was about 0.5cm long. Note the lateral compression of the body, lack of carapace, and the enormous compound eye. (eye on other side of head not visible). (m) The unusual amphipod *Caprella equilibra,* an animal highly modified for clinging to algae, hydroids, and other substrates. Caprellid amphipods are from 1 to 32 mm long.   (c) *After D. Krauss.* (d) *Courtesy of W. Lang.* (h) *From: Comparative Morphology of Recent Crustacea by McLaughlin. Copyright © 1980 by W. H. Freeman and Company. Used with permission.* (i) *After Harger.* (j) *After Pimental.* (k) *After Stebbing.* (l) *Photo by J. A. Pechenik.* (m) *After Light.*

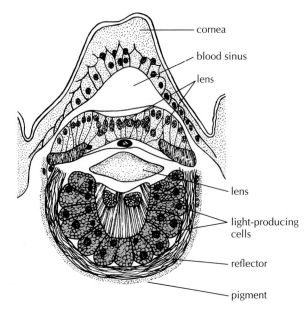

cornea

blood sinus

lens

lens

light-producing cells

reflector

pigment

Figure 18.26

A complex photophore of the euphausiid *Meganyctiphanes norvegica*. Such organs produce light intracellularly and are distributed in species-specific patterns on the appendages and ventrally on the thorax and abdomen. *From J. A. Colin Nichol, in* The Biology of Marine Animals, *2d ed. Copyright © 1967 Pitman Publishing, Division Longman Group UK, Ltd., London. Reprinted by permission.*

350 species. Stomatopods are rather large (up to about 35 cm long), bottom-dwelling, violent carnivores. Unlike those of the euphausiids, the *abdominal* appendages of stomatopods bear conspicuous gills (Fig. 18.25h). Stomatopods resemble flattened shrimp but, not being decapods, have eight pairs of thoracic legs like euphausiids. The second pair are large and extremely powerful, modified for smashing into hard-shelled prey, such as bivalves, gastropods, and crabs, or for spearing fish and other soft-bodied prey. Stomatopods deal with their prey with impressive speed; the specialized raptorial legs have been reported to strike their victims with speeds of up to 1,000 cm per second. In this respect, stomatopods resemble the terrestrial preying mantids and thus are commonly known as "mantis shrimp." The stomatopods are unique among crustaceans in having a jointed head, enabling the anterior and posterior portions to move independently and adding to the preying mantid appearance.

Most stomatopods are tropical, but some species do very well in temperate areas. They typically live in burrows—in rock, coral, or mud—in shallow water.

Two of the largest orders of nondecapod malacostracan crustaceans are the Isopoda and Amphipoda. The order Isopoda contains at least 10,000 species, about as many species as described in the Decapoda. Most isopods are marine, although both freshwater and terrestrial species (including the familiar "pill bugs" or "sow bugs") occur. Unlike the decapods, isopods have no carapace. Moreover, they have only a single pair of maxillipeds, in contrast to the three pairs found in decapods, and they have uniramous first antennae, as opposed to the decapods' biramous first antennae. Isopods tend to be small, about 0.5 cm to 3.0 cm in length. Compound eyes, if present at all, are not on movable stalks. Isopods are characteristically flattened dorso-ventrally (Fig. 18.25i) and accomplish gas exchange by means of flattened pleopods. Thus, respiratory appendages are associated with the abdomen. A number of terrestrial isopod species possess a system of tracheae. This is another example of convergent evolution, in which two or more groups of animals have independently evolved similar adaptations in response to similar selective pressures. In this case, as in the insects and arachnids, selection has favored internalization of the respiratory system as a means of deterring water loss in a terrestrial environment.

In contrast to the isopods, members of the order Amphipoda tend to be flattened laterally (Fig. 18.25l). Again, no carapace is present, individuals possess only a single pair of maxillipeds, and the compound eyes are sessile. About 6,000 amphipod species have been described, mostly from salt water. However, both freshwater and terrestrial species also exist. In contrast to the isopods, amphipod gills are found on the thorax, attached to the pereopods. In some species, a portion of the protopodite (the **coxa**) of each of several pairs of anterior appendages is elaborated into a large, flattened sheet (Fig. 18.25k), contributing significantly to the flattened appearance of the amphipod body.

Class Branchiopoda

Class Branchio · poda
(G: gill foot)
brank-ē-ō-pō´-dah

Branchiopods are a diverse group of small, primarily freshwater crustaceans. On each thoracic appendage, the **coxa** (one of the basal segments of the typical crustacean appendage) is modified to form a large, flattened paddle; this paddle functions in gas exchange and locomotion, giving rise to the name of the class (*branchio* = G: a gill; *pod* = G: foot). Most species are filter feeders, although a few are carnivorous. The bodies of most branchiopods are at least partially enclosed in a bivalved carapace (Fig. 18.27a–c). However, some species (the fairy shrimp) completely lack a carapace (Fig. 18.27d). Fairy shrimp include the well-known brine shrimp (*Artemia salina*),

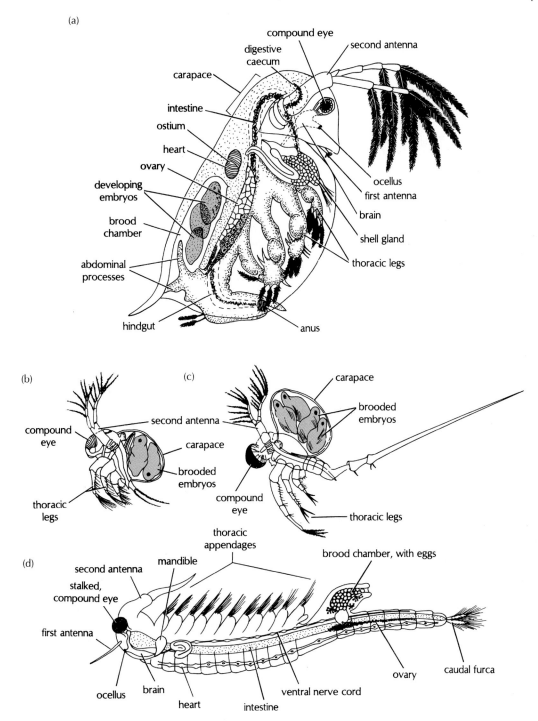

Figure 18.27

Branchiopod diversity. (a) *Daphnia pulex* (female), the freshwater water flea. Except for the head, most of the body lies within a laterally flattened, bivalve like carapace. The long, biramous second antennae are the primary locomotory appendages of cladocerans. The compound eyes are paired and sessile (not stalked). Other cladocerans: (b) *Podon intermedius* and (c) *Bythotrephes longimanus*. The carapace is greatly reduced in size,

serving primarily as a brood chamber. Cladocerans range from a few hundred microns to nearly 2 cm in length. (d) *Artemia salina*, the brine shrimp, in normal swimming orientation. Brine shrimp may attain lengths of about 1 cm, and possess a pair of stalked, compound eyes.   (a) *After Pennak, after Claus.* (b,c) *From G. Hutchinson, A Treatise on Limnology, Vol. 2. Copyright © 1967 John Wiley & Sons, New York. Reprinted by permission of the Estate of George Hutchinson.* (d) *After Brown; after Lochhead.*

which are found in waters whose salinities range from about 0.1 to 10.0 times the salt concentration of open-ocean seawater. The fertilized eggs of this species are commonly marketed as "sea monkeys."

Of the approximately 800 branchiopod species so far described, at least 50% are contained within the order Cladocera, the water fleas. Cladocerans, including the familiar genus *Daphnia* (Fig. 18.27a), dominate the zooplankton

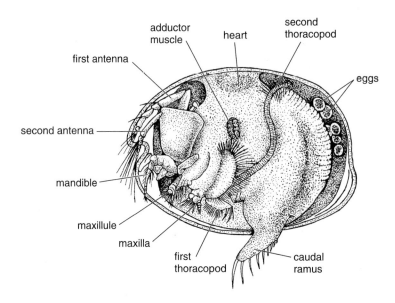

Figure 18.28
A marine ostracod seen in lateral view, with the left valve removed to show internal structure.

of freshwater lakes, although a few species are marine. All species are microscopic, with most of the body contained within a bivalved carapace. Protruding from this carapace is the head, bearing a pair of large, biramous second antennae that are used to propel the animal through the water. The first antennae and the second maxillae are much reduced. The thoracic region bears five to six pairs of appendages, which generate feeding and respiratory currents; food particles are filtered from the water by fine setae on these appendages. There are no abdominal appendages. The head bears, in addition to the second antennae, a single, huge, compound eye, formed by the fusion of the ancestral compound eyes from each side of the head. The eye is not stalked but can be rotated in various directions by associated musculature.

Members of another branchiopod group (the tadpole shrimp) completely lack second antennae; only a vestige remains. Clearly, the branchiopod body plan is rather plastic.

Class Ostracoda

Class Ostra · coda
(G: a shell)
os-truh-cō′-dah

Ostracods are small (rarely more than a few millimeters across) but widespread crustaceans, common in both

marine and freshwater habitats. Remarkably, a few species are terrestrial. Ostracods are mostly head; the rest of the body is greatly reduced, bearing at most two pairs of appendages (fewer than in any other crustacean), and showing no sign of segmentation externally. The body is completely encased within a partially calcified, bivalved carapace (Fig. 18.28). Most of the approximately 6,650 species are free-living, although some ostracods are commensal with other crustaceans or with certain echinoderms; there are no parasitic ostracods. Another 10,000 ostracod species are known only as fossils, with an extensive record extending back some 530 million years.

A few extant species are completely planktonic, but most ostracods leave the bottom only periodically, using their first or second antennae, or sometimes both pairs of antennae, for swimming. A number of species are completely benthic bottom-dwellers, in some cases burrowing within the sediment. Ostracods include carnivorous, suspension-feeding, scavenging, and herbivorous species.

Class Copepoda

Class Cope · poda
(G: oar foot)
cō-peh-pō′-dah

Most of the approximately 8,500 species in the class Copepoda are marine and feed on unicellular, free-floating, photosynthesizing protists called **phytoplankton** (*phyto* = G: plant; *plankton* = G: that which is forced to wander). Copepods are invariably small, usually less than 1 mm to 2 mm long. Some copepod species occur in freshwater lakes and ponds, while terrestrial species live in soil or in moist surface films in humid environments. Perhaps two-thirds of all copepod species are planktonic in the ocean. Because of their small size, such copepods are, to a large extent, at the mercy of currents. Together with other animals of limited locomotory capability (relative to movement of water currents), these copepods form a major component of the **zooplankton** (*zoo* = G: animal). Most other copepod species are specialized for life in or on substrates, forming a major component of the **meiobenthos** (i.e., the community of small animals living in association with sediment). Locomotion of planktonic copepods is accomplished primarily by the actions of a pair of biramous second antennae, while benthic species use their thoracic appendages to walk over surfaces.

Copepods are among the most abundant animals on earth and are also among the most important herbivores of the ocean. In part, they collect phytoplankton through the activities of the first and second maxillae (Fig. 18.29a), although the details of food capture are complex and incompletely understood (Research Focus Box 18.2).[8] Copepods are at the base of the oceanic food chain in another respect as well: They are a major food source for primary carnivores, including fish of commercial importance.

Most free-living copepods have a single, median eye on the head (Fig. 18.29a,b,d). The eye usually consists of three lens-bearing ocelli; two of the units look forward and upward, and the third ocellus is directed downward. Yet, exceptions do exist; some species have a pair of eyes, placed laterally and with conspicuous lenses (Fig. 18.29c), while other species lack eyes entirely. Compound eyes are never encountered among the copepods.

In contrast to many other crustaceans, copepods lack gills and abdominal appendages. However, the other tagma (head and thorax) do bear appendages (Fig. 18.30). The structure and function of copepod appendages vary substantially with species and lifestyle and, often, with sex as well. One or both first antennae may be hinged in the male (Fig. 18.29a[3]), functioning to capture females for mating, in addition to performing the standard sensory function. In addition, one or both of the male's fifth thoracic appendages may terminate in a claw (Fig. 18.29a[2]), which is used to hold the female during mating.

Although a free-living, suspension-feeding existence is the norm for planktonic copepods, a number of planktonic species are carnivorous, and many benthic species, living on sediment or in the spaces between sand grains,

scrape off food particles from solid substrates. Still other copepods, about 25% of all described species, parasitize a variety of vertebrates and invertebrates. As might be expected, the head appendages and other body parts are modified, sometimes extravagantly, as reflections of these different lifestyles (Fig. 18.31).

Class Pentastomida

Class Penta · stomida
(G: five mouths)
pen-tah-stō´-mid-ah

The taxonomic affinities of pentastomids ("tongue worms") have long been uncertain. All pentastomids are internal parasites of vertebrates, and most characteristics of potential diagnostic utility have been lost; most aspects of adult external morphology—and of the sensory, digestive, excretory, and reproductive systems—have become highly modified as adaptations for an exclusively endoparasitic existence (Fig. 18.32). The larval stages of pentastomids also are parasitic, with an associated disappearance or extreme reduction of virtually all potentially revealing ontogenetic characteristics. The fertilized egg exhibits spiral cleavage, so at least the protostome affiliation of the group is clear. But a consensus for placing pentastomids among the Crustacea—and even among the Arthropoda—has emerged only recently; for many years, the Pentastomida was considered a separate phylum.

In support of their affiliation with arthropods, pentastomids show such arthropod characteristics as an exoskeleton that is periodically molted, striated musculature arranged metamerically, a spacious hemocoel, and larval stages often possessing three pairs of appendages. However, pentastomids also show some decidedly unarthropod characteristics: The larval legs are not jointed, and the adult is legless; the only adult appendages are four pairs of anterior, chitinous hooks. Moreover, the pentastomid cuticle is not chitinous. But the weight of available evidence argues strongly for pentastomid membership within the Arthropoda. Perhaps the most compelling evidence of their arthropod affinity is found in the unusual morphology of their sperm, the characteristics of which suggest that pentastomids are most closely related to a group of marine crustaceans, the Branchiura, which are external parasites (**ectoparasites**) of fish (see *Taxonomic Detail* at the end of this chapter, p. 412). Detailed studies of cuticle ultrastructure that have accumulated over the past 15 years or so, together with recent analyses of DNA sequences coding for 18S ribosomal RNA support the idea that pentastomids are highly modified crustaceans.

---

8. See *Topics for Further Discussion and Investigation*, no. 11.

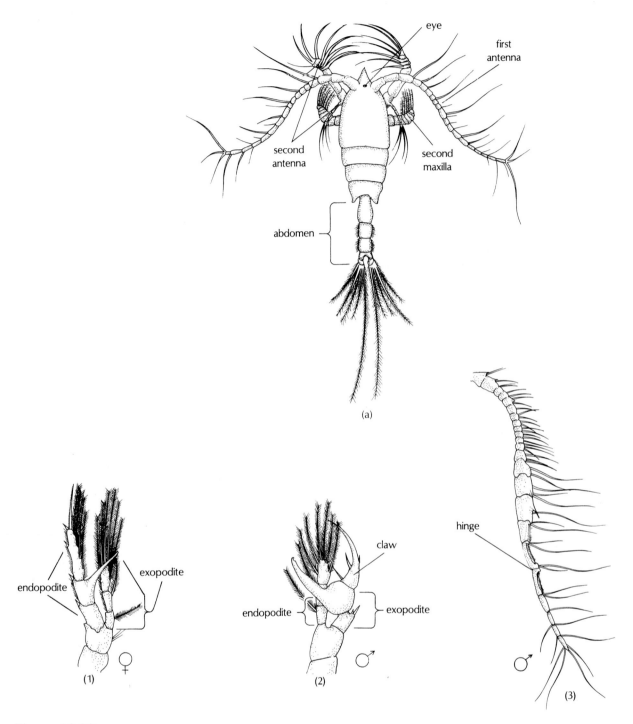

(a)

Figure 18.29
Diversity among the free-living copepods. (a) *Euchaeta prestandreae,* a typical, actively swimming copepod. Copepod appendages: The fifth right thoracic legs of a female (1) and a male (2) copepod, *Centropages typicus,* illustrating the clawed appendage

of the male. The right first antenna of the male (3) illustrates the hinge. (b) Free-living copepod carrying two clusters of fertilized eggs on the abdomen. (c) A harpacticoid copepod, *Tisbe furcata.* (d) A cyclopoid copepod, *Sapphirina angusta.*   (a) *After McConnaughey*

Adult pentastomids dwell mainly in the lungs and nasal passages of reptiles, although some species instead specialize in the same areas within amphibians, birds, and even some mammals, including dogs and horses. All species feast on the host's blood, clinging to host tissues

with sharp anterior hooks. Variations in hook morphology provide one of the few useful criteria for classifying pentastomid species. The parasites are dioecious (i.e., there are separate males and females), and embryogenesis occurs within a shelled capsule. The nascent, encysted

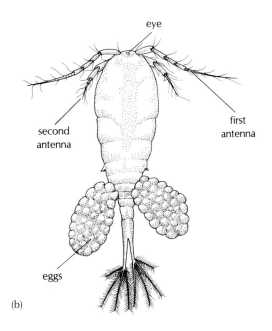

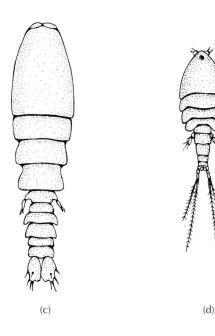

(b)

eye

second
antenna

first
antenna

eggs

(c) (d)

and Zottoli; after Brady, 1883. Challenger Reports, Vol. 8. from C. B. Wilson, 1932, "The Copepods of the Woods Hole Region". Smithsonian Institution Bulletin 158. (b) From Pimentel. (c,d) Reprinted from C. B. Wilson, "The Copepods of the Woods Hole

Region", 1932, from the Bureau of American Ethnology (Washington, DC: Smithsonian Institution Press:, pages 197, 343, 354, by permission of the publisher. Copyright 1932.

parasite usually leaves the host in nasal secretions and saliva. It develops no further unless eaten by the proper intermediate host, usually another vertebrate. Some insects act as obligate intermediate hosts for a few species. If taken inside the correct intermediate host, the larvae emerge, tear through the gut wall, and develop further within species-specific tissues, undergoing up to nine molts before reaching a stage infective to the final, definitive host. As with other internal parasites, pentastomids cannot reach sexual maturity in the intermediate host; that is, after all, why these hosts are termed "intermediate." If the intermediate host is ingested by the right definitive host, the juvenile migrates back up to the pharynx and thence to the desired location in the respiratory system, using its piercing claws.

In at least one pentastomid species, the intermediate host can be bypassed. In this single documented case, the larvae can develop and become infective within the definitive host (Norwegian reindeer) and then, in female hosts, penetrate placental tissue to infect the next generation of reindeer directly. This finding provides one of the few examples of likely recent evolution within a parasite life cycle; perhaps the intermediate host will eventually be eliminated from the life cycle of this species. How this pentastomid species is able to develop to adulthood entirely within the definitive host when an intermediate host is absolutely required for other parasitic species in this phylum (and most other phyla) is unknown.

Each pentastomid species is highly specific for particular hosts, but the host is never seriously harmed. People occasionally become infected by eating intermediate

stages in uncooked meat, but the parasites cannot reach adulthood in humans and die within a few weeks, after causing only minor irritation.

Since all extant pentastomids parasitize vertebrates, one might expect that pentastomids evolved (presumably from crustacean ancestors) only after the evolution of vertebrates. However, what appear to be fully convincing pentastomids were described in 1994 from Cambrian rocks formed some 500 million years ago, well before the advent of vertebrates, or even of terrestrial arthropods. One can but wonder about the original pentastomid hosts—they must have been aquatic and invertebrate—and about why modern pentastomids parasitize only vertebrates.

Class Cirripedia

Class Cirri · pedia
(L: hairy foot)
sēr-i-pē´-dē-ah

Cirripedes are more commonly billed as the barnacles. The approximately 1,000 species in this class are exclusively marine and show a greater departure from the basic crustacean body plan than the members of any other class. Unlike most other crustaceans, all cirripedes are exclusively sedentary organisms, permanently affixed to, or burrowed into, living (including whales) or nonliving substrates. Barnacles attached to moving or floating

Research Focus Box 18.2

Copepod Feeding

Cowles. T. J., R. J. Olson, and S. W. Chisholm. 1988. Food selection by copepods: Discrimination on the basis of food quality. *Marine Biol.* 100:41.

Over the past 40 to 50 years, the feeding biology of herbivorous marine copepods has been the subject of a growing number of studies, in recognition of the great role these small planktonic crustaceans play in marine food chains that lead to commercially important fish species. Until about 1970, most biologists believed that copepods passively collected food particles from the surrounding water, using the second antennae to generate feeding currents and setae of the first and second maxillae as sieves. This mechanism implied that copepods indiscriminately ingest all particles large enough to be captured by the "sieve." However, research in the past 20 years has indicated that both marine and freshwater planktonic crustaceans deliberately can accept some algal cells and reject others. *How* planktonic crustaceans make these discriminations among food particles is not yet clear. Cowles, Olson, and Chisholm (1988)

have demonstrated that at least one copepod species can distinguish between foods on the apparent basis of differences in nutritional value.

To study whether herbivorous copepods (*Acartia tonsa*—about 500 µm long) can recognize differences in nutritional value requires using two diets identical in all physical characteristics and apparently differing only in intracellular chemistry. The researchers achieved this by growing the marine diatom *Thalassiosira weissflogii* (about 14 µm in diameter) at two different nitrogen concentrations (provided as nitrate). The resulting cells were identical in cell size and in average carbon content, but they differed in individual nitrogen content, chlorophyll content, protein content, and amino acid content; the nutrient-deprived cells were lower in each of these biochemical measures (Focus Table 18.4).

Focus Table 18.4   Chemical Composition of the Diatom *Thalassiosira weissflogii* Cultured at Two Nutrient Levels

Nutrient Level	Cell Diameter (µm)	Carbon (pg cell^{-1})	Nitrogen (pg cell^{-1})	C:N	Chl *a* (pg cell^{-1})	Protein (pg cell^{-1})	Free Amino Acids (pg cell^{-1})
Low	14.1 ± 2.3	202.4 ± 29.8	11.8 ± 2.6	17.2 ± 3.7	0.65 ± 0.5	41.2 ± 9.0	735.5
High	14.0 ± 2.1	179.2 ± 25.0	16.5 ± 3.9*	10.9 ± 2.0*	1.85 ± 0.38*	93.6 ± 30.0*	1537.6

Note: Values are means ±2 SE. n = 12, except for protein, where n = 7, and free amino acids, where n = 1. pg = picograms.

*Significant difference (p<0.05) between cell types, paired comparison t-test (df = 2n–2).

After collecting copepods with plankton nets and acclimating them to laboratory conditions for several days, the researchers determined copepod feeding rates (cells ingested per five hours) on each of the two chemically distinct algal diets. If the initial number of algal cells per milliliter of seawater in the test container, the final number of cells per milliliter after five hours of feeding, the volume of seawater in the container, and the number of copepods put into the container are all known, then feeding rates are easily determined; if cells disappear, it can only be through ingestion. In five of the eight experiments reported in the paper by Cowles et al. (1988), copepods fed substantially more

quickly on the cells grown in the high-nutrient medium (Focus Fig. 18.2). These high-nutrient cells thus seemed to somehow stimulate the copepods to feed more actively. Other experiments reported in this paper indicated that the stimulatory agent is a chemical released by the diatom. In all eight experiments, copepods ingested protein at a significantly faster rate when feeding on the high-nutrient cells, even when eating comparable numbers of cells per hour (Focus Fig. 18.3); this is possible due to the higher protein content of the cells grown in the high-nutrient medium.

What would happen if the copepods were offered a choice of diets? Could they actually distinguish between

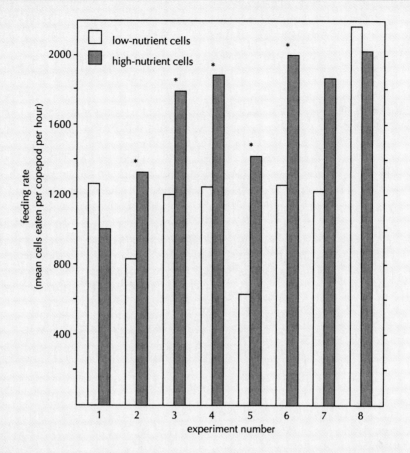

Focus Figure 18.2

The rate at which the copepod *Acartia tonsa* fed on cells of the diatom *Thalassiosira weissflogii* when those cells were cultured either under high-nutrient or low-nutrient conditions. The height of each bar represents the mean of three determinations. Asterisks signify significant differences between mean feeding rates.

cells differing in nutrient composition and *preferentially* choose to eat the most nutritionally valuable ones? How would you design such an experiment? If the two types of cells are physically identical, how could researchers determine how many of each type the copepods ate?

It turns out that when cells of the two nutrient types are exposed to light and then darkness, they fluoresce at markedly different wavelengths. If a sample of the feeding mixture is placed—before and after the five-hour feeding study—into a complex machine called a "flow cytometer," the machine will examine each cell's fluorescence pattern and then sort the sample cell by cell into a high-nutrient or low-nutrient category. The flow cytometer thus makes it possible to distinguish between the cells, even though they look identical. In this experiment, many more high-nutrient cells disappeared from the medium during feeding than would have been predicted from their abundance in the

mixture at the start of the study; these nutritionally superior cells seem to have been eaten preferentially, certainly not in direct proportion to their abundance (Focus Figure 18.4).

How can a copepod possibly tell the difference between one algal cell and the next if the cells are morphologically indistinguishable? The researchers suggest that either (1) the copepod recognizes a nutritionally more valuable cell from a distance and preferentially captures such a cell, or (2) the copepod captures both cell types at equal rates but then preferentially *ingests* the high-nutrient cells, rejecting the others. In both cases, the copepod is presumed capable of perceiving a subtle chemical gradient surrounding each cell and behaving accordingly. Certainly, diatoms and other unicellular algae are known to produce a variety of chemical exudates. But can a copepod really be so sensitive to what are likely very subtle differences in chemical gradients surrounding individual algal cells?

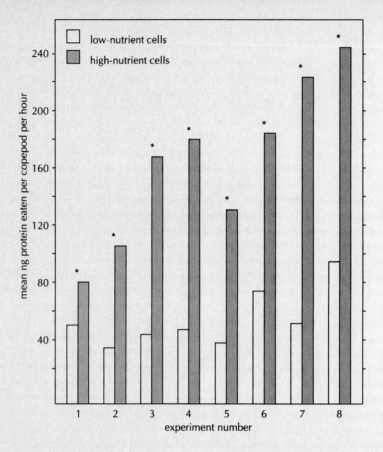

Focus Figure 18.3

The rate at which the copepod *Acartia tonsa* ingested protein when fed diatoms (*Thalassiosira weissflogii*) reared under either high-nutrient or low-nutrient conditions. The height of each bar represents the mean of three replicates. Asterisks signify significant differences between mean protein ingestion rates. (ng = nanograms.)

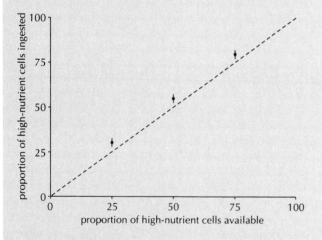

Focus Figure 18.4

Preference for high-nutrient algal cells by copepods (*Acartia tonsa*) fed mixtures of high- and low-nutrient cells. If copepods were eating cells in proportion to their abundance, data points would fall on the dotted line. Points above the dotted line indicate a preference for high-nutrient cells. Vertical bars attached to each point indicate one standard deviation about the mean. Each point represents the mean of three replicates.

Could the copepod instead be responding to small differences in cell surface texture or buoyancy that were undetected by the researchers? And does the ability to somehow make this fine distinction between cells of different chemical composition make a difference—that is, do the copepods grow or reproduce significantly better as a consequence of this selective feeding behavior?

Whatever the eventual answers to these and related questions, copepod feeding rates clearly are not easily predicted by knowing only how many algal cells are available in the water. Thus, copepod growth and reproductive rates in the field are difficult to predict with confidence, and such information is needed to predict successfully the number of carnivores (including various edible fish species) that a given copepod population can support. Copepods obviously are capable of far more behavioral sophistication than might be expected from an animal barely visible to the unaided eye.

Figures and table from T. J. Cowles, R. J. Olson, and S. W. Chisholm, "Food Selection by Copepods: Discrimination on the Basis of Food Quality" in Marine Biology, 100:41–49, 1988. Copyright © 1988 Springer-Verlag, Heidelberg, Germany. Reprinted by permission.

(a)   (b)

Figure 18.30

(a) Scanning electron micrograph of a free-living cyclopoid copepod in dorsal view. The total length is about 1 mm. Note the point of articulation between the fourth and fifth thoracic segments, and the short first antennae. (b) Ventral view of same animal, showing head appendages. *(a,b) Courtesy of C. Bradford Calloway.*

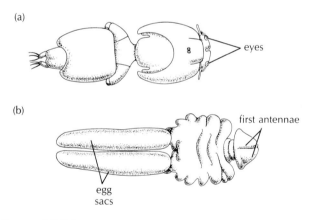

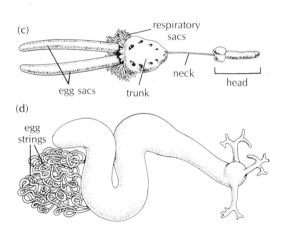

Figure 18.31

Parasitic copepods. (a) *Caligus curtus,* female, found on the outside surface of various marine fish, including cod, pollock, and halibut. Total length is 8–12 mm. (b) *Chondracanthopsis nodosus,* female, taken from the gills of redfish. The body is about 7 mm long. (c) *Rebelula bouvieri,* female, taken from the flesh of a marine fish. The egg sacs alone are 30–40 mm long. (d) *Lernaea branchialis,* a parasite in the gills of flounder. The body of the copepod may reach 40 mm in length, while the egg strings may be several hundred mm long.

Reprinted from C. B. Wilson, "The Copepods of the Woods Hole Region", 1932, from the Bureau of American Ethnology Washington, DC: Smithsonian Institution Press:, pages 197, 343, 354, by permission of the publisher. Copyright 1932.

substrates often are conspicuously stalked (Fig. 18.33c). Other species may be cemented directly to the substrate at the "basis." In keeping with a nonmotile existence, the head is greatly reduced, the first antennae are much reduced, and the second antennae are completely absent. Most species live within a thick calcium carbonate protective shell (Fig. 18.33), which they secrete. Because of this shell, barnacles were classified as molluscs until about 150 years ago. Indeed, the barnacle shell is secreted by the "mantle" tissue, and the space between the inner wall of the shell and the animal itself is called the "mantle cavity," as in molluscs.

Barnacles have attracted a reasonable amount of attention as foulers of ship bottoms; even a moderate

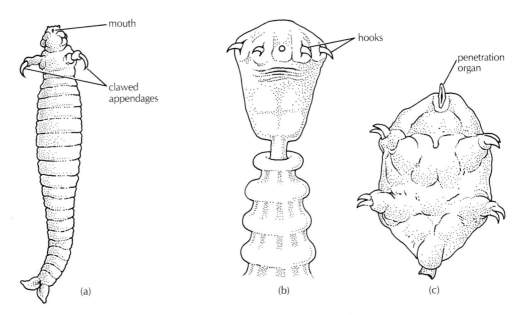

Figure 18.32

Representative pentastomids. The adults are parasitic in the nasal passages of snakes and lizards, while cockroaches, bats, raccoons, muskrats, and armadillos commonly serve as intermediate hosts for the larval stages. (a) *Raillietiella mabuiae*, entire individual. (b) Anterior end of *Armillifer annulatus*. (c) Intermediate larval stage of *Porocephalus crotali*. (a,b) *From Baer, 1951. Ecology of Animal Parasites. Champaign: University of Illinois Press.* (c) *From E. R. Noble and G. A. Noble, Parasitology: The Biology of Animal Parasites, 5th ed. Philadelphia: Lea & Febiger, 1982. After Penn. Courtesy of* The Journal of Parasitology.

encrustation of barnacles can greatly reduce ship speed and fuel efficiency. The happy consequence of barnacle encrustation is that research on the biology of cirripedes has not been difficult to justify.

The shell of barnacles is composed of numerous plates, including the **carina, rostrum, scuta,** and **terga** (Fig. 18.33a–c). The rostrum represents the side of the shell at which the body of the barnacle attaches to the mantle. Certain of the scuta and the terga are movable. Thus, the scuta and terga can occlude the opening at the top of the shell when the animal withdraws, and they can move apart when the feeding appendages are to be protruded. The feeding appendages, called **cirri,** are modified thoracic appendages (Fig. 18.33a); these are used by free-living barnacles to filter food particles from the water. Recent studies indicate that members of at least some species feed actively when water velocities moving past the barnacles are low but switch to passive filtering when water velocities exceed a particular threshold (Fig. 18.34).

Barnacles lack abdominal segments, gills, and a heart; the blood circulates through sinuses entirely by movements of the body. The circulatory system is open, as it is in all arthropods.

A number of cirripedes live in burrows within calcareous substrates (e.g., shell, coral) or as parasites within the bodies of other animals, including other crustaceans. As might be expected, such specializations in habitat and lifestyle are reflected by major morphological modifications, particularly by the presence of root-like absorptive structures (Fig. 18.33d,e).[9] The life cycles of parasitic barnacles are often bizarre and certainly unique among crustaceans, as discussed in the next section.

Other Features of Arthropod Biology

Reproduction and Development

Sexual reproduction is the rule among arthropods. Fertilization is internal in most species, but it is external in some. Most species are **dioecious** (i.e., have separate sexes), although sedentary and parasitic arthropod species are often hermaphroditic, and some free-living species exhibit various degrees of asexual reproduction. **Parthenogenesis—** that is, production of offspring from unfertilized eggs—is commonly encountered among the Insecta and Branchiopoda and in some freshwater copepods. Indeed, males have never been found in some species of these groups.

Marine species often have a free-living larval stage in the life history (Fig. 18.35). A free-living **nauplius** larva is typical of several diverse groups of crustaceans, including the copepods, ostracods, branchiopods, and cirripedes.

---

9. See *Topics for Further Discussion and Investigation*, no. 4.

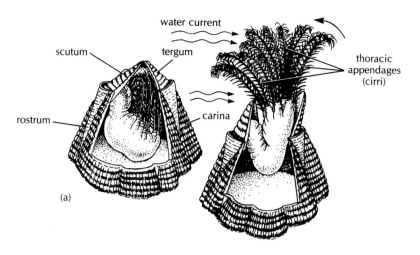

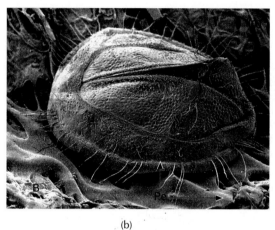

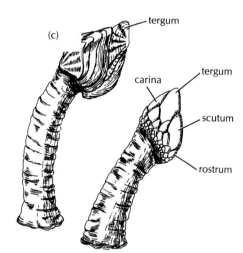

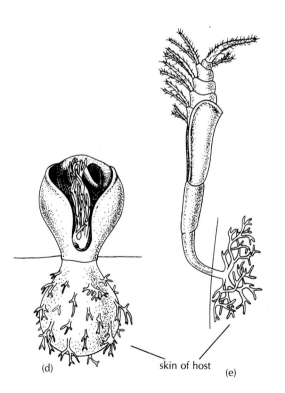

water current

scutum   tergum

thoracic
appendages
(cirri)

rostrum

carina

(a)

(b)

(c)

tergum

tergum

carina

scutum

rostrum

skin of host

(d)   (e)

Figure 18.33

(a) The free-living barnacle *Balanus* sp., an "acorn barnacle." The scutum and tergum of the barnacle shell can move apart, allowing the animal to extend its cirri in the water currents and filter out food particles. The scuta and terga form a type of operculum, protecting the animal from predators, salinity stress, and desiccation stress once the thoracic appendages have been withdrawn. (b) Juvenile of the barnacle *Balanus improvisus* 4–6 hours after the free-swimming larval stage attached to the surface and shed its larval cuticle. The plates of the adult shell are already conspicuous. This individual is about 680 μm (micrometers) in longest dimension. (c) Stalked gooseneck barnacles; *Lepas* sp. on the left and *Pollicipes* sp. on the right. Barnacles encrust a variety of substrates, including ship bottoms, turtle shells, and whales. (d,e) Parasitic barnacles, shown in the skin of a shark (d) and an annelid (e).   (a) *From Wells, Lower Animals. Copyright © 1968 McGraw-Hill, Inc., New York. Reprinted by permission of McGraw-Hill, Inc. and Geo. Weidenfeld & Nicholson Ltd., London. (b) From H. Glenner and J. T. Høeg. 1993. Marine Biology. 117:431–439. Springer-Verlag, New York. (c) After Pimentel. (d,e) From Baer, 1951. Ecology of Animal Parasites. Champaign: University of Illinois Press.*

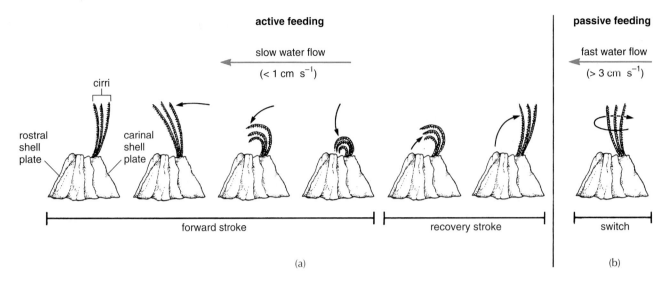

active feeding

slow water flow
(< 1 cm s⁻¹)

passive feeding

fast water flow
(> 3 cm s⁻¹)

Figure 18.34
Feeding by the common barnacle *Semibalanus balanoides.*
(a) Active capture of food particles by cirri when water is flowing
slowly. (b) Under conditions of faster water flow, the cirri
reverse their orientation and filter particles from the water

passively. Barnacles switched to passive feeding when mean water
velocity exceeded 3.10 cm · sec⁻¹ (centimeters per second).

(a) From C. C. Trager et al., in Marine Biology, 105:117–127, 1990. Copyright © 1990
Springer-Verlag, New York. Reprinted by permission.

Although the adult body plan of barnacles has become highly modified from the basic crustacean pattern, the typical nauplius larva remains unchanged; even highly modified parasitic species produce typical nauplii. In all species, the nauplius has a characteristic triangular shape, a good crustacean carapace, and a median eye composed of three ocelli (Fig. 18.35a,b). Periodically, the nauplius larva molts and subsequently adds or modifies appendages and gets larger. After going through several naupliar stages, the nauplius metamorphoses into a morphologically distinct larval stage. Among the Copepoda, the final naupliar stage undergoes a transition to a **copepodite** form, and then the individual goes through five copepodite stages before the final adult body plan is attained. Among the barnacles, the larva proceeds through several naupliar stages and then metamorphoses into a remarkable **cypris** stage (Fig. 18.35c). The cypris larva, or **cyprid,** is housed in a bivalved, noncalcified carapace. The thoracic appendages with which the larva swims become the filtering appendages of the adult barnacle. A pair of sensory antennae are conspicuous at the animal's anterior end. When the cyprid of nonparasitic species locates a suitable substrate, it secretes a glue from anterior cement glands and the animal then becomes permanently attached to the substrate by its head. Dramatic internal reorganization and production of the adult body enclosure quickly ensue.

Parasitic barnacles also have free-living, perfectly normal cyprids as the terminal larval stage, but their life cycles are often strikingly modified from the moment of substrate selection. A short summary will illustrate just how modified the life cycle can become. Female cyprids of several well-studied species (e.g., members of the genus *Sacculina*) attach to a prospective crustacean host and

metamorphose into another form (the **kentrogon**), specialized for piercing through the host cuticle. The entire kentrogon does not enter the host, however; rather, it injects only a part of itself through the cuticle, and this material then develops extensively within the host. Eventually, a large, specialized reproductive mass, containing the ovary, emerges on the host's lower abdominal surface (Fig. 18.36a). Once a small aperture develops on this external female gonad, the gonad becomes attractive to male cyprids, which walk into the opening and attach to the female. Within an hour, a spiny, amorphous mass of tissue is discharged through one of the male cyprid's antennae. This small mass of tissue, which is best thought of as a juvenile male, somehow migrates within the female mantle cavity to one of two receptacle ducts and eventually implants at the distal end of a waiting receptacle chamber (Fig. 18.36b). The dwarf male then begins differentiating gonadal tissue, and little else. Only one male is apparently ever found in a single receptacle, so each female supports and nourishes up to two males within her reproductive tract for their entire lives. The female gonad does not mature unless at least one of the two receptacle chambers houses a male. Eggs can then be fertilized and free-living nauplii released into the surrounding water.

Not all crustaceans produce free-living nauplius larvae, although all pass through a naupliar stage as they develop. Decapod crabs and crayfish, for instance, brood their embryos (externally, beneath the abdomen) to a more advanced stage, releasing **zoeae** larvae (Fig. 18.35d). The zoeal stages are characterized by a pair of large compound eyes and a carapace from which two spines project ventrally and dorsally. Decapods typically go through a number of zoeal stages before metamor-

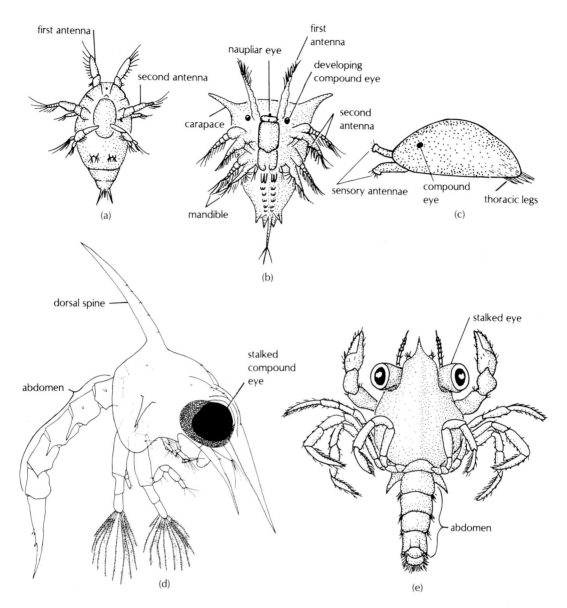

Figure 18.35
Crustacean larval stages. (a) Copepod nauplius. (b) Barnacle
nauplius. (c) Barnacle cyprid. (d) Decapod zoea: *Portunis sayi*, stage
IV. (e) Decapod megalopa larva.   (b) *After Korschelt.* (d) *Courtesy of I. P. Williams.*
(e) *After Hardy.*

phosing into a **megalopa** stage, which looks much like a
crab except that the abdomen is not tucked up under the
thorax (Fig. 18.35e). Other crustacean groups produce
other types of larval stages, always recognizable as being
arthropod. Freshwater crustaceans often lack free-living
larval stages, probably for reasons discussed in Chapter 1;
freshwater copepods, however, often have free-swimming
naupliar stages in the life history.

Isopods and amphipods also lack larval stages. Young
emerge as miniatures of the adult, after a period of protection by the adult female. The lack of free-living larval stages
in aquatic species may be a preadaptation for a terrestrial

existence. Not surprisingly, isopods and amphipods are the
only groups of malacostracans to have accomplished major
radiations into terrestrial or semiterrestrial habitats.

Although internal fertilization is sporadically distributed among marine arthropods, it is the rule among
terrestrial species. Indeed, internal fertilization in marine
forms must have been a prerequisite, a preadaptation, for
the invasion of land. Many arthropods, including insects,
scorpions, and arachnids transfer sperm to females indirectly, by means of specialized containers (**spermatophores**) that are passed to the female in a variety of
ways, as described in Chapter 25 (pp. 511–513).

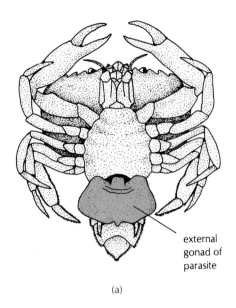

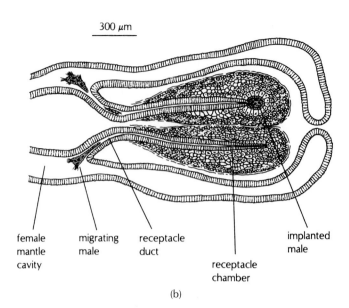

(a)   (b)

Figure 18.36

(a) External gonad of the rhizocephalan *Sacculina carcini*, parasitic on (and in) the green crab *Carcinus maenas*. The crab is shown in ventral view. Internally, the parasite ramifies throughout the crab's tissues. (b) Diagrammatic cross section through a portion of the external gonad of a female rhizocephalan barnacle, *Sacculina carcini*. As shown, the female has two narrow receptacle ducts opening into the mantle chamber. A dwarf male, expelled through one antenna of a cyprid larva, is shown entering the

lower of the two ducts. The upper duct already houses one male at its distal end; another male, shown about to enter this upper duct, will never reach the end: The passageway is blocked by the cuticle shed previously by the first male to enter.   (a) *From K. Rhode, 1982. Ecology of Marine Parasites. New York: Queensland Press; after Boas.* (b) *From J. T. Høeg, Philosophical Transactions of the Royal Society, 317:47–63, 1987. Copyright © 1987 The Royal Society, London. Reprinted by permission.*

The fertilized eggs of terrestrial species require some form of protection, especially from desiccation, and are often provided with sufficient food to fuel most or all of their prejuvenile development. The nutritional requirements of the fertilized eggs of terrestrial species can often be met only if the female has access to a high-protein diet during the period of egg formation, or **oogenesis** (*oo* = G: egg; *genesis* = G: birth). Hence, many female insects require a blood meal to mature their eggs prior to **oviposition** (i.e., discharge and placement of eggs). A number of insects, notably the wasps, meet the nutritional needs of their larvae by placing their eggs in or adjacent to the eggs of other insect species or within the bodies of other adult insects, which are then devoured by the developing young from the inside out (Research Focus Box 18.1).[10] Some insects deposit their eggs in plants, which respond by forming protective galls. The eggs are inserted into these various substrates through a long tube, called an **ovipositor,** typically protruding from the abdomen. Harvestmen (class Arachnida) are similarly equipped, and for a similar purpose. In some insects (e.g., bees), the ovipositor has been modified to form a stinger.

During development, insects pass through several larval stages, called **instars.** This is conspicuously true for insect species that undergo a **metamorphosis** from a larval to a distinctly different adult body plan. In some species, this transition is gradual, and the different in-

stars are called **nymphs** (Fig. 18.37a,b). Aquatic nymphs are sometimes referred to as **naiads.** Dragonflies, grasshoppers, and cockroaches, for example, develop in this manner and are said to be **hemimetabolous** (*hemi* = G: half; *metabolo* = G: change) (Fig. 18.38a). In most other insect species, the change to adult form is radical and abrupt, and termed **holometabolous** (*holo* = G: whole; *metabolo* = G: change). The feeding, immature stages are termed **larvae** (Fig. 18.37c,d). After passing through several larval instars of ever-increasing size, a morphologically distinct, nonfeeding pupal stage is formed. The **pupa** then undergoes extensive internal and external reorganization to form the adult morph. Butterflies provide what is probably the most familiar example of holometabolous development (Fig. 18.38b). Wasps and ants are other noteworthy examples, with the adults laboriously tending the helpless larvae and pupae. Holometabolous development characterizes about 88% of all extant insect genera, as opposed to less than 50% of insect genera known from 250 million year old fossils; life histories with ecologically distinct larval stages and a dramatic transition to adulthood clearly have had a selective advantage over those exhibiting more gradual development to the adult stage.

Adults of **apterygote** species, such as silverfish, lack wings (*a* = G: without; *ptero* = G: wing), and in fact are believed to be direct descendants of wingless ancestors; thus,

10. See *Topics for Further Discussion and Investigation,* no. 4.

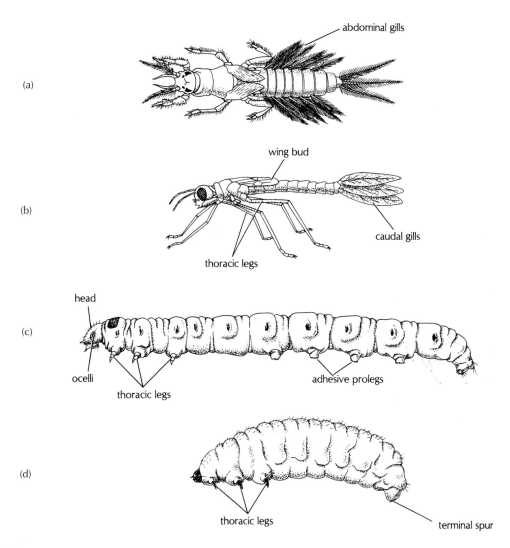

Figure 18.37

Insect development. The nymphs show a gradual transition to the adult form, exhibiting hemimetabolous development. In contrast, the larval stages of moths, beetles, and related species show little resemblance to the adult; development to adulthood is radical and abrupt, and is termed holometabolous. (*a*) Mayfly nymph, *Ephemera varia*. (b) Damselfly nymph. (c) Moth larva, *Bellura* sp. (d) Beetle larva *Donacia* sp.   (a) *After Pennak: after Needham.* (b) *After Sherman and Sherman.* (c,d) *After McCafferty.*

they primitively do not exhibit a pronounced metamorphosis as they develop. Instead, immatures simply get larger with each succeeding molt, and the body plan resembles that of the final adult at each stage. Such development is termed **ametabolous** (i.e., without change). Some other wingless insect species, such as fleas and lice, have evolved from winged ancestors and lost the wings secondarily; these species do metamorphose during development.

Holometabolous development is particularly characterized by a distinctly insect phenomenon, the formation of **imaginal discs.** At the completion of cleavage, small groups of up to about 30 cells give rise to discrete "discs" (spheres, actually), which are destined to differentiate into very well-defined adult epidermal structures, such as the eyes, the antennae, or the wings. Throughout larval development, however, these imaginal discs remain quiescent, or they divide and grow at a very slow rate compared to the rest of the larva. The cells of one disc are distinguishable from those of other discs found elsewhere in the body only by their position and a small number of structural details. Nevertheless, the eventual fate of the cells in a disc is fixed, and at metamorphosis a particular disc always will give rise to the same structure, even if surgically transplanted elsewhere in the body. The orientation of the structure within the body may be incorrect following such disc transplantation, but the structure itself will be perfectly formed. Imaginal discs long have been utilized by developmental biologists to probe the manner in which the expression of genes in individual cells is controlled.

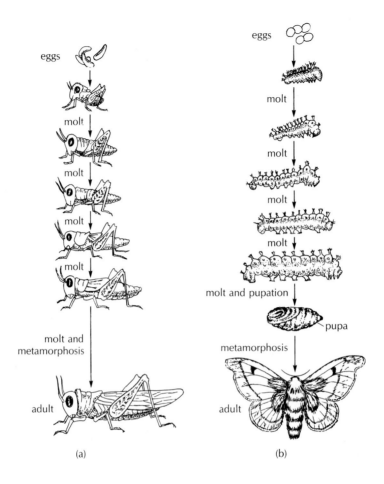

Figure 18.38
(a) Hemimetabolous development of a grasshopper.
(b) Holometabolous development in the silkworm moth.

Figure from General Endocrinology, by C. Donnell Turner, Copyright 1948 by Saunders College Publishing and renewed 1976 by C. Donnell Turner, reproduced by permission of the publisher.

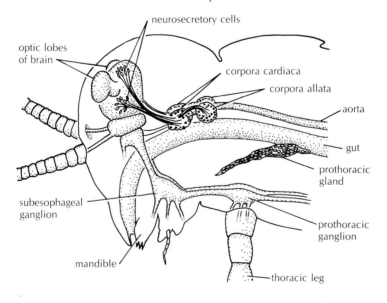

Figure 18.39
Anterior end of an insect, showing the location of the brain hormone, juvenile hormone, and ecdysone secretory centers. Neurosecretory cells in the brain secrete a brain hormone (PTTH) that stimulates the prothoracic glands to secrete ecdysone. The corpora allata secrete juvenile hormone. The corpora cardiaca innervate the corpora allata and are also neurosecretory. One of their major roles is the regulation of heartbeat rate. *From Wells, Lower Animals. Copyright © 1968 McGraw-Hill, Inc., New York. Reprinted by permission of McGraw-Hill, Inc. and Geo. Weidenfeld & Nicholson Ltd., London.*

Molting and metamorphosis during insect development is under complex environmental and hormonal control. The molting process—in particular, the resorption of some of the old cuticle and the development of new cuticle—is triggered by a steroid called **ecdysone,** but ecdysone production is itself regulated in a surprisingly complex manner. Neurosecretory cells in the brain secrete a **prothoracicotropic hormone (PTTH),** a polypeptide that activates (*tropic* = G: stimulate) a pair of glands in the anterior portion of the thorax, the **prothoracic glands (PG)** (Fig. 18.39). The PG, in turn, typically secrete an ecdysone precursor that is rapidly converted to ecdysone, the molting hormone, by particular enzymes in the hemolymph or target tissues; the ecdysone in turn is converted to a final, active form (20-hydroxyecdysone) by cells in target tissues, triggering the molting process. The PG thus seem to perform the same role as the Y-organ in crustaceans. Adult insects do not molt, and the prothoracic glands degrade during pupal development or shortly after metamorphosis to adulthood is completed.

The extent to which morphological differentiation occurs during molting depends upon whether another hormone, **juvenile hormone (JH),** is present in the blood at certain critical periods (**gates**). JH, a sesquiterpene lipid, is produced by yet another gland, the paired **corpora allata,** located just behind the brain. Experiments have demonstrated that high quantities of JH in the blood generally inhibit differentiation. In particular, the switch from larval to pupal development is inhibited by JH; insects metamorphose only after the corpora allata cease JH production and all circulating JH has been destroyed by specific enzymes in the hemolymph. The normal sequence of insect development depends upon pulses of JH secretion being critically timed to coincide with gates of sensitivity of target tissues to the hormone. JH also plays other important roles during development—for example, in determining caste in social insects and in stimulating yolk deposition during oogenesis in females and, at least in some species, sperm development in males. A number of intriguing behavioral patterns also are known to be under hormonal control. Curiously, peptide and protein hormones first identified as endocrine products in humans have recently been found in the insect nervous system; at least some of these hormones seem to control aspects of insect physiology and behavior. Recent studies indicate that juvenile hormone precursors (including methyl farnesoate and farnesoic acid) are synthesized by crustacean mandibular organs, which may be homologous with insect corpora allata; the role played by these compounds in crustacean biology is still unclear, although it seems they may regulate aspects of female gametogenesis.

A number of insects enter into a resting state (**diapause**) at some point in their development, as an adaptation for withstanding adverse conditions, such as cold winters. Entrance into the diapause state is under hormonal control (JH secretion again plays a role here) and is often triggered (and later released) by changing day length (**photoperiod**) and/or changing temperature. The stage at which diapause occurs is species dependent: In some species, an early embryonic stage enters diapause; in other species, either a larval instar, the pupal instar, or even the adult typically enters diapause. Diapause is also commonly encountered among copepods and branchiopods.

Digestion

The arthropod gut is divisible into three areas: foregut, midgut, and hindgut. All free-living species exhibit a distinct and separate mouth and anus, and in all species, food must be moved through the digestive tract by muscular activity rather than ciliary activity, since the lumen of the foregut and hindgut is lined with cuticle. Digestion is generally extracellular. Nutrients are distributed to the tissues through the hemal system.

Class Merostomata

Members of the class Merostomata are carnivorous. The mouth leads into an esophagus and thence into a **gizzard.** Both the esophagus and the gizzard are lined with cuticle, which is molted periodically along with the rest of the exoskeleton. The gizzard is equipped with chitinous teeth for grinding ingested food. A valve prevents passage of undigestible material from the gizzard into the stomach. A pair of elongated pouches (**hepatic,** or digestive, **ceca**) extend laterally from the stomach. Most food digestion and nutrient absorption occur in these hepatic ceca. Wastes pass through the rectum and out an anus, located ventrally at the base of the caudal spine.

Class Arachnida

Most arachnid species are carnivores. Because arachnids lack mandibles, the chelicerae must tear and grind the food prior to ingestion. In addition, glands associated with the oral cavity release enzymes into the prey, and the food is thus predigested externally. Digestive enzymes contributed by glands in the chelicerae or pedipalps may also participate in this process. After sufficient time has elapsed, the solubilized tissue is moved into the foregut through a muscular pump and then travels from there into the stomach; with few exceptions, arachnids do not ingest food in particulate form. Digestion and absorption take place primarily in the extensive tubular outgrowths of the stomach wall, the digestive diverticula. Undigested material and wastes pass through an intestine and exit from a posterior anus.

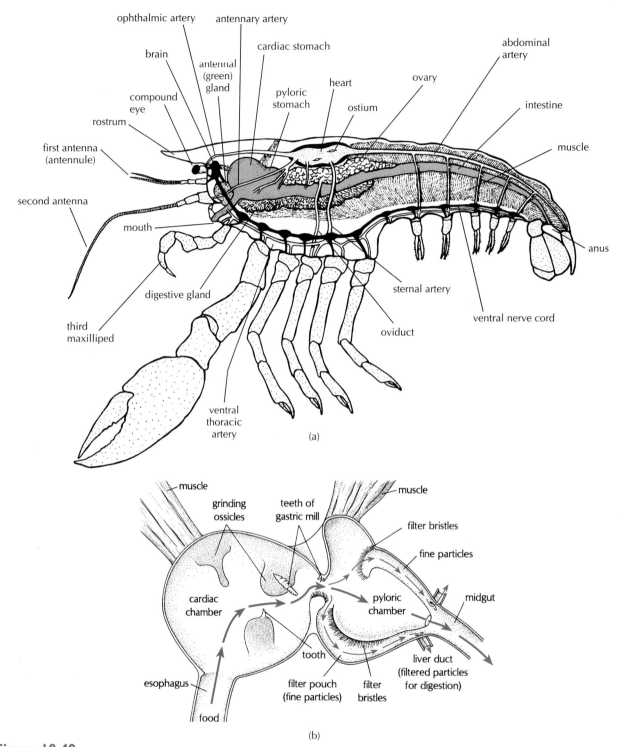

Figure 18.40

(a) Internal anatomy of a female crayfish. (b) The digestive system of a malacostracan. *(a) From Charles F. Lytle and J. E. Wodsedalek, General Zoology Laboratory Guide, 10th ed. Copyright © 1987 Wm. C. Brown Communications, Inc. Modified with permission of Times Mirror Higher Education Group, Inc., Dubuque, Iowa. All Rights Reserved. (b) After Hickman; after Yonge.*

Classes Chilopoda and Diplopoda

Most species of chilopod (centipede) are predaceous carnivores on other invertebrates and on some smaller vertebrates, using their modified maxillipeds to hold and poison prey. The digestive tract is usually a straight tube. Diplopods (millipedes), on the other hand, are usually herbivores, feeding on both living and, especially, dead or decaying plant material or the juices of living plants; some carnivorous species also exist. A **peritrophic**

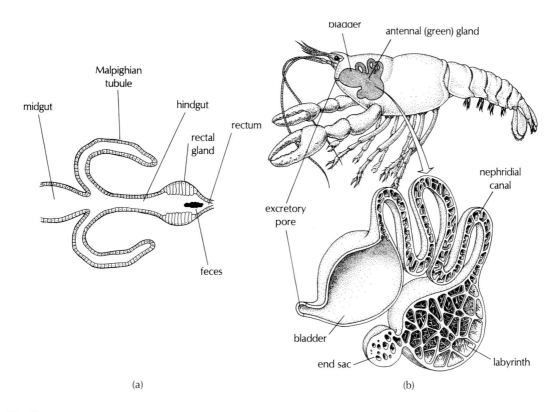

Figure 18.41
(a) Malpighian tubule system from an insect. (b) Antennal gland of the crayfish. In the enlargement, the gland has been "untangled" for clarity.   (b) *From Purves and Orians*, Life: The Science of Biology, *2d ed. Copyright © 1983 Sinauer Associates, Sunderland MA. Reprinted by permission.*

membrane lines the midgut of millipedes, presumably to protect against abrasion. Food becomes enclosed by this membrane as it moves through the gut, and new peri-trophic membrane is then secreted. As in the centipedes, the millipede gut is essentially a linear tube.

Class Insecta

Insects feed on a variety of nutritive sources, including plant and animal tissues and fluids. The esophagus is highly muscularized and serves as a pump, moving food into a crop for storage and/or preliminary digestion. A **proventriculus** is generally present, functioning as a valve to regulate the passage of food and, in some species, to grind ingested food. Digestion and absorption occur in the midgut and its associated gastric ceca. Symbiotic bacteria and protozoans harbored in the gastric ceca of many species participate in the digestive process. The walls of the midgut are often lined by peritrophic membrane, as in the Diplopoda. This membrane is discarded and renewed periodically. Before wastes reach the anus, most of the water is resorbed from the fecal material by the rectal glands of the hindgut (Fig. 18.20).

Subphylum Crustacea

In the crustaceans, food is ground and strained in a muscular foregut consisting of a large **cardiac stomach** and a smaller **pyloric stomach** (Fig. 18.40). The food is ground by the chitinous, toothed ridges of a **gastric mill.** Stiff setae often prevent food particles from passing farther until they have reached the proper consistency. Food passes on to the midgut, which is associated with an extensive array of digestive ceca in which the food is digested, absorbed, and stored. The digestive tubules often form a distinct organ, the **hepatopancreas,** or liver.

Excretion

Class Merostomata

Members of the Merostomata bear four pairs of excretory **coxal glands** adjacent to the gizzard. The glands empty into a common chamber, which then leads through a coiled tubule and into a bladder. Salts are resorbed in the bladder as needed, the final urine being discharged through pores at the base of the last pair of walking legs.

Class Arachnida

Coxal glands are also common among arachnids. Here they are spherical sacs resembling annelid nephridia. Wastes are collected from the surrounding blood of the hemocoel and discharged through pores on from one to several pairs of appendages. Recent evidence suggests that the coxal glands may also function in the release of pheromones.

Some arachnid species have Malpighian tubules instead of, or in addition to, the coxal glands. In some of these species, however, the Malpighian tubules seem to function in silk production rather than excretion. The excretory role of Malpighian tubules in insects has already been discussed (p. 362). Arachnids possess in addition to the coxal glands and Malpighian tubules, a number of strategically placed cells called **nephrocytes,** which phagocytize waste particles. The major waste product of protein metabolism among arachnids is guanine, a purine biochemically related to uric acid.

Classes Chilopoda and Diplopoda

The major excretory structures in the Chilopoda and Diplopoda are Malpighian tubules. Although some uric acid is produced, the major waste product of centipedes is ammonia.

Class Insecta

The major excretory structures of insects are Malpighian tubules (Fig. 18.41a), as discussed earlier (p. 362). Several other mechanisms also may be involved in insect waste elimination. In particular, evidence indicates that some wastes are incorporated into the cuticle, to be shed at ecdysis. Many insects also possess nephrocytes, as found in arachnids.

Insects eliminate a significant fraction of their nitrogenous wastes as water-insoluble uric acid and related compounds. These compounds are eliminated from the body in nearly dry form, as most of the water in the urine is resorbed in transit through the rectum. This is an outstanding physiological adaptation for terrestrial existence.

Taxonomic Summary

Phylum Arthropoda
    Subphylum Trilobitomorpha
        Class Trilobita—the trilobites
    Subphylum Chelicerata
        Class Merostomata—horseshoe crabs
        Class Arachnida—spiders, mites, ticks, scorpions
        Class Pycnogonida (= Pantopoda)—sea spiders
    Subphylum Uniramia
        Class Chilopoda—centipedes
        Class Diplopoda—millipedes
        Class Insecta
          Subclass Apterygota—the wingless insects
          Subclass Pterygota—the winged insects
    Subphylum Crustacea
        Class Branchiopoda—brine (fairy) shrimp, clam shrimp, water fleas
        Class Ostracoda—the ostracods
        Class Copepoda—the copepods
        Class Pentastomida
        Class Cirripedia—the barnacles
        Class Malacostraca
          Order Isopoda—pillbugs, woodlice
          Order Amphipoda—sand fleas
          Order Euphausiacea—euphausiids (krill)
          Order Stomatopoda—stomatopods
          Order Decapoda—crabs, lobsters, shrimp, hermit crabs

Subphylum Crustacea

In crustaceans, nitrogenous wastes are generally removed by diffusion across the gills—for those species that have gills. Most crustaceans release ammonia, although some urea and uric acid are also produced. The so-called excretory organs may be more involved with resorption of salts and elimination of water than with discharge of nitrogenous wastes in freshwater species. In some species, these excretory organs are called **antennal glands** or **green glands** because of their location near the antennal segments (Fig. 18.41) and their color. In other species, the organs are found near the maxillary segments and are termed **maxillary glands.** These excretory organs are structurally similar to the coxal glands of chelicerates. Fluid collects within the tubules from the surrounding blood of the hemocoel, and this primary urine is modified substantially by selective reabsorption and secretion as it moves through the excretory system and rectum.

At least some terrestrial crab species reprocess their urine at the gills, resorbing additional ions and adding substantial amounts of ammonia before the urine is finally discharged; the urine reaches the gills by trickling downward into the gill chamber from the nephropores (G: kidney openings).

Blood Pigments

The bloods of many arthropod species lack respiratory pigments. This is particularly true of terrestrial species, in which gas exchange is accomplished primarily through tracheae. In other species, both hemocyanin (HCy) and hemoglobin (Hb) are found. Both pigments are sporadically distributed, even among the members of a given class.

Topics for Further Discussion and Investigation

1. Arthropod affinities and origins are far from certain, even at the highest taxonomic levels. Arthropods may have derived from annelid-like ancestors. However, embryological studies and studies of limb structure and function in various arthropod and polychaete species suggest that insects, millipedes, and centipedes may have had a very different evolutionary origin from that of other arthropod groups, such as the Crustacea and Arachnida. Indeed, a number of authorities give the Uniramia (including the insects, millipedes, and centipedes) phylum status. What is the evidence in favor of an independent origin for these arthropods, and what difficulties does this scheme present in terms of the numbers of different arthropod characteristics that must be presumed to have evolved independently through convergent evolution?

Ballard, J. W. O., G. J. Olsen, D. P. Faith, W. A. Odgers, D. M. Rowell, and P. W. Atkinson. 1992. Evidence from 12S ribosomal RNA sequences that onychophorans are modified arthropods. *Science* 258:1345.

Cisne, J. L. 1974. Trilobites and the evolution of arthropods. *Science* 186:13.

Eernisse, D. J., J. S. Albert, and F. E. Anderson. 1992. Annelida and Arthropoda are not sister taxa: A phylogenetic analysis of spiralian metazoan morphology. *Syst. Biol.* 41:305.

Emerson, M. J., and F. R. Schram. 1990. The origin of crustacean biramous appendages and the evolution of Arthropoda. *Science* 250:667–69.

Evans, H. E. 1959. Some comments on the evolution of the Arthropoda. *Evolution* 13:147.

Gupta, A. P. 1979. *Arthropod phylogeny.* New York: Van Nostrand Reinhold.

Hessler, R. R., and W. A. Newman. 1975. A trilobitomorph origin for the Crustacea. *Fossils and Strata* 4:437.

Manton, S. M. 1967. The polychaete *Spinther* and the origin of the Arthropoda. *J. Nat. Hist.* 1:1.

Manton, S. M. 1973. Arthropod phylogeny—A modern synthesis. *J. Zool., London* 171:111.

Shear, W. A. 1992. End of the "Uniramia" taxon. *Nature* 359:477.

Tiegs, O. W., and S. M. Manton. 1958. The evolution of the Arthropoda. *Biol. Rev.* 33:255.

2. Living in the terrestrial environment poses many difficulties for invertebrates. Not least among these is that the air is dry and that temperatures fluctuate considerably, even on a day-to-day or hour-to-hour basis. Investigate some of the physiological and behavioral adaptations that enable terrestrial and semiterrestrial arthropods to tolerate dehydrating conditions and marked shifts in environmental temperature.

a. References concerning temperature regulation:

Dorsett, D. A. 1962. Preparation for flight by hawkmoths. *J. Exp. Biol.* 39:579.

Duman, J. G. 1977. Environmental effects on antifreeze levels in larvae of the darkling beetle, *Meracantha contracta. J. Exp. Zool.* 201:333.

Edney, E. B. 1953. The temperature of woodlice in the sun. *J. Exp. Biol.* 30:331.

Heinrich, B. 1974. Thermoregulation in endothermic insects. *Science* 185:747.

Heinrich, B., and H. Esch. 1994. Thermoregulation in bees. *Amer. Sci.* 82:164.

Kingsolver, J. G. 1987. Predation, thermoregulation, and wing color in pierid butterflies. *Oecologia (Berlin)* 73:301.

Kugal, O., B. Heinrich, and J. G. Duman. 1988. Behavioural thermoregulation in the freeze-tolerant arctic caterpillar, *Gynaephora groenlandica J. Exp. Biol.* 138:181.

Morgan, K. R. 1985. Body temperature regulation and terrestrial activity in the ectothermic beetle *Cicindela tranquebarica. Physiol. Zool.* 58:29.

Southwick, E. E., and R. A. Moritz. 1987. Social control of air ventilation in colonies of honey bees, *Apis mellifera. J. Insect Physiol.* 33:623.

Vogt, F. D. 1986. Thermoregulation in bumblebee colonies. I. Thermoregulatory versus brood-maintenance behaviors during acute changes in ambient temperature. *Physiol. Zool.* 59:55.

Wilkens, J. L., and M. Fingerman. 1965. Heat tolerance and temperature relationships of the fiddler crab, *Uca pugilator,* with reference to its body coloration. *Biol. Bull.* 128:133.

Willmer, P. G. 1982. Thermoregulatory mechanisms in Sarcophaga. *Oecologia (Berlin)* 53:382.

b.   References concerning water balance:

Cohen, A. C., R. B. March, and J. D. Pinto. 1981. Water relations of the desert blister beetle *Cysteodemus armatus* (Leconte) (Coleoptera: Meloidae). *Physiol. Zool.* 54:179.

Combs, C. A., N. Alford, A. Boynton, M. Dvornak, and R. P. Henry. 1992. Behavioral regulation of hemolymph osmolarity through selective drinking in land crabs, *Birgus latro* and *Gecarcoidea lalandii. Biol. Bull.* 182:416.

Dresel, E. I. B., and V. Moyle. 1950. Nitrogenous excretion in amphipods and isopods. *J. Exp. Biol.* 27:210.

Edney, E. B. 1966. Absorption of water vapour from unsaturated air by *Arenivaga* sp. (Polyphgidae, Dictyoptera). *Comp. Biochem. Physiol.* 19:387.

Hamilton, W. J., III, and M. K. Seely. 1976. Fog basking by the Nambi desert beetle, *Onymacris unguicularis. Nature (London)* 262:284.

Noble-Nesbitt, J. 1970. Water uptake from subsaturated atmospheres: Its site in insects. *Nature (London)* 225:753.

Seely, M. K., and W. J. Hamilton III. 1976. Fog catchment sand trenches constructed by tenebrionid beetles, *Lepidochora,* from the Namib Desert. *Science* 193:484.

Standing, J. D., and D. D. Beatty. 1978. Humidity behaviour and reception in the sphaeromatid isopod *Gnorimosphaeroma oregonensis* (Dana). *Canadian J. Zool.* 56:2004.

Willmer, P. G., M. Baylis, and C. L. Simpson. 1989. The roles of colour change and behavior in the hygrothermal balance of a littoral isopod, *Ligia oceanica. Oecologia (Berlin)* 78:349.

Wolcott, D. L. 1991. Nitrogen excretion is enhanced during urine recycling in two species of terrestrial crab. *J. Exp. Zool.* 259:181.

3.   Invertebrates are basically poikilothermic; that is, in the absence of behavioral modification (including flight), body temperatures generally follow those of the surrounding air or water rather closely. This poses a particular problem for terrestrial and shallow-water invertebrates when temperatures fall below freezing. Investigate the mechanisms used by arthropods to prevent freezing of tissues during periods of cold weather.

Horwarth, K. L., and J. G. Duman. 1982. Involvement of the circadian system in photoperiodic regulation of insect antifreeze proteins. *J. Exp. Zool.* 219:267.

van der Laak, S. 1982. Physiological adaptations to low temperature in freezing-tolerant *Phylodecta laticollis* beetles. *Comp. Biochem. Physiol.* 73A:613.

Lee, R. L., Jr. 1989. Insect cold-hardiness: To freeze or not to freeze? *BioScience* 39:308.

Lee, R. E., Jr., C. Chen, M. H. Meacham, and D. L. Denlinger. 1987. Ontogenetic patterns of cold-hardiness and glycerol production in *Sarcophaga crassipalpis. J. Insect Physiol.* 33:587.

Lee, R. E., Jr., J. M. Strong-Gunderson, M. R. Lee, K. S. Grove, and T. J. Riga. 1991. Isolation of ice nucleating active bacteria from insects. *J. Exp. Zool.* 257:124.

Turnock, W. J., R. J. Lamb, and R. P. Bodnaryk. 1983. Effects of cold stress during pupal diapause on the survival and development of *Mamestra configurata* (Lepidoptera: Noctuidae). *Oecologia (Berlin)* 56:185.

Tursman, D., J. G. Duman, and C. A. Knight. 1994. Freeze tolerance adaptations in the centipede *Lithobius forficatus. J. Exp. Zool.* 268:347.

Yingst, D. R. 1978. The freezing resistance of the arctic subtidal isopod, *Mesidotea entomon. J. Comp. Physiol.* 125:165.

4.   Symbiotic relationships (including parasitism) are commonly encountered among the Crustacea, Insecta, and Arachnida. For one of these groups, discuss the morphological, behavioral, and/or physiological adaptations for the symbiotic lifestyle.

A general treatment can be found in any parasitology text, such as Noble, E. R., and G. A. Noble. 1982. *Parasitology: The Biology of Animal Parasites,* 5th ed. Philadelphia: Lea & Febiger. Or Cheng, T. C. 1986. *General Parasitology,* 2d ed. New York: Academic Press.

Christensen, A. M., and J. J. McDermott. 1958. Life-history and biology of the oyster crab, *Pinnotheres ostreum* Say. *Biol. Bull.* 114:146.

Day, J. H. 1935. The life-history of *Sacculina. Q. J. Microsc. Sci.* 77:549.

DeVries, P. J. 1990. Enhancement of symbioses between butterfly caterpillars and ants by vibrational communication. *Science* 248:1104.

Gotto, R. V. 1979. The association of copepods with marine invertebrates. *Adv. Marine Biol.* 16:1.

Høeg, J. T. 1987. Male cypris metamorphosis and a new male larval form, the trichogon, in the parasitic barnacle *Sacculina carcini* (Crustacea: Cirripedia: Rhizocephala). *Phil. Trans. Royal Soc. London (B)* 317:47.

Letourneau, D. K. 1990. Code of ant-plant mutualism broken by parasite. *Science* 248:215.

McClintock, J. B., and J. Janssen. 1990. Chemical defense in a pelagic antarctic amphipod via pteropod abduction: A novel symbiosis. *Nature (London)* 346:462.

Mitchell, R. 1968. Site selection by larval water mites parasitic on the damselfly *Cercion hieroglyphicum* Brauer. *Ecology* 49:40.

Moyse, J. 1983. *Isadascus bassindalei* gen. nov., sp. nov. (Ascothoracica: Crustacea) from northeast Atlantic with a note on the origin of barnacles. *J. Mar. Biol. Assoc. U.K.* 63:161.

Price, P. W. 1972. Parasitoids utilizing the same host: Adaptive nature of differences in size and form. *Ecology* 53: 190.

Salt, G. 1968. The resistance of insect parasitoids to the defense reactions of their hosts. *Biol. Rev.* 43:200.

Takasu, K., and Y. Hirose. 1991. The parasitoid *Ooencyrtus nezarae* (Hymenoptera: Encyrtidae) prefers hosts parasitized by conspecifics over unparasitized hosts. *Oecologia* 87:319.

5.   Investigate form and function in spider webs.

Craig, C. L. 1986. Orb-web visibility: The influence of insect flight behaviour and visual physiology on the evolution of web designs within the Araneoidea. *Anim. Behav.* 34:54.

Denny, M. 1976. The physical properties of spiders' silk and their role in the design of orb webs. *J. Exp. Biol.* 65:483.

Köhler, T., and F. Vollrath. 1995. Thread biomechanics in the two orb-weaving spiders *Araneus diadematus* (Araneae, Araneidae) and *Uloborus walckenaerius* (Araneae, Uloboridae). *J. Exp. Zool.* 271:1.

Nentwig, W. 1983. Why do only certain insects escape from a spider's web? *Oecologia* (*Berlin*) 53:412.

Nentwig. W. 1983. The non-filter function of orb webs in spiders. *Oecologia* (*Berlin*) 58:418.

Opell, B. D. 1990. Material investment and prey capture potential of reduced spider webs. *Behav. Ecol. Sociobiol.* 26:375.

Palmer, J. M., F. A. Coyle, and F. W. Harrison. 1982. Structure and cytochemistry of the silk glands of the mygalomorph spider *Antrodiaetus unicolor* (Araneae, Antrodiaetidae). *J. Morphol.* 174:269.

Popock, R. I. 1895. Some suggestions on the origin and evolution of web spinning in spiders. *Nature* (*London*) 51:417.

Rypstra, A. L. 1982. Building a better insect trap: An experimental investigation of prey capture in a variety of spider webs. *Oecologia* (*Berlin*) 52:31.

Seibt, U., and W. Wickler. 1990. The protective function of the compact silk nest of social *Stegodyphus* spiders (Araneae, Eresidae). *Oecologia* (*Berlin*) 82:317.

Smith, R. B., and T. P. Mommsen. 1984. Pollen feeding in an orb-weaving spider. *Science* 226:1330.

Witt, P. N., and C. F. Reed. 1965. Spider-web building. *Science* 149:1190.

6.   How do insects fly?

Alexander, R. M. 1979. *The invertebrates.* New York: Cambridge University Press, 407–16.

Alexander, R. M. 1995. Springs for wings. *Science* 268:50.

Betts, C. R. 1986. Functioning of the wings and axillary sclerites of Heteroptera during flight. *J. Zool., London* (*B*) 1:283.

Boettiger, E. G., and E. Furshpan. 1952. The mechanics of flight movements in Diptera. *Biol. Bull.* 102:200.

Borst, A., and S. Bahde. 1988. Visual information processing in the fly's landing system. *J. Comp. Physiol.* (*A*) 163:167.

Miyan, J. A., and A. W. Ewing. 1985. Is the "click" mechanism of dipteran flight an artifact of CCl_4 anaesthesia? *J. Exp. Biol.* 116:313.

Roeder, K. D. 1951. Movements of the thorax and potential changes in the thoracic muscles of insects during flight. *Biol. Bull.* 100:95.

Weis-Fogh, R. 1975. Unusual mechanisms for the generation of lift in flying animals. *Sci. Amer.* 233:80.

Wells, M. 1968. *Lower Animals.* New York: McGraw-Hill, 101–15.

7.   Hermit crabs are an active marine group of decapod crustaceans whose abdomen is unprotected by a carapace. Instead, they protect their soft parts by taking up residence inside empty gastropod shells. How do hermit crabs choose their shells, and what selective pressures probably have molded these choices?

Abrams, P. A. 1987. An analysis of competitive interactions between three hermit crab species. *Oecologia* (*Berlin*) 72:233.

McClintock, T. S. 1985. Effects of shell condition and size upon the shell choice behavior of a hermit crab. *J. Exp. Marine Biol. Ecol.* 88:271.

Mercando, N. A., and C. F. Lytle. 1980. Specificity in the association between *Hydractinia echinata* and sympatric species of hermit crabs. *Biol. Bull.* 159:337.

Mesce, K. A. 1982. Calcium-bearing objects elicit shell selection behavior in a hermit crab. *Science* 215:993.

Vance, R. 1972. Competition and mechanism of coexistence in three sympatric species of intertidal hermit crabs. *Ecology* 53:1062.

Wilber, T. P., Jr. 1990. Influence of size, species, and damage on shell selection by the hermit crab *Pagurus longicarpus. Marine Biol.* 104:31.

8.   Through the process of natural selection, many arthropods have developed remarkable abilities to disguise themselves, either by blending in with their surroundings or by imitating other species. Investigate the adaptive value of mimicry and camouflage among the Arthropoda.

Blest, A. D. 1963. Longevity, palatability, and natural selection in five species of New World Saturuiid moth. *Nature* (*London*) 197:1183.

Cusson, M., and J. A. McNeil. 1989. Involvement of juvenile hormone in the regulation of pheromone release activities in a moth. *Science* 243:210.

Eisner, T., K. Hicks, M. Eisner, and D. S. Robson. 1978. "Wolf-in-sheep's-clothing" strategy of a predaceous insect larva. *Science* 199:790.

Finday, R., M. R. Young, and J. A. Finday. 1983. Orientation behaviour in the Grayling butterfly: Thermoregulation or crypsis? *Ecol. Entomol.* 8:145.

Greene, E. 1989. A diet-induced developmental polymorphism in a caterpillar. *Science* 243:643.

Greene, E., L. J. Orsak, and D. W. Whitman. 1987. A tephritid fly mimics the territorial displays of its jumping spider predators. *Science* 236:310.

Kettlewell, H. B. D. 1956. Further selection experiments on industrial melanism in the Lepidoptera. *Heredity* 10:287.

Körner, H. K. 1982. Countershading by physiological colour change in the fish louse *Anilocra physodes* L. (Crustacea: Isopoda). *Oecologia* (*Berlin*) 55:248.

Mather, M. H., and B. D. Roitberg. 1987. A sheep in wolf's clothing: Tephritid flies mimic spider predators. *Science* 236:308.

Platt, A., R. Coppinger, and L. Brower. 1971. Demonstration of the selective advantage of mimetic *Limentis* butterflies presented to caged avian predators. *Evolution* 25:692.

9.   Much communication among arthropods is accomplished by means of chemicals. Investigate the mechanism and/or the adaptive significance of chemical communication among the Arthropoda.

Adams, T. S., and G. G. Holt. 1987. Effect of pheromone components when applied to different models on male sexual behaviour in the housefly, *Musca domestica. J. Insect Physiol.* 33:9.

Boeckh, J., H. Sass, and D. R. A. Wharton. 1970. Antennal receptors: Reactions to female sex attractant in *Periplaneta americana*. *Science* 168:589.

Butler, C. G., D. J. C. Fletcher, and D. Walter. 1969. Nest-entrance marking with pheromones by the honey bee *Apis mellifera* L. and by a wasp, *Vespula vulgaris*. *Anim. Behav.* 17:142.

Dahl, E., H. Emanuelsson, and C. von Mecklenburg. 1970. Pheromone transport and reception in an amphipod. *Science* 170:739.

Eisner, T., and Y. C. Meinwald. 1965. Defensive secretion of a caterpillar (*Papilio*). *Science* 150:1733.

Fitzgerald. T. D. 1976. Trail marking by larvae of the eastern tent caterpillar. *Science* 194:961.

Linn, C. E., Jr., M. G. Campbell, and W. L. Roelofs. 1987. Pheromone components and active spaces: What do moths smell and where do they smell it? *Science* 237:650.

Mafra-Neto, A., and R. T. Cardé. 1994. Fine-scale structure of pheromone plumes modulates upwind orientation of flying moths. *Nature* 369:142.

McAllister, M. K., and B. D. Roitberg. 1987. Adaptive suicidal behaviour in pea aphids. *Nature* (*London*) 328:797.

Myers, J., and L. P. Brower. 1969. A behavioural analysis of the courtship pheromone receptors of the queen butterfly, *Danaus gilippus berenice*. *J. Insect Physiol.* 15:2117.

Nault, L. R., M. E. Montgomery, and W. S. Bowers. 1976. Ant-aphid association: Role of aphid alarm pheromone. *Science* 192:1349.

Pierce, N. E., and S. Eastseal. 1986. The selective advantage of attendant ants for the larvae of a Lycaenid butterfly, *Glaucopsyche lygdamus*. *J. Anim. Ecol.* 55:451.

Price, P. W. 1970. Trail odors: Recognition by insects parasitic on cocoons. *Science* 170:546.

Rust, M. K., T. Burk, and W. J. Bell. 1976. Pheromone-stimulated locomotory and orientation responses in the American cockroach *Periplaneta americana*. *Anim. Behav.* 24:52.

Schneider, D. 1969. Insect olfaction: Deciphering system for chemical messages. *Science* 163:1031.

Zimmer-Faust, R. K. 1993. ATP: A potent prey attractant evoking carnivory. *Limnol. Oceanogr.* 38:1271.

10. It is becoming clear that all invertebrates possess some form of immune response; that is, they have the ability to distinguish self from nonself at the cellular level. Investigate the ability of arthropods to make this distinction.

Berg, R., I. Schuchmann-Feddersen, and O. Schmidt. 1988. Bacterial infection induces a moth (*Ephestia kuhniella*) protein which has antigenic similarity to virus-like particle proteins of a parasitoid wasp (*Venturia canescens*). *J. Insect Physiol.* 34:473.

Brehélin, M., and J. A. Hoffmann. 1980. Phagocytosis of inert particles in *Locusta migratoria* and *Galleria mellonella*: A study of ultrastructure and clearance. *J. Insect Physiol.* 26:103.

Briggs, J. D. 1958. Humoral immunity in lepidopterous larvae. *J. Exp. Zool.* 138:155.

Chisholm, J. R. S., and V. J. Smith. 1992. Antibacterial activity in the haemocytes of the shore crab, *Carcinus maenas*. *J. Marine Biol. Assoc. U.K.* 72:529.

Dunn, P. E. 1990. Humoral immunity in insects. *BioScience* 40:738.

Edson, K. M., S. B. Vinson, D. B. Stoltz, and M. D. Summers. 1981. Virus in a parasitoid wasp: Suppression of the cellular immune response in the parasitoid's host. *Science* 211:582.

Ennesser, C. A., and A. J. Nappi. 1984. Ultrastructural study of the encapsulation response of the American cockroach, *Periplaneta americana*. *J. Ultrastr. Res.* 87:31.

Götz, P., G. Enderlein, and I. Roettgen. 1987. Immune reactions of *Chironomus* larvae (Insecta: Diptera) against bacteria. *J. Insect Physiol.* 33:993.

Ratcliffe, N. A., C. Leonard, and A. F. Rowley. 1984. Prophenoloxidase activation: Nonself recognition and cell cooperation in insect immunity. *Science* 226:557.

Salt, G. 1968. The resistance of insect parasitoids to the defense reactions of their hosts. *Biol. Rev.* 43:200.

Sloan, B., C. Yocum, and L. W. Clem. 1975. Recognition of self from non-self in crustaceans. *Nature* (*London*) 258:521.

White, K. N., and N. A. Ratcliffe. 1982. The segregation and elimination of radio- and fluorescent-labelled marine bacteria from the haemolymph of the shore crab, *Carcinus maenas*. *J. Mar. Biol. Assoc. U.K.* 62:819.

11. Small planktonic crustaceans, such as copepods and cladocerans, are near the base of the food chain in aquatic ecosystems; that is, the growth rate and reproductive potential of many fish depend largely on the size of the herbivorous zooplankton population available as food. In turn, the size of the herbivorous zooplankton population depends on the amount of phytoplankton available, and the rate at which the phytoplankton can be captured, ingested, and converted into new zooplankton biomass and offspring. Consequently, considerable effort has gone into studying zooplankton feeding biology. What factors are involved in the collection of phytoplankton by crustacean zooplankton?

Anraku, M., and M. Omori. 1963. Preliminary survey of the relationship between the feeding habit and structure of the mouth parts of marine copepods. *Limnol. Oceanogr.* 8:116.

Butler, N. M., C. A. Suttle, and W. E. Neill. 1989. Discrimination by freshwater zooplankton between single algal cells differing in nutritional status. *Oecologia* (*Berlin*) 78:368.

Costello, J. H., J. R. Strickler, C. Marvasé, G. Trager, R. Zeller, and A. J. Freise. 1990. Grazing in a turbulent environment: Behavioral response of a calanoid copepod, *Centropages hamatus*. *Proc. Nat. Acad. Sci. USA* 87:1648.

DeMott, W. R. 1989. Optimal foraging theory as a predictor of chemically mediated food selection by suspension feeding copepods. *Limnol. Oceanogr.* 34:140.

Gerritsen, J., and K. G. Porter. 1982. The role of surface chemistry in filter feeding by zooplankton. *Science* 216:1225.

Gophen, M., and W. Geller. 1984. Filter mesh size and food particle uptake by *Daphnia*. *Oecologia* (*Berlin*) 64:408.

Hamner, W. M., P. P. Hamner, S. W. Strand, and R. W. Gilmer. 1983. Behavior of antarctic krill, *Euphausia superba*: Chemoreception, feeding, schooling, and molting. *Science* 220:433.

Huntley, M. E., K.-G. Barthel, and J. L. Star. 1983. Particle rejection by *Calanus pacificus*: Discrimination between similarly sized particles. *Marine Biol.* 74:151.

Poulet, S. A., and P. Marsot. 1978. Chemosensory grazing by marine calanoid copepods (Arthropoda: Crustacea). *Science* 200:1403.

Price, H. J., G.-A. Paffenhöfer, and J. R. Strickler. 1983. Modes of cell capture in calanoid copepods. *Limnol. Oceanogr.* 28:116.

Richman, S., and J. N. Rogers. 1969. The feeding of *Calanus helgolandicus* on synchronously growing populations of the marine diatom *Ditylum brightwelli*. *Limnol. Oceanogr.* 14:701.

Turner, J. T., P. A. Tester, and J. R. Strickler. 1993. Zooplankton feeding ecology: A cinematographic study of animal-to-animal variability in the feeding behavior of *Calanus finmarchicus*. *Limnol. Oceanogr.* 38:255.

12. What is the evidence indicating that color changes in arthropods are regulated by hormones?

Brown, F. A., Jr. 1935. Control of pigment migration within the chromatophores of *Palaemonetes vulgaris*. *J. Exp. Zool.* 71:1.

Brown, F. A., Jr. 1940. The crustacean sinus gland and chromatophore activation. *Physiol. Zool.* 13:343.

Brown, F. A., Jr., and H. E. Ederstrom. 1940. Dual control of certain black chromatophores of *Crago*. *J. Exp. Zool.* 85:53.

Fingerman, M. 1969. Cellular aspects of the control of physiological color changes in crustaceans. *Amer. Zool.* 9:443.

McWhinnie, M. A., and H. M. Sweeney. 1955. The demonstration of two chromatophorotropically active substances in the land isopod, *Trachelipus rathkei*. *Biol. Bull.* 108:160.

Pérez-González, M. D. 1957. Evidence for hormone-containing granules in sinus glands of the fiddler crab *Uca pugilator*. *Biol. Bull.* 113:426.

13. What are the roles of predation and competition in regulating the sizes of arthropod populations?

Bertness, M. D. 1989. Intraspecific competition and facilitation in a northern acorn barnacle population. *Ecology* 70:257.

Chew, F. S. 1981. Coexistence and local extinction in two pierid butterflies. *Amer. Nat.* 118:655.

El-Dessouki, S. A. 1970. Intraspecific competition between larvae of *Sitona* spp. (Coleoptera, Curculionidae). *Oecologia* (Berlin) 6:106.

Enders, F. 1974. Vertical stratification in orb-web spiders (Araneidae, Araneae) and a consideration of other methods of coexistence. *Ecology* 55:317.

Frank, J. H. 1967. The insect predators of the pupal stage of the winter moth, *Operophtera brumata* (L.) (Lepidoptera: Hydiomenidae). *J. Anim. Ecol.* 36:375.

MacKay, W. P. 1982. The effect of predation of western widow spiders (Araneae: Theridiidae) on harvester ants (Hymenoptera: Formicidae). *Oecologia* (Berlin) 53:406.

Moore, N. W. 1964. Intra- and interspecific competition among dragonflies (Odonata). *J. Anim. Ecol.* 33:49.

Wise, D. H. 1983. Competitive mechanisms in a food-limited species: Relative importance of interference and exploitative interactions among labyrinth spiders (Araneae: Araneidae). *Oecologia* (Berlin) 58:1.

14. Based upon lectures and your reading in this text, what factors contribute to making arthropods the most abundant animals (with the possible exception of the nematodes) on the earth?

15. Based upon lectures and your reading in this text, what functions does the blood of insects serve, now that gas exchange is achieved by means of tracheae rather than by a circulatory system?

16. Growing evidence indicates that the morphology of some aquatic arthropods is influenced by predation. Discuss the evidence supporting this statement, and the adaptive value and costs of the response.

Barry, M. J. 1994. The costs of crest induction for *Daphnia carinata*. *Oecologia* 97:278.

Black, A. R. 1993. Predator-induced phenotypic plasticity in *Daphnia pulex*: Life history and morphological responses to *Notonecta* and *Chaoborus*. *Limnol. Oceanogr.* 38:986.

Dodson, S. I. 1974. Adaptive change in plankton morphology in response to size-selective predation: A new hypothesis of cyclomorphosis. *Limnol. Oceanogr.* 19:721.

Dodson, S. I. 1989. The ecological role of chemical stimuli for the zooplankton: Predator-induced morphology in *Daphnia*. *Oecologia* (Berlin) 78:361.

Lively, C. M. 1986. Predator-induced shell dimorphism in the acorn barnacle *Chthamalus anisopoma*. *Evolution* 40:232.

Morgan, S. G. 1989. Adaptive significance of spination in estuarine crab zoeae. *Ecology* 70:464.

O'Brien, W. J., J. D. Kettle, and H. P. Riessen. 1979. Helmets and invisible armor: Structures reducing predation from tactile and visual planktivores. *Ecology* 60:287.

Stenson, J. A. E. 1987. Variation in capsule size of *Holopedium gibberum* (Zaddach): A response to invertebrate predation. *Ecology* 68:928.

Walls, M., and M. Ketola. 1989. Effects of predator-induced spines on individual fitness in *Daphnia pulex*. *Limnol. Oceanogr.* 34:390.

Zaret, T. M. 1969. Predation-balanced polymorphism of *Ceriodaphnia cornuta* Sars. *Limnol. Oceanogr.* 14:301.

17. Many plants have evolved chemical defenses against herbivore feeding. Discuss the manner in which at least some insect species circumvent these defenses or even appropriate the plant chemicals for their own uses.

Dussourd, D. E., and T. Eisner. 1987. Vein-cutting behavior: Insect counterploy to the latex defense of plants. *Science* 237:898.

Eisner, T., J. S. Johanessee, J. Carrel, L. B. Hendry, and J. Meinwald. 1974. Defensive use by an insect of a plant resin. *Science* 184:996.

Kearsley, M. J. C., and T. G. Whitham. 1992. Guns and butter: A no cost defense against predation for *Chrysomela confluens*. *Oecologia* 92:556.

Peterson, S. C., N. D. Johnson, and J. L. LeGuyader. 1987. Defensive regurgitation of allelochemicals derived from host cyanogenesis by eastern tent caterpillars. *Ecology* 68:1268.

18. Discuss the evidence that bees use memory in navigating to and from their hives.

Dyer, F. C. 1987. Memory and sun compensation by honeybees. *J. Comp. Physiol.* (A) 160:621.

Gould, J. L. 1986. The locale map of honeybees: Do insects have cognitive maps? *Science* 232:861.

19. Most biologists agree that insect wings originally must have evolved in response to pressures not associated with flight. Only after the basic wing structure evolved could the further evolution of wings as organs of flight become possible. Compare and contrast the different hypotheses accounting for the origin of insect wings. Is the evidence supporting any one hypothesis more compelling than that supporting the other hypotheses?

Douglas, M. W. 1981. Thermoregulatory significance of thoracic lobes in the evolution of insect wings. *Science* 211:84.

Heinrich, B. 1993. *The Hot-blooded Insects: Strategies and Mechanisms of Thermoregulation.* Cambridge, Mass.: Harvard Univ. Press, 104–13.

Kingsolver, J. G., and M. A. R. Koehl. 1985. Aerodynamics, thermoregulation, and the evolution of insect wings: Differential scaling and evolutionary change. *Evolution* 39:488.

Kukalova-Peck, J. 1978. Origin and evolution of insect wings and their relation to metamorphosis, as documented by the fossil record. *J. Morphol.* 156:53.

Marden, J. H., and M. G. Kramer. 1994. Surface-skimming stoneflies: A possible intermediate stage in insect flight evolution. *Science* 266:427.

20. Many small aquatic crustaceans make extensive daily vertical migrations, often ascending dozens of meters (or more), only to return to greater depths a few hours later. Discuss the adaptive value of this behavior. What are the advantages and limitations of each hypothesis you discuss?

Dini, M. L., and S. R. Carpenter. 1991. The effect of whole-lake fish community manipulations on *Daphnia* migratory behavior. *Limnol. Oceanogr.* 36:370.

Enright, J. T. 1977. Diurnal vertical migration: Adaptive significance and timing. Part I. Selective advantage: A metabolic model. *Limnol. Oceanogr.* 22:856.

Gliwicz, M. Z. 1986. Predation and the evolution of vertical migration in zooplankton. *Nature (London)* 320:746.

McLaren, I. A. 1974. Demographic strategy of vertical migration by a marine copepod. *Amer. Nat.* 108:91.

Neill, W. E. 1990. Induced vertical migration in copepods as a defense against invertebrate predation. *Nature* 345:524.

Ohman, M. D., B. W. Frost, and E. B. Cohen. 1983. Reverse vertical migration: An escape from invertebrate predators. *Science* 220:1404.

Orcutt, J. D., Jr., and K. G. Porter. 1983. Diel vertical migration by zooplankton: Constant and fluctuating temperature effects on life history parameters of *Daphnia. Limnol. Oceanogr.* 28:720.

Taxonomic Detail

Phylum Arthropoda

This phylum probably contains up to five million species, of which about one million have been described so far.

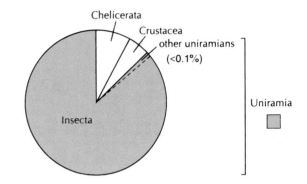

Subphylum Chelicerata

The approximately 75,000 species in this subphylum are divided among three classes.

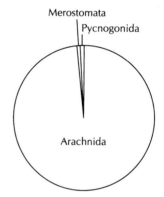

Class Merostomata.
Only four living species.

Order Xiphosura.
Limulus—the horseshoe crabs. All species are marine bottom-dwellers, growing up to 60 cm long, with a life span of up to 19 years.

Order Eurypterida.

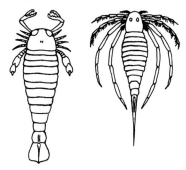

From E. N. K. Clarkson, Invertebrate Palaeontology & Evolution, 2d ed. Copyright © 1986 Chapman & Hall, Div. International Thomson Publishing Services Ltd., U. K. Reprinted by permission.

This group went extinct some 230 million years ago, and was maximally abundant about 350 to 450 million years ago. Originating in the sea, where members probably both swam and walked along the bottom, eurypterids later invaded estuarine and freshwater environments, and some even may have been amphibious, spending part of their time on land. Most preserved individuals are less than 20 cm long, but members of some species attained lengths exceeding 2 m, making them the largest arthropods that ever lived. Many biologists believe that arachnids evolved from eurypterid ancestors.

Class Arachnida.

The approximately 74,000 species are divided among 11 orders and about 550 families.

Order Scorpiones.

Diplocentrus, Centruroides—the scorpions. This group contains the largest of all arachnids, with individuals of some species reaching lengths of 18 cm. All species are terrestrial and carnivorous. There are 1,500 species divided among nine families.

Order Uropygi.

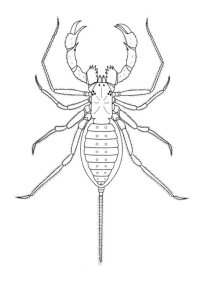

From James Nybakken, The Diversity of Invertebrates. Copyright © 1996 Times Mirror Higher Education Group, Inc., Dubuque, Iowa. All Rights Reserved. Reprinted by permission.

Mastigoproctu—the whip scorpions. Eighty-five species, one family.

Order Amblypygi.

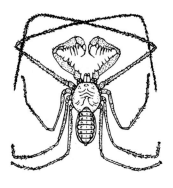

The tail-less whip scorpions.

Order Araneae.

The spiders. This group contains about half of all arachnid species. All are predators, primarily on insects. About 36,000 described species (with many more species yet to e described) in three suborders (one minor, omitted here); about 90 families.

Suborder Orthognatha.

Bothriocyrtum, Actinopus—trapdoor spiders; *Aphonopelma, Acanthoscurria*—tarantulas. Over 1,200 species divided among 11 families. Mostly tropical.

Suborder Labidognatha. Seventy-four families, about 35,000 species.

Family Symphytognathidae. This family contains the smallest of all known spider species (about 0.3 mm in body length). The group occurs only in New Zealand, Australia, and the neotropics.

Family Argyronetidae. *Argyronecta*—water spiders. All live submerged in freshwater, coming to the surface periodically to capture air on fine abdominal hairs. They then submerge and brush the air into a dome-shaped underwater web for storage.

Family Lycosidae. *Lycosa*—wolf spiders. Most species hunt on the ground, and only a few species make webs. Males follow a silken thread to locate the female, which leads to an elaborate courtship ritual as in many other arachnids. Up to 3,000 species.

Family Araneidae. *Araneus, Argiope*—orb weavers. Between 3,000 and 4,000 species.

Family Theridiidae. Comb-footed spiders. About 2,000 species. *Latrodectus* (the black widow spider) produces a venom that is highly toxic to humans (usually not deadly) and other vertebrates.

Family Salticidae. Jumping spiders. This is one of the largest of all spider families, containing nearly 5,000 species. Jumping spiders have the most highly developed of all arachnid eyes and use them to pounce on prey from a considerable distance. Salticids use silk primarily for nest building rather than for prey capture.

Family Thomisidae. Crab spiders. About 1,500 species.

Order Ricinulei.

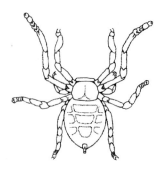

The 33 species in this order form a relatively inconspicuous group of tropical and subtropical eyeless arachnids. One family.

Order Pseudoscorpiones.

Chelifer. The 2,000 species in this group resemble the true scorpions, but they lack the long tail and stinger, and most are only 1 mm to 7 mm long. Most species are terrestrial, but some live intertidally, among seaweed stranded at high tide. All feed on the body juices of other invertebrates. Many hitch rides on insects, a traveling behavior termed "phoresy." Most species manufacture silken nests, spun from the tips of the chelicerae, and many indulge in elaborate courtship rituals. Twenty-two families.

Order Solifugae (= Solpugida).

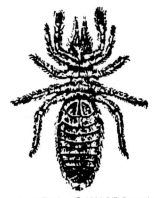

Engemann/Hegner, Invertebrate Zoology © 1981 3/E, Figure depicting Arachnoidea, Solpugida Figure 10-46C. Reprinted by permission of Prentice Hall, Inc., Upper Saddle River, NJ.

The solifugids (or solpugids)—wind scorpions (because they run "like the wind"). Individuals can be up to 7 cm long. Unlike other arachnids, these bear huge chelate chelicerae that point forward. The more than 900 species are found mostly in tropical and subtropical areas other than Australia and New Zealand. All are fast-moving predators. Members of many species eat termites. Twelve families.

Order Opiliones.

Leiobunum—the harvestmen, "daddy longlegs." Between 4,500 and 5,000 known species. Individuals range from less than a millimeter to nearly 22 mm in body length. Legs often very long and thin. These arachnids prey on other arthropods and on gastropods, or scavenge on fruit, vegetables, and decomposing animals. Twenty-eight families.

Order Acari.

Oppia, Dermatophagoides—the mites; *Dermacentor, Ixodes*—the ticks. About 30,000 species have been described to date, and some researchers estimate that another million species await discovery and description. Few mites are longer than 1 mm. The largest members of the order are the ticks, which can reach lengths of 5 mm to 6 mm; when fully fed, they can reach lengths of nearly 30 mm. Many species are major agricultural pests, and others transmit debilitating or deadly viral diseases to humans and livestock. Others help to decompose and recycle organic material. Nearly 400 families.

Class Pycnogonida (= Pantopoda).

Nymphon—the sea spiders. All 1,000 or so species are marine, living from shallow water to depths of 6,800 m. Most are external parasites or sucking predators. If the nauplius-like larva leaves the egg before development is completed, it grows as a parasite of cnidarians. Eight to nine families.

Subphylum Uniramia

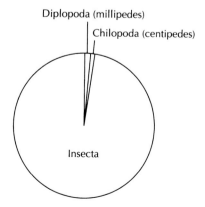

Diplopoda (millipedes)

Chilopoda (centipedes)

Insecta

Class Chilopoda.

The centipedes. All approximately 2,500 species are predators: The first pair of trunk appendages are modified into a pair of poison fangs, which are used to paralyze prey before consumption. Most species are tropical, although many are temperate. *Scutigera,* the common household centipede, is unusual among centipedes in possessing large, multifaceted (compound) eyes, apparently independently evolved from those of insects. The smallest centipedes are about 4 mm long, while members of the largest species, *Scolopendra gigantea,* reach lengths of nearly 30 cm. Twenty families.

Class Diplopoda.

The millipedes. Nearly 10,000 species have been described to date. Each millipede segment typically bears a pair of poison glands that secrete a volatile and irritating, or even deadly, liquid; among a variety of other organic constituents, the secretions usually contain hydrogen cyanide. At least one species can spray organisms a meter away. Most species are scavengers, relatively few are carnivores, and a few others are vegetarians. Approximately 120 families.

Class Symphyla.

Scutigerella, Symphella. These small (1–8 mm long) terrestrial arthropods live in damp habitats, such as beneath rotting wood, or in moist soil or leaf litter. Individuals usually have 10 to 12 pairs of legs. Most of the 160 species are vegetarians, and at least one species is an agricultural pest. Two families.

Class Pauropoda.

These are extremely small (usually less than 1.5 mm long) terrestrial arthropods that live in leaf litter and soil in forests. Each individual has nine pairs of legs. Most species are fast-moving and feed on fungi. Fewer than 500 species have been described. Five families.

Class Insecta (= Hexapoda).

At least 750,000 species have been described to date, and probably several times that number await discovery. Insects are found in every known terrestrial and freshwater habitat; they even occur in the Antarctic, living as ectoparasites on seals and sea birds.

Subclass Apterygota.

Primitive wingless (*a* = G: without; *pteron* = G: wings) insects, contained within a single order. The ancestors of this group are also believed to have been wingless. Four orders.

Order Thysanura.

Silverfish, bristletails. The approximately 600 species in this group are typically very fast runners. A number of species associate only with ants, termites, and birds' nests. A recently discovered fossilized thysanuran has been judged to be nearly 400 million years old. Seven families.

Order Collembola.

Spring tails. A group of small insects, no more than several millimeters long, possessing a characteristic abdominal jumping organ. Spring tails are common in various freshwater, marine, and terrestrial habitats. Five families, containing some 2,000 species.

Subclass Pterygota.

These are the winged (*pteron* = G: wing) insects. Although in many species the wings have been secondarily lost, all members of this group are descended from winged ancestors. Twenty-seven orders.

Order Ephemeroptera.

Ephemera, Ephemerella—mayflies. Over 2,000 species have been described from all over the world. Adults are nonfeeding and short-lived: Few live longer than a few days, and some live only a few hours. Fertilized eggs are deposited exclusively into freshwater, where the larvae develop for up to several years, passing through as many as 55 distinct stages (instars). Ephemeropterans are the only insects to have wings before adulthood: The penultimate larval stage (the "subimago"), which lasts only a few days at most, is also winged. Individuals of most species bear two pairs of wings. Like members of the following order, these insects cannot fold the wings flat against the body when at rest, believed to be a primitive condition. Nineteen families.

Order Odonata.

The damselflies (suborder Zygoptera) and dragonflies (suborder Anisoptera). These species are winged (two pairs), but, as with the Ephemeroptera, the wings cannot be folded along the abdomen when the flies are resting. The gilled larvae (nymphs) develop in freshwater, where they are major predators and also key food resources for fish. The largest dragonflies *look* intimidating—they are about 10 cm long—but they do not bite or sting. Adults eat only other insects. Twenty-five families.

Order Blattaria.

Periplaneta, Blatella, Blatta, Blaberus, Blatteria, Cryptocercus—cockroaches. Most of the nearly 3,700 species in this order are tropical; a few species are house pests. Although many species are omnivorous, others specialize on such diets as wood and live in such unlikely habitats as deserts, caves, or ant nests. Some species (members of the primitive family Cryptocercidae) harbor in their hindguts protozoan symbionts closely related to those inhabiting the hindguts of termites. Many species lack wings. One wingless Australian species, *Macroparesthia rhinoceros*, weighs 20 gm despite being only 6.5 cm long. Another large cockroach species, *Blaberus giganteus*, found in Mexico and South America, grows to over 6 cm in body length. Five families.

Order Mantodea.

Mantids, preying mantids. All 1,800 species prey on other insects. The nymphs resemble small adults and develop without a true metamorphosis. Adults generally possess two pairs of wings, although females may lack both pairs. Eight families.

Order Isoptera.

Termites ("white ants"). These insects are believed to have evolved from primitive wood-eating cockroaches. Like those cockroaches, termites house symbiotic protozoa or bacteria that digest cellulose and release the nutrients to the insect host. A single termite colony may contain over one million individuals in below-ground colonies or in above-ground earthen colonies as large as a house. Termites are remarkably destructive of wooden structures (homes, trees), but are extremely important in nutrient and energy recycling, especially in tropical areas. All species are eusocial, and individual termites fall into certain castes that define their social rank and life's work: workers, males, soldiers, and queens. The approximately 2,000 described termite species are distributed among six families.

Order Grylloblattaria.

The 12 species of wingless omnivores contained in this order are found only in cold environments, including glaciers and ice caves. One family.

Order Orthoptera.

Crickets, katydids, grasshoppers, locusts, monkey hoppers (referring to their agility in trees and shrubs). This is a large group of some 20,000 species, some members of which grow to over 11 cm long, with wingspans exceeding 22 cm. Many species "stridulate," producing a species-specific song by rubbing specialized portions of their wings (*not* the hind legs) together. Members of one genus, *Mecopoda*, are commonly caged as long-lived pets in China and Japan. Sixty-one families.

Order Phasmida (= Phasmatoptera).

Carausius, Diapheromera—walking sticks (stick insects); *Phyllium*—walking leaves (leaf insects). The approximately 2,500 species are infamous for their ability to almost perfectly mimic the stems and leaves of the plants on which they feed. Most species can alter their body coloration with each molt to better match the color pattern of their surroundings. The largest stick insects reach about 33 cm in length. Eleven families.

Order Dermaptera.

Earwigs. Most earwigs (about 99% of the 1,500 species) are free-living herbivores or carnivores, but about 20 species are exclusively parasitic or commensal on bats and rodents. Most species are tropical, and many lack wings. Eleven families.

Order Embiidina.

Embiids. This group contains about 200 species of tropical or near-tropical silk-spinning insects that live mostly in narrow galleries lined with silk, in soil, in wood, or in leaf litter. Males usually have wings, but the females are always wingless. Embiids primarily eat decaying plant matter. Eight families.

Order Plecoptera.

Stoneflies. These insects are distributed worldwide, except for Antarctica. The adult life span is typically short—just long enough for mating and egg laying; most of the life cycle is spent in the larval stage, usually in freshwater. Stonefly nymphs closely resemble those of mayflies (order Ephemeroptera), but have only two tails instead of three and bear gills on the thorax instead of on the abdomen. Fifteen families.

Order Psocoptera.

Book lice, bark lice. Despite the connotation of their common name, these insects are not parasites. Most feed on algae, mold, lichens, pollen, or dead insects. Individuals are small, usually 1 mm to 6 mm long. Most of the nearly 2,600 species are found in leaf litter, under bark, on leaves, under stones, and in caves, particularly in the tropics. Some species are annoying to humans, thriving on stored foods in pantries and cupboards. Thirty-seven families.

Order Anoplura.

Sucking lice; crab lice (transmitted venereally among humans); *Pediculus*—human body lice. These 500 species of blood-sucking ectoparasites are small, wingless insects that never exceed about 4 mm in length. The abdomen swells extensively to accommodate a large volume of blood. All parasitize mammals, including such diverse animals as aardvarks, camels, monkeys, llamas, seals, ungulates, and humans. Livestock infestations can be extremely debilitating. Human head and body lice transmit typhus. Fifteen families.

Order Mallophaga.

Chewing lice, biting lice. All are small (less than 5–6 mm), wingless parasites of birds and mammals. Eleven families.

Order Thysanoptera.

Thrips. All 5,000 species are small (rarely larger than 5.0 mm, with some as small as 0.5 mm). Some species are winged; some are not. Most species feed on various parts of plants and may transmit diseases among the plants on which they feed. Five families.

Order Hemiptera.

The true bugs. Many of the 35,000 species in this large group are major agricultural pests or transmitters of diseases; many other species are beneficial. Most species feed on various portions of plants, but some prey on other arthropods, and a few are ectoparasites of vertebrates. Seventy-four families.

Order Homoptera.

This large order (some 45,000 species have been described to date) includes the cicadas, aphids, mealy bugs, spittle bugs, jumping plant lice, and leaf hoppers. All species feed on plants and often require specific plant hosts. The group includes many agricultural pest species. Fifty-five families.

Family Aphididae. Aphids. Many of the 3,500 aphid species ("plant lice") are severe agricultural pests and vectors of serious plant diseases. The life cycle usually involves several asexual generations: Females hatch in the spring and deposit eggs, which develop parthenogenetically into more females, which continue to reproduce parthenogenetically for several more generations. The last generation of the summer develops into males and females; these mate and deposit fertilized eggs that overwinter (lay dormant) until the following spring.

Family Cicadidae. Cicadas. These large insects have two pairs of wings. The male usually has sound-producing organs at the base of the abdomen, although some species produce sound using the wings. Eggs are deposited in trees; nymphs and adults feed on tree sap.

Family Cercopidae. Spittlebugs (= froghoppers). About 23,000 species have been described to date. Adults are herbivorous and usually require particular host species. Females lay eggs in plant tissue, and the developing nymphs usually produce a conspicuous, foamy white mass, which protects them from predation and desiccation. Nymphs of other species instead secrete a calcareous tubular house on the host plant. Some species stunt or otherwise damage pine trees and clover.

Superorder Holometabola.

The nine orders in this group include most insect families and most insect species. All exhibit complete, dramatic metamorphosis.

Order Neuroptera.

Dobson flies, lacewings, ant lions, and snake flies (so-named from the snake-like movements and shape of the prothorax). This is a small but diverse and widespread group of rather primitive holometabolous insects. The larvae are aquatic and secrete silken cocoons from the Malpighian tubules. Twenty-one families.

Order Coleoptera.

The beetles. This is the largest of all insect orders, including over 339,000 described species. Nearly 70% of these species are contained within only seven families. Some families contain fewer than a dozen species, but most contain hundreds or thousands, and a few contain more than 30,000 species each. In addition to the families about to be discussed, coleopterans include whirligig beetles, ladybugs (ladybirds), click beetles (which click while jumping in the air), Japanese beetles, and waterpenny beetles. Most beetles have two pairs of wings, with the front pair serving only as a protective sheath for the rear pair, which are used for flying. Many species produce sounds in various ways, including stridulation (rubbing various specialized body parts together). One hundred fifty-three families.

Family Carabidae. This group of some 30,000 species includes the colorful tiger beetles and the remarkable bombardier beetles, which explosively discharge a severe irritant as a potent defense measure. Most species are carnivores, even as larvae; the larvae digest their prey before ingestion and then slurp up the liquid food.

Family Ptiliidae. This is a small group of about 430 species. Many feed on fungal spores. A few highly specialized species live only in ant colonies, feasting on the excretory products of ant larvae.

Family Staphylinidae. Rove beetles. The approximately 30,000 species in this family live mostly among leaf litter. Many species live in ant or termite nests and feed on fungal spores and hyphae, but most are carnivorous.

Family Scarabaeidae. Dung beetles, scarab beetles. This group contains about 25,000 species. Members of most species eat dung, although some feed on fungi, flowers, and grasses. Some species are serious pests on golf courses, ruining the greens. Larvae often destroy crops by eating the roots.

Family Buprestidae. Jewel beetles (= metallic beetles). Adults often have a distinct metallic coloration. The larvae often do serious damage to shrubs and trees, especially fruit trees. About 15,000 species have been described.

Family Lampyridae. *Photinus, Photurus.* The fireflies. The 2,000 species are characterized by specialized bioluminescent organs at the tip of the abdomen that produce a light signal that attracts mates. The larvae are ground-dwelling predators of other terrestrial invertebrates, including snails, slugs, caterpillars, and earthworms.

Family Dermestidae. These are small beetles (1–12 mm long) with a wide range of tastes, feeding on such diverse foods as pollen, nectar, carpets, upholstery, grains, dead and decaying vertebrates, and dead insects. Members of the genus *Dermestes* are routinely used to help clean vertebrate skeletons for display or study. Other species destroy prized insect collections. The larvae are especially damaging. The group contains about 850 species.

Family Tenebrionidae. This group includes some 18,000 species, many wingless, which feed mostly on plant material. Several genera, including *Tribolium* and *Tenebrio,* commonly infest stored foods. *Tribolium* (the flour beetle), in particular, has been used widely in ecological studies of population growth.

Family Cerambycidae. Timber beetles. The larvae of these 35,000 species bore into plant tissue, living or dead. Adults mostly feed on pollen and nectar and are therefore often seen on flowers.

Family Curculionidae. The weevils. This large group of some 50,000 described species includes major agricultural pests (e.g., rice weevils, cotton-boll weevils). Most species feed on various parts of flowering plants.

Family Chrysomelidae. Leaf beetles. This group contains about 35,000 species. All adults and many larvae feed on plant leaves. The larvae of some species feed underground, on plant roots. Many species are agricultural pests.

Order Strepsiptera.

This group of holometabolous insects contains about 500 species. The females all are wingless, often legless, endoparasites of other insects, including bees, wasps, thysanurans, and cockroaches. Females spend their entire adult lives within the host's body, often with only the head protruding between a pair of adjacent host abdominal segments. Males are winged (although the front pair of wings are greatly reduced in size) and free-living, and they soon locate a female and fertilize her eggs. Hundreds or thousands of very active six-legged larvae, less than 300 µm long, escape from the parental host. The larvae must then locate and bore into the larval stage of the host insect; larvae that mature as males then leave the host and fly off, while those that mature as females remain forever within the host. Eight families.

Order Mecoptera.

Scorpion flies, snow fleas. These common forest insects feed on nectar or eat other insects. The abdomen of males ends in an upward, pointed curve, resembling a scorpion's stinger; nevertheless, the flies do not sting. Eight to nine families.

Order Siphonaptera.

Fleas, jiggers. Approximately 1,750 species of these wingless, holometabolous, biting, and blood-sucking insects have been described. Adults are parasitic—usually ectoparasitic—on warm-blooded animals, usually mammals (especially rodents). The larvae are typically not parasitic and pupate within silken cocoons. Because adults frequently jump from one host to another, fleas are excellent vehicles for transferring diseases among hosts. In particular, fleas are vectors for bubonic plague (the black death). Certain flea species are also obligate intermediate hosts for the common tapeworm of dogs and cats. Fleas lack compound eyes. Fifteen families.

Order Diptera.

The true flies. This immense group of some 100,000 to 150,000 species contains such beloved insects as mosquitoes, gnats, black- and greenflies, no-see-ums, botflies, fruit flies, dung flies, and houseflies. Unlike other so-called flies, adult dipterans exhibit a posterior pair of club-shaped reduced wings (halteres, used for balancing during flight) and only one pair of flying wings. The members of this group occur worldwide and some can breed successfully in such unlikely places as oil seeps, hot springs, and the sea floor. The larvae show a terrific diversity of feeding patterns, including leaf-mining, predation, detritus feeding, and ecto- and endoparasitism. The larvae of many species lack legs and are known as "maggots." Many species transmit diseases, such as malaria, typhoid, yellow fever, and dysentery, and many other species are important agricultural pests. On the other hand, many other dipteran species eat or parasitize various insect pests, pollinate flowers, or destroy certain weeds; these dipterans thus are undeniably beneficial. One hundred sixty-two families.

Family Chironomidae. Midges. Approximately 5,000 species of ubiquitous, nonbiting, flying insects. A number of species are marine. The larvae are generally aquatic.

Family Tipulidae. Crane flies. With over 13,000 described species, this is the largest dipteran family.

Family Chaoboridae. Phantom midges (e.g., *Chaoborus*). The larvae are aquatic and commonly prey on larval mosquitoes, serving as natural control agents. Only about 75 species have been described.

Family Culicidae. Mosquitoes (e.g., *Culex, Anopheles, Aedes*). The female adult proboscis is modified for piercing; females require a blood meal before egg laying. Mosquitoes play major roles in transmitting such devastating diseases as malaria, yellow fever, and filariasis. The larvae are aquatic. About 3,000 species have been described.

Family Simuliidae. Blackflies (buffalo gnats). Females are blood-sucking parasites that can inflict a memorable bite. A swarm of adults can kill livestock and even humans. One species is essential in the transmission of river blindness (Africa and Central America). The larvae develop in streams.

Family Tabanidae. Deerflies, horseflies. The females are bloodsuckers and often large-bodied ones. The larvae are aquatic predators on various other aquatic invertebrates. Deerflies transmit anthrax. More than 3,000 species have been described.

Family Tephritidae. Fruit flies. Over 4,000 species are known. Larvae mostly feed on fruits, such as apples and cherries, making the larvae major agricultural pests.

Family Drosophilidae. Vinegar flies. The best-known genus in this group of some 1,500 species is *Drosophila*, widely used in evolutionary, genetic, and developmental studies.

Family Muscidae. This group includes houseflies (*Musca domestica*), which transmit typhoid, anthrax, dysentery, and conjunctivitis; the cattle face fly (*M. autumnalis*), which is commonly seen clustering around the heads of cattle; and the tsetse fly (*Glossina*, sometimes placed in a separate family, the Glossinidae), which transmits sleeping sickness and other similar diseases caused by trypanosomes. Larvae generally feed on decaying animal and plant matter or on feces.

Family Calliphoridae. Blowflies (the "housefly" of the western and southwestern United States); screwworm. Most larvae (maggots) develop on decaying animals. Some species preferentially deposit their eggs on open sores of living animals, rather than on the carcasses of dead ones.

Family Oestridae. Botflies. Most of the 65 described species resemble bees. The larvae are endoparasites of mammals, including sheep, cattle, and other livestock. Horse botflies (about 45 species) are sometimes placed in another family.

Family Bombylliidae. Bee flies. Many of the approximately 4,000 species closely resemble bees or wasps. Although adults generally feed on nectar, the larvae are always parasitic on developmental stages of other insects or on other insect parasites; as such, bee-fly larvae control many insect pest populations, including locusts and tsetse flies.

Order Trichoptera.

Caddisflies. About 7,000 species have been described to date. Adults resemble small moths but feed exclusively on liquids. Some species never get longer than about 2 mm. The larvae and pupae are generally aquatic (mostly in freshwater, although some species develop in salt marshes), but the developmental stages of some species are fully terrestrial. Larvae typically feed on algae, fungi, and bacteria. Approximately 40 families.

Order Lepidoptera.

Moths and butterflies. This enormous group of insects contains nearly 140,000 described species. The females of some species are wingless, and many species are strictly nocturnal (active only at night). The larvae typically feed on plants (leaves, stems) or plant products (fruits, seeds), but some prey on other insects. There are 137 families (with some families containing fewer than a dozen species each).

Family Noctuidae. Cutworms, armyworms. This is the largest lepidopteran family, with about 25,000 described species. The larvae of many species are major agricultural pests, feeding on plants and fruits.

Family Cyclotornidae. This group of only five Australian species is interesting despite its small size. The larvae are initially ectoparasitic on ants. As the larvae get older, they drop off the host back at the nest, where they then provide the ants with nectar and feed on the ants' larvae.

Family Pieridae. The 2,000 species in this group often show highly specific feeding preferences, and several species have been much studied by ecologists interested in plant-insect coevolution. Some species are pests, particularly those feeding on legumes and crucifers.

Family Pyralidae. This major (about 20,000 species) group of moths contains many agricultural pest species.

Family Bombycidae. This group of only about 100 Asiatic moth species includes one of the most famous of all lepidopteran species, the well-known silkworm *Bombyx mori.* The silkworm, in addition to its long-standing commercial importance, has played a major role in the development of molecular biology; the first messenger RNA molecules to be isolated in large quantities from any eukaryote were the mRNAs coding for the silk protein of *B. mori.*

Family Saturniidae. *Hyalophora cecropia*—giant silk moths. This group of about 1,000 species includes the largest of all moths, with wingspreads up to 25 cm. Some species are economic pests on various trees, while others produce a commercially valuable silk. Females are well known for using pheromones to attract mates from long distances.

Family Sphingidae. Sphinx moths. This group of some 850 species includes the well-studied tobacco hornworm caterpillar *Manduca sexta,* a serious pest of tobacco and tomato plants.

Order Hymenoptera.

This group of some 130,000 holometabolous insect species includes the familiar ants, bees, and wasps. Many species form functionally complex societies. Eighty-nine families.

Suborder Symphyta.

The sawflies. The caterpillars (larvae) of all species feed on terrestrial plant tissues and often specialize on particular plant species or groups. Adults of some species prey on other insects. Fourteen families.

Suborder Apocrita.

Wasps, ants, and bees. Adults are often nectar feeders, although members of some species suck the body juices from other arthropods. The larvae are usually legless and blind. Many of these white grubs and maggots feed in or on the bodies of a host arthropod or its larvae; others develop within plant galls, fruits, or seeds. Seventy-five families.

Family Ichneumonidae. These wasps are mostly parasitoids, living freely as adults but developing at the expense of an arthropod host, usually an insect but sometimes a spider, bee, or pseudoscorpion; the larva feeds on the host and eventually kills it. At least 15,000 species have been described to date, although perhaps three times as many species await discovery. Only about 5% of species are ensocial, and only a few species sting people.

Family Formicidae. *Formica, Myrmica, Solenopsis*—the ants. Adults typically feed on fungi or nectar, or they prey on other terrestrial arthropods. At least 9,500 species are known, each of which forms complex social groups of dozens to thousands of cooperating individuals per colony. Most ant colonies include members of at least three distinct castes: workers (wingless, sterile females), males, and queens. Probably another 20,000 ant species remain to be described, primarily from tropical rain forests.

Family Apidae. *Apis*—honeybees; *Bombus*—bumblebees. This is one of eight bee families, the entire group encompassing perhaps 20,000 species. Only about 5% of the species, including the honeybees, are eusocial. Unlike most other bees, the members of this family do not dig burrows, but rather nest in cells of wax or resin, sometimes supplementing the structure with other materials, such as bark, mud, or even vertebrate feces. Bees are major flower pollinators, and many flowers, including orchids, depend on bees for pollination; both adults and larvae subsist on nectar and pollen. Not all bee species bear stingers. The so-called "killer bees" now working their way northward from Central and South America are honeybees; they are unusually aggressive, but they may produce more honey than most other bees. Honeybees are well known for their complex dances, through which workers communicate the location of desirable flowers.

Subphylum Crustacea

This subphylum contains approximately 45,000 species.

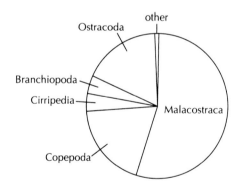

Class Cephalocarida.

Hutchinsoniella. The nine species in this group are thought to be among the most primitive of living crustaceans. All are marine bottom-dwellers living in soft sediments intertidally to depths exceeding 1,500 m. No species exceeds 3.7 mm in length. All are suspension feeders, collecting food particles using spines on the legs and circulating water through the legs by the rhythmic beating of the limbs.

Class Malacostraca.

The over 20,000 described species are distributed among 12 orders.

Superorder Syncarida.

Anaspides, Bathynella—the syncarids. The 150 species contained in this group are mostly freshwater, elongated crustaceans lacking a carapace. A number of species are interstitial, living in the spaces between sand grains. A few other species live only in burrows made by freshwater crayfish. Six families.

Superorder Hoplocarida.

Order Stomatopoda.

Squilla, Pseudosquilla, Gonodactylus—mantis shrimp. The 350 stomatopod species are active, hard-hitting predators living in crevices and holes in hard substrate or in extensive burrows of their own making in soft substrates. Some species are fished commercially. Twelve families.

Superorder Peracarida.

The over 11,000 described species are distributed among seven orders.

Order Thermosbaenacea.

Thermosbaena. Small (about 3 mm long) crustaceans living only in saline hot springs, at water temperatures of up to 43°C. Other members of the order live in caves, in cold, freshwater. Two families.

Order Mysidacea.

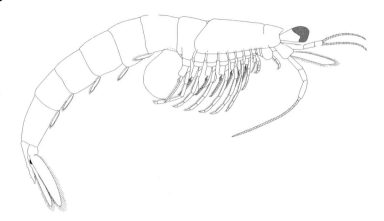

Mysis—mysid shrimp. Most of the 780 species in this group are marine or estuarine, but several dozen inhabit freshwater environments, including lakes, wells, and caves. Many species are heavily preyed on by various fish. Six families.

From: Comparative Morphology of Recent Crustacea by McLaughlin. Copyright © 1980 by W. H. Freeman and Company. Used with permission.

Order Cumacea.

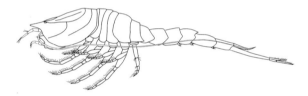

The cumaceans. These are exclusively marine, living in soft substrates, usually at less than about 200 m depth. The almost 1,000 species are distributed among eight families.

Order Tanaidacea.

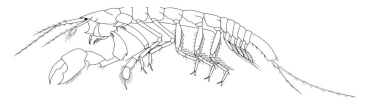

The tanaids. The 550 species are all marine, occurring from shallow to abyssal depths. They are among the most commonly encountered malacostracans in the deep sea. Sixteen families.

Order Isopoda.

Asellus, Ligia, Idotea, Paracerceis, Sphaeroma—the isopods; woodlice; pillbugs. This large group of some 10,000 species occurs in marine, estuarine, freshwater, interstitial, and terrestrial habitats. Some species are blind cave dwellers. Others parasitize fish, cephalopods, or other crustaceans, including other isopods. *Bathynomus giganteus,* a deep-sea species, attains a length of 42 cm (nearly 1.5 ft) and a width of about 15 cm. Most species are considerably smaller. *Limnoria,* although less than 0.5 cm long, is a very destructive, marine, wood-boring isopod. Many species are blind. In some, males are greatly reduced and live in a pouch within the female or attached externally to her antennae. Isopods are among the most common deep-sea malacostracans and are the dominant terrestrial malacostracans. One hundred families.

Order Amphipoda.

Amphipods dominate the small (1–50 mm) malacostracan fauna in freshwater and in shallow coastal waters in temperate regions. Some of the approximately 6,000 species are common in underground streams and in caves. About 120 families.

Suborder Gammaridea.

Ampelisca, Corophium, Gammarus; Orchestia, Talorchestia (beachhoppers and sand fleas)—the gammarid amphipods. The over 4,700 species are distributed among some 91 families. Most individuals are only 1 mm to 15 mm long, although one abyssal species reaches a length of 25 cm. Nearly one-third of the species live in freshwater. Another 3,350 species are marine, mostly living intertidally or in shallow water, and a few species are terrestrial. Gammarid amphipods are major food sources for many fish species and

for some marine mammals. Amphipod concentrations can reach 73,000 individuals per m^2. Gammarids exhibit a variety of lifestyles. Various species are herbivores, carnivores, deposit feeders, commensals, or external parasites (of invertebrates and fishes). Some species display complex mating behavior.

Suborder Hyperiidea.

Hyperia, Phronima—the hyperid amphipods. All 300 or so species are marine and planktonic and occur in all ocean waters. Despite the relatively small number of species, hyperid amphipods are important in marine food chains: as key food items for marine mammals, sea birds, and large fish and as carnivores of other invertebrates. These amphipods are typically associated with gelatinous animals, such as salps, jellyfish, and siphonophores. They often exhibit extensive diurnal vertical migrations (presumably with their hosts), often traveling over 1,000 m vertically in a 24-hour period. *Phronima* can apparently kill the gelatinous host and live inside the hollowed-out test, propelling it by beating its pleopods.

Suborder Caprellidea.

Caprella—the caprellid amphipods; whale lice; skeleton shrimp. This group contains about 200 strange-looking crustaceans. Individuals of most species are sedentary, clinging to a substrate (such as macroalgae) or to a host (echinoderm, cetacean). The body colors closely match the surroundings and can be changed with each molt. Whale lice (family Cyamidae) are exclusively ectoparasitic on whales and dolphins, especially in or near open wounds, but most caprellids are free-living.

Suborder Ingolfiellidea.

Most of the 27 species (two families) in this group are restricted to freshwater and brackish-water caves and subterranean freshwater habitats. Some species are marine, living in the deepest reaches of the sea, while others are interstitial, specialized for life between grains of sand.

Superorder Eucarida.

The approximately 10,100 species are distributed among three orders (one minor, omitted here).

Order Euphausiacea.

Euphausia, Meganyctiphanes—krill, euphausiids (= euphausiaceans). The 85 described species in this order have a worldwide distribution and constitute a major food of baleen whales, many fishes, squid, birds, and true shrimp. All species are marine and are particularly common in the open ocean. Extensive daily vertical migrations of up to 400 m each way are well documented. Two families.

Order Decapoda.

This group of some 10,000 species, distributed over some 100 families, comprises the shrimp, crabs, hermit crabs, and lobsters.

Infraorder Penaeidea.

The penaeids and sergestids, with about 350 species. Five families.

Family Penaeidae. *Penaeus*—the penaeid shrimp. Most commercially important shrimp are included in this family of some 250 species. The species are especially common in estuaries.

Family Sergestidae. *Lucifer, Sergestes*—the sergestid shrimp. Some species are fished commercially in India and Southeast Asia.

Infraorder Caridea.

The carideans, with about 1,600 species. Twenty-two families.

Family Palaemonidae. *Macrobrachium, Palaemonetes.* These 450 shrimp species are common in marine, estuarine, and freshwater habitats.

Family Alpheidae. *Alpheus*—snapping shrimp. Approximately 425 species.

Family Pandalidae. *Pandalus*—pandalid shrimp. About 115 species are known, including some of commercial importance.

Family Crangonidae. *Crangon.* These small (less than 1 cm long), omnivorous shrimp are common in temperate and tropical waters. One species is fished commercially by Europeans. Approximately 140 species.

Infraorder Astacidea.

The Astacids, with 12 families.

Family Nephropidae. *Homarus*—the American lobster. Adult lobsters may reach nearly 60 cm (2 ft) in length. Very tasty.

Family Cambaridae. *Cambarus*—the freshwater shrimp (crayfish). These are commonly dissected in laboratory courses. None have free-living larvae.

Family Astacidae. *Astacus*—freshwater shrimp (crayfish). Thirteen species, none with free-living larvae.

Family Parastacidae. *Cherax*–freshwater shrimp (crayfish). Several species support aquaculture operations in The People's Republic of China, Australia, and The USA.

Infraorder Palinura.

About 130 species, five families.

Family Palinuridae. *Panulirus*—the tropical (spiny) lobster, commercially important in southern waters; *Palinurus*—the European spiny lobster.

Infraorder Anomura.

The anomurans, with nearly 1,600 species in 13 families.

Family Upogebiidae. *Upogebia.* Worldwide, mud-burrowing species that typically live in shallow water. All 30 species are marine.

Family Callianassidae. *Callianassa*—mud shrimp. The nearly 170 different species are all marine and live in Y- or U-shaped burrows in mud, typically in shallow water.

Family Diogenidae. *Cancellus, Clibanarius*—hermit crabs. The 350 described species are mostly tropical and often very colorful.

Family Paguridae. *Pagurus*—hermit crabs. The over 550 species are all marine. Hermit crabs have vulnerable abdomens, which they typically protect by living within empty gastropod shells that they carry with them as they walk. Fossil evidence indicates that hermit crabs have inhabited gastropod shells for a good 113 to 150 million years. Several dozen species deliberately place sea anemones on their shells, gaining added protection from predators, and a few deep-water hermit crab species live in chitinous, snail-like "shells" secreted for the crabs by the anemones themselves. A few species are sessile, living in calcareous tubes built by sedentary polychaetes and vermetid gastropods, or within calcareous pits in coral colonies. Recent molecular data indicate that king crabs (*Paralithodes*) and their relatives evolved from hermit crab ancestors and thus also belong within this family of remarkable animals.

Family Lithodidae. *Lithodes, Paralithodes*—king crabs. King crabs are among the largest and tastiest crustaceans; they support a substantial fishery in the North Pacific. Recent molecular analyses of DNA coding for certain mitochondrial rRNA sequences indicate that king crabs and their relatives evolved from hermit crab ancestors and thus should be moved to the family Paguridae.

Family Galatheidae. *Galathea.* Marine bottom-dwelling crabs, especially common in the tropics. Some of the 258 described species are blind and are common at abyssal depths.

Family Porcellanidae. *Petrolisthes, Porcellana*—porcelain crabs. Most of the 225 species are marine; they are commonly symbiotic with sponges, tube-dwelling polychaetes, and sea anemones.

Family Hippidae. *Emerita, Hippa*—sand crabs, mole crabs. These small, marine crabs have a worldwide distribution. They burrow into sand on wave-swept beaches and emerge briefly to strain food from the water each time the water recedes.

Infraorder Brachyura.

The brachyurans, with about 4,500 species distributed among nearly 50 families.

Family Majidae. *Libinia, Maja*—spider crabs. A Japanese species (*Macrocheira kaempteri*) may reach nearly 4 m width (across the legs). The group is entirely marine and has a worldwide distribution, except at the poles. Individuals are found to depths of about 2,000 m and also occur in shallow water.

Family Cancridae. *Cancer*—the cancer crabs. This marine group includes numerous species of commercial importance. They are also important in marine food chains, as major predators of other invertebrates and of fish.

Family Portunidae. *Callinectes, Carcinus* (the green crab), *Ovalipes, Portunus*—the swimming crabs. The 230 species in this group are common in shallow marine and estuarine habitats. *Callinectes sapidus* (the blue crab) is a commercially important food species along the southeastern Atlantic coast.

Family Xanthidae. *Menippe, Panopeus, Rhithropanopeus*—mud crabs. This is the largest of all crab families, containing about 1,000 species. Most species are marine, but some extend into freshwaters. Some species are exclusively symbiotic with corals.

Family Grapsidae. *Sesarma.* Most species are marine or estuarine, although some live in rivers, and a few tropical species are fully terrestrial.

Family Gecarcinidae. *Cardisoma, Gecarcinus.* These brachyuran crabs are terrestrial for most of their adult lives, although they must migrate back to the sea to reproduce. *C. guanhummi* can weigh 500 g and boasts a carapace up to 11 cm across.

Family Pinnotheridae. *Pinnixa, Pinnotheres*—pea crabs. These are small (typically less than 1 cm) marine crabs, exclusively parasitic on, or commensal with, echinoderms, bivalves, or tube-dwelling polychaetes. About 225 species.

Family Hapalocarcinidae. The coral gall crabs. This small group (27 species) is exclusively tropical, and its members are all tiny; individuals typically measure only a few millimeters across the carapace. The female lives within a lump of coral tissue produced by hard corals as the coral slowly grows around and eventually encloses the patient crab, which then lives protected but imprisoned.

Family Ocypodidae. *Uca*—the fiddler crabs. The several hundred species in this group are mostly semiterrestrial, living intertidally in estuaries and salt marshes.

Class Branchiopoda.

Most of the approximately 800 species in this group live in freshwater, with only a few species living in the sea. Four orders.

Order Notostraca.

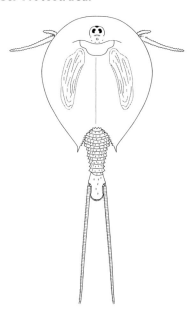

From *Comparative Morphology of Recent Crustacea* by *McLaughlin*. Copyright © 1980 by W. H. Freeman and Company. Used with permission.

Triops—tadpole shrimp. The nine species in this order generally inhabit temporary lakes. The eggs withstand considerable dehydration. One family.

Order Cladocera.

Daphnia, Bosmina, Polyphemus, Moina, Podon, Evadne—water fleas. Individuals are usually smaller than 3 mm. This order contains about half of the approximately 800 branchiopod species. Nine families.

Order Conchostraca.

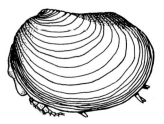

From R. Pennak, *Freshwater Invertebrates of the U.S., 3d ed.* Copyright © 1978 John Wiley & Sons, Inc., New York. Reprinted by permission.

Clam shrimp. This group is believed to have given rise to the cladocerans. The fewer than 200 species are common in temporary freshwater ponds, and the eggs withstand considerable dehydration. As with cladocerans, embryos are brooded within the female carapace, which completely encloses the adult conchostracan body. Individuals can grow to lengths of 2 cm. Five families.

Order Anostraca.

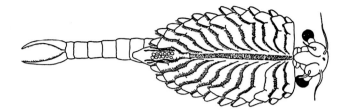

From R. Pennak, Freshwater Invertebrates of the U.S., 3d ed. Copyright © 1978 John Wiley & Sons, Inc., New York. Reprinted by permission.

Branchinecta; Eubranchipus; Streptocephalus—the fairy shrimp; *Artemia*—the brine shrimp. This group contains about 200 species, which grow to 10 cm in length. These animals live in both temporary and permanent bodies of water, covering a wide range of salinities. Some groups, such as members of the genus *Artemia,* are found worldwide except for the poles, while others occur mainly, or only, at the poles. Members lack a carapace, and as many as 19 of the body segments bear swimming appendages. Eight families.

Class Ostracoda.

Cypridina, Cypris, Gigantocypris, Pontocypris—seed shrimp. Most of the 6,650 ostracod species are only a few millimeters long, but females of *Gigantocypris* can exceed 3 cm. Most species are marine, estuarine, or freshwater, but a few occur in moist terrestrial habitats. Ostracods exhibit a variety of lifestyles, but none are parasitic. Marine species are found from shallow to abyssal depths. Ostracods are mostly divided among the two major subclasses Myodocopa and Podocopa. Forty-three families.

Class Mystacocarida.

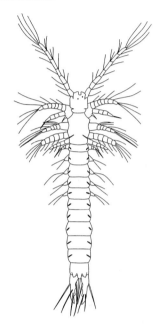

From: Comparative Morphology of Recent Crustacea by McLaughlin. Copyright I© 1980 by W. H. Freeman and Company. Used with permission.

The eight known species are exclusively marine and very small: Most are less than 0.5 mm long, and none are larger than 1 mm. All species are interstitial, living in the spaces between sand grains intertidally or in shallow water.

Class Copepoda.

The approximately 8,500 species are distributed among six orders.

Order Calanoida.

Calanus, Euchaeta, Eurytemora, Centropages, Diaptomus (occurs in freshwater only), *Candacia, Bathycalanus* (the largest of free-living copepods, up to 17 mm long), *Acartia.* These animals are mostly marine, and in fact they dominate the marine plankton; they are extremely important in marine food chains leading to many important fish species. Copepods occur in all ocean waters, from the surface to depths exceeding 5,000 m. Many planktonic species perform extensive vertical migrations, swimming to the surface at night and moving into deeper water at dawn. The members of one family (Diaptomidae) are found exclusively in freshwater. The order contains over 1,900 species, distributed among 37 families.

Order Harpacticoida.

Euterpina, Psammis, Tisbe. These small copepods (less than 2.5 mm long) are found in marine, estuarine, and freshwaters. Most species live in or on the bottom, many living interstitially in the spaces between sand grains. Some freshwater species live in damp moss. A few species are commensal in the baleen of certain whale species, and a few spend their lives swimming in the plankton. Most species occur only in shallow water, but a few are restricted to the deep sea. The order contains 34 families, housing about 2,250 species.

Order Cyclopoida.

These copepods are found in marine and freshwater habitats. Most species are free-living, but some are commensal, and others are parasitic. Members of the genus *Mytilicola* are intestinal parasites of bivalves. Others are ectoparasites on gill filaments of fish. Some species are intermediate hosts for the parasitic nematode *Dracunculus medinensis* (p. 185). The over 3,000 species are distributed among 12 to 16 families.

Order Monstrilloida.

Monstrilla. All are marine, bizarre, often highly degenerate crustaceans, and all are parasites, as larvae, on various invertebrates. The adults are free-living but nonfeeding, lacking mouthparts or gut; the adult stage of the life cycle is therefore short. Adults also lack a second antenna. The approximately 80 species are distributed among only two families.

Order Caligoida.

Caligus, Salmincola (the salmon gill maggot). Although the nauplii larvae are typical and free-living, the adults are exclusively external parasites of marine fish, commonly affixed to the gills. Eleven families.

Class Branchiura.

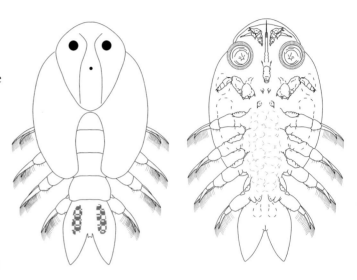

From L. Margolis and K. Zabata, "Guide to the Parasites of Fishes of Canada, Part II: Crustacea." in Fisheries & Oceans. Copyright © Canada Communication Group, Hull, Quebec, Canada. Communication Group, Hull, Quebec, Canada. Reprinted with the permission of the Minister of Supply and Services Canada, 1995.

Argulus. All 125 species are ectoparasitic on freshwater or marine fish, even in the larval stages. The animals are typically smaller than 2 cm. One family.

Class Pentastomida.

The approximately 100 species are distributed among two orders and seven families.

Order Cephalobaenida.

Cephalobaena, Reighardia. The life cycles of the species in this order are largely undocumented. This order contains one of the few pentastomid species living in a nonreptilian host; members of the genus *Reighardia* parasitize only birds, including gulls. Two families.

Order Porocephalida.

This order includes the largest of all pentastomids, members of one species reaching lengths of 9 cm. Five families with about 65 species, most of which parasitize reptiles.

Family Linguatulidae. *Lingulata.* This family is unusual in that all its members parasitize mammals rather than reptiles. Developmental stages occur in the glandular tissues of rabbits, horses, cows, sheep, and pigs, while adulthood is reached only in canines. Developmental stages can persist for several weeks in humans, who become infected by eating raw glands from slaughtered mammalian hosts. In humans, the nymphs migrate out of the gut and take up residence in the nasopharynx, or sometimes the eye, before dying a few weeks later.

Class Tantulocarida.

A recently discovered group of only five known species, all ectoparasitic on deep-water crustaceans. The members resemble copepods, except thoracic legs are absent.

Class Remipedia.

This group was first recognized in 1983. The eight species described so far are restricted to tropical, underwater marine caves. The long body with its abundant lateral appendages resembles that of a polychaete. Remipedians may be the closest of all animals to the ancestral crustacean condition.

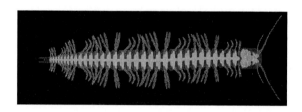

Class Cirripedia.

All nearly 1,000 species are entirely sedentary as adults and are found only in marine and estuarine waters. The approximately 25 families are distributed among four orders. This class includes the barnacles.

Order Acrothoracica.

Most species lack calcareous protective plates, burrowing instead into mollusc shells, coral skeletons, or limestone. The females are fully developed, but the males are little more than inconspicuous sacs of sperm living adjacent to females. A few species are found at depths of 600 m to 1,000 m, but most are found in shallow water. Three families.

Order Ascothoracica.

All 30 species are marine ecto- or endoparasites of cnidarians or echinoderms. The body is enclosed within a bivalved carapace or sac. These animals occur over a wide depth range, from shallow water to abyssal depths.

Order Thoracica.

Balanus, Lepas, Chthamalus, Verruca, Elminius—the true barnacles. Most cirripedes are placed in this order. All of the approximately 800 species are marine; species in this group are found from the intertidal zone to the deepest reaches of the sea. Individuals of most species are enclosed within a complex of calcareous plates. A wide variety of lifestyles is exhibited: Most species are sessile suspension feeders, but others are parasitic in sharks, polychaetes, or corals, and a few are pelagic. Some species live attached to the outsides of fish, jellyfish, turtles, or whales, and others are commensal with anthozoans, crustaceans, or sponges or in the mantle cavity of certain bivalves. This order contains the oldest known cirripedes, with a fossil record extending back over 400 million years. Seventeen families.

Order Rhizocephala.

Sacculina, Lernaeodiscus, Loxothylacus. Of the 230 species in this order, all but three or four are exclusively marine; the few misfits occur only in freshwater. Regardless of habitat, all species are internal parasites of other crustaceans, especially of crabs and other decapods. The female gonad of the parasite protrudes conspicuously from the host, usually ventrally, while the large absorptive portion of the parasite remains inside. Males are small and inconspicuous, little more than masses of sperm cells. Species occur from shallow waters to depths exceeding 4,000 m. Seven families.

Some General References About the Arthropods

Ali, M. F., and E. D. Morgan. 1990. Chemical communication in insect communities: A guide to insect pheromones with special emphasis on social insects. *Biol. Rev.* 65:227–47.

Bliss, D. E., ed. 1982–1987. *The Biology of Crustacea*, vols. 1–9. New York: Academic Press.

Boardman, R. S., A. H. Cheetham, and A. J. Rowell, eds. 1987. *Fossil Invertebrates.* Palo Alto, Calif.: Blackwell Scientific, 205–69.

Coddington, J. A., and H. W. Levi. 1991. Systematics and evolution of spiders (Araneae). *Ann. Rev. Ecol. Syst.* 22:565–92.

Davies, R. G. 1988. *Outlines of Entomology*, 7th ed. London: Chapman & Hall.

Foelix, 1982. *Biology of Spiders.* Cambridge, Mass.: Harvard Univ. Press.

Harrison, F. W., and R. F. Foelix, eds. in press. *Microscopic Anatomy of the Invertebrates, Vol. 8: Chelicerate Arthropoda.* New York: Wiley-Liss.

Harrison, F. W., A. G. Humes, and E. E. Ruppert, eds. 1992. *Microscopic Anatomy of the Invertebrates, Vol. 9: Crustacea.* New York: Wiley-Liss.

Harrison, F. W., and A. G. eds. 1992. *Microscopic Anatomy of the Invertebrates, Vol. 10: Decapod Crustacea.* New York: Wiley-Liss.

Harrison, F. W., and M. Locke eds. in press *Microscopic Anatomy of the Invertebrates, Vol. 11: Insecta.* New York: Wiley-Liss.

Harrison, F. W., and M. E. Rice, eds. 1993. *Microscopic Anatomy of the Invertebrates, Vol. 12: Onychophora, Chilopoda, and Lesser Protostomata.* New York: Wiley-Liss.

Heinrich, B. 1993. *The Hot-Blooded Insects: Strategies and Mechanisms of Thermoregulation.* Cambridge, Mass.: Harvard Univ. Press.

Hölldobler, B., and E. O. Wilson. 1990. *The Ants.* Cambridge, Mass.: Harvard Univ. Press.

Papaj, D. R., and A. C. Lewis. 1993. *Insect Learning: Ecological and Evolutionary Perspectives.* New York: Routledge, Chapman and Hall, Inc.

Parker, S. P., ed. 1982. *Classification and Synopsis of Living Organisms*, vol. 2. New York: McGraw-Hill, 71–728.

Polis, G. A., ed. 1990. *The Biology of Scorpions.* Stanford, Calif.: Stanford Univ. Press.

Rosmoser, W. S. 1981. *The Science of Entomology*, 2d ed. New York: Macmillan Publ. Co.

Savory, T. 1977. *Arachnida*, 2d ed. New York: Academic Press.

Schram, F. R. 1986. *Crustacea.* New York: Oxford Univ. Press.

Schultz, J. W. 1987. The origin of the spinning apparatus in spiders. *Biol. Rev.* 62:89–113.

Weygoldt, P. 1969. *The Biology of Pseudoscorpions.* Cambridge, Mass.: Harvard Univ. Press.

Wilson, E. O. 1971. *The Insect Societies.* Cambridge, Mass.: Harvard Univ. Press.

Winston, M. L. 1987. *The Biology of the Honeybee.* Cambridge, Mass.: Harvard Univ. Press.

Witt, P. N., and J. S. Rovner, eds. 1982. *Spider Communication: Mechanisms and Ecological Significance.* Princeton, N.J.: Princeton Univ. Press.

19

Two More Phyla of Uncertain Affiliation: Tardigrades and Onychophorans

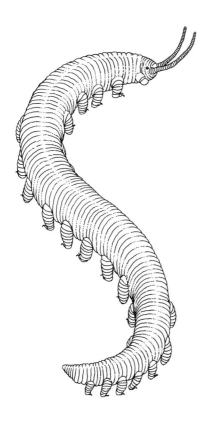

Introduction and General Characteristics

One would think that a group as stereotyped, and with as long a fossil record, as the Arthropoda would have a relatively uncontroversial phylogeny. The following features are considered arthropod characteristics: external, jointed exoskeleton; tracheae; compound eyes; Malpighian tubules; mandibles; heart with ostia. Nevertheless, there is considerable controversy regarding these animals' evolutionary history. In large part, the controversy centers on how many times these arthropod characteristics have independently evolved.

All of the arthropod classes listed in Chapter 18 are already represented in the earliest fossil records of about 600 million years ago. The group's ancestry is difficult to trace because arthropods most certainly are derived from soft-bodied ancestral forms poorly represented as fossils. Many biologists have long believed that arthropods are derived from annelids, or at least from annelid-like ancestors. In particular, both arthropods and annelids demonstrate a clearly segmented coelomic space during embryogenesis; absence of this characteristic in the arthropod adult could easily be a modification of the ancestral condition. The existence of animals possessing some characteristics of arthropods and some characteristics typical of other groups, or at least atypical for arthropods, presents tantalizing phylogenetic implications. The animals in the two phyla discussed in this chapter—Tardigrada and Onychophora—show this intriguing combination of arthropod and nonarthropod characteristics.

Phylum Tardigrada

Members of the phylum Tardigrada (the water bears) have clear arthropod affinities, but don't quite seem to fit in as *bona fide* members of that phylum (or of any other phylum). About 750 tardigrade species have been described. All tardigrades are quite small, ranging between about 50 μm (micrometers) and 1,200 μm in length. The typical tardigrade is only about 0.5 mm (millimeters) long. Most species live in surface films of freshwater on terrestrial plants, especially mosses and lichens. Some marine species have been described, many of them living in the spaces between sand grains (i.e., **interstitially**), at depths of up to nearly 5,000 m (meters). Although small, water bears may occur in impressively dense aggregations, up to several million per square meter of substrate surface. A few species are commensal or parasitic with other invertebrates.

In 1994, the first known fossil tardigrades were reported, from early Cambrian rocks in Siberia formed some 530 million years ago.

Like the arthropods, tardigrades possess a complex, chitinous cuticle that is periodically molted and that not only covers the outside of the body, but also lines the foregut and hindgut. Moreover, tardigrades have recently been shown to possess arthropod-like striated muscles. Because the tardigrades are small and the cuticle is highly permeable to water and gases, they are restricted to moist habitats. Gas exchange occurs across the general body surface; tardigrades have no specialized respiratory structures. Like the arthropods, tardigrades lack motile cilia. All tardigrades possess four pairs of clawed appendages, with which the animals lumber over the substrate in bearish fashion (Fig. 19.1), but the appendages are never jointed.

The tardigrade nervous system is organized in annelid/arthropod manner, with a paired ventral nerve cord (Fig. 19.2). Also, several tardigrade glands look suspiciously like Malpighian tubules, although their function in tardigrades has yet to be convincingly demonstrated. The mouthparts are a pair of stylets, which are mostly used to pierce plant cells; a few species are carnivorous. Tardigrades have no specialized larval stages; offspring develop as miniature adults.

Several tardigrade characteristics are decidedly nonarthropod. As already mentioned, the appendages are not jointed. Moreover, the cuticle is never calcified. Tardi-

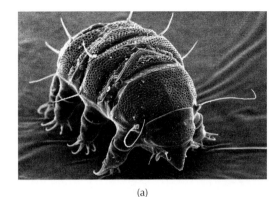

(a)

(b)

Figure 19.1

Scanning electron micrographs of (a) *Echiniscus spiniger* and (b) *Macrobiotus hufelandi,* two tardigrades.   (a,b) *Courtesy of D. R. Nelson.*

grade embryological development may display at least one typically deuterostome characteristic (enterocoely—something never encountered among arthropods), although this is no longer certain. Moreover, it is not clear that the capacious body cavity is a hemocoel; it may well be a pseudocoel. If tardigrades are indeed coelomates, that coelom is reduced to a small pouch surrounding the gonads. In fact, some workers link tardigrades to the nematodes and other pseudocoelomate aschelminths, based upon constancy of cell numbers in the tardigrade cuticle and the ultrastructural organization of the pharynx. Moreover, like nematodes and rotifers, tardigrades exhibit **cryptobiosis,** a bizarre ability to dehydrate and reduce metabolic rate to withstand extreme environmental conditions of low-temperature and desiccation stress.[1] A tardigrade may live more than 10 years—more than 100 years in some instances!—in this cryptobiotic state; the total life span, including episodes of cryptobiosis, may thus be many decades in some species, although life spans of less than one year are more common.

---

1. See *Topics for Further Discussion and Investigation,* at the end of the chapter.

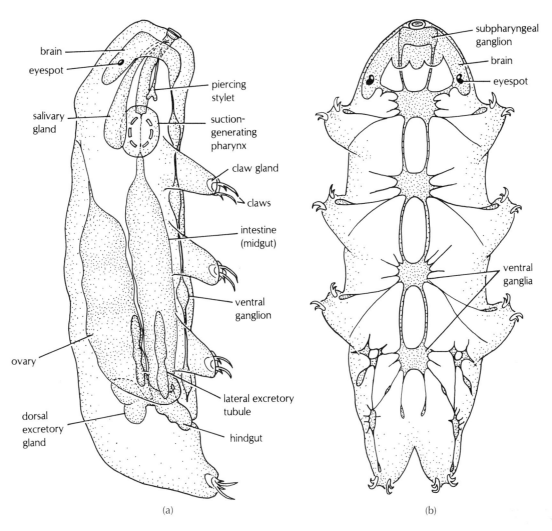

Figure 19.2

(a) Internal anatomy of a typical tardigrade. Note that the nerve cord is ventrally located, as in annelids and arthropods. The claw glands secrete the claws. (b) Detail of the tardigrade nervous system. Note the large ganglia associated with each pair of appendages.   (a,b) *From Pennak, Fresh-Water Invertebrates of the United States, 2d ed. Copyright © 1978 John Wiley & Sons, New York. Reprinted by permission of John Wiley & Sons, Inc.*

Phylum Onychophora

Phylum Onycho · phora
(G: claw bearer)
on-ē-koh´-fer-ah

Members of the second phylum of uncertain affiliation—the phylum Onychophora—possess some characteristics that are clearly annelid in nature, some that are clearly arthropod in nature, and none that affiliate them with any other animal group. All members are free-living and are clearly protostomous coelomates. More than 100 species have been described, with *Peripatus* being the best-known genus (Fig. 19.3a).

All onychophorans are terrestrial, and most are found in moist habitats in tropical environments and in southern temperate regions (e.g., New Zealand). Indeed, they appear to be restricted to such environments largely because they possess a thin, nonwaxy cuticle that is not effective in deterring evaporative loss of body water. Perhaps this danger of dehydration is also a factor in making onychophorans exclusively nocturnal (active only at night). Some species are carnivores (feeding particularly on various smaller arthropods), some are herbivores, and some are omnivores. Onychophoran predators attack their prey from a distance by shooting a proteinaceous glue from specialized oral protuberances. Once the victim is sufficiently entangled in the glue, the onychophoran bites through any protective coverings, secreting into the tissue substances

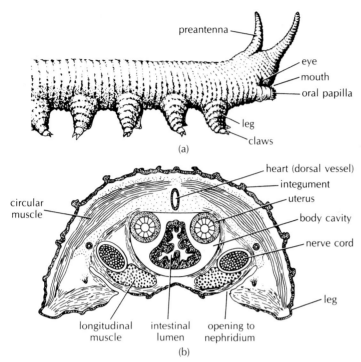

Figure 19.3
(a) The onychophoran *Peripatus,* in lateral view. (b) Same animal, seen in diagrammatic cross section.   (a,b) *Sherman/Sherman, The Invertebrates: Function and Form, 2/e, © 1976, pp. 111, 112, 236, 172, 169, 45, 9, 15. Reprinted by permission of Prentice Hall, Upper Saddle River, New Jersey.*

that kill the prey and partially liquify the tissues (Research Focus Box 19.1). Onychophorans also eject glue as a defense against predators.

Individual onychophorans may grow to 15 cm (centimeters) in length. Onychophorans possess the following annelid-like characteristics: (1) body wall musculature is smooth and composed of longitudinal, circular, and diagonal elements (Fig. 19.3b); (2) a single pair of feeding appendages (jaws) is present; (3) no appendages are jointed; (4) a hydrostatic skeleton plays a role in locomotion (as described later); (5) one pair of nephridia is found in most segments; (6) light receptors are ocelli rather than compound eyes; and (7) the outer body wall is deformable. The following characteristics are arthropod-like: (1) the jaw musculature is striated; (2) the cuticle contains chitin; (3) the main body cavity is a hemocoel, not a true coelom; (4) gas exchange is achieved by spiracles opening into a tracheal system (the spiracles cannot be closed, however—another factor preventing the onychophorans from invading drier habitats); (5) mouth appendages are mandible-like; (6) the heart bears ostia; (7) legs are extended by hemocoelic pressure rather than by direct muscular contraction (as also found for the legs of arachnids and merostomes and the maxillae of butterflies); (8) excretory organs closely resemble the green glands of crustaceans; and (9) adhesive defense secretions are reminiscent of those produced by

repugnatorial glands of the centipedes and millipedes (myriapods). The onychophoran nervous system is of the annelid/arthropod type: segmented, with a pair of ventral nerve cords.

A few characteristics separate the onychophorans from both the arthropods and the annelids. The head appendages consist of a pair of antennae, a pair of jaws, and a pair of oral papillae. These are followed by a series of unjointed walking legs that are quite unlike the parapodia of polychaete annelids, both structurally and functionally. Blood pigment is lacking.

Studies of several onychophoran species have shown onychophoran locomotion to be unique. Propulsion is generated directly by the musculature of the limbs themselves, with the body remaining rigid, serving as an anchor point against which the limbs can operate. Of course, in the absence of a solid skeleton, body rigidity is a function of the body wall musculature. The limbs (about 20 pairs are typically found) project ventrolaterally, elevating the body above the ground. This is quite different from polychaete parapodia, which project laterally, leaving the body surface in direct contact with the substrate.

As in polychaetes, waves of limb activity pass down the length of the onychophoran body, and several waves are generally progressing concurrently (Fig. 19.4). In the preparatory stroke for any given onychophoran limb, the tip of the leg is raised above the substrate and extended forward. On the backstroke (power stroke), the tip of the leg is applied to the substrate, and the leg is held straight and stiff. The tip of the limb remains stationary while the limb musculature contracts, sweeping the body past the point of contact between the limb and the substrate. As the body progresses past the stationary end of the leg, the leg shortens until it is essentially perpendicular to the substrate and then elongates as the portion of the body wall above the limb continues to move forward. Because of these well-timed changes in the limb length, the body itself shows little undulation, changing position neither laterally nor vertically with respect to the ground. If the legs could not be seen, the body would appear to be gliding. The jointed legs of arthropods also show such a shortening during movement, but the mechanism by which this is accomplished is necessarily quite different.

As so far described, the hydrostatic skeleton appears to have a rather passive role in onychophoran locomotion. However, a more active involvement has been demonstrated through analysis of film footage. As the animal moves, the body's length and width are continually changing; indeed, body dimensions rarely remain constant for more than a few seconds. Changes in body length are correlated with changes in speed and gait (i.e., the manner of walking) (Fig. 19.4). These alterations of body dimensions are brought about through the actions of the circular and longitudinal musculature, interacting

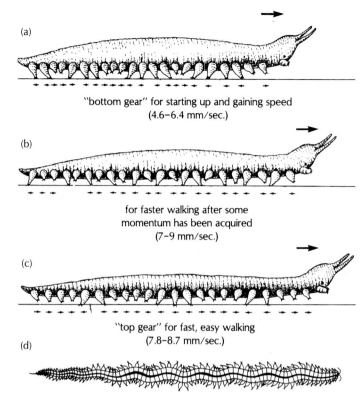

(a)

"bottom gear" for starting up and gaining speed
(4.6–6.4 mm/sec.)

(b)

for faster walking after some
momentum has been acquired
(7–9 mm/sec.)

(c)

"top gear" for fast, easy walking
(7.8–8.7 mm/sec.)

(d)

Figure 19.4

(a–c) The three most common gaits observed during locomotion of the onychophoran *Peripatopsis sedgwicki*. Shifts in walking speeds are associated with changes in body length and width. Sufficient speed must be obtained using gaits (a) and (b) before the animal can switch to gait (c) for prolonged, rapid walking. (d) Locomotion of the polychaete *Nereis diversicolor* (Annelida: Polychaeta), shown for comparison. The locomotory appendages (parapodia) project laterally, and waves of limb activity pass down the length of the body. Note that, for each segment, the left parapodium and right parapodium are not in the same phase of the activity cycle. Among polychaetes, waves of longitudinal muscle contraction aid in generating forward thrust.   *(a,c) From Manton, in* Journal of the Linnean Society (Zoology), *41:529, 1950. Copyright © Academic Press, Ltd., London. Reprinted by permission. (d) From Gray, in* Journal of Experimental Biology, *16:9, 1939. Copyright © 1939 Company of Biologists, Colchester, England. Reprinted by permission.*

through a constant-volume, internal hydrostatic skeleton (the hemocoel). Because onychophorans have no internal septa, contractions in one part of the body can bring about rapid distension, or rigidity, in any other part of the body. Changes in the animal's speed are brought about through changes in body length, limb length, the distance to which the legs reach forward on each stroke, and the amount of time required between the initiation of the forward swing and the completion of the power stroke. Speed changes appear to be achieved primarily by altering the gait.

Onychophoran locomotion is somewhat similar to that found within the Polychaeta, in that outfoldings of the body wall (projecting ventrally in the onychophoran and laterally in the polychaete) are used to

generate thrust and in that a hydrostatic skeleton is implicated. However, there are important differences. First, the onychophoran limb is stiffened not by internal, rigid acicula, but rather by the intrinsic musculature of the limbs themselves and by hydrostatic forces generated by contraction of the body or limb musculature against the fluid-filled hemocoel. Moreover, the onychophoran body does not undulate; all thrust is generated by the limbs directly. In contrast, among the polychaetes, contractions of the longitudinal body wall musculature may be used to generate waves of body undulation (Fig. 19.4d). These undulations transmit additional thrust against the substrate through the stationary parapodia. Indeed, when polychaetes are really making headway, most of the progress is attributable to contractions of the body wall musculature rather than to the parapodial elements directly. An additional difference between the locomotory mechanics of the two groups is that the polychaete limb does not change length during the propulsive cycle as does the onychophoran limb. These locomotory differences support the contention that if onychophorans arose from annelids or annelid-like ancestors, they probably did not arise from a polychaete ancestor.

The evolutionary history of the Onychophora is unknown. Fossils of marine, onychophora-like animals from the mid-Cambrian period (approximately 550 million years ago) have been described, but it is impossible to tell whether these fossilized animals were more like arthropods or more like annelids. Existence of these fossils suggests, at least, that the evolution of onychophoran-like limbs predated movement of the group to land; limb development can thus be viewed as a preadaptation for a terrestrial existence in this group. But little else about the evolutionary history of onychophorans is even hinted at in examination of other animals, either living or extinct. Certainly, there has been no consensus about the degree of relationship between the Onychophora and the Arthropoda. Indeed, scientists who believe that the insects and myriapods (including the centipedes and millipedes) evolved independently of the Chelicerata and Crustacea often place the Onychophora together with these terrestrial arthropods in a separate phylum, the Uniramia, so-called in recognition of the fact that all members have only uniramous appendages. Evidence against this view seems to be accumulating rapidly, however; both morphological and molecular evidence argue increasingly that arthropods had a single origin, and molecular data in particular place the Onychophora squarely within the phylum Arthropoda as a highly modified—not primitive—member of that phylum. By the next edition of this book, onychophorans may comprise a new class within the Arthropoda, their enigmatic evolutionary status apparently resolved at last.

Research Focus Box 19.1

The Energetics of Onychophoran Feeding

St J. Read, V. M., and R. N. Hughes. 1987. Feeding behaviour and prey choice in *Macroperipatus torquetus* (Onychophora). *Proc. Royal Soc. London B* 230:483.

Most animals feed only on certain types of foods or on foods within a certain size range. Ecologists have long been interested in documenting the relative costs and benefits of feeding on particular foods in an effort to understand these species-specific food preferences. One explanation for a predator specializing on prey of particular size is based on the feeding-efficiency argument that natural selection should favor feeding on prey that provide the greatest caloric content for the amount of energy expended in prey capture and ingestion. Carnivorous onychophorans seem well suited to feeding-efficiency studies: They are slow-moving for their size (about 4 cm per minute) and rarely travel more than a meter from their burrows, which means that experiments can easily be performed in the laboratory and that prey capture must account for a good fraction of the animal's total daily energy expenditure. In addition, onychophorans typically eat only one victim per night and ingest the soft tissues completely, facilitating measurements of energy intake. Also, the primary cost of food collection is the proteinaceous glue needed to entangle the prey—a single, measurable entity.

Read and Hughes (1987) examined the energetic efficiency of food capture for a single species of onychophoran (*Macroperipatus torquatus*) found in the rain forests of Trinidad. To estimate the net amount of energy gained from feeding, the biologists first needed to determine the cost of prey capture, as estimated from the amount of glue expelled in capturing prey of different sizes. To do this, they weighed individual prey and put each into a small box with onychophorans. Once an onychophoran attacked and completely subdued its intended victim, the researchers scraped any excess glue from the bottom of the box and weighed it along with the glue-entangled prey to assess the amount of glue used in the attack. They also weighed each predator to determine how the use of glue varied with predator size. A single attack typically used up 40% to 50%, and as much as about 80%, of an individual's available glue supply.

Equating the amount of glue released with the cost of feeding may overestimate the cost of prey capture: In eating

the prey, the predator also consumes some of the glue, so this protein is reclaimed, not lost. On the other hand, ignoring the energetic costs of *producing* the glue and of physically locating and consuming the predator may underestimate the cost of prey capture. On balance, the procedure may well produce a very reasonable estimate of capture costs, assuming that the two inaccuracies are approximately equal in magnitude.

Since the onychophoran must produce glue in order to eat, the amount of time required for an individual to replenish its glue supply after feeding determines how often the animal can indulge. To estimate the rate of glue replenishment, the biologists first had to determine how much glue a fully loaded onychophoran possesses and then how long it takes the onychophoran to produce that much glue. How would you go about making these determinations? The researchers chose a simple but clever approach. They induced individual onychophorans to squirt all of their glue at preweighed pieces of aluminum foil, and then they reweighed the foil. The onychophorans were then allowed to restore their glue supplies for different amounts of time, and the extent of restocking was again assessed by force-firing the animals at preweighed aluminum foil. The researchers found that these onychophorans needed more than five weeks to fully reload (Focus Fig. 19.1).

To measure the energetic benefits of feeding on arthropod prey of different sizes, Read and Hughes weighed prey before they were offered to an onychophoran of known weight, and then reweighed what remained of the prey after the onychophoran had concluded its meal; weight loss reflects food consumption. Typically, the remains consisted only of the prey's external cuticle.

A priori, it might seem advantageous for an onychophoran to attack any animal it can capture, but such is not the case. For one thing, more glue is needed to subdue a larger animal, so the cost of predation rises for larger prey. The larger animal also has a greater chance of escaping with some of the predator's precious glue and providing nothing in return. In addition, the predator has a gut of finite capac-

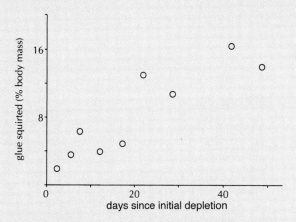

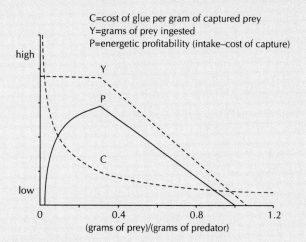

Focus Figure 19.1
Rate at which the onychophoran *Macroperipatus torquatus* replenishes its supply of glue. The animals were forced to expel all of their glue on day zero. The amount of glue held on subsequent days is shown here as a percentage of the animal's body mass.

Focus Figure 19.2
The relationship between yield of prey and cost of capture for prey and predators of different sizes. Individual data points have been omitted for clarity.

ity; it does not make energetic sense to expel extra glue to capture something too big to fully ingest. Finally, the onychophoran that exhausts its glue supplies attacking large prey has a long wait until it can capture its next meal; the animal also becomes more vulnerable to predation, since it relies on its glue stores for defense as well as food capture.

Focus Figure 19.2 summarizes some of the key relationships concerning the profitability of prey capture. The *X* axis shows the weight of the prey relative to the weight of the predator. A value of 1.0 means that the predator and prey are equal in weight. A value of 0.5 indicates that the prey weighs half as much as the predator.

As shown in Focus Figure 19.2 (curve *C*), the amount of glue that must be expended to capture each gram of prey declines exponentially for larger and larger prey. Although it takes more glue to capture a larger animal, the amount of glue needed does not increase directly with the size of the prey; it increases more slowly. For example, if an onychophoran weighing 3 g (grams) requires 0.25 g of glue to subdue an animal half its size and 0.30 g of glue to subdue an animal of its own size, then the relative amount of glue required in the first case (with the smaller prey) is 0.25 g glue/1.5 g prey = 0.17 g glue used per gram of prey, but only 0.3 g glue/3.0 g prey = 0.1 g glue used per gram of prey in the second case, with the larger prey.

Curve *P* (Focus Fig. 19.2) shows the difference between the amount of energy extracted by a predator feeding on prey of different sizes and the approximate energy content of the glue expended in prey capture. The net energy obtained declines dramatically for heavier prey because of the predator's limited gut capacity; more glue is needed to capture larger prey, and excess prey tissue (curve *Y*) simply cannot be ingested. Curve *P* shows that it is energetically most advantageous for an onychophoran to feed on prey about 0.2 to 0.6 times its own weight. Attacking much larger or much smaller prey is not nearly as efficient.

Additional laboratory studies showed that these onychophorans did indeed grow more slowly when reared exclusively on very small or very large prey and that, when given a choice, they generally did not capture the smallest prey. It seems reasonable to assume that natural selection has encouraged a predilection for feeding on prey that will return the greatest, or at least a substantial, amount of the energy invested in prey capture.

Figures from V. M. St. J. Read and R. N. Hughes, "Feeding Behaviour and Prey Choice in Macroperipatus torquatus (Onychophora)" in Proceedings of the Royal Society of London B 230:483–506, 1987. Copyright © 1987 The Royal Society, London.

Taxonomic Summary

Phylum Tardigrada
Phylum Onychophora

Topic for Further Discussion and Investigation

Investigate the tardigrades' tolerance of low temperature and desiccation.

Crowe, J. H. 1972. Evaporative water loss by the tardigrades under controlled relative humidities. *Biol. Bull.* 142:407.

Crowe, J. H., and A. F. Cooper. 1971. Cryptobiosis. *Sci. Amer.* 225:30.

Pigón, A., and B. Weglarska. 1955. Rate of metabolism in tardigrades during active life and anabiosis. *Nature (London)* 176:121.

Pollock, L. W. 1975. Tardigrada. *Reproduction of Marine Invertebrates*, vol. II. Giese, A. C., and J. S. Pearse, eds. New York: Academic Press, 43–54.

Westh, P., and H. Ramløv. 1991. Trehalose accumulation in the tardigrade *Adorybiotus coronifer* during anhydrobiosis. *J. Exp. Zool.* 258:303.

Taxonomic Detail

Phylum Tardigrada

The approximately 765 species are distributed among three classes.

Class Heterotardigrada.
Although most species live in terrestrial or freshwater habitats, many are marine, often living interstitially in the spaces between sand grains. Five families. About 340 species.

> **Family Echiniscidae.** *Echiniscus* (a genus containing nearly 25% of all tardigrade species). All members of the family are terrestrial or freshwater inhabitants, and a few species occur only in South America at altitudes exceeding 1,000 m. Many species are brightly colored: red, yellow, or orange.

Class Mesotardigrada.
Thermozodium esakii. The single species in this class is known only from hot springs (65°C) in Japan.

Class Eutardigrada.
This class includes about 425 species from marine, freshwater, and terrestrial habitats. Many of the marine species have been described over the past several years. Eutardigrades are divided between two orders based on differences in claw morphology.

Order Parachela. *Macrobiotus, Minibiotus, Hipsibus.* At least seven families, containing about 420 species.

Order Apochela. *Milnesium, Limmenius.* These tardigrades are exclusively terrestrial. Some species are carnivorous, eating rotifers, nematodes, and other tardigrades inhabiting the mosses and lichens on which they live.

Phylum Onychophora

The 70 species are distributed among two families.

> **Family Peritopsidae.** *Peripatoides, Peripatopsis. Peripatoides* occurs only in Australia and New Zealand, while other family members live in parts of Africa, South America, and New Guinea.

> **Family Peripatidae.** *Peripatus.* These best-known onychophorans are widespread in subtropical lands.

Some General References About the Tardigrades and Onycophorans

Harrison, F. W., and M. E. Rice, eds. 1993. *Microscopic Anatomy of the Invertebrates, Vol. 12: Onychophora, Chilopoda, and Lesser Protostomata.* New York: Wiley-Liss.

Parker, S. P., ed. 1982. *Classification and Synopsis of Living Organisms, Vol. 2,* (Tardigrada and Onychophora). New York: McGraw-Hill, 729–30, 731–39.

Thorpe, J. H., and A. P. Covich, eds. 1991. *Ecology and Classification of North American Freshwater Invertebrates.* New York: Academic Press, 501–21.

Wright, J. C., P. Westh, and H. Ramløv. 1992. Cryptobiosis in Tardigrada. *Biol. Rev.* 67:1–29.

20

The Lophophorates (Phoronids, Brachiopods, Bryozoans) and Entoprocts

Introduction and General Characteristics

The four phyla discussed in this chapter—Phoronida, Brachiopoda, Bryozoa, and Entoprocta—have long had uncertain phylogenetic relationships to other animal phyla and to each other. We begin with the first three phyla, as they have in common one major anatomical feature that may be homologous in the three groups: the **lophophore.** The lophophore is a circumoral (i.e., around the mouth) body region characterized by a circular or *U*-shaped ridge around the mouth. This ridge bears either one or two rows of ciliated, hollow tentacles. The internal space within the lophophore and its tentacles is always a coelomic cavity, and the anus always lies outside the circle of tentacles; both characteristics have come to be important parts of the definition of a lophophore. The lophophore of all species functions as a food-collecting organ and as a surface for gas exchange. All members are sessile or sedentary suspension feeders, employing the lophophore cilia to capture phytoplankton and small planktonic animals, and no lophophorate ever has a distinct head. Moreover, the pattern of water flow created by the tentacular cilia is the same in all lophophorate species: Water is pulled down into the center of the lophophore by the action of lateral cilia and expelled between adjacent tentacles after frontal cilia have removed food particles (Fig. 20.1b).

The lophophore's coelomic cavity, called the **mesocoel,** is physically separated by a septum from the larger, primary coelomic cavity, the **metacoel** (Fig. 20.1a). In some groups, this septum is perforated or incomplete, so

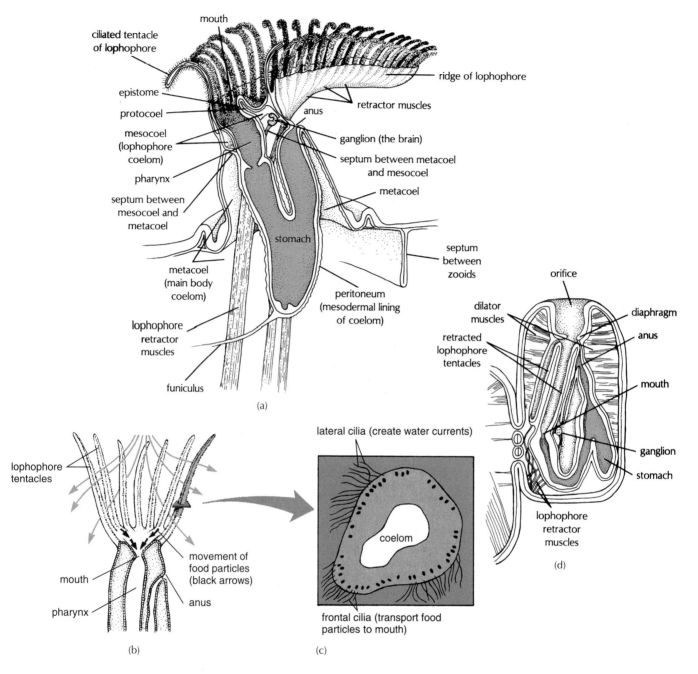

Figure 20.1

(a) Longitudinal section through the body of a lophophorate (the bryozoan *Cristatella* sp.). Note that the mouth opening lies within the ring of lophophore tentacles, but that the anus does not. (b) Movement of water through a lophophore. (c) Cross section through one lophophore tentacle, showing location and function of ciliary traits. (d) Illustration of a polypide retracted into a cystid. Anatomical details are discussed further in the section on phylum Bryozoa.   *(a) From Hyman; after Cori. (b,c) From P. G. Willmer, Invertebrate Relationships. Copyright © 1990 Cambridge University Press, New York. Reprinted by permission. (d) From Hyman, after Marcus.*

the two coelomic cavities are interconnected. A third coelomic cavity, the **protocoel,** may be present anterior to the mesocoel. The protocoel is distinct but quite small in the adults of some lophophorate groups, but it is particularly conspicuous during the larval stage of phoronids. The absence of a conspicuous protocoel in the remaining lophophorate species appears related to the evolutionary reduction of the head.

One feature common to all lophophore-bearing animals is the possession of very simple gonads, arising from a portion of the mesodermal lining of the main (trunk) coelomic compartment—that is, from the metacoel. Additionally, all species possess U-shaped digestive tracts, in which the anus (when present) terminates near the mouth (Fig. 20.1). Finally, all species secrete some form of protective covering about the body.

Despite these similarities, there is no agreement about the likely origins of any of the three lophophorate phyla or about the most likely primitive lophophorate condition. The three groups differ markedly with respect to their circulatory and excretory systems, the nature of the protective coverings they secrete, and important details of development. Indeed, lophophorate developmental patterns are unlike those reported for any other coelomate group. Features characteristic of both protostome and deuterostome development occur within the three lophophorate phyla, as do features that characterize neither group. Cleavage is basically radial, as in deuterostomes, and cell fates are not fixed—at least in the species so far studied—until after the first several cleavages. Such indeterminate cleavage is a typical deuterostome characteristic. However, coelom formation is not distinctly enterocoelous (as it would be in a typical deuterostome) or schizocoelous (as it would be in a typical protostome). Both modes of coelom formation have been observed among lophophorates, and in at least some phoronid species the coelom forms as a split (schizocoely) in endodermal tissue (enterocoely!) rather than mesodermal tissue. To further add to the intrigue, the bryozoan "coelom" forms by neither of these mechanisms! It arises not during embryogenesis, but only following the completion of larval life; indeed, there is presently some controversy over whether bryozoans are truly coelomates.

In at least one lophophorate group (the Phoronida), the mouth forms from the blastopore, a distinctly protostome characteristic, and in two of the groups (the Phoronida and the Brachiopoda), typical protostome nephridia are found in larvae and adults. On the other hand, the tripartite coelom typically encountered among lophophorates is considered a uniquely deuterostome feature. On balance, morphological and embryological criteria link the lophophorates with deuterostomes. But this view is countered by recent (1995) molecular data that place lophophorates securely with the protostomes, and make their bizarre combination of developmental characteristics all the more intriguing; the evolutionary origins of lophophorates, their relationship to members of other phyla, and the precise evolutionary relationships among the lophophorate phyla remain ambiguous, to say the least.

Nearly all lophophorate species are marine, except for a few bryozoan species that are commonly encountered in freshwater. None of the species is terrestrial, in any sense of the word; the sedentary, suspension-feeding lifestyle that all lophophorates exhibit is not a likely preadaptation for life on land. Approximately 4,500 lophophorate species are alive today, but lophophorate diversity was far greater during the Paleozoic and Mesozoic eras, approximately 135 to 500 million years ago; indeed, 7 to 10 times as many species flourished in past millenia, and are present now only as much-studied fossils.

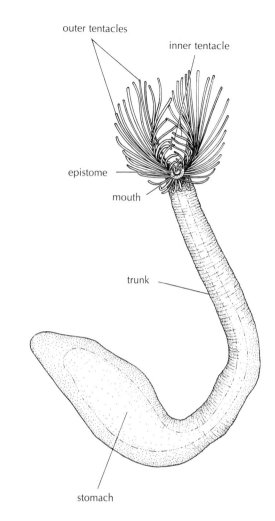

Figure 20.2

A phoronid, *Phoronis architecta,* removed from its tube.   *From Hyman; after Wilson.*

Phylum Phoronida

Members of the phylum Phoronida (for-ō´-ni-dah) are the easiest of the lophophorate phyla to discuss, as all members conform closely to the same basic body plan. Members of the other two lophophorate phyla are not nearly so cooperative. Only about one dozen phoronid species are known, and all of these are marine.

Most phoronids live in permanent, chitinous tubes implanted in muddy or sandy sediments or attached to solid surfaces. A few species burrow into hard, calcareous substrates; even so, these species secrete a chitinous tube within the burrow. Adult phoronids do not move from place to place, although they can move within their tubes and, if artificially removed from these tubes, can burrow back into the sediment. A giant nerve fiber permits the animal to withdraw rapidly into its tube or burrow upon provocation.

Phoronids typically are about 12 cm long and have the form of an elongated, cylindrical sac (Fig. 20.2). Appendages, except for the anterior lophophore, are lacking.

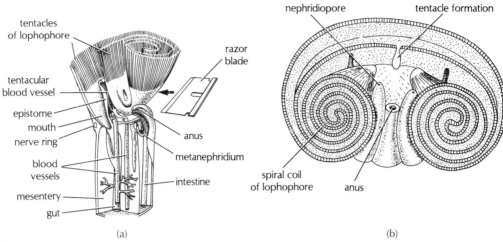

Figure 20.3
Phoronid anatomy. (a) Lateral view of *Phoronis australis,* cut vertically to reveal internal anatomy. (b) Looking down on the lophophore, after the tentacles have been chopped off in the plane indicated by the razor blade in (a). The tentacles would be coming out of the page if they had not been chopped off. The

lophophore is basically U-shaped, although each end of the U is curled to form a spiral in this species.   *From Hyman; after Shipley and Benham.*

A flap of tissue called the **epistome** covers the mouth (*epi* = G: around; *stoma* = G: mouth). The epistome is hollow, as it contains a remnant of the embryonic protocoel.

The lophophore is the phoronids' only prominent external structure. It consists of a conspicuous ring of tentacles, usually deeply indented to form a *U*-shape, and a less conspicuous, ciliated food groove (Fig. 20.3). Ciliary activity drives water into the ring of tentacles from the top of the lophophore and outward through the narrow spaces between the tentacles, as with other lophophorates. In this manner, suspended food particles can be captured by the tentacular cilia and mucus, transferred to the cilia of the food groove, and conducted to the mouth for ingestion.

Internally, phoronids possess a pair of metanephridia, with the ciliated nephrostomes collecting coelomic fluid and the nephridiopores discharging urine near the anus (Fig. 20.3*b*). A blood circulatory system with hemoglobin (contained within blood corpuscles) is present in all species. There is no distinct heart, but the major trunk blood vessel is contractile. Blood circulates largely through a series of interconnected, discrete vessels. Each tentacle of the lophophore is serviced by a single, small vessel, through which the blood both ebbs and flows.

Many biologists feel that the phoronids are closest to the primitive lophophorate condition, although most workers would not consider other lophophorates to have evolved from phoronids *per se.* Phoronid-like animals may have given rise to all living deuterostomes.

Phylum Brachiopoda

Phylum Brachio · poda
(G: arm foot)
brak-ē-ō-pō´-dah

The external appearance of brachiopods, which include the so-called "lampshells," is quite unlike that of the phoronids. Brachiopods superficially resemble the bivalved molluscs, in that the body is protected externally by a pair of convex, calcified shells (Fig. 20.4) that are coated with a thin layer of organic periostracum. Indeed, up until about 100 years ago, brachiopods were considered *bona fide* members of the phylum Mollusca. Only about 350 brachiopod species currently exist, but nearly 20,000 species, extending back nearly 600 million years, are represented in the fossil record. This phylum has clearly seen better times (Fig. 20.5).[1]

The lophophore of brachiopods is similar to that of the phoronids except that it is drawn out into two arms (Fig. 20.6), increasing the effective surface area for food collection and gas exchange. Also, as in the Phoronida, one or two pairs of metanephridia serve as excretory organs. A blood circulatory system is also present, with the blood being circulated by the action of a well-developed heart and from one to several contractile vessels associated with the main dorsal vessel. Subsequent to leaving this main vessel, the blood moves somewhat haphazardly

---

1. See *Topics for Further Discussion and Investigation,* no. 2, at the end of the chapter.

Figure 20.4

Photograph of an inarticulate brachiopod, *Lingula* sp. (right), and four specimens of an articulate brachiopod, the lampshell *Terebratella* sp. (left). *Lingula* sp. has a long muscular pedicle, which anchors the animal in a burrow in the sediment. In contrast, lampshells generally have a short, nonmuscular pedicle that attaches to rocks. The pedicle of *Terebratella* sp. is seen protruding from an opening in the ventral shell valve in one of the specimens. In a few articulate and inarticulate brachiopod species, the pedicle is completely lost; the valves cement directly to solid substrate.   *Photograph by J. Pechenik and L. Eyster.*

among animals has long puzzled zoologists; it is found only in brachiopods, sipunculans, priapulids, and a few species of polychaetes, groups with no close phylogenetic connection. Hemerythrin is not found among any other lophophore-bearing animals.

Most brachiopods live permanently attached to a solid substrate or firmly implanted in sediment. Attachment is generally achieved by means of a stalk, called the **pedicle** (L: little foot), which protrudes posteriorly through a notch or hole in the ventral shell valve (Figs. 20.4 and 20.6*b,c*). The stalk is quite long and flexible in some species, perhaps serving to keep the body up above the substrate and in a zone of greater water flow.[2] This could benefit the brachiopod both in terms of increased gas exchange and increased rate of food capture. The pedicle is often muscular and hollow, housing an extension of the main coelomic cavity (the metacoel), through which coelomic fluid can circulate.

Brachiopod shells are composed of a protein matrix plus either calcium carbonate or calcium phosphate, and are secreted by two lobes of tissue referred to as the **mantle tissue.** This mantle tissue is by no means homologous with the shell-secreting molluscan tissue of the same name. Like the "mantle" of barnacles, the terminology simply reflects past ideas about the relationships between animal groups, ideas that have changed substantially over the past hundred years or so; both barnacles and brachiopods were once classified as molluscs. The brachiopod body is oriented so that the shell valves are ventral and dorsal. This is quite unlike the situation encountered among the bivalved molluscs, in which the shell valves are to the left and right sides of the body. Brachiopod shells are typically less than 10 cm in any dimension.

In some brachiopod species, comprising the class Inarticulata, the shell valves are held together entirely by adductor muscles. In members of the only other brachiopod class, the Articulata, the shell valves are hinged and articulate, as in molluscan bivalves; that is,

through a system of interconnected blood sinuses. The blood contains no oxygen-binding pigments; a blood pigment is found in the coelomic fluid, but it is hemerythrin rather than hemoglobin. Hemerythrin, although it does contain iron, is structurally and functionally quite different from hemoglobin. Hemerythrin distribution

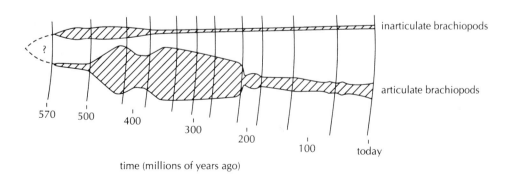

time (millions of years ago)

Figure 20.5

The occurrence of brachiopods in the fossil record. Decreases in width correspond to comparable decreases in the number of species. The distinction between articulate and inarticulate

species is given in Table 20.1.   *From Boardman, et al., Fossil Invertebrates. Palo Alto: Blackwell Scientific Publications, Inc., 1987. Reprinted by permission.*

---

2. See *Topics for Further Discussion and Investigation*, no.3.

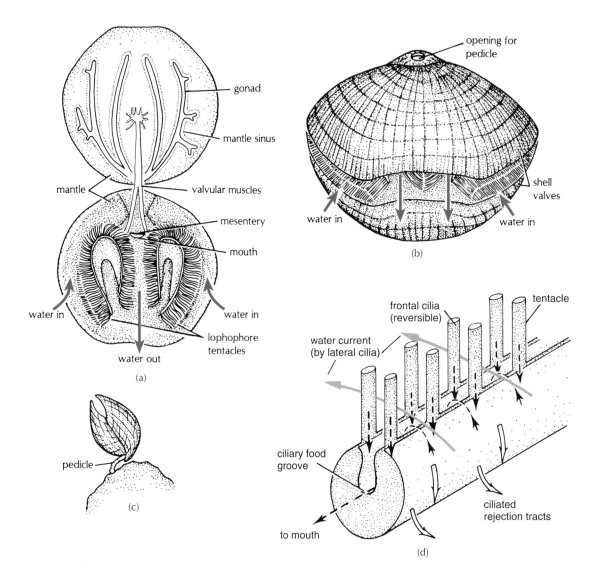

Figure 20.6

(a) Brachiopod with valves opened to show orientation of lophophore. (b) Path of water flow through the lophophore of a brachiopod. (*Terebratella* sp.) in the process of feeding. Note the opening through which the pedicle protrudes in life. (c) The brachiopod *Hemithyris psittacea* in its normal feeding posture. (d) Detail of one part of a lophophore arm. Dashed arrows show paths of captured food particles in transit to the mouth; open arrows show particles being rejected. Blue arrows indicate water flow through adjacent tentacles.   (a) *Beck/Braithwaite, Invertebrate Zoology, Laboratory Workbook, 3/e,* © *1968. Reprinted by permission of Prentice Hall, Upper Saddle River, New Jersey.* (b) *After Hyman; after Blochman.* (c) *After Hyman.* (d) *Modified after Russell-Hunter.*

the shell margins possess a series of interlocking teeth and sockets that prevent substantial sliding of one valve relative to the other.

In both classes, the shell valves are brought together by contracting adductor muscles, as in bivalved molluscs (Fig. 20.7). However, there is no springy bivalve-style hinge ligament to force open the valves when the adductors relax. Instead, separating the valves from each other is an active process, dependent upon the contraction of an opposing set of muscles, the **diductor muscles.** The shell thus acts as a complete skeletal system; not only do the shell valves protect the soft body parts, but they also serve as the vehicles through which the two muscle groups, the adductors and the diductors, antagonize each other (Fig. 20.7).

Members of the two brachiopod classes differ with respect to their shells' chemical composition and the morphology of their digestive tracts (Table 20.1). The shells of articulate brachiopods are all strengthened with calcium carbonate. In contrast, those of inarticulate species usually contain calcium phosphate. The digestive tract of the inarticulates is always U-shaped, with a mouth and separate terminal anus. The digestive tract of articulate brachiopods, on the other hand, terminates

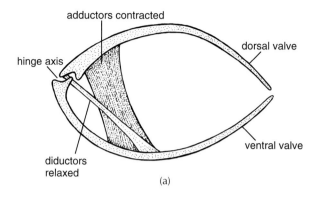

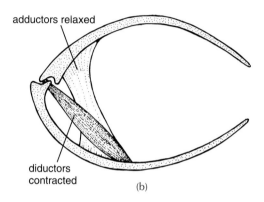

Figure 20.7

The use of adductor and diductor muscles in opening and closing the shell valves of a typical hinged brachiopod. (a) Contracting the adductor muscles while relaxing the diductor muscles closes the valves and stretches the diductor muscles. (b) Contracting the diductor muscles while relaxing the adductor muscles levers the valves apart and stretches the adductor muscles, preparing them for another contraction.   *From Boardman, et al.,* Fossil Invertebrates. *Palo Alto: Blackwell Scientific Publications, Inc., 1987. Reprinted by permission.*

blindly (Fig. 20.8). Thus, the articulate brachiopods, sophisticated in so many other respects, make do without an anus. Lastly, the lophophores of some articulate brachiopod species contain a rigid, calcified, internal support never encountered within the Inarticulata.

Phylum Bryozoa (= Ectoprocta; = Polyzoa)

Phylum Bryo · zoa
(G: moss animals)
brī-ō-zō′-ah

All three names for the phylum Bryozoa are widely used in the literature. The name "Bryozoa" (moss animal) refers to the finely branched appearance of many common species. "Ectoprocta" emphasizes that the anus lies

Table 20.1   Comparison of the Two Major Brachiopod Classes

Characteristic	Articulata	Inarticulata
Shell	Always calcium carbonate	Usually calcium phosphate
	Tooth and socket hinge	No articulating hinges
Digestive tract	Ends blindly	Open at mouth and anus
Rigid internal support	Present in some species	Never present

outside the ring of tentacles, as in all lophophorates (*ecto* = G: outside; *proct* = G: anus). Finally, "polyzoa" (i.e., multiple animals) refers to the colonial nature of all bryozoan species: They always form clusters of asexually produced, physically interconnected units. The name Bryozoa is used in this text, as it seems to be the term of choice in recent publications.

All bryozoans secrete a house around the body. The contents of the house—that is, the lophophore, gut, nerve ganglia, and most of the musculature—are referred to as the **polypide** (Fig. 20.9*a*). The house itself, plus the body wall that secretes it, constitutes the **cystid,** while the secreted, nonliving part of the house is termed the **zooecium** (*zoo* = G: animal; *oecus* = G: house). This rather confusing terminology evolved from the mistaken idea that the body wall of the house and the contents of the house were two separate individuals. That misconception was corrected quite some time ago, but the terminology had been used so widely that it has persisted. The body wall is actually attached to the zooecium, so the bryozoan is essentially glued to its house. One entire individual (cystid and polypide) is termed a **zooid.**

The bryozoan body wall has a unique developmental potential: The entire zooid can be generated, or regenerated, from the body wall of the cystid. Such generation and regeneration is, in fact, an integral part of the bryozoan life cycle.

During the life of an individual zooid, the entire polypide periodically degenerates into a dark-pigmented, spherical mass known as a **brown body.** In most species, a new polypide is subsequently produced from the cystid. In some species, the brown body remains conspicuously present in the coelomic space of the new polypide. In other species, the brown body is engulfed by the digestive system of the regenerating polypide and is then discharged unceremoniously through the anus. Bryozoan zooids may go through four or more such cycles of degeneration and rebirth during their lives. In some instances, brown-body formation also may provide a mechanism for bypassing unfavorable environmental conditions. Brown-body

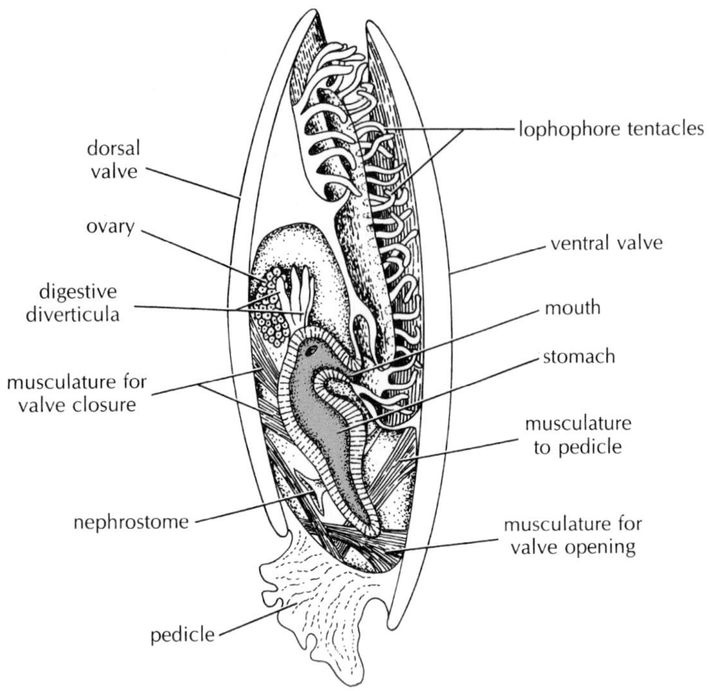
dorsal valve
ovary
digestive diverticula
musculature for valve closure
nephrostome
pedicle
lophophore tentacles
ventral valve
mouth
stomach
musculature to pedicle
musculature for valve opening

Figure 20.8
Diagrammatic illustration of the internal anatomy of an articulate brachiopod. Note the blind-ending intestine and the complex musculature operating the shell valves. Also note the conspicuous nephrostome. *After Harmer and Shipley.*

formation may also be a mechanism for eliminating insoluble waste products, by disposing of the entire polypide—a rather remarkable method of garbage disposal, to say the least. Bryozoans lack nephridia.

The septum dividing the metacoel from the mesocoel is very incomplete in bryozoans, so the coelomic fluid of the body cavity is continuous with that of the lophophore and tentacle cavities. The bryozoan lophophore is strikingly similar to that of phoronids, except that it can be retracted within the zooecium for protection and protruded for feeding and gas exchange. It protrudes through an **orifice** in the zooecium (Fig. 20.9). In many species, a muscular **diaphragm** is positioned just beneath this orifice. Protrusion of the lophophore is accomplished indirectly, by increasing the hydrostatic pressure within the main body cavity and dilating the diaphragm. The means of achieving the temporary increase in pressure within the body cavity differ among bryozoan species, as discussed later in the chapter.

Bryozoans show a great variety of external morphology, far greater than that encountered within other lophophorate groups. Moreover, all bryozoans are colonial; that is, a single individual reproduces itself asexually to form a contiguous grouping of genetically identical individuals, with up to about two million zooids in a single

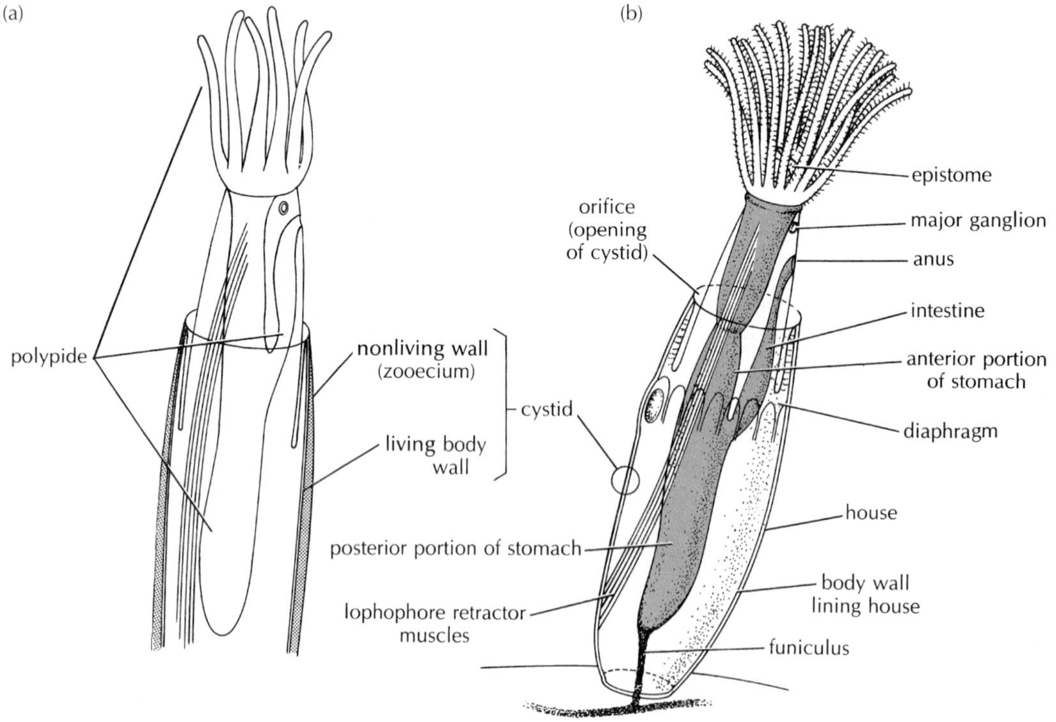
(a)
polypide
nonliving wall (zooecium)
living body wall
cystid
(b)
orifice (opening of cystid)
epistome
major ganglion
anus
intestine
anterior portion of stomach
diaphragm
house
posterior portion of stomach
lophophore retractor muscles
body wall lining house
funiculus

Figure 20.9
(a) Diagrammatic illustration demonstrating terminology of the polypide, cystid, and zooecium. (b) The freshwater bryozoan (class Phylactolaemata) *Fredericella,* showing details of the internal anatomy. *From Hyman; after Allman.*

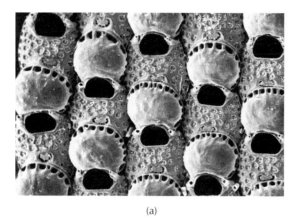

(a)

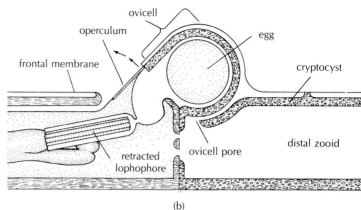

(b)

Figure 20.10

(a) Scanning electron micrograph of the marine bryozoan *Fenestrulina malusii.* Each zooid is associated with an ovicell, in which a single embryo is brooded after fertilization, as indicated

in (b). The arrows adjacent to the operculum indicate the direction of opercular movement.   (a,b) *From Claus Nielsen, in Ophelia, 9:209–341, 1971. Copyright © 1971 Ophelia Publications, Helsingor, Denmark.*

colony (Fig. 20.10*a*). Each of these individual zooids is extremely small, usually less than 1 mm each, although an entire colony can exceed 0.5 m in length or circumference. Since the colonies grow by adding new modules, rather than by increasing the size of each unit, the aging and growth of bryozoans seem not to be subject to the sorts of physiological rules and limitations imposed on most other animals (Research Focus Box 20.1).

Bryozoan colonies show a considerable variety of species-specific geometrical patterns. Colonies may be erect and branching or flat and encrusting. Great diversity of form is encountered within each of these two basic patterns. Many people have erect bryozoans in their homes without knowing it: Colonies of these marine species are dried, dyed, and sold commercially as "air ferns." No wonder they never need to be watered!

In many bryozoans, thick mesenchymal cables—called **funicular cords**—form tissue connections between the individual colony members. A single cable is termed a **funiculus** (Figs. 20.1*a* and 20.9*b*). The funicular system of a single zooid may extend across the coelomic cavity from the stomach to a pore in the body wall, providing a mechanism for possible direct transfer of nutrients among adjacent zooids.[3] The funicular system may be homologous with the circulatory systems of other metazoans, including other lophophorates.

Most bryozoans (about 4,000 species) are marine, but about 50 species are restricted to freshwater. These 50 bryozoan species constitute the only nonmarine lophophorates. Approximately 15,000 additional marine bryozoan species are known only from the fossil record. Like that of brachiopods, the bryozoan fossil record is long and interesting (Fig. 20.11).

Bryozoans are divided among two major classes and one minor class, based largely on differences in the morphology of the lophophore, mechanism of lophophore protrusion, chemical composition of the external body covering, and the presence or absence of an epistome and of body wall musculature.

Class Phylactolaemata

All members (about 50 species) of the class Phylactolaemata (fī-lak´-tō-lē-mah´-tah) are found in freshwater, and most (but not all) freshwater bryozoans are members of this class. The zooids of a given colony are morphologically identical; that is, the colonies are **monomorphic.** Some species secrete a chitinous outer covering, while others produce thick, gelatinous surroundings (Fig. 20.12*a*). In some species, the diameter of a single colony can exceed 50 cm. Colonies develop on a variety of submerged, solid surfaces, including shells, rocks, and the leaves and branches of freshwater vegetation. Most species are permanently affixed to these substrates and are incapable of locomotion. However, the colonies of a very few species are capable of slowly moving from place to place, using the single muscular "foot" shared by all of the zooids in the colony.

The lophophore is U-shaped in all phylactolaemate species, as in phoronids. A flap of tissue, termed the **epistome** as in phoronids, hangs over the mouth. A protocoel is evident within the epistome of some species. As in the phoronids, the body wall contains both circular and longitudinal muscles. A pronounced, hollow funiculus extends from the stomach into the coelomic cavity shared by all of the zooids (Fig. 20.9).

---

3. See *Topics for Further Discussion and Investigation*, no. 6.

Research Focus Box 20.1

Colonial Metabolism

Hughes, D. J., and R. N. Hughes. 1986. Metabolic implications of modularity: Studies on the respiration and growth of *Electra pilosa*. *Phil. Trans. Royal Soc. London B* 313:23–29.

As most animals grow larger, their weight-specific metabolic rate (typically measured as oxygen consumed per hour per gram of tissue) declines; although a fully grown whale certainly consumes oxygen faster than a baby whale, the full-grown whale consumes oxygen more slowly *per unit of body weight*. Such a decline in weight-specific metabolic rate with increasing body volume results in most animals accumulating biomass at progressively slower rates as they grow.

Does coloniality offer any escape from this limitation on growth rate, or must colonial animals like bryozoans also grow more slowly with time? To answer this question, Hughes and Hughes (1986) measured metabolic rates and growth rates of the encrusting bryozoan *Electra pilosa*. Oxygen consumption rates of 37 colonies of widely different sizes were determined by placing individual colonies in small volumes of seawater and measuring the rate at which seawater oxygen concentration declined, at constant temperature. Oxygen concentrations were measured electronically.

At the end of each measurement, the researchers scraped each colony from its substrate, dried it at 60°C, and weighed it to the nearest milligram. This total weight would include inert calcium carbonate. To determine the actual amount of metabolizing tissue (i.e., **biomass**) present,

Hughes and Hughes then cremated the samples at 500°C for six hours to combust all the organic material to carbon dioxide (CO_2) and water (H_2O). The researchers then reweighed the samples; the amount of weight lost in the cremation process reflects tissue weight, since only the inorganic "ash" is left behind. Tissue weights of the colonies tested in this experiment ranged from less than 0.1 mg to over 10 mg.

This "ashing" procedure enabled the researchers to examine how weight-specific oxygen consumption changed as the *E. pilosa* colonies grew. In contrast to the clear inverse relationship seen with noncolonial animals, weight-specific oxygen consumption did not decline with growth of the bryozoan colony (Focus Fig. 20.1). Moreover, monitoring the growth rates of 76 individual bryozoan colonies over a period of two weeks showed that, as colony size increased, so did the rate at which zooids on the growing edge (the peripheral zooids, located along the circumference of the colony) reproduced asexually (Focus Fig. 20.2). Thus, colonies of *E. pilosa* actually grew more rapidly as they aged, not more slowly.

In short, encrusting bryozoans appear to have a decided advantage over noncolonial organisms when competing for space on solid substrates, such as rocks and macroalgae: By

---

Lophophore protraction is brought about by the contraction of muscles in the deformable body wall.[4] Because the coelomic fluid is essentially incompressible, this contraction increases the pressure within the metacoel and, since the septum between the two main body cavities is incomplete, squeezes coelomic fluid from the metacoel into the mesocoel of the lophophore. Contraction of the body wall musculature thus inflates the lophophore tentacles and increases the hydrostatic pressure within the metacoel as well. Once the orifice diaphragm is opened by the contraction of specialized dilator muscles, the lophophore is forced out from the cystid by the elevated hydrostatic pressure within the metacoel. Retractor muscles extending from the body wall to the lophophore (Figs. 20.9 and 20.13) bring the lophophore back within the zooecium.

Perhaps the most intriguing feature unique to the phylactolaemates is the formation of **statoblasts.** These

structures, produced seasonally, can withstand considerable desiccation and thermal stress. The statoblasts consist of a cell mass enclosed within a bivalved protective capsule of species-specific morphology. They are produced along the funiculus of each zooid (Fig. 20.13), often in great numbers, and are released from the zooecium through a pore or, more typically, upon degeneration of the polyp in the late fall. When environmental conditions improve the following spring, the two valves of the statoblast separate along a preformed suture line, and a polypide soon emerges. Colony formation ensues by standard means, through asexual budding of additional zooids. Phylactolaemates produce two types of statoblasts: **floatoblasts,** which are buoyant by means of gas-filled cells, and **sessoblasts,** which are firmly cemented to a solid substrate. Floatoblasts may be dispersed considerable distances by water, wind, or animals.

---

4. See *Topics for Further Discussion and Investigation*, no.4.

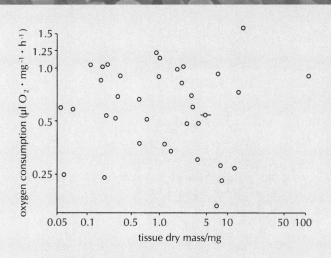

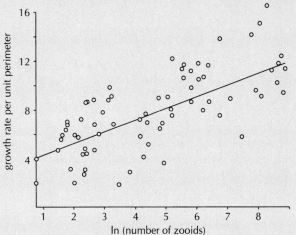

Focus Figure 20.1

The effect of colony size (measured as dry tissue weight) on weight-specific oxygen consumption (measured as microliters of oxygen consumed per milligram dry tissue weight per hour) in the encrusting bryozoan. *Electra pilosa*. No obvious relationship is apparent; certainly there is no indication that weight-specific oxygen consumption declines with increasing colony size.

Focus Figure 20.2

The effect of colony size on the division rate of zooids along the edge of a colony in the encrusting bryozoan, *Electra pilosa*. Increasing colony size is represented by the increasing number of zooids along the X-axis (plotted as natural log). In general, zooids along the edge of the colony divided faster in larger colonies.

proliferating an endless series of genetically identical modules, *E. pilosa* avoids the slowdown in growth inevitable in organisms that grow by increasing their body volume. Indeed, growth actually accelerates with increased colony size, possibly because the feeding, nondividing zooids nearer the colony's center contribute nutrients to those actively dividing zooids along the edge. Can you design an experiment to

determine whether the increased division rates of these edge zooids are, in fact, brought about by increased movement of nutrients from more central zooids?

Figures from D. J. and R. N. Hughes, "Metabolic Implications of Modularity: Studies on the Respiration and Growth of Electra pilosa" in Transactions of the Royal Society of London. Copyright © 1986 The Royal Society, London. Reprinted by permission.

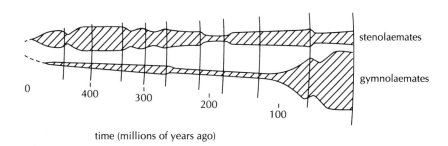

time (millions of years ago)

Figure 20.11

The occurrence of calcified bryozoans in the fossil record. Increases in width correspond to proportional increases in the number of species. The distinction between gymnolaemates and stenolaemates is given on pp. 437–440. *From Boardman, et al., Fossil Invertebrates. Palo Alto: Blackwell Scientific Publications, Inc., 1987. Reprinted by permission.*

Statoblast formation by phylactolaemate bryozoans corresponds to the formation of gemmules and resting eggs by freshwater sponges and tardigrades, respectively. Clearly, formation of resting (i.e., **diapause**) stages is a common adaptation to the vagaries of freshwater existence.

Class Gymnolaemata

Most extant bryozoan species are assigned to the class Gymnolaemata (jim´-nō-lē-mah´-tah) (Fig. 20.12*b–e*), although they have dominated the phylum for only the past 70 million years or so (Fig. 20.11). The

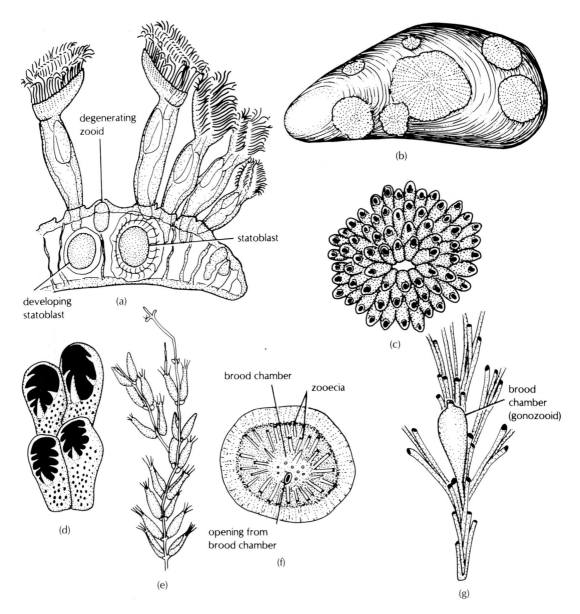

Figure 20.12
Bryozoan diversity. Phylactolaemata: (a) *Cristatella* colony. This species is found commonly encrusting lily pads in freshwater ponds. Note the gelatinous, creeping sole. The degenerating zooid is forming a brown body. Gymnolaemata: (b) Mussel shell valve encrusted with colonies of *Cryptosula,* shown in detail in frame (c). (d) *Electra pilosa.* Note the conspicuous spines guarding the frontal opening. (e) *Bowerbankia gracilis,* with tentacles withdrawn in several zooids. Stenolaemata: (f) *Lichenopora* colony. Note the communal brood chamber. (g) *Crisia ramosa,* showing tubular zooecium.　 *(a) After Hyman. (b,c) From Hyman; after Rogick. (d) From R. I. Smith, Key to Marine Invertebrates of the Woods Hole Region. Copyright © 1964 Marine Biological Laboratory, Woods Hole, MA. (After Rogick and Croasdale.) Reprinted by permission. (e) From Smith; after Rogick and Croasdale. (f) From Hyman; after Hincks. (g) After Harmer and Shipley.*

gymnolaemates are primarily marine. They are especially fascinating animals, even by bryozoan standards, displaying a broad range of morphological and functional diversity. Bryozoans are heavily preyed upon by turbellarians, polychaetes, insect larvae, crustaceans, arachnids (mites), gastropods, asteroids, and fish. To a great extent, the story of gymnolaemate evolution is one of increasing the degree to which the polypide is pro-

tected from these predators. In general, this is achieved by strengthening the zooecia, which is not so simply accomplished among bryozoans. Recall that lophophore protrusion depends upon a change of shape of the animal, causing an elevation in coelomic pressure, and that the body wall is joined to the zooecium. The trick, then, is to strengthen the zooecium without losing the ability to generate elevated pressures within the metacoel.

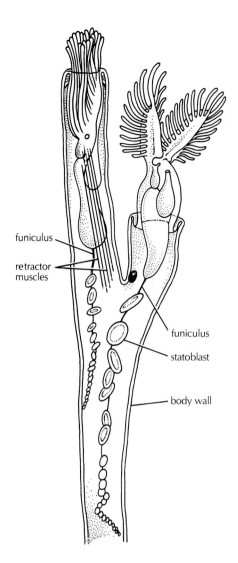

funiculus

retractor
muscles

funiculus

statoblast

body wall

Figure 20.13
Statoblast formation in the freshwater bryozoan *Plumatella repens*.
Note the association of the statoblasts with the funiculus and the
lack of boundaries between adjacent zooids.   *After Harmer and Shipley.*

In a few gymnolaemate species, the feeding zooids
are borne on stolons—that is, on tubular extensions of
the body wall. The stolons may be upright (**erect**) or flat
against the substrate. In most gymnolaemate species,
however, stolons are absent. Instead, zooids are contiguous, with the zooecium of one zooid supported by the
zooecium of adjacent zooids (Figs. 20.10*a* and 20.12*c*),
thereby strengthening the entire colony.

The colonies of most marine gymnolaemate species
are composed of a large array of small, box-like or ellip-
tical houses, typically encrusting on any submerged,
solid surface. Zooids are arranged in species-specific
patterns, determined by the pattern of asexual budding.
The zooecia themselves are characterized by species-
specific morphology and are therefore of great taxonomic value.

The Gymnolaemata are contained within two orders:
the Ctenostomata (tēn´-ō-stō-mah´-tah) and the
Cheilostomata (kēl´-ō-stō-mah´-tah). The ctenostomes
are characterized by a flexible, chitinous zooecium, and
lophophore protraction is achieved in much the same
way as in the phylactolaemates. Muscle contractions
draw the zooecium inward, and the resulting increase in
hydrostatic pressure of the coelomic fluid forces the
lophophore out through the orifice.

But most gymnolaemate bryozoans are cheilostomes, all of which calcify their zooecia to varying degrees. In many species, the orifice is sealed by a hinged,
calcareous **operculum** when the polypide is fully withdrawn within the zooecium (Figs. 20.10*b* and 20.14).
Calcium carbonate is also liberally deposited between the
chitinous cuticle of the zooecium and the epidermis of
the cystid of most species. The **frontal membrane,** however, often remains uncalcified, and muscles attach from
the calcified body wall to this flexible surface (Fig.
20.14*a*, top); lophophore protraction is achieved by contracting these muscles, which increases the coelomic
pressure. The body wall musculature is dramatically reduced or, more frequently, completely absent. Although
calcification in these species certainly strengthens the
colony structurally and probably provides some protection from predators, the zooid remains vulnerable
through its uncalcified frontal membrane.

Often, the frontal membrane is partially shielded by
spines projecting from the frontal margins of the zooecium (Fig. 20.12*d*) or, more rarely, from the frontal
membrane itself. Nevertheless, the frontal membrane remains the Achilles' heel of the zooid.

In some cheilostome species, the frontal membrane
remains uncalcified and flexible, but a calcareous shelf,
called the **cryptocyst,** is secreted beneath it (Fig. 20.14*a*,
middle). Muscles pass to the frontal membrane through
small holes in the cryptocyst. Thus, the lophophore can
still be protracted by muscles pulling downward on the

Gymnolaemates differ from the phylactolaemates in
a number of major respects. In phylactolaemates, the
walls of the cystid are very incomplete, so adjacent
zooids lack morphological boundaries (Fig. 20.13). Although each zooid protracts its lophophore through its
own orifice, the polypides of a phylactolaemate colony
essentially share a common metacoel. In contrast, the
zooids of gymnolaemates are morphologically distinct
individuals, although small **pore plates** allow the exchange of coelomic fluid between neighboring zooids
(Fig. 20.14*b*). The two classes also differ with respect to
lophophore morphology. Whereas the lophophore of
phylactolaemate bryozoans has the deeply invaginated
U-shape seen in phoronids, that of the gymnolaemates
has a circular appearance.

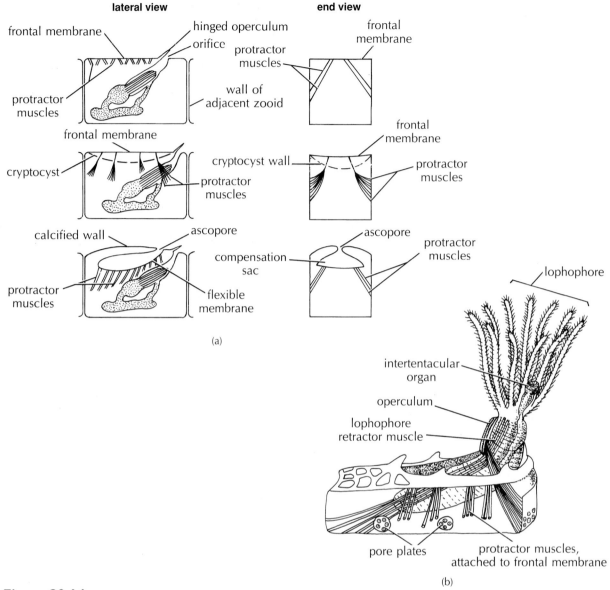

lateral view

frontal membrane
hinged operculum
orifice
protractor muscles
protractor muscles
wall of adjacent zooid

end view

frontal membrane

frontal membrane
cryptocyst
protractor muscles

cryptocyst wall
frontal membrane
protractor muscles

calcified wall
ascopore
compensation sac
protractor muscles
flexible membrane

ascopore
protractor muscles

(a)

lophophore

intertentacular organ
operculum
lophophore retractor muscle

pore plates
protractor muscles, attached to frontal membrane

(b)

Figure 20.14
(a) Functional diversity among the gymnolaemates, showing different degrees of protection of the frontal membrane.

(b) *Electra pilosa,* a species with a well-exposed frontal membrane, shown with lophophore protruded from the zooecium. (a) *After Clark; after Harmer.* (b) *After Clark; after Marcus.*

frontal membrane, but the polypide is protected within the zooecium by the calcified cryptocyst. The frontal membrane itself, however, remains vulnerable to attack.

A different means of protecting the polypide has evolved in yet a third group of cheilostomes. In these species, the entire zooecium, including the frontal membrane, calcifies. All of the soft tissues are thus completely protected within the zooecium, but the frontal membrane can no longer serve as the vehicle through which internal hydrostatic pressure can be elevated. Instead, these species possess a new, uncalcified membrane that separates the metacoel from the calcified frontal surface. A space exists between the calcified frontal wall and this new membrane, forming a sac—specifically, a **compensation sac** or **ascus** (Fig. 20.14*a*, bottom). The compensation sac opens to the outside through a single pore, the **ascopore,** in the frontal surface (*ascus* = G: sac). Muscles extend from the calcified sides of the zooecium to the undersurface of the compensation sac. Contraction of these muscles pulls the compensation sac tissue inward, elevating the pressure within the coelomic cavity and forcing the lophophore out through the orifice of the zooecium. A vacuum does not form within the compensation sac during this process because seawater is drawn into the compensation sac through the ascopore.

The zooids described so far are the feeding and reproductive members of the colony, the **autozooids.** All gymnolaemate colonies contain some nonfeeding members as well. Thus, all gymnolaemate colonies are **polymorphic,** with individuals being morphologically or physiologically specialized for different functions, even though all individuals in the colony have a single genotype. The nonfeeding colony members, called **heterozooids,** take a variety of forms. The stolons and holdfasts of stoloniferous species are composed of a series of short heterozooids. Strangely enough, the pore plates permitting the exchange of coelomic fluid between adjacent zooids are also highly specialized heterozooids in some species. Other heterozooids, found only among the cheilostomes, are specialized for protecting and cleaning the colony. One such heterozooid is little more than a highly modified operculum, drawn out into a long, movable bristle. The continual sweeping movements made by these **vibracula** (Fig. 20.15*d*) presumably discourage invertebrate larvae from attaching to the colony and keep the colony surface free of debris. Other heterozooids form immobile, protective spines.

Probably the most intriguing heterozooids are the **avicularia.** Avicularia, in their least modified form, have the external appearance of a normal autozooid (Fig. 20.15*c*). Internally, the avicularia consist of little more than well-developed muscle fibers extending across a capacious coelom to the vicinity of the ventral, hinged operculum. The polypide is much reduced or completely absent. The operculum can be closed suddenly and with considerable force, discouraging, mutilating, or even killing potential predators. The avicularia of some species are highly modified for their tasks. In the most extreme form, the avicularia consist primarily of a modified opercular system, now referred to as a **mandible** (Fig. 20.15*a–c*). These avicularia may be mounted on a stalk, permitting some rotation. Some of these very specialized avicularia closely resemble a bird's head, a resemblance that gave rise to their name (*aves* = L: bird; thus, *avicularia* = little bird) (Fig. 20.15*a,b*).

In at least some species, adaptive shifts in zooid morphological development are induced by contact with potential predators and competitors.[5]

Class Stenolaemata

A small number of bryozoans are sufficiently dissimilar from the phylactolaemates and gymnolaemates to warrant placement into a third class, the Stenolaemata (Stē′-nō-lē-mah′-tah) The members of this class are all marine, and the living species belong to a single order, the Cyclostomata (sī′-clō-stō-mah′-tah) Cyclostome zooids are always tubular and erect, and the zooecia are completely calcified (Fig. 20.12*f,g*). Not surprisingly, the cystid is nonmuscular, as in most cheilostomes. A thin, cylindrical membrane divides the main coelomic space into two compartments, one of which contains the polypide. Lophophore protraction is accomplished by the sideways, muscular displacement of this membrane. As in all bryozoans, the lophophore is drawn back into the zooecium by powerful lophophore retractor muscles.

Other Features of Lophophorate Biology

Reproduction

Asexual reproduction is most characteristic of the Bryozoa, in keeping with their habit of colony formation. Asexual reproduction by budding or fission is encountered in only one or two phoronid species, and reproduction is exclusively sexual in brachiopods. Brachiopods are also the exception when it comes to sexuality, with most species being dioecious. In contrast, most phoronids and bryozoan colonies are hermaphroditic. Gametes are formed by simple gonads—really just a cluster of germ cells in the mesodermal lining—within the metacoel. Phoronids and brachiopods usually discharge their gametes through the nephridia. Bryozoans, however, lack nephridia; instead, their sperm are released through the lophophore tentacles, and eggs are released through a special pore located between two of the tentacles. Lophophorates never copulate.

Phylum Phoronida

In phoronid reproduction, sperm emitted through the nephridiopores of one individual are captured from the seawater by another, neighboring individual. Most species develop into a characteristic, ciliated, feeding larval stage called an **actinotroch** (Fig. 20.16*a*). Metamorphosis to adult form involves a rapid and dramatic "turning inside out" of the larval body.

Phylum Brachiopoda

Among brachiopods, sperm and eggs are discharged through the nephridiopores, and fertilization occurs in the surrounding seawater. Free-swimming, ciliated larvae are produced (Fig. 20.16*b*).

Phylum Bryozoa

Although all bryozoan colonies are hermaphroditic, an individual zooid may be of single sex. Sperm are released into the zooid's coelomic cavity and exit from openings in the

---

5. See *Topics for Further Discussion and Investigation,* no. 7.

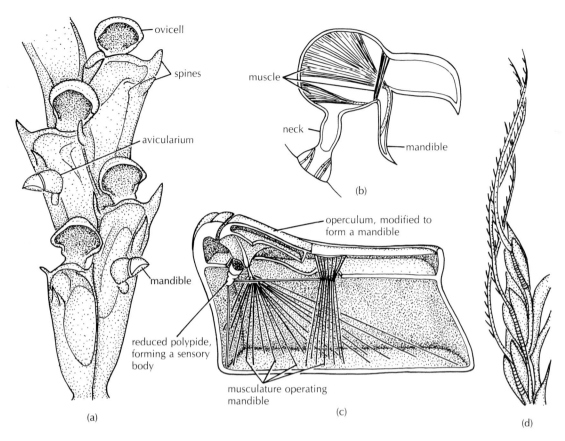

Figure 20.15

(a) *Bugula* sp., an erect, marine bryozoan with ovicells and highly modified avicularia. (b) Detail of the avicularia shown in (a). (c) Relatively unmodified avicularium of *Flustra foliacea*, showing more clearly the relationship between the avicularium and a normal zooid. (d) *Caberea ellisi*, showing three members of a colony, each possessing a single, long vibraculum.   (a) *From Hyman; after Rogick and Croasdale.* (b) *From Laverak and Dando, in* Essential Invertebrate Zoology, *2d ed. Copyright © 1987 Blackwell Scientific, Oxford, England. Reprinted by permission.* (c) *From Lars Silen, in Woollacott and Zimmer, eds.,* The Biology of Bryozoa. *Copyright © 1977 Academic Press, Inc. Reprinted by permission.* (d) *From Smith.*

lophophore tentacles. Neighboring individuals collect the sperm from the surrounding seawater. Once the eggs have been fertilized, a period of brood protection generally follows. Some species brood their embryos within the metacoel, but a variety of other brooding sites are found within the phylum. Most cheilostomes possess specialized brooding chambers, called **ovicells,** at one end of the zooid (Fig. 20.10). As the embryos develop within the various brood chambers, the polypide of the parent usually degenerates. This presumably reflects a transformation of parental tissues into nutrition for the offspring. The polypide of the parent may later be regenerated from the cystid.

In some brooding cheilostome species, the maternal zooid contributes substantial nutrients to the developing embryo in the ovicell. Because of this "placental brooding" system, the volume of the final larva as it escapes from the ovicell may be 15 to 500 times greater than the volume of the newly fertilized egg.

One or, more rarely, several nonfeeding, ciliated **coronate larvae** eventually emerge from each ovicell (Fig. 20.16c). After a limited period of dispersal away from the parent colony, the larvae attach to a substrate through the eversion of a sticky **adhesive sac,** and then

they metamorphose to adult form (Fig. 20.17). During metamorphosis, all larval tissues move to the interior of the animal and are destroyed through a combination of phagocytosis and autolysis. Only the outer body wall remains intact, and this becomes the cystid. The first zooid of a colony is termed an **ancestrula.** The rest of the colony is subsequently generated via asexual budding.

The fertilized eggs of a few, nonbrooding species develop into feeding larval stages called **cyphonautes larvae** (Fig. 20.16d). The cyphonautes larva is enclosed within a pair of triangular, chitinous valves and is ciliated on the marginal surfaces not covered by these "shells." The shell valves are lateral, as in bivalved molluscs. The cyphonautes larva is equipped with a complete digestive tract, enabling it to feed and remain in the plankton for long periods of time, perhaps as long as several months. It eventually attaches to a substrate by everting an adhesive sac, and metamorphosis to the ancestrula quickly follows.

Digestion

The digestive tract of all phoronids and bryozoans is U-shaped, with a mouth and separate anus. The anus always lies

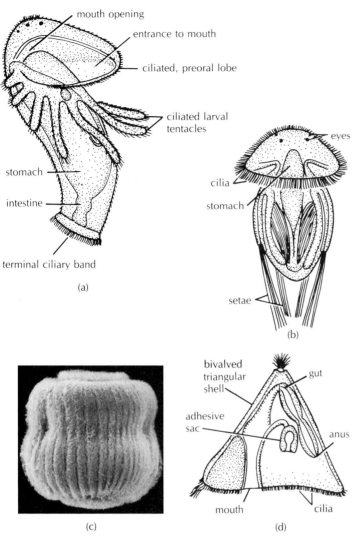

(a)

(b)

(c)   (d)

Figure 20.16

(*a*) Actinotroch larva of a phoronid. (*b*) Larva of the brachiopod *Argyrotheca* sp. (*c*) Scanning electron micrograph of the coronate larva of the bryozoan *Bugula neritina*. (*d*) Cyphonautes larva of a bryozoan.   *(a) From Hyman; after Wilson. (b) From Hyman; after Kowalevsky. (c) Courtesy of R. M. Woollacott and C. G. Reed. (d) After Hardy.*

outside the circle of lophophore tentacles, by definition. Digestion occurs within the relatively large stomach, and it has both extracellular and intracellular components. Brachiopods may (Inarticulata) or may not (Articulata) possess a U-shaped, one-way digestive system. In either case, the stomach connects to a large **digestive gland,** within which food material is broken down. Digestion is believed to be primarily intracellular.

Nervous System

The nervous system takes the form of a ring in all lophophorate species. This ring is found at the base of the lophophore in phoronids, encircling the esophagus in brachiopods, and adjacent to the pharynx in bryozoans.

The ring may or may not be distinctively ganglionated. From the nerve ring, nerve fibers innervate the tentacles and musculature. An epidermal nerve network is commonly found in the body wall. Discrete sense organs are rare among lophophorates. Although a balance organ (**statocyst**) has been described in one species of inarticulate brachiopod, the sensory apparatus of other lophophorate species consists of scattered mechanoreceptor and chemoreceptor cells and/or setae.

Phylum Entoprocta (= Kamptozoa)

Phylum Ento · procta (= Kampto · zoa)
(G: within anus [G: flexible animal])

The Entoprocta is a phylum of benthic, solitary or colonial animals restricted to marine environments. About 140 species have been described, and all are suspension feeders. Like the lophophorates, entoprocts collect food particles using an anterior organ bearing numerous ciliated tentacles (Fig. 20.18). However, since the anus lies within this ring of tentacles (*ento* = G: within; *procta* = G: anus), the organ is not, by definition, a lophophore, and entoprocts are therefore not lophophorates. Moreover, the pattern of water flow is opposite that exhibited by lophophorates: Among entoprocts, water flows into the center of the tentacular circle laterally, between the tentacles, and then passes upwards and outwards through the center (Fig. 20.18b). In several other respects, however, colonial entoprocts superficially resemble bryozoans, and some workers feel that bryozoans may, in fact, have evolved from entoprocts. Like bryozoan zooids, entoprocts are small, typically less than a few millimeters. Like bryozoans, entoprocts lack blood vessels and, as with most lophophorates, the entoproct digestive tract is U-shaped (Fig. 20.18). However, unlike bryozoan development, entoproct cleavage is determinate and spiral, and some species develop into perfectly formed protostome-like trochophore larvae.

Entoprocts have a small body cavity of sorts, but how it forms is still unclear. Consequently, some workers consider the entoprocts to be acoelomate, while others believe them to be pseudocoelomate. They are indeed a troublesome group at present, with phylogenetic relationships that are particularly difficult to ascertain. Many species form extensive colonies through asexual replication, as do bryozoans, but entoproct colonies are never polymorphic; all individuals in an entoproct colony look the same and are capable of feeding and sexual reproduction. All entoprocts are apparently hermaphroditic.

Figure 20.17
Four ancestrulae of *Bugula neritina,* newly metamorphosed from four coronate larvae.   *Courtesy of R. M. Woollacott and C. G. Reed.*

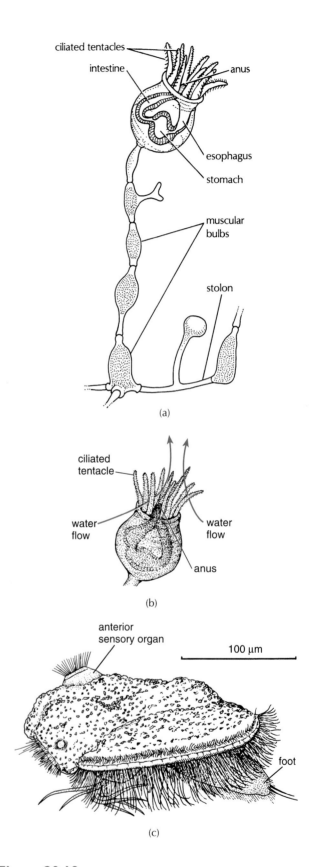

Figure 20.18
(a) An entoproct, *Anthropodaria* sp. (b) Flow of water past the feeding tentacles of an entoproct. (c) Larval stage of the entoproct *Loxosomella harmeri,* viewed from the left side.   (a) *From Hyman; after Nasonov.* (c) *From Claus Nielsen, in Ophelia, 9:209–341, 1971. Copyright © 1971 Ophelia Publications, Helsingor, Denmark.*

Taxonomic Summary

Phylum Phoronida
Phylum Brachiopoda—the lampshells
    Class Inarticulata—the inarticulate brachiopods
    Class Articulata—the articulate brachiopods
Phylum Bryozoa (= Ectoprocta; = Polyzoa)—the moss animals
    Class Phylactolaemata
    Class Gymnolaemata
        Order Ctenostomata
        Order Cheilostomata
    Class Stenolaemata
        Order Cyclostomata
Phylum Entoprocta (= Kamptozoa)

Topics for Further Discussion and Investigation

1. For a long time, entoprocts and ectoprocts (bryozoans) were grouped in a single phylum. Discuss the anatomical and functional similarities and differences between bryozoans and entoprocts, and the evidence for and against a close evolutionary relationship between these two groups.

Hyman, L. H. 1951. *The invertebrates. Vol. III, Acanthocephala, Aschelminthes and Entoprocta.* New York: McGraw-Hill.

Mariscal, R. N. 1965. The adult and larval morphology and life history of the entoproct *Barentsia gracilis* (M. Sars, 1835). *J. Morphol.* 116:311.

Nielsen, C. 1977. The relationships of Entoprocta, Ectoprocta and Phoronida. *Amer. Zool.* 17:149.

Nielsen, C. 1977. Phylogenetic considerations: The protostomian relationships. *Biology of bryozoans.* Woollacott, R. M., and R. L. Zimmer, eds. New York: Academic Press, 519–34.

2. Discuss the potential role of bivalved molluscs in bringing about the dramatic evolutionary decline of brachiopods.

Gould, S. J., and C. B. Calloway. 1980. Clams and brachiopods—ships that pass in the night. *Paleobiology* 6:383.

Stanley, S. M. 1968. Post-Paleozoic adaptive radiation of infaunal bivalve molluscs; a consequence of mantle fusion and siphon formation. *J. Paleontol.* 42:214.

3. Investigate the role of water currents in the feeding biology of lophophorates.

Eshleman, W. P., and J. L. Wilkins. 1979. Brachiopod orientation to current direction and substrate position (*Terebratalia transversa*). *Canadian J. Zool.* 57:2079.

Gilmour, T. 1979. Ciliation and function of food-collecting and waste-rejecting organs of lophophorates. *Canadian J. Zool.* 56:2142.

LaBarbera, M. 1977. Brachiopod orientation to water movement. I. Theory, laboratory behavior, and field orientations. *Paleobiology* 3:270.

Strathmann, R. R. 1973. Function of lateral cilia in suspension-feeding of lophophorates (Brachiopoda, Phoronida, Ectoprocta). *Marine Biol.* 23:129.

4. Based upon your readings in this book, compare and contrast the operation of the bryozoan lophophore with the operation of the nemertine proboscis.

5. Some species of branching bryozoans superficially resemble some colonial hydrozoan colonies. Based upon your readings in this book, what are some of the similarities and dissimilarities between these two animal groups?

6. Discuss the evidence that nutrients are shared among the zooids of a bryozoan colony.

Best, M. A., and J. P. Thorpe. 1985. Autoradiographic study of feeding and the colonial transport of metabolites in the marine bryozoan *Membranipora membranacea. Marine Biol.* 84:295.

Carle, K. J., and E. E. Ruppert. 1983. Comparative ultrastructure of the bryozoan funiculus: A blood vessel homologue. *Z. Zool. Syst. Evol.-forsch.* 21:181.

7. Investigate the manner in which gymnolaemate bryozoan colonies compete for space on solid substrates. Compare and contrast these mechanisms with those mediating protection from predators.

Harvell, C. D. 1986. The ecology and evolution of inducible defenses in a marine bryozoan: Cues, costs, and consequences. *Amer. Nat.* 128:810.

Harvell, C. D., and D. K. Padilla. 1990. Inducible morphology, heterochrony, and size hierarchies in a colonial invertebrate monoculture. *Proc. Nat. Acad. Sci.* 87:508.

Jackson, J. B. C., and L. W. Buss. 1975. Allelopathy and spatial competition among coral reef invertebrates. *Proc. Nat. Acad. Sci.* 72:5160.

Shapiro, D. F. 1992. Intercolony coordination of zooid behavior and a new class of pore plates in a marine bryozoan. *Biol. Bull.* 182:221.

Tzioumis, V. 1994. Bryozoan stolonal outgrowths: A role in competitive interactions? *J. Marine Biol. Assoc. U.K.* 74:203.

Taxonomic Detail

Phylum Phoronida

Phoronis. The 10 species in this phylum are found intertidally to depths of about 400 m. The smallest species are less than about 0.5 cm long, and the largest are about 50 cm long.

Phylum Brachiopoda

This phylum contains approximately 350 species distributed between two classes.

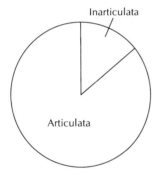

Class Inarticulata.
The approximately 50 species are contained within two orders.

Order Lingulida.
Lingula. Individuals are up to 44 mm in shell length. All species use their shells to form burrows in sediment, intertidally to depths of about 125 m. Some species are eaten by humans (in Australia and Japan). Fossils over 400 million years old closely resemble species of present-day *Lingula.* One family.

Order Acrotretida.
Crania. Some of the species in this group range from the intertidal zone to depths exceeding 7,600 m. Two families.

Class Articulata.
The approximately 300 species are contained within two orders.

Order Rhynchonellida.
These brachiopods live permanently attached to solid substrates from about 6 m depth to over 3,000 m, in both tropical and cold waters. Approximately 30 species, distributed among four families.

Order Terebratulida.
This group contains most living brachiopod species. The approximately 250 species are distributed among 12 families.

Family Terebratellidae. *Terebratella.* Shells are up to 91 mm long. One species, *Magadina cumingi,* uses its pedicle to move up and down in the substrate, unlike any other species of articulate brachiopod. Species occur from the intertidal zone to depths exceeding 4,000 m.

Phylum Bryozoa

This phylum contains approximately 4,000 species distributed among three classes.

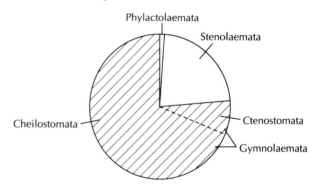

Class Phylactolaemata.
These bryozoans are found only in freshwater. The approximately 50 species are distributed among four families.

Family Plumatellidae. *Plumatella.* Members of this family produce both sessile and free-floating, dispersive statoblasts.

Family Lophopodidae. *Pectinatella.* Colonies form large, gelatinous masses up to 0.5 m across, and the lophophores of some species bear up to 120 tentacles. All species produce floating statoblasts.

Family Cristatellidae. *Cristatella.* The polypides are confined to the upper surface of the rounded, unbranched colony. The colony's muscular lower surface enables the entire colony to move slowly, perhaps 10 cm per day, over a solid substrate. All species produce only floating statoblasts.

Class Stenolaemata.

Crisia, Tubulipora. All species are marine and exhibit tubular, calcified zooids. Reproduction is unique among bryozoans in that each fertilized embryo may divide asexually, numerous times, giving rise to 100 or more genetically identical embryos within each reproductive individual (gonozooid). The approximately 900 species are distributed among about 20 families.

Class Gymnolaemata.

This is the most diverse group of bryozoans and contains the greatest number of species (approximately 3,000), most of which are marine. Two orders.

Order Ctenostomata.

Zooid walls are not calcified but instead are membranous or gelatinous. Avicularia, opercula, and ovicells are lacking. The approximately 250 species are distributed among about 20 families.

Family Nolellidae.
Nolella, Aethozoon. Members of the genus *Aethozoon* have zooids up to 8 mm long, the largest of any bryozoan species.

Family Paludicellidae.
Paludicella. All species are restricted to freshwater.

Family Monobryozoontidae.
Monobryozoon. The smallest bryozoan colonies are found in this group: The entire colony consists of a single, feeding autozooid. As new zooids are produced, they break free of the parent, which soon degenerates. All three species are marine and live in sandy sediments.

Family Hypophorellidae.
Hypophorella. The branched stolons bore holes into sedentary polychaete tubes or into the calcareous walls of other bryozoans. All species are marine.

Family Penetrantiidae.
These species secrete phosphoric acid, thereby boring into the calcareous shells of molluscs and barnacles. All species are marine.

Family Vesiculariidae.
Bowerbankia, Zoobotryon. All species are marine, forming branching, typically erect colonies. Lophophores bear only 8 to 10 tentacles each. Members of one genus, *Zoobotryon,* attain colony lengths of up to 0.5 m.

Order Cheilostomata.

Most species are marine and form highly polymorphic colonies with specialized reproductive zooids, attachment zooids, avicularia, vibracula, and ovicells. The zooids are often rectangular, and the walls are always at least partly calcified. This group contains the greatest diversity of bryozoan morphology. The approximately 2,750 species are distributed among about 70 families.

Suborder Anasca.

Approximately 30 families. Most or all of each zooid's frontal surface is membranous and virtually transparent.

Family Membraniporidae.
Membranipora. These species lack avicularia and ovicells. Species may be encrusting, often forming almost circular colonies, or erect. All species produce feeding cyphonautes larvae with bivalved shells.

Family Electridae.
Electra. Colonies lack avicularia or ovicells, and most species are encrusting. All species produce cyphonautes larvae.

Family Microporidae.
Micropora. A number of the species in this group are free-living. One, *Selenaria maculata,* walks on the long setae of its avicularia at speeds of over 1.5 cm per minute.

Family Bugulidae.
Bugula, Dendrobeania. These species form bushy, erect colonies. The zooids are only lightly calcified, and the entire frontal surface is membranous, so the internal organs are easily visible through a dissecting microscope. Colonies usually possess well-formed "bird's head" avicularia on stalks. *Bugula* is one of the largest and most widespread of all bryozoan genera.

Suborder Ascophora.

This is the largest and most diverse group of all living bryozoans. The frontal wall demonstrates a great variety of form and function. The lophophore is always operated through a compensation sac (= ascus). About 50 families.

Family Watersiporidae.
Watersipora. The members of this group are all encrusting and lack avicularia and ovicells; zooids brood embryos internally.

Family Hippoporinidae.
Pentapora. All species have ovicells, and many have avicularia. Colonies of *Pentapora* may become larger than 1 m in circumference, making them the largest of all bryozoan colonies.

Family Schizoporellidae.
Schizoporella.

Phylum Entoprocta

Most of the approximately 140 species are marine, although a few live in freshwater. Four families.

Family Loxosomatidae. *Loxosoma, Loxosomella.* This family contains all of the solitary species; all other entoprocts are colonial. Some species are mobile, using a muscular, distal expansion of the stalk. All of the at least 110 species are marine.

Family Barentsiidae. *Barentsia, Urnatella.* All species are colonial. The few freshwater entoprocts are all placed in the genus *Urnatella.* Some species produce cysts, which tolerate adverse environmental conditions. Approximately 30 species.

Some General References About the Lophophorates

Boardman, R. S., A. H. Cheetham, and A. J. Rowell, eds. 1987. *Fossil Invertebrates.* Palo Alto, Calif.: Blackwell Scientific, 445–549.

Harrison, F. W. and R. W. Woollacott. 1996. *Microscopic Anatomy of Invertebrates,* Vol. 13. Lophophorates and Entoprocta. Wiley-Liss, NY.

Hyman, L. H. 1959. *The Invertebrates, Vol. 5. Smaller Coelomate Groups.* New York: McGraw-Hill.

James, M.A., A.D. Ansell, M.J. Collins, G.B. Currey, L.S. Peck, and M.C. Rhodes. 1992. *Biology of Living Brachiopods.* Adv. Marine Biol. 28: 176–387.

Parker, S. P., ed. 1982. *Classification and Synopsis of Living Organisms,* Vol. 2, Phoronids, Bryozoans, and Brachiopods. New York: McGraw-Hill, 741, 743–69, 773–80.

Ryland, J. S. 1970. *Bryozoans.* London: Hutchinson Univ. Library.

Thorpe, J. H., and A. P. Covich, eds. 1991. *Ecology and Classification of North American Freshwater Invertebrates.* New York: Academic Press, 481–99 (bryozoans).

Woollacott, R. M., and R. Zimmer, eds. 1977. *Biology of Bryozoans.* New York: Academic Press.

Some General References About the Entoprocts

Harrison, F. W. and R. W. Woollacott. 1996. *Microscopic Anatomy of Invertebrates,* Vol. 13. Lophophorates and Entoprocta. Wiley-Liss, NY.

Hyman, L. H. 1951. *The Invertebrates, Vol. 3. Acanthocephala, Aschelminthes and Entoprocta.* New York: McGraw-Hill.

Nielson, C. 1964. Studies on Danish Entoprocta. *Ophelia* 1:1–76.

Parker, S. P., ed. 1982. *Classification and Synopsis of Living Organisms,* vol. 2. New York: McGraw-Hill, 771–72.

21

The Echinoderms

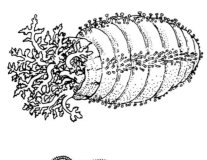

Introduction and General Characteristics

....

Phylum Echino · dermata
(G: spine skin)
ē-kīn´-ō-der-mah´-tah

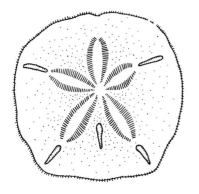

The phylum Echinodermata includes the sea lilies, feather stars, brittle stars, sea stars, sand dollars, sea urchins, sea biscuits, and sea cucumbers.

Nearly all of the approximately 6,000 echinoderm species living today are marine; a few species are estuarine, but none live in freshwater. An additional 13,000 or so species, distributed among approximately 20 classes, are known from the fossil record; most of those classes have no living representatives. Although their free-living larval stages are bilaterally symmetrical, most adult echinoderms show a basic five-point (pentamerous) radial symmetry. As radially symmetrical animals, they lack cephalization. Thus, adult echinoderms generally do not have anterior and posterior ends. Instead, body surfaces are designated as being either **oral** (bearing the mouth) or **aboral** (not bearing the mouth).

Most echinoderms possess a well-developed internal skeleton composed largely (up to 95%) of calcium carbonate, with smaller amounts of magnesium carbonate (up to 15%), even lesser amounts of other salts and trace metals, and a small amount of organic material. The components of the echinoderm skeleton are individually manufactured within specialized cells originating from embryonic mesoderm. This is in sharp contrast to the method of shell production in molluscs and other invertebrate groups, in which minerals are deposited into an extracellular protein matrix.

The major unifying characteristic of the phylum Echinodermata is the presence of what is known as the

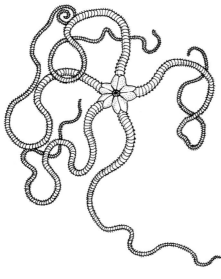

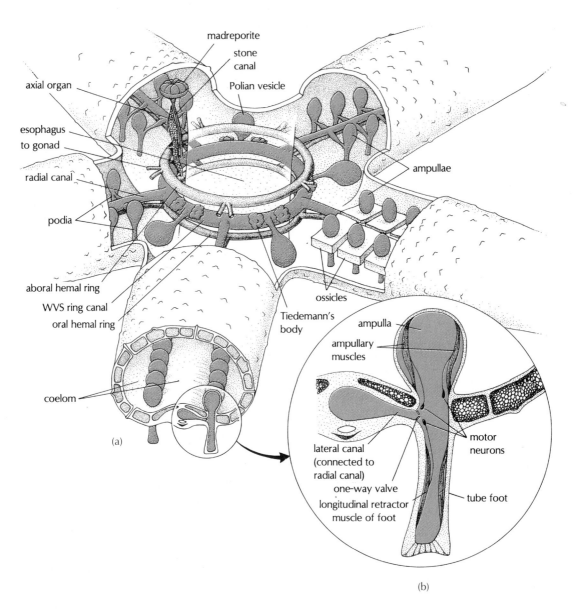

Figure 21.1
(a) Diagrammatic illustration of the echinoderm water vascular
system. The hemal system is also shown. (b) Detail of individual
tube foot.   (a) *Modified after Fretter and Graham, and others.* (b) *Modified after
Fretter and Graham; after Smith.*

water vascular system (often abbreviated as the **WVS**).
The WVS consists of a series of fluid-filled canals derived primarily from one of three pairs of coelomic
compartments (the hydrocoel) that form during embryonic development. These canals lead to thin-walled
tubular structures called **podia,** or **tube feet** (*podium* =
G: a foot) (Figure 21.1a.) The podia are best visualized
as tubular extensions of the WVS that penetrate the
echinoderm body wall and skeleton in particular regions, known as **ambulacral zones,** or, in some groups,
ambulacral grooves (*ambulacr* = L: walk). The system
of internal WVS canals is generally linked to the outside seawater through a sieveplate called the

madreporite, which leads down a **stone canal** (so
named because it is reinforced with spicules or plates
of calcium carbonate) and then to a **ring canal,** which
forms a ring around the esophagus in all but a few
echinoderm species (Fig. 21.1*a*). Accessory fluid-
storage structures called **Polian vesicles** and **Tiede-
mann's bodies** are often associated with the ring canal.
Five (or some multiple thereof) **radial canals** radiate
symmetrically from the ring canal. Pairs of bulb-
shaped **ampullae** usually connect to these radial
canals, with each ampulla servicing a single tube foot.
Both ampullae and tube feet are supported by a system
of calcareous ossicles, the **ambulacral ossicles.**

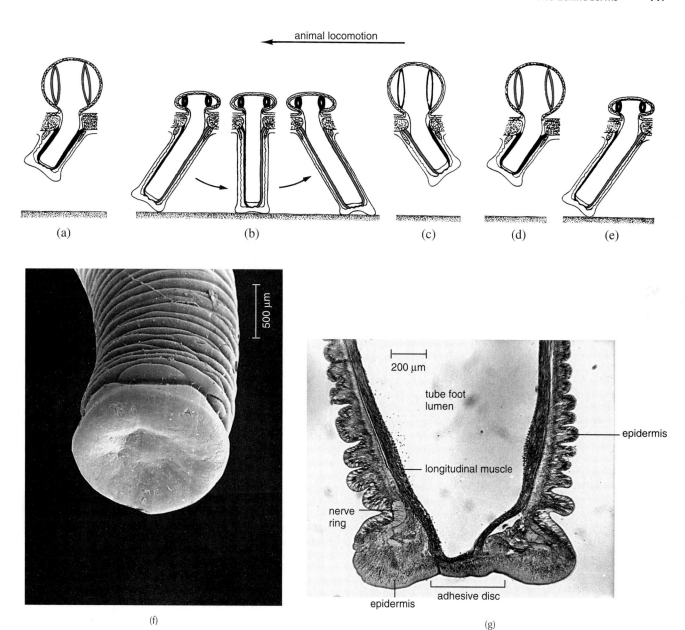

animal locomotion

(a)　　　(b)　　　(c)　　　(d)　　　(e)

500 μm

200 μm

tube foot
lumen

epidermis

longitudinal muscle

nerve
ring

epidermis

adhesive disc

(f)　　　(g)

Figure 21.2

Tube foot locomotion over a solid substrate. Blue shading indicates muscles about to contract to produce the next step in the sequence. (a) Ampullar muscles about to contract, elongating the tube foot by hydraulic action. (b) Base of tube foot applied to substrate, movement achieved by contracting longitudinal muscles on one side of tube foot. (c–e) Tube foot being prepared for the next step; note that fluid is pumped back into the ampulla as longitudinal muscles are contracted. (f) Distal portion of tube foot from the seastar *Marthasterias glacialis,* showing terminal adhesive disk. (g) Longitudinal section through distal end of a tube foot from the seastar *M. glacialis.*   *(f,g) Courtesy of P. Flammang. From P. Flammang et al., 1994. Biol. Bull, 187:35–47.*

Tube feet often lack circular muscles (Fig. 21.1b) and so cannot extend themselves. In most species, fluid is pumped into the tube foot by contraction of the ampulla, extending the foot hydraulically (Fig. 21.2). A one-way valve at the juncture of the ampulla and radial canal ensures that fluid flows from ampulla to tube foot, rather than to the radial canal, when the ampulla contracts. The tube foot retracts when longitudinal muscles in each tube foot contract.

A single echinoderm may possess more than 2,000 tube feet. Locomotion often requires the coordinated protraction and retraction of all those feet.[1] Tube feet

---

1. See *Topics for Further Discussion and Investigation,* no. 5, at the end of the chapter.

attach to solid substrates largely through ionic interactions, although suction is probably involved as well. Evidence of a duo-gland adhesion system also has been reported recently for several sea star species. One or more gland cells on each tube foot apparently secrete an adhesive that binds that tube foot to a substrate; adjacent gland cells then apparently release another chemical that somehow breaks those bonds. Similar duo-gland adhesive systems have been described from other animals as disparate as flatworms and scaphopod molluscs.

The inner surface of each tube foot is well ciliated, so the fluid in the WVS circulates. The thin-walled podia can thus function effectively as respiratory structures as well as locomotory structures, and the fluid of the WVS, together with that of the other coelomic compartments, serves as the primary circulatory medium. Tube feet also may be the primary sites of excretion (by simple diffusion), and in at least some groups, may function in chemoreception and food collection as well.

Specialized excretory organs are never found among adult echinoderms, although a cilia-driven nephridial system occurs in the larvae. A true heart is also absent.

Associated with the echinoderm WVS is a peculiar system of tissues and organs known as the **hemal system.** A major component of this hemal system is a spongy **axial organ** that lies adjacent to the stone canal of the WVS (Fig. 21.1a). The axial organ is housed in its own coelomic compartment, the **axial sinus,** and connects to two **hemal rings,** one oral and the other aboral. From the aboral hemal ring, strands of tissue, each contained within a coelomic, **perihemal canal,** extend outward to the gonads. Another series of strands radiates from the oral hemal ring to the tube feet.

The hemal system's functional significance has long been uncertain. Direct connections between the hemal system and the WVS, or between the hemal and digestive systems, do not appear to exist in most species. Recent studies with asteroids (sea stars) and holothurians (sea cucumbers) suggest that the hemal system functions in transporting nutrients from the coelomic fluid to the gonads. This conclusion is based in part on determining the sites of uptake of ^{14}C-labeled food material; radioactivity appeared sequentially in the digestive system, hemal system, and finally, in the gonads. In addition, concentrations of carbohydrates, lipids, proteins, and amino acids in hemal fluid may be 10 times greater than in other echinoderm fluids, indicating a high nutritional content. How nutrients move from the digestive system to the hemal system remains uncertain.

Morphological studies suggest that the axial organ of asteroids and echinoids may have an excretory function, although this has yet to be demonstrated experimentally.

The axial organ also may be involved in producing intriguing cells called **coelomocytes,** which are found in nearly all echinoderm tissues and body fluids, including the coelomic fluid. These coelomocytes are involved in recognizing and phagocytosing foreign material, including bacteria[2]; synthesizing pigments and collagen (for connective tissue); transporting oxygen (some coelomocytes contain hemoglobin) and nutritive material; and digesting food particles. They also play a role in wound repair. Most echinoderms have great, and sometimes extraordinary, regenerative capabilities.[3]

At least 85 echinoderm species are known to be toxic or venomous, although few are deadly to humans.

Class Crinoidea

Class Crin · oidea
(G: lily-like)
krī-noy´-dē-ah

The class Crinoidea is the oldest of the extant echinoderm classes, with a fossil record extending back nearly 600 million years; its members show many characteristics that, based upon studies of fossils, appear to be primitive. ("Primitive" characteristics are those that appear to show the least change from the presumed ancestral condition; the word does not imply lack of complexity.) The Crinoidea is comprised of the stalked crinoids (the sea lilies, of which about 80 species now exist, all in deep water) and the nonstalked, motile comatulid crinoids (the feather stars), of which about 550 living species are known. Crinoids were far more successful long ago, and indeed they make up the majority of fossilized echinoderms. Today, the most diverse assemblage of crinoid species is found on the Great Barrier Reef surrounding Australia, where more than 50 different species co-occur in some areas. All crinoids are suspension-feeders.

The sea lilies, although a small group at present, were quite numerous 300 to 500 million years ago.[4] Most fossil crinoids were, as the sea lilies remain today, permanently attached to the substrate by a **stalk** (Fig. 21.3). The stalk is flexible, being composed of a series of calcareous discs (**columnals**) stacked one on top of the other and held together by connective tissue.

The feeding and reproducing part of the animal is situated at the top of the stalk. The digestive system is tubular, consisting of a mouth, intestine, and terminal anus, and is confined entirely to a calyx/tegmen complex; the **calyx,** a cup-shaped structure containing the complete digestive system (Fig. 21.3a), is covered by a lid-forming

---

2. See *Topics for Further Discussion and Investigation*, no. 2.
3. See *Topics for Further Discussion and Investigation*, no. 1.
4. See *Topics for Further Discussion and Investigation*, no. 7.

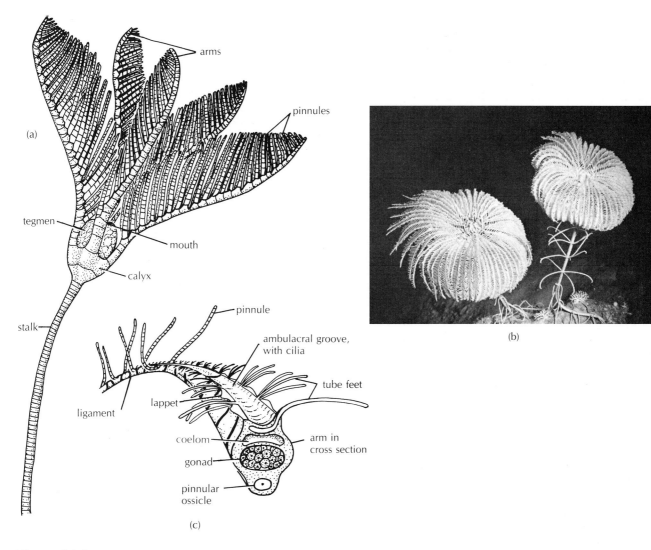

Figure 21.3
(a) Diagrammatic illustration of a stalked crinoid. (b) The stalked crinoid *Cenocrinus asterias*. (c) Detail of a crinoid arm. *(a) From Hyman,* The Invertebrates, *Vol. III. Copyright 1951 McGraw-Hill Book Company, New York. Reprinted by permission. (b) Courtesy of J. E. Miller, Harbor Branch Oceanographic Institution, Inc.*

membrane (the **tegmen**) that bears the mouth. The mouth, and therefore the oral surface in general, is always directed away from the base of the calyx. Crinoids are unique among echinoderms in having the oral surface on the upper half of the body, a clear adaptation for suspension feeding. Both the calyx and tegmen generally bear numerous protective calcareous plates on the outer surface.

From 5 to more than 200 arms (usually in some multiple of 5) extend outward from the sea lily calyx, and all bear tube feet (podia). These arms, like the stalk, consist of a series of jointed calcareous ossicles, so that the arms can bend.

Two rows of tubular **pinnules** extend outward from each arm, one row on each side of the ambulacral groove, which runs the length of the arm (Fig. 21.3c). Thin, elongated podia, grouped in triplets, also flank the ambulacral groove of each arm. These podia are studded with mucus-secreting glands. Crinoids collect food by extending the arms, pinnules, and tube feet into the surrounding water current,[5] the arms and pinnules being moved by contractions of extensor muscles and ligaments. As food particles in the water contact the podia, the particles are entangled in mucus and then flicked into the ambulacral grooves. The food particles are then

---

5. See *Topics for Further Discussion and Investigation,* no. 3.

transported to the mouth by ambulacral cilia. Note that crinoid tube feet have no locomotory function. Their use is limited to food collection, gas exchange, and, probably, the elimination of nitrogenous wastes by diffusion.

The sea lily stalk contains no musculature. Yet the animal can reorient its body quickly and repeatedly for more efficient food capture in response to temporary changes in current speeds and directions. Sea lilies accomplish this reorientation by rapidly altering the stiffness of the connective tissues holding the columnal discs of the stalk together. The transition between solid and fluid states in the connective tissue is under direct nervous control, and it takes less than one second to accomplish. This so-called "smart" connective tissue is so far known only among echinoderms.

Crinoid tube feet are never associated with ampullae, distinguishing the crinoids from most other echinoderms; crinoids protract their podia by contracting muscles in the radial canals. Crinoids also lack a madreporite, although stone canals opening into the coelom are numerous. The WVS opens to the outside through a large number of ciliated tubes penetrating the tegmen.

Feather stars, also referred to as **comatulids** after the name of the single order in which all are placed (order Comatulida), resemble the sea lilies from the calyx upward. However, in place of a long stalk, a series of jointed, flexible appendages called **cirri** occur near the base of the body (Fig. 21.4). The cirri are used to grasp solid substrates during periods of resting and feeding, which, for a comatulid, is most of the time. Comatulids are often observed perching atop sponges, corals, and other structures, living and nonliving; in this way, the feather stars extend their feeding appendages into the faster-moving water above the local substrate, increasing the frequency of food capture. Food collection is as described for stalked crinoids.

Feather stars locomote either by "snowshoeing" atop soft sediments using their cirri, or by swimming short distances above the substrate using forceful downward movement of the arms. A number of arms distributed uniformly around the calyx beat downward in unison; another group of arms then beats downward while the first group of arms makes its recovery stroke. Swimming thus involves a highly coordinated series of arm movements. The ability to move gives the comatulids a means of escaping or avoiding predators that is not available to the stalked crinoids.

Class Stelleroidea

> *Class Steller · oidea*
> (L: a star)
> stel-er-oy´-dē-ah

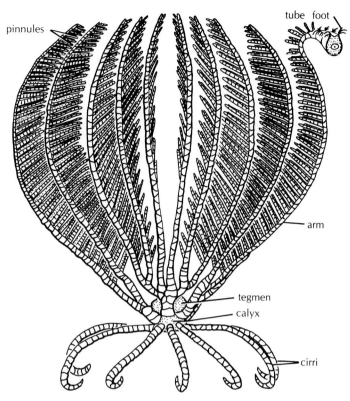

Figure 21.4
Diagrammatic illustration of a comatulid crinoid.   *Modified from Hyman; after Clark.*

The class Stelleroidea contains all other armed echinoderms—namely, the brittle stars (ophiuroids) and sea stars (asteroids). As the class and common names imply, these animals differ from the crinoids in lacking stalks and in having the arms arrayed star-like around a flattened body. Despite the pronounced morphological differences between brittle stars and sea stars that will be discussed shortly, early fossil remains indicate a sufficiently close evolutionary relationship to warrant grouping these animals in this one class. Asteroids and ophiuroids also share a peculiar arrangement of mitochondrial DNA (a conspicuous multigene inversion), supporting other evidence that members of the two classes are closely related. In this taxonomic arrangement, brittle stars and sea stars are placed in separate subclasses within a single class.

Subclass Ophiuroidea

> *Subclass Ophiur · oidea*
> (G: snake-like)
> ōf´-ē-yor-oy´-dē-ah

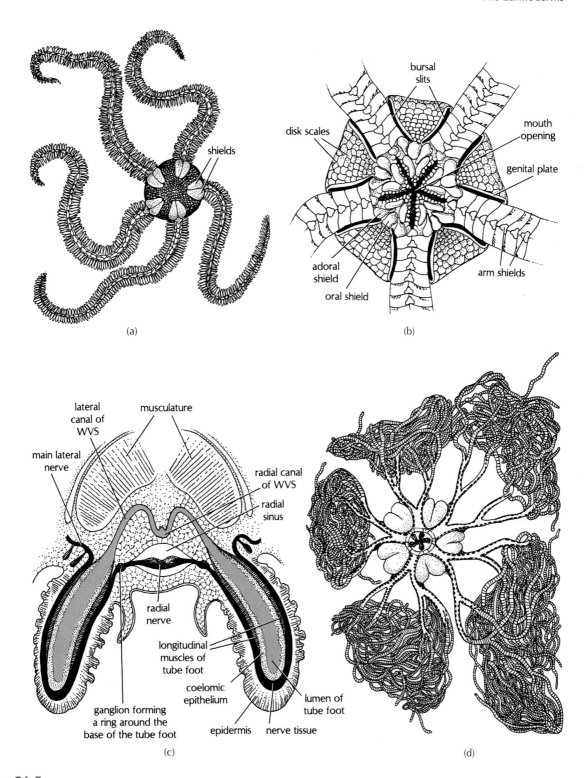

Figure 21.5

(a) An ophiuroid, *Ophiothrix fragilis,* seen from the aboral surface. (b) Oral view of the disc of *Ophiomusium sp.,* showing the location of the mouth opening and bursal slits. (c) Cross section through an arm of *Ophiothrix* sp. Note the absence of ampullae. (d) The basket star, *Gorgonocephalus* sp.   (a) *After Kingsley.* (b) *From Hyman, The Invertebrates, Vol. III. Copyright © 1951 McGraw-Hill Book Company, New York. Reprinted by permission.* (c) *From Hyman; after Cuenot.* (d) *After Pimentel.*

The approximately 2,100 species in the subclass Ophiuroidea are motile, as are all living echinoderms other than the stalked crinoids (sea lilies). As with the crinoids, arms extend from a central body, are built of jointed, vertebral ossicles (**vertebrae**), and are quite flexible (Fig. 21.5); the subclass is named in recognition of the snakelike movements made by these arms during locomotion. Even so, tube feet often play a role in ophiuroid movement, particularly in very young individuals, in species that remain small as adults, and in burrowing species.

The ophiuroids generally possess five long arms, radiating symmetrically from a small central disc (generally only a few centimeters in diameter). In some species (the basket stars), each arm branches several to many times (Fig. 21.5d). The discs of basket stars may reach 10 cm in diameter, and the outstretched arms may be about 1 m across! The common name for a typical ophiuroid is "brittle star," reflecting the tendency of the arms to detach from the central disc when provoked.[6] It is not uncommon to find 50% or more of the individuals in an ophiuroid population regenerating at least one arm, a process that requires several months to complete.

The oral surface of the ophiuroid body disc is often covered with a thin layer of minute, calcareous scales, while the body's aboral surface usually bears a number of protective calcareous plates, or **shields** (Fig. 21.5a). The arms are similarly encased in a series of endoskeletal ossicles. The bulk of each arm, however, consists of a series of thick, articulating, calcareous discs (the vertebral ossicles), with the tube feet (Fig. 21.5c) penetrating these "vertebrae" to the outside through a series of minute holes.

The ophiuroid digestive system, like that of the crinoids, is generally confined to the central disc. Unlike most other members of the Echinodermata, however, the ophiuroids possess only a single opening to the digestive system: A mouth is present, but an anus is lacking.

One final similarity between the crinoids and ophiuroids is that the podia are generally not operated by ampullae. In most other respects, the ophiuroid WVS follows the typical echinoderm pattern described earlier, except that a single ophiuroid may possess numerous madreporites, all of which open on the oral surface.

One feature found among most ophiuroids that distinguishes them from all other echinoderms is the occurrence of slit-shaped infoldings on the oral surface along the arm margins, adjacent to the arm shields. These 10 invaginations are known as **bursae** (Fig. 21.5b), and they project well into the coelomic space in the central disc. Seawater is constantly circulated through the bursae, presumably for gas exchange and, perhaps, for waste elimination. This circulation of external fluid is accomplished by cilia and, in some species, by muscular contractions. The bursae may also function in reproduction: In many species, the bursae serve as brood chambers within which the embryos develop.

Many ophiuroids are deposit feeders, ingesting sediment and assimilating the organic fraction. They also capture small animals in the sediment and ingest them individually. Some other brittle star species are suspension feeders, filtering food particles from the water, while still other species function as carnivores or scavengers. Ophiuroids typically hide under rocks or in crevices during the day, emerging to feed only at night; feeding aggregations of several thousand individuals per square meter have been reported from some shallow-water habitats. Many species live in association with other invertebrates, especially sponges and sessile cnidarians. A few tropical ophiuroid species recently have been shown to exhibit diurnal alterations in body coloration that are mediated by light-sensitive chromatophores.

Subclass Asteroidea

Subclass Aster · oidea
(G: star-like)
as-ter-oy´-dē-ah

Approximately 1,600 species of sea stars (often misleadingly termed "starfish") have been described from the living fauna, making the Asteroidea the second largest group (behind the Ophiuroidea) in the Echinodermata. The asteroids and the ophiuroids are superficially similar in that the members of both groups possess arms and a basically star-shaped body (Fig. 21.6). However, the arms of the sea stars are not distinct from the central body disc, and they do not generally play an active, direct role in locomotion. Moreover, there are important morphological and functional differences in the digestive system and WVS. Finally, adult asteroids tend to be larger than adult ophiuroids. Few sea stars are smaller than several centimeters in diameter; individuals of most species are about 15 cm to 25 cm in diameter, and some are considerably larger.

Locomotion in sea stars is slow and is accomplished by the highly coordinated activities of all the tube feet that radiate out along the oral surface of each arm. The WVS is essentially as described at the beginning of this chapter, with the madreporite opening on the aboral surface. The tube feet lie in distinct **ambulacral grooves** on the oral surface (Fig. 21.6b). Such grooves are not encountered among ophiuroids (or in any other group of echinoderms, other than the crinoids). Each tube foot is individually operated by an ampulla and generally terminates in a small suction cup at the distal end (Fig. 21.7). A given tube foot is extended through contraction of the associated ampulla, and it is swung forward or backward by contraction of the longitudinal musculature on one side or the other of the podium. The **ambulacral ossicles** apparently support the podia during these movements. Once the terminal portion of the podium contacts the substratum, slight contractions of the longitudinal muscles pull the central portion of the distal end of the podium upward, creating suction. Suction-cup locomotion seems best adapted for movement over firm substrates; the tube feet of asteroid species that move over or burrow into soft substrates do not terminate in suction cups.

The sea star mouth is directed downward, opens into a very short esophagus, and passes upward into a lower stomach (the **cardiac stomach**) (Fig. 21.8). This "lower" stomach is confined to the central body disc and is chiefly

---

6. See *Topics for Further Discussion and Investigation*, no. 1.

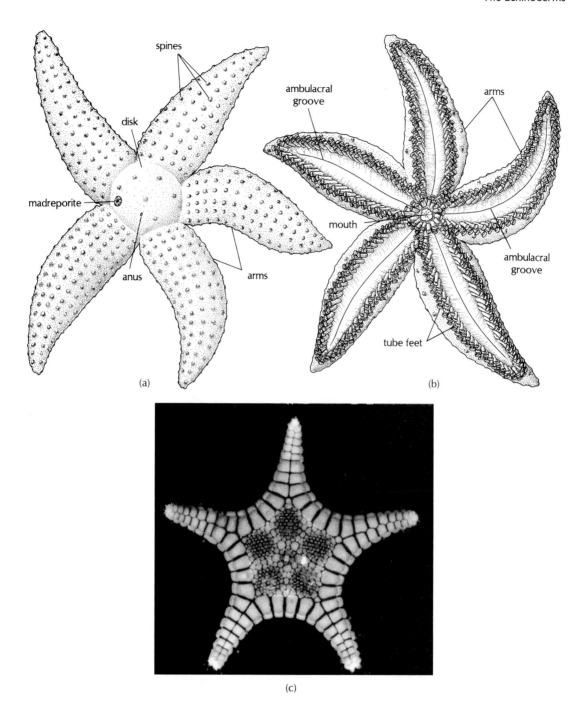

Figure 21.6

(a) A sea star, *Asterias vulgaris,* in aboral view. (b) Oral view of *Asterias vulgaris,* showing the tube feet in ambulacral grooves. (c) A deep-water sea star, *Rosaster alexandri.* Note the conspicuous tube feet extending from the five slender arms.

(c) *Courtesy of J. E. Miller, Harbor Branch Oceanographic Institution, Inc.* (a) *After Hyman.* (b) *After Sherman and Sherman.* (c) *Courtesy of J. E. Miller. Harbor Branch Oceanographic Institution, Inc.*

responsible for the digestion of food. Above the lower stomach is an upper or **pyloric stomach,** branches of which radiate out into each arm as **pyloric caeca.** The great surface area of the pyloric caeca, achieved through outfolding of the tissue, is in keeping with the primary functions of these organs: secretion of digestive enzymes and absorption of digested nutrients. The pyloric caeca are also primary storage sites for assimilated food. The anus lies on the aboral surface, nearly in line with the mouth.

Asteroids typically are predators on large invertebrates, including sponges, gastropods, polychaetes, bivalves, and other echinoderms.[7] A few species consume small fish. During feeding on large prey, the cardiac stomach of some species is actually protruded out of the body disc through

7. See *Topics for Further Discussion and Investigation,* no. 3.

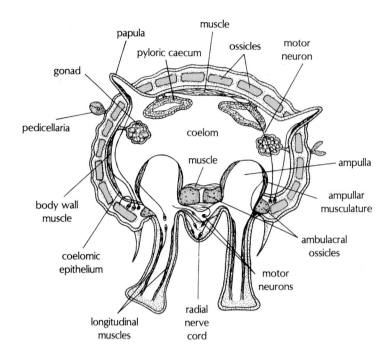

Figure 21.7
Cross section through the arm of an asteroid. Note that the podia protrude through minute pores in the ambulacral ossicles.

Other anatomical features, including the papulae and pedicellariae, are discussed later in the chapter.   *From Hyman; after J. E. Smith.*

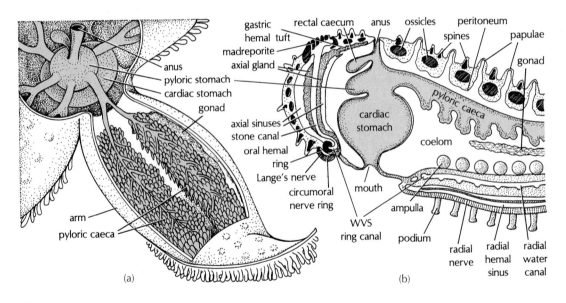

Figure 21.8
(a) Diagrammatic illustration of the digestive system of a sea star.
(b) Diagrammatic illustration of a sea star digestive system,
viewed laterally.   *(b) After Chadwick.*

the mouth and placed in contact with the prey's soft tissues. The cardiac stomach can be protruded through spaces as narrow as 1 mm to 2 mm (e.g., the gap between the shell valves of a nearly closed bivalve). Digestion in such species is frequently external, the resulting nutrient broth being transferred to the pyloric stomach by means of the ciliated channels of the cardiac stomach. No other echinoderms feed in

this manner. On the other hand, if the prey is sufficiently small, the stomach will be retracted while holding the victim, and digestion will proceed internally. Alternatively, the small prey may be ingested directly at the mouth, without the stomach having to be protruded at all. Some suspension-feeding asteroid species have also been described.

(a)

(b)

Figure 21.9
Autotomy in the asteroid, *Pycnopodia helianthoides*. Coelomic fluid from an autotomizing individual was injected into the intact specimen shown. The animal was photographed (a) 25 seconds and (b) 60 seconds later. *(a,b) Courtesy of V. Mladenov, from Mladenov et al., 1989. Biological Bulletin 176:169–175. Permission of Biological Bulletin.*

Many sea stars will sever, or **autotomize,** some of their arms if subjected to physical disturbance. This appears to be a form of escape, leaving potential predators with only a nutritious souvenir of the encounter. The response seems to be chemically mediated, since coelomic fluid from an autotomizing sea star induces autotomy if injected into another sea star, as shown in Figure 21.9. The lost arms are eventually regenerated.

The sea star's calcareous skeleton takes the form of discrete rods, crosses, and plates, which are embedded in connective tissue (Fig. 21.10). Ossicles other than those supporting the ambulacral grooves often bear outwardly directed calcareous spines that can be moved from side to side by muscles connecting to the underlying ossicles. The spines and outer surface of the skeletal ossicles are covered by a fairly thick cuticle that is secreted by the underlying, ciliated epidermis.

Asteroids possess two other types of appendages in addition to spines and podia. Thin, noncalcified outfoldings of the outer body wall serve a respiratory function. These structures, called **papulae,** are found protruding between ossicles and are connected directly with the main coelomic cavity. The second type of appendage is much more dynamic. These appendages, called **pedicellariae,** consist of two (sometimes three) calcium carbonate ossicles (**valves**), whose ends can be moved together or apart by muscles (Fig. 21.11). The two jaws are supported by a nonmovable, basal ossicle. The pedicellariae generally function in the removal of unwanted organisms and debris that contact the surface of the animal, and they also have been shown to capture living prey (including small fish!) in several asteroid species. Pedi-

cellariae are found in only one other echinoderm class, the Echinoidea, which includes the sea urchins and sand dollars.

Class Concentricycloidea

Class Concentri · cycloidea
L: concentric ring
kon-sen´-trē-sī-kloy´-dē-ah

"Many a flower was born to blush unseen. . . ."
(From Sir Thomas Gray, "*Elegy Written in a Country Churchyard*")

The last three classes of the Echinodermata remaining to be discussed consist of species that lack arms.

A marked departure from other echinoderm body plans was discovered in 1986 in a species collected from wood submerged in over 1,000 m of water off the coast of New Zealand. A second species now has been collected from wooden panels deliberately planted in the Bahamas, at a depth of about 2,000 m. Each animal is circular, flat, and without arms, and more resembles a jellyfish (medusa) or a flower than an echinoderm (Fig. 21.12). Recent studies show that the sperm of these newly discovered animals—with the nucleus and mitochondrion drawn out into long, thin threads and the elongated acrosome divided into numerous distinct segments—are unlike those known from any other echinoderms.

These so-called sea daisies, or concentricycloids, however, are indeed echinoderms and perhaps are best

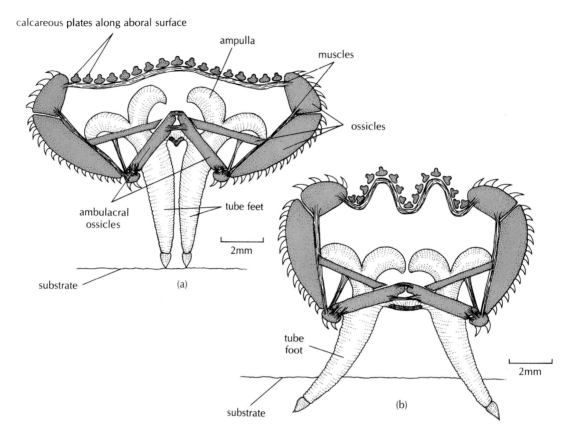

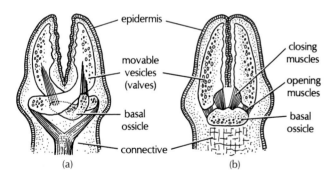

Figure 21.10
Transverse section through the arm of the asteroid *Astropecten irregularis*, showing the arrangement of calcareous ossicles and plates. The overlying cuticle has been omitted for clarity. In (a), the animal is walking over a substrate. In (b), the animal is burrowing. Note the changes in orientation of the ossicles, made possible by the elastic connective tissue that joins adjacent plates.
From D. Heddle, in Symposium of the Zoological Society of London, 20:125. Copyright © 1967 Zoological Society of London. Reprinted by permission.

Figure 21.11
Two types of pedicellariae found on the body of *Asterias* sp. In (a), the movable ossicles (valves) cross, as in the blades of a pair of scissors. In (b), the valves are straight and nearly parallel, and operate in the manner of forceps. In both types of pedicellariae, the basal ossicle supports the valves but does not itself move.
(a, b) *After Hyman.*

thought of as highly modified, armless sea stars. On the other hand, some biologists recently have argued that concentricycloids are at least as likely to have evolved from ophiuroid ancestors. In any event, their status as a separate class is controversial. The individuals so far discovered are all smaller than 1 cm in diameter and have been found exclusively in the crevices of submerged wood recovered from the deep sea.

Like most other echinoderms, sea daisies have a spacious coelom, a water vascular system complete with unmistakable tube feet serviced by individual ampullae (Fig. 21.12), and a calcareous endoskeleton of distinct ossicles. Unlike the tube feet of any other known echinoderm, however, those of the sea daisies are arranged in a single circle along the periphery of the animal and are connected to a double-ringed water vascular system (Fig. 21.12), for which the class is named. The two concentric water vascular rings are connected at intervals by short radial canals. In all other echinoderms, radial canals bearing the tube feet radiate from a single ring canal encircling the esophagus, as discussed earlier.

The sea daisy body is supported by a series of overlapping skeletal plates (ossicles) arranged in concentric rings, and the daisy's "petals" (Fig. 21.12) are actually calcareous spines. The oral surface of one species is covered by a thin sheet of tissue, the **velum,** further increasing the animal's superficial resemblance to a hydrozoan medusa. In that species, the adults have no mouth or gut; they are presumed to subsist on dissolved organic matter. Members of the second species have a large mouth and a

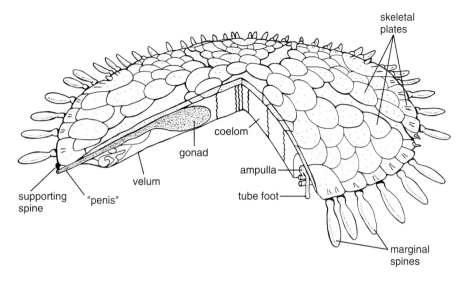

Figure 21.12

A male concentricycloid, *Xyloplax medusiformis*, with a portion cut away to show internal features.   *From F. W. E. Rowe, in Proceedings of the* Royal Society, London B. *233:431–459. Copyright © 1988 The Royal Society, London. Reprinted by permission.*

well-defined stomach, which is probably eversible and is presumed to function as in asteroids. The stomach ends blindly, as in ophiuroids; there is no anus. Neither species has any specialized gas exchange organs. Most details of the biology of these species have yet to be described.

Class Echinoidea

Class Echin · oidea
(G: spine-like)
ek-in-oy′-dē-ah

The Echinoidea include the sea urchins, heart urchins, and sand dollars, somewhat less than 1,000 species in total. The class is perhaps best represented by the sea urchins, which possess large numbers of long, rigid, calcium carbonate spines. The Greek word "*echinus*" means, literally, "a hedgehog." The spines serve for protection and, in some species, are actively involved in locomotion. Most sea urchins are free-living, roaming individuals, but a number of species bore into rock.

Echinoid spines attach to the underlying skeleton via ball-and-socket joints and can be declined rapidly in various directions by contracting specialized muscle fibers that connect between the ball (**tubercle**) and the spine (Fig. 21.13). The spines are often thin and sharp, but they are thick and blunt in a few species (Fig. 21.15*a,b*). They may function in bracing the animal when it wedges into crevices, in gathering and manipulating food, and in defense. Toxins may be extruded through the spines or from glands associated with the spines in some species, as many an unwary, warm-water

vacationer has discovered. If broken or damaged, the spines are replaced or repaired within a month or two.

The ossicles comprising most echinoid skeletons are flat and fused together, so ossicles cannot move relative to one another. In most species, the skeleton thus forms a solid, inflexible **test,** a feature that sets the typical echinoid apart from most other echinoderms. As the individual echinoid grows, the test is enlarged simultaneously in all directions much in the way that vertebrates enlarge their skulls during development to adulthood; the process involves the selective resorption and deposition of calcareous material by cells located in the sutures between ossicles.

Tube feet are widely distributed on the body, protruding through five double rows of pores in the **ambulacral plates** of the test (Fig. 21.14a). These pores are most easily seen if a denuded, empty, sea urchin test is held before a bright light. The body areas containing tube feet (i.e., the ambulacral plates) are distributed symmetrically about the body in strips extending orally/aborally. These regions are separated from each other by distinct **interambulacral** areas that are devoid of tube feet. The tube feet of echinoderms are especially well developed and generally bear suction-cup ends; the podia commonly function in locomotion, as in the Asteroidea.

Pedicellariae are also prominent echinoid appendages. However, they are generally borne on stalks and, unlike asteroid pedicellariae, are often equipped with calcareous support rods and bear three opposing jaws rather than two (Fig. 21.14c). Globular forms of pedicellariae, found in most urchin species, discharge poisons for defense (Fig. 21.14d). Some echinoid species

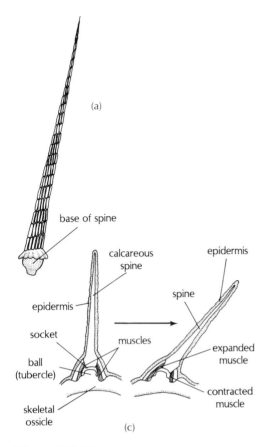

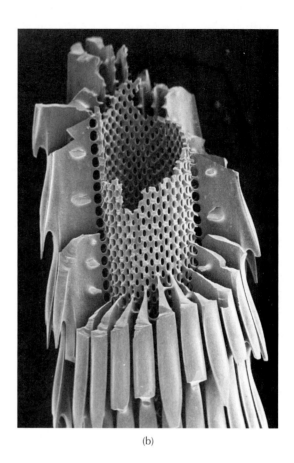

Figure 21.13

(a) A spine of the tropical sea urchin Diadema antillarus. (b) A spine of *Diadema setosum,* as seen with the scanning electron microscope. The spine has been broken to reveal the complexity of the internal construction. (c) Movement of a sea urchin spine.

(a,c) After Hyman. (b) Courtesy of K. Märkel from Burkhardt et al., 1983. Zoomorph. 102:189–203. Courtesy © Springer-Verlag Co.

have thin-walled outfoldings surrounding the mouth. These small appendages, called **gills,** presumably function in gas exchange.

Echinoids may be either regular or irregular (Fig. 21.15). **Regular urchins** have an almost perfect, spherical symmetry. All sea urchins fall into this group. Other echinoids are classified as **irregular** and display varying degrees of bilateral symmetry. This bilateral symmetry may be associated with a lifestyle of burrowing through sand, mud, or gravel. In association with the burrowing habit, the tube feet of irregular urchins tend to lack terminal suckers. In all irregular echinoids, ambulacral areas (and thus the tube feet) are restricted to the oral and aboral surfaces, rather than extending in an unbroken line from the oral to the aboral surfaces. The ambulacral areas on the aboral surface form a conspicuous five-pointed pattern, resembling the petals of a flower. The heart urchins have distinct anterior and posterior ends, with the mouth located anteriorly and the anus posteriorly. The spines of heart urchins are much more numerous than those of regular urchins, but they are also much shorter. Sand dollars (Fig. 21.15c) also bear very short spines, most likely

an adaptation for burrowing. Unlike the heart urchins, sea biscuits, and regular urchins, in which the aboral surface is convex, the test of most sand dollars is greatly flattened to form a very thin disc.

The echinoid systems for feeding and digestion differ significantly from those of all other echinoderm species. A complex system of ossicles and muscles, called **Aristotle's lantern,** surrounds the esophagus in all regular echinoids and in some irregular species as well (Fig. 21.16). The teeth of Aristotle's lantern can be protruded from the mouth and moved in various directions to scrape food, especially algae, from solid substrates. Detritus-feeding deep-sea urchins use the lantern to scoop mud. Some urchin species also are known to routinely consume small bivalved molluscs and other invertebrates. Echinoids lacking Aristotle's lantern generally feed on small organic debris, which is often collected by means of modified tube feet, spines, and/or external ciliary tracts.[8]

The echinoid stomach is not protrusible. Indeed, no echinoid has a true stomach, the esophagus leading instead

8. See *Topics for Further Discussion and Investigation,* no. 3.

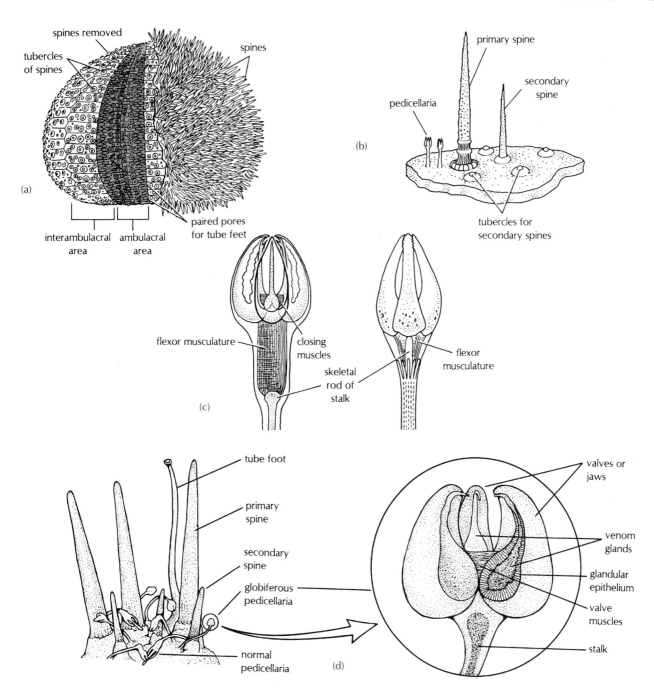

Figure 21.14

(a) A sea urchin, *Echinus esculentus,* with the spines removed from a portion of the test to reveal the ambulacral and interambulacral zones. Small pores, through which tube feet protrude in the living urchin, are seen in the ambulacral plates. Tubercules, upon which the spines pivot, are conspicuous on the interambulacral plates. The sea urchin test is comprised of five double rows of ambulacral plates and five double rows of interambulacral plates, for a total of 20 rows of plates. (b) Schematic illustration of an urchin test with appendages. (c) Pedicellariae from the echinoids *Strongylocentrotus droebochienis* (left) and *Eucidaris* sp. (right), showing the stalks and three-part jaws. (d) Globiferous pedicellaria, showing poison glands.   (b) *After Jackson.* (c) *From Hyman; after Mortensen.* (d) *From F. E. Russell, in J. H. S. Blaxter, F. S. Russell and M. Yonge, Eds., Advances in Marine Biology, 21:60-217, 1984. Copyright © 1984 Academic Press. Reprinted by permission of Harcourt Brace & Company Limited, London, England.*

into a very long, convoluted intestine (Fig. 21.17), where the food is both digested and absorbed (**assimilated**). The anus is located aborally and is surrounded by a series of plates comprising the **periproct** (G: around the anus).

Assimilated food passes into the echinoid coelomic fluid. The echinoid coelomic space is immense, particu-larly in the regular urchins (Fig. 21.17). Coelomic fluid is the principal transporter of both food and wastes; that is, the coelomic fluid is the primary circulatory fluid. The inner surface of the mesodermal lining of the coelomic cavity is ciliated, maintaining a constant movement of the circulatory medium.

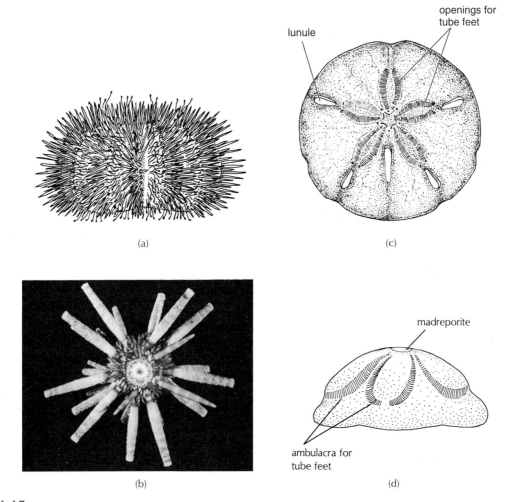

(a)

(b)

(c)

(d)

Figure 21.15

(a) A spiny urchin, a regular echinoid; see also Figure 21.17.
(b) *Eucidaris tribuloides,* a tropical pencil urchin; the common name refers to the animal's unusually thick, cylindrical spines. (c) A sand dollar, an irregular echinoid. (d) A sea biscuit, another irregular echinoid. Regular echinoids (the sea urchins) are characterized by pentamerous symmetry, a globular test, and long spines. Irregular echinoids (including the heart urchins, sea biscuits, and sand dollars) have a somewhat to very flattened test and relatively short spines, and tend toward bilateral symmetry; the periproct with anus is displaced posteriorly, and the mouth may be displaced anteriorly. The ambulacra of irregular species often resemble flower petals in outline as in (c) and (d).   (a,c,d) *After Brown.* (b) *Courtesy of J. E. Miller, Harbor Branch Oceanographic Institution, Inc.*

The WVS of echinoids follows the archetypical pattern closely, with a single madreporite opening aborally, as in the asteroids (Fig. 21.17). The Japanese (and people living near the Mediterranean Sea) prize echinoid gonads as an edible delicacy, for which they are willing to pay more than $100 per pound. As a result of overfishing and pollution, local Japanese urchin populations no longer meet the culinary demand; canned echinoid gonads are increasingly imported from the United States.

Class Holothuroidea

The class Holothuroidea (hol´-ō-thur-oy´-dē-ah) contains over 1,000 species, making it only slightly larger than the Echinoidea. Holothurians and echinoids resemble each other in lacking arms. In certain respects, the typical holothurian may be regarded as a flexible echinoid, whose morphological modifications reflect adaptation for a different lifestyle. Transforming an echinoid into a typical holothurian would first require removing

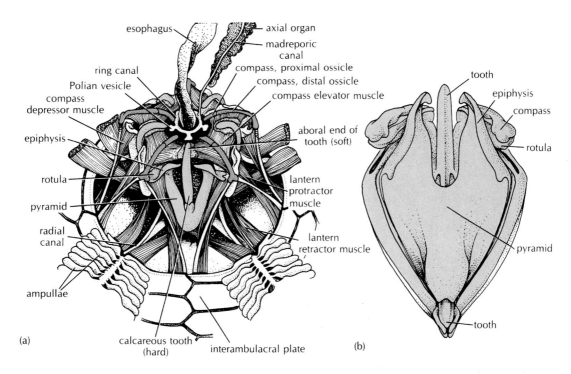

esophagus — **axial organ**
madreporic canal
ring canal — **compass, proximal ossicle**
Polian vesicle — **compass, distal ossicle**
compass depressor muscle — **compass elevator muscle**
epiphysis — **aboral end of tooth (soft)**
rotula — **lantern protractor muscle**
pyramid
radial canal — **lantern retractor muscle**
ampullae
(a) — **calcareous tooth (hard)** — **interambulacral plate**

tooth
epiphysis
compass
rotula
pyramid
(b) — **tooth**

Figure 21.16

(a) Aristotle's lantern and its associated complex musculature, from the sea urchin *Arbacia punctulata*. (b) Detail of the lantern ossicles, with the musculature omitted for clarity. Lantern morphology differs considerably among echinoids and is an important tool for species identification. The lantern consists of five bulky pyramids that support the five teeth; a series of bars (epiphyses) running along the aboral ends of the pyramids; a series of five, thin compass ossicles aborally; and a series of five similar pieces, the rotulas, lying below the compass. Protractor muscles push the teeth outward; retractor muscles move the teeth apart and draw them back into the test. The teeth are very hard at the tips but soft at the aboral end; new tooth material is formed continually at the aboral end, compensating for tooth wear distally.   (a,b) *After Brown.*

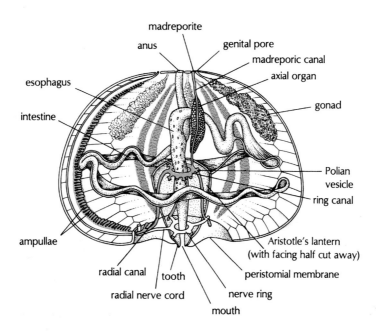

madreporite
anus — **genital pore**
madreporic canal
esophagus — **axial organ**
intestine — **gonad**
Polian vesicle
ring canal
ampullae — **Aristotle's lantern (with facing half cut away)**
radial canal — **peristomial membrane**
tooth — **nerve ring**
radial nerve cord
mouth

Figure 21.17

A sea urchin, *Arbacia punctulata*, seen laterally in diagrammatic section. Note the large coelomic space contained within the test.
After Brown; after Petrunkevitch.

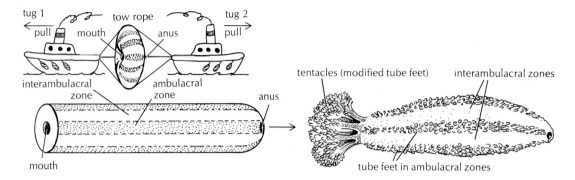

Figure 21.18
Relationship between the basic body plans of the echinoids and the holothurians. The holothurian shown at the end of the sequence is *Cucumaria frondosa.*   *After Hyman.*

all spines and pedicellariae and then discarding Aristotle's lantern. (This is a thought experiment, not something that echinoderms actually do!) Next, the ossicles of the test would have to be separated from each other and greatly reduced in size. This mutual detachment and reduction in size of the ossicles would render the body wall stretchable, since it would now be composed largely of connective tissue. From this point to a finished holothurian would be largely a matter of stretching the imaginary animal, increasing the distance between the oral and aboral surfaces (Fig. 21.18).

Thus, holothurians are typically soft-bodied, bilaterally symmetric, vermiform creatures with distinct anterior and posterior ends and with podia generally confined to distinct ambulacral strips (as in the Echinoidea) (Fig. 21.18). The calcareous ossicles, so conspicuous in other echinoderm classes, are microscopic in holothurians and embedded in the body wall; these ossicles are often exquisitely shaped (Fig. 21.19). Calcareous ossicles compose up to 80% of the total dry weight of the body wall in some species, while a few species lack ossicles entirely. The outer body wall is often warty and dark colored. Some species actually resemble very closely the fruit from which they derive their common name, "sea cucumber." Adults range from several centimeters to over 1 m in length. Strangely enough, cephalization is not pronounced among holothurians, despite the presence of a distinct anterior end.

A holothurian's oral tube feet are modified as large, often feathery tentacles at the anterior end; these tentacles can be protracted from the mouth and used to capture food (Figs. 21.18 and 21.20). Each tentacle may be operated by a single, large ampulla. In some species, the tentacles are coated with sticky mucus to trap food particles from suspension, but most sea cu-

cumber species are deposit feeders, ingesting sediment and extracting the organic component. Some holothurians have been estimated to pass over 130 kg of substrate through their digestive systems per year. As deposit feeders, sea cucumbers thrive on the fine mud ooze that characterizes the deep-sea floor; they make up more than 90% of the biomass in some abyssal habitats. Another group of deposit-feeders, the brittle stars (class Ophiuroidea), are also plentiful on the surface of such soft, deep-sea sediments.

The ambulacral tube feet of surface-living sea cucumber species generally bear suckers and are used in locomotion and attachment, as in the echinoids and asteroids.

The holothurian digestive system resembles that of the echinoids, except that it is greatly elongated (Fig. 21.20a). The WVS follows the typical echinoderm pattern, with the ring canal forming a ring about the esophagus, as usual. The ring canal is supported, however, by a calcareous ring, which may have an evolutionary origin in common with the Aristotle's lantern of echinoids. The madreporite generally lies free in the coelomic cavity (Fig. 21.20), so the WVS does not appear to be directly connected to the outside. The holothurian coelomic space is very large, as in the echinoids and asteroids, and the coelomic fluid is the primary circulatory medium. A few holothuroids also possess an extensive hemal system, with pulsatile hearts.

In contrast to other echinoderms, most holothurians have a body wall that contains well-developed layers of both circular and longitudinal musculature (Fig. 21.19a) and is considered by some to be an edible delicacy.

Clearly, holothurians have the characteristics necessary for operating a hydrostatic skeleton: a large, fluid-filled, constant-volume body cavity; a deformable body wall; and an appropriate musculature. It should not be surprising to learn, then, that members of many holothurian species

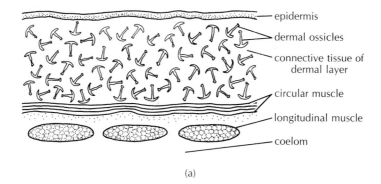

(a)

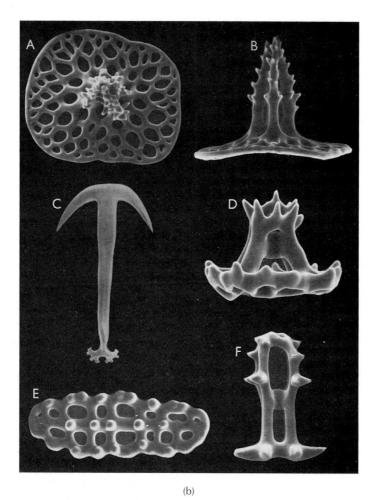

(b)

Figure 21.19

(a) Schematic cross section of the body wall of a typical holothurian, showing the arrangement of the musculature and location of the ossicles. Separation and size reduction of the ossicles permits holothurians to undergo major shape changes; the body wall, musculature, and coelom form a functional hydrostatic skeleton. (b) Scanning electron micrograph of the microscopic ossicles removed from the body walls of several holothurian species. Ossicle morphology plays a major role in species identification. (*A,B*) *Eostichopus regalis* (Cuvier), dorsal and lateral views; (c) *Euapta lappa* (Müller); (d) *Holothuria* (*Cystipus*) *occidentalis* Ludwig; (e) *Holothuria* (*Cystipus*) *pseudofossor* Deichmann; (f) *Holothuria* (*Semperothuria*) *surinamensis* Ludwig. These ossicles range from 60 μm to 400 μm (microns) in longest dimension.   (b) *Courtesy of Harbor Branch Oceanographic Institution, Inc. Fort Pierce, Florida.*

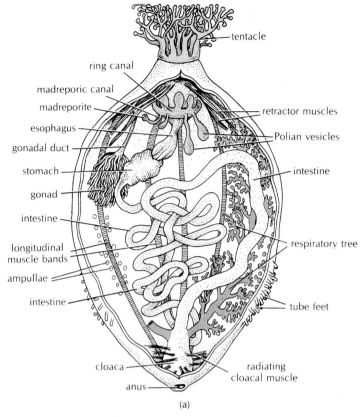

(a)

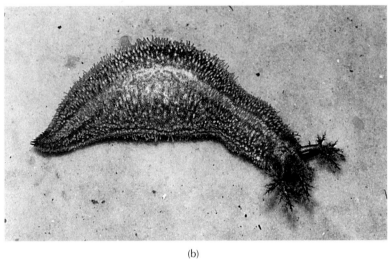

(b)

Figure 21.20
(a) Internal anatomy of the holothurian *Thyone briareus.* Note the extensive respiratory trees (only one shown), large coelomic space, and long intestine. (b) A mud-dwelling, burrowing holothurian, *Thyonella gemmata,* from Florida. Note the expanded tentacles, which are modified tube feet.   (a) *After Hyman; After Coe.*
(b) *Courtesy of Ralph Buschbaum.*

burrow in sand and mud (Fig. 21.21a). The tube feet of such species are often much reduced, and some of the more specialized burrowing species lack tube feet entirely. Locomotion in these burrowing species is accomplished in part by using the tentacles to push substrate away, but it primarily results from waves of contraction of the circular and longitudinal muscles, earthworm style.

Holothurians are the only echinoderms to possess truly specialized, internal respiratory structures, called **respiratory trees.** Most holothurian coeloms hold a pair of these highly branched, muscular structures. The respiratory trees connect to the cloaca, which pumps water into the trees. Water is expelled through the cloaca by contraction of the respiratory tree tubules themselves.

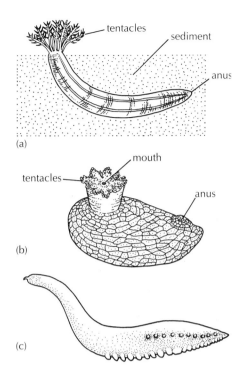

Figure 21.21

Holothurian diversity. (a) A burrowing holothurian, *Leptosynapta inhaerens,* in sand. This species is only a few centimeters long; has a very thin, flexible body wall; lacks respiratory trees; and except for the tentacles, lacks podia. (b) A highly modified holothurian, *Psolus fabricii.* The ventral surface is flattened, forming a "creeping sole." Podia are restricted to the flattened surface; the dorsal surface is covered with protective, calcareous scales. (c) A deep-sea holothurian, *Psychropotes longicauda,* with the tube feet modified as legs. Holothurians comprise one of the dominant elements of the deep-sea macrofauna. The specimen illustrated was obtained from a depth of perhaps 800 m; the body may reach nearly 0.5 m in length. Other deep-sea species, such as those in the genus *Scotoanassa,* are gelatinous and extremely fragile, and live several meters *above* the bottom.   (a,b) *Modified from Fretter and Graham, 1976. A Functional Anatomy of Invertebrates. Orlando: Academic Press, Inc. (c) From Marshall, Deep-Sea Biology. Copyright © 1979 Garland STPM Press, New York. Reprinted by permission.*

I have saved the best for last. Many holothurian species respond to a variety of physical and environmental factors by eviscerating. In some species, this is limited to the expulsion of incredibly sticky and/or toxic structures called **Cuvierian tubules,** which are attached to the respiratory trees and apparently used only for discouraging potential predators. In many other species, true **evisceration** occurs, in which the entire digestive system may be expelled, along with the respiratory trees and gonads. All lost body parts are eventually reformed, reflecting the substantial regenerative capabilities possessed by most echinoderms.[9]

Other Features of Echinoderm Biology

Reproduction and Development

Asexual reproduction is often encountered among echinoderms. Among asteroids and ophiuroids, the central disc separates into two pieces, and each piece proceeds to reform the missing arms and organs. At least one asteroid species reproduces asexually from pieces of the arms alone and asexual replication of larvae has been reported for several asteroid species. Holothurians also exhibit asexual replication; in a few species, the adult body routinely breaks in half, transversely, with each half-cucumber regenerating its missing parts. Asexual reproduction is unknown among crinoids and echinoids.

Most echinoderm species reproduce only sexually, and the sexes are usually separate. All echinoderms except holothurians and crinoids bear multiple gonads. In the asteroids, at least one pair of gonads extends into each arm. Concentricycloids possess 10 gonads (five pairs), which lie in the capacious coelomic cavity. In ophiuroids, from one to many gonads empty into each bursa. Echinoids usually have five gonads. The gametes of crinoids develop in tissues of the arms or pinnules, not in true gonads. Holothurians are unique among the Echinodermata in commonly possessing a single gonad.

In most echinoderm species, gametes are liberated into the surrounding seawater, so fertilization is typically external. Concentricycloids seem to be a particularly notable exception to this general rule of external fertilization.

9. See *Topics for Further Discussion and Investigation,* no. 1.

Research Focus Box 21.1

Influence of Pollutants

Canicatti, C., and M. Grasso, 1988. Biodepressive effect of zinc on humoral effector of the *Holothuria polii* immune response. *Marine Biol.* 99:393.

All animals, including the invertebrates, can distinguish between self and nonself. Among echinoderms, the immune response is mediated both by (1) humoral factors, which agglutinize (clump) or lyse (break apart) materials recognized as nonself, and (2) coelomocytes, which produce the lysins and also directly phagocytize or encapsulate foreign materials. Canicatti and Grasso (1988) set out to assess the effects of heavy metal pollution on the functioning of the self/nonself surveillance system. Heavy metals are major constituents of industrial effluents. Although numerous studies have considered the effects of various organic and inorganic pollutants on echinoderm survival, feeding, growth, reproduction, and development, the study by Canicatti and Grasso is one of the first to examine effects on the echinoderm immune response.

To conduct their study, the investigators drained the coelomic fluid from adult sea cucumbers (*Holothuria polii*) that were about 33 cm long. They then centrifuged the fluid to separate the liquid portion from the cellular constituents, including the coelomocytes. The pelleted cells were then opened by sonication, producing a "coelomocyte lysate." The ability of the centrifuged supernatant and coelomocyte lysate to agglutinate and lyse foreign cells was then assessed with and without the addition of zinc, cadmium, or mercury. Rabbit red blood cells (erythrocytes) served as foreign agents, to test the immune response.

Of the various heavy metals tested, only zinc affected the immune response, and the effect varied strikingly with concentration. Zinc concentrations of 1 mM (millimolar) or higher significantly depressed the lytic abilities of both the liquid and cellular fractions of the coelomic fluid (Focus Fig. 21.1). In contrast, lower zinc concentrations actually increased the lytic activity of the fluid fraction above control levels (Focus Fig. 21.1). Zinc concentrations up to 4 mM, the highest level tested, had no effect on the ability of coelomic fluid to agglutinate rabbit red blood cells; effects were limited to lytic activity.

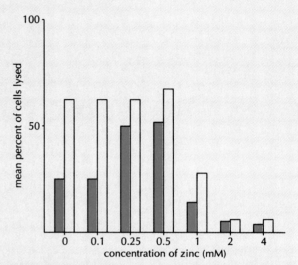

Focus Figure 21.1
Effect of dissolved zinc on the lytic activity of the liquid (shaded bars) and cellular (open bars) components of sea cucumber (*Holothuria polii*) coelomic fluid. Activity was tested against rabbit red blood cells, and each bar represents the mean of five replicates.

The data clearly indicate the ability of zinc pollution to alter a holothurian's immune recognition system. This raises a number of compelling questions: Would the researchers have seen the same depressive effect on the immune response if intact animals, rather than isolated cells and cellular products, were exposed to the same zinc levels? Do *organic* pollutants, such as fuel oils and pesticides, have comparable effects? Of the metals tested in this study, why did only zinc have an effect, and why did low concentrations of zinc *improve* the ability to recognize or attack foreign cells? How might suppression of the immune response affect the holothurian's survival in the field? Finally, does zinc have similar effects on the immune systems of other animals?

Figure from C. Canicatti and M. Grasso, "Biodepressive Effect of Zinc on Humoral Effector of the Holothuria polii Immune Response" in Marine Biology, 99:393–396, 1988. Copyright © 1988 Springer-Verlag, Heidelberg, Germany. Reprinted by permission.

The ducts leading out from the male testes continue beyond the margins of the circular body and are stiffened by peripheral, supportive spines (Fig. 21.12). These ducts may thus serve as copulatory organs, perhaps permitting internal fertilization.

With the exception of the concentricycloids, distinctive ciliated larval stages characterize each class, as illustrated in Figures 21.22 and 21.23. A delicate, internal, calcareous skeleton supports the larval arms in echinoids and ophiuroids. Echinoderm metamorphosis to the adult body plan is often dramatic, in terms of speed and the

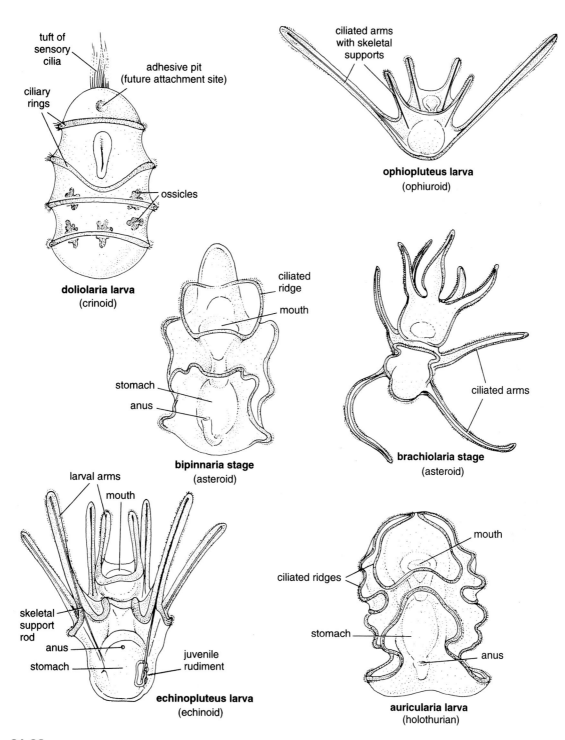

Figure 21.22
Larval forms typical of the various extant echinoderm classes.

complexity and magnitude of the morphological reorganization.[10] However, a number of species in most echinoderm classes exhibit severe reductions in the development of larval structures—particularly those associated with food collection—and an acceleration in the rate at which adult structures form, so embryos become adults more directly and often more quickly. Such modifica-

tions are often associated with some form of parental brood care. These evolutionary departures from the standard larval morphology seem to have evolved independently a number of times even within the various echinoderm classes. Echinoderms are thus providing particularly good experimental material for studying the underlying molecular mechanisms through which

10. See *Topics for Further Discussion and Investigation.* no. 6.

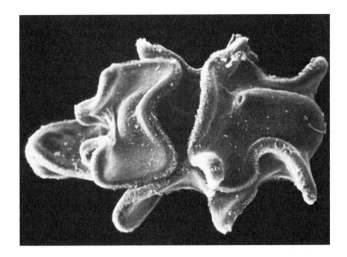

Figure 21.23
Scanning electron micrograph showing 10-day-old bipinnaria larva of the crown-of-thorns sea star, *Acanthaster planci*. This larva is about 750 μm long. *Courtesy of R. R. Olson.*

developmental patterns have become modified by selective forces. The development of prosobranch gastropods would also be well worth studying in this regard, but so far most of the action has been focused on the echinoderms and ascidians (Chapter 24).

Nervous System

Echinoderms do not have a centralized brain; nor are distinct ganglia generally found in them. Instead, the nervous system is composed of three diffuse nerve networks. An **ectoneural** system receives sensory input from the epidermis. This system, highly developed in all but the crinoids, consists of a ring around the esophagus with five associated nerves radiating outward. In species with arms, these radial nerve cords extend down each arm to the tube feet, ampullae (where present), and pedicellariae (where present). A second, **hyponeural** system is exclusively concerned with motor function. It, too, consists of a circumoral nerve ring with five associated radial nerves, but it lies deeper within the animal's tissues. The hyponeural system is well developed only in ophiuroids and, to a lesser extent, in asteroids. In crinoids, the major nerve network is an **entoneural** system (Fig. 21.24), associated with the animal's aboral end. From a central mass in the calyx/tegmen complex, nerves radiate down the stalk of the crinoid to the cirri and up into each arm. The entoneural system is inconspicuous or entirely absent in the other echinoderm classes.

Taxonomic Summary

Phylum Echinodermata
    Subphylum Crinozoa
        Class Crinoidea—the sea lilies and feather stars
    Subphylum Asterozoa
        Class Stelleroidea
            Subclass Ophiuroidea—the brittle stars
            Subclass Asteroidea—the sea stars
        Class Concentricycloidea—the sea daisies
    Subphylum Echinozoa
        Class Echinoidea—the sea urchins, heart urchins, and sand dollars
        Class Holothuroidea—the sea cucumbers

Topics for Further Discussion and Investigation

1. Investigate the factors controlling the loss and regeneration of body parts among echinoderms.

Anderson, J. M. 1965. Studies on visceral regeneration in sea stars. III. Regeneration of the cardiac stomach in *Asterias forbesi* (Desor). *Biol. Bull.* 129:454.

Carnevali, M. D. C., E. Lucca, and F. Bonasoro. 1993. Mechanisms of arm regeneration in the feather star *Antedon mediterranea*: Healing of wound and early stages of development. *J. Exp. Zool.* 267:299.

Dobson, W. E. 1985. A pharmacological study of neural mediation of disc autotomy in *Ophiophragmus filograneus* (Lyman) (Echinodermata: Ophiuroidea). *J. Exp. Marine Biol. Ecol.* 94:223.

Lawrence, J. M., T. S. Klinger, J. B. McClintock, S. A. Watts, C.-P. Chen, A. Marsh, and L. Smith. 1986. Allocation of nutrient resources to body components by regenerating *Luidia clathrata* (Say) (Echinodermata: Asteroidea). *J. Exp. Marine Biol. Ecol.* 102:47.

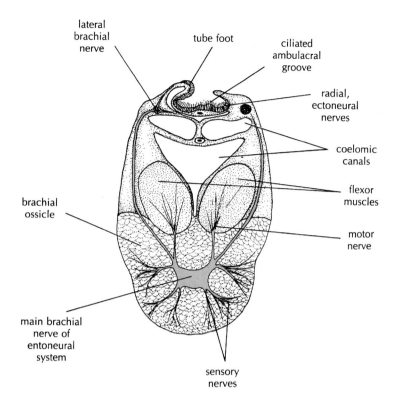

Figure 21.24
Cross section through the arm of a crinoid, showing the
arrangement of the nervous system.   *From Hyman; after Hamann.*

Mladenov, P. V., R. H. Emson, L. V. Colpit, and I. C. Wilkie. 1983.
Asexual reproduction in the West Indian brittle star *Ophiocomella
ophiactoides* (H. L. Clark) (Echinodermata: Ophiuroidea). *J. Exp.
Marine Biol. Ecol.* 72:1.

Mladenov, P. V., S. Igdoura, S. Asotra, and R. D. Burke. 1989.
Purification and partial characterization of an autotomy-
promoting factor from the sea star *Pycnopodia helianthoides. Biol.
Bull.* 176:169.

Sides, E. M. 1987. An experimental study of the use of arm
regeneration in estimating rates of sublethal injury on brittle-stars.
J. Exp. Marine Biol. Ecol. 106:1.

Smith, G. N., Jr. 1971. Regeneration in the sea cucumber
Leptosynapta. I. The process of regeneration. *J. Exp. Zool.* 177:319.

Wilkie, I. C. 1978. Arm autotomy in brittle stars (Echinodermata:
Ophiuroidea). *J. Zool., London* 186:311.

Zeleny, C. 1903. A study of the rate of regeneration of the arms in
the brittle star *Ophioglypha lacertosa. Biol. Bull.* 6:12.

**2.   What is the role of coelomocytes in mediating the
echinoderm immune response?**

Bang, F. B. 1982. Disease processes in seastars: A Metchnikovian
challenge. *Biol. Bull.* 162:135.

Canicatti, C., and A. Quaglia. 1991. Ultrastructure of *Holothuria
polii* encapsulating body. *J. Zool., London* 224:419.

Dan-Sohkawa, M., J. Suzuki, S. Towa, and H. Kaneko. 1993. A
comparative study on the fusogenic nature of echinoderm and
nonechinoderm phagocytes *in vitro. J. Exp. Zool.* 267:67.

Hilgard, H. R., and J. H. Phillips. 1968. Sea urchin response to
foreign substances. *Science* 161:1243.

Kaneshiro, E. S., and R. D. Karp. 1980. The ultrastructure of
coelomocytes of the sea star *Dermasterias imbricata. Biol. Bull.*
159:295.

Service, M., and A. C. Wardlaw. 1984. Echinochrome-A as a
bactericidal substance in the coelomic fluid of *Echinus esculentus*
(L.). *Comp. Biochem. Physiol.* 79B:161.

Yui, M. A., and C. J. Bayne. 1983. Echinoderm immunology:
Bacterial clearance by the sea urchin *Strongylocentrotus purpuratus.
Biol. Bull.* 165:473.

**3.   Discuss the role of tube feet, spines, pedicellariae,
cilia, and water flow in the collection of food by echinoderms.**

Burnett, A. L. 1960. The mechanism employed by the starfish
Asterias forbesi to gain access to the interior of the bivalve *Venus
mercenaria. Ecology* 41:583.

Chia, F. S. 1969. Some observations on the locomotion and
feeding of the sand dollar, *Dendraster excentricus* (Eschscholtz). *J.
Exp. Marine Biol. Ecol.* 3:162.

De Ridder, C., M. Jangoux, and L. De Vos. 1987. Frontal
ambulacral and peribuccal areas of the spatangoid echinoid
Echinocardium cordatum (Echinodermata): A functional entity in
feeding mechanism. *Marine Biol.* 94:613.

Ellers, O., and M. Telford. 1984. Collection of food by oral surface
podia in the sand dollar, *Echinarachnius parma* (Lamarck). *Biol.
Bull.* 166:574.

Emson, R. H., and J. D. Woodley. 1987. Submersible and laboratory observations on *Asteroschema tenue*, a long-armed euryaline [*sic*] brittle star epizoic on gorgonians. *Marine Biol.* 96:31.

Emson, R. H., and C. M. Young. 1994. Feeding mechanism of the brisingid starfish *Novodinia antillensis. Marine Biol.* 118:433.

Fankboner, P. V. 1978. Suspension-feeding mechanisms of the armoured sea cucumber *Psolus chitinoides* Clark. *J. Exp. Marine Biol. Ecol.* 31:11.

Ghiold, J. 1983. The role of external appendages in the distribution and life habits of the sand dollar *Echinarachnius parma* (Echinodermata: Echinoidea). *J. Zool., London* 200:405.

Hendler, G., and J. Miller. 1984. Feeding behavior of *Asteroporpa annulata,* a gorgonocephalid brittle star with unbranched arms. *Bull. Marine Sci.* 34:449.

Hendler, G. 1982. Slow flicks show star tricks: Elapsed-time analysis of basketstar (*Astrophyton muricatum*) feeding behavior. *Bull. Marine Sci.* 32:909.

Holland, N. D., A. B. Leonard, and J. R. Strickler. 1987. Upstream and downstream capture during suspension feeding by *Oligometra serripinna* (Echinodermata: Crinoidea) under surge conditions. *Biol. Bull.* 173:552.

Lasker, R., and A. C. Giese. 1954. Nutrition of the sea urchin, *Strongylocentrotus purpuratus. Biol. Bull.* 106:328.

LaTouche, R. W. 1978. The feeding behavior of the feather star *Antedon bifida* (Echinodermata: Crinoidea). *J. Marine Biol. Assoc. U.K.* 58:877.

Leonard, A. B. 1989. Functional response in *Antedon mediterranea* (Lamarck) (Echinodermata: Crinoidea): The interaction of prey concentration and current velocity on a passive suspension-feeder. *J. Exp. Marine Biol. Ecol.* 127:81.

Macurda, D. B., and D. L. Meyer. 1974. Feeding posture of modern stalked crinoids. *Nature (London)* 247:394.

Mauzey, K. P., C. Birkeland, and P. K. Dayton. 1968. Feeding behavior of asteroids and escape responses of their prey in the Puget Sound region. *Ecology* 49:603.

Meyer, D. L. 1979. Length and spacing of the tube feet in crinoids (Echinodermata) and their role in suspension-feeding. *Marine Biol.* 51:361.

O'Neill, P. L. 1978. Hydrodynamic analysis of feeding in sand dollars. *Oecologia* 34:157.

4. Seawater contains fairly high concentrations (up to 3×10^3 g carbon/liter) of dissolved organic matter (DOM), especially in shallow coastal waters. To what extent are echinoderms capable of meeting their nutritional needs through the uptake of DOM directly from seawater?

Ferguson, J. C. 1980. The non-dependency of a starfish on epidermal uptake of dissolved organic matter. *Comp. Biochem. Physiol.* 66A:461.

Fontaine, A. R., and F. S. Chia. 1968. Echinoderms: An autoradiographic study of assimilation of dissolved organic molecules. *Science* 161:1153.

Hammond, L. S., and C. R. Wilkinson. 1985. Exploitation of sponge exudates by coral reef holothuroids. *J. Exp. Marine Biol. Ecol.* 94:1.

Manahan, O. T., J. P. Davis, and G. C. Stephens. 1983. Bacteria-free sea urchin larvae: Selective uptake of neutral amino acids from seawater. *Science* 220:204.

Shilling, F. M., and D. T. Manahan. 1990. Energetics of early development for the sea urchins *Strongylocentrotus purpuratus* and *Lytechinus pictus* and the crustacean *Artemia* sp. *Marine Biol.* 106:119–27.

Stephens, G. C., M. J. Volk, S. H. Wright, and P. S. Backlund. 1978. Transepidermal accumulation of naturally occurring amino acids in the sand dollar, *Dendraster excentricus. Biol. Bull.* 154:335.

5. How are individual echinoderm tube feet operated and integrated into the total behavior of the animal?

Binyon, J. 1964. On the mode of functioning of the water vascular system of *Asterias rubens* L. *J. Marine Biol. Assoc. U.K.* 44:577.

Kerkut, G. A. 1953. The forces exerted by the tube feet of the starfish during locomotion. *J. Exp. Biol.* 30:575.

Lavoie, M. E. 1956. How sea stars open bivalves. *Biol. Bull.* 111:114.

Polls, I., and J. Gonor. 1975. Behavioral aspects of righting in two asteroids from the Pacific coast of North America. *Biol. Bull.* 148:68.

Prusch, R. D., and F. Whoriskey. 1976. Maintenance of fluid volume in the starfish water vascular system. *Nature (London)* 262:577.

Smith, J. E. 1947. The mechanics and innervation of the starfish tube foot-ampulla system. *Phil. Trans. Royal Soc. B* 232:279.

Thomas, L. A., and C. O. Hermans. 1985. Adhesive interactions between the tube feet of a star fish, *Leptasterias hexactis,* and substrata. *Biol. Bull.* 169:675.

6. What major morphological changes take place as echinoderms metamorphose from larval to adult form?

Cameron, R. A., and R. T. Hinegardner. 1978. Early events in sea urchin metamorphosis, description and analysis. *J. Morphol.* 157:21.

Emlet, R. B. 1988. Larval form and metamorphosis of a "primitive" sea urchin, *Eucidaris thouarsi* (Echinodermata: Echinoidea: Cidaroida), with implications for developmental and phylogenetic studies. *Biol. Bull.* 174:4.

Hardy, A. 1965. Pelagic larval forms. *The open sea: Its natural history.* Boston: Houghton Mifflin, 178–98.

Hendler, G. 1978. Development of *Amphioplus abditus* (Verrill) (Echinodermata: Ophiuroidea). II. Description and discussion of ophiuroid skeletal ontogeny and homologies. *Biol. Bull.* 154:79.

Mladenov, P. V. M., and F. S. Chia. 1983. Development, settling behaviour, metamorphosis and pentacrinoid feeding and growth of the feather star *Florometra serratissima. Marine Biol.* 73:309.

Smiley, S. 1986. Metamorphosis of *Stichopus californicus* (Echinodermata: Holothuroidea) and its phylogenetic implications. *Biol. Bull.* 171:611.

7. The echinoderms have an extensive fossil record, in keeping with their ancient origin, marine habitat, and solid skeleton. In fact, a number of classes are known only from the fossil record, there having been no living representatives of these classes for tens of thousands of years. Most of these extinct echinoderms were sessile animals, permanently attached to a substrate. What are the

morphological similarities and differences between these extinct species and the present-day sea lilies (class Crinoidea)?

Boardman, R. S., A. H. Cheetham, and A. J. Rowell, eds. 1987. *Fossil Invertebrates.* Palo Alto, Calif.: Blackwell Scientific Publications.

Clarkson, E. N. K. 1986. *Invertebrate Paleontology and Evolution,* 2d ed. Boston: Allen and Unwin.

Hyman, L. H. 1955. *The Invertebrates,* vol. IV. New York: McGraw-Hill.

Taxonomic Detail

Phylum Echinodermata

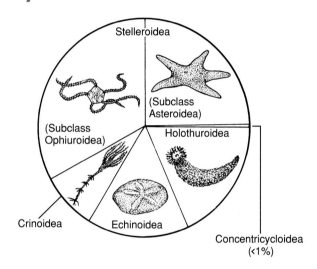

Subphylum Crinozoa

Representatives of most classes are extinct and are known only as fossils; existing species are contained within a single class. All are marine.

Class Crinoidea.
The sea lilies and feather stars. The approximately 620 crinoid species are distributed among five orders, most of which contain only one to three families and 25 or fewer species.

Order Millericrinida.
Ptilocrinus. These stalked sea lilies lack cirri and are mostly restricted to depths of 2,000 m or more. One family.

Order Cyrtocrinida.
Holopus. These sea lilies live attached to the substrate with a very short stalk, or none at all, and lack cirri. They are restricted to intermediate depths (several hundred meters) in the Caribbean. One family.

Order Bourgueticrinida.
Bathycrinus, Rhizocrinus. These stalked sea lilies lack cirri and have 5 or 10 unusually short arms. The group includes the deepest known crinoids (members of the genus *Bathycrinus*), which extend to depths of nearly 10,000 m. Three families.

Order Isocrinida.
Cenocrinus, Neocrinus, Metacrinus. These sea lilies have long stalks (up to 1 m). Many species live permanently attached to hard substrates, anchored at the base of their stalks. The other species attach to firm substrates using prehensile cirri found at intervals along the stalk; these cirri can release their grip on the substrate, allowing the animals to then walk or, in some species, swim to other sites. Most species are confined to depths of several hundred meters or more; all are located in the Caribbean and tropical Indo West Pacific. Three families.

Order Comatulida.
The feather stars. *Antedon, Comantheria, Comanthina, Florometra, Heliometra.* This highly diverse and colorful group of crinoids contains about 88% of all crinoid species. They are widely distributed, being especially common both on coral reefs and in polar waters; representatives occur at all depths, from less than 1 m to the abyss. One of the largest feather star species, *Comanthina schlegelii,* has up to 200 arms, with the actual number being highly variable among individuals.

Subphylum Asterozoa

The approximately 3,600 species, distributed among two classes, are all marine.

Class Stelleroidea.
All but a few asterozoan species are included in this class. Three subclasses.

Subclass Somasteroidea.
The members of this subclass share characteristics with both ophiuroids and asteroids, indicating a close relationship between these two major groups. Most representatives are known only as fossils.

Subclass Ophiuroidea.
The brittle stars and basket stars. The approximately 2,100 species are distributed among three orders.

Order Phrynophiurida.
Five families.

Family Gorgonocephalidae. *Gorgonocephalus.* This group, which includes the basket stars (species with highly branched arms), is distributed from the tropics to the poles and from shallow water to the abyss. Individuals often occur in dense aggregations, perched on solid substrates with the oral surface facing up and the arms fully extended into the surrounding water; these animals are suspension feeders. Including the arms, some individuals grow to diameters exceeding 70 cm. About 100 living species are known.

Order Ophiurida.

Amphiura, Amphipholis, Ophiura, Ophiomusium, Ophiocoma, Ophioderma, Ophiacantha, Ophiactis, Ophiopholis, Ophiothrix. This is the largest order of brittle stars, including over 2,000 described species distributed among 11 families. Representatives are found in all oceans, at all depths.

Subclass Asteroidea.

The sea stars. Nearly 1,600 species have been described, distributed among five orders.

Order Platyasterida.

Luidia. This is the most primitive extant group of asteroids, with a fossil record extending back nearly 500 million years. The stars in this order were formerly placed in the now abandoned order Phanerozonida. Most species are found on sandy substrates in shallow, tropical waters. At least some species are euryhaline, capable of living in water of substantially reduced osmotic concentration. The free-swimming larvae produced by some species of *Luidia* are among the largest of all known invertebrate larvae, attaining lengths of 2 cm or more, and seem capable of remaining in the water for more than a year before metamorphosing; the larvae are unusual in reproducing more larvae by asexual fission while still planktonic. Some zoologists consider *Luidia* an aberrant member of another order, the Paxillosida. One family.

Order Paxillosida.

Astropecten; Ctenodiscus—mud stars (they ingest sediment). This large order of nearly 400 described species is composed of primitive stars living in or on sediment. Not surprisingly for animals that do not walk on solid substrates, the tube feet lack suckers. Many species are exclusively abyssal, to depths of nearly 8,000 m. *Luidia* may belong to this order (see order Platyasterida). Five families.

Order Valvatida.

Goniaster, Linckia, Odontaster, Oreaster. Some species grow to diameters of only about 20 mm, while others reach diameters of nearly 50 cm. The order contains one of the largest asteroid families, the Goniasteridae, which contains about 300 described species. Species in the genus *Linckia*

show remarkable regenerative powers, even in comparison with most other asteroids: An entire animal can develop from a single arm. Nine families.

Order Spinulosida.

Acanthaster—the crown-of-thorns sea star, a voracious predator on coral, with toxic spines and 9 to 23 arms; *Echinaster; Henricia*—blood stars; *Pteraster; Solaster*—sun stars, with as many as 15 arms. Twelve families.

Order Forcipulata.

Heliaster—another group of sun stars, with up to 50 arms per animal; *Zoroaster*—rat tail stars. Four families.

> **Family Asteriidae.** *Asterias, Leptasterias, Pisaster, Pycnopodia.* This is the largest of all asteroid families, containing more than 300 species, most of which live on hard substrates. Members of the genus *Pycnopodia* possess as many as 40,000 tube feet distributed among as many as 25 arms; these unusually large stars can attain diameters exceeding 1 m.

Order Brisingida.

Brisinga, Novodinia. Sea stars in this order are common in the deep sea. Members of the genus *Brisinga* may exceed 1 m in diameter, including the arms; these stars possess unusually long tube feet, which they use for suspension feeding. Members of the genus *Novodinia* have been observed capturing prey using pedicellariae.

Class Concentricycloidea.

Xyloplax—the sea daisies. These deep-water echinoderms seem related to the asteroids and are possibly most closely related to members of the asteroid order Valvatida; their status as a separate class is controversial. Only a few species have been described to date, all associated with submerged wood. One family.

Subphylum Echinozoa

Two classes.

Class Echinoidea.

The urchins, sea biscuits, and sand dollars. Thirteen orders, many with few surviving species.

Order Cidaroida.

Eucidaris—pencil urchins. These widespread urchins are an ancient group, with fossilized representatives extending back over 300 million years. Each ambulacral plate is penetrated by a single tube foot, which sets these urchins apart from all others. All are nonselective scavengers. The nearly 140 species that have been described are distributed among two families.

Order Echinothuroida.

These are all deep-water species (1,000–4,000 meters depth), with a delicate, highly flexible test. They may reach 70 cm in diameter. All live on soft sediments, often in large aggregations. All species studied to date develop as highly modified, nonfeeding larvae, and are all contained in one family.

Order Diadematoida.

Diadema, Echinothrix. These are common in warm, shallow waters, particularly on coral reefs. All representatives have extremely long, sharply pointed spines. The spines are not venomous, but you will sincerely regret stepping on them. Four families.

Order Arbacioida.

Arbacia. One of the several dozen species in this group (*A. punctulata*) has been used widely in research by developmental biologists. One family (Arbaciidae).

Order Temnopleuroida.

Lytechinus, Tripneustes. This group contains over 100 species, mostly in shallow tropical waters. One Pacific species (in the genus *Toxopneustes*) is venomous, although not deadly to humans; the pedicellariae are unusually large and capable of piercing the skin. Two families.

Order Echinoida.

Echinus, Paracentrotus, Echinometra, Strongylocentrotus, Heterocentrotus. Several species of *Strongylocentrotus* have long been used in studies of developmental processes. This order includes the only known suspension-feeding urchin, *Derechinus horridus;* the animal remains stationary on rocks in the subantarctic, and grows only in height, projecting a large surface area into the water for particle capture. Four families, based largely on morphological differences in the pedicellariae.

Order Holectypoida.

Echinoneus. This order contains some of the most widely distributed shallow-water tropical echinoid species. One family.

Order Clypeasteroida.

Clypeaster—sea biscuits; *Dendraster, Echinarachnius, Mellita*—sand dollars. These echinoids possess numerous but very short spines with which they burrow into sand and mud substrates. All species deposit-feed and display varying degrees of bilateral symmetry, as in heart urchins. Because of this bilateral rather than radial symmetry, the members of this order, together with those in the order Spatangoida, are often referred to as **irregular urchins.** Nine families.

Order Spatangoida.

Brissopsis, Echinocardium, Meoma, Moira, Spatangus—heart urchins; *Pourtalesia*—bottle urchins. These deposit-feeding echinoids show a distinct bilateral symmetry, as in sand dollars, and spines and tube feet on different parts of the test are specialized for either burrowing or food collection. As with the other irregular urchins (contained in the order Clypeasteroida), the spines of heart and bottle urchins are very short, with more the appearance of fur than of spines. Fourteen families, including a number of poorly known deep-water groups.

Class Holothuroidea.

The sea cucumbers. About 1,100 species have been described, distributed among six orders.

Order Dendrochirotida.

Most of the 400 or so species in this order are restricted to shallow waters, and all possess oral tentacles that are elaborately branched. The tentacles trap food particles from suspension in the surrounding water and push them into the mouth. Seven families.

Family Placothuriidae.   *Placothuria.* These are strange cucumbers with many primitive features. The body is U-shaped and enclosed completely in a test of large, overlapping plates. Living representatives are known only off the coast of New Zealand.

Family Psolidae.   *Psolus.* This is a widespread group, with representatives extending from the intertidal zone to depths of about 2,800 m. As an adaptation for life on hard substrates, they have a distinct, flattened, ventral surface that lacks ossicles. The rest of the body is enveloped by a test of overlapping plates. About 80 species are known.

Family Cucumariidae.   *Cucumaria.* The more than 150 described species in this group have thick-walled but flexible bodies that contain numerous small and separate calcareous ossicles. The group is especially well represented in temperate and cold waters, but it is distributed worldwide and at all depths.

Order Aspidochirotida.

Holothuria, Stichopus, Parastichopus, Thelenota. Most species are nonselective deposit feeders. Some are commercially important as a food source in the Far East, being fished and sold as "beche-de-mer" or "trepang"; only the body wall is eaten. Some individuals reach lengths of 1 m to 2 m, with correspondingly impressive diameters, maintaining a cucumber physique. More than a third of the species in this order are restricted to deep waters, and at least a few (such as *Bathyplotes natans*) are good swimmers. Three families.

Order Elasipodida.

Psychropotes, Pelagothuria. The hundred or so deposit-feeding species in this group are widespread in deep waters and, in fact, may constitute about 95% of the total biomass in some abyssal habitats. They are remarkable in a number of respects: None have respiratory trees; the bodies are typically gelatinous and extremely fragile; all show pronounced bilateral symmetry; and many species swim, either some or all of the time. Five families.

Order Apodida.

Chiridota, Euapta, Leptosynapta. The several hundred species in this order are characterized by the complete absence of tube feet (*a* = G: without; *poda* = G: feet) and respiratory trees. Some representatives also lack body wall ossicles. In all, the body wall is very thin and worm-like. Most species live in shallow water, burrowed in sand or sequestered under rocks. The sticky tentacles, which are not serviced by ampullae, exhibit characteristic feeding movements in which they take turns carefully and deliberately wiping food particles into the mouth. Many species possess minute, attached, internal sacs (vibratile urnae) within the coelom that remove foreign particles from coelomic fluid. Three families.

Order Molpadiida.

Molpadia, Caudina. These deposit-feeding cucumbers are mostly confined to shallow waters, although some species live at depths to about 8,000 m. All have respiratory trees. Four families.

Some General References About the Echinoderms

Binyon, G. 1972. *Physiology of Echinoderms.* New York: Pergamon Press.

Boardman, R. S., A. H. Cheetham, and A. J. Rowell, eds. 1987. *Fossil Invertebrates.* Palo Alto, Calif.: Blackwell Scientific, 550–611.

Gage, J. D., and P. A. Tyler. 1991. *Deep-Sea Biology: A Natural History of Organisms at the Deep-Sea Floor.* New York: Cambridge University Press.

Harrison, F. W., and F. S. Chia, eds. 1994. *Microscopic Anatomy of Invertebrates, Vol. 14. Echinoderms.* New York: Wiley-Liss.

Hyman, L. H. 1955. *The Invertebrates, Vol. IV: Echinoderms.* New York: McGraw-Hill.

Lawrence, J. M. 1987. *A Functional Biology of Echinoderms.* London: Croom Helm.

Nichols, D. 1962. *The Echinoderms.* London: Hutchinson University Library.

22

The Chaetognaths

Introduction and General Characteristics

Phylum Chaeto · gnatha
(G: bristle jaw)
kēt-og-nath´-ah

Chaetognaths are all free-living, marine carnivores. The average chaetognath is only a few centimeters long, and even the largest individuals are no more than 15 cm long. Apparently, the ratio between surface area and volume is sufficiently large that gas exchange and excretion requirements can be met by diffusion across the general body surface; chaetognaths bear no specialized respiratory or excretory organs, and they lack a blood circulatory system. The inner lining of the adult body cavity is ciliated, however, so that gas exchange and nutrient transport are achieved by circulation of the coelomic fluid.

Typically, the nearly transparent chaetognath lies motionless in the water, sinking slowly until something edible comes its way. The chaetognath then darts forward and grasps its prey with two rows of long, curved, stiff spines adjacent to the mouth (Fig. 22.1). Two rows of short teeth on either side of the mouth aid in holding prey during ingestion. These teeth also can puncture the prey exoskeleton and body wall; recent studies indicate that toxic secretions may exit from the vestibular pits, or from pores along the vestibular ridge, and enter prey through these punctures (Research Focus Box 22.1).

Scattered fan-like clusters of external cilia are sensitive to vibration, enabling chaetognaths to detect the presence and location of the copepods and fish larvae that constitute their principal foods. The large, grasping spines also may serve as mechanoreceptors.

Chaetognaths bear a pair of small eyes, each composed of five pigment-cup ocelli. The ocelli are oriented in several different directions within a single eye, so the

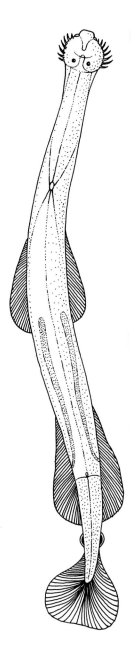

Research Focus Box 22.1

Chaetognath Prey Capture

Thuesen, E. V., K. Kogure, K. Hashimoto, and T. Nemoto. 1988. Poison arrowworms: A tetrodotoxin venom in the marine phylum Chaetognatha. *J. Exp. Marine Biol. Ecol.* 116:249–256.

For many years, biologists have suspected that prey capture among chaetognaths is facilitated by paralytic toxins, but the animals' small size made the hypothesis impossible to demonstrate: If the entire animal is only perhaps a few centimeters long, how does one isolate, and then hope to characterize, what are likely to be minute amounts of toxic secretions? Thuesen et al. (1988) developed techniques sufficiently sensitive to conclusively demonstrate that at least six chaetognath species do indeed produce a potent neurotoxin, one that paralyzes victims by blocking sodium channels in cell membranes.

To isolate any toxins, the researchers collected adult chaetognaths, decapitated them, and then extracted potential toxins by bathing the heads in a solution of 0.1% acetic acid. Remaining tissues were discarded, and the extract was concentrated by evaporation. To bioassay for the presence of neurotoxins, the researchers used tissue cultures of mouse nerve cells (neuroblastomas). The approach was both clever and simple, and it made use of the known physiological effects of two other chemicals: ouabain and veratridine.

Ouabain disengages the sodium-potassium pump that maintains the normal resting potential of all eukaryotic cells. With the pump operating normally, sodium ions are actively expelled while potassium ions are simultaneously carried in. In the presence of ouabain, the pump is deactivated, and sodium ions flood into cells along their concentration gradient, a response that is further amplified by the second chemical, veratridine. Thus, when the two chemicals are added to the nerve cells in tissue culture, sodium ions flood into the cells, and the cells quickly swell and die. In the presence of a neurotoxin that blocks sodium channels, however, the cells should be spared since sodium ions will not be able to move across the cell membrane. Thuesen et al. could therefore show that the chaetognath head contains neurotoxins by demonstrating that the extract blocks the effects of ouabain and veratridine.

This was precisely the effect shown by head extracts of the six chaetognath species tested; in contrast, extracts of headless chaetognaths did not save the cells from the effects of ouabain and veratridine.

Do all chaetognaths produce neurotoxins? The chaetognaths tested in the Thuesen et al. study included both benthic and pelagic species from three different families and a wide range of geographical distributions (Focus

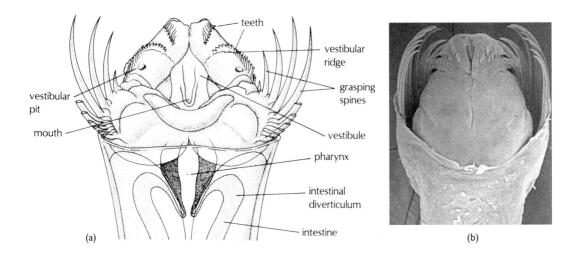

(a)   (b)

Figure 22.1

(a) The anterioir end of *Parasagitta elegans*, in ventral view. The eyes are located dorsally and are not visible in this orientation. The vestibular pits and the papillae of the vestibular ridges are probably secretory. The spines and teeth of the head are normally covered by a hood, which is retracted only for prey capture. (b) Scanning electron micrograph of *Sagitta setosa*, anterior end.   (a) From Hyman; after Ritter-Zahony. (b) Courtesy of Q. Bone et al., 1983. Journal of the Marine Biological Association of the United Kingdom, *Vol. 63:929. Reprinted with the permission of Cambridge University Press.*

Focus Table 22.1 Habitat and Geographical Distribution of Chaetognaths Shown to Use Neurotoxins in Prey Capture

Family and Species	Habitat	Geographical Distribution
Eukrohniidae		
Eukrohnia hamata	Open ocean	Polar waters, deep tropical waters
Sagittidae		
Parasagitta elegans	Coastal	Arctic waters
Flaccisagitta scrippsae	Open ocean	North Pacific
F. enflata	Open ocean	Tropical waters
Aidanosagitta crassa	Coastal	Japan
Spadellidae		
Spadella angulata	Benthic	Japan, Southeast Asia

From E. V. Thuesen, K. Kogure, K. Hashimoto, and T. Nemoto. 1988. Poison arrowworms: A tetrodotoxin venom in the marine phylum Chaetognatha. *J. Exp. Marine Biol. Ecol.* 116:249–56. Copyright © 1988 Elsevier Science Publishers, Amsterdam. Reprinted by permission.

Table 22.1), suggesting that the phenomenon is widespread. Secretion of neurotoxins certainly would explain the ability of many chaetognaths to capture larval fish as large as themselves.

Where is the toxin produced and secreted? The researchers suggest that the most likely sites are the papillae of the vestibular ridges adjacent to the mouth (Fig. 22.1), since venom secreted here would have ready access to piercing wounds made by the teeth. How could you demonstrate whether this is indeed the secretion site?

chaetognath has a wide visual field; in fact, several ocelli point downward, so the chaetognath actually sees through its own, transparent body. The ocelli are probably not image forming, but they may enable the animal to detect motion and changes in light intensity.[1]

Chaetognaths are commonly called "arrow worms," since the body is substantially longer than it is wide and lacks appendages, although it does bear one or two pairs of delicate, **lateral fins** and a terminal **caudal fin** (Fig. 22.2). The lateral fins probably serve to reduce the rate at which the animal sinks while motionless, by increasing body surface area, and to stabilize the body when the animal swims. The caudal fin generates forward thrust for swimming and also must aid in flotation. The entire body is covered by a thin cuticle secreted by the underlying epidermis.

The body cavity of adult chaetognaths is compartmentalized. One septum isolates the head compartment from that of the trunk. A second septum divides the trunk compartment from that of the tail (Fig. 22.2).

Most chaetognath species are planktonic, spending their entire lives being passively transported by ocean currents. Although, as planktonic animals, arrow worms cannot swim against any substantial water current, many species can and do undertake extensive vertical migrations (Fig. 22.3), sometimes traveling hundreds of meters daily. Many other planktonic animals, including pteropods (Gastropoda), copepods (Crustacea), and a variety of invertebrate larval stages display similar patterns of behavior. Generally, the animals migrate downward during the day and swim upward at night, although the timing and extent of the migrations often differ significantly among species and among different developmental stages of a given species. No single rationale for such **diurnal vertical migrations** has been demonstrated conclusively

---

1. See *Topics for Further Discussion and Investigation*, no. 1, at the end of the chapter.

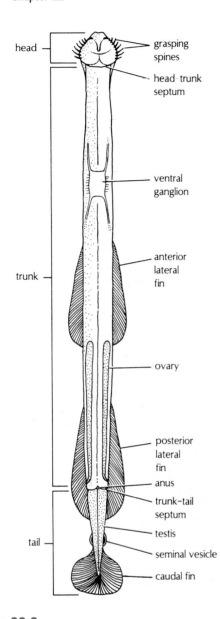

Figure 22.2
Parasagitta elegans, seen in ventral view.   *From Hyman; after Ritter-Zahony.*

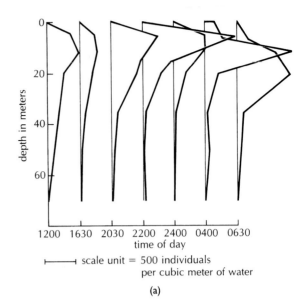

(a)

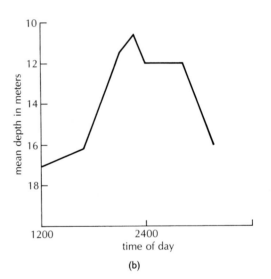

(b)

Figure 22.3
Diurnal vertical migration by the chaetognath *Sagitta elegans.*
(a) Changes in numbers of individuals per unit volume of seawater at different depths over time. Note that a pronounced population peak begins to form toward the surface in the early evening and that the average depth of this peak shifts downward toward early morning. (b) The average depth for the population over time, from the data in (a). The population clearly migrates upward toward evening and downward in the early morning.
(a,b) From Pearr, in Ecology, 54:300, 1973. Copyright © 1973 Ecological Society of America. Reprinted by permission.

in laboratory or field experiments, although several reasonable adaptive benefits have been proposed: (1) avoiding visual predators, (2) increasing the energetic efficiency of feeding and digestion, (3) reducing competitive interactions both within and among species, (4) improving physiological well-being by experiencing varied temperatures and salinities, and (5) achieving dispersal by taking advantage of vertical differences in speeds and directions of water currents.[2] The most convincing evidence presented to date (field studies on certain planktonic crustaceans) best supports the antipredation hypothesis.

One family of chaetognaths (the Spadellidae) is entirely benthic. These arrow worms use specialized **adhesive papillae** to form temporary attachments to solid substrates, usually rocks and/or macroalgae. In the typical feeding posture the attachment is made posteriorly, with the rest of the body held elevated above the substrate. When the need arises, the attachment can be broken and the animal can dart off to a new location.

Only about 100 chaetognath species are known (Fig. 22.4). This phylum is nevertheless an important one, ecologically and economically. Local concentrations as

---

2. See *Topics for Further Discussion and Investigation,* no. 2.

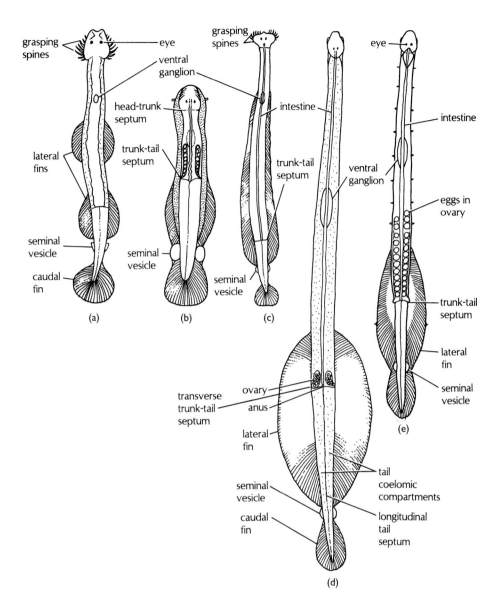

Figure 22.4
Chaetognath diversity. Note the differences in the number and extent of the lateral fins and in the external appearance of the seminal vesicles. (a) *Sagitta macrocephala*. (b) *Sagitta cephaloptera*.

(c) *Eukrohnia fowleri*. (d) *Krohnitta subtilis*. (e) *Krohnitta pacifica*.
(d,e) *From Hyman; after Tokioka*.

high as several hundred chaetognaths per cubic meter of seawater have been reported. As significant predators of fish embryos, fish larvae, and copepods, chaetognaths must be important components of oceanic food chains and, in particular, must play a major role in determining the size of herring and other commercially exploitable fish populations.

Chaetognath bioluminescence was described for the first time in 1994, for one deep water species rarely found at depths shallower than 700 m. This species luminesces using the same chemical substrate found in other marine animals as disparate as cnidarians, squid, and crustaceans. Bioluminescence is known from no other chaetognath species.

Chaetognaths are also intriguing for their enigmatic place in the phylogenetic scheme. Their relationship to other animal groups has long been uncertain. The adult body cavity is lined with mesodermally derived peritoneum and is thus a true coelom, by definition. Studies of chaetognath embryology establish the arrow worms as deuterostomous coelomates in that cleavage is basically radial and indeterminant (i.e., cell fates are not irrevocably fixed following the first cell division) and the site of the blastopore gives rise to the anus; as in other deuterostomes, the mouth arises elsewhere (recall that *deuterostome* = G: second mouth). Moreover, the embryonic coelom arises from an archenteron, although, in detail, the method of coelom formation by chaetognaths differs

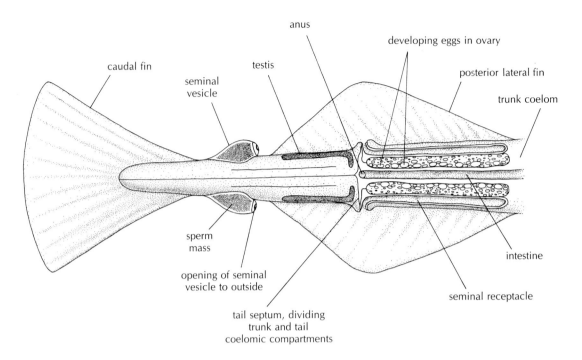

Figure 22.5
Ventral view showing reproductive system of *Parasagitta elegans*. Chaetognaths are simultaneous hermaphrodites, possessing both male and female gonads. The sperm are packed into spermatophores within the seminal vesicles. Sperm received from another individual are stored in the seminal receptacles.
From Frank Brown, Selected Invertebrate Types. Copyright © John Wiley & Sons, Inc., New York. Reprinted by permission of Mrs. Frank Brown.

significantly from the basic deuterostome plan. As discussed in Chapter 2, the standard deuterostome coelom arises from a symmetrical outpouching of the archenteron. In chaetognaths, on the other hand, the coelom is formed by invagination of the archenteron. Nevertheless, coelom formation is clearly enterocoelous, rather than schizocoelous, in nature.

In many other respects, however, chaetognaths show very weak affinities with other deuterostomes. Unique among deuterostomes is the absence of a ciliated larval stage in the life history; the morphology of the young chaetognath closely resembles that of the adult. The body cavity and musculature of adult chaetognaths are also unlike those of other deuterostomes. In fact, the only major, conspicuous morphological similarity with other deuterostomes is the division of the adult body cavity into three distinct compartments. The trunk and tail compartments are clearly illustrated in Figure 22.5; the head contains a small, additional coelomic space. For many years, it was believed that the embryonic coelom was obliterated during subsequent development, with the adult body cavity forming later and by an entirely different mechanism. However, recent studies indicate that the embryonic coelomic cavities are simply compressed for a time and later reopen in the juvenile; in this respect, at least, chaetognath development is not as strange as had been thought. Still, if chaetognaths are indeed valid deuterostomes, their evolutionary relationship to echinoderms and other deuterostome groups is obscure, to say the least.

The musculature of chaetognaths bears some resemblance to that encountered within the pseudocoelomate phylum Nematoda, and some workers have in fact allied the chaetognaths with pseudocoelomates. No circular muscles are found among the members of either phylum. The musculature of the chaetognath body wall is exclusively longitudinal and, like that of the nematodes, is arranged in discrete bundles. Alternating contraction of the ventral and dorsal musculature of the chaetognath trunk and tail provides the thrust for locomotion, the two sets of muscles antagonizing each other through the hydrostatic skeleton of the fluid-filled, compartmentalized body cavity. Unlike nematodes, however, chaetognaths do not generate sinusoidal waves of activity. Instead, muscle contractions are sporadic, so the animal darts forward intermittently. Unlike pseudocoelomates, chaetognaths are not eutelic (i.e., chaetognaths grow primarily through increases in cell numbers, not increases in individual cell size).

No unambiguous chaetognaths are preserved as fossils, so nothing about this group's evolutionary origins can be learned from the fossil record. Chaetognath embryology clearly marks this phylum as deuterostome. However, like the lophophorates, chaetognaths must have diverged from the major line of deuterostome evolution very early in the evolutionary history of metazoans. The present resemblance between adult chaetognaths and adult pseudocoelomates appears to reflect convergent evolution rather than any direct phylogenetic relationship.

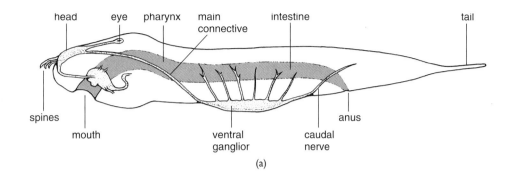

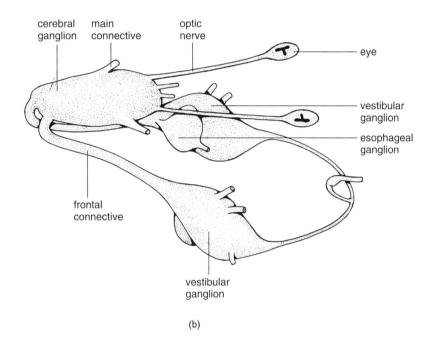

Figure 22.6
(a) Nervous system of a typical chaetognath. (b) Detail of nervous system in the head of a typical chaetognath. *(a,b) From T. Goto and M. Yoshida, in Nervous Systems in Invertebrates, M. A. Ali, ed. Copyright © 1987 Plenum Publishing Corporation, New York. Reprinted by permission.*

Other Features of Chaetognath Biology

Reproduction

All chaetognaths are simultaneous hermaphrodites (Fig. 22.5), with the male gonads generally maturing earlier than the female gonads. The ovaries are found in the coelomic compartment of the trunk, while the testes occupy the coelomic compartment of the tail. Self-fertilization occurs at least occasionally in some species. There is no copulation during cross-fertilization; instead, sperm transfer is indirect, being mediated through an often mutual exchange of sperm-filled containers called **spermatophores.** The spermatophores are manufactured within a conspicuous pair of seminal vesicles and are then attached to the outside of the recipient's body. Once

sperm escape from the spermatophores, they migrate along the body to the opening of the female reproductive tract. Fertilization is internal, after which the fertilized eggs are released and develop into miniatures of the adult. Chaetognaths do not have a morphologically distinct larval stage in their life history.

Digestion

The chaetognath gut is linear and open at each end (Fig. 22.4). Digestion is extracellular and occurs within the intestine. A distinct stomach is lacking.

Nervous System

The chaetognath nervous system is fairly complex (Fig. 22.6). The anterior part of the intestine is encircled by a ring of nervous tissue bearing several ganglia. Posteriorly, there is a conspicuous ventral ganglion. Sensory and motor nerves extend from this and associated ganglia to the various light, tactile, and chemosensory systems; to the musculature of the trunk, tail, spines, and digestive tract; and to the cerebral ganglion located in the head.

Topics for Further Discussion and Investigation

1.   Discuss the evidence suggesting that chaetognaths can orient to light and to vibrations.

Feigenbaum, D., and M. R. Reeve. 1977. Prey detection in the Chaetognatha: Response to a vibrating probe and experimental determination of attack distance in large aquaria. *Limnol. Oceanogr.* 22:1052.

Goto, T., and M. Yoshida. 1983. The role of the eye and CNS components in phototaxis of the arrow worm, *Sagitta crassa* Tokioka. *Biol. Bull.* 164:82.

Newbury, T. K. 1972. Vibration perception by chaetognaths. *Nature* (London) 236:459.

2.   What environmental cues appear to regulate the cycle of vertical migration of chaetognaths? Discuss the likely adaptive benefits and costs associated with this migratory behavior.

Pearre, S., Jr. 1973. Vertical migration and feeding in *Sagitta elegans* Verrill. *Ecology* 54:300.

Taxonomic Detail

Phylum Chaetognatha

The 70 to 100 chaetognath species are distributed among two orders within a single class. The division into orders is based on the presence (order Phragmophora) or absence (order Aphragmophora) of ventral, transverse muscles.

Class Sagittoidea.
Two orders.

Order Phragmophora.
Paraspadella, Spadella—all members of these genera are bottom-dwellers, mostly in shallow waters; *Eukrohnia*—species in this genus are planktonic.

Order Aphragmophora.
Parasagitta, Sagitta. The group contains 50% or more of all chaetognath species. Its members are all planktonic and compose a major fraction of the animal biomass in the world's oceans.

Some General References About the Chaetognaths

Alvarino, A. 1965. Chaetognaths. *Oceanogr. Marine Biol. Ann. Rev.* 3:115–94.

Ghirardelli, E. 1968. Some aspects of the biology of the chaetognaths. *Adv. Marine Biol.* 6:271–375.

Hyman, L. H. 1959. *The Invertebrates, Vol. 5. Smaller Coelomate Groups.* New York: McGraw-Hill.

Parker, S. P., ed. 1982. *Classification and Synopsis of Living Organisms,* vol. 2. New York: McGraw-Hill, 781–83.

23

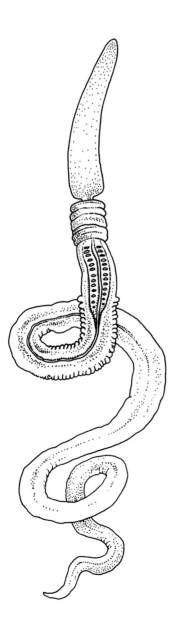

The Hemichordates

Introduction and General Characteristics

> *Phylum Hemi · chordata*
> (G: one-half string)
> hem-ē-kor-dah´-tah

The hemichordates are a small group of marine worms with an apparent and intriguing relationship with two major phyla: the Echinodermata and our own phylum, the Chordata. As typical deuterostomes, cleavage is radial, coelom formation is enterocoelic, and the coelom forms as three distinct compartments (protocoel, mesocoel, and metacoel). Although hemichordates lack a notochord, excluding them from membership in the phylum Chordata, they do exhibit two other chordate characteristics: pharyngeal gill slits and, in some species, a dorsal, hollow nerve cord. As of 1994, 18S rDNA sequence analyses suggest a closer link between hemichordates and echinoderms than between hemichordates and chordates, while other molecular-phylogenetic analyses link hemichordates more closely to vertebrate chordates. Most of the 85 hemichordate species are contained within a single class, the Enteropneusta.

Class Enteropneusta

> *Phylum Entero · pneusta*
> (G: gut breathing)
> en-ter-op-nū´-stah

The enteropneusts are common inhabitants of shallow water, with most species forming mucus-lined burrows in sandy or muddy sediment. The body is long and narrow, and it is divided into three distinct regions

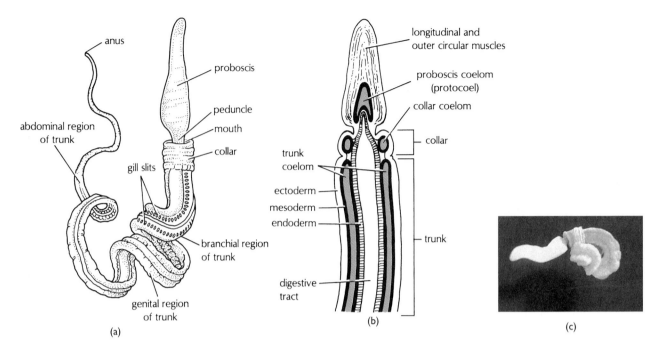

Figure 23.1
(a) The acorn worm, *Saccoglossus kowalevskii,* removed from its burrow to show the basic body structure. (b) Schematic illustration of the coelomic compartments in an enteropneust. (c) Individual of *S. kowalevskii* collected intertidally from Nahant, MA.   *Photo by J. Pechenik.*

corresponding to the tripartite compartmentation of the coelom (Fig. 23.1). The anteriormost section, called the **proboscis,** houses a single coelomic chamber, the **protocoel** (*proto* = G: first; *coel* = G: cavity). This anterior section of the body is generally conical in shape and has given rise to the common name for enteropneusts: "acorn worms." The proboscis is highly muscular and has major responsibility for burrowing and food collection. Indeed, the proboscis is the only truly active body part; enteropneusts are sedentary animals, rarely moving from place to place as adults. Locomotion is largely restricted to movements within burrows (Fig. 23.2).

The proboscis is followed by a narrow **collar** region, containing a pair of coelomic chambers derived from the embryonic mesocoel (*meso* = G: middle). The enteropneust mouth opens on the collar's ventral anterior surface. The bulk of the body, termed the **trunk,** contains a pair of coelomic compartments derived from the embryonic metacoel. External ciliation of the trunk may aid locomotion within burrows. The total body length is typically about 8 cm to 45 cm, although the members of one South American enteropneust species attain lengths of up to 2.5 m!

Many hemichordate species are **deposit feeders,** ingesting sediment, extracting the organic constituents, and extruding a coil (**casting**) of mucus-bound, organically deprived sediment from the anus (Fig. 23.3). Other species are **suspension feeders;** planktonic organisms and detritus adhere to mucus on the proboscis and are then conveyed along ciliated tracts to the mouth.

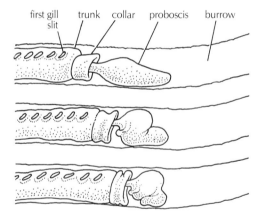

Figure 23.2
Use of the proboscis in locomotion within the burrow by an acorn worm, *Saccoglossus horstii.*   *Adapted from R. B. Clark, 1964.*
Dynamics in Metazoan Evolution. New York: Oxford University Press. After Burdon-Jones.

The enteropneust digestive tract is tubular, consisting of a mouth (located behind the proboscis, on the collar—see Figures 23.1 and 23.4) leading into an esophagus (which compacts ingested particles into a mucus-bound rope), a pharynx, an intestine (the main site of digestion and absorption), and a terminal anus. Food is moved through the gut primarily by the action of ciliated cells lining the inner wall of the digestive tract.

An anterior extension of the pharynx forms a **buccal tube,** or **stomochord** (G: gut chord), within the collar of

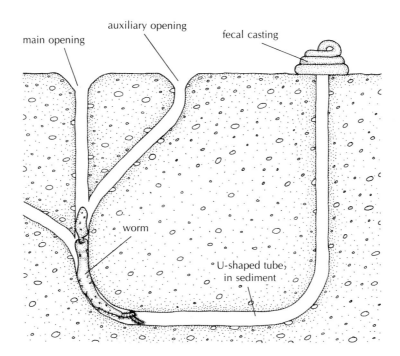

Figure 23.3
The U-shaped burrow of a typical enteropneust. Note the coil of
fecal material that has been deposited at the burrow's posterior end.
Modified from Hyman; after Stiasny and after Burdon-Jones.

the animal (Figs. 23.1b and 23.4a). For some time, this
tube was thought to be a notochord, and the hemichordates were thus included within our own phylum, the
Chordata. The two structures are no longer considered
homologous; that is, the enteropneust buccal tube and
the chordate notochord are presumed to have had independent origins.

The pharynx opens to the outside through a series of
lateral, paired gill slits (more than a hundred in some
species). Cilia lining these gill slits beat in coordinated
fashion so that water is drawn in at the mouth and discharged through the slits (Fig. 23.4). This flow of water is
believed to serve for gas exchange, and the meaning of the
name "enteropneust" then becomes apparent: The animals essentially breathe through a portion of their gut.

Acorn worms possess a true blood circulatory system. The blood, which lacks pigment, is circulated
through dorsal and ventral blood vessels and associated
blood sinuses by pulsations of the muscular blood vessels
themselves; a distinct heart is never found. Evaginations
of the blood sinuses are pronounced in the pharynx, so
the gill slits are highly vascularized, in keeping with their
likely function in respiration.

The enteropneust nervous system is echinoderm-like
in that much of it is in the form of an epidermal nerve
network. However, although there is clearly no brain, the
nerve network is consolidated in some body regions to
form longitudinal nerve cords; a midventral and a mid-
dorsal nerve cord are particularly well developed. Only
the dorsal nerve cord extends into the collar region;

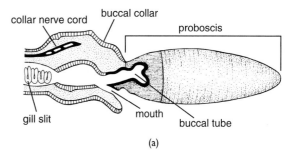

(a)

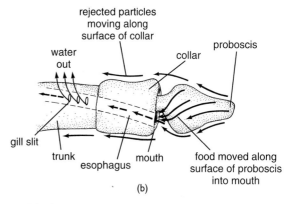

(b)

Figure 23.4
(a) Longitudinal section through anterior portion of an enteropneust,
showing internal features. The proboscis is drawn intact (not
sectioned). (b) Diagrammatic illustration of the respiratory and
feeding pathways of a typical enteropneust. Fine food particles are
trapped in mucus and transferred by cilia along the proboscis to the
mouth, which is located in the collar; rejected particles are moved
posteriorly over the collar surface. Ciliary activity draws water in at
the mouth, through the pharynx, and out at the gill slits. The general
body surface also plays a major role in gas exchange. *From George C.
Kent,* Comparative Anatomy of the Vertebrates, *6th ed. Copyright © 1987 The C. V.
Mosby Company, St. Louis, Missouri. Reprinted by permission of the author. (b) After
Russell-Hunter.*

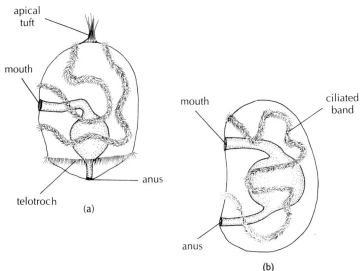

Figure 23.5

(a) The tornaria larva of *Balanoglossus* sp. The tornaria bears a conspicuous terminal ring of cilia (the telotroch), which is lacking in the auricularia larva of holothurian echinoderms (b). In other respects, ciliation patterns for larvae of the two groups are similar.

(a,b) *Adapted from Hardy, 1965. The Open Sea: Its Natural History. Boston: Houghton Mifflin Company.*

moreover, this dorsal, collar nerve cord lies well below the epidermis and, in some species, is hollow (Fig. 23.4a). Such a nerve cord may well be homologous with the dorsal, hollow nerve cord of vertebrates and other chordates.

Acorn worms are **dioecious** (sexes are separate), with the gonads housed in the trunk, and fertilize their eggs externally in the seawater. A free-living larval form is found among a number of enteropneust species (Fig. 23.5). This planktonic, **tornaria** larva is equipped with a series of sinuous, ciliated bands reminiscent of those encountered among the echinoderms.

Class Pterobranchia

> *Class Ptero · branchia*
> (G: feather (or wing) gill)
> ter-ō-brank´-ē-ah

A small number of hemichordates are markedly dissimilar from the enteropneusts and are placed in a separate class, the Pterobranchia. Pterobranchs have been collected primarily by dredging in fairly deep water, especially in the Antarctic; until recently, they were considered inaccessible to the average invertebrate zoologist. However, over the past 20 years or so, several shallow-water species have been described, particularly from the clear and inviting waters around Bermuda.

Unlike the enteropneusts, pterobranchs possess ciliated, anterior tentacles and a U-shaped gut. No pterobranch has more than one pair of pharyngeal gill slits, and some species have none. Also unlike enteropneusts, members of most pterobranch species occupy rigid tubes that they secrete (Fig. 23.6). Tube secretions are produced by glands on an anterior **cephalic shield** (Fig. 23.6). The cephalic shield also serves as an attachment organ and can be used as a muscular proboscis to crawl within the tube or even on solid substrates adjacent to the tube; at the slightest provocation, the animal withdraws into its tube with a rapid contraction of its muscular, highly extensible stalk (Fig. 23.6).

As illustrated in Figure 23.6, a number of pterobranch species are colonial. Each individual, called a **zooid,** is only a few millimeters long, and colonies rarely exceed 20 mm in diameter. The zooids within one colony are produced by asexual budding, so all are descended from a single larva.

Current thinking suggests that pterobranchs may be direct descendants of an ancient, but now extinct group of colonial, often planktonic, marine animals called "graptolites."[1] Graptolites are very abundant as fossils; they flourished in the oceans about 300 to 500 million years ago. Some zoologists view pterobranchs, in turn, as likely ancestors of the Enteropneusta just discussed. Others would link pterobranchs more closely to lophophorates (Chapter 20) than to enteropneusts. The U-shaped gut and the structure of the anterior region of the pterobranch body—complete with ciliated tentacles containing extensions of the mesocoel and surrounding the mouth but not the anus—does seem remarkably phoronid-like (p. 426). In addition, the path of water flow, mechanism of particle capture, and mechanism of particle rejection are decidedly lophophore-like.[2] The enigmatic evolutionary position of pterobranchs has focused far greater attention on them than might be expected, given the animals' dimensions, their general inaccessibility, and the small size of the phylum. The rather recent discovery of shallow-water pterobranch species increases the likelihood that their evolutionary position will be clarified in time. Current (1995) 18S rDNA molecular data support a close link between enteropneusts and pterobranchs.

---

1. See *Topics for Further Discussion and Investigation,* no. 2, at the end of this chapter.

2. See *Topics for Further Discussion and Investigation,* no. 4.

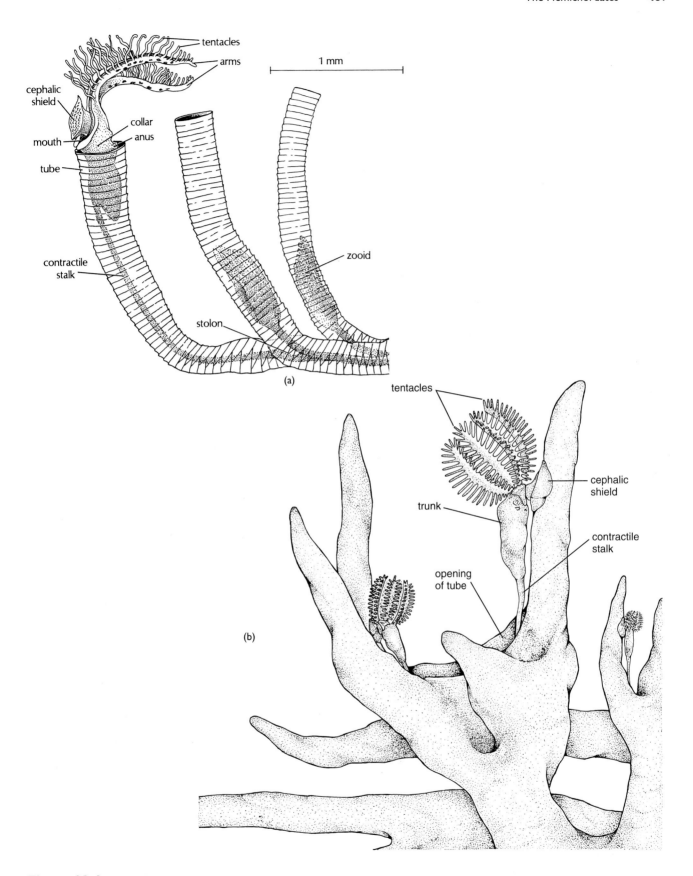

Figure 23.6

(a) A colony of the pterobranch *Rhabdopleura* sp. (b) Colony of *Cephalodiscus* sp., showing several individuals climbing about on their elaborate tubes. Each zooid (excluding the stalk) is only about 2 mm long.   (b) *From S. M. Lester in* Marine Biology, *85:263–268, 1985. Copyright © 1985 Springer-Verlag, Heidelberg, Germany. Reprinted by permission.*

Taxonomic Summary

Phylum Hemichordata
  Class Enteropneusta—the acorn worms
  Class Pterobranchia

Topics for Further Discussion and Investigation

1. What morphological features of pterobranchs ally these animals with the enteropneusts?

Barrington, E. J. W. 1965. *The biology of the Hemichordata and Protochordata.* Edinburgh, Scotland: Oliver & Boyd.

Hyman, L. H. 1959. *The invertebrates.* Vol. V, *Smaller coelomate groups.* New York: McGraw-Hill.

2. Discuss the evidence for and against a likely relationship between living pterobranchs and the extinct graptolites. What additional sorts of information might convince you one way or the other?

Bates, D. E. B., and N. H. Kirk. 1985. Graptolites, a fossil case-history of evolution from sessile, colonial animals to automobile superindividuals. *Proc. Royal Soc. London B* 228:207.

Dilly, P. N. 1988. Tube building by *Cephalodiscus gracilis. J. Zool., London* 216:465.

3. Some time ago, a separate class (the Plancto-sphaeroidea) was established to hold a single species of what was thought to be a planktonic hemichordate. It was subsequently determined that these organisms were unusual larval stages of a benthic adult. How would you determine whether a newly discovered animal was an adult or a larva? If the animal was a larval form, how could you determine the species to which it belonged?

4. Compare and contrast the feeding biology of pterobranchs and lophophorates.

Halanych, K. M. 1993. Suspension feeding by the lophophore-like apparatus of the pterobranch hemichordate *Rhabdopleura normani. Biol. Bull.* 185:417.

Taxonomic Detail

Phylum Hemichordata

The approximately 85 species are distributed among two classes.

Class Enteropneusta.

Balanoglossus, Ptychodera, Saccoglossus—acorn worms. The approximately 70 species are distributed among four families.

Class Pterobranchia.

Cephalodiscus, Rhabdopleura. The 15 species are distributed among three families.

Some General References About the Hemichordates

Barrington, E. J. W. 1965. *The Biology of Hemichordata and Protochordata.* San Francisco: W. H. Freeman.

Bullock, T. H. 1946. The anatomical organization of the nervous system of the Enteropneusta. *Q. J. Microsc. Sci.* 86:55–111.

Harrison, F. W. and E. E. Ruppert, eds. 1996. *Microscopic Anatomy of Invertebrates*, Vol. 15. Hemichordata, Chaetognatha, and the Invertebrate Chordates. Wiley-Liss, NY.

Hyman, L. H. 1959. *The Invertebrates, Vol. 5: Smaller Coelomate Groups.* New York: McGraw-Hill.

Parker, S. P., ed. 1982. *Classification and Synopsis of Living Organisms*, vol. 2. New York: McGraw-Hill, 819–21.

24

The Nonvertebrate Chordates

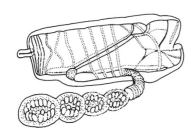

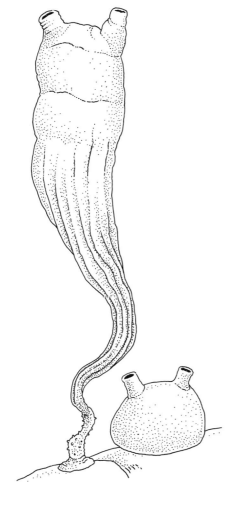

Introduction and General Characteristics

> *Phylum Chordata*
> (G: string)
> kor-dah´-tah

In addition to the 48,000 more familiar species contained in the subphylum Vertebrata, the phylum Chordata includes some 1,400 invertebrate species in the subphyla Urochordata and Cephalochordata. The nonvertebrate chordates are all marine. In common with their vertebrate relatives, the invertebrate chordates generally show, at some point in their life histories, the following characteristics: pharyngeal gill slits; a hollow, dorsal nerve cord; and a notochord. The **notochord** (G: back string) consists of a linear series of cells, each of which contains a large, fluid-filled vacuole. In the vertebrates, the notochord is eventually replaced by cartilaginous tissue or by calcareous vertebrae during **ontogeny** (i.e., development). About 90% of the invertebrate chordate species are contained within the subphylum Urochordata.

Subphylum Urochordata

> *Subphylum Uro · chordata*
> (G: tail string)
> ūr´-´ō-kor-dah´-tah

The Urochordata is one of the few major taxonomic groups to contain no parasitic species; all urochordates feed by straining small particles (especially phytoplankton) from the surrounding water. The method of generating the water current from which these food particles are obtained differs dramatically among the different urochordate classes. Members of this subphylum are commonly referred to as "tunicates," for reasons soon to become apparent, and are distributed among three classes: Ascidiacea, Larvacea, and Thaliacea. None of these animals have left a fossil record.

Class Ascidiacea

Class: *Ascidia · cea*
G: a little bag
a-sid-ē-ā´-sē-ah

The class Ascidiacea contains at least 90% of all described urochordate species; its members are found throughout the world's oceans, in both shallow and deep water. Most adult ascidians, commonly known as "sea squirts," live attached to solid substrates. Some species live anchored in soft sediments. With few exceptions, ascidian adults are **sessile** (i.e., incapable of locomotion). The body is bag-like and covered by a secretion of the epidermal cells (Fig. 24.1). This secreted, protective **test,** also called the **tunic,** is composed largely of protein and a polysaccharide (tunicin) that closely resembles cellulose. Amoeboid cells, blood cells, and in some species, blood vessels are found within the tunic. In some species, the tunic also bears numerous calcareous spines, hairs, or spicules (Fig. 24.2). Although lacking true nerves and muscles, the tunic of a few species has recently been found to contain elongated, multipolar cells capable of both nervous conduction and contraction.[1] Tunic coloration varies from nearly transparent in some species to dramatically pigmented in others; reds, browns, greens, and yellows are common.

The ascidian digestive tract is especially noteworthy in that the pharynx is of unusually great diameter and is perforated with numerous slits called **stigmata;** stigmata morphology differs substantially among species and is therefore an important taxonomic characteristic. The perforated pharynx thus forms a large **pharyngeal (or branchial) basket** (Fig. 24.1). The outer margin of the pharynx is lined by a double ring of ciliated cells—the **peripharyngeal band** (*peri* = G: around)—and the general surface and the stigmata of the pharynx are also ciliated; these cilia create a flow of water through the ascidian for feeding and gas exchange. Water is drawn into the ascidian through a **buccal (oral) siphon,** which is studded with sensory receptors on the outer surface and sensory tentacles on the inner surface. Incoming water is

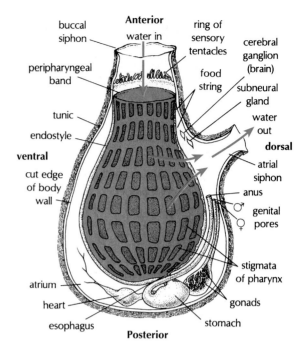

Figure 24.1

Diagrammatic illustration of the anatomy of a typical ascidian. A thin sheet of mucus, produced by the endostyle, coats the basketlike pharynx. All incoming water must pass through this mucous coating before reaching the atrium, so that food particles are strained out. Cilia move the sheet dorsally, where it compacts to form a discrete mucus-food string, ready for ingestion and digestion. The water current is generated by the action of ciliated cells along the margins of the stigmata.
Modified after Bullough and other sources.

also apparently sampled by a **subneural gland** located beneath the brain; the subneural gland's precise function is still uncertain.

Extending along the ventral surface of the pharynx is an elongated gland called the **endostyle** (Fig. 24.1). This endostyle is generally thought to be homologous with the vertebrate thyroid gland, whose iodine-rich secretions regulate important aspects of development and metabolism. Among urochordates, the endostyle secretes a thin sheet of iodine-containing mucus across the pharynx; water passing through the stigmata must pass also through this mucous sheet. Food particles as small as approximately 1 µm (micrometer) thus can be filtered from the inflowing water. Pharyngeal cilia continually move the particle-laden sheet of mucus dorsally (away from the endostyle) to form a mucus-food string. The string is then moved posteriorly (away from the siphons) and enters the esophagus and then the stomach for digestion. Digestion is extracellular, and nutrients are absorbed

1. Mackie, G. O., and C. L. Singla. 1987. Impulse propagation and contraction in the tunic of a compound ascidian. *Biol. Bull.* 173:188.

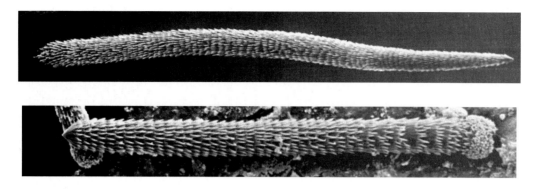

Figure 24.2

Calcareous spicules isolated from the test of the solitary, warmwater ascidian *Herdmania momus*. The spicules were soaked in a mild base to remove adhering tissue. Each spicule bears at least 100 rows of overlapping spines, as illustrated.   *From G. Lambert and C. C. Lambert, Journal of Morphology 192:145–59, figs. 9a & b. Copyright © 1987 Alan R. Liss, Inc. Reprinted by permission of Wiley-Liss, A Division of John Wiley & Sons, Inc.*

in the intestine. Food is moved through the gut by cilia lining the digestive organs, and solid wastes are discharged through the anus.

The water that has been relieved of its particles (and oxygen) during passage through the stigmata enters a chamber, termed the **atrium,** enclosed by the tunic (Fig. 24.1). Water leaves the atrium through an **atrial siphon,** taking with it feces, excretory products, respired carbon dioxide, and any gametes that may have been discharged from the gonad. The buccal and atrial siphons can be closed off by contracting sphincter muscles ringing the incurrent and excurrent openings.

The ascidian body wall (not the tunic) contains both circular and longitudinal muscles. The animal is thus capable of making major shape changes, although it cannot, with few exceptions, move from place to place. Rapid muscle contractions cause a characteristic squirting behavior, important in rejecting particles and dislodging anything stuck in the esophagus.

Although many ascidian species are **solitary** (i.e., individuals are physically separated from each other), a large number of species are **colonial.** As with other colonial species (e.g., hydrozoans and bryozoans), colonial ascidians are produced by asexual budding from a sexually produced founder and remain connected to the "parent." Commonly, the individuals in an ascidian colony have separate buccal siphons and separate mouths, but they share a single atrial siphon—the epitome of communal living (Fig. 24.3).

A small, tubular heart is located adjacent to the stomach. The heart is notable primarily because it reverses the direction of its pumping many times each hour; the opening to the heart through which circulatory fluid enters will be the opening through which fluid exits several minutes later, and what was a vein becomes an artery and vice versa.[2] The circulatory fluid itself is peculiar in that it contains **amoebocyte cells,** which accumulate and store excretory wastes, and **morula cells,** which accumulate vanadium, sulfuric acid, or both; the concentration of vanadium in morula cells may be nearly 10 million times higher than its concentration in seawater! In many species, the morula cells deposit these substances in the tunic.[3] The ascidian circulatory fluid contains no oxygen-carrying respiratory pigments, so the blood's oxygen-carrying capacity is determined entirely by the solubility of oxygen in the fluid.

The only diagnostic chordate characteristic encountered among adult ascidians is the presence of **pharyngeal gill slits** (stigmata). To see your chordate reflection in the ascidians, you must look to the larval stage, which is known as a **tadpole** because of its superficial resemblance to the developmental stage of frogs (Fig. 24.4). The tadpole heart and digestive system are restricted to the "head" region. The larvae are not capable of feeding, and the digestive tract becomes functional only following metamorphosis to the adult form. The tadpole's tail contains a conspicuous, dorsal, hollow nerve cord, a stiff notochord, and longitudinal muscles extending the length of the tail. The notochord can be bent but can be neither elongated nor shortened. The longitudinal muscles on one side of the tail can thus antagonize the longitudinal muscles on the other side of the tail, enabling the tadpole to swim by flexing the tail from side to side. In the absence of the notochord, contractions of the longitudinal musculature in the tail would merely cause the tail to buckle or shorten. Swimming of tadpole larvae is utterly dependent upon the skeletal function subserved by the notochord.[4]

---

2. See *Topics for Further Discussion and Investigation,* no. 1, at the end of the chapter.

3. See *Topics for Discussion and Investigation,* no. 2.

4. See *Topics for Further Discussion and Investigation,* no. 8.

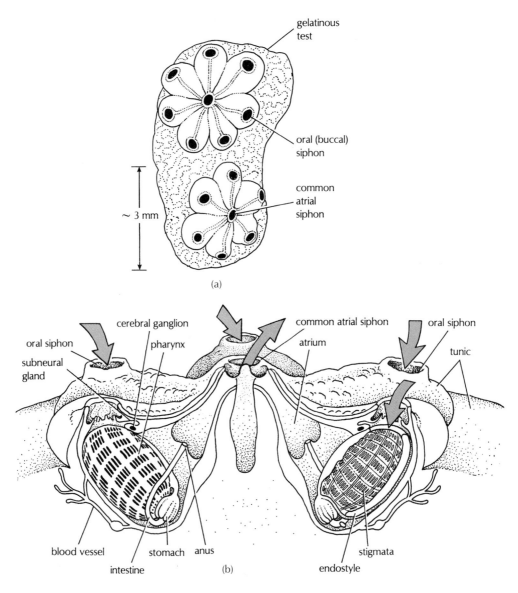

Figure 24.3
(a) A colonial ascidian, *Botryllus violaceus*. As many as a dozen individuals may share a single excurrent opening.
(b) Diagrammatic section through a colony, showing the anatomy of the individual ascidians. Note that the circulatory system extends into the tunic; this is also true in solitary species.

(a) *After Milne-Edwards.* (b) *After Delage and Herouard.*

Attachment to a substrate by means of anterior suckers and a subsequent dramatic metamorphosis to adulthood generally follow a free-swimming larval life of less than one day. In particular, metamorphosis includes the resorption of the larval notochord and tail. Muscular activity thus plays a significant, but very short-lived, role in ascidian locomotion.

Class Larvacea (= Appendicularia)

The members of the class Larvacea (lar-vā´-sē-ah) may have evolved from ascidian ancestors through the process of **neoteny,** in which the rates at which **somatic** (i.e.,

nongonadal) body structures differentiate are slowed down relative to the rate of differentiation of reproductive structures.[5] Thus, the animal becomes sexually mature while retaining the larval morphology. The former adult morphology is thereby deleted from the life cycle, and the new adult morph gradually becomes increasingly better adapted to its new lifestyle through natural selection. Adult larvaceans rarely exceed 5 mm to 6 mm in body length, although a few species are said to reach lengths of 100 cm. Larvaceans are encountered in most areas of the ocean, at all latitudes.

The larvacean heart and respiratory, digestive, and reproductive systems are confined to the head

---

5. See *Topics for Further Discussion and Investigation,* no. 10.

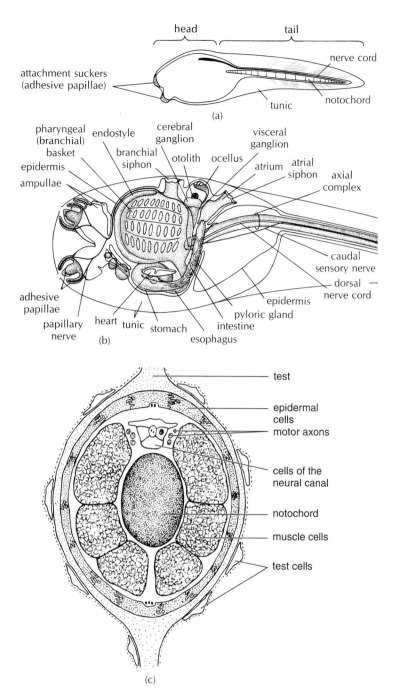

head | tail

nerve cord

attachment suckers
(adhesive papillae)

tunic

notochord

(a)

pharyngeal
(branchial)
basket

endostyle

cerebral
ganglion

visceral
ganglion

epidermis

branchial
siphon

otolith

ocellus

atrium

atrial
siphon

axial
complex

ampullae

adhesive
papillae

papillary
nerve

heart

tunic

stomach

esophagus

intestine

pyloric gland

epidermis

dorsal —
nerve cord

caudal
sensory nerve

(b)

test

epidermal
cells

motor axons

cells of the
neural canal

notochord

muscle cells

test cells

(c)

Figure 24.4

(a) Diagrammatic illustration of an ascidian tadpole larva, with the digestive system omitted for clarity. (b) Internal anatomy of an ascidian tadpole. (c) Diagrammatic cross section through the tail of an ascidian tadpole larva (*Ciona intestinalis*). Note the dorsal nerve cord above the conspicuous notochord, which is surrounded by a substantial musculature. Compare with the

Cephalochordata cross section in Figure 24.10.   (b) *From Cloney, in American Zoologist, 22:817, 1982. Copyright © 1982 American Society of Zoologists. Reprinted by permission. © From Q. Bone, et al., "Diagrammatic cross-section of mid-region of tail . . . in Ciona" in Journal of the Marine Biological Association of the United Kingdom, Vol. 72. Copyright © 1992 Cambridge University Press, New York. Reprinted by permission.*

(Fig. 24.5a), as in the ascidian tadpole larva. Larvaceans, however, are feeding individuals. The mechanism of food collection differs considerably from that encountered among the Ascidiacea. Unlike the ascidian tadpole, the larvacean secretes a gelatinous house around itself (Fig. 24.5b). This house plays a role in both locomotion and food collection; the undulation of the larvacean tail—

which, like that of the ascidian tadpole, contains a notochord—drives water through the house for locomotion and feeding. Water exits through a narrow-diameter opening that is partially occluded by a fine-mesh filter. The orientation of this opening determines the direction of movement of the house and animal resulting from the expulsion of fluid, as summarized poetically for members

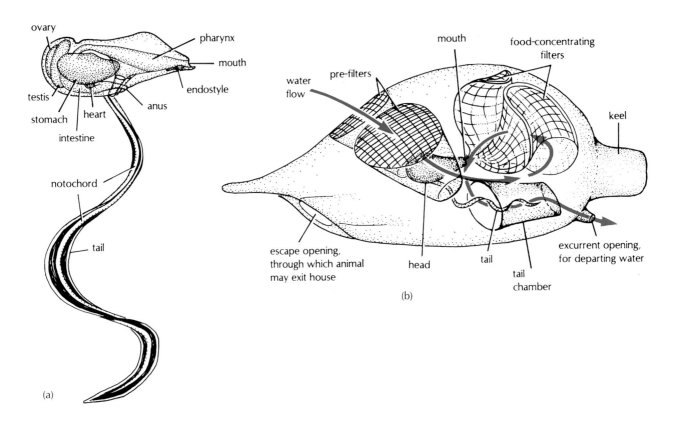

Figure 24.5

(a) Schematic illustration of a larvacean, removed from its gelatinous house. (b) The larvacean *Oikopleura* sp., generating water currents within its complex, disposable house. The house is about 5 cm long. (a) *After Pimentel.* (b) *After Hardy: after Lohmann.*

of the larvacean genus *Oikopleura* by Sir Walter Garstang:

> *Oikopleura,* masquerading as a larval Ascidian,
> Spins a jelly-bubble-house about his meridian:
> His tail, doubled under, creates a good draught,
> That drives water forward and sucks it in aft.

A typical individual with a trunk length of about 1.2 mm can move about 35 ml of water through its house each hour (about 0.8 liter per day), and it can jet along at about 1 cm • sec[1] (centimeter per second).

Particles as small as 0.1 μm are filtered out as water passes through a mucous food-concentrating filter before leaving the house. The concentrated food suspension is then swept through a second, pharyngeal mucous filter and ingested. Very coarse particles, which might damage the delicate house, are screened out from the incoming water current by a coarser mesh positioned across the incurrent opening. The larvacean abandons its house periodically as the meshes become clogged and as the house becomes littered with feces. Before abandoning the old house, the larvacean secretes the material for a new one; it then inflates the new house within seconds of leaving the old one. The food-laden filters of the abandoned house may be important food sources for open-ocean fishes: A single larvacean may form and abandon 15 houses daily.

Class Thaliacea

The members of the class Thaliacea (thal-ē-ā´-s´ē-ah) are mostly free-living, planktonic individuals, but they achieve their mobility through modification of the adult ascidian body plan rather than through exploitation of the tadpole morph.

Thaliaceans, including the three orders of animals known commonly as pyrosomes, salps, and doliolids, are planktonic and nearly transparent. They commonly attain concentrations of hundreds to thousands of individuals per m³ (cubic meter) in subtropical continental shelf waters. The buccal and atrial siphons are at opposite ends of the body, and bands of circular muscle are generally highly developed. These two features account for most thaliaceans' substantial locomotory capabilities.

The most primitive (least modified) of the thaliaceans, members of the genus *Pyrosoma,* are much like colonial ascidians except that the buccal and atrial siphons are at opposite ends of the body. Like colonial ascidians, feeding is accomplished by means of cilia, the pharyngeal basket is well developed, and water is discharged through a common exit (Fig. 24.6a). Locomotion of *Pyrosoma* spp. is also achieved through ciliary activity.

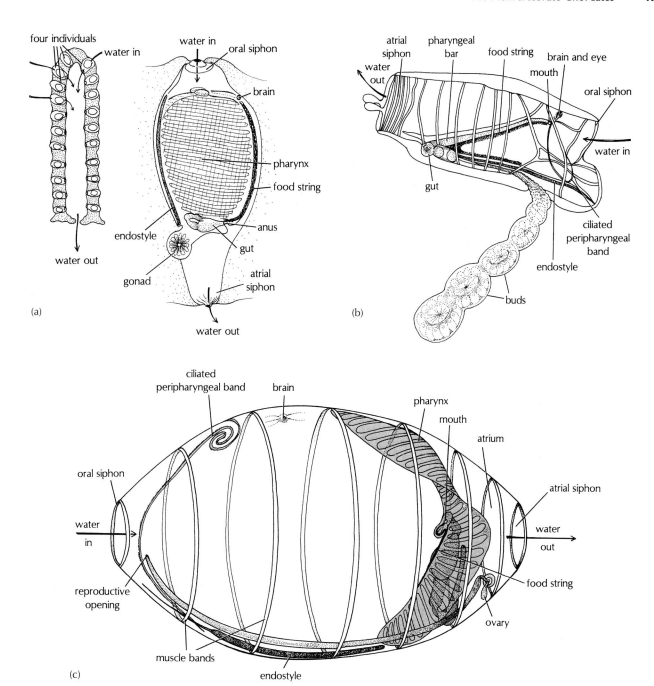

Figure 24.6

(a) *Pyrosoma atlanticum.* The general shape of a typical colony is
shown on the left; the general morphology of an individual colony
member is shown on the right. Although individuals are only a
few millimeters long, a single colony may exceed 3 m in length.
Pyrosome anatomy is much like that of ascidians, except that the
atrial and oral siphons are directly opposite each other. Tinted
areas indicate sheets of mucus. (b) Diagrammatic illustration of a
salp, *Cyclosalpa affinis.* The pharyngeal basket is reduced to a thin
bar. The cilia on the pharyngeal bar function only in compacting

the mucous net into a food string; locomotion and feeding are
accomplished through contraction of the well-developed
muscular bands. (c) Diagrammatic illustration of a doliolid,
Doliolina intermedia. Body lengths rarely exceed 4–5 mm. Note
the conspicuous rings of musculature that generate locomotory
currents. Feeding currents are produced by cilia.   (a,b) *From
N. J. Berrill, The Tunicata. Copyright © The Ray Society, London, England. Reprinted by
permission. (c) Reprinted from Tokioka and Berner, in Pacific Science, 12:317–326,
1958, by permission of University of Hawaii Press, Honolulu, Hawaii.*

Other thaliaceans move through the water by closing off the buccal aperture and contracting the thick circumferential bands of muscle underlying the test. This contraction deforms the test. The resulting decrease in internal volume forces water out of the atrial siphon (since water is incompressible), and the animal is jet-propelled forward. With a single such contraction, one doliolid only about 5 mm long can achieve instantaneous velocities of about 25 cm (50 body lengths) per second, albeit for only a fraction of a second. Following the contraction, the aperture of the atrial siphon is closed while the buccal aperture is opened, and the circular muscles relax. The volume of the animal increases as the elastic test quickly regains its resting shape. Water is thus drawn into the buccal siphon, and the animal is pulled forward as it prepares for the next power stroke. Note that the bands of circular muscle are antagonized by the test itself, rather than by longitudinal musculature.

The evolution of thaliaceans from ascidian-like ancestors appears to be a story of decreasing reliance on cilia and increasing development of musculature. In doliolids, the pharyngeal basket is reduced to a flattened plate, although cilia and stigmata are still conspicuous (Fig. 24.6c). The details of feeding have yet to be documented for these animals, but pharyngeal cilia seem to play a major role in food collection. Among the salps, however, the pharyngeal basket is reduced to a slender **branchial bar.** Water flows through a mucous bag suspended between the endostyle, branchial bar, and peri pharyngeal bands; stigmata are absent (Fig. 24.6b). In salps (Fig. 24.7), muscular contractions thus play the dominant role in both feeding and locomotion.

Although most thaliacean species occur primarily as individuals that rarely exceed 5 cm in length, species in the genus *Pyrosoma* are entirely colonial and may attain lengths of 8 m to 10 m. Most thaliaceans occur only in warm waters.[6]

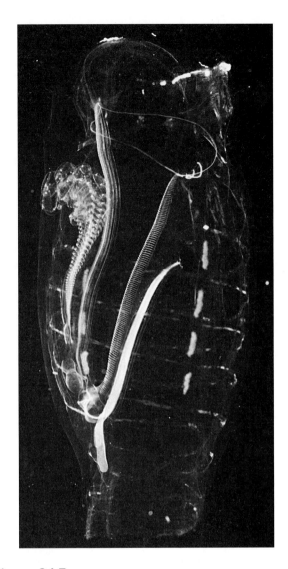

Figure 24.7
Salp photographed in the open ocean.   *Courtesy of G. R. Harbison and J. Carleton.*

Other Features of Urochordate Biology

Reproduction

Reproductive biology differs considerably among the various urochordates. Fertilization is almost always external in solitary ascidians and, probably, in all larvaceans, but it is almost always internal in colonial ascidians and thaliaceans. Gametes or larvae are typically discharged through the atrial siphon or, primarily in larvaceans and some colonial ascidians, through rupture of the body wall. Ascidians commonly exhibit asexual reproduction by budding, interspersed with bouts of sexual reproduction; most ascidians are simultaneous hermaphrodites. Most larvaceans are also hermaphroditic, but reproduction is exclusively sexual. Since asexual reproduction is not found within the class Larvacea, larvaceans are always solitary (i.e., there are no colonial species).

Thaliacean reproduction is both sexual and asexual, as with ascidians, but among the Thaliacea, the two forms of reproduction are allocated between two types of individuals. The union of sperm and egg gives rise to an asexually reproducing morph called the **oozooid.** The oozooid produces no gametes. Instead, individuals called **blastozooids** are budded off from the oozooid (Fig. 24.8).[7] These blastozooids develop hermaph roditic gonads, and their gametes ultimately give rise to the next generation of oozooids.

---

6. See *Topics for Further Discussion and Investigation,* no. 5.

7. See *Topics for Further Discussion and Investigation,* no. 9.

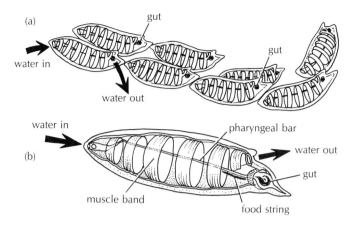

Figure 24.8

In the life cycle of many salps, chains of blastozooids (a) are budded off from a solitary oozoid (b). The blastozooids are released from the parent while still aggregated. After swimming for a time as a group, the individual blastozooids separate and reach sexual maturity. Note that the blastozooids are produced asexually from the oozoid, so that all of the individuals in an aggregate are genetically identical to each other and to the parental oozoid.

(a,b) *After Hardy.*

Asexual replication combined with generation times as short as one or two days permit rapid buildups of thaliacean populations. Doliolids have been found in groups of more than 1,000 individuals per cubic meter of seawater, over perhaps a hundred square kilometers of ocean. Larvaceans have been reported at even higher concentrations, sometimes exceeding 20,000 individuals per cubic meter of seawater despite their exclusively sexual reproduction.

Excretory and Nervous Systems

Specialized excretory organs are rare among the urochordates. In most species, wastes are apparently removed by diffusion across general body surfaces and through the activities of scavenging amoebocytes. In several species within the ascidian genus *Molgula,* however, excretory organs have been described.[8] These ductless, blind-ending **renal sacs** accumulate nitrogenous wastes and deposit them as solid concretions composed primarily of uric acid and related compounds. These concretions accumulate within the renal sacs throughout the life of the ascidian.

A conspicuous cerebral ganglion is located in the urochordate body wall and innervates the muscles and sensory cells. Its role in ascidian biology seems minimal, since surgical removal of the cerebral ganglion causes little apparent change in activity. The brain's involvement in coordinating the activities of larvaceans and thaliaceans

is likely to be much greater, since muscular activity plays a greater role in these animals' lives. In addition, salps have a prominent, horseshoe-shaped photoreceptor associated with the cerebral ganglion.

Subphylum Cephalochordata

Subphylum Cephalo · chordata
(G: head string)
sef´-al-´ō-cor-dah´-tah

Cephalochordates are small (less than 10 cm long), laterally flattened, marine invertebrates that have much in common both with urochordates and with members of our own subphylum, the Vertebrata. Comparisons of gene sequences coding for 18S rRNA indicate that cephalochordates are indeed our closest nonvertebrate relatives. Their fossil record extends back to the mid-Cambrian, about 535 million years ago. Like vertebrates, cephalochordates have a distinct notochord and a dorsal, hollow nerve cord (Figs. 24.9 and 24.10), and coelomic body cavities that form by enterocoely. Specific homeobox regulatory gene sequences expressed during development of the cephalochordate nerve cord appear homologous with sequences expressed during development of the vertebrate hindbrain, suggesting that the vertebrate brain has indeed evolved from an extensive portion of the cephalochordate nerve cord.[9]

But cephalochordate feeding biology most closely resembles that of urochordates: Microscopic food particles are captured on mucous sheets as water flows through gill slits in the ciliated pharynx. The captured food then enters the gut as a mucous string, to be digested enzymatically and by cellular phagocytosis. As in urochordates, the mucus is secreted by an endostyle and the pharynx functions in gas exchange as well as in food collection, with the pharyngeal gill bars being serviced by a *bona fide* circulatory system; colorless blood is circulated through the gill and to the rest of the body tissues by muscular contractions of the blood vessel walls. Wastes are eliminated through a well-formed nephridial system located in the pharyngeal region. The paired excretory organs have a combination of protonephridial and metanephridial characteristics.[10]

As with doliolids and ascidians, water flow is created by cilia on the pharynx, not by muscular activity. Water flows into the cephalochordate through a buccal opening, which is fringed with sensory tentacles, and exits through an atrial opening located anterior to the anus (Fig. 24.9). The buccal tentacles bear chemosensory and

---

8. See *Topics for Further Discussion and Investigation,* no. 3.

9. Holland, P. W. H., et al. 1992. *Development* 116:653–61.
10. Ruppert, E. E., and P. R. Smith. 1988. The functional organization of filtration nephridia. *Biol. Rev.* 63:231.

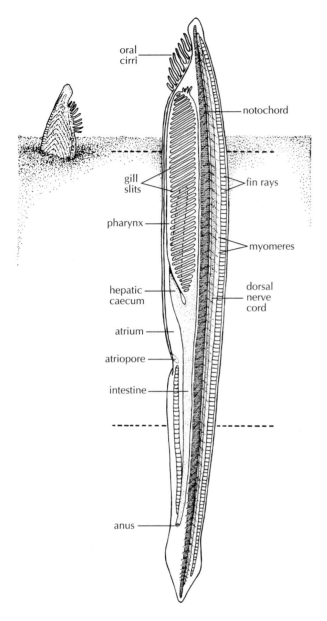

oral
cirri

notochord

gill
slits

fin rays

pharynx

myomeres

hepatic
caecum

dorsal
nerve
cord

atrium

atriopore

intestine

anus

Figure 24.9

The lancelet *Branchiostoma* sp. burrowed into substrate, with its internal anatomy showing through the nearly transparent body. Dotted lines indicate regions from which the cross sections of Figure 24.10 were taken.   *After Lytle/Wodsedalek, 1987. Dubuque, Iowa: Times Mirror Higher Education Group, Inc.*

mechanoreceptor cells, and they also serve to prevent very large particles from entering the pharynx.

Unlike that of urochordates, the cephalochordate notochord persists throughout life. But cephalochordates are clearly not vertebrates: The nerve cord does not elaborate anteriorly to form a brain; there is no protective cranium about the anterior part of the nervous system (not surprisingly, since there is nothing much to protect); the notochord extends the entire length of the animal, well into the head (hence, the name "cephalochordate"; *cephalo* = G: head); and the notochord is never replaced by a vertebral column. Specialized sensory organs are limited to the buccal tentacles and to a series of simple, pigmented light receptors (*ocelli*) distributed along the nerve cord.

Adult cephalochordates are nearly transparent and have fishlike bodies, earning them the common name of "lancelets" (little spears). The body being somewhat pointed at both ends, the members of this subphylum are also called by the general term "amphioxus," meaning "sharp at both ends" (*amphi* = G: on both sides; *oxys* = G: sharp). The body is covered neither by a test nor by scales, and adults spend most of their time feeding in sand, with the head protruding above the substrate (Fig. 24.9). When they need to find another feeding area or to escape predators, they move by using strong contractions of paired, lateral muscles. The muscles are arranged in a longitudinal row of about 60 distinct, V-shaped segments (**myomeres**) that are readily visible through the thin, nearly transparent epidermis. Muscle groups on opposite sides of the body (Fig. 24.10) antagonize each other through the flexible but incompressible notochord, as in larvaceans and ascidian tadpole larvae; the backbone serves this same function in fish. Unlike most urochordates, lancelets are almost always dioecious rather than hermaphroditic. Eggs are fertilized externally, and the embryos, after passing through a vertebrate-like "neurula" stage, develop to short-lived, free-swimming larvae.

About two dozen lancelet species have been described, some of which are eaten by people in parts of China.

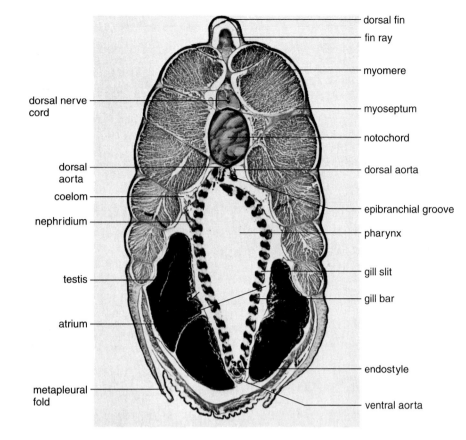

(a)

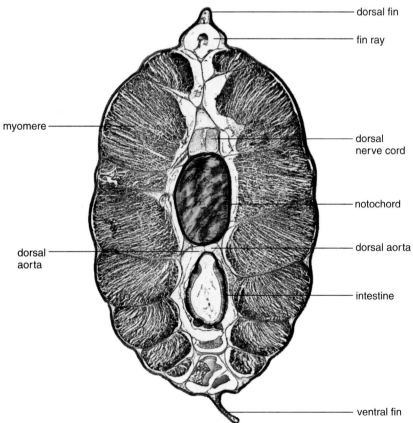

(b)

Figure 24.10
Cross sections taken through (a) the anterior region and (b) the posterior region of a male lancelet. Approximate areas from which the sections were taken are shown in Figure 24.9 (a,b).

© Carolina Biological Supply Company/Phototake.

Taxonomic Summary

Phylum Chordata
  Subphylum Urochordata
    Class Ascidiacea
    Class Larvacea (= Appendicularia)
    Class Thaliacea
        Order Pyrosomatida
        Order Doliolida
        Order Salpida
    Subphylum Cephalochordata
    Subphylum Vertebrata

Topics for Further Discussion and Investigation

1. Investigate the significance of heartbeat reversal in urochordates and the mechanism by which it is accomplished.

Herron, A. C. 1975. Advantages of heart reversal in pelagic tunicates. *J. Marine Biol. Assoc. U.K.* 55:959.

Jones, J. C. 1971. On the heart of the orange tunicate, *Ecteinascidia turbinata* Herdman. *Biol. Bull.* 141:130.

Kriebel, M. E. 1968. Studies on cardiovascular physiology of tunicates. *Biol. Bull.* 134:434.

Ponec, R. J. 1982. Natural heartbeat patterns of six ascidians and environmental effects on cardiac function in *Clavelina huntsmani*. *Comp. Biochem. Physiol.* 72A:455.

2. Investigate the adaptive significance of vanadium and acid accumulation in the tests of ascidians.

Davis, A. R., and A. E. Wright. 1989. Interspecific differences in fouling of two congeneric ascidians (*Eudistoma olivaceum* and *E. capsulatum*): Is surface acidity an effective defense? *Marine Biol.* 102:491.

Stoecker, D. 1978. Resistance of a tunicate to fouling. *Biol. Bull.* 155:615.

Young, C. M. 1986. Defenses and refuges: Alternative mechanisms of coexistence between a predatory gastropod and its ascidian prey. *Marine Biol.* 91:513.

3. Discuss the evidence for functional excretory systems among ascidians.

Das, S. M. 1948. The physiology of excretion in *Molgula* (Tunicata; Ascidiacea). *Biol. Bull.* 95:307.

Heron, A. 1976. A new type of excretory mechanism in tunicates. *Marine Biol.* 36:191.

Saffo, M. B. 1988. Nitrogen waste or nitrogen source? Urate degradation in the renal sac of molgulid tunicates. *Biol. Bull.* 175:403.

4. Discuss the similarities and differences in the feeding mechanisms of urochordates and bivalves.

Deibel, D. 1986. Feeding mechanism and house of the appendicularian *Oikopleura vanhoeffeni*. *Marine Biol.* 93:429.

Deibel, D., and G.-A. Paffenhöfer. 1988. Cinematographic analysis of the feeding mechanism of the pelagic tunicate *Doliolum nationalis*. *Bull. Marine Sci.* 43:404.

MacGinitie, G. E. 1939. The method of feeding of tunicates. *Biol. Bull.* 77:443.

Young, C. M., and L. F. Braithwaite. 1980. Orientation and current-induced flow in the stalked ascidian *Styela montereyensis*. *Biol. Bull.* 159:428.

5. Discuss the potential ecological significance of planktonic urochordates in the open ocean.

Bruland, K. W., and M. W. Silver. 1981. Sinking rates of fecal pellets from gelatinous zooplankton (salps, pteropods, doliolids). *Marine Biol.* 63:295.

Deibel, D. 1988. Filter feeding by *Oikopleura vanhoeffeni*: Grazing impact on suspended particles in cold ocean waters. *Marine Biol.* 99:177.

Flood, P. R., D. Deibel, and C. C. Morris. 1992. Filtration of colloidal melanin from sea water by planktonic tunicates. *Nature* 355:630.

Harbison, G. R., and R. W. Gilmer. 1976. The feeding rates of the pelagic tunicate *Pegea confederata* and two other salps. *Limnol. Oceanogr.* 21:517.

Madin, L. P. 1974. Field observations on the feeding behavior of salps (Tunicata: Thaliacea). *Marine Biol.* 25:143.

Morris, C. C., and D. Deibel. 1993. Flow rate and particle concentration within the house of the pelagic tunicate *Oikopleura vanhoeffeni*. *Marine Biol.* 115:445.

Wiebe, P. H., L. P. Madin, L. R. Haury, G. R. Harbison, and L. M. Philbin. 1979. Diel vertical migration by *Salpa aspera* and its potential for large-scale particulate organic matter transport to the deep-sea. *Marine Biol.* 53:249.

6. About 20 ascidian species invariably house algal symbionts. Evaluate the contributions made by these algae to their ascidian hosts.

Olson, R. R. 1986. Photoadaptations of the Caribbean colonial ascidian-cyanophyte symbiosis *Trididemnum solidum. Biol. Bull.* 170:62.

7. Investigate self/nonself recognition among ascidians.

Kingsley, E., D. A. Briscoe, and D. A. Raftos. 1989. Correlation of histocompatibility reactions with fusion between conspecifics in the solitary urochordate *Styela plicata. Biol. Bull.* 176:282.

Peddie, C. M., and V. J. Smith. 1994. Mechanism of cytotoxic activity by hemocytes of the solitary ascidian, *Ciona intestinalis. J. Exp. Zool.* 270:335.

Raftos, D. A., and E. L. Cooper. 1991. Proliferation of lymphocyte-like cells from the solitary tunicate, *Styela clava,* in response to allogeneic stimuli. *J. Exp. Zool.* 260:391.

Rinkevich, B., and I. L. Weissman. 1987. A long-term study on fused subclones in the ascidian *Botryllus schlosseri:* The resorption phenomenon (Protochordata: Tunicata). *J. Zool., London* 213:717.

Rinkevich, B., Y. Saito, and I. L. Weissman. 1993. A colonial invertebrate species that displays a hierarchy of allorecognition responses. *Biol. Bull.* 184:79.

Wright, R., and E. L. Cooper. 1984. Protochordate immunity—II. Diverse hemolymph lectins in the solitary tunicate *Styela clava. Comp. Biochem. Physiol.* 79B:269.

8. Compare the locomotory mechanism of ascidian tadpoles with that of nematodes.

9. Compare the reproductive biology of thaliaceans with that of the scyphozoans.

10. Argue either for or against the proposition that larvaceans evolved from ascidian tadpole larvae by neoteny.

Berrill, N. J. 1955. *The Origin of Vertebrates.* Oxford: Oxford Univ. Press.

Bone, Q. 1992. On the locomotion of ascidian tadpole larvae. *J. Marine Biol. Assoc. U.K.* 72:161.

Garstang, W. 1928. The morphology of the Tunicata, and its bearing on the phylogeny of the Chordata. *Q. J. Micros. Sci. 72:51.*

Holland, L. Z., G. Gorsky, and R. Fenaux. 1988. Fertilization in *Oikopleura dioica* (Tunicata, Appendicularia): Acrosome reaction, cortical reaction and sperm-egg fusion. *Zoomorphology* 108:229.

Wada, H., and N. Satoh. 1994. Details of the evolutionary history from invertebrates to vertebrates, as deduced from the sequences of 18S rDNA. *Proc. Nat. Acad. Sci. USA* 91:1801.

Taxonomic Detail

Phylum Chordata
Subphylum Urochordata

The approximately 1,400 species are distributed among three classes.

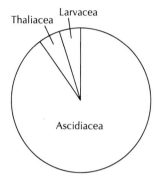

Class Ascidiacea.
The approximately 1,250 species are distributed among three orders, each of which has worldwide representation.

Order Asplousobranchia.
Didemnum, Diplosoma, Clavelina, Distaplia, Trididemnum. All species form colonies by asexual budding; in some species, colony members are encased in a common gelatinous tunic. Didemnid colonies often contain symbiotic unicellular algae; they are also capable of locomotion, through gradual extension of the tunic in finger-like projections. Some species contain chemicals of possible value in cancer treatment, and at least one species in this order (*Clavelina minata,* from Japan) exhibits intracellular bioluminescence. Three families.

Order Phlebobranchia.
Ciona, Ascidia, Phallusia, Ecteinascidia (colonial). Most species are solitary, rather than colonial, and are especially common in shallow water.

Order Stolidobranchia.
Boltenia, Halocynthia, Molgula, Polyandrocarpa (colonial), *Botryllus* (colonial). Representatives of this order occur worldwide, often in shallow water. Most species are solitary, with some solitary individuals reaching lengths of about 15 cm. Some members live interstitially in sand, and a few species are restricted to abyssal depths. Some species develop without a free-living larval stage. Three families.

Class Larvacea (= Appendicularia).

Oikopleura, Fritillaria. Larvaceans occur in all oceans and are particularly common near the water's surface. Individuals can reach about 9 cm in length, although most are smaller than 1 cm. Some species of *Oikopleura* bioluminesce when poked or shaken. The class contains about 70 species.

Class Thaliacea.

All species are colonial in at least part of the life cycle. About 70 species, distributed among three orders.

Order Pyrosomida.

Pyrosoma. This group of colonial urochordates is best represented in warm water. All individuals exhibit striking bioluminescence, probably through the chemical activities of bacterial symbionts. Colonies commonly range from several centimeters to several meters in length.

Order Doliolida.

Doliolum. Representatives live in all oceans, although they are most common in warmer waters. Doliolids are the fastest moving of all thaliaceans, despite their small size (usually only a few millimeters long).

Order Salpida.

Salpa. Representatives live in all the world's oceans, although they are especially abundant in warm waters. Some species are found at depths exceeding 1,500 m, although most salps live much closer to the surface. Salps reproduce rapidly; some species have generation times shorter than one day. Individuals may grow to lengths exceeding 24 cm. About 30 species have been described.

Subphylum Cephalochordata (= Acrania)

Branchiostoma—amphioxus, or lancelets. All 23 described species occur in shallow water and live partially buried in sandy substrates. Some species tolerate estuarine conditions, although most are fully marine. Two families, two genera.

Some General References About the Nonvertebrate Chordates

Alexander, R. M. 1975. *The Chordates.* New York: Cambridge Univ. Press.

Alldredge, A. 1976. Appendicularians. *Sci. Amer.* 235:94–102.

Barrington, E. J. W. 1965. *The Biology of Hemichordata and Protochordata.* San Francisco: W. H. Freeman.

Berrill, N. J. 1955. *The Origin of Vertebrates.* London: Oxford University Press.

Berrill, N. J. 1961. Salpa. *Sci. Amer.* 204:150–60.

Conklin, E. G. 1932. The embryology of amphioxus. *J. Morphol.* 54:69–151.

Goodbody, I. 1974. The physiology of ascidians. *Adv. Marine Biol.* 12:1–149.

Harrison, F. W. and E. E. Ruppert, eds. 1996. *Microscopic Anatomy of Invertebrates*, Vol. 15. Hemichordata, Chaetognatha, and the Invertebrate Chordates. Wiley-Liss, NY.

Millar, R. H. 1971. The biology of ascidians. *Adv. Marine Biol.* 9:1–100.

Parker, S. P., ed. 1982. *Classification and Synopsis of Living Organisms*, vol. 2. New York: McGraw-Hill, 823–30.

25

Invertebrate Reproduction and Development—An Overview

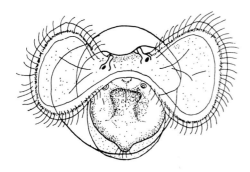

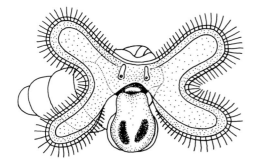

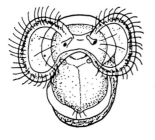

Introduction

The continued existence of a species depends upon the ability of individuals of that species to reproduce. Nearly every behavioral, morphological, or physiological adaptation of a species may be presumed to contribute to its reproductive success, either directly or indirectly. In a sense, then, all organisms live to reproduce. Moreover, no matter how the process of reproduction begins or proceeds, differentiation (genetically controlled specialization of cells) is always involved; much of what is presently understood about the control of gene expression comes from the study of invertebrate development. The topic of reproduction and development thus seems ideal for uniting all of the phyla in the consideration of a single aspect of invertebrate biology. As you read through this chapter, I hope you will be able to take pride in recognizing many terms that were new to you when you began this book.

Invertebrates show a great diversity of reproductive and developmental patterns, a diversity that far surpasses that encountered among the vertebrates. Most vertebrates fertilize internally and show some degree of parental care for the developing young. All vertebrates are deuterostomes, and thus cleavage is basically radial and indeterminate, and the mouth does not form from the blastopore. Variations on the basic deuterostome theme occur, of course, largely because of differing amounts of yolk in the eggs. The diversity of developmental patterns observed among invertebrates, however, is much more than variations on a theme. Among invertebrates, there are radical differences in

1. expression of sexuality;
2. site of fertilization (if present);
3. pattern of cell division;
4. stage at which cell fates become determined;

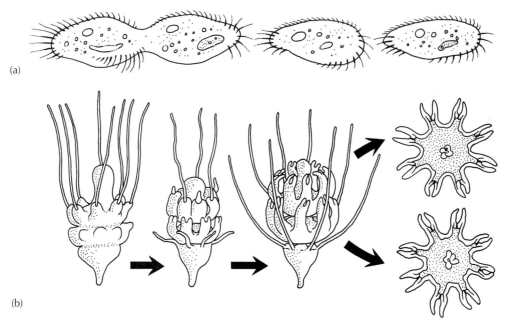

Figure 25.1
(a) Binary fission in a ciliated protozoan. (b) Strobilation in the scyphozoan *Stomolophus meleagris*, culminating in the release of numerous ephyrae. (b) *From D. R. Calder in* Biological Bulletin, *162: 149, 1982. Copyright © 1982 Biological Bulletin. Reprinted by permission.*

5. number of distinct tissue layers formed;
6. mechanism through which mesoderm (if any) is formed;
7. extent to which a body cavity develops;
8. mechanism through which a body cavity develops; and
9. origin of the mouth and anus (when present).

In this chapter, we survey invertebrate patterns of reproduction and development, identify some of the groups most closely associated with these different patterns, and consider the ecological significance—that is, the likely adaptive benefits—of the various patterns discussed. Cleavage patterns, modes of coelom formation, and other features of early metazoan embryology have already been summarized in Chapter 2. The goal of this chapter is to put reproduction and development into ecological and evolutionary contexts.

Asexual Reproduction

Reproduction among invertebrates may be either sexual or asexual. Sexual reproduction always involves the union of genetic material contributed by two genomes. Asexual reproduction, on the other hand, is reproduction in the absence of fertilization (i.e., without a union of gametes). The timing of both sexual and asexual reproductive events is controlled by both environmental and internal factors.[1]

Asexual reproduction is often a process of exact replication; in such instances, barring mutation, asexually produced offspring are genetically identical to the progenitor. This form of asexual reproduction (**ameiotic;** i.e., without meiosis) can add no genetic diversity to a population. On the other hand, through asexual reproduction, a single individual can contribute to a potentially rapid increase in population size, excluding potential competitors and flooding the population with a particularly successful genotype.

Asexual reproduction need not involve egg production by a female. In sponges (poriferans), hydrozoans, scyphozoans, bryozoans, thaliaceans, and some ascidians and protozoans, for example, asexual reproduction is accomplished through the budding of new individuals from preexisting individuals (Fig. 25.1). Among the Protozoa, replication is often achieved through binary fission. In the Trematoda, asexual reproduction takes the form of ameiotic replication of larval stages, greatly increasing the probability that any one genotype will locate a suitable host. Similarly, larvae of many fungus-feeding gall midges (class Insecta) asexually produce more genetically identical larvae within themselves, increasing the likelihood that a given genotype will eventually locate a suitable fungus. In some other groups, such as the Anthozoa, Ctenophora, Turbellaria, Rhynchocoela, Polychaeta, Asteroidea, and Ophiuroidea, body parts may be detached from the adult (or even the larvae, in some as-

---

1. See *Topics for Further Discussion and Investigation,* nos. 1 and 2, at the end of the chapter.

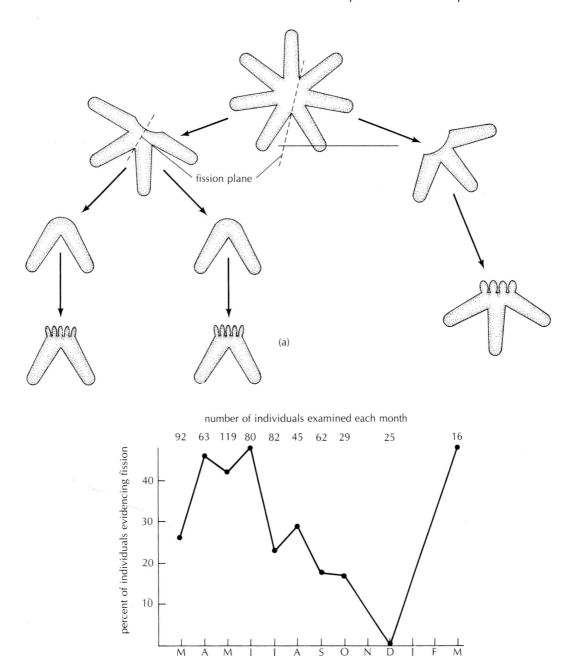

Figure 25.2

(a) Asexual reproduction in the asteroid *Nepanthia belcheri*. This species routinely undergoes fission, in which six- or seven-armed adults divide to form primarily two- or three-armed individuals, which then proceed to regenerate the full complement of arms. (b) Percent frequency of fission varied between 0% and nearly 50% of the population over the course of one year. The number of individuals examined each month is shown at the top of the graph.

From Ottesen and Lucas in Marine Biology, *69:223, 1982. Copyright © 1982 Springer-Verlag, Heidelberg, Germany. Reprinted by permission.*

teroid species) and left behind to regenerate into new, morphologically complete individuals (Fig. 25.2).

Egg production is intimately involved in the ameiotic, asexual reproduction of many other invertebrate species. Among selected arthropods and rotifers, asexual reproduction takes the form of **parthenogenesis,** in which eggs develop to adulthood in the absence of fertilization. As simple as this sounds, there can be unusual complications. In some mites and ticks (class Arachnida), for example, the females cannot oviposit unless they first mate with a male, even though the eggs never become fertilized and the male makes no genetic

Table 25.1 Summary of Invertebrate Reproductive Patterns. (+ = present; – = absent; ? = unknown; – ? = probably absent; + ? = probably present; internal fertilization = within the body of the animal, but not necessarily within the reproductive tract.)

Taxonomic Group	Reproductive Mode		Spermato-phore	Sexuality		Larval Stage	Fertilization	
	Asexual	Sexual		Dioe-cious	Hermaphro-ditic		Internal	External
Protozoa	+	+	–	Have mating strains		none	+	–
Porifera	+	+	–	+(rare)	+	parenchymula	+	–
Cnidaria								
Scyphozoa	+	+	–	+	+(rare)	planula	+(rare)	+
Hydrozoa	+	+	–	+	+(rare)	planula, actinula	+	+
Anthozoa	+	+	–	+	+(rare)	planula	+	+
Ctenophora	+(rare)	+	–	+(rare)	+	cydippid	+(rare)	+
Platyhelminthes								
Turbellaria	+	+	–	+(rare)	+	Müller's	+	–
Trematoda	+	+	–	–	+	miracidium, cercaria	+	–
Cestoda	+	+	–	+(rare)	+	none	+	–
Rhynchocoela	+	+	–	+	+(rare)	pilidium	+(rare)	+
Nematoda	+	+	–	+	+(rare)	none	+	–
Rotifera								
Seisonidea	–	+	–	+	–	none	+	–
Bdelloidea	+	–	–	+	–	none	–	–
Monogononta	+	+	–	+	–	none	+	–
Loricifera	?	+	?	+	–	Higgins	?	?
Kinorhyncha	–	+	+	+	–	none	+	+
Gastrotricha	+	+	+	–	+	none	+	–
Priapulida	–	+	–	+	–	nameless (benthic)	+	+
Nematomorpha	–	+	–	+	–	nameless	+	–
Pentastomida	–	+	–	+	–	nameless (parasitic)	+	–
Acanthocephala	–	+	–	+	–	acanthor	+	–
Gnathostomulida	–	+	–	–	+	none	+	– ?
Annelida								
Polychaeta	+	+	+	+	+(rare)	trochophore	–	+
Oligochaeta	+	+	–	–	+	none	–	+
Hirudinea	–	+	+	–	+	none	+	–
Mollusca								
Gastropoda	+(rare)	+	+	+	+	veliger	+	+(rare)
Bivalvia	–	+	–	+	+	trocophore, veliger	+(rare)	+
Polyplacophora	–	+	–	+	+(rare)	trochophore	+(rare)	+
Cephalopoda	–	+	+	+	+(rare)	none	+	–
Scaphopoda	–	+	–	+	–	trochophore	–	+
Apalcophora	–	+	– ?	+	+	trochophore	+?	+?
Monoplacophora	–	+	–	+	–	veliger?	–?	+?

contribution to the offspring. This is termed **pseudogamy** (false marriage). Something similar occurs in many beetles, except that males do not exist in some of these species. In such cases, the females mate with males of an allied species; although no union of gametes occurs, the eggs will not develop in the absence of contact with sperm.

In other groups of invertebrates, asexual reproduction may involve meiosis, so that pairing and segregation of chromosomes occurs, and new genetic combinations

Table 25.1—*Continued*

Taxonomic Group	Reproductive Mode		Spermato-phore	Sexuality		Larval Stage	Fertilization	
	Asexual	Sexual		Dioe-cious	Hermaphro-ditic		Internal	External
Arthropoda								
Merostomata	−	+	−	+	−	trilobite	−	+
Arachnida	−	+	+	+	−	none	+	−
Chilopoda	−	+	+	+	−	none	+	−
Diplopoda	−	+	+	+	−	none	+	−
Insecta	+	+	+	+	+(rare)	larva, pupa	+	−
Crustacea	+	+	+	+	+	nauplius, cyprid, zoea, megalopa	+	+(rare)
Sipuncula	+(rare)	+	−	+	+(rare)	trochopore, pelagosphera	−	+
Echiura	−	+	−	+	−	trochophore	+(rare)	+
Pogonophora	−	+	+	+(rare)	−	?	?	?
Tardigrada	+	+	−	+	−	none	+	−
Onychophora	−	+	+	+	−	none	+	−
Bryozoa	+	+	−	+(rare)	+	coronate, cyphonautes	+	+(rare)
Phoronida	+	+	+	+(rare)	+	actinotroch	+	+(rare)
Brachiopoda	−	+	−	+	+(rare)	nameless	+	+(rare)
Chaetognatha	−	+	+	−	+	none	+	−
Echinodermata								
Asteroidea	+	+	−	+	+(rare)	bipinnaria, brachiolaria	+(rare)	+
Ophiuroidea	+	+	−	+	+	ophiopluteus	+	+
Echinoidea	−	+	−	+	+(rare)	echinopluteus	+(rare)	+
Holothuroidea	−	+	−	+	+(rare)	auricularia, doliolaria	−	+
Crinoidea	−	+	−	+	−	doliolaria	+	+
Urochordata								
Ascidiacea	+	+	−	+(rare)	+	tadpole	+	+
Larvacea	−	+	−	+(rare)	+	none	− ?	+ ?
Thaliacea	+	+	−	−	+	none	+	−
Hemichordata								
Enteropneusta	+(rare)	+	−	+	−	tornaria	−	+
Cephalochordata	−	+	−	+	+(rare)	nameless	−	+

can be generated despite the lack of genetic input by a second individual. This occurs in some protozoans and nematodes, both parasitic and free-living, but it is most commonly encountered among arthropods, especially insects and arachnids. As seen in Table 25.1, asexual re-production is quite common among invertebrates. In fact, reproduction without fertilization is the primary re-productive mode in many species. Note that, with few exceptions, asexual reproduction requires the presence of only a single individual.

Table 25.2 Forms of Sexuality Encountered Among Invertebrates

Term	Form
Dioecious (gonochoristic)	♂ or ♀
Simultaneous hermaphrodite	♂
Sequential hermaphrodite	+ ♀
Protandric	♂ → ♀
Protogynous (relatively rare)	♀ → ♂

Sexual Reproduction

Patterns of Sexuality

Although many invertebrates reproduce asexually, sexual reproduction, which requires the fusion of haploid gametes, is quite common. Two individuals are usually involved in bringing this about. Moreover, the genetic composition of the offspring is always dissimilar to that of either parent. The two parents are usually of different sexes, in which case the species is said to be **dioecious** or **gonochoristic.** Alternatively, a single individual may be both male and female, either simultaneously (**simultaneous hermaphroditism**) or in sequence (**sequential hermaphroditism**) (Table 25.2).

Hermaphroditism is common among invertebrates, as shown in Table 25.1. The East Coast oyster, *Crassostrea virginica,* is a good example of a species exhibiting sequential hermaphroditism. The young oyster matures as a male, later becomes a female, and may change sex every few years thereafter. Most sequential hermaphrodites change sex only once and usually change from male to female. This is known as **protandric hermaphroditism,** or **protandry** (*prot* = G: first; *andros* = G: male).

In contrast to species that change sex as they age, many invertebrates, including most ctenophores and cestodes, are simultaneous hermaphrodites (Fig. 25.3). Self-fertilization is rare among simultaneous hermaphrodites, although it can occur, as in the Cestoda and within some barnacle species. An advantage of simultaneous hermaphroditism is that a meeting between any two mature individuals can result in a successful mating. This is especially advantageous in sessile animals, such as barnacles (Crustacea: Cirripedia). Indeed, the benefits of simultaneous hermaphroditism are so conspicuous that one wonders why such reproductive patterns are rare among vertebrates. Possibly the reproductive systems and behaviors of most advanced vertebrates have become so complex and specialized that two-sexed individuals are simply not feasible. This is perhaps unfortunate; many of the inequalities

commonly encountered among human societies would be unlikely in a society of simultaneous hermaphrodites.

The advantages of sequential hermaphroditism are less evident. Age-dependent sex changes can deter self-fertilization (a rather extreme form of inbreeding) in hermaphroditic species. In addition, it may be energetically more efficient to be one sex when small and the other sex when larger. Certainly, a single spermatozoon requires less structural material and nutrients than does a single ovum and is therefore less costly to produce. The total cost of reproduction, however, will depend upon the total number of gametes produced, the cost of each individual gamete, and the amount of energy expended in obtaining a mate and protecting the young; it may or may not be cheaper to be a male, depending upon the species.

Gamete Diversity

Invertebrate gametes show considerable diversity of structure and, often, of function. Some invertebrates produce a percentage of eggs that are incapable of being fertilized and/or of sustaining development following fertilization. These **nurse eggs** are ultimately consumed by neighboring embryos (Fig. 25.4). Nurse eggs are especially common among the Gastropoda. Sperm also show a high degree of functional diversity. Many invertebrate species produce only a percentage of normal sperm; that is, sperm that have a haploid DNA content and are capable of fertilizing an egg and promoting subsequent development of the embryo. These are **eupyrene** sperm. The other sperm can play no direct role in development because they have either an excess number of chromosomes or too few. In the extreme case, the abnormal sperm are without chromosomes entirely; that is, they are **apyrene.** Apyrene sperm production is especially common among gastropods and insects (Fig. 25.5). The functional significance of apyrene and other atypical sperm remains speculative at the present time.[2]

Considerable morphological diversity exists among even normal invertebrate sperm (Fig. 25.6), and the arrangement of microtubules within the axoneme (see Chapter 3, pp. 28–29) often departs radically from the usual 9 + 2 arrangement (Fig. 25.7). Indeed, the sperm of some species lack flagella entirely, in which case the sperm are either incapable of any movement or move in amoeboid fashion. Aberrant sperm are encountered especially often within the Arthropoda.

Getting the Gametes Together

All sexual development begins with the fertilization of a haploid egg; the trick, then, is to get the eggs and sperm together. Invertebrates demonstrate a variety of ways of bringing this union about. On land and in freshwater, fer-

2. See *Topics for Further Discussion and Investigation,* no. 6.

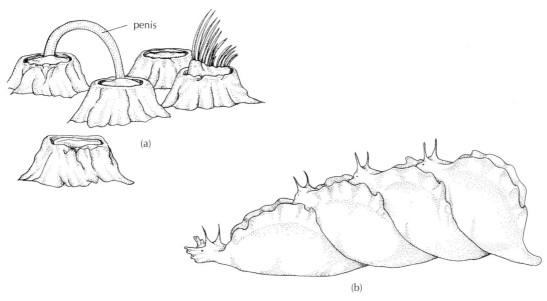

(a)

(b)

Figure 25.3

(a) Copulation among the Cirripedia. Barnacles are simultaneous hermaphrodites, a decided advantage for individuals that are incapable of locomotion as adults; any two adjacent individuals are potential mates. (b) Each sea hare, *Aplysia brasiliana* (Gastropoda: Opisthobranchia), bears a penis at the right side of the head and a vaginal opening posteriorly. Mutual insemination is common. The individuals in the middle of the chain illustrated are functioning simultaneously as males and females.   (a) *After Barnes and Hughes.* (b) *From Purves and Orians,* Life: The Science of Biology, *2d ed. Copyright © 1983 Sinauer Associates, Sunderland MA. Reprinted by permission.*

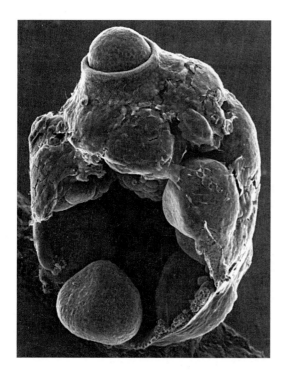

Figure 25.4

Ingestion of nurse eggs during prosobranch gastropod development. The marine species *Searlesia dira* is shown in the process of ingesting a single nurse egg. A portion of the body wall has been torn away, revealing a number of previously ingested nurse eggs. Each nurse egg is approximately 230 μm (micrometers) in diameter.   *Courtesy of Brian Rivest, from Rivest, 1983. J. Exp. Marine Biol. Ecol. 69:217. Elsevier Science Publishers, Physical Sciences and Engineering Division.*

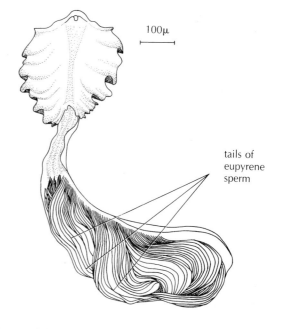

100μ

tails of eupyrene sperm

Figure 25.5

A giant, apyrene sperm of the marine gastropod *Cerithiopsis tubercularis*, with thousands of eupyrene sperm embedded in its tail. This form of apyrene sperm, called spermatozeugmata, may function in the transport of eupyrene sperm to the egg.   *From V. Fretter and A. Graham,* A Functional Anatomy of Invertebrates. *Copyright © 1976 Academic Press. Reprinted by permission of Harcourt Brace & Company Limited, London, England.*

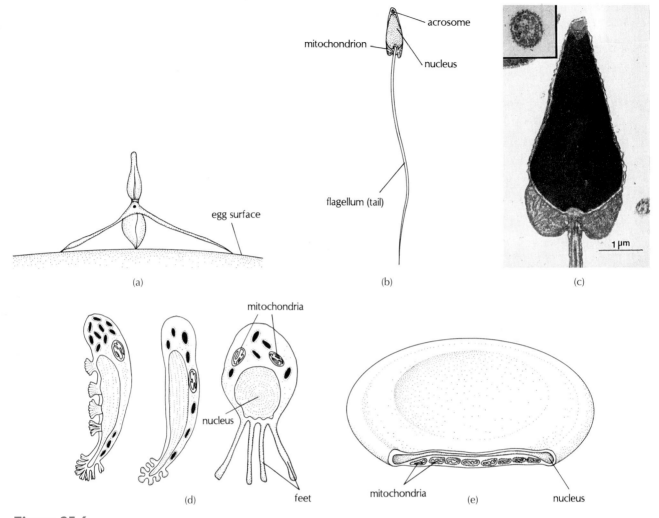

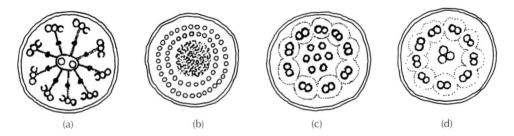

Figure 25.6

Gamete diversity. (a) The tripodlike sperm of the crustacean *Galathea* sp. The sperm is about to penetrate the egg. (b) Sea urchin sperm. (c) Transmission electron micrograph of a sea urchin sperm (head end only). (d) The aflagellate sperm of gnathostomulid worms (interstitial acoelomates). These sperm use the small feetlike processes to move. (e) The disc-shaped sperm of the insect *Eosentomon transitorium.*   (a) *From Kume and Dan,*

1968. Invertebrate Embryology. Washington D. C. : National Science Foundation. After Kortzoff. (c) Courtesy of K. J. Eckelbarger. From Eckelbarger et al., 1989. Biological Bulletin 176:257–71. Permission of Biological Bulletin. (d) From Bacetti & Afzelius, "Biology of the Sperm Cell" in Monographs in Developmental Biology, No.10. Copyright © 1976 S. Karger, AG, Basel, Switzerland. (e) From Bacetti and Afzellius; after Bacetti, et al.

Figure 25.7

(a) Cross section through the sperm tail of a sea urchin. Note the 9 + 2 arrangement of microtubules in the axoneme. (b) Cross section through the axoneme of an insect sperm, *Parlatoria oleae.* (c) Axoneme of a caddisfly sperm, *Polycentropus* sp. The microtubules have a 9 + 7 arrangement. (d) 9 + 3 arrangement of microtubules in the axoneme of sperm from the spider *Pholeus phalangioides.*   *From Bacetti & Afzelius, "Biology of the Sperm Cell" in* Monographs in Developmental Biology, *No. 10. Copyright © 1976 S. Karger, AG, Basel, Switzerland.*

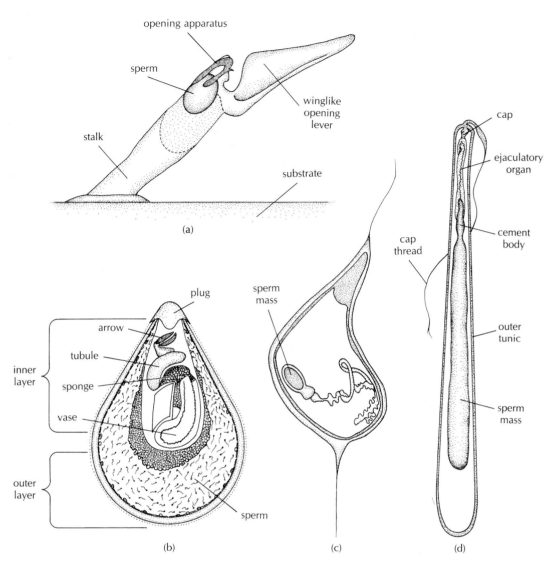

Figure 25.8

(a) The spermatophore of a scorpion. (b) Longitudinal section through the spermatophore of a tick (Arachnida). After transfer to the female, the outer layers of the spermatophore elongate over a one-minute period. The tip of the inner section of the spermatophore then opens, and within one additional second, the arrow, tubule, sponge, and vase are shot out from the spermatophore, discharging sperm. (c) Spermatophore of a vermetid gastropod, tentatively identified as *Dendropoma platypus.* (d) A cephalopod spermatophore.   (a) *From Barnes, 1980.* Invertebrate Zoology, *4th ed. Orlando, Florida: W. B. Saunders College Publishing. After Angermann.* (b) *Reprinted from Feldman-Muhsam, in* Journal of Insect Physiology, *29.449, 1983 with kind permission from Elsevier Science Ltd., The Boulevard, Langford Lane, Kidlington OX5 1GB, U. K.* (c) *From Hadfield and Hopper in* Marine Biology, *57:315, 1980. Copyright © 1980 Springer-Verlag, Heidelberg, Germany. Reprinted by permission.*

tilization of the egg is, with few exceptions, internal, for reasons presented in Chapter 1 (these environments are generally too dry or osmotically stressful for survival of exposed gametes). Internal fertilization may be accomplished in several ways. Males of many invertebrate species are equipped with a penis, through which sperm are transferred directly into the female's genital opening. In the case of hypodermic impregnation, as encountered among some turbellarians, leeches, gastropods, and rotifers, sperm are forcefully injected through the female's body wall. In both cases, sperm transfer is said to be **direct.**

Males of other species lack any such copulatory organs, and yet internal fertilization may still occur. Several means of achieving such **indirect sperm transfer** are encountered among invertebrates.[3] Commonly, the sperm are packaged in containers of varying complexity. These sperm-filled containers, called **spermatophores** (literally, "sperm carriers"), are secreted by specialized glands found only in the male. Among terrestrial invertebrates, spermatophores are typically employed by pulmonate gastropods, onychophorans, and terrestrial arthropods, including the insects, arachnids, centipedes, and millipedes (Fig. 25.8*a,b*). The spermatophore is transferred

---

3. See *Topics for Further Discussion and Investigation,* no. 3.

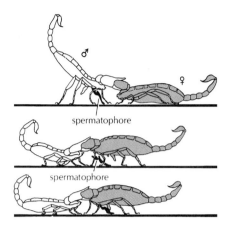

Figure 25.9
Exchange of spermatophores in scorpions. The male is shown on the left. After the male deposits the spermatophore, he guides the female over it.   *From Angerman, in Ziteschrift fur Tierpsychologie, 14:276, 1957. Copyright © 1957 Verlag Paul Parey, Berlin. Reprinted by permission.*

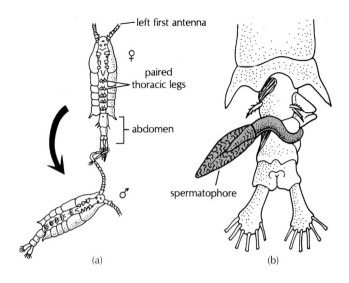

Figure 25.10
Spermatophore exchange in a marine copepod, *Centropages typicus* (as seen in ventral view). (a) The male grabs the female, using the hinged right first antenna. The male then swings the female around and grasps her around the abdomen, using his clawed right fifth thoracic appendage. (b) The spermatophore is shown in position after its transfer to the female.   *From Blades in Marine Biology, 40:57, 1977. Copyright © 1977 Springer-Verlag, Heidelberg, Germany. Reprinted by permission.*

from the male to the female in diverse ways. In some pseudoscorpions (Arthropoda: Arachnida), for example, no actual mating occurs. Males may deposit spermatophores onto suitable substrates without a female being present; the sperm capsules are located by the female either chemotactically or, in a few species, by following silk threads secreted by the male. Once the female has located a spermatophore, she inserts it into her genital opening, discharging the sperm from the container. This mechanism seems to be admirably suited for achieving internal fertilization in a species in which the proximity of one individual to another frequently prompts physical attack and even cannibalism.

The pseudoscorpion example to the contrary, internal fertilization generally requires a high degree of cooperation between pairs of individuals, and it is often preceded by elaborate courtship displays. This is typically true when fertilization is accomplished by copulation, but it is also common when fertilization is achieved through use of spermatophores. In what appear to be the more highly evolved pseudoscorpions, for example, males again deposit spermatophores on the ground but only in the presence of females and only after a complex mating dance. The male then physically guides the female and positions her over the stalk of the attached spermatophore, whereupon she takes up the sperm packet through her genital opening. The two partners then quickly separate from each other. The female removes the capsule a short time later, after it has been emptied by osmotically generated pressures and the sperm have been safely stored within her.

This mode of sperm transfer closely resembles that observed in true scorpions. Here, too, the stalked spermatophore is deposited only in the presence of the female and its deposition is preceded by an intricate dance, during which time the male searches for a suitable substrate on which to cement the sperm-filled container.

The scorpion spermatophore is quite complex, possessing a mechanically operated sperm-ejection lever (Fig. 25.8a). Again, the male positions the female so that her genital opening is over the tip of the capsule (Fig. 25.9). The capsule is then inserted into the genital opening just far enough to operate the lever of the spermatophore; the ejected sperm are then taken up and stored for later use.

In some other arachnids, the spermatophore's exact functional significance is dramatically apparent. The males of some species forcefully subdue a female, open her genital pore, deposit a spermatophore on the ground, pick it up with the chelicerae, insert the spermatophore into the female's genital opening, close the opening, and leave. Spermatophore transfer in centipedes and millipedes follows a somewhat similar script, except that the female voluntarily picks up and inserts the sperm packet following an often elaborate mating dance. Clearly, the spermatophore is a functional substitute for a copulatory organ in transporting spermatozoa to the female.

In bushcrickets and many other terrestrial arthropod species the female feeds on a specialized part of the spermatophore, both during and after sperm transfer. In such cases the "nuptial gift" occupies the female's attention for some time, increasing the likelihood that sperm transfer will be completed before the female removes the spermatophore, and also contributes nutrients that can increase female fecundity and offspring quality; either outcome will clearly increase the fitness of the donor male.

Spermatophores also are used by many marine and some freshwater invertebrates. Spermatophores are known

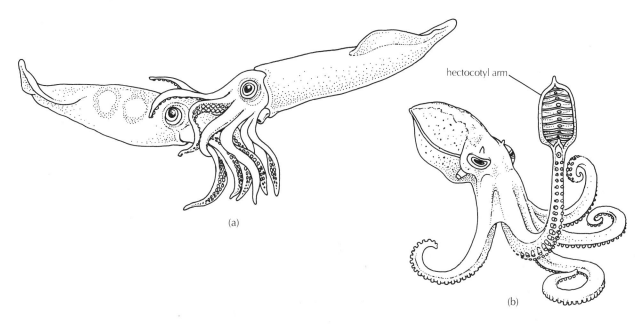

Figure 25.11

(a) Mating in the squid *Loligo*. (b) *Octopus lentus*, showing hectocotyl arm. The end of the arm forms a broad, cuplike depression that holds the spermatophores after the male has removed them from his mantle cavity.   (a) *After Barnes; after R. F. Sisson.* (b) *After Huxley; after Verrill.*

to occur in monogonont rotifers, polychaetes, oligochaetes, leeches, gastropods (Fig. 25.8c), cephalopods (Fig. 25.8d), crustaceans (Fig. 25.10), phoronids, pogonophorans, and chaetognaths. In some species, the spermatophores are simply discharged into the sea and reach a female by chance. Species employing this mode of sperm transfer always live in close association (i.e., communally), so the loss of sperm is not so great as might be supposed. Floating spermatophores have been reported among the Gastropoda, Polychaeta, Phoronida, and Pogonophora. More commonly, males deliver the spermatophores directly.

Among the cephalopods, the complexity of the spermatophore, and of its mode of transfer to the female, rivals that encountered among terrestrial invertebrates. Males often use their chromatophores to perform species-specific color displays for the female as a prelude to sperm transfer. The spermatophore is a large, cylindrical mass of sperm, incorporating a complex osmotically or mechanically activated sperm discharge mechanism (Fig. 25.8d). Large numbers of spermatophores are stored in a pouch, called **Needham's sac,** opening into the mantle cavity. The spermatophores of some species attain lengths of 1 m and contain up to approximately 10^{10} sperm! Typically, at the appropriate moment, the male grabs one or more spermatophores from Needham's sac and inserts them into the female's mantle cavity, adjacent to the genital opening. The organ used to insert the spermatophores is an often highly modified arm called the **hectocotylus** (Fig. 25.11). In some cephalopod species, the hectocotylus breaks off from the male follow-

ing its insertion into the female's mantle cavity. Before the reproductive biology of these species was understood, the disconnected hectocotylus was thought to be a parasitic worm, a not unreasonable assumption.

The evolutionary movement of invertebrates to land and freshwater from the sea may well have been accompanied not by the development of entirely new systems of fertilization, but rather by the modification of already existing ones. Indeed, reproduction through the use of spermatophores or through copulation may be considered a preadaptation for life on land; development of spermatophores may well have been one of the preadaptations that made the evolutionary transition from salt water to land or freshwater possible.

In marine animals, internal fertilization also may be accomplished in the absence of any form of physical copulation or sophisticated sperm packaging. Because the concentration of dissolved salts in seawater closely approximates that of most cells and tissues, sperm can be discharged freely into the seawater and transported to the female by water currents. Those species of bryozoans, echinoderms, bivalves, sponges, and cnidarians that have internal fertilization commonly employ this mechanism.

Once an egg has been fertilized internally, the embryos may develop within the female's body until released as miniatures of the adult—as, for example, in some brooding gastropods, bivalves, and ophiuroids.[4] Alternatively, the fertilized eggs may be packaged in groups within egg capsules or egg masses, which are then either protected by the female or affixed to, or buried within, a

4. See *Topics for Further Discussion and Investigation*, no. 8.

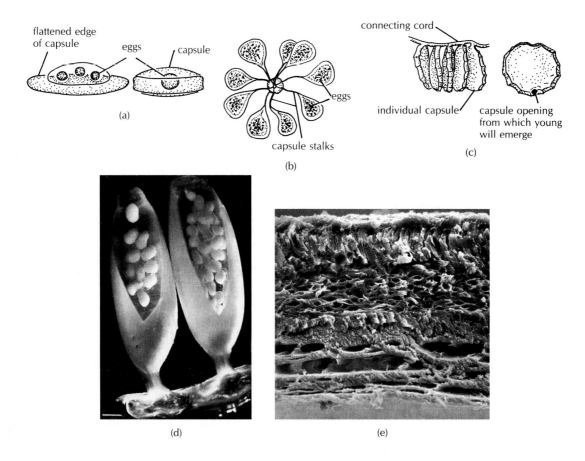

Figure 25.12

Representative egg capsules of marine prosobranch gastropods. (a) Planktonic egg capsules of two periwinkles, *Littorina littorea* and *Tectarius muricatus*. (b) Cluster of egg capsules of *Turritella*. (c) Necklace of egg capsules deposited by the whelk *Busycon carica*. Each capsule is about the size of a quarter. (d) Egg capsules of the marine gastropod *Nucella lapillus*. Part of the capsule wall has been cut away, revealing numerous yolky embryos within each capsule. The capsules are about 1 cm (centimeter) tall. Offspring emerge from the top of each capsule as miniature snails, having passed through the veliger stage within the capsules. (e) Cross section (scanning electron micrograph) of the egg capsule wall of *Nucella* (= *Thais*) *lima,* showing the complex, multilayered construction. The wall is approximately 55 μm thick.

(a–c) Illustrations by George F. Sandstrom from Seashells of North America. © 1968 Western Publishing Company, Inc. Used by permission. (d) Courtesy of R. Stöckman-Bosbach. From Stöckman-Bosbach et al., 1989. Marine Biology 102:283–89. Permission of Springer-Verlag. (e) From Pechenik, unpublished micrograph.

substrate and abandoned. These encapsulating structures are especially complex among members of the Gastropoda (Fig. 25.12).

Among marine invertebrates, union of gametes also may be accomplished in the absence of the structurally and behaviorally complex mechanisms generally associated with internal fertilization. In the ocean, fertilization may in fact be achieved by the coordinated release of eggs and sperm into the surrounding seawater. Such **external fertilization** is common in the marine environment, as seen in Table 25.1. The percentage of eggs that become fertilized varies with water current velocity and direction, position of individuals within a spawning aggregation, the size of the aggregation, egg diameter, the degree to which eggs chemically attract sperm, and the extent to which spawning is coordinated within any given population.[5]

Larval Forms

The product of an external fertilization generally develops into a free-living, swimming larva, an individual that grows and differentiates entirely in the water, as a member of the plankton. Larval forms are also frequently produced by species that have internal fertilization, the larvae either emerging from the females after a period of brooding, or from egg capsules or egg masses. The larvae of most invertebrate species are ciliated, the cilia serving for locomotion and, in species with feeding larvae, for food collection as well. External ciliation is incompatible with the chitinous exoskeleton of larval arthropods; among such larvae, locomotion and food collection must be achieved using specialized appendages (Table 25.3). External ciliation is also lacking during the development

---

5. See *Topics for Further Discussion and Investigation,* no. 9.

Table 25.3 Representative Larval Invertebrates

Phylum	Characteristic Larva	Adult	Page Reference
Porifera	amphiblastula	sponge	74
Cnidaria (= Coelenterata)	strobilating scyphistoma / ephyra	jellyfish (Scyphozoa)	86
	planula	hydroid (Hydrozoa)	93
		anemone (Anthozoa)	95
Platyhelminthes	Müller's larva	flatworm (Turbellaria)	132

continued on next page

of nematodes, which also are enclosed within a complex cuticle. Urochordates and chaetognaths constitute the final exceptions to the general rule of marine invertebrate larvae being ciliated.

The likely adaptive benefits of free-living larval forms for aquatic species that are slow moving or sessile as adults are (1) dispersal and genetic exchange between geographically separated populations of the same species, (2) rapid recolonization of areas following local extinctions, and (3) lack of direct competition with adults for food or space during development. The last benefit is also of particular significance for terrestrial insects.

Table 25.3—*Continued*

Phylum	Characteristic Larva	Adult	Page Reference
Platyhelminthes (continued)	miracidium → redia → cercaria →	fluke (Trematoda)	137
Rhynchocoela (= Nemertea)	pilidium →	ribbon worm	170
Annelida	trochophore → setigerous larva →	polychaete worm (Polychaeta)	304
Sipuncula	→ pelagosphaera →	sipunculan	323

Table 25.3—*Continued*

Phylum	Characteristic Larva	Adult	Page Reference
Mollusca			

trochophore

chiton (Polyplacophora)   268

scaphopod (Scaphopoda)   268

veliger   snail (Gastropoda)   268

veliger   clam (Bivalvia)   268

continued on next page

Table 25.3—*Continued*

Phylum	Characteristic Larva	Adult	Page Reference
Arthropoda			380

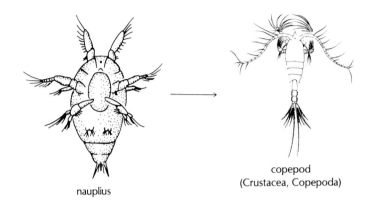

nauplius   copepod
(Crustacea, Copepoda)

380

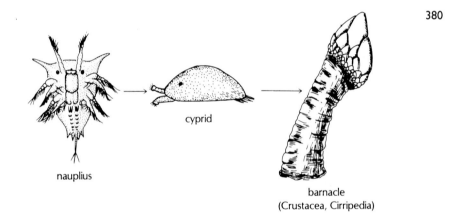

nauplius   cyprid   barnacle
(Crustacea, Cirripedia)

382

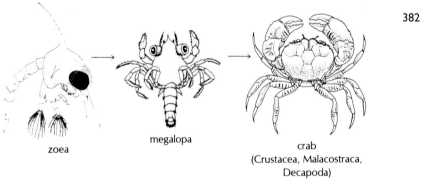

zoea   megalopa   crab
(Crustacea, Malacostraca,
Decapoda)

Table 25.3—*Continued*

Phylum	Characteristic Larva	Adult	Page Reference
Arthropoda (continued)			384

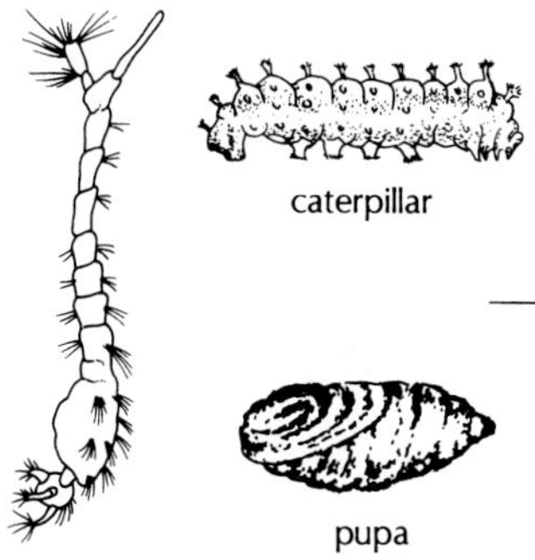

caterpillar

beetle

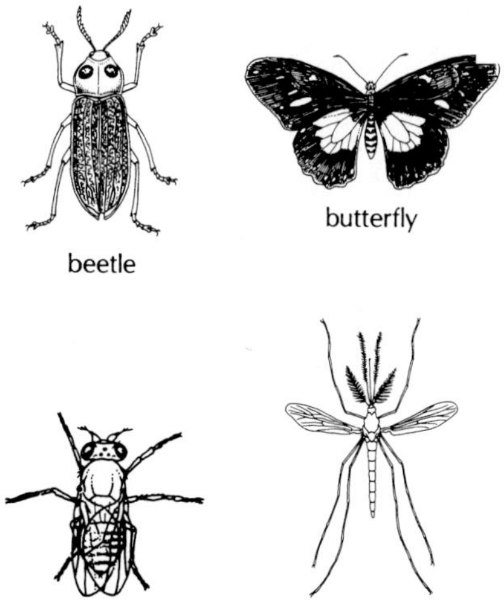

butterfly

nymph

pupa

fly

midge

(Insecta)

| Bryozoa | | | 437 |

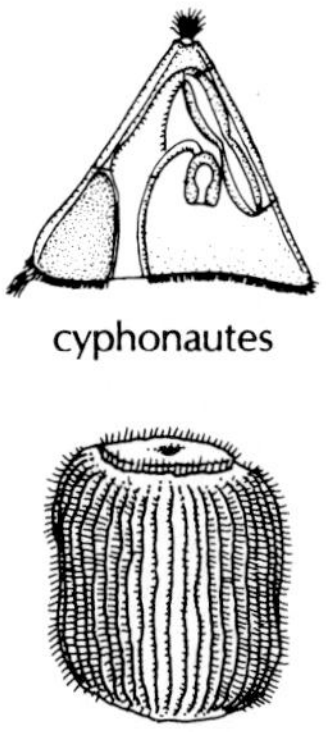

cyphonautes

coronate larva

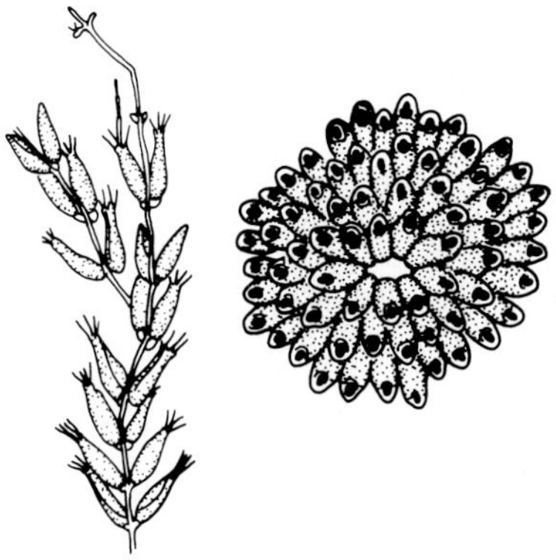

bryozoans

| Phoronida | | | 437 |

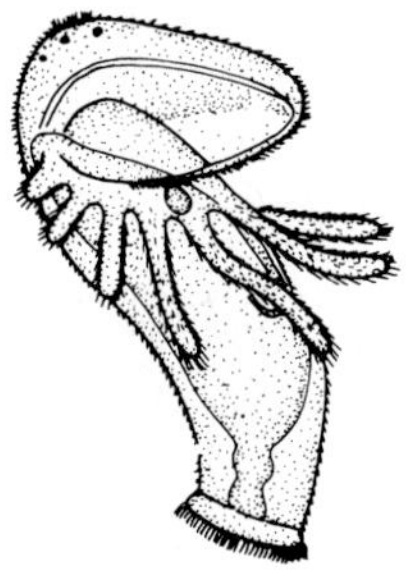

actinotroch

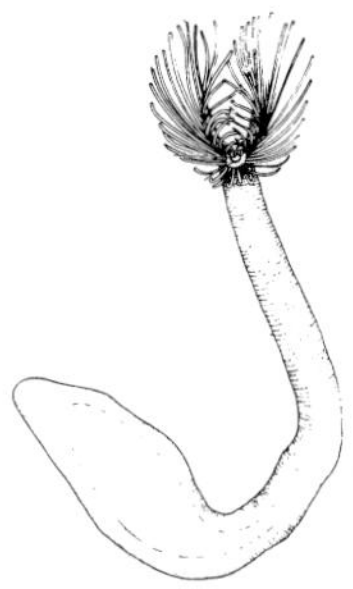

phoronid

continued on next page

Table 25.3—*Continued*

Phylum	Characteristic Larva	Adult	Page Reference
Brachiopoda	brachiopod larva	lampshell	437
Echinodermata	doliolaria	sea lily (Crinoidea)	465
	bipinnaria brachiolaria	sea star (Asteroidea)	465
	echinopluteus	sand dollar sea urchin (Echinoidea)	465

Table 25.3—*Continued*

Phylum	Characteristic Larva	Adult	Page Reference
Echinodermata (continued)	ophiopluteus	brittle star (Ophiuroidea)	465
	auricularia	sea cucumber (Holothuroidea)	465
Hemichordata	tornaria	acorn worm	486
Chordata, Urochordata	tadpole	tunicate, sea squirt (Ascidiacea)	491 496

Through the course of evolution by natural selection, larval forms have become increasingly well adapted to their own niches and may little resemble the adults of their own species, either morphologically or physiologically (Fig. 25.13). Larvae and adults may be considered ecologically distinct organisms—often exploiting entirely different habitats, lifestyles, and food sources—that just happen to have a genome in common. This latter point is crucial. Despite their ecological dissimilarity, the success of the one form in its stage of the life cycle determines the very existence of the other. Reproductive patterns comprising two or more ecologically distinct phases are termed **complex life cycles.**

The transition between phases of a complex life cycle often takes the form of an abrupt morphological, physiological, and ecological revolution termed a **metamorphosis** (Fig. 25.14). The greater the degree of difference between the adult and larval lifestyles, and the greater the degree of adaptation to those different lifestyles, the more dramatic the metamorphosis.[6] Complex life cycles are believed to be the original condition for marine invertebrates. Such life cycles seem generally to have been lost in association with the invasion of land and freshwater, but they have been reevolved in at least one group, the Insecta. The percentage of insect species exhibiting **holometabolous development** (i.e., development involving a conspicuous metamorphosis) has increased from about 10% (325 million years ago) to about 63% (200 million years ago) to about 90% (presently). The adaptive benefits of complex life histories must indeed be considerable.

The larvae of some invertebrates are **lecithotrophic** (*lecitho* = G: yolk; *trophy* = G: feeding); that is, they subsist on nutrient reserves supplied to the egg by the parent and are independent of the outside world for food. Lecithotrophic development is common among most groups of invertebrates, particularly among marine invertebrates living at high latitudes or in very deep water. It is much less common in shallow-water environments of the tropics and subtropics. In such shallow-water habitats, the larval stages typically develop functional guts and feed upon other members of the plankton, both plant (phytoplankton) and animal (zooplankton). Such larvae are said to be **planktotrophic.** Although there would seem to be many advantages to producing such planktotrophic larvae—for example, parents need to supply only minimal nutrients per egg, so a greater number of offspring can be produced from any given amount of stored yolk—many species have apparently switched from planktotrophic to lecithotrophic development in the course of their evolution. The selective pressures responsible for such shifts in developmental mode are difficult to document convincingly, but certainly lecithotrophic larvae are independent of seasonal and spatial variations in phytoplankton concentration and also may benefit from more rapid development to the

metamorphic stage.[7] In any event, once feeding larvae are lost from the life history of any particular species, they typically appear to be lost from that lineage forever.

Dispersal As a Component of the Life-History Pattern

Most freshwater, marine, or terrestrial animals have a dispersal stage at some point in their life histories. For those marine invertebrates that exploit the properties of seawater by living a sedentary or even a sessile adult existence, dispersal is typically accomplished by a planktonic larval stage, as previously discussed. How much dispersal occurs depends on how long the larval form can be maintained and on the speed and direction of water currents in which the larvae live (Fig. 25.15). In general, greater dispersal is associated with greater genetic homogeneity between geographically separated adult populations, greater ability to recolonize areas after localized extinction events, reduced rates of speciation, and greater species longevity (Fig. 25.16). In recent times, larvae of many aquatic invertebrate groups have been dispersed enormous distances artificially by ocean-going ships, which take on ballast water in shallow coastal areas and release it when docking thousands of miles away. The long-term ecological consequences of such ship-mediated dispersal are likely to be substantial, as in the case of the freshwater zebra mussel, which has already spread throughout much of North America after its probable introduction from Russia in 1986.

Freshwater invertebrates must deal with an often ephemeral habitat. Some groups, such as the sponges, tardigrades, bryozoans, and a number of crustaceans and rotifers, often avoid the need to disperse away from unfavorable conditions by forming resistant stages of arrested development. Such stages take the form of gemmules in sponges, resting eggs in crustaceans and rotifers, and statoblasts in bryozoans. Rotifers, tardigrades, and many protozoans also may enter into a cryptobiotic state during periods of dehydration. When environmental conditions improve, these various resting stages become revitalized. Wind-borne dispersal to new habitats is also common for these different diapause forms.

For terrestrial invertebrates, a sedentary adult existence is rare, due to the air being dry and of low density. Not surprisingly then, dispersal of terrestrial species is generally achieved by the adults, and the developing larvae are the stay-at-homes. The major exceptions to this generalization are found among the "suspension-feeding" arachnids—that is, the web-building spiders. In many species, shortly after the young spiders emerge from the silken cocoon built for them by the mother, they climb to the top of the nearest twig or blade of grass and allow

---

6. See *Topics for Further Discussion and Investigation,* no. 5.

7. See *Topics For Further Discussion and Investigation,* no. 10.

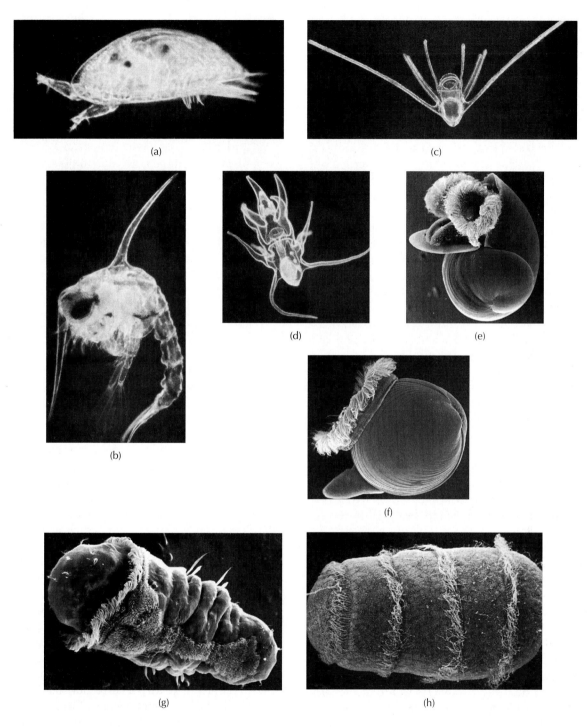

Figure 25.13

Invertebrate larvae. (a) Barnacle cyprid. (b) Zoea of a decapod crustacean. (c) Ophiopluteus of *Ophiothrix fragilis*, an ophiuroid. (d) Brachiolaria of the asteroid *Asterias vulgaris*. (e) Veliger of the opisthobranch *Rostanga pulchra.*(f) Veliger of the bivalve *Lyrodus pedicellatus,* showing shell, foot, and ciliated velum. Shell length is about 330 μm. (g) Advanced polychaete trochophore larva. Note that setae have already developed on two of the segments. (*h*) Doliolaria of the crinoid *Florometra serratissima.*   (a–d) *Courtesy of D. P. Wilson/Eric and David Hosking.* (e) *Courtesy of F. S. Chia, from Chia, 1978. Marine Biology 46:109. Permission of Springer-Verlag.* (f) *Courtesy of C. Bradford Calloway.* (g) *Courtesy of F. S. McEuen, from McEuen, 1983. Marine Biology 76:301. Permission of Springer-Verlag.* (h) *Courtesy of Philip V. Mladenov, from Mladenov, P. V., and F. S. Chia, 1983. Marine Biology 73:309. Permission of Springer-Verlag.*

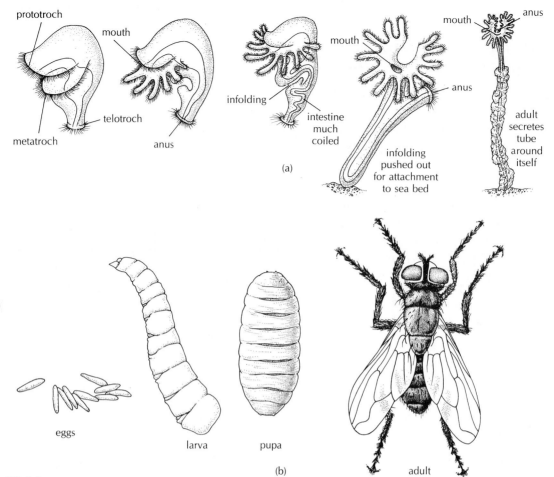

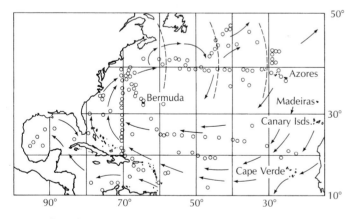

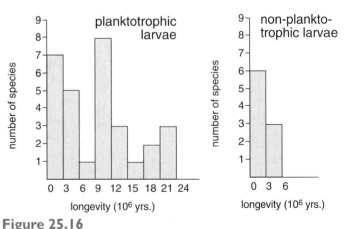

Figure 25.14
Examples of invertebrate metamorphosis. (a) Metamorphosis of an actinotroch larva into an adult phoronid. Note the dramatic reorientation of the digestive tract. (b) Stages in the development of the common housefly. This animal undergoes two distinct metamorphoses; one from the larval to the pupal stage, and another from the pupa to the adult. (a) *Adapted from Hardy, 1965.* The Open Sea: Its Natural History. *Boston: Houghton Mifflin Company.* (b) *After Engelmann and Hegner; after Packard.*

Figure 25.15
Dispersal of larvae of the tropical prosobranch gastropod *Cymatium parthenopeum* throughout the Atlantic Ocean. Circles indicate stations at which larvae of this species were collected. Arrows indicate surface circulation patterns. The adults are restricted to shallow-water habitats on the east coast of the United States and the west coast of Europe. *From Pechenik et al., in Science 224:1097, 1984. Copyright © 1984 American Association for the Advancement of Science, Washington, D.C.*

Figure 25.16
Relationship between mode of larval development and species duration in the fossil record, for 40 species in the gastropod genus *Nassarius*. Developmental mode is deduced from morphological characteristics of the premetamorphic shell, which is retained at the apex of juvenile and adult shells as the individual grows. No species with nonplanktotrophic larvae (limited dispersal potential) persisted beyond 4.5 million years, whereas many species with planktotrophic larvae (extensive dispersal potential) have fossil records extending from 9–24 million years. *Reprinted from "Relationship between species longevity and larval ecology in nassariid gastropods" by G. Gili and J. Martinell, 1994, Lethaia Volume 27, pages 291-299, by permission of Scandinavian University Press.*

themselves to be taken by the wind. This is termed **ballooning.** Air currents may disperse baby spiders for hundreds of miles (Fig. 25.17).

The degree to which the dispersing individual selects the appropriate habitat for the sedentary phase of the life cycle obviously determines the probability that a given individual will indeed survive to reproductive maturity. Among marine invertebrates with free-living larvae, selective pressures acting against random metamorphosis to adult form and habitat must have been substantial. The dispersing larvae of many species are highly selective about where they will metamorphose and can delay their metamorphosis for varying periods of time (up to many months for some species of molluscs, echinoderms, polychaetes and arthropods) if they fail to encounter the appropriate cues. The cues that trigger metamorphosis are often associated with some component of the adult environment: the adult food source or prey species, for example, or quite commonly, adults of the same species (Table 25.4).[8]

The extent to which larvae actually delay their metamorphosis in the field and the conditions under which they delay their metamorphosis have been difficult to explore. It is becoming clear, however, that there may be unanticipated costs associated with delaying metamorphosis in terms of postmetamorphic survival or growth rate (Research Focus Box 25.1); in such cases, the adaptive benefits of prolonging the competent period of larval development, so obvious and convincing in theory, may not be realized in practice.

Among the terrestrial insects, the larval stages are the sedentary eating machines, and the adults are the dispersal stages. Not surprisingly, the larvae of many species have become highly adapted for life on specific hosts, and the adults, in turn, often have evolved adaptations promoting the placement of eggs in the habitats supporting the best larval growth and survival.

Figure 25.17

Young spiders ballooning. . . . A warm draft of rising air blew softly through the barn cellar. The air smelled of the damp earth, of the spruce woods, of the sweet springtime. The baby spiders felt the warm updraft. One spider climbed to the top of the fence. Then it did something that came as a great surprise to Wilbur. The spider stood on its head, pointed its spinnerets into the air, and let loose a cloud of fine silk. The silk formed a balloon. As Wilbur watched, the spider let go of the fence and rose into the air.

"Good-bye!" it said, as it sailed through the doorway.

"Wait a minute!" screamed Wilbur. "Where do you think you're going?"

But the spider was already out of sight. . . . The air was soon filled with tiny balloons, each balloon carrying a spider.

Wilbur was frantic. Charlotte's babies were disappearing at a great rate.   *Illustration and text from Charlotte's Web, by E. B. White, pictures by Garth Williams. Copyright, 1952, by E. B. White; text copyright renewed © 1980 by E. B. White; illustrations copyright renewed © 1980 by Garth Williams.*

Table 25.4 Gregarious Metamorphosis by Barnacle Larvae

	Control	B. balanoides	B. crenatus	E. modestus
Total number of larvae in experiment	120	120	120	120
Average number metamorphosing per dish (± one standard error about the mean)	0.4 ± 0.31	9.5 ± 0.81	3.7 ± 0.93	1.6 ± 0.85
Average percent metamorphosed (± one standard error about the mean)	3.3 ± 2.5%	79.2 ± 6.7%	30.8 ± 7.7%	13.3 ± 7.0%

Source: Data from Knight-Jones. 1953. *J. Exp. Biol.* 30:584.

Note: Groups of 12 cyprid larvae of *Balanus balanoides* were placed in each of 10 dishes of seawater (controls) or in dishes containing seawater plus the adults and shells of *B. balanoides* or of two other barnacle species, *B. crenatus* or *Elminius modestus.* The number of metamorphosed individuals was assessed after 24 hours.

8. See *Topics for Further Discussion and Investigation,* no. 4.

Research Focus Box 25.1

Costs of Delaying Metamorphosis in a Marine Bryozoan

Woollacott, R. M., J. A. Pechenik, and K. M. Imbalzano. 1989. Effects of duration of larval swimming period on early colony development in *Bugula stolonifera* (Bryozoan: Cheilostomata). *Marine Biol.* 102:57–63.

Aquatic invertebrate larvae typically must develop for a time in the plankton before they become competent to metamorphose. This is true for species with either planktotrophic or lecithotrophic development. Competent larvae often metamorphose in response to specific chemical or physical cues typically associated with conditions appropriate for juvenile and adult development. In the absence of those cues, competent larvae of many species delay their metamorphosis, thereby increasing the likelihood of metamorphosing into a suitable adult habitat. However, if delaying metamorphosis imposes costs, such as reduced juvenile survival, reduced juvenile growth rates, or reduced fecundity, the benefits of delaying metamorphosis will not be fully realized. Woollacott et al. (1989) set out to deter-

mine whether delaying metamorphosis affects rates of postmetamorphic development in the common East Coast bryozoan *Bugula stolonifera*. Like most bryozoans, this species broods larvae to an advanced stage of development; the larvae are competent to metamorphose within a few hours of their release into the plankton (Research Focus Fig. 25.1). Conveniently, adults are readily stimulated to release larvae by keeping them in the dark for a time and then exposing them to bright light. Within a few days after attaching to a substrate, the larvae complete their transformation into a polyp (the "ancestrula") and begin to feed (Fig. 20.17, Chapter 20). Within another few days, each ancestrula buds off the next polyp—the second zooid in the growing colony—which itself soon begins to feed.

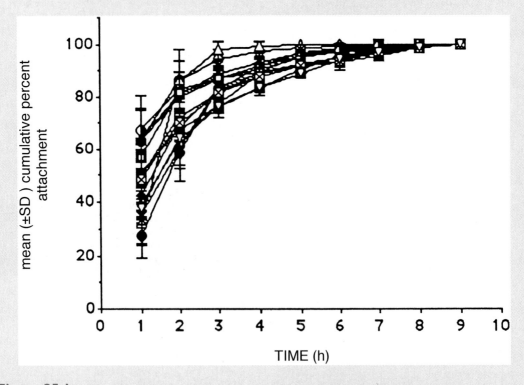

Focus Figure 25.1

Percent attachment of *Bugula stolonifera* larvae in the presence of continuously available plastic substrate at 20°C. Data are shown for 14 experiments conducted over a two-year period. Each point is the mean of 9–10 replicates, with error bars representing one standard deviation about the mean. Although there is considerable variability both within and among

experiments, most of the larvae were clearly competent to metamorphose within 3–4 hours after their release from parental brood chambers.   *From R. M. Woollacott, J. A. Pechenik, and K. M. Imbalzano, "Effects of duration of larval swimming period on early colony development in Bugula stolonifera (Bryozoa: Cheilostomata)" in* Marine Biology, *102:57–63, 1989. Copyright © Springer-Verlag, New York. Reprinted by permission.*

Woollacott et al. collected sexually mature bryozoan colonies from shallow water along the coast of Cape Cod, Massachusetts, and they then triggered larval release in the laboratory 2 to 7 days later. Larvae were transferred to a large glass jar within an hour of their release from the parent colonies. Two, 6, 8, and 10 hours later, the biologists added rectangular pieces of translucent plastic, a material previously shown to stimulate substantial attachment and metamorphosis. The substrate was removed after one hour, and the development of all attached individuals was subsequently followed for the next six days (6 d). In particular, the researchers determined the number of individuals developing to the feeding ancestrula stage of development after 3 d and the number of individuals that had developed a second feeding zooid by 6 d. All experiments were conducted at 20°C, with 9 to 10 replicates for each larval swimming period tested. Juveniles were fed on a mixture of several phytoplankton species.

Delaying metamorphosis depressed colony growth rates. The effects were seen for individuals delaying their metamorphosis for as little as 6 h, but the results were especially dramatic for individuals delaying their metamorphosis for 10 h.

In 10 of the 14 experiments conducted, the bryozoans developed more slowly to the ancestrula stage if the larvae had been kept swimming for 10 h, in comparison with colonies developing from larvae that had been kept swimming for only 2 h (Research Focus Fig. 25.2). Delaying metamorphosis also increased the time required for the second zooid to form and begin feeding.

In this bryozoan species, delaying metamorphosis for more than about 6 h at 20°C clearly compromised the postmetamorphic rate of development. This could lead to a reduced ability to compete for space with other sedentary organisms, including other bryozoans of this or other species, and a longer time to attain reproductive maturity. How might you go about designing a field experiment to test the ecological consequences of this effect? And how do you suppose the effect is mediated? For example, does delaying metamorphosis somehow compromise juvenile feeding ability? What other factors could account for the observed reduction of developmental rate, and how would you test the relative importance of these factors?

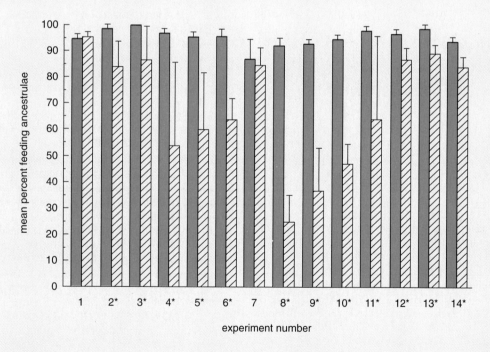

Focus Figure 25.2

The influence of delayed metamorphosis on rate of colony development by the bryozoan *Bugula stolonifera*. Solid bars represent data for larvae kept swimming for only two hours (at 20°C) before adding plastic substrate; striped bars represent data for larvae kept swimming for 10 hours. Each bar shows mean percent (N = 9–10 replicates per treatment) of individuals reaching the feeding ancestrula stage within three days of metamorphosis. Error bars show one standard deviation about the mean. An * indicates a significant difference between means in particular experiments (P < 0.05). *From J. A. Pechenik in Ophelia, 32:63–94, 1990. Copyright © 1990 Ophelia Publications, Helsingor, Denmark. Reprinted by permission.*

Even though much of the existing diversity of reproductive pattern has not found its way into the preceding discussion, invertebrates clearly show a dazzling variety of ways to reproduce. Many of the patterns that seem strange to us—external fertilization, simultaneous and sequential hermaphroditism, and parthenogenesis, for example—are actually very common reproductive modes. Why are there so many different ways to reproduce?[9] The question is much discussed in the literature of evolutionary ecology but challenging to answer definitively. For one thing, the adaptive significance of any particular pattern is often difficult to demonstrate. Moreover, the selective pressures responsible for presently observed reproductive patterns may be quite different today from what they were hundreds of thousands of years ago. Lastly, the role of historical accident in shaping the reproductive patterns of various invertebrate groups is hidden among evolutionary interrelationships that are often unclear and that may never be satisfactorily unraveled. Nevertheless, discussions regarding the remarkable diversity of reproductive patterns encountered among invertebrates surely will continue long into the future, and our understanding surely will improve as new data on the costs and benefits associated with different patterns of reproduction and development are accumulated for more and more species, and as the mechanisms through which reproductive patterns evolve become clearer.

Topics for Further Discussion and Investigation

1. What are the roles of temperature and light in regulating the cycles of invertebrate reproductive activity?

Davison, J. 1976. *Hydra hymanae:* Regulation of the life cycle by time and temperature. *Science* 194:618.

De March, B. G. 1977. The effects of photoperiod and temperature on the induction and termination of reproductive resting stage in the freshwater amphipod *Hyallela azteca* (Sanssure). *Canadian J. Zool.* 55:1595.

Fell, P. E. 1974. Diapause in the gemmules of the marine sponge, *Haliclona loosanoffi,* with a note on the gemmules of *Haliclona oculata. Biol. Bull.* 147:333.

Hardege, J. D., H. D. Bartels-Hardege, E. Zeeck, and F. T. Grimm. 1990. Induction of swarming of *Nereis succinea. Marine Biol.* 104:291.

Hayes, J. L. 1982. Diapause and diapause dynamics of *Colias alexandra* (Lepidoptera: Pieridae). *Oecologia* (Berlin) 53:317.

Jokiel, P. L., R. Y. Ito, and P. M. Liu. 1985. Night irradiance and synchronization of lunar release of planula larvae in the reef coral *Pocillopora damicornis. Marine Biol.* 88:167.

Pearse, J. S., and D. J. Eernisse. 1982. Photoperiodic regulation of gametogenesis and gonadal growth in the sea star *Pisaster ochraceus. Marine Biol.* 67:121.

Rose, S. M. 1939. Embryonic induction in *Ascidia. Biol. Bull.* 77:216.

Schierwater, B., and C. Hauenschild. 1990. A photoperiod determined life-cycle in an oligochaete worm. *Biol. Bull.* 178:111.

Stross, R. G., and J. C. Hill. 1965. Diapause induction in *Daphnia* requires two stimuli. *Science* 150:1462.

Vowinckel, C. 1970. The role of illumination and temperature in the control of sexual reproduction in the planarian *Dugesia tigrina* (Girarad). *Biol. Bull.* 138:77.

Wayne, N. L., and G. D. Block. 1992. Effects of photoperiod and temperature on egg-laying behavior in a marine mollusk, *Aplysia californica. Biol. Bull.* 182:8.

West, A. B., and C. C. Lambert. 1976. Control of spawning in the tunicate *Styela plicata* by variations in a natural light regime. *J. Exp. Zool.* 195:263.

Williams-Howze, J., and B. C. Coull. 1992. Are temperature and photoperiod necessary cues for encystment in the marine benthic harpacticoid copepod *Heteropsyllus nunni* Coull? *Biol. Bull.* 182:109.

2. What aspects of reproductive activity appear to be under chemical control among invertebrates?

Adamo, S. A., and R. Chase. 1990. The "love dart" of the snail *Helix aspersa* injects a pheromone that decreases courtship duration. *J. Exp. Zool.* 255:80.

Crisp, D. J. 1956. A substance promoting hatching and liberation of young in cirripedes. *Nature* 178:263.

Engelmann, F. 1959. The control of reproduction in *Diploptera punctata* (Blattaria). *Biol. Bull.* 116:406.

Forward, R. B., Jr., and K. J. Lohmann. 1983. Control of egg hatching in the crab *Rhithropanopeus harrisii* (Gould). *Biol. Bull.* 165:154.

Gibson, G. D., and F-S. Chia. 1994. A metamorphic inducer in the opisthobranch *Haminaea callidegenita:* Partial purification and biological activity. *Biol. Bull.* 187:133.

Golden, J. W., and D. L. Riddle. 1982. A pheromone influences larval development in the nematode *Caenorhabditis elegans. Science* 218:578.

Golding, D. W. 1967. Endocrinology, regeneration and maturation in *Nereis. Biol. Bull.* 133:567.

Kanatani, H., and M. Ohguri. 1966. Mechanism of starfish spawning. I. Distribution of active substance responsible for maturation of oocytes and shedding of gametes. *Biol. Bull.* 131:104.

Katona, S. K. 1973. Evidence for sex pheromones in planktonic copepods. *Limnol. Oceanogr.* 18:574.

Kelly, T. J., and L. M. Hunt. 1982. Endocrine influence upon the development of vitellogenic competency in *Oncopeltus fasciatus. J. Insect Physiol.* 28:935.

Kupfermann, I. 1967. Stimulus of egg laying: Possible neuroendocrine function of bag cells of abdominal ganglia of *Aplysia californica. Nature* (London) 216:814.

Marthy, H., J. R. Hauser, and A. Scholl. 1976. Natural tranquilizer in cephalopod eggs. *Nature* 261:496.

---

9. See *Topics for Further Discussion and Investigation*, nos. 8 and 11.

Morse, D. E., N. Hooker, H. Duncan, and L. Jensen. 1979. γ-Aminobutyric acid, a neurotransmitter, induces planktonic abalone larvae to settle and begin metamorphosis. *Science* 204:407.

Painter, S. D., M. G. Chong, M. A. Wong, A. Gray, J. G. Cormier, and G. T. Nagle. 1991. Relative contributions of the egg layer and egg cordon to pheromonal attraction and the induction of mating and egg-laying behavior in *Aplysia. Biol. Bull.* 181:81.

Reynolds, S. E., P. H. Taghert, and J. W. Truman. 1979. Eclosion hormone and bursicon titres and the onset of hormonal responsiveness during the last day of adult development in *Manduca sexta* (L.). *J. Exp. Biol.* 78:77.

Shorey, H. H., and R. J. Bartel. 1970. Role of a volatile sex pheromone in stimulating male courtship behaviour in *Drosophila melanogaster. Anim. Behav.* 18:159.

Sullivan, C. H., and D. B. Bonar. 1984. Biochemical characterization of the hatching process of *Ilyanassa obsoleta. J. Exp. Zool.* 229:223.

Takeda, N. 1979. Induction of egg-laying by steroid hormones in slugs. *Comp. Biochem. Physiol.* 62A:273.

Truman, J. W., and P. G. Sokolove. 1972. Silk moth eclosion: Hormonal triggering of a centrally programmed pattern of behavior. *Science* 175:1491.

Wheeler, D. E., and H. F. Nijhout. 1983. Soldier determination in *Pheidole bicarinata:* Effect of methoprene on caste and size within castes. *J. Insect Physiol.* 29:847.

Wigglesworth, V. B. 1934. The physiology of ecdysis in *Rhodnius prolixus* (Hemiptera). II. Factors controlling moulting and "metamorphosis." *Quart. J. Microsc. Sci.* 77:191.

Xu, R. A., and M. F. Barker. 1990. Annual changes in the steroid levels in the ovaries and the pyloric caeca of *Sclerasterias mollis* (Echinodermata: Asteroidea) during the reproductive cycle. *Comp. Biochem. Physiol.* 95A:127.

Zimmer-Faust, R. K., and M. N. Tamburri. 1994. Chemical identity and ecological implications of a waterborne, larval settlement cue. *Limnol. Oceanogr.* 39:1075.

3.   Discuss the morphological and behavioral adaptations for indirect sperm transfer.

Blades, P. I. 1977. Mating behavior of *Centropages typicus* (Copepoda: Calanoida). *Marine Biol.* 40:57.

Feldman-Muhsam, B. 1967. Spermatophore formation and sperm transfer in ornithodoros ticks. *Science* 156:1252.

Hsieh, H.-L., and J. L. Simon. 1990. The sperm transfer system in *Kinbergonuphis simoni* (Polychaeta: Onuphidae). *Biol. Bull.* 178:85.

Legg, G. 1977. Sperm transfer and mating in *Ricinoides hanseni* (Ricinulei: Arachnida). *J. Zool., London* 182:51.

Reeve, M. R., and M. A. Walter. 1972. Observations and experiments on methods of fertilization in the chaetognath *Sagitta hispida. Biol. Bull.* 143:207.

Wedell, N. 1994. Dual function of the bushcricket spermatophore. *Proc. Royal Soc. London B* 258:181.

Weygoldt, P. 1966. Mating behavior and spermatophore morphology in the pseudoscorpion *Dinocheirus tumidus* Banks (Cheliferinea: Chernetidae). *Biol. Bull.* 130:462.

Yund, P. O. 1990. An in situ measurement of sperm dispersal in a colonial marine hydroid. *J. Exp. Zool.* 253:102.

4.   How are the cues used by dispersal stages related to the habitat requirements of the sedentary or sessile stage in a complex life cycle?

Brewer, R. H. 1976. Larval settling behavior in *Cyanea capillata* (Cnidaria: Scyphozoa). *Biol. Bull.* 150:183.

Chew, F. S. 1977. Coevolution of pierid butterflies and their cruciferous food plants. II. The distribution of eggs on potential food plants. *Evolution* 31:568.

Cohen, L. M., H. Neimark, and L. K. Eveland. 1980. *Schistosoma mansoni:* Response of cercariae to a thermal gradient. *J. Parasitol.* 66:362.

Grosberg, R. K. 1981. Competitive ability influences habitat choice in marine invertebrates. *Nature* 290:700.

Highsmith, R. C. 1982. Induced settlement and metamorphosis of sand dollar (*Dendraster excentricus*) larvae in predator-free sites: Adult sand dollar beds. *Ecology* 63:329.

Hunte, W., J. R. Marsden, and B. E. Conlin. 1990. Habitat selection in the tropical polychaete *Spirobranchus giganteus* III. Effects of coral species on body size and body proportions. *Marine Biol.* 104:101.

Hurlbut, C. J. 1993. The adaptive value of larval behavior of a colonial ascidian. *Marine Biol.* 115:253.

Jensen, R. A., and D. E. Morse. 1984. Intraspecific facilitation of larval recruitment: Gregarious settlement of the polychaete *Phragmatopoma californica* (Fewkes). *J. Exp. Marine Biol. Ecol.* 83:107.

Knight-Jones, E. W. 1953. Laboratory experiments on gregariousness during settling in *Balanus balanus* and other barnacles. *J. Exp. Biol.* 30:584.

MacInnes, A. J., W. M. Bethel, and E. M. Cornfield. 1974. Identification of chemicals of snail origin that attract *Schistosoma mansoni* miracidia. *Nature (London)* 248:361.

Olson, R. 1983. Ascidian-*Prochloron* symbiosis: The role of larval photoadaptations in midday larval release and settlement. *Biol. Bull.* 165:221.

Pawlik, J. R., C. A. Butman, and V. R. Starczak. 1991. Hydrodynamic facilitation of gregarious settlement of a reef-building tube worm. *Science* 251:421.

Raimondi, P. T. 1988. Settlement cues and determination of the vertical limit of an intertidal barnacle. *Ecology* 69:400.

Scheltema, R. S. 1961. Metamorphosis of the veliger larvae of *Nassarius obsoletus* (Gastropoda) in response to bottom sediment. *Biol. Bull.* 120:92.

Turner, E. J., R. K. Zimmer–Faust, M. A. Palmer, M. Luckenbach, and N. D. Pentcheff. 1995. Settlement of oyster (*Crassostrea virginica*) larvae: Effects of water flow and a water-soluble chemical cue. *Limnol. Oceanogr.* 1579.

Wallace, R. L. 1978. Substrate selection by larvae of the sessile rotifer *Ptygura beauchampi. Ecology* 59:221.

Williams, K. S. 1983. The coevolution of *Euphydryas chalcedona* butterflies and their larval host plants. III. Oviposition behavior and host plant quality. *Oecologia (Berlin)* 56:336.

Wood, E. M. 1974. Some mechanisms involved in host recognition and attachment of the glochidium larva of *Anodonta cygnea* (Mollusca: Bivalvia). *J. Zool.* 173:15.

5. Describe the morphological changes that take place during metamorphosis in one group of marine invertebrates.

Atkins, D. 1955. The cyphonautes larvae of the Plymouth area and the metamorphosis of *Membranipora membranacea* (L.). *J. Marine Biol. Assoc. U.K.* 34:441.

Berrill, N. J. 1947. Metamorphosis in ascidians. *J. Morphol.* 81:249.

Bickell, L. R., and S. C. Kempf. 1983. Larval and metamorphic morphogenesis in the nudibranch *Melibe leonina* (Mollusca: Opisthobranchia). *Biol. Bull.* 165:119.

Bonar, D. B., and M. G. Hadfield. 1974. Metamorphosis of the marine gastropod *Phestilla sibogae* Bergh (Nudibranchia: Aeolidacea). I. Light and electron microscopic analysis of larval and metamorphic stages. *J. Exp. Marine Biol. Ecol.* 16:227.

Cameron, R. A., and R. T. Hinegardner. 1978. Early events in sea urchin metamorphosis, description and analysis. *J. Morphol.* 157:21.

Cloney, R. A. 1977. Larval adhesive organs and metamorphosis in ascidians. *Cell Tissue Res.* 183:423.

Cole, H. A. 1938. The fate of the larval organs in the metamorphosis of *Ostrea edulis*. *J. Marine Biol. Assoc. U.K.* 22:469.

Emlet, R. B. 1988. Larval form and metamorphosis of a "primitive" sea urchin, *Eucidaris thouarsi* (Echinodermata: Echinoidea: Cidaroida), with implications for developmental and phylogenetic studies. *Biol. Bull.* 174:4.

Factor, J. R. 1981. Development and metamorphosis of the digestive system of larval lobsters, *Homarus americanus* (Decapoda: Nephropidae). *J. Morphol.* 169:225.

Lang, W. H. 1976. The larval development and metamorphosis of the pedunculate barnacle *Octolasmis mülleri* (Coker, 1902) reared in the laboratory. *Biol. Bull.* 150:255.

Reed, C. G., and R. M. Woollacott. 1982. Mechanisms of rapid morphogenetic movements in the metamorphosis of the bryozoan *Bugula neritina* (Cheilostomata, Cellularioidea). I. Attachment to the substratum. *J. Morphol.* 172:335.

Stricker, S. A. 1985. An ultrastructural study of larval settlement in the sea anemone *Urticina crassicornis* (Cnidaria, Actiniaria). *J. Morphol.* 186:237.

6. Discuss the possible adaptive significance of the apyrene sperm of insects and suggest ways to test the various hypotheses.

Silberglied, R. E., J. G. Shepherd, and J. L. Dickinson. 1984. Eunuchs: The role of apyrene sperm in Lepidoptera? *Amer. Nat.* 123:255.

7. What impact do environmental pollutants have on the reproductive biology of marine invertebrates?

Bellam, G., D. J. Reish, and J. P. Foret. 1972. The sublethal effects of a detergent on the reproduction, development, and settlement in the polychaetous annelid *Capitella capitata*. *Marine Biol.* 14:183.

Bigford, T. E. 1977. Effects of oil on behavioral responses to light, pressure and gravity in larvae of the rock crab *Cancer irroratus*. *Marine Biol.* 43:137.

Heslinga, G. A. 1976. Effects of copper on the coral-reef echinoid *Echinometra mathaei*. *Marine Biol.* 35:155.

Holland, D. L., D. J. Crisp, R. Huxley, and J. Sisson. 1984. Influence of oil shale on intertidal organisms: Effect of oil shale extract on settlement of the barnacle *Balanus balanoides* (L.). *J. Exp. Marine Biol. Ecol.* 75:245.

Hunt, J. W., and B. S. Anderson. 1989. Sublethal effects of zinc and municipal effluents on larvae of the red abalone *Haliotis rufescens*. *Marine Biol.* 101:545.

Kobayashi, N. 1980. Comparative sensitivity of various developmental stages of sea urchins to some chemicals. *Marine Biol.* 58:163.

Macdonald, J. M., J. D. Shields, and R. K. Zimmer-Faust. 1988. Acute toxicities of eleven metals to early life-history stages of the yellow crab *Cancer anthonyi*. *Marine Biol.* 98:201.

Muchmore, D., and D. Epel. 1977. The effects of chlorination of wastewater on fertilization in some marine invertebrates. *Marine Biol.* 19:93.

Short, J. W., S. D. Rice, C. C. Brodersen, and W. B. Stickle. 1989. Occurrence of tri-*n*-butyltin-caused imposex in the North Pacific marine snail *Nucella lima* in Auke Bay, Alaska. *Marine Biol.* 102:291.

8. Many aquatic invertebrates release gametes or larvae into the environment, whereas many others show varying degrees of protection and care for their developing young. What are the costs and benefits that might explain either (a) why so many species have evolved some form of direct parental care or (b) why so many species have not?

Gillespie, R. G. 1990. Costs and benefits of brood care in the Hawaiian happy face spider *Theridion grallator* (Araneae, Theridiidae). *Amer. Midl. Nat.* 123:236.

Menge, B. A. 1975. Brood or broadcast? The adaptive significance of different reproductive strategies in two intertidal seastars, *Leptasterias hexactis* and *Pisaster ochraceus*. *Marine Biol.* 31:87.

Milne, I. S., and P. Calow. 1990. Costs and benefits of brooding in glossiphoniid leeches with special reference to hypoxia as a selection pressure. *J. Anim. Ecol.* 59:41.

Pechenik, J. A. 1979. Role of encapsulation in invertebrate life histories. *Amer. Nat.* 114:859.

Thorson, G. 1950. Reproductive and larval ecology of marine bottom invertebrates. *Biol. Rev.* 25:1.

9. External fertilization is common among marine invertebrates. How can one assess fertilization success in the field and the biological and physical factors that contribute to such success?

Babcock, R. C., C. N. Mundy, and D. Whitehead. 1994. Sperm diffusion models and *in situ* confirmation of long-distance fertilization in the free-spawning asteroid *Acanthaster planci*. *Biol. Bull.* 186:17.

Benzie, J. A. H., and P. Dixon. 1994. The effects of sperm concentration, sperm:egg ratio, and gamete age on fertilization success in crown-of-thorns starfish (*Acanthaster planci*) in the laboratory. *Biol. Bull.* 186:139.

Denny, M. W., and M. F. Shibata. 1989. Consequences of surf-zone turbulence for settlement and external fertilization. *Amer. Nat.* 134:859.

Grosberg, R. K. 1991. Sperm-mediated gene flow and the genetic structure of a population of the colonial ascidian *Botryllus schlosseri*. *Evolution* 45:130.

Levitan, D. R., M. A. Sewell, and F-S. Chia. 1992. How distribution and abundance influence fertilization success in the sea urchin *Strongylocentrotus franciscanus*. *Ecology* 73:248.

Pennington, J. T. 1985. The ecology of fertilization of echinoid eggs: The consequences of sperm dilution, adult aggregation, and synchronous spawning. *Biol. Bull.* 169:417.

Young, C. M., P. A. Tyler, J. L. Cameron, and S. G. Rumrill. 1992. Seasonal breeding aggregations in low-density populations of the bathyal echinoid *Stylocidaris lineata*. *Marine Biol.* 113:603.

Yund, P. O. 1990. An *in situ* measurement of sperm dispersal in a colonial marine hydroid. *J. Exp. Zool.* 253:102.

10. In many marine invertebrate groups, species apparently have switched from having a feeding (planktotrophic) larva to a nonfeeding (lecithotrophic) larva. What selective pressures likely account for such a shift, what morphological changes accompany such a shift, and through what developmental mechanisms have such shifts been accomplished?

Amemiya, S., and R. B. Emlet. 1992. The development and larval form of an echinothurioid echinoid, *Asthenosoma ijimai,* revisited. *Biol. Bull.* 182:15.

Jeffrey, W. R. 1994. A model for ascidian development and developmental modification during evolution. *J. Marine Biol. Assoc. U.K.* 74:35.

McEdward, L. R. 1992. Morphology and development of a unique type of pelagic larva in the starfish *Pteraster tesselatus* (Echinodermata: Asteroidea). *Biol. Bull.* 182:177.

Raff, R. A. 1987. Constraint, flexibility, and phylogenetic history in the evolution of direct development in sea urchins. *Devel. Biol.* 119:6.

Sinervo, B., and L. R. McEdward. 1988. Developmental consequences of an evolutionary change in egg size: An experimental test. *Evolution* 42:885.

Strathmann, R. R. 1978. The evolution and loss of feeding larval stages of marine invertebrates. *Evolution* 32:894.

Strathmann, R. R., L. Fenaux, and M. F. Strathmann. 1992. Heterochronic developmental plasticity in larval sea urchins and its implications for evolution of nonfeeding larvae. *Evolution* 46:972.

Wray, G. A., and R. A. Raff. 1991. The evolution of developmental strategy in marine invertebrates. *Trends Ecol. Evol.* 6:45.

11. The widespread occurrence of sexual reproduction has long fascinated evolutionary biologists, partly because, as a result of meiosis, each adult gets to contribute only half of its genes to the next generation and partly because male offspring make no direct contribution to future reproduction; parthenogenesis would seem to be the more efficient reproductive pattern. What factors might select for sexual reproduction over parthenogenetic reproduction, and how can the various hypotheses be tested?

Bell, G. 1982. *The Masterpiece of Nature: The Evolution and Genetics of Sexuality*. Berkeley, Calif.: Univ. California Press.

Case, T. J., and M. L. Taper. 1986. On the coexistence and coevolution of asexual and sexual competitors. *Evolution* 40:366.

Ghiselin, M. T. 1974. *The Economy of Nature and the Evolution of Sex*. Berkeley, Calif.: Univ. Calif. Press.

Jaenike, J. 1978. An hypothesis to account for the maintenance of sex within populations. *Evol. Theory* 3:191.

Lively, C. M., and S. G. Johnson. 1994. Brooding and the evolution of parthenogenesis: Strategy models and evidence from aquatic invertebrates. *Proc. Royal Soc. London B* 256:89.

Lloyd, D. 1980. Benefits and handicaps of sexual reproduction. *Evol. Biol.* 13:69.

Michod, R. E., and T. W. Gayley. 1992. Masking of mutations and the evolution of sex. *Amer. Nat.* 139:706.

Some General References About Invertebrate Reproduction and Development

Adiyodi, K. G., and R. G. Adiyodi, eds. 1983–1993. *Reproductive Biology of Invertebrates. Vol. I: Oogenesis, Oviposition, and Oosorption; Vol. II: Spermatogenesis and Sperm Function; Vol. III: Accessory Sex Glands; Vol. IV A and B: Fertilization, Development, and Parental Care; Vol. V: Sexual Differentiation and Behaviour; Vol. VI A and B: Asexual Propagation and Reproductive Strategies.* New York: Wiley Interscience.

Chia, F. S., and M. E. Rice. 1971. *Settlement and Metamorphosis of Marine Invertebrate Larvae*. New York: Elsevier/North Holland Inc.

Giese, A. C., and J. S. Pearse eds. 1974–1979. *Reproduction of Marine Invertebrates. Vol. I: Acoelomate and Pseudocoelomate Metazoans; Vol. II: Entoprocts and Lesser Coelomates; Vol. III: Annelids and Echiurans; Vol. IV: Molluscs: Gastropods and Cephalopods; Vol. V: Molluscs: Pelecypods and Lesser Classes.* New York: Academic Press.

Giese, A. C., J. S. Pearse, and V. B. Pearse. 1987. *Reproduction of Marine Invertebrates. Vol. IX: General Aspects: Seeking Unity in Diversity.* Palo Alto, Calif.: Blackwell Scientific Publications.

Gilbert, L. I., and E. Frieden, eds. 1981. *Metamorphosis: A Problem in Developmental Biology,* 2d ed. New York: Plenum Press.

Jägersten, G. 1972. *Evolution of the Metazoan Life Cycle*. New York: Academic Press.

McEdward, L. R., ed. 1995. *Ecology of Marine Invertebrate Larvae*. New York, CRC Press.

Reverberi, G. 1971. *Experimental Embryology of Marine and Fresh-Water Invertebrates*. Amsterdam: North-Holland Publ. Co.

Wilbur, K. M., ed. 1983. *The Mollusca, Vol. 3: Development*. New York: Academic Press.

Glossary of Frequently Used Terms

Definitions of more specialized terms can be found by using the Index.

A

aboral The part of the body farthest from the mouth.

acoelomate Lacking a body cavity between the gut and the outer body wall musculature.

annulation External division of a worm-shaped body into a series of conspicuous rings.

apomorphic A modified ("derived") character state.

archenteron A cavity that eventually becomes the digestive tract of the adult or larva; formed during the development of a deuterostome embryo.

asexual reproduction Reproduction that does not involve the fusion of gametes; reproduction without fertilization.

B

benthic Living on or within a substrate.

benthos The aquatic animals, plants, and algae living on or within a substrate.

binary fission Asexual division of one organism into two nearly identical organisms.

bioluminescence Biochemical production of light by living organisms.

biramous Two-branched.

blastocoel The internal cavity commonly formed by cell division early in embryonic development, prior to gastrulation.

brooding Parental care of developing young.

budding A form of asexual reproduction in which new individuals develop from a portion of the parent, as in all bryozoans and in many protozoans, cnidarians, and polychaetes.

C

cephalization Concentration of nervous and sensory systems in one part of the body, which becomes known as the "head."

chromatophore A pigment-containing cell that can be used by an animal to vary its external coloration.

cilium A thread-like locomotory organelle containing a highly organized array of microtubules; shorter than a flagellum.

cirri Among ciliated protozoans, a group of cilia that function as a single unit; among barnacles, the thoracic appendages, which are modified for food collection; among crinoids, the prehensile appendages located aborally that are used for walking and for clinging to solid substrates.

coelom An internal body cavity lying between the gut and the outer body wall musculature that is lined with derivatives of the embryonic mesoderm.

colony A cluster of genetically identical individuals formed asexually from a single colonizing individual.

conjugation A temporary physical association in which genetic material is exchanged between two ciliate protozoans.

convergent evolution The process whereby similar characteristics are independently evolved by different groups of organisms in response to similar selective pressures.

cuticle A noncellular, secreted body covering.

cyst A secreted covering that protects many small invertebrates, including some protozoans, rotifers, and nematodes, from environmental stresses, such as desiccation and overcrowding.

D

deposit feeding Ingesting substrate (sand, soil, mud) and assimilating the organic fraction.

desiccation Dehydration.

dioecious Characterized by having separate sexes; that is, an individual is either male or female, but never both. "Gonochoristic" means the same thing.

diploblastic Possessing only two distinct tissue layers during embryonic development.

E

ectoderm An embryonic tissue layer; forms epidermal, nervous, and sensory organs and tissues.

encystment The secretion of a protective outer covering that permits some small invertebrates to withstand exposure to extreme environmental stresses, such as desiccation and overcrowding.

endoderm An embryonic tissue layer; forms wall of digestive system.

enterocoely Formation of a coelom through outpocketing of the inner portion of the archenteron in some animals (deuterostomes).

estuary A partially enclosed body of water influenced by both tidal forces and by freshwater input from the land.

eutely Species-specific constancy of cell numbers or nuclei; growth occurs through increased cell size rather than increased cell number.

exoskeleton A system of external levers and joints that permits pairs of muscles to act against, or antagonize, each other; the exoskeleton is also protective.

F

filiform Thread-like.

filter feeder An organism that filters food particles from the surrounding fluid.

flagellum A thread-like locomotory organelle containing a highly organized array of microtubules; longer than a cilium and often bearing numerous lateral projections.

flame cell Flagellated cell associated with protonephridia, as in flatworms, rotifers, and some polychaetes.

G

gametes The sex cells involved in fertilization.

gastrovascular canals Fluid-filled canals opening at the mouth of cnidarians and ctenophores that function in gas exchange and in the distribution of nutrients.

gastrulation Creation of a new tissue layer by the movement of cells in the early embryo (blastula).

gill A structure specialized for gas exchange in aquatic animals.

gonochoristic Having separate sexes; that is, an individual is either male or female, but never both. "Dioecious" means the same thing.

H

hermaphrodite A single individual that functions as both male and female, either simultaneously or in sequence.

homology Having identical evolutionary origins and developing through identical developmental pathways.

homoplasy Independent evolution of similar or identical character states through convergence or parallel evolution.

hydrostatic skeleton A constant-volume, fluid-filled cavity that permits muscles to be restretched after contraction, often through the mutual antagonism of muscle pairs.

I

interstitial Living in the spaces between sand grains.

intertidal Living in the area between high and low tides and, thus, alternately exposed to the air and to the sea.

introvert A tubular, eversible extension of the head, bearing the mouth at its tip.

L

larva A free-living developmental stage in the life history of many invertebrate species.

M

meiofauna Small, interstitial animals living between sand grains.

mesenteries Infoldings of gastroderm and mesoglea extending into the gastrovascular cavity of cnidarians; sheets of peritoneum from which the digestive tract is suspended in coelomates.

mesoderm An embryonic tissue layer that gives rise to certain tissues and organs of the adult, including the muscles and gonads.

mesoglea Gelatinous layer found between the epidermis and the gastrodermis of cnidarians.

mesohyl Nonliving, middle gelatinous layer of sponges; living cells are often found within it.

metamerism Serial repetition of organs and tissues, including the body wall, nervous and sensory systems, and musculature.

metamorphosis A dramatic transformation of morphology and function occurring over a short period of time during development.

metanephridium An organ open to the body cavity through a ciliated funnel (nephrostome) and involved in excretion or in the regulation of water balance or salt content.

metazoan A multicellular animal.

microtubules Tubulin-containing cylinders characteristic of cilia and flagella.

monoecious Characterized by the presence of both sexes in a single individual, either in sequence or sequentially; hermaphroditic.

monophyletic groups Groups that contain an ancestor and all the descendants of that ancestor.

N

nematocyst Cnidarian cell type that explosively emits long threads specialized for defense and food capture.

O

ocellus A simple, pigment-containing photoreceptor found in a variety of unrelated invertebrates.

osmosis Diffusion of water across a semipermeable (permeable to water but not to solute) membrane along a concentration gradient.

P

paraphyletic groups Groups that do not include all descendants of a common ancestor.

parasitism A close association between species in which one member benefits at the expense of the partner.

parthenogenesis Development of an unfertilized egg into a functional adult.

pelagic organism An organism that lives above the bottom in the open sea.

peristalsis Progressive waves of muscular contraction passing down the length of an organism or organ system.

peritoneum The mesodermal lining of the body cavity of coelomates.

phagocytosis The process through which food particles are surrounded by cell membrane and incorporated into the cell cytoplasm, forming a food vacuole.

plankton Animals (zooplankton) and unicellular algae (phytoplankton) that have only limited locomotory capabilities and are therefore distributed by water movements.

plesiomorphic An ancestral ("primitive") character state.

preadaptation A trait that is adaptive only in a new set of physical or biological circumstances.

primitive Believed to closely resemble the ancestral form; possessing characteristics resembling ancestral characteristics.

proboscis A tubular extension at the anterior of an animal, generally used for locomotion or food collection; may or may not be directly connected to the gut.

protandric hermaphroditism A pattern of sexuality in which a single individual functions as male and then female in sequence.

pseudocoel An internal body cavity lying between the outer body wall musculature and the gut; not lined with mesoderm and generally formed by persistence of the embryonic blastocoel.

pseudopodia Amorphous protrusions of cytoplasm involved in the locomotion and feeding of amoebae and related protozoans.

R

radial cleavage A form of early cell division in which all cleavage planes are perpendicular, so daughter cells come to lie directly in line with each other.

S

schizocoely Coelom formation accomplished by a split in the mesoderm during embryonic development of some animals (protostomes).

sedentary Bottom-dwelling and capable of only limited locomotion.

septa Peritoneal (mesodermal) sheets separating adjacent segments, as in annelids, or body divisions, as in chaetognaths.

sessile Bottom-dwelling and generally incapable of locomotion.

sexual reproduction Reproduction involving fusion of gametes.

spermatophore A container of sperm transferred from one individual to another during mating.

spicules Calcareous or silicious formations present in the tissues of some organisms and generally serving protective or supportive functions.

spiral cleavage Pattern of cell division in which cleavage planes are at 45° to the animal-vegetal axis of the egg.

statocyst A sense organ that informs the bearer of orientation of the body to gravity.

suspension feeder An animal that feeds on particulates suspended in the surrounding medium; this may be accomplished by filtering or by other means.

synapomorphic Uniquely shared, derived (modified) character states; describes a group of species that can be defined as different from all others because they share some unique homologous character.

T

test Any hard external covering; may be secreted by the animal or constructed from surrounding materials.

triploblastic Exhibiting three distinct tissue layers during embryonic development.

V

vector Any organism that transmits parasites from one host species to another.

vermiform Worm shaped—that is, soft-bodied and substantially longer than wide.

Z

zooid A single member of a colony.

zooplankton The animal component of the plankton, having only limited locomotory powers.

Index